Electromagnetic Waves

Electromagnetic Waves

Carlo G. Someda

CRC Press is an imprint of the
Taylor & Francis Group, an **informa** business
A TAYLOR & FRANCIS BOOK

CRC Press
Taylor & Francis Group
6000 Broken Sound Parkway NW, Suite 300
Boca Raton, FL 33487-2742

ISBN-13: 978-0-8493-9589-5 (hbk)
Library of Congress Card Number 2005043549

Publisher's Note

The publisher has gone to great lengths to ensure the quality of this reprint but points out that some imperfections in the original copies may be apparent.

Library of Congress Cataloging-in-Publication Data

Someda, Carlo G.
Electromagnetic waves / Carlo G. Someda.--2nd ed.
p. cm.
Includes bibliographical references and index.
ISBN 0-8493-9589-5
1. Electromagnetic waves. I. Title.

QC661.S66 2006
539.2--dc22 2005043549

Visit the Taylor & Francis Web site at
http://www.taylorandfrancis.com

and the CRC Press Web site at
http://www.crcpress.com

To Gian Carlo Corazza and Enrique A.J. Marcatili
who taught me how to mix
Maxwell's equations and practical life

To my grandchildren, Matilde and Edoardo

Contents

Preface

Is there a real need to publish, in 1997, a new book on electromagnetic waves? Clearly, the answer to this question will come from the market, not from the author. What the author can do is just to try to explain why he accepted to take this challenge.

In the late 1970's, UTET, one of the most important and prestigious Italian publishers, decided to prepare a series of volumes in electrical and electronic engineering, written by Italian authors. The overall purpose was to provide a complete cross-section of how these disciplines were taught in Italian universities. Francesco Barozzi, a very distinguished professor, then head of the Electrical and Electronic Engineering Department at the University of Trieste, was appointed as the editor in chief. He was a charming gentleman, with an immense, ubiquitous background in humanities and arts, which he never showed o . With a friendly smile, he always took an extremely calm approach to any kind of problem, making everybody feel at ease. To be selected by him as one of the authors was so rewarding, that my decision to prepare a volume for the UTET series was taken in less than one second.

Writing the book was not equally fast: it was published in 1986. After that, for the rst few years its presence on the market was timid, but later sales increased to a very satisfactory level, which is now steadily maintained. Of course, this statement must be calibrated with respect to the limited size of the potential market: unfortunately, there is just one country where electromagnetic waves are taught in Italian!

There are, on the other hand, several scientists around the world, who can read Italian. One of them, professor Michael C. Sexton, from University College Cork, Ireland, saw my book during one of his frequent visits to Padua in relationship to his research in the eld of plasma physics and fusion engineering. He was the rst who suggested to write an English version. Shortly after, Dr. Richard J. Black, who was then teaching at the Ecole Polytechnique in Montreal, Canada, and therefore was familiarizing himself with technical literature in the Romance languages, expressed the same opinion. They said "there is good stu in there." I do not know precisely what they meant. On my part, I can say, comparing the contents of my book with those of many others, that I was fortunate to work, in a relatively short time interval, in three elds: microwaves, lasers, and ber optics. What I tried to do in teaching, and consequently in writing my book, was to strengthen the links, and unify the background. The reader will judge whether I achieved this goal.

The role Professor Sexton played was not only to trigger my interest in an English version of my book. He also contacted Chapman & Hall. The project was approved by the publisher under the condition that some changes were made, with respect to the Italian edition. The suggestions made to this purpose by anonymous referees matched exactly my own opinions. Indeed, my experiences as a student indicated that in the Latin countries there is a long tradition — maybe even a philosophical attitude — to teach "by concepts." Conversely, in English-speaking countries one teaches — and learns — "by examples." As a result, the English book could not be a mere translation, but had to be enriched with examples, and with more problems.

As a net result, this book is very large. Maybe it is too thick, but I would not know what to cut. For this reason, I am very grateful to the publisher, who never put on me any pressure in order to reduce the market risks that this unusual size entails.

C.G. Someda

Acknowledgments

The list of those towards whom I became indebted, during the preparation of this book, has no upper limits. Let me begin by repeating two names that I already mentioned in the Preface: those of Professor Michael C. Sexton and Dr. Richard J. Black. Their contributions range from suggesting the material to be added to the Italian version, and choosing examples and problems, to polishing my English, and adding to the references. In one word, their roles are quite similar to those of co-authors. It was very generous on their part to leave only to me the honor to sign this book.

Next, I wish to thank an entire team of graduate students, post-doctoral fellows, and research assistants, who spent so much of their time and energy doing the initial translations, and then reading the manuscript over and over again, looking for no other reward but to see me happy and thankful. They were so many, that the list of their names would be too long, and the probability of forgetting someone would be too high. I am sure they understand what I mean, and recognize themselves in these lines.

I want to mention explicitly, and to thank enthusiastically, Ms. Barbara Sicoli, who polished with professionality and patience the final version of the manuscript, and Ms. Flavia Bergamin, who did the computer processing of almost all the figures.

Finally, let me thank UTET, the Publisher of the Italian text, for the generous permission to publish the English version as a totally independent book, and Chapman & Hall, for being so patient during the long preparation of the manuscript, and for encouraging me when the goal looked out of my reach.

Preface to the Second Edition

The rst edition of my "Electromagnetic Waves" is now seven years old - not so old, in a eld where basic knowledge has a long record of stability, and technological breakthroughs occur on a time scale of several years, if not decades. Why a second edition, then? In fact, the main motivation does not come from the evolution of the contents, but from facts that took place in the book company arena. The publisher of the rst edition, Chapman & Hall, was acquired by another group - a prominent one - long ago, shortly after publishing my book. What I learnt during the following years was that, when similar events occur, regardless of the size and of the prestige of the acquiring group, redistributing the titles of the acquired company is a very complicated issue, which does not necessarily entail happy endings for all authors. It turned out that, without anybody being speci cally guilty for that, the highly positive, I would almost say enthusiastic, reviews that my book had enjoyed when it appeared, were not followed by a marketing campaign of comparable intensity. Loss of enthusiasm can be contagious. Just to mention one of the negative fallouts, I never completed the exercise manual. Sales declined.

The reader can imagine, then, how grateful I am to the Taylor & Francis Group, and to its CRC Press branch in particular, for proposing a second edition at the right moment. The rst one was almost sold out, and I was facing a bifurcation: either restart with new motivations, or abandon the whole project. I believe that the whole team wants now to show that we made the right decision.

For what refers to the contents, the list of changes is very short. I have added, here and there, a few de nitions, which are not new, but whose importance I had underestimated in the past; just to quote an example, antenna e ective height, in Chapter 12. The reference list has been updated, as a consequence of the wide set of rst-class books that were published in these seven years. The text of several problems has been edited, to remove some ambiguities or to complete, here and there, insu cient data. The discovery of little imperfections in the problems occurred while their solutions were being prepared; this entails that the manual should be on the market soon after the book. Last but not least, several tens of misprints and minor mistakes that, in spite of automatic correctors etc., appeared in the rst edition (an extremely frustrating experience) have been taken care of. This is a signi cant improvement, although, I am afraid, typographical perfection remains a dream. I am very grateful to an innumerable list of colleagues, friends, students, who had an active role in identifying these errors.

To conclude, let me thank collectively all those who contributed to the second edition. Two persons, however, deserve more than anonymous thanks, as they have been instrumental in preparing the manuscript. One is Mr. Emanuele Zattin, who, while completing his education as a computer engineer, has developed a terri c experience as a professional electronic editor. It took all his skills to recover the previous les, and make them compatible with state-of-the-art software. The second one is a junior colleague of mine, Antonio Daniele Capobianco. Without Tony's continuous dedication and unlimited patience, preparation of this manuscript would have been an endless, painful process, not only for me, but for many other people.

C.G. Someda

CHAPTER 1

Basic equations for electromagnetic elds

1.1 Introduction: Experimental laws

Electromagnetic waves form a chapter of mathematical physics which can be organized as an axiomatic theory. Indeed, all the fundamental concepts, as well as many notions of technical interest, can be deduced from a small set of postulates. This type of approach is convenient in terms of conceptual economy and compactness. On the other hand, there is a danger in it, shared in general by all axiomatic theories: connections with physical intuition may become very weak, especially in the beginning. However, in our case, we expect this drawback to be mitigated, since we take it for granted that the reader has already been exposed to a more elementary description of electromagnetic phenomena, at least in their time-independent or slowly-varying versions. Our rst concern will be to show how the postulates of the axiomatic theory can be linked to what the reader knows already about slowly-varying elds, although postulates, by de nition, do not require justi cation.

For this purpose, let us brie y review the basic laws of slowly-varying electromagnetic elds. We will take them simply as experimental data.

1.1.1 Conservation of electric charge

The theory to be presented here deals only with phenomena on a macroscopic scale, where consequences of the discrete nature of the electric charge are irrelevant. Therefore, we shall model charge in terms of a function of spatial coordinates P and time t, which we denote as (P, t) and refer to as *electric charge density*. Throughout this book, the phrase "charge density" will mean the so-called *free* charge density, i.e., the local imbalance between densities of positive and negative charges, even though, on a microscopic scale, charges of opposite signs cannot be located exactly at the same points. In order not to make the text unnecessarily cumbersome, the term "function" (of spatial coordinates and/or time) will also implicitly include the case of a generalized function. Those cases where this is not acceptable will be outlined explicitly. A typical example of generalized function, that we will use soon, is a point charge $q(P_0) = q\ (P\quad P_0)$, where $(\mathbf{r})$ stands for a three-dimensional Dirac delta function.

The in nitesimal charge contained, at the time t, in the in nitesimal volume dV_P, centered at the point P, is expressed by $(P, t)\, dV_P$. In SI (or rationalized MKS) units (e.g., see Weast, 1988-89), is expressed in coulombs per cubic meter (C/m^3). Consequently, electric currents, as e ects of moving electric charges, will be modeled in terms of a vector $\mathbf{J}\ (P, t)$ in three-dimensional (3-D) real space, which depends on the coordinates of the point P and on the time, t. We shall refer to $\mathbf{J}\ (P, t)$ as the *electric current density* (expressed in amperes per square meter, A/m^2). It is de ned so that the in nitesimal current that ows across the incremental surface dS_P, passing through P, is expressed by $\mathbf{J}\ (P, t) \quad \mathbf{dS}_P$, the vector $\mathbf{dS}_P$ being normal to dS_P in P.

The macroscopic law of *electric charge conservation* is then expressed, with the symbols de ned above, by the relation:

$$\frac{d}{dt} \int_V \quad dV = \quad \int_{S_v} \mathbf{J} \quad \mathbf{dS}_V \tag{1.1}$$

which holds for any region V in real 3-D space, bounded by a closed surface S_V. The vector $\mathbf{dS}_V$ is oriented, at each point of S_V, toward the region complementary to V (i.e., in common language, "toward the exterior" of region V).

Conventions on the signs and on the orientations, that we introduced with this procedure, entail that the vector $\mathbf{J}$ is oriented in the sense in which positive charges move. This is purely a mathematical fact, without any physical link to the sign of those charges that are actually present and moving in each speci c case.

1.1.2 *Lorentz force*

Experiments show that a point charge q (i.e., a nite charge within a negligibly small volume) moving at an instantaneous velocity $\mathbf{v}$ (m/s) in a region where an electromagnetic eld is present experiences a force (expressed in newtons, N):

$$\mathbf{F} = q(\mathbf{E} + \mathbf{v} \quad \mathbf{B}) \quad , \tag{1.2}$$

which is called the *Lorentz force.* Eq. (1.2) can be taken as the de nition of two vectors, functions of coordinates in 3-D space and of time: $\mathbf{E}$, called the electric eld intensity, or, for short, the *electric eld*, expressed in volts per meter (V/m); and $\mathbf{B}$, called the *magnetic induction*, expressed in teslas (T).

1.1.3 *Faraday's law of electromagnetic induction*

The two space- and time-dependent vectors $\mathbf{E}$ and $\mathbf{B}$ are linked by the following integral relationship between the ux of the vector $\mathbf{B}$ and the circulation of the vector $\mathbf{E}$:

$$\oint_\ell \mathbf{E} \quad \mathbf{d\ell} = \quad \frac{d}{dt} \int_S \mathbf{B} \quad \mathbf{dS} \quad , \tag{1.3}$$

where S is any regular two-faced surface, within the domain where $\mathbf{E}$ and $\mathbf{B}$ are de ned; ℓ is the boundary of S. The orientations of the normal to S (i.e., of $\mathbf{dS}$) and of ℓ (i.e., of $\mathbf{d}\boldsymbol{\ell}$) are linked by the right-hand screw rule. Eq. (1.3), usually referred to as *Faraday's law of electromagnetic induction*, provides the rst connection between electric and magnetic phenomena, showing that a time-varying magnetic induction eld is always accompanied by an electric eld.

1.1.4 Ampere's circuital law

Eq. (1.3) is not the only connection which exists between electric and magnetic phenomena. Experiments show that, in the presence of moving charges, i.e., of an electric conduction or convection current, there are always accompanying magnetic phenomena. Also this link is expressed quantitatively by a relationship between a ux and a circulation. More precisely, *in the static case*, i.e., if the current density vector, $\mathbf{J}$, is not a function of time, it is:

$$\oint_\ell \mathbf{H}\ \mathbf{d}\boldsymbol{\ell} = \int_S \mathbf{J}\ \mathbf{dS} \quad , \tag{1.4}$$

where the symbols S and ℓ have again the meanings explained after Eq. (1.3). Eq. (1.4) can be taken as the de nition of the space- and time-dependent vector $\mathbf{H}$, called *magnetic eld intensity*, or, for short, *magnetic eld*, and expressed in amperes per meter (A/m).

It is simple to show that, for time-dependent quantities, Eq. (1.4) cannot be valid in general. Indeed, let us consider two surfaces, S_1 and S_2, having a common boundary ℓ, and let S_v be the closed surface de ned as the union of S_1 and S_2. Then, Eqs. (1.4) and (1.1) would imply that the total charge inside S_v can not vary in time. However, as S_v can be any surface in the region where the elds are de ned, then the charge density, , can not vary in time, at any point. This restriction is physically untenable. Consequently, in the time-dependent regime it is necessary to replace Eq. (1.4) by another relation, which must reduce to Eq. (1.4) in the static case.

A priori, there are no logical reasons why we should change either the right-hand side of Eq. (1.4), or its left-hand side. However, a change on the left-hand side would clearly imply a di erent de nition of the eld $\mathbf{H}$. Now, there is experimental evidence that, in the static case and within a given material, the vectors $\mathbf{B}$ and $\mathbf{H}$, if de ned as above, satisfy a simple relation, called the constitutive relation of the material (see Section 1.3). In the time-dependent case, experience again shows that it is possible, and useful, to preserve such a relation essentially unaltered. Consequently, we shall leave the left-hand side of Eq. (1.4) unchanged, in the dynamic regime, and modify the right-hand side, with the addition of a term that can account for a time-dependent charge density. These requirements are satis ed by the so-called *displacement current density*, i.e., the space- and time-dependent vector $\partial \mathbf{D}\ /\partial t$. It has the same physical dimensions as the conduction or convection current density $\mathbf{J}$, i.e., (A/m^2), but its physical nature is di erent, as it does not consist of moving electric charges. It is de ned by the following relationship:

$$\oint_{\ell} \mathbf{H} \; d\boldsymbol{\ell} = \int_{S} \left(\mathbf{J} + \frac{\partial \mathbf{D}}{\partial t} \right) \; d\mathbf{S} \quad , \tag{1.5}$$

where it is assumed that the surface S does not vary with time. The vector $\mathbf{D}$, whose partial derivative with respect to time is the displacement current density, is referred to as the *electric displacement* or the *electric induction* (expressed in C/m^2).

Eq. (1.5) is referred to as *Ampere's circuital law* (sometimes referred to as the Biot-Savart law); it provides another link between electric and magnetic phenomena completely independent of Faraday's law of induction Eq. (1.3).

1.2 Maxwell's equations and the charge continuity equation

In the previous section, experimental background laws were quickly recalled in their *integral form*, i.e., written for physical quantities that are observable on a nite length scale, and expressed mathematically by volume, surface or line integrals. On the other hand, the axiomatic theory to be developed throughout this book becomes simpler, and more compact, if it starts from a *local* description of electromagnetic phenomena. Local relations between quantities de ned in the previous section will now play the logical role of *axioms*.

Two of these local relations, commonly referred to as *Maxwell's equations*, read:

$$\nabla \quad \mathbf{E} = \frac{\partial \mathbf{B}}{\partial t} \quad , \tag{1.6}$$

$$\nabla \quad \mathbf{H} = \mathbf{J} + \frac{\partial \mathbf{D}}{\partial t} \quad . \tag{1.7}$$

If, in particular, we assume that the vectors in Eqs. (1.6) and (1.7) are time-harmonic, i.e., vary sinusoidally in time with an angular frequency ω, then, we may exploit the one-to-one and onto mapping between the set of time-harmonic vectors in real 3-D space, and the 3-D complex-vector space, illustrated in the next chapter, but probably known to the reader from elementary circuit theory, and summarized as $\mathbf{a} = \mathrm{Re}(\mathbf{a}\, e^{j\omega t})$. Thus, we can replace Eqs. (1.6) and (1.7) by the following equations, where the vectors are space-dependent, time-independent complex functions:

$$\nabla \quad \mathbf{E} = j\omega \mathbf{B} \quad , \tag{1.8}$$

$$\nabla \quad \mathbf{H} = \mathbf{J} + j\omega \mathbf{D} \quad . \tag{1.9}$$

In the following, we will use "Maxwell's equations" to refer to either Eqs. (1.6) and (1.7), or to Eqs. (1.8) and (1.9), depending on the context, without danger of ambiguity.

To develop the deductive theory, it is necessary to combine Maxwell's equations with a third postulate, known as the *equation of charge continuity*. It relates the electric charge density with the conduction or convection current density, as follows:

$$\nabla \quad \mathbf{J} = \quad \frac{\partial}{\partial t} \quad . \tag{1.10}$$

In the time-harmonic case, the one-to-one and onto mapping that we just recalled enables us to replace Eq. (1.10) with:

$$\nabla \quad \mathbf{J} = \quad j\omega \quad . \tag{1.11}$$

We attributed the character of postulates to Maxwell's equations and to the charge continuity equation. Therefore, from a strictly logical viewpoint, there is no need to provide a physical justi cation for them. Nevertheless, such a justi cation is straightforward given the experimental fundamentals that were summarized in Section 1.1. If the vector functions of space and time **E**, **B**, **H**, **D**, **J** and the scalar function are regular, in the sense that they have partial derivatives that are continuous to all the orders implied in the calculations, then, applying the de nitions of curl and divergence, Eqs. (1.6), (1.7) and (1.10) can be derived, respectively, from Faraday's law of electromagnetic induction Eq. (1.3), Ampere's circulation law Eq. (1.5), and charge conservation Eq. (1.1).

We shall not overlook the case where one or more of the above-listed quantities is not regular throughout its domain of de nition. Indeed, we will see, in Section 1.6, that some of the electromagnetic vectors can be discontinuous across surfaces where the materials which occupy their domain of de nition have step discontinuities. In order to "justify" Maxwell's equations and the charge continuity equation in such circumstances, we will accept, in our theory, material discontinuities only if they satisfy the following two requirements, which agree fully with physical intuition:

1) the number of discontinuities over any nite interval in space is nite;
2) each discontinuity is of the rst type, i.e., it can be interpreted as the limit of a continuous transition (distributed over a characteristic length , and having derivatives with respect to spatial coordinates up to the order one needs) as tends to zero (see Figure 1.1).

Under these assumptions, the curl and divergence vector operators, de ned in Appendix A, and used as described above, lead always to vector functions that admit a right limit and a left limit everywhere. Where these two limits do not coincide, an integration (over a surface, in the case of a curl, over a volume, in that of a divergence) leads to a relation *between continuous quantities*, as required by the macroscopic laws Eqs. (1.1), (1.3) and (1.4).

The subject of discontinuities is a delicate topic. We will come back to it, shortly, in Section 1.5.

1.3 Constitutive relations

In Section 1.1, we anticipated that the vectors **B** and **H** are not independent of each other. As the reader should know from previous studies of electrostatic

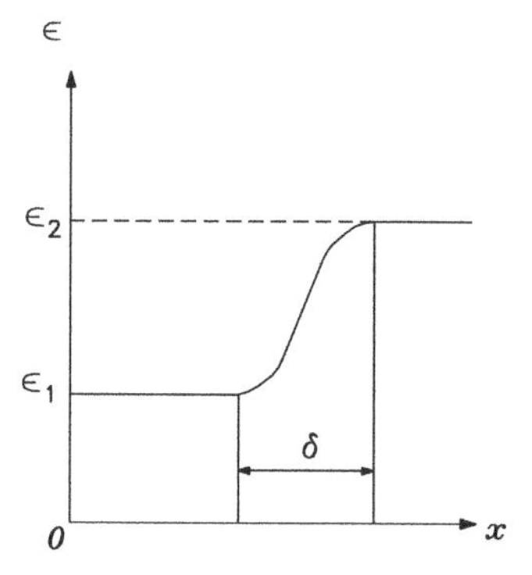

Figure 1.1 *A step discontinuity in a parameter of the medium is seen as the limiting case of a smooth transition.*

and magnetostatic phenomena, in general the vectors de ned so far are related not only by Maxwell's equations and the charge continuity equation, but also by other relations, which express properties of the medium where these vectors are de ned. They are referred to as the *constitutive relations.*

The mathematics of electromagnetism is drastically simpli ed in all those cases in which the constitutive relations are *linear*, i.e., when some, among those vectors, can be expressed as linear functions of the others. In this book, we will restrict ourselves only to materials belonging to this class, which we call linear media. This implies a loss in generality. The reader is certainly aware that, in basic physics as well as in elementary electrical engineering, there are very important problems (for example, ferromagnetism) where the assumption of linearity is totally inapplicable, even as a rst approximation. In our framework, consequences of this loss in generality are not too restrictive. Indeed, a very broad set of electromagnetic phenomena can be explained in a satisfactory manner under the assumption of linearity. Furthermore, in many nonlinear media which are exploited at radio, microwave or optical frequencies (semiconductors, ferrites, nonlinear crystals, etc.), e.m. wave propagation can often be studied with a linearized model (the so-called small-signal approximation). For all those cases that do not belong to this set, the reader should consult more specialized textbooks (see Section 1.11).

Among *linear media*, the simplest to describe and, at the same time, those of major practical interest, are *isotropic* media. This term means that, when such media are at rest with respect to an inertial reference frame, the displacement vector $\mathbf{D}$, and the conduction or convection current density vector, $\mathbf{J}$, are parallel to the electric eld vector, $\mathbf{E}$, while the magnetic induction vector, $\mathbf{B}$, is parallel to the magnetic eld vector, $\mathbf{H}$. In mathematical terms:

$$\mathbf{D} = \mathbf{E} \ , \tag{1.12}$$

$$\mathbf{J} = \mathbf{E} \ , \tag{1.13}$$

$$\mathbf{B} = \mathbf{H} \ . \tag{1.14}$$

Eqs. (1.12), (1.13) and (1.14) de ne the following scalar quantities: (i) the *permittivity* (or dielectric constant) , expressed in farads per meter (F/m), (ii) the *conductivity* , expressed in siemens (or inverse ohms) per meter (S m^{1} ^{1} m^{1}), and (iii) the *magnetic permeability* , expressed in henries per meter (H/m).

For *time-invariant* media, , and , in the preceding relations, are constant; therefore, for time-harmonic vectors, the above mentioned one-to-one and onto mapping in the 3-D space of complex vectors allows us to write:

$$\mathbf{D} = \mathbf{E} \ , \tag{1.15}$$

$$\mathbf{J} = \mathbf{E} \ , \tag{1.16}$$

$$\mathbf{B} = \mathbf{H} \ . \tag{1.17}$$

Let us stress that going back from Eqs. (1.15), (1.16) and (1.17) to Eqs. (1.12), (1.13) and (1.14), is a legitimate operation if and only if , and are *constant with respect to the angular frequency* ω. Often, it turns out to be convenient to use Eqs. (1.15)–(1.17) as de nitions of , and also in the very broad class of media in which these parameters are functions of ω. But then, strictly speaking, the right-hand sides of Eqs. (1.12), (1.13) and (1.14) should be replaced by a *convolution integral* between the inverse-Fourier transforms of the quantities which appear in the right-hand side of Eqs. (1.15), (1.16) and (1.17). We may then look at Eqs. (1.12), (1.13) and (1.14) as *approximations*, which are acceptable only when the spectral width, in the frequency domain, of all the quantities that one is dealing with, is narrow enough. We will return to this point brie y in Chapter 5.

The constitutive relation in a *non-ideal dielectric medium* (i.e., in a medium with conductivity $\neq 0$), in the time-harmonic regime, can be expressed in a more compact way, by de ning a *total current density* vector (A/m), as the sum of the conduction (or convection) current density and of the displacement current density:

$$\mathbf{J}_t = \mathbf{J} + \frac{\partial \mathbf{D}}{\partial t} \ , \tag{1.18}$$

whose corresponding complex vector is:

$$\mathbf{J}_t = \mathbf{J} + j\omega\,\mathbf{D} \ . \tag{1.19}$$

Then, in a linear isotropic medium, from Eqs. (1.15) and (1.16), we get the following constitutive relation:

$$\mathbf{J}_t = (\ + j\omega \ )\,\mathbf{E} \ . \tag{1.20}$$

We can further elaborate on this relation, to make it formally identical to that which applies to an ideal dielectric ($= 0$) where the right-hand side of Eq. (1.20) is reduced to $j\omega$ $\mathbf{E}$. We introduce the so-called *complex permittivity*:

$$_c = \ \ j \frac{}{\omega} \ . \tag{1.21}$$

Table 1.1 *Index of refraction and test wavelength, for selected materials.*

Material	n	*Wavelength*
Air	1.000294	589 nm
Distilled water	1.333	589 nm
Aqueous sucrose (59% soln)	1.544	589 nm
Fused silica	1.46	633 nm
Glass (zinc crown)	1.517	589 nm
Glass (extra dense int)	1.92	589 nm
Gallium Arsenide	3.37	1.15 m
" "	3.34	10.6 m
Indium Antimonide (InSb)	3.95	10.6 m
Copper	0.16 j3.37	633 nm
Gold	0.16 j3.21	633 nm
Silver	0.067 j4.05	633 nm

See Weast (1988-89), p. E-383, for wavelength dependence.

Using this de nition, Eq. (1.20) can be written as:

$$\mathbf{J}_t = j\omega \ _c \mathbf{E} \quad . \tag{1.22}$$

The usefulness of the last equation will become evident later, when we discuss how to elaborate on Maxwell's equations and in particular Eq. (1.9).

For *passive* media, i.e., media that can not contribute energy to a wave propagating therein, we have 0. Thus, the quantity expressed by Eq. (1.21) belongs to the fourth quadrant of the complex plane. A widely used notation is $_c = (' \ j '') \ _0$, where $_0 = (1/36 \ ) \ 10 \ ^9$ F/m is the electric permittivity of free-space. The quantity $= \arctan(''/ ')$ is called the *loss angle* of the dielectric; the corresponding $''/ '$ is called the loss tangent.

Table 1.1 shows values of the refractive index $n = (_r \ _r)^{1/2}$, where $_r = / _0$ is the relative magnetic permeability, and $_r = / _0$ is the relative dielectric permittivity, for some materials which are of particular signi cance at optical or infrared frequencies. In a few cases, where the loss tangent is very signi cant in practical applications, the relative dielectric permittivity is de ned as $_r = _c/ _0$, and, consequently, the refractive index is complex. For more extensive tables, the reader is referred to other books, like Weast (1988-89) for a general coverage, Yariv and Yeh (1984) for a more extensive list of optical materials and acousto-optical materials, and Guenther (1990) for various thin- lm materials.

As for the magnetic permeability, , in isotropic media of technical interest for e.m. waves, it is not signi cantly di erent from the free-space value, $_0 = 4 \quad 10 \ ^7$ H/m. (For a list of examples, see again Guenther, 1990). Signi cant examples for another class of media, to be de ned now, will be given in Chapter 6.

Before closing this section, let us brie y mention *linear anisotropic media*,

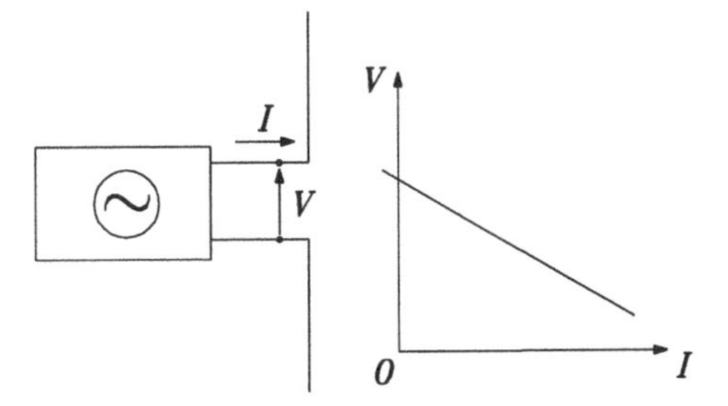

Figure 1.2 *Schematic representation of a transmitting antenna.*

in which linear relations, analogous to Eqs. (1.15)–(1.17) still hold, but are expressed in terms of *dyadic operators* (Collin, 1991; Morse and Feshbach, 1953; Nye, 1985) which are not proportional to the identity dyadic, instead of in terms of scalars:

$$\mathbf{D} = {}^{-}\mathbf{E} \; , \tag{1.23}$$

$$\mathbf{J} = {}^{-}\mathbf{E} \; , \tag{1.24}$$

$$\mathbf{B} = {}^{-}\mathbf{H} \; . \tag{1.25}$$

We will devote the whole of Chapter 6 to such media and to their applications. However, many of the theorems to be proved in Chapter 3 are valid also for anisotropic media, and that will be pointed out in all cases where it is indeed the case. Where, on the other hand, the opposite is not explicitly stated, in the following, it will be understood that we are referring just to isotropic media.

Finally, it is worth mentioning that there are media, not without practical importance (for instance, in propagation in non-inertial reference frames), which are again linear, but still more complex than the anisotropic ones. They are the so-called linear *bi-anisotropic media*, whose characteristic is that each of the vectors on the left of Eqs. (1.23)–(1.25) is a linear function of *both* vectors **E** and **H**. For problems of e.m. wave propagation in such media, the reader should consult specialized texts.

1.4 Imposed currents

We have written the fundamental equations; now, we should specify which terms are known, and which are unknown. The concept of "known terms" is straightforward in a mathematical sense; but if we refer to physical reality, then we see that it is, to a large extent, an abstraction, as none of the quantities involved in our equations can be thought of as being rigorously predetermined, independent of the others. Let us try to clarify this statement with an example.

Figure 1.2 represents a transmitting antenna, i.e., a metallic wire fed by a generator, and surrounded by a dielectric. If we are given the voltage-current characteristic of the generator (Figure 1.2), then, from elementary circuit theory, we are led to believe that the generator imposes a well de ned, single-valued current in the wire. But, in reality, the electromagnetic eld which surrounds the wire "reacts" on the antenna, that is, it a ects the current distribution in the wire. In turn, the eld depends on the environment around the antenna. Strictly speaking, thus, the choice of a quantity like the current in the wire as a known term, is not justi able.

On the other hand, a *mathematical* distinction between known and unknown terms is absolutely necessary, to proceed further. The way out of this dilemma consists of promoting to the rank of "known terms" those, among the physical quantities involved in a speci c problem, whose distribution can either be "guessed" to a good approximation before commencing the calculation of the other quantities, or is easier than the others to determine experimentally. Usually, this is the case for quantities which are restricted to a region in space much smaller than that in which we are interested in knowing the whole eld. Usually, the electric current density satis es these requirements, being de ned only within the regions in space where the "sources" of the e.m. eld are located. In this spirit, and with these limitations, we will use the term *"imposed" current density* for known quantities, ($\mathbf{J}_0$ or, in the harmonic-regime complex-vector notation, $\mathbf{J}_0$).

In introductory classes, the notion of *imposed electric eld*, $\mathbf{E}_0$, and that of imposed current density, $\mathbf{J}_0$, are often presented as perfectly equivalent to each other, and interchangeable through Ohm's law. This equivalence exists as long as one speci es unambiguously the conductivity of the medium in which these quantities are de ned, so that:

$$\mathbf{J}_0 = \quad \mathbf{E}_0 \ . \tag{1.26}$$

But this approach is probably not the best one for electrodynamics. Indeed, Maxwell's equations (1.6) and (1.7) show that the elds $\mathbf{E}$ and $\mathbf{H}$ do not change, if a given conduction current density is replaced by a convection (or displacement) current density, as long as they are numerically identical. Eq. (1.26), in electrodynamics, might be misleading. The choice of a current density, rather than of an electric eld, as a known term, is encouraged by the fact that many procedures for solving Maxwell's equations apply only to elds in *homogeneous media*, i.e., elds de ned in media where the constitutive relations are constant in space. If we take an imposed electric eld as known term, then we need to assume that in the source-containing region (in the example, inside the antenna) Eq. (1.26) holds, with a well-de ned value for the conductivity . This could be (and normally is) in contrast with the requirement of a homogeneous medium.

Consequently, hereafter, unless the opposite is stated, we will use Maxwell's equations in the form:

$$\nabla \quad \mathbf{E} = \quad \frac{\partial \mathbf{B}}{\partial t} \ , \tag{1.27}$$

$$\nabla \quad \mathbf{H} = \mathbf{J} + \frac{\partial \mathbf{D}}{\partial t} + \mathbf{J}_0 \quad , \tag{1.28}$$

with the addition of the continuity equation:

$$\nabla \ (\mathbf{J} + \mathbf{J}_0) = \quad \frac{\partial}{\partial t} \quad . \tag{1.29}$$

$\mathbf{J}_0$ (called the *imposed current density*) is a known term, whereas **E**, **D**, **J**, **H** and **B** are the unknowns, related among themselves by the constitutive relations of Section 1.3.

For the complex representation of the time-harmonic case, we can write Maxwell's equations in the form:

$$\nabla \quad \mathbf{E} = \quad j\omega \mathbf{B} \quad , \tag{1.30}$$

$$\nabla \quad \mathbf{H} = \mathbf{J} + j\omega \mathbf{D} + \mathbf{J}_0 \quad , \tag{1.31}$$

with the addition of the continuity equation:

$$\nabla \ (\mathbf{J} + \mathbf{J}_0) = \quad j\omega \quad . \tag{1.32}$$

Again, $\mathbf{J}_0$ is known, while **E**, **H**, **D**, **B** and **J** are unknowns, related by the constitutive relations.

We note in particular that in *linear media*, both isotropic and anisotropic, it is straightforward to eliminate the vectors **B**, **D** and **J** from Eqs. (1.30) and (1.31), or from Eqs. (1.27) and (1.28). This results in a system of *two linear equations in two unknowns*, **E** and **H**.

Hereinafter, the term *e.m. eld* will refer to a pair of point functions (or generalized functions) $\{\mathbf{E}, \mathbf{H}\}$ satisfying Eqs. (1.30) and (1.31).

The mathematical problem is now well de ned. On the other hand, there is no guarantee that its solution e ectively matches physical reality. This basic target can be reached only if the choice of the known term in the preceding equation, $\mathbf{J}_0$, is a realistic one.

1.5 Divergence equations

Before tackling the mathematical problem of solving Maxwell's equations, i.e., of evaluating an e.m. eld once its sources are known, it is useful to derive further relations between the quantities which we de ned so far. These new relations are not independent of those that we took as postulates in the previous sections; thus they do not modify the balance between equations and unknowns. Nevertheless, they may be useful in that they present the same information in a di erent way. For example, they may provide useful tests to discriminate spurious solutions from acceptable ones, when one solves Maxwell's equations with numerical methods.

Let us start from Maxwell's equations, for example in the form in Eqs. (1.27) and (1.28), or in the equivalent form for complex vectors, and combine each of these with the vector identity of Eq. (C.5), $\nabla \quad \nabla \quad \mathbf{a} = 0$. From Eq. (1.27) we have:

$$\nabla \quad \left(\frac{\partial \mathbf{B}}{\partial t}\right) = 0 \quad . \tag{1.33}$$

The operators ∇ and $\partial/\partial t$ commute whenever $\mathbf{B}$ satis es well-known requirements of regularity (partial di erentiability twice, and continuity of these derivatives). It can be shown that the same is true also in the case of generalized functions of the type illustrated in Section 1.1. Therefore, in Eq. (1.33) we can always put $\nabla \quad \partial/\partial t = \partial/\partial t\,(\nabla \quad)$. Then, since whatever e.m. eld of technical interest is identically zero before an initial instant, we can conclude that:

$$\nabla \quad \mathbf{B} = 0 \quad . \tag{1.34}$$

For the case of the complex-vector representation of a time-harmonic vector, we obviously obtain a similar relation:

$$\nabla \, \mathbf{B} = 0 \quad . \tag{1.35}$$

In the case of linear isotropic media, substituting Eq. (1.17) into Eq. (1.35) we immediately get $\nabla \,(\quad \mathbf{H}) = 0$. If moreover is constant in space (i.e., if the medium is *homogeneous*), we obtain:

$$\nabla \, \mathbf{H} = 0 \quad . \tag{1.36}$$

The property in Eqs. (1.34), or (1.35), i.e., the fact that the magnetic induction eld is a solenoidal eld, is linked, as the reader should know from time-independent magnetic elds, to the fact that magnetic monopoles do not exist.[†]

Let us now apply again the identity Eq. (C.5), $\nabla \quad \nabla \quad \mathbf{a} = 0$ to Eq. (1.28), taking account of the charge continuity equation (1.29). With straightforward operations, identical to the preceding ones, we obtain:

$$\nabla \, \mathbf{D} = \quad . \tag{1.37}$$

In this relation between the displacement vector and the charge density, the reader familiar with time-independent elds will recognize the di erential form of *Gauss' law.*

Using the complex representation of time-harmonic variables, we get:

$$\nabla \, \mathbf{D} = \quad . \tag{1.38}$$

That identity, like the others that will be cited later in this text, are reported in Appendix C, and were derived under the assumption that one deals with a *regular* function of spatial coordinates. The interested reader can extend the proof of these identities to generalized functions, provided these satisfy the conditions given in Section 1.1, i.e., are de ned as the limiting case of regular functions.

[†] No magnetic charges have been found so far in matter as we know it. While a development in the theory of elementary particles suggested that the universe ought to contain at least a few magnetic monopoles left over from the "big bang," not one magnetic monopole has yet been detected, and it is evident that if they exist at all they are exceedingly rare, e.g., see Section 11.2 of Purcell (1985).

In the case of linear isotropic media, (i.e., if the constitutive relation is given by Eq. (1.15)) as well as *homogeneous* (i.e., if is independent of the spatial coordinates) from Eq. (1.38) it follows that:

$$\nabla \ \mathbf{E} = - \quad . \tag{1.39}$$

In the time-harmonic case, we can develop a further result, making use of the complex permittivity in Eq. (1.21). Let us start from the complex form of the second Maxwell's equation which, in a linear isotropic medium, can be written as:

$$\nabla \quad \mathbf{H} = j\omega \ _c \mathbf{E} + \mathbf{J}_0 \quad . \tag{1.40}$$

Again applying the identity $\nabla \ \nabla \quad \mathbf{a} = 0$, we immediately see that the divergence of the entire right-hand side of Eq. (1.40) is zero. If the medium is *homogeneous* ($_c$ = constant), this result can be put in the following form:

$$\nabla \ \mathbf{E} = \frac{\nabla \ \mathbf{J}_0}{j\omega \ _c} \quad . \tag{1.41}$$

This means that, inside a homogeneous medium, the divergence of the electric eld can di er from zero only either where the ow lines of the imposed current are open, or at the boundary of the medium. Comparing Eqs. (1.39) and (1.41) we reach another physical interpretation of this fact, via the equality:

$$\nabla \ \mathbf{J}_0 = \quad j\omega \frac{\ _c}{\ } \quad , \tag{1.42}$$

which corresponds to the following, rather intuitive reasoning: in the time-harmonic regime, and within a homogeneous medium, the free-charge density (which is also varying sinusoidally in time), is nonzero only at points at which the *imposed* current ux lines are open. If there are no sources with open lines of current, the ux lines of the total current (conduction plus displacement), within a homogeneous medium, are necessarily closed; they can begin or end only at boundaries which separate di erent media.

It is worth mentioning that, for historical reasons, in many texts the divergence equations, normally in the forms of Eqs. (1.35) and (1.38), are also referred to as Maxwell's equations, like Eqs. (1.30) and (1.31). We prefer to use distinct names, to underline the di erent weightings of the four equations in the axiomatic theory. While Maxwell's curl equations are essential ingredients in the set of fundamental postulates, the divergence equations are not independent of the former.

1.6 Continuity conditions

When dealing with static or slowly-varying e.m. elds, readers were told that one can obtain continuity conditions for the tangential components of the vectors **E** and **H**, across an abrupt discontinuity of either , or , or , by suitably handling the integral relations in Eqs. (1.3) and (1.4). Similarly, it is well known that the integral form that can be given, via Gauss' theorem

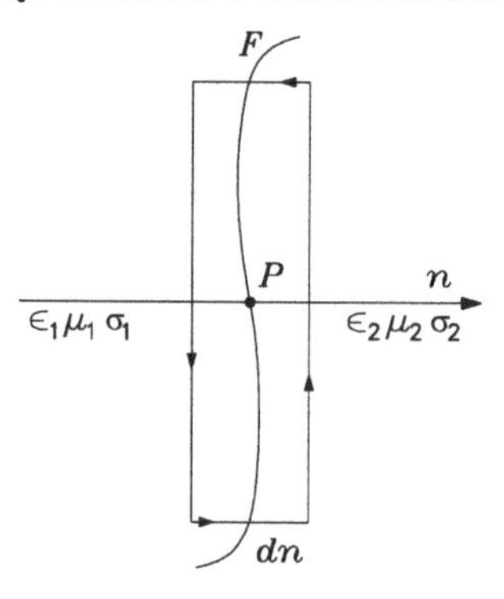

Figure 1.3 *Derivation of the continuity conditions.*

(see Appendix A), to the divergence equations (1.34) and (1.37), can yield continuity conditions for the normal components of the vectors **B** and **D**, again at a sharp discontinuity either in or in .

The aim of this section is to see under which assumptions the continuity conditions on the above-mentioned vectors can be derived from Maxwell's equations, at abrupt changes in the parameters of the medium. Without any possible exception, our fundamental postulates entail that there can not be any discontinuity for the elds inside a medium where the parameters vary continuously.

Let P be any point on the boundary between two di erent media, F, that we suppose to be a regular surface (see Figure 1.3). The restrictions imposed in Section 1.2, where we established under which assumptions generalized functions were admitted in our theory, entitle us to state now that, along the normal to F at P, the right limit and the left limit of $\nabla\ \ \mathbf{E}$ and $\nabla\ \ \mathbf{H}$ exist, and that both of them are nite. The same can be said for the corresponding complex vectors. If (and only if) the right and the left limits coincide, then $\nabla\ \ \mathbf{E}$ and $\nabla\ \ \mathbf{H}$ are regular at P, and the same is true for the right-hand sides of the corresponding Maxwell's equations. Therefore, at that point (or better, in a region that contains only such points) Stokes' theorem (see Appendix A) holds. Let us apply it to a path of the type illustrated in Figure 1.3, around an area dS, in nitesimal in that the length dn of the sides normal to F tends to zero. It leads us to the following *continuity conditions*:

$$\mathbf{E}_{\tan 1} = \quad \mathbf{E}_{\tan 2} \ \ (\mathbf{E}_{\tan 1} = \mathbf{E}_{\tan 2}) \quad , \tag{1.43}$$

$$\mathbf{H}_{\tan 1} = \quad \mathbf{H}_{\tan 2} \ \ (\mathbf{H}_{\tan 1} = \mathbf{H}_{\tan 2}) \quad , \tag{1.44}$$

where the subscript "tan" denotes the vector component tangential to the surface F at P, while the subscripts 1 and 2 denote values on sides 1 and 2, respectively, of the surface F.

If, on the contrary, at a point P the right and the left limits of the curl do not coincide, then the curl is de ned there only as a generalized function. In that case, also the quantity on the other side of the corresponding Maxwell's equation is a generalized function, which takes an infinite value at P. Let us check whether such a behavior is compatible with the assumption that we made in Section 1.2, of interpreting the abrupt discontinuity of a parameter of the medium as the limit of a sequence of smooth transitions, parameterized by a characteristic length , which is gradually reduced to zero. Suppose that, for $\rightarrow 0$, we have $|\mathbf{B}(P)| \rightarrow \infty$. This implies, via the constitutive material relations, $|\mathbf{H}(P)| \rightarrow \infty$; and furthermore, that happens only at points P which are located on F. Consequently, partial derivatives of $\mathbf{H}$ with respect to some coordinates must also diverge to infinity around P; this is in contrast to what we said previously, regarding the right and left limits of $\nabla \quad \mathbf{H}$ being both finite. A similar proof "by contradiction" holds, if we suppose that, for $\rightarrow 0$, the vector $|\mathbf{D}|$ diverges.

In contrast, such a proof does not apply to the imposed current density ($\mathbf{J}_0$), whose magnitude may tend to infinity without any untenable consequences for other electromagnetic quantities. It is not even applicable to the conduction current density $\mathbf{J}$, provided we accept that there may be media with infinite conductivity, an abstraction whose usefulness will be apparent often, in subsequent chapters. In all these cases, the current density (either imposed or not) becomes a two-dimensional distribution, i.e., a generalized function of the coordinate perpendicular to the surface F.

We can conclude that Eq. (1.43) *is always valid* (postponing to Section 3.5 a discussion of the consequences of introducing "magnetic currents"), while, on the contrary, Eq. (1.44) is valid *on all surfaces on which there are no surface currents* (either imposed or not).

Let us discuss now continuity of the normal components of the vectors $\mathbf{B}$ and $\mathbf{D}$ in the divergence equations (1.34) and (1.37), or in the corresponding equations for complex vectors. We know that right and left limits exist for $\nabla \quad \mathbf{B}$ and $\nabla \quad \mathbf{D}$. Then, applying Gauss' divergence theorem (see Appendix A), we obtain the following relation, where the subscript n indicates the component normal to F, whereas the meaning of the subscripts 1 and 2 remains the same as above:

$$B_{\mathrm{n}1} = B_{\mathrm{n}2} \qquad (B_{\mathrm{n}1} = B_{\mathrm{n}2}) \quad , \tag{1.45}$$

which is *always* valid; and:

$$D_{\mathrm{n}1} = D_{\mathrm{n}2} \qquad (D_{\mathrm{n}1} = D_{\mathrm{n}2}) \quad , \tag{1.46}$$

which holds at *all points of F where there are no free surface charges*, i.e., there is no current density represented by a generalized function of the coordinate normal to F.

1.7 The wave equation. The Helmholtz equation

In a time-invariant, linear, isotropic medium (leaving for Chapter 6 the case of an anisotropic medium), the constitutive relations enable us to modify

Maxwell's equations in the time domain, as follows:

$$\boldsymbol{\nabla} \quad \mathbf{E} \quad = \quad \frac{\partial \mathbf{H}}{\partial t} \ , \tag{1.47}$$

$$\boldsymbol{\nabla} \quad \mathbf{H} \quad = \quad \frac{\partial \mathbf{E}}{\partial t} + \quad \mathbf{E} + \mathbf{J}_0 \quad . \tag{1.48}$$

This system of two rst-order linear partial di erential equations (PDE's) is not separated: each equation involves partial derivatives of both unknowns. An elementary approach to their solution may start with separation of the two unknowns, at the cost of increasing the order of the equations. This operation is very simple when , and are constants, i.e., for a *homogeneous medium.* First, we calculate the curl of both sides of Eq. (1.48); on the right-hand side, we make use of the fact that the operators $\boldsymbol{\nabla}$ and $\partial/\partial t$ commute, under our assumptions on regularity of the unknown functions. Then, we replace the curl of $\mathbf{E}$ in the second term by Eq. (1.47), and nally obtain an equation in only one unknown, $\mathbf{H}$, namely:

$$\boldsymbol{\nabla} \quad \boldsymbol{\nabla} \quad \mathbf{H} + \quad \frac{\partial^2 \mathbf{H}}{\partial t^2} + \quad \frac{\partial \mathbf{H}}{\partial t} = \boldsymbol{\nabla} \quad \mathbf{J}_0 \quad . \tag{1.49}$$

It contains partial derivatives of second order with respect to both time and the spatial coordinates. Recalling the identity in Eq. (A.7) for the Laplacian of a vector, $\nabla^2 \mathbf{a} = \nabla(\nabla \quad \mathbf{a}) \quad \boldsymbol{\nabla} \quad \boldsymbol{\nabla} \quad \mathbf{a}$, and taking account of the fact that, as we saw in Section 1.5, in a homogeneous medium we always have $\boldsymbol{\nabla} \quad \mathbf{H} = 0$, Eq. (1.49) can be rewritten as follows:

$$\nabla^2 \mathbf{H} \quad \frac{\partial^2 \mathbf{H}}{\partial t^2} \quad \frac{\partial \mathbf{H}}{\partial t} = \quad \boldsymbol{\nabla} \quad \mathbf{J}_0 \quad . \tag{1.50}$$

This is easier to handle than Eq. (1.49), for reasons which will be explained later.

If instead we calculate the curl of both sides of Eq. (1.47) and then we replace the right-hand side with Eq. (1.48), we obtain the following equation where $\mathbf{E}$ is the only unknown:

$$\boldsymbol{\nabla} \quad \boldsymbol{\nabla} \quad \mathbf{E} + \quad \frac{\partial^2 \mathbf{E}}{\partial t^2} + \quad \frac{\partial \mathbf{E}}{\partial t} = \quad \frac{\partial \mathbf{J}_0}{\partial t} \quad . \tag{1.51}$$

Apart from physical dimensions, Eq. (1.51) di ers from Eq. (1.49) only on the right-hand side. We may obtain an equation whose left-hand side is formally the same as that of Eq. (1.50) only if there is no free charge, i.e., when Eq. (1.39) reduces to $\nabla \quad \mathbf{E} = 0$. We then have:

$$\nabla^2 \mathbf{E} \quad \frac{\partial^2 \mathbf{E}}{\partial t^2} \quad \frac{\partial \mathbf{E}}{\partial t} = \quad \frac{\partial \mathbf{J}_0}{\partial t} \quad . \tag{1.52}$$

As anticipated earlier, separation of the two unknowns has increased the order of these di erential equations. Consequently, the system consisting of Eqs. (1.50) and (1.52) has more solutions than Maxwell's equations. However,

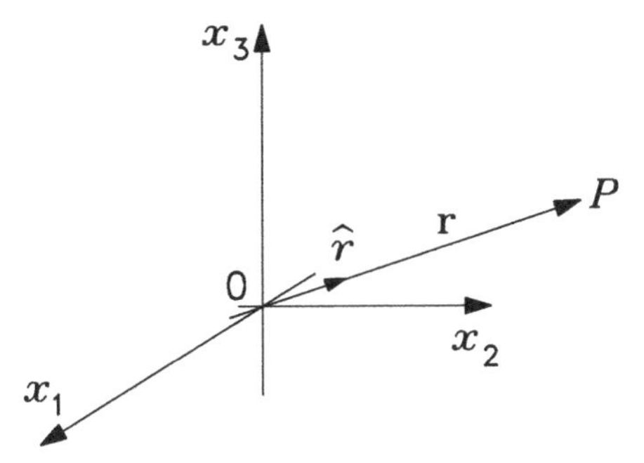

Figure 1.4 *Reference frame adopted in the 3-D real space.*

solutions of the latter only have the physical signi cance of e.m. elds. Therefore, whenever Eqs. (1.50) and (1.52) are used, caution is necessary to avoid spurious solutions; usually, it is enough simply to solve only one of the two Eqs. (1.50) and (1.52), verify that the solution obtained in this way satis es the conditions $\boldsymbol{\nabla}\ \mathbf{H} = 0$ (or $\boldsymbol{\nabla}\ \mathbf{E} = 0$), and then calculate the other eld via one of Maxwell's equations.

The physical sense of Eq. (1.50) or, equivalently, of Eq. (1.52), is simple to illustrate in a region where there are no sources. To make this argument as simple as possible, let us rst suppose that the medium in this region is an ideal dielectric, i.e., has zero conductivity ($= 0$). The two equations become then of the following form:

$$\nabla^2 \mathbf{a} \qquad \frac{\partial^2 \mathbf{a}}{\partial t^2} = 0 \quad , \tag{1.53}$$

that is usually called the *vector wave equation* (or the D'Alembert's vector equation). The *scalar* wave equation, that di ers from Eq. (1.53) only because the unknown vector $\mathbf{a}$ is replaced by an unknown scalar , is one of the most studied equations in mathematical physics (Morse and Feshbach, 1953). Its solutions, referred to as *waves* or *wave functions*, can have quite di erent physical dimensions and meanings, depending on the context. By analogy with terminology, which has been in use in other branches of physics for about two centuries, the expression *electromagnetic wave* is often taken as synonymous with *electromagnetic eld*, in the fast time-varying regime. We will accept that, in spite of the fact that the validity of Eq. (1.53) is subject to very limiting assumptions, that were introduced along the way in order to obtain it from Maxwell's equations.

This terminology may be justi ed in intuitive terms, insofar Eq. (1.53) is satis ed by any regular vector function of the following type, corresponding to the intuitive notion of propagation:

$$\mathbf{a}\ (P, t) = \mathbf{a}\ (\mathbf{r} \quad \hat{\mathbf{r}}\, ct) \quad , \tag{1.54}$$

where $\mathbf{r} = P \quad 0 = (x_1 \hat{a}_1 + x_2 \hat{a}_2 + x_3 \hat{a}_3)$ denotes the position vector of the point

$P = (x_1, x_2, x_3)$, with respect to an arbitrarily chosen origin 0 (Figure 1.4); $\hat{\mathbf{r}} = \mathbf{r}/|\mathbf{r}|$ denotes the unit vector in the direction of $\mathbf{r}$. We set:

$$c = 1/\sqrt{} \quad . \tag{1.55}$$

This quantity has the dimensions of *velocity*; for reasons that will be outlined later, it is referred to as the *speed of light* in the medium characterized by the parameters and . In particular, in free-space, where $_0 = 4 \quad 10 \;^7\,\mathrm{H/m}$ and $_0 = (1/36\ ) \quad 10 \;^9\,\mathrm{F/m}$, we have $c = c_0 = 1/\sqrt{\ _0\ _0} = 3 \quad 10^8\,\mathrm{m/s}$.[‡] Eq. (1.54) describes, as we just said, a *propagation* phenomenon, i.e., a function that travels unchanged, in the direction $\mathbf{r}$, with velocity c.

Eqs. (1.50) and (1.52) are very similar to those which describe oscillations. This suggests that the term $\partial \mathbf{a}\ /\partial t$ represents a *damping* of the traveling function, as it propagates in space.

As for the time-harmonic regime, we leave it to the reader to prove, similarly to what we have just carried out, that the complex vectors, in a homogeneous medium, satisfy the following equations:

$$\nabla^2\mathbf{E} + \omega^2 \quad \mathbf{E} \quad j\omega \quad \mathbf{E} = j\omega\ \mathbf{J}_0 \quad , \tag{1.56}$$

$$\nabla^2\mathbf{H} + \omega^2 \quad \mathbf{H} \quad j\omega \quad \mathbf{H} = \quad \nabla \quad \mathbf{J}_0 \quad . \tag{1.57}$$

Using the complex permittivity, the left-hand sides of these two equations become not only more compact, as the third term gets encompassed into the second one, but also formally independent of whether the medium is lossless ($= 0$) or lossy ($\neq 0$). Let:

$$^2 = \quad \omega^2 \quad _c \quad , \tag{1.58}$$

and let us establish that $\ = \ + j\ $ is that root of $\ ^2$ which lies in the rst quadrant of the complex plane (see Figure 1.5). We conclude then that, in a source-free region, both $\mathbf{E}$ and $\mathbf{H}$ satisfy the equation:

$$\nabla^2\mathbf{a} \quad ^2\mathbf{a} = 0 \quad , \tag{1.59}$$

usually referred to as the *homogeneous vector Helmholtz equation.*

Like the wave equations (1.50) and (1.52), Eqs. (1.56) and (1.57) do not form a system equivalent to Maxwell's equations. Indeed, the increase in order of the derivatives has introduced spurious solutions. To retain only the correct solutions, one should proceed as we said after deriving Eqs. (1.50) and (1.52).

1.8 Magnetic vector potential

As we saw in the previous section, to eliminate one unknown from Maxwell's equations one has to pay the cost of increasing the equation order. Nevertheless, decreasing the number of unknowns in the problem is still a useful, if not

[‡] In contrast to the one-dimensional case (where $\mathbf{a}$ is a function of only one spatial coordinate and of time), in the two- and three-dimensional cases Eq. (1.54) is not the general integral of Eq. (1.53).

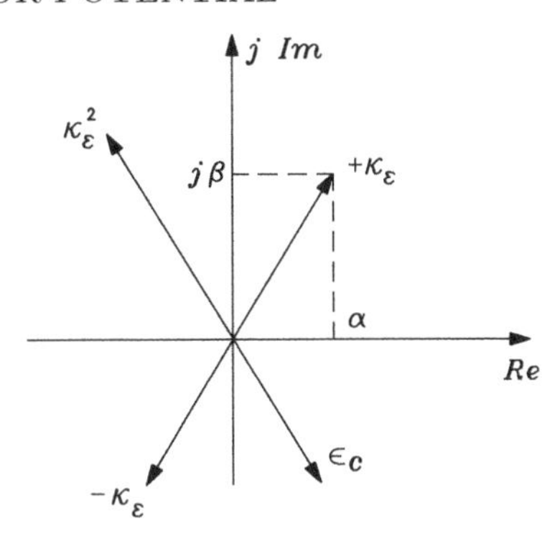

Figure 1.5 *The complex permittivity $_c$, the quantity ^{2}, and its square roots, on the complex plane.*

a necessary, step to begin with. To this aim, a widely used approach consists of introducing auxiliary variables, which are usually referred to as *scalar and vector electromagnetic potentials.* They are not necessarily measurable with physical instruments. Still, they are useful, because they allow one to obtain more compact mathematical formulations of the problem, where the number of unknowns is smaller than in Maxwell's equations. This section and the following ones will be devoted to the most popular potentials in e.m. wave theory. To avoid unnecessarily cumbersome algebra, we will only develop the time-harmonic case, in terms of complex vectors.

First, let us prove that, in a linear isotropic medium, there exists a vector function of spatial coordinates, **A**, called the *magnetic vector potential,* such that both the electric eld **E** and the magnetic eld **H** can be calculated from **A**, and from the known term $\mathbf{J}_0$. The starting point is the divergence equation (1.35). A well-known theorem from vector analysis tells that, as a consequence, there exists a vector **A**, such that:

$$\mathbf{H} = (1/\ )\ \nabla\ \ \mathbf{A}\ \ . \tag{1.60}$$

Now, Maxwell's equation:

$$\nabla\ \ \mathbf{H} = j\omega\ _c\mathbf{E} + \mathbf{J}_0 \tag{1.61}$$

can be easily rewritten with **E** on the left-hand side; hence, this vector, as well, can be expressed in terms of **A** and $\mathbf{J}_0$, as we intended to prove. In particular, if the medium is magnetically homogeneous ($=$ constant), the following expression holds:

$$\mathbf{E} = \frac{1}{j\omega\ \ _c}(\nabla\ \ \nabla\ \ \mathbf{A}\ \ \ \ \mathbf{J}_0)\ \ . \tag{1.62}$$

If the medium is also electrically homogeneous ($_c =$ constant), it is easy to nd a di erential equation which **A** must satisfy. Substituting Eqs. (1.60) and (1.62) in the other Maxwell's equation:

$$\nabla \quad \mathbf{E} = \quad j\omega \quad \mathbf{H} \quad , \tag{1.63}$$

we in fact obtain:

$$\nabla \quad (\nabla \quad \nabla \quad \mathbf{A} + \quad {}^2\mathbf{A} \quad \mathbf{J}_0) = 0 \quad , \tag{1.64}$$

where $\ ^2$ is the quantity de ned by Eq. (1.58).

Eq. (1.64) shows that we have been able again to reduce the number of unknowns (from the two vectors **E** and **H** to the vector **A**) only. However, also in this new approach we have paid the price of increasing the order of the partial di erential equation, from one to three. This increase can be easily restricted to two, rather than three, provided the region where the eld is de ned is a *simply connected domain.* In this case, in fact, another well-known theorem from vector analysis (see any text on mathematical physics, e.g., Boas, 1983, or on vector calculus, e.g., Marsden and Tromba, 1988, or Schey, 1973) ensures that the expression in round brackets on the left-hand side of Eq. (1.64), whose curl is zero, is a *conservative eld.* Therefore, it can be set equal to the gradient of a scalar function. Just for a matter of dimensions, it is convenient to put:

$$\nabla \quad \nabla \quad \mathbf{A} + \quad {}^2\mathbf{A} \quad \mathbf{J}_0 = \quad j\omega \quad {}_c \nabla\varphi \quad , \tag{1.65}$$

where $\varphi(\mathbf{r})$ is a *completely arbitrary* scalar function of the spatial coordinates. We will refer to it as the *scalar electric potential* associated with the vector magnetic potential $\mathbf{A}(\mathbf{r})$.§

The di erential equation for **A** has been reduced to second order. Moreover, replacing from Eq. (1.65) into Eq. (1.62), we may obtain a new expression for the electric eld **E**:

$$\mathbf{E} = \quad j\omega\mathbf{A} \quad \nabla\varphi \quad . \tag{1.66}$$

This reminds us of the fundamental equation of electrostatics $\mathbf{E} = \quad \nabla\varphi$, at least as far as the dimensions are concerned. Incidentally, Eq. (1.66) allows us to verify that the physical dimensions of the magnetic vector potential are:

$$[\mathbf{A}] = [\mathbf{E}]/[\omega] = \mathrm{V} \quad \mathrm{s} \quad \mathrm{m} \quad {}^1 = \mathrm{Wb} \quad \mathrm{m} \quad {}^1. \tag{1.67}$$

The reader may be surprised by the complete arbitrariness of the scalar function φ; but this can be fully justi ed, as follows. Let us calculate the divergence of both sides of Eq. (1.65). Using the vector identity (C.5), $\nabla \quad \nabla \quad \mathbf{a} = 0$, we obtain:

$$\nabla \quad \mathbf{A} = \quad (1/\omega^2 \quad {}_c)\, \nabla \quad \mathbf{J}_0 \quad (1/j\omega)\, \nabla^2\varphi \quad . \tag{1.68}$$

§ If the region where the e.m. eld is de ned is not simply connected, Eq. (1.64) can be satis ed also by vectors that do not satisfy Eq. (1.65). However, this is not a major drawback, in our context. Consider, for example, a coaxial cable of in nite length: we can subdivide the region between the conducting cylinders into two (or more) simply-connected regions. In each of them, one is permitted to pass to Eq. (1.65). Fields calculated in this fashion can then be matched by continuity, at the interfaces between neighboring regions. Therefore, in the following we will consider that Eq. (1.65) is *always* satis ed, without further distinguishing (unless the opposite is stated) between simply and non-simply connected regions.

Before writing Eq. (1.65), the function $\mathbf{A}$ always appeared within the argument of a curl operator; we mentioned its *existence*, but we never spoke of uniqueness. Until that point, we were allowed to modify arbitrarily the divergence of $\mathbf{A}$, without inducing any change in the elds $\{\mathbf{E}, \mathbf{H}\}$ given by Eqs. (1.62) and (1.60). But now, as Eq. (1.68) shows, once the function φ has been speci ed in a unique way, then $\nabla \ \mathbf{A}$ is also determined in a unique way. A classical theorem of vector analysis, known as Helmholtz's theorem, says that a vector eld is uniquely determined, once its curl and its divergence are both uniquely de ned in all its domain of de nition. We conclude then that any choice of φ automatically eliminates the arbitrariness that was still contained until then in the vector function $\mathbf{A}$. Note that this statement can not be reversed. Indeed, if one chooses the divergence of $\nabla \ \mathbf{A}$, one does not get a uniquely de ned φ, because Eq. (1.68) has to be looked at as a di erential equation in an unknown φ.

Let us now discuss brie y some of the most classical choices, or *gauges*, of the function φ, and their main implications.

1.8.1 $\varphi = 0$ *gauge*

At rst sight, one may think that $\varphi = 0$ is the simplest choice. Let $\mathbf{A}_0$ denote the corresponding vector magnetic potential. Eq. (1.65) reduces to:

$$\nabla \quad \nabla \quad \mathbf{A}_0 + \ {}^2\mathbf{A}_0 = + \quad \mathbf{J}_0 \quad , \tag{1.69}$$

whereas Eq. (1.68) becomes:

$$\nabla \ \mathbf{A}_0 = \quad (1/\omega^2 \ _c) \nabla \ \mathbf{J}_0 \quad . \tag{1.70}$$

In the following, and particularly in Chapters 4 to 7, we will see that, in cartesian coordinates, it is fairly straightforward to solve the vector Helmholtz equation (1.59) via its scalar components. This is not generally the case, on the contrary, for Eq. (1.69), since the operator $\nabla \quad \nabla$ is very cumbersome to express by cartesian components. Eq. (1.70) indicates that Eq. (1.69) can be reduced to a non-homogeneous Helmholtz equation if $\nabla \ \mathbf{J}_0 = 0$ everywhere. In this case, we have $\nabla \quad \nabla \quad \mathbf{A}_0 = \quad \nabla^2 \mathbf{A}_0$. Otherwise, the gauge $\varphi = 0$ appears to be rather impractical. Sources whose divergence is not vanishing everywhere are indeed a matter of practical concern; an antenna whose length is signi cantly shorter than the wavelength at which it operates is an important example in this sense (see Chapter 12).

On the other hand, note that, with this gauge, Eq. (1.66) becomes:

$$\mathbf{E} = \quad j\omega \, \mathbf{A}_0 \quad . \tag{1.71}$$

This indicates that the boundary conditions for the vector magnetic potential (which, in general, are di cult to link to physical intuition) are simple to express, in the $\varphi = 0$ gauge, by direct proportionality with the boundary conditions for $\mathbf{E}$. These, on the contrary, are very often quite intuitive (see, for example, all the problems where the boundary is a perfect conductor).

1.8.2 Coulomb gauge

There are at least two motivations which induce us to test another gauge. One is the similarity between Eq. (1.66) and the expression $\mathbf{E} = \quad \nabla\varphi$ that holds in electrostatics. The other one is the desire to explore what happens if we have $\nabla\ \mathbf{A} = 0$. Both suggest choosing φ as a solution of the equation which is satis ed by the electrostatic potential, i.e., the *Poisson equation*:

$$\nabla^2\varphi = \quad / \quad , \tag{1.72}$$

which, using Eq. (1.42), can also be written as:

$$\nabla^2\varphi = \frac{\nabla\ \mathbf{J}_0}{j\omega\ _c} \quad . \tag{1.73}$$

Eq. (1.68) shows that this is equivalent to imposing that the magnetic vector potential be a *solenoidal* eld, even when (and this makes a remarkable di erence with the previous gauge) the imposed currents have nonzero divergence:

$$\nabla\ \mathbf{A} = 0 \quad . \tag{1.74}$$

This transforms Eq. (1.65) into a nonhomogeneous Helmholtz equation:

$$\nabla^2\mathbf{A} \qquad {}^2\mathbf{A} = \qquad \mathbf{J}_0 + j\omega\ \ _c\nabla\varphi \quad . \tag{1.75}$$

As we said before, a Helmholtz equation often entails a signi cant simpli - cation. However, in this case, it turns out that there is a price to pay: we must rst solve the Poisson equation (1.72), because φ appears on the right-hand side of Eq. (1.75). This di culty is avoided by the Lorentz gauge, that we shall discuss below.

Eq. (1.72) has an in nite number of solutions; they can be obtained one from another using the rule:

$$\varphi' = \varphi + \varphi_0 \quad , \tag{1.76}$$

where φ_0 is any solution of *Laplace's equation*:

$$\nabla^2\varphi_0 = 0 \quad . \tag{1.77}$$

For any change of φ expressed by Eq. (1.76) there is a corresponding change of $\mathbf{A}$. Indeed, suppose that some suitable conditions (to be discussed in Section 3.3) assure the uniqueness of the e.m. eld corresponding to given sources. Then, any pair of potentials $(\varphi, \mathbf{A})$ must yield the same eld $\{\mathbf{E}, \mathbf{H}\}$. The transformation rule for $\mathbf{A}$ is then easy to derive from Eq. (1.66), imposing $\mathbf{E} = \quad j\omega\mathbf{A} \quad \nabla\varphi = \quad j\omega\mathbf{A}' \quad \nabla\varphi'$, with φ' given by Eq. (1.76). We get:

$$\mathbf{A}' = \mathbf{A} \qquad \frac{1}{j\omega}\ \nabla\varphi_0 \quad . \tag{1.78}$$

Eqs. (1.76) and (1.78) de ne a *Coulomb gauge transformation*.

1.8.3 Lorentz gauge

We may note that, for both gauges considered so far, the di erential equation for **A** contains, in addition to second partial derivatives with respect to the coordinates, the term $^2\mathbf{A}$; in the time domain, it would correspond to a second order di erential operator with respect to time. Also, in Section 1.7 we showed that an identical term appears in the equations for **E** and **H** obtained directly from Maxwell's equations. The physical meaning of this term is, as we pointed out there, that elds travel at a nite propagation velocity. This comment induces us to investigate what happens if we introduce a similar term, $^2\varphi$, in the equation for the scalar electric potential, subtracting it from the left-hand side of Eq. (1.68).

The *Lorentz gauge* consists in fact of choosing for φ one solution of the nonhomogeneous scalar Helmholtz equation:

$$\nabla^2\varphi \quad {}^2\varphi = \quad - = \frac{\nabla \ \mathbf{J}_0}{j\omega \quad c} \quad . \tag{1.79}$$

Eq. (1.68) then provides the following relation between **A** and φ, which is referred to as the *Lorentz condition*:

$$\nabla \ \mathbf{A} = \quad j\omega \quad {}_c\varphi \quad . \tag{1.80}$$

Using Eq. (1.80), Eq. (1.65) also becomes a nonhomogeneous Helmholtz equation, which (note the di erence with respect to the Coulomb gauge) is totally uncoupled from Eq. (1.79), namely:

$$\nabla^2\mathbf{A} \quad {}^2\mathbf{A} = \quad \mathbf{J}_0 \quad . \tag{1.81}$$

It is worth mentioning that the formal identity between Eqs. (1.79) and (1.81) has a very profound meaning in the theory of special relativity (Becker, 1962).

The Lorentz gauge is not unique. It is straightforward to verify that Eqs. (1.76) and (1.78) constitute a *Lorentz gauge transformation*, provided the symbol φ_0 stands now for any solution of the homogeneous Helmholtz equation, $\nabla^2\varphi_0$ $^2\varphi_0 = 0$. Note that, if **A** and φ satisfy the Lorentz condition, then whatever pair $(\mathbf{A}', \varphi')$ obtained by the above-mentioned transformation also satis es the same condition.

Finally, let us point out that if (and only if) we have $\nabla \ \mathbf{J}_0 = 0$ in all of the region of interest, then $\varphi = 0$ is a Lorentz gauge and, incidentally, also a Coulomb gauge. This fortunate case is anything but rare in practice, as we will see in several of the following chapters.

1.9 Fitzgerald electric vector potential

In spite of the large freedom in the choice of the scalar electric potential φ, which allows one to adapt the vector magnetic potential **A** to the various problems, the following chapters will show many examples of problems where it is preferable to use, for the same purposes, di erent variables from those de ned in Section 1.8. This topic will be discussed in this and in the following section.

The almost perfect symmetry in the two Maxwell's equations induces us to try to develop a "dual" theory, with respect to that of the previous section. But this requires rst a limiting assumption. In fact, only if we suppose that at every point in the region of interest it is:

$$\nabla \ \mathbf{J}_0 = 0 \quad , \tag{1.82}$$

then we also have $\nabla \ \mathbf{E} = 0$, because of Eq. (1.41), and we thus have a starting point which is the dual of that of the previous section, Eq. (1.40). In the present case, then, there exists a vector $\mathbf{F}$ such that:

$$\mathbf{E} = \ \frac{1}{c} \nabla \ \ \mathbf{F} \quad . \tag{1.83}$$

This is referred to as the *Fitzgerald electric vector potential.*

As we indicated at the end of the last section, the assumption (1.82) applies quite frequently in practice. The scope of applicability of this new potential $\mathbf{F}$ is therefore rather broad. From $\mathbf{F}$ we can obtain not only $\mathbf{E}$, but also the magnetic eld $\mathbf{H}$, whose expression can be deduced from Maxwell's equation (1.63), and, in the case of a homogeneous medium ($_c$ = constant), reads:

$$\mathbf{H} = (1/j\omega \ \ _c) \nabla \ \ \nabla \ \ \mathbf{F} \quad . \tag{1.84}$$

Let us look now for a di erential equation for $\mathbf{F}$. Referring to a homogeneous medium only, note that, from Eq. (1.82), it follows that there exists a vector function, $\mathbf{K}_0$, such that:

$$\mathbf{J}_0 = \nabla \ \ \mathbf{K}_0 \quad . \tag{1.85}$$

Then, substituting Eqs. (1.83), (1.84) and (1.85) in the second Maxwell's equation (1.61), we get:

$$\nabla \ \ (\nabla \ \ \nabla \ \ \mathbf{F} + \ ^2\mathbf{F} \ \ j\omega \ \ _c\mathbf{K}_0) = 0 \quad . \tag{1.86}$$

When the region of interest is a simply connected domain, from this third-order equation we can obtain a second order equation by putting:

$$\nabla \ \ \nabla \ \ \mathbf{F} + \ ^2\mathbf{F} \ \ j\omega \ \ _c\mathbf{K}_0 = \ \ j\omega \ \ _c\nabla \quad , \tag{1.87}$$

where the *scalar magnetic potential* is an arbitrary function. It makes it possible to replace Eq. (1.84) by the following expression:

$$\mathbf{H} = \ \ j\omega\mathbf{F} \ \ \nabla \ \ + \mathbf{K}_0 \quad , \tag{1.88}$$

which lends itself to the determination of the physical dimensions of the electric vector potential:

$$[\mathbf{F}] = [\mathbf{H}]/[\omega] = \mathrm{A} \ \ \mathrm{s} \ \ \mathrm{m} \ \ ^1 = \mathrm{coulomb} \ \ \mathrm{meter} \ \ ^1. \tag{1.89}$$

The term $\mathbf{K}_0$¶ makes this equation asymmetrical with respect to Eq. (1.66);

¶ As for $\mathbf{K}_0$, we only know its curl, $\mathbf{J}_0 = \nabla \times \mathbf{K}_0$; but it is well known that, based on Helmholtz theorem, it is possible to express $\mathbf{K}_0$ in terms of a volume integral involving $\mathbf{J}_0$.

this is just a consequence of the fact that the known terms, in Eqs. (1.62) and (1.84), are not symmetrical either.

The only gauge of practical interest for is the *Lorentz's gauge*, which is in this case a solution of the *homogeneous* Helmholtz equation:

$$\nabla^2 \qquad {}^2 \quad = 0 \quad . \tag{1.90}$$

The reader can show, as an exercise, that the following *Lorentz condition* holds, provided one exploits the residual degree of freedom left by Eq. (1.85), and sets $\boldsymbol{\nabla}\ \mathbf{K}_0 = 0$:

$$\boldsymbol{\nabla}\ \mathbf{F} = \quad j\omega \quad {}_c \qquad . \tag{1.91}$$

In particular, note that, for the scalar magnetic potential, $= 0$ is *always* a Lorentz gauge. However, as indicated by Eq. (1.88), the choice $= 0$ leads to direct proportionality between **H** and **F** only if $\mathbf{K}_0 \quad 0$ (i.e., only if the entire region of interest does not contain sources, because Eq. (1.82) together with the present restriction imply $\mathbf{J}_0 \quad 0$). Proportionality with **H** can be exploited to write the boundary conditions for **F**.

For any Lorentz gauge, also di ering from $= 0$, Eq. (1.87) becomes the following non-homogeneous Helmholtz equation:

$$\nabla^2\mathbf{F} \qquad {}^2\mathbf{F} = \quad j\omega \quad {}_c\mathbf{K}_0 \quad , \tag{1.92}$$

where it should be recalled that $\mathbf{K}_0$ is *the* vector with zero divergence satisfying Eq. (1.85).

1.10 Hertz vector potential

In the preceding section, we introduced a vector potential that can be used only when the restriction in Eq. (1.82) is satis ed. Thus, it is applicable only when the eld sources are electric charges moving along closed loops. There is also considerable interest in the opposite case, i.e., in all those phenomena whose sources can be represented as electric charges that oscillate in time around points of stable equilibrium. Such a case lends itself to be modeled by the following expression, which is always mathematically valid in the time-harmonic regime, but has a clear physical meaning only for oscillating charges:

$$\mathbf{J}_0 = j\omega\,\mathbf{P}_0 \quad . \tag{1.93}$$

The vector $\mathbf{P}_0$ is referred to as the *imposed electric polarization.*

We would certainly be permitted to use the magnetic vector potential **A**. We might then substitute Eq. (1.93) in Eq. (1.64), and then follow, from that point forth, the same procedure as in Section 1.8. But the problem is greatly simpli ed, under the formal viewpoint, if we introduce a new vector , referred to as the *Hertz vector potential*, de ned as follows:

$$\quad = \quad (1/j\omega \quad {}_c)\,\mathbf{A} \quad . \tag{1.94}$$

Its dimensions are:

$$[\quad] = [\mathbf{A}]/[\omega]\,[\quad {}_c] = \mathrm{V} \quad \mathrm{s} \quad \mathrm{m} \quad {}^1 \quad \mathrm{s} \quad \mathrm{m}^2 \quad \mathrm{s} \quad {}^2 = \mathrm{V} \quad \mathrm{m}. \tag{1.95}$$

From Eqs. (1.64) and (1.93) it follows that satis es:

$$\nabla \left(\nabla \ \nabla \ + {}^2 \ \frac{\mathbf{P}_0}{c} \right) = 0 \quad . \tag{1.96}$$

This can also be reduced to a second-order partial di erential equation with the introduction of an arbitrary scalar function . It is advisable to give the same dimensions (Volts) as the electric scalar potential introduced in Section 1.8, letting:

$$\nabla \ \nabla \ + {}^2 \ \frac{\mathbf{P}_0}{c} = \ \nabla \quad . \tag{1.97}$$

Once again, the arbitrariness of allows choices. The most interesting one is a Lorentz gauge, i.e., a solution of an equation of the same type as Eq. (1.79), whose right-hand side can be expressed in terms of the vector $\mathbf{P}_0$, as the reader may prove as an exercise:

$$\nabla^2 \ \ {}^2 = \frac{\nabla \ \mathbf{P}_0}{c} \quad . \tag{1.98}$$

One then may show, using the procedure of Section 1.8, that for a Lorentz gauge the link between the Hertz vector potential and the corresponding scalar electric potential is simply:

$$= \ \nabla \quad . \tag{1.99}$$

Eq. (1.97) reduces to a Helmholtz equation:

$$\nabla^2 \ \ {}^2 = \frac{\mathbf{P}_0}{c} \quad . \tag{1.100}$$

The e.m. eld vectors, expressed in terms of a Hertz vector potential in the Lorentz gauge, are:

$$\mathbf{E} = \nabla(\nabla \ ) \ \ {}^2 \ = \nabla \ \nabla \ \ \frac{\mathbf{P}_0}{c} \ , \tag{1.101}$$

$$\mathbf{H} = j\omega \ {}_c \nabla \quad . \tag{1.102}$$

Regardless of its applications (which are much more extensive in areas such as quantum electronics or solid-state physics, than in electromagnetic propagation), the Hertz vector deserves to be introduced to the reader. Indeed, it shows how an accurate choice of dimensions may lead to a remarkable formal simpli cation of the main equations. The reader interested in the topic should see (e.g., in Jackson, 1975) how further compactness can be gained, if the Hertz potential theory is written in the Gaussian CGS system of units, in which ${}_0 = \ {}_0 = 1$.

1.11 Further applications and suggested reading

The table of contents of this book clearly shows that the author was biased, when he chose them, by his telecommunications-oriented educational background and professional interests. Electromagnetic (e.m.) waves, herein, are regarded mainly as signal carriers, irrespective of whether they propagate freely through unbounded media, or are guided along transmission lines or similar structures.

The word "telecommunications" (Meyers, 1989, and references therein) is to be taken in a broad sense, which encompasses, for example, *radar* (Meyers, 1989; Skolnik, 1980; Taylor, 1995), as a system based on e.m. signals. In the same sense, *remote sensing* (Barrett and Curtis, 1992; Cracknell and Hayes, 1991; Elachi, 1987), which can be looked at as one modern evolution of the same principles radar is based on, may be included in the telecommunications world.

To be fair, the reader should be informed that e.m. waves have found very successful applications in many areas of science and technology, that have little in common with telecommunications, except for the frequency ranges, which often overlap. Many subjects that are touched on only marginally in this book, can be expanded, in order to link e.m. waves with non-telecommunications applications. It will be one of the main purposes of the nal section in each chapter, to provide hints in these directions. For this chapter, being basic and general, the range of possible applications becomes arbitrarily broad. In practice, it is absolutely impossible to make an exhaustive list of the topics where Maxwell's equations, as well as the other equations that we derived from them, play a crucial role. We will restrict ourselves just to the very fundamental subjects.

One of the most classical non-telecom exploitations of e.m. wave theory is the design of *optical instruments*. Only simple objects, like lenses and mirrors, can be explained, and just to some extent, in terms which do not require Maxwell's equations. At a slightly more sophisticated level, such as, for example, a Fresnel lens or a di raction grating, the e.m. approach becomes a necessity. It is not surprising therefore that, in the history of technology, optics was probably the rst eld where the theory of rapidly varying e.m. elds found practical applications (Born and Wolf, 1980; Guenther, 1990).

Other non-telecom subjects, which have generated extremely deep interest during the last decades, include *industrial*, *scienti c* and *medical* (ISM) applications of e.m. waves. The most widely used industrial application is probably heating, of either metallic materials (at frequencies which are usually below the MHz range), or dielectric materials (from a few MHz to a few GHz). Very popular examples of the latter set are domestic microwave ovens. Medical applications, based on interaction of e.m. radiation with biological tissues, include diagnosis and therapy. Introductory reading for newcomers to these elds can be found in Thuerry (1992). It is important to point out that these areas enjoy substantial independence from telecommunications, under the regulatory viewpoint.

As for scienti c applications, they are virtually unlimited. We restrict ourselves to mention just what is, probably, the widest application of *e.m. potentials* outside the eld of telecommunications, namely, the *interaction of radi-*

ation with matter. This title includes, just to quote some examples: plasma physics (Chen, 1984) and magnetohydrodynamics, which nd formidable technical application in controlled nuclear fusion aiming at new energy production plants; particle accelerators for fundamental research in physics, where high-frequency e.m. elds, traveling together with charged particles, increase the particle kinetic energy; devices where the inverse process takes place, i.e., microwave tubes, where energetic electrons are exploited to generate high-frequency e.m. waves; quantum electronics, the branch of applied physics which deals with lasers (Siegman, 1986; Svelto, 1989), microwave masers (Siegman, 1971), and similar devices; solid-state physics (Ashcroft and Mermin, 1976; Kittel, 1996), where fundamental properties of materials (typically, semiconductors) are measured through interactions between e.m. elds and carriers (electrons and holes) in the material itself.

The list of references herein should be taken just as a small set of examples. In most cases, other excellent textbooks, omitted here because of space limits, are perfectly equivalent to those that we quote (see also Jackson, 1975; Panofsky and Phillips, 1962; Ramo *et al.*, 1994; Dobbs, 1993; Louisell, 1960).

Finally, let us mention that *electromagnetic compatibility* (EMC) (or the minimization of *electromagnetic interference* (EMI)) can be looked at as the borderline between telecom and non-telecom applications of e.m. waves. It is a fast growing branch of applied science. As implied by its name, EMC is the discipline which studies how di erent systems and subsystems, which make use of elds and waves at overlapping frequencies, can share physical space, without mutually impairing their operation. As an introduction to this subject, the reader is referred to Paul (1992), and for a detailed reference, to Perez (1995).

Problems

1-1 In rectangular coordinates (x_1, x_2, x_3), let $\mathbf{E} = E_0\,\hat{x}_1$ exp(j x_3), $\mathbf{H} =$ $(E_0/\ \ )\,\hat{x}_2$ exp(j x_3), where   and   are two real constants.

a) Show that, in a homogeneous, losselss, source-free medium, these vectors satisfy the divergence equations for any value of   and .

b) Find the values of   and   for which they satisfy Maxwell's equations, in a medium of the the previously speci ed class.

c) Verify the physical dimensions of the expressions found for   and .

1-2 Calculate the magnitude of the displacement current density in free space (= $_0$) in the following situations:

an electric eld of 1 V/m at a frequency $f = 100\,\mathrm{MHz}$;

an electric eld of 100 kV/m at a frequency $f = 60\,\mathrm{Hz}$;

an electric eld of 1 kV/m at a frequency $f = 10^{15}\,\mathrm{Hz}$.

1-3 The current owing, in opposite directions, in two coaxial cylinders of

diameters $2a$ and $2b$, increases linearly with time, giving rise, in the region $a < r < b$, to an azimuthal time-dependent magnetic eld, i.e., a eld whose components in cylindrical coordinates (r, φ, z) are $H_r{=}H_z{=}\ 0$, $H_\varphi{=}\ H_0\, tr\ ^1$ (for 0 t T).
Assuming $=\ _0$, calculate the corresponding electric eld. (*Hint*: assume that all elds are independent of z).

1-4 Write the two equations which express in integral form the same information as the divergence equations, Eq. (1.34) and Eq. (1.37). Analyze their physical dimensions. Show that Eqs. (1.45) and (1.46) are implicit in these integral relationships.

1-5 Show that the Lorentz condition, Eq. (1.91), is valid provided one sets $\boldsymbol{\nabla}\ \mathbf{K}_0 = 0$.

1-6 Derive Eq. (1.98).

1-7 Show that in a rectangular coordinate frame each scalar component of Eq. (1.53) is of the following type:

$$\nabla^2 a \qquad \frac{\partial^2 a}{\partial t^2} = 0 \quad ,$$

which is usually called the *scalar wave equation.* Show that in a one-dimensional case, i.e., if a depends only one spatial coordinate, x, and on time, then the previous equation becomes of the following type:

$$\frac{\partial^2 a}{\partial x^2} \quad \frac{1}{v^2}\frac{\partial^2 a}{\partial t^2} = 0 \quad ,$$

which is usually called the *vibrating string equation.* Show, by inspection, that its general integral is of the form $a\ (x, t) = f(x \quad vt) + g(x + vt)$, where $f(u)$ and $g(u)$ are two completely arbitrary functions of a single variable, u, satisfying the only requirement that they have continuous second derivatives.

1-8 Explain what distinguishes a *voltage* between two points, P, Q, i.e., the line integral of the electric eld along a path from Q to P, and a *potential di erence* between the same points, de ned in terms of any of the scalar electric potentials de ned in Section 1.8.
(*Hint.* A time-harmonic electric eld satisfying Maxwell's equation is not, in general, a conservative eld.)

1-9 Let φ be an electric scalar potential in the Coulomb gauge. Show that $\varphi' = \varphi + C$, where C is a constant, yields the same electromagnetic eld. Does the same rule apply to a Lorentz gauge?

1-10 Find the physical dimensions of the scalar magnetic potential, de ned by Eq. (1.87).

1-11 Find the physical dimensions of the vector $\mathbf{P}_0$ (the imposed electric polarization) de ned by Eq. (1.93).

1-12 Consider the following electric and magnetic elds (representing a plane wave in a good conductor, see Section 4.12), de ned with respect to a rectangular coordinate frame (x_1, x_2, x_3):

$$\mathbf{E} = \mathbf{E}_0 \, e^{\;\; Sx_1}$$

$$\mathbf{H} = \frac{S\hat{x}_1 \quad \mathbf{E}}{j\omega} \, ,$$

where

$$S = (1 + j) \, (\omega \quad /2)^{1/2} \, ,$$

and $\mathbf{E}_0$ is a real constant complex vector lying in the $\{x_2, x_3\}$ plane. Show that in a homogeneous medium (constant , ,) they satisfy exactly Maxwell's equation (1.63), but only approximately Maxwell's equation $\nabla \quad \mathbf{H} = j\omega \;\; _c \mathbf{E}$, under the assumption ω . What is the physical meaning of this assumption?

1-13 Find the physical dimensions of the quantity indicated by S in the previous problem. Why is the inverse of its real part usually referred to as the *penetration depth* in the good conductor?

1-14 Find an imposed current density $\mathbf{J}_0$ such that the elds of Problem 1-12 satisfy Maxwell's equation $\nabla \quad \mathbf{H} = j\omega \;\; _c \mathbf{E} + \mathbf{J}_0$ *exactly.*

1-15 Show that the imposed current density of Problem 1-14 satis es $\nabla \quad \mathbf{J}_0 = 0$.

1-16 Calculate the time-harmonic vectors, in the time domain, which correspond to the complex vectors of Problems 1-12 and 1-14.

1-17 In rectangular coordinates (x, y, z), in a source-free region where the medium is homogeneous and lossless, consider a magnetic vector potential $\mathbf{A} = A_z(x, y) \, \hat{a}_z \, \exp(\quad z)$, where is a complex constant and $A_z(x, y)$ does not depend on z.

a) Show that the corresponding eld is transverse magnetic (TM), i.e., its magnetic eld is orthogonal to the z axis.

b) Show that for this $\mathbf{A}$ to be a Lorentz gauge, the corresponding scalar electric potential must be of the type $\varphi = \quad A_z \exp(\quad z)/(j\omega \quad)$. Examine whether must satisfy some relationship with other parameters.

c) Show that in general the corresponding electric eld has non-zero components both along z and in the plane orthogonal to z.

1-18 In rectangular coordinates (x, y, z), in a source-free region where the medium is homogeneous and lossless, consider an electric vector potential $\mathbf{F} = F_z(x, y)\,\hat{a}_z \exp(\quad z)$, where is a complex constant and $F_z(x, y)$ does not depend on z.

a) Show that the corresponding eld is transverse electric (TE), i.e., its electric eld is orthogonal to the z axis.

b) Show that for this $\mathbf{F}$ to be a Lorentz gauge, the corresponding scalar magnetic potential must be $\quad = \quad F_z \exp(\quad z)/(j\omega \quad)$.

c) Show that in general the corresponding magnetic eld has non-zero components both along z and in the plane orthogonal to z.

1-19 Modify the scalar electric potential of Problem 1-17 as follows: $\varphi = A_z\,[\exp(\quad z) + \exp(\; z)]/(j\omega \quad)$.

a) Show that this is again a Lorentz gauge, provided satis es a suitable relationship.

b) Find the corresponding magnetic vector potential.

1-20 An in nitely long, non-magnetic ($\;=\;$ $_0$) lossless dielectric rod, of radius a and permittivity $_1 \neq \; _0$, is surrounded by air (permittivity $\; = \; _0$).

a) Which components of the vectors $\mathbf{E}, \mathbf{H}$ in a cylindrical reference frame are continuous at $r = a$?

b) Which components are not continuous, and what relationship do they satisfy?

c) Express all the previous continuity or discontinuity conditions in terms of the cartesian components of the vectors.

1-21 How is the Hertz vector potential related to the magnetic vector potential $\mathbf{A}$, for time-harmonic elds?

1-22 For a completely arbitrary time dependence of the elds (provided that all the required derivatives exist), write, for the Hertz vector potential:

the partial di erential equation that it satis es;

the electric and magnetic elds in terms of the Hertz vector potential;

the relationships between the imposed electric polarization vector $\mathbf{P}_0$, the imposed current density and the imposed charge density.

1-23 Consider the following magnetic eld, referred to rectangular coordinates (x, y, z) (the eld of the so-called TE_{10} mode of a rectangular waveguide, see Chapter 7): for $0 < x < a$, $0 < y < b$, $H_x = A \sin(\; x/a) \exp(\quad z)$, $H_y = 0$, $H_z = A(\quad /a) \cos(\; x/a) \exp(\quad z)$; elsewhere, $H_x = H_y = H_z = 0$. Find the surface current densities which are required at $x = 0$, $x = a$, $y = 0$, $y = b$, to sustain such a eld.

References

Ashcroft, N.W. and Mermin, N.D. (1976) *Solid State Physics.* Holt, Rinehart and Winston, New York.

Barrett, E.C. and Curtis, L.F. (1992) *Introduction to Environmental Remote Sensing.* 3rd ed., Chapman & Hall, London.

Becker, R. (1962) *Theorie der Elektrizitat. Band II.* Teubner, Stuttgart.

Boas, M.L. (1983) *Mathematical Methods in the Physical Sciences.* 2nd ed., Wiley, New York.

Born, M. and Wolf, E. (1980) *Principles of Optics.* 6th ed., Pergamon Press, Oxford.

Chen, F.F. (1984) *Introduction to Plasma Physics and Controlled Fusion.* 2nd ed., Plenum Press, New York.

Collin, R.E. (1991) *Field Theory of Guided Waves.* 2nd ed., IEEE Press, Pistacaway, New Jersey.

Cracknell, A.P. and Hayes, L. (1991) *Introduction to Remote Sensing.* Taylor and Francis, London.

Dobbs, E.R. (1993) *Electromagnetic Waves.* Chapman & Hall, London.

Elachi, C. (1987) *Introduction to the Physics and Techniques of Remote Sensing.* Wiley, New York.

Guenther, R.D. (1990) *Modern Optics.* Wiley, New York.

Jackson, J.D. (1975) *Classical Electrodynamics.* 2nd ed., Wiley, New York.

Kittel, C. (1996) *Introduction to Solid State Physics.* 7th ed., Wiley, New York.

Louisell, W.H. (1960) *Coupled Mode and Parametric Electronics.* Wiley, New York.

Marsden, J.E. and Tromba, A.J. (1988) *Vector Calculus.* 3rd ed., Freeman, New York.

Meyers, R.A. (1989) *Encyclopedia of Telecommunications.* Academic, New York.

Morse, P.M. and Feshbach, H. (1953) *Methods of Theoretical Physics.* McGraw-Hill, New York.

Nye, J.F. (1985) *Physical Properties of Crystals.* Clarendon Press, Oxford.

Panofsky, W.K.H. and Phillips, M. (1962) *Classical Electricity and Magnetism.* 2nd ed., Addison-Wesley, Reading, MA.

Paul, C.R. (1992) *Introduction to Electromagnetic Compatibility.* Wiley, New York.

Perez, R. (1995) *Handbook of Electromagnetic Compatibility.* Academic Press, San Diego.

Purcell, E.M. (1985) *Electricity and Magneticism.* 2nd ed., Berkeley physics course, Vol. 2, McGraw-Hill, New York.

Ramo, S., Whinnery, J.R. and van Duzer, T. (1994) *Fields and Waves in Communication Electronics.* 3rd ed., Wiley, New York.

Schey, H.M. (1973) *Div, Grad, Curl and All That.* Norton, New York.

Siegman, A.E. (1971) *An Introduction to Lasers and Masers.* McGraw-Hill, New York.

Siegman, A.E. (1986) *Lasers.* University Science Books, Mill Valley, California.

Skolnik, M.I. (1980) *Introduction to Radar Systems.* 2nd ed., McGraw-Hill, New York.

Svelto, O. (1989) *Principles of Lasers.* 3rd ed., translated from Italian and edited by D.C. Hanna, Plenum, New York.

Taylor, J.D. (1995) *Introduction to Ultra-Wideband Radar Systems.* CRC Press, Boca Raton, Florida.

Thuerry, J. (1992) *Microwaves: Industrial, Scienti c, and Medical Applications.* Artech House Microwave Library, E.H. Grant (Ed.), Artech House, Boston.

Weast, R.C. (Ed.), (1988-89) *CRC Handbook of Physics and Chemistry.* 75th ed., CRC Press, Boca Raton, Florida.

Yariv, A. and Yeh, P. (1984) *Optical Waves in Crystals: Propagation and Control of Laser Radiation.* Wiley, New York.

CHAPTER 2

Polarization

2.1 Introduction

As we said at the beginning of the previous chapter, we take it for granted that the reader has been previously exposed to the fundamentals of circuit and network theory. In particular, we assume that the reader is familiar with the representation of time-harmonic (i.e., varying sinusoidally in time) *scalar* quantities by means of complex numbers, i.e., with the so-called Steinmetz method (or procedure). This assumption is based on typical electrical engineering and/or electronics curricula, and also applies to curricula in physics in many universities. Those readers for whom this assumption does not apply can grasp the fundamentals of Steinmetz procedure in the next section. However, it is highly recommended that they read books that treat this subject in depth, and especially that they familiarize themselves with the Steinmetz method by solving many problems on circuits. For those readers who, on the contrary, feel con dent that they know the procedure, the rst part of this chapter may look rather tedious. Nevertheless, we recommend that they read carefully at least the second part, beginning with Section 2.5, as otherwise some subtle details in the following chapters might be di cult to grasp.

Throughout this book, substantial bene t can be gained if we assume, whenever possible, a purely monochromatic sinusoidal time dependence of the electromagnetic quantities that we are interested in, such as elds, potentials, etc. However, in contrast to circuit theory, here most of our variables are vectors, in three-dimensional (3-D) real space. We must rst ask ourselves how the Steinmetz method can be generalized to time-harmonic vectors. As we will see in the next section, this entails some additional di culty. The rest of this chapter is then devoted to demonstrating why these complications are indeed worth the time and e ort they require. The net result will indeed be, that the complex representation of a time-harmonic vector can convey, in compact forms that lend themselves to very elegant analytical and geometrical interpretations, information on an important physical feature of the time-harmonic vector itself, namely, its polarization. Applications are found in all the following chapters, ranging from plane waves to waveguides, from beams to antennas, from di raction to coherence.

2.2 Steinmetz representation of time-harmonic vectors

In real 3-D space, let us introduce a completely arbitrary cartesian orthogonal reference frame, $\{x_1, x_2, x_3\}$. Let $\{\hat{a}_1, \hat{a}_2, \hat{a}_3\}$ be the unit-vectors of this frame. We call *time-harmonic*, with angular frequency ω, a vector whose components in that reference frame vary sinusoidally in time, with angular frequency ω.

Let then $S_v(\omega)$ be the set of all time-harmonic vectors of angular frequency ω:

$$S_v(\omega) \quad \left\{ \sum_{n=1}^{3} \hat{a}_n a_n \ \cos(\omega t + \varphi_n) \ ; \ a_n \in \mathcal{R} \ ; \ \varphi_n \in \mathcal{R} \ ; \ a_n \quad 0 \ ; \ 0 \quad \varphi_n < 2 \quad \right\} . \tag{2.1}$$

It is easy to show that $S_v(\omega)$ satis es the axioms that de ne a vector space over the real eld.

We then de ne a *complex vector* as a linear combination of time-independent vectors, each of which belongs to the real 3-D space, where the scalar coef- cients are arbitrary *complex* numbers. Let $\mathcal{C}_3$ be the set of all the complex vectors de ned in this way:

$$\mathcal{C}_3 \quad \left\{ \sum_{n=1}^{3} \hat{a}_n a_n \ e^{j\varphi_n} \ ; \ a_n \in \mathcal{R} \ ; \ \varphi_n \in \mathcal{R} \ ; \ a_n \quad 0 \ ; \ 0 \quad \varphi_n < 2 \quad \right\} . \tag{2.2}$$

Then, it is easy to show that $\mathcal{C}_3$ is also a vector space over the real eld. The relation:

$$\mathbf{a} \ (t) = \sum_{n=1}^{3} \hat{a}_n a_n \ \cos(\omega t + \varphi_n) = \mathrm{Re} \left\{ \sum_{n=1}^{3} \hat{a}_n a_n \ e^{j\varphi_n} \ e^{j\omega t} \right\} = \mathrm{Re} \left\{ \mathbf{a} \, e^{j\omega t} \right\} \tag{2.3}$$

is a one-to-one and onto mapping, , between $S_v(\omega)$ and $\mathcal{C}_3$. This relationship forms the *complex representation of time-harmonic vectors*, and is also referred to as the vectorial Steinmetz method.

Let us de ne, in $\mathcal{C}_3$, a *scalar product* as a straightforward extension of two well-known operations, the inner product of two real vectors, and the product of complex numbers, i.e.:

$$\mathbf{a} \quad \mathbf{b} = \left(\sum_{n=1}^{3} \hat{a}_n a_n \ e^{j\varphi_n} \right) \quad \left(\sum_{n=1}^{3} \hat{a}_n b_n \ e^{j \quad _n} \right) = \sum_{n=1}^{3} a_n b_n \ e^{j(\varphi_n + \quad _n)} \quad . \tag{2.4}$$

With this de nition of scalar product, $\mathcal{C}_3$ is not a unitary space, according to the de nition given in classical textbooks on linear algebra (e.g., Halmos, 1958), because some axioms are violated. Some consequences, in particular those that involve the de nitions of parallel and orthogonal complex vectors, will be outlined in the next section. Another consequence that we wish to point out is that the mapping in Eq. (2.3) *is not an isomorphism*, as there is no one-to-one matching between scalar products in the two vector spaces.

The scalar products in the two spaces are related to each other as follows:

$$\frac{1}{2}\,\mathrm{Re}\{\mathbf{a}\quad\mathbf{b}\ \} = \overline{\mathbf{a}\quad\mathbf{b}}\quad , \tag{2.5}$$

where the bar above the factors on the right-hand side indicates a time average over a period $T = 2\ \ /\omega$. We may then call the mapping in Eq. (2.3) a *pseudo-isomorphism.*

2.3 Parallel and orthogonal complex vectors

Two complex vectors, **A** and **B**, are said to be *parallel* when there exists a complex number c, such that the relationship:

$$\mathbf{A} = c\mathbf{B} \tag{2.6}$$

is satis ed.

We want to preserve the validity, for complex vectors, of the following relation, which is well known to hold for real vectors and for real numbers:

$$|\mathbf{A}| = |c|\ |\mathbf{B}|\quad . \tag{2.7}$$

Note that, in order to do that, it is necessary to de ne the *modulus* of a complex vector in the following way:

$$|\mathbf{A}| = \sqrt{\mathbf{A}\quad\mathbf{A}\quad}\quad . \tag{2.8}$$

Another notion that we want to preserve from the space of real vectors is that the *projection* of a complex vector onto its own direction must coincide with its modulus. Then, to be consistent, the *projection* of a complex vector, **A**, onto the direction of another, completely arbitrary complex vector, **C**, must be de ned as follows:

$$A_c = \mathbf{A}\quad\mathbf{C}\ /|\mathbf{C}| = (\mathbf{C}\quad\mathbf{A}\ )\ /|\mathbf{C}|\quad . \tag{2.9}$$

Consequently, two complex vectors **A** and **C** are said to be *orthogonal* to each other when the following relationship:

$$\mathbf{A}\quad\mathbf{C}\ \ = \mathbf{A}\quad\mathbf{C} = 0 \tag{2.10}$$

is satis ed. Note that in general Eq. (2.10) does not imply $\mathbf{A}\quad\mathbf{C} = 0$.

These results are probably surprising, and the reader might believe that they lead, sooner or later, to inconsistencies. On the contrary, we will see that these new rules are instrumental in some steps. For example, in Chapter 4 they will help us to correctly calculate the per-unit-area power carried by an arbitrarily polarized plane wave.

2.4 Properties of time-harmonic vectors

Any element in the set de ned by Eq. (2.1) can be written in the following way:

$$\begin{aligned} \mathbf{A} &= \left(\sum_{n=1}^{3} \hat{a}_n a_n \, \cos\varphi_n \right) \cos\omega t \quad \left(\sum_{n=1}^{3} \hat{a}_n a_n \, \sin\varphi_n \right) \sin\omega t \\ &= \mathbf{A}' \cos\omega t \quad \mathbf{A}'' \sin\omega t \quad , \end{aligned} \tag{2.11}$$

where the two vectors:

$$\mathbf{A}' = \sum_{n=1}^{3} \hat{a}_n a_n \, \cos\varphi_n \quad , \qquad \mathbf{A}'' = \sum_{n=1}^{3} \hat{a}_n a_n \, \sin\varphi_n \tag{2.12}$$

are *real and time-independent.* Eq. (2.11) implies that, at any time t, the vector $\mathbf{A}$ belongs to the plane de ned by the two vectors in Eq. (2.12). Hence, for any time-harmonic vector originating at a given point O, the trajectory described vs. time by its terminal point is a plane curve. Furthermore, as the two vectors in Eq. (2.12) are linearly combined with time-periodic coe cients, $\cos\omega t$ and $\sin\omega t$, that trajectory has to be a closed curve. Let us show that in fact Eq. (2.11) is the *parametric equation of an ellipse.* Indeed, looking for the equation of that trajectory in any orthogonal reference frame $\{x', y'\}$ on the $\{\mathbf{A}', \mathbf{A}''\}$ plane (see Figure 2.1), we may write $\mathbf{A} = x' \, \hat{a}_{x'} + y' \, \hat{a}_{y'}$, where $\hat{a}_{x'}$ and $\hat{a}_{y'}$ are obviously the unit vectors along the axes. Then, as a consequence of Eq. (2.11), we have:

$$x' = a'_{x'} \cos\omega t \quad a''_{x'} \sin\omega t \quad , \tag{2.13}$$

$$y' = a'_{y'} \cos\omega t \quad a''_{y'} \sin\omega t \quad . \tag{2.14}$$

We now multiply Eq. (2.13) by $a'_{y'}$, and Eq. (2.14) by $a'_{x'}$, and subtract. Then, we multiply Eq. (2.13) by $a''_{y'}$, and Eq. (2.14) by $a''_{x'}$, and subtract. Finally, we square the two expressions, and sum, making use of the identity $\sin^2 \omega t + \cos^2 \omega t = 1$. The nal result is:

$$(x' a'_{y'} \quad y' a'_{x'})^2 + (x' a''_{y'} \quad y' a''_{x'})^2 = (a'_{x'} \, a''_{y'} \quad a'_{y'} \, a''_{x'})^2 \quad , \tag{2.15}$$

which is the standard form for the equation of an ellipse on the $\{x', y'\}$ plane.

This result is usually expressed saying that a generic time-harmonic vector is *elliptically polarized.*

It is usually convenient to express the ellipse in terms of the coordinate system corresponding to what are known as its *principal axes.* The principal axes are labelled x and y in Figure 2.1, and are parallel to the major and minor axes of the ellipse, respectively. In this reference frame, the instantaneous vector (2.11) can be expressed in the following form:

$$\mathbf{A} = \mathbf{F}_x \, \cos(\omega t + \varphi) \quad \mathbf{F}_y \, \sin(\omega t + \varphi) \quad , \tag{2.16}$$

where $\mathbf{F}_x$ is a real vector parallel to the x axis, and $\mathbf{F}_y$ is a real vector parallel to the y axis.

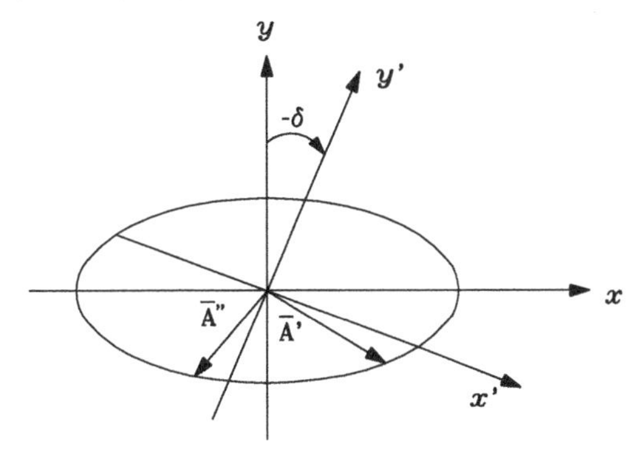

Figure 2.1 *The polarization ellipse, its principal axes, and a generic cartesian reference frame.*

To nd the principal axes, when a vector is given in the form Eq. (2.11), is perfectly equivalent to nding the phase angle φ in Eq. (2.16). The problem is much simpler to tackle if one makes use of the corresponding complex vector, and therefore its solution is left to the next section.

2.5 Properties of the complex vectors

From Eq. (2.11) it follows that:

$$\mathbf{A} = \mathrm{Re}\left[(\mathbf{A}' + j\,\mathbf{A}'')\, e^{j\omega t}\right] \quad . \tag{2.17}$$

Comparing this with Eq. (2.3), we see that the complex vector **A**, which corresponds, according to the Steinmetz procedure, to the time-harmonic vector **A**, is connected to the two time-independent real vectors that appear in Eq. (2.11) by the simple relationship:

$$\mathbf{A} = \mathbf{A}' + j\,\mathbf{A}'' \quad . \tag{2.18}$$

If we need to express the vector (2.18) in a generic orthogonal cartesian frame $\{x', y'\}$, we may write:

$$\mathbf{A} = A_{x'}\,\hat{a}_{x'} + A_{y'}\,\hat{a}_{y'} \quad , \tag{2.19}$$

and the complex numbers $A_{x'}, A_{y'}$ are the components of **A** in this frame, which is real, so that we may write:

$$\begin{aligned} A_{x'} &= \mathbf{A} \quad \hat{a}_{x'} = \mathbf{A} \quad \hat{a}_{x'} \quad , \\ A_{y'} &= \mathbf{A} \quad \hat{a}_{y'} = \mathbf{A} \quad \hat{a}_{y'} \quad . \end{aligned} \tag{2.20}$$

If we want to introduce the reference frame of the principal axes, then, combining Eq. (2.16) with Eq. (2.3), we get:

$$\mathbf{A} = A_x\hat{a}_x + A_y\hat{a}_y = |\mathbf{F}_x|\, e^{j\varphi}\, \hat{a}_x \quad j\,|\mathbf{F}_y|\, e^{j\varphi}\, \hat{a}_y \quad . \tag{2.21}$$

This result shows that there is another way to de ne the principal axes of the ellipse represented by Eq. (2.15). It is the basis in which the ratio between the cartesian components of the complex vector **A** is purely imaginary, that is, the ratio:

$$p = j\,\frac{A_y}{A_x} \quad , \tag{2.22}$$

is *real*. On the other hand, in a generic orthogonal basis, the ratio:

$$p' = j\,\frac{A_{y'}}{A_{x'}} \quad , \tag{2.23}$$

is a complex number.

As is well known, a rotation of a real reference frame by an angle (see again Figure 2.1) entails the following transformation for the vector components:

$$\begin{pmatrix} A_{x'} \\ A_{y'} \end{pmatrix} = \begin{pmatrix} \cos & \sin \\ \sin & \cos \end{pmatrix} \begin{pmatrix} A_x \\ A_y \end{pmatrix} \quad . \tag{2.24}$$

Then, comparing Eq. (2.22) and Eq. (2.23), we see that a simple way to nd the principal axes, starting from the vector expressed in a generic reference frame, is to impose that the rotation yield a real value for Eq. (2.22), i.e., to impose $p = p$. We leave it as an exercise for the reader to prove that the nal result is the following rule:

$$\tan 2 \;= j\,\frac{p' \quad p'}{1 \quad p'p'} \quad . \tag{2.25}$$

This point will be exploited in the next two sections.

2.6 Linear polarization ratio

The complex number in Eq. (2.23) is referred to as the *linear polarization ratio* of either the time-harmonic vector **A**, or the corresponding complex vector **A**, in the generic orthogonal reference frame $\{x', y'\}$.

When the reference frame is rotated by a generic angle ″, then the components are transformed according to Eq. (2.24), with replaced by ″. It is easy to check that the corresponding transformation rule for the linear polarization ratio is:

$$p'' = \frac{p' + j\tan \;''}{1 + j\,p'\tan \;''} \quad . \tag{2.26}$$

Eq. (2.26) is a *bilinear* relationship between p'' and p'. The reader may prove, as an exercise, that an invariant under this transformation is *the sign of the real part* of the linear polarization ratio. This suggests that this sign has a meaning which is intrinsic to the vector **A**, independent of the reference frame. Indeed, we may observe that:

if $\mathrm{Re}(p'') < 0$, i.e., when the sign in front of the last term in Eq. (2.21) is a plus, then the time-harmonic vector **A** rotates *clockwise* for an observer watching at the positive face of the $\{x, y\}$ plane — we refer to this as a *left-handed* elliptical polarization;

if $\mathrm{Re}(p'') > 0$, i.e., when the sign in front of the last term in Eq. (2.21) is a minus, then the time-harmonic vector $\mathbf{A}$ rotates *counterclockwise* for an observer watching at the positive face of the $\{x, y\}$ plane — we refer to this as a *right-handed* elliptical polarization;

nally, if $\mathrm{Re}(p'') = 0$, i.e., if $\mathbf{F}_y = 0$ in Eqs. (2.21) and (2.16), then the ellipse described by the point $P = O + \ \mathbf{A}\ (t)$ degenerates into a segment — we refer to this as a *linear polarization.*

In particular, when $p'' = \quad 1$, then $|\mathbf{F}_x| = |\mathbf{F}_y|$ in Eqs. (2.21) and (2.16). The ellipse becomes a circle, and we refer to these cases as *circular polarizations*, either right- or left-handed, obviously depending on the sign of p''.

2.7 Circular polarization ratio

Consider the following pair of complex vectors:

$$\hat{R} = (\hat{a}_x \quad j\,\hat{a}_y)/\sqrt{2} \quad , \qquad \hat{L} = (\hat{a}_x + j\,\hat{a}_y)/\sqrt{2} \quad . \tag{2.27}$$

From what we saw in the previous section, they represent two *circularly polarized* time-harmonic vectors, a right-handed and a left-handed one, respectively. These two complex vectors:

belong to the (x, y) plane;

have moduli equal to unity:

$$\hat{R} \quad \hat{R} \quad = 1 \quad , \qquad \hat{L} \quad \hat{L} \quad = 1 \quad , \tag{2.28}$$

so they can be called unit vectors;

are orthogonal to each other, according to the de nition in Eq. (2.10):

$$\hat{R} \quad \hat{L} \quad = 0 \quad . \tag{2.29}$$

Therefore, they form an orthonormal basis, which we refer to as the *rotating frame.* Any complex vector $\mathbf{A}$ belonging to the plane of Figure 2.1 can be expanded into this basis, as:

$$\mathbf{A} = A_R\,\hat{R} + A_L\,\hat{L} \quad . \tag{2.30}$$

The components of $\mathbf{A}$ in this basis, the rotating frame, are related to those in the principal axis frame $\{x, y\}$ by:

$$\begin{aligned} A_R &= \mathbf{A} \quad \hat{R} \quad = \frac{1}{\sqrt{2}}\,(A_x + j\,A_y) \quad , \\ A_L &= \mathbf{A} \quad \hat{\mathrm{L}} \quad = \frac{1}{\sqrt{2}}\,(A_x \quad j\,A_y) \quad . \end{aligned} \tag{2.31}$$

The complex number:

$$q = \frac{A_L}{A_R} \tag{2.32}$$

is called the *circular polarization ratio* of the time-harmonic vector $\mathbf{A}$, or of the corresponding complex vector $\mathbf{A}$.

All the de nitions that we gave in this section originated from quantities that were expressed in the principal axis reference frame, $\{x, y\}$. If one starts from a completely arbitrary real orthogonal frame, $\{x', y'\}$, rotated by an angle with respect to the principal axes, one gets, analogously:

$$q' = \frac{A'_L}{A'_R} = q\, e^{j2} \quad . \tag{2.33}$$

Hence, the quantity $|q|$ is independent of the reference frame, and therefore has an intrinsic meaning. This is recon rmed by the relationship which links the circular polarization ratio to the linear polarization ratio de ned in the previous section, namely:

$$q' = \frac{1 \quad p'}{1 + p'} \quad . \tag{2.34}$$

This is again a bilinear transformation, which maps the loci $|q'| =$ constant on the q' complex plane into curves on the p' complex plane along which the sign of $\mathrm{Re}(p')$ remains constant. In more detail, we have:

the line $\mathrm{Re}(p') = 0$ is mapped onto the circle $|q'| = 1$ (locus of *linear polarizations*);

the half-plane $\mathrm{Re}(p') < 0$ is mapped onto the region $|q'| > 1$, where *left-handed elliptical polarizations* are located. In particular, $q' = \infty$ corresponds to left-handed circular polarization;

the half-plane $\mathrm{Re}(p') > 0$ is mapped onto the region $|q'| < 1$, where *right-handed elliptical polarizations* are located. In particular, $q' = 0$ corresponds to right-handed circular polarization.

2.8 Stokes parameters

In the last two sections, we stressed that some properties of the linear and circular polarization ratios are independent of the reference frame one adopts on the plane of the ellipse, Eq. (2.15). Those remarks show how to proceed further, searching for quantities that characterize polarization, and are invariant with respect to a change in the reference frame, or at least, if they can not be constant, change in simple ways. Among many possibilities that were suggested in the past decades, the so-called *Stokes parameters* became quite popular, being both elegant and useful at the same time.

In order to link their de nition to what we have seen so far, let us start from Eq. (2.15), which describes the elliptical trajectory of the point $P = O + \mathbf{A}$, in the generic frame $\{x', y'\}$ (see Figure 2.1), where the components of the complex vector $\mathbf{A}$ are expressed by Eq. (2.20).

From now on, the notation that we have been using so far may be simpli ed, just because we are looking for quantities that are insensitive to a change in the reference frame. Therefore, let us write the components of the complex vector $\mathbf{A}$ as follows, omitting primes, although in general we are *not* using the

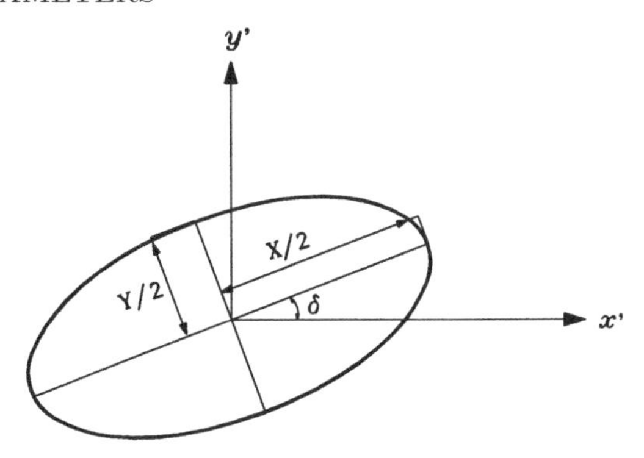

Figure 2.2 *Geometrical illustration of the symbols* X, Y *and* *(orientation angle).*

principal axes as the reference system:

$$A_{x'} = b_x \, e^{j\varphi_x}, \quad A_{y'} = b_y \, e^{j\varphi_y}, \quad \varphi = \varphi_x \quad \varphi_y, \quad \mathbf{A} \text{ as } \mathrm{Re}(\mathbf{A}\, e^{j\omega t}), \tag{2.35}$$

where the parameters b_x, b_y, φ_x and φ_y are positive real. The equation of the ellipse, Eq. (2.15), can then be rewritten, after a little manipulation, as follows:

$$\left(\frac{x'}{b_x}\right)^2 + \left(\frac{y'}{b_y}\right)^2 \quad 2\frac{x'}{b_x}\frac{y'}{b_y} \cos\varphi = \sin^2\varphi \quad . \tag{2.36}$$

From this, one sees that the angle between the major axis of the ellipse and the x' axis, referred to as the *orientation angle* (see Figure 2.2), is given by:

$$\tan 2 \quad = \frac{2b_x b_y}{b_x^2 \quad b_y^2} \cos\varphi \quad , \tag{2.37}$$

while the ratio between the lengths Y and X of, respectively, the minor and major axes of the ellipse is related to the parameters of Eq. (2.36) as follows:

$$\frac{Y}{X} = \quad \tan \quad , \quad = \frac{1}{2} \arcsin\left[\frac{2b_x b_y}{b_x^2 + b_y^2} \sin\varphi\right] \quad , \quad | \; | \quad \frac{}{4} \quad , \tag{2.38}$$

where: (i) in line with the previous de nitions, the plus and minus signs apply for the right-handed and left-handed polarizations respectively, and (ii) the quantity is referred to as the *eccentricity angle.*

The four *Stokes parameters* are de ned as follows:

$$\begin{aligned}
s_1 &= A_{x'} A_{x'} \quad A_{y'} A_{y'} = b_x^2 \quad b_y^2 \quad , \\
s_2 &= A_{x'} A_{y'} + A_{x'} A_{y'} = 2b_x b_y \cos\varphi \quad , \\
s_3 &= \quad j(A_{x'} A_{y'} \quad A_{x'} A_{y'}) = 2b_x b_y \sin\varphi \quad , \\
s_0 &= A_{x'} A_{x'} + A_{y'} A_{y'} = b_x^2 + b_y^2 \quad .
\end{aligned} \tag{2.39}$$

All the information concerning the polarization ellipse of Figure 2.1 is concentrated in the two real numbers and , plus the sign in Eq. (2.38). One may then wonder why we introduced a set of four real parameters. The answer may be given as follows. First of all, squaring the expressions in Eq. (2.39), and summing, we nd:

$$s_1^2 + s_2^2 + s_3^2 = s_0^2 \quad , \tag{2.40}$$

which shows that in any case the four Stokes parameters are not independent. A usual way to express this link in words is to refer to $\mathbf{S} \quad (s_1, s_2, s_3)$ as the *Stokes vector*, and to state that $|\mathbf{S}| = s_0$. Furthermore, if we want to keep information on polarization of a time-harmonic vector separate from that on vector *amplitude*, we may use *normalized amplitudes*, i.e., *impose*:

$$s_0 = 1 \quad . \tag{2.41}$$

In this way, the number of independent Stokes parameters goes down to two, with a further ambiguity in the sign of the third one. This result matches precisely our statement after Eq. (2.39). However, let us emphasize that we were entitled to make the normalization (2.41) *only because* we were dealing with a *purely monochromatic, time-harmonic eld.* The de nition of the Stokes parameters has a broader scope of validity. For example, it may be used for narrow-band signals, i.e., for time-harmonic carriers modulated by slowly-varying envelopes, a subject with which we will deal in Chapter 5. Another very important example of application of Stokes parameters beyond the scope of the previous sections in this chapter is the representation of *incoherent* (or partially coherent) elds, that is, elds whose slowly varying envelopes are random functions of time. In this case, the products of the type $A'_j A'_j$ that appear in Eq. (2.39) have to be replaced by time averages, and therefore the results are strongly dependent on the assumptions one makes about the correlation properties of the eld.

In all these cases, the di erence between s_0 and unity becomes a measure of the "*depolarization*" of the eld $\mathbf{A}$. We advise the reader to return to these comments and reconsider them after studying the last chapter in this book which deals with statistical properties of e.m. elds.

Once the Stokes parameters have been determined in a completely arbitrary reference frame, then it is straightforward to identify, starting from them, the polarization ellipse de ned with respect to the same frame. In fact, from Eq. (2.39), one nds:

$$b_x^2 = \frac{s_0 + s_1}{2} \quad , \quad b_y^2 = \frac{s_0 \quad s_1}{2} \quad , \quad \varphi = \arctan \frac{s_3}{s_2} \quad , \tag{2.42}$$

or, alternatively:

$$2 \quad = \arctan \frac{s_2}{s_1} \quad , \quad 2 \quad = \arcsin \frac{s_3}{s_0} \quad , \quad X^2 + Y^2 = s_0 \quad . \tag{2.43}$$

If the reference frame is rotated by an angle on the plane of the ellipse, then, as the reader may check as an exercise, two parameters (namely, s_3 and s_0) are invariant, while the other two obey the simple transformation rule:

$$s_1' \quad = \quad s_1 \cos 2 \quad + s_2 \sin 2 \quad ,$$

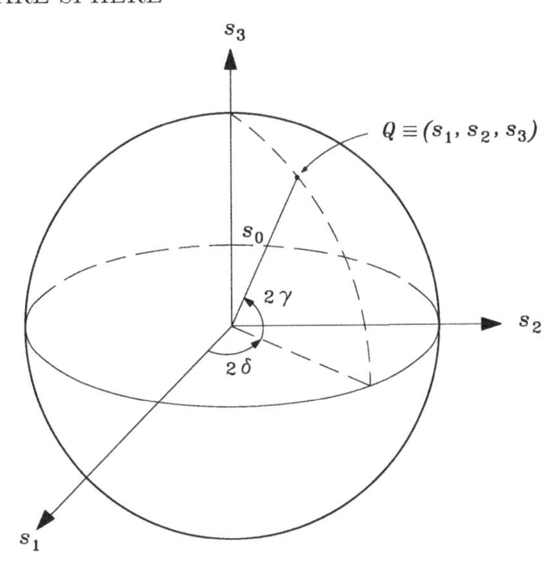

Figure 2.3 *Geometrical representation of the Stokes parameters: the Poincare sphere.*

$$s_2' = s_1 \sin 2 \qquad s_2 \cos 2 \qquad , \tag{2.44}$$

which is typical of a rotation of the $\{s_1, s_2\}$ plane by an angle 2 . The invariance of s_3 with respect to this kind of basis transformation is connected to the fact that s_3 measures the content of circular polarization in the vector $\mathbf{A}$, as the reader can infer from the de nition of s_3 itself, and as will be recon rmed in the next section.

Many of the most interesting properties of Stokes parameters become simple and evident if we adopt a geometrical construction, which will in fact be dealt with in the next section.

2.9 The Poincare sphere

The relationship in Eq. (2.40) between the Stokes parameters lends itself to an immediate geometrical interpretation. If we introduce, in real 3-D space, three mutually orthogonal axes labeled $\{s_1, s_2, s_3\}$, then Eq. (2.40) can be read as the equation of a spherical surface of radius s_0 centered at the origin. This is usually referred to as the *Poincare sphere* (see Figure 2.3).

For a purely monochromatic time-harmonic eld, as noted in the previous section, one may separate information on polarization from that on amplitude, normalizing the eld so that $s_0 = 1$. In general, as the reader will be able to appreciate more easily after reading the nal chapter of this book, the radius of the Poincare sphere will convey information regarding the *degree of polarization* of the eld one is dealing with. Unit-radius will imply completely polarized eld, i.e., the state of polarization is purely deterministic and known at any time without any uncertainty. At the opposite extreme, the radius will go to zero for a completely *unpolarized* eld, i.e., for a eld whose state of polarization varies randomly in time, with a correlation time shorter than the response time of the instrumentation that measures it, i.e., shorter than the time scale over which the quantities in Eq. (2.39) are averaged.

The reader can now easily grasp the simple relationship that exists between the *position* of a point on the Poincare sphere and the *parameters of the ellipse* traced out by the point $P = O + \mathbf{A}$. Indeed, the Eqs. in (2.43) indicate that *latitude* and *longitude* on the Poincare sphere equal 2 and 2 respectively, where and are respectively the ellipse eccentricity and orientation angles. Consequently:

the upper hemisphere ($s_3 > 0$) is the locus of right-handed polarizations, and the lower hemisphere ($s_3 < 0$) is the locus of left-handed polarizations;

the equator ($s_3 = 0$) is the locus of *linear* polarizations;

the two poles ($s_1 = s_2 = 0$) represent *circular* polarizations, right-handed at the north pole, and left-handed at the south pole;

mirror symmetry with respect to the plane of the equator re ects that, in real 3-D space where the vector $\mathbf{A}$ is de ned, a clockwise rotation becomes counterclockwise, and vice-versa, if one reverses the sign of the coordinate axis orthogonal to the plane of the ellipse.

We think it might now be simpler for the reader to understand why the Stokes parameters transform according to Eq. (2.44) when the reference frame on the plane of the ellipse is rotated by an angle . Purely circular polarizations are completely insensitive to such a change in the reference frame. Linear polarizations remain linear, but their orientation angle, i.e., the angle between the reference frame and the direction of polarization, changes when the axes are rotated. The periodicity in such a change must be , and that is the source of the factor of two in front of in Eq. (2.44). Similarly, implications for elliptical polarizations can be deduced in a straightforward manner.

2.10 Evolution of polarization in a linear medium: Jones matrix

All the de nitions that we gave so far, and all the operations that were described in this chapter, dealt with polarization as a *local* property. They referred to a vector which was de ned at a given point in real 3-D space. On the other hand, it is self-explanatory why, in a book on electromagnetic waves, it is of interest to study how the polarization of a vector wave evolves as it propagates. This subject can be dealt with in depth, grasping the physical meaning of the equations one is writing, only after one familiarizes oneself with the contents of the next three chapters. Nevertheless, it appears to be useful to present here, in a rather abstract form, a general formalism, which is consistent with what we saw in this chapter, and will be applicable to the subjects to be dealt with later on.

Consider two points in real 3-D space, P_{in} and P_{out}, both situated on the same path along which a wave is propagating. P_{in} is "upstream," P_{out} is "downstream". As we saw in the previous sections, polarization at each point is completely characterized by a complex vector with two nonvanishing components. For simplicity, let us suppose that the two vectors at P_{in} and P_{out}, $\mathbf{A}_{\text{in}}$ and $\mathbf{A}_{\text{out}}$, respectively, lay in parallel $z =$ constant planes. If the medium between P_{in} and P_{out} is *linear*, then the two vectors $\mathbf{A}_{\text{in}}$ and $\mathbf{A}_{\text{out}}$ must be linearly related to each other. If we adopt the formalism of Section 2.5, and use *the same reference frame* $\{x, y\}$ (not necessarily the principal axes) at the two points, then the relationship between the vectors can be expressed in matrix form, as:

$$\begin{pmatrix} A_{\text{out}x} \\ A_{\text{out}y} \end{pmatrix} = \begin{pmatrix} J_{11} & J_{12} \\ J_{21} & J_{22} \end{pmatrix} \begin{pmatrix} A_{\text{in}x} \\ A_{\text{in}y} \end{pmatrix} \quad . \tag{2.45}$$

The matrix denoted as $(\mathbf{J})$ in Eq. (2.45) is referred to as the *Jones matrix.* If one changes the reference frame, rotating it by an angle   (see again Figure 2.1), then the Jones matrix in the new system, $(\mathbf{J}')$, is related to that in the old system by:

$$(\mathbf{J}') = (\mathbf{R})\,(\mathbf{J})\,(\mathbf{R})^{\ \ 1} \quad , \tag{2.46}$$

where $(\mathbf{R})$ is the rotation matrix that appears in Eq. (2.24).

The physical reasons why polarization may change as a wave propagates along a linear system are not simple to understand at this stage. As previously mentioned, they will be clari ed in the following chapters. However, it is rather intuitive that, independent of the causes, the two components of the vector $\mathbf{A}_{\text{in}}$ may experience either a *di erential loss (or gain, in an active medium)*, or *a di erential phase shift.* A simple example of the rst kind is an *ideal polarizer*, whose Jones matrix, in a reference frame where the y axis is along the blocking direction, is:

$$(\mathbf{J}_0) = \begin{pmatrix} 1 & 0 \\ 0 & 0 \end{pmatrix} \quad . \tag{2.47}$$

A simple example of the second kind is a so-called *quarter-wave retardation plate*, i.e., a device which introduces a delay of $/2$ radians on one component (say the one along y) with respect to the other. Then (except for an unde ned factor of unit modulus in front of the matrix, which represents an irrelevant phase shift, equal for all components), the Jones matrix reads:

$$(\mathbf{J}_{\ /2}) = \begin{pmatrix} 1 & 0 \\ 0 & j \end{pmatrix} \quad . \tag{2.48}$$

When two or more linear systems are cascaded, along the path of a given wave, then it is evident from the de nition that the Jones matrix of the whole system is the product, in the appropriate order, of the Jones matrices of the individual components. Some of the problems at the end of this chapter deal with these subjects. Another point, which is probably much less obvious, and therefore must be emphasized here, is that the Jones matrix does not always lend itself to the inverse operation, namely, to reconstruct the input vector polarization from that of the output vector. An example in this sense is provided by the ideal polarizer: the matrix in Eq. (2.47) cannot be inverted, as it is singular. Indeed, this corresponds to the physical fact that the output of an ideal polarizer is, by de nition, linearly polarized in a given direction, irrespective of the input eld.

Similarly, in general one can not infer that the Jones matrix remains invariant when a wave travels through the same device in the opposite direction, i.e., when the roles of the points P_{in} and P_{out} are interchanged. Such a statement will have to be checked against the so-called *reciprocity theorem*, to be proved in the next chapter.

2.11 Further applications and suggested reading

For further details regarding the complex (phasor) representation of time harmonic vectors, we refer to any text on mathematical physics and Section 1.3 of Yariv and Yeh (1984) for some pertinent comments as well as Section 1.4 of Azzam and Bashara (1987). For an in-depth treatment of polarized light including the Poincare sphere and Jones matrices, we refer to Azzam and Bashara (1987) — see also the introductory level optics treatment on pp. 48–49 and 544–545 of Guenther (1990). We also refer to Kliger *et al.* (1990) and Chapter 5 of Yariv and Yeh (1984) for further applications in optics as well as Someda and Stegeman (1992) and references therein for advanced waveguide theory applications, and Galtarossa and Menyuk (2005) for a thorough coverage of the now popular subject of polarization mode dispersion (PMD) in optical bers.

Problems

2-1 Derive the rule expressed by Eq. (2.25).

2-2 Show that the sign of the real part of the linear polarization ratio is an invariant under the transformation de ned by Eq. (2.26).

2-3 Show that, if the reference frame is rotated by an angle on the plane of the polarization ellipse, two Stokes parameters (s_3 and s_0) are invariant, while the other two obey the transformation rule expressed by Eq. (2.44).

2-4 Consider the following vector eld which corresponds to the magnetic eld of the so-called TE_{10} mode of a rectangular waveguide (see Chapter 7) and is de ned with respect to rectangular coordinates (x, y, z) for $0 < x < a$, $0 < y < b$ as $H_x = A \sin(\ x/a) \exp(\ \ z)$, $H_y = 0$, $H_z = A(\ \ /a) \cos(\ x/a) \exp($

a) First assume that is real. Show that in this case the vector is linearly polarized at any point, and nd its direction of polarization as a function of the x coordinate.

b) Next assume that $\ = j \ $ is imaginary. Calculate the linear polarization ratio and the circular polarization ratio of the vector as a function of x. Find the values of x, if any, for which the vector is linearly polarized, and those for which it is circularly polarized. Distinguish between right-hand and left-hand circular polarizations, for a given sign of .

2-5 Refer to the vector de ned in the previous problem, and let (see Chapter 7, Eq. (7.24)):

$$= j \quad = j\omega \sqrt{\quad_0 \quad_0}\, \sqrt{1 \quad (\omega_c/\omega)^2} \quad .$$

Discuss the shape and the orientation of the polarization ellipse as a function of ω, over the range $\omega_c < \omega < 2\omega_c$, at the points $x = a/4$ and $x = 3a/4$.

2-6 Calculate the Stokes parameters, vs. ω, for the vector of the previous problem, assuming that the constant A is real. Study the corresponding curve on the Poincare sphere.

2-7 Consider the following vector eld, **E**, which corresponds to the electric eld of the so-called TE_{11} mode of a circular waveguide (see Chapter 7) and is de ned within the circle of radius a centered at the origin, on the $z = 0$ plane of a cylindrical coordinate frame, (r, φ, z) as $E_r = E_0\, J_1(1.84\, r/a)\, (1/r)\, \sin\varphi$, $E_\varphi = (1.84/a)\, E_0\, J_1'(1.84\, r/a)\, \cos\varphi$, $E_z = 0$, where E_0 is a constant which is in general complex. (For the de nition of the Bessel function J and of its derivative J', see Appendix D).

a) Show that this vector is linearly polarized at any point where it is de ned.
b) Find how its direction of polarization changes along a circle $r =$ constant, for example the circle $r = a/2$.

2-8 Write in component form, using cylindrical and cartesian coordinates, a vector, $\mathbf{E}_1$, such that the vector $\mathbf{E} + \mathbf{E}_1$ is circularly polarized at any point, where **E** is the vector de ned in the previous problem.

2-9 Consider the following vector eld, **E**, which is de ned in all 3-D space with respect to a spherical coordinate system (r, ϑ, φ) and corresponds to the electric eld of the short electric dipole (see Chapter 12):

$$\mathbf{E} = \frac{1}{j\omega\ _0}\frac{\mathcal{M}}{2}\left(\quad + \frac{1}{r}\right)\frac{e^{\quad r}}{r^2}\cos\vartheta\,\hat{r}$$

$$+ \frac{1}{j\omega\ _0}\frac{\mathcal{M}}{4}\left(\ ^2 + \frac{\quad}{r} + \frac{1}{r^2}\right)\frac{e^{\quad r}}{r}\sin\vartheta\,\hat{\vartheta} \quad ,$$

where $\quad = \omega\sqrt{\ _0\ _0}$, and $\mathcal{M}$ is a constant. For simplicity, retain only the term of order $1/r^2$ in the r-component, and the term of order $1/r$ in the ϑ-component. Find the linear polarization ratio and the axes of the polarization ellipse, as a function of ϑ, for $\omega = 2\quad 10^6$ rad/s, $r = 300\,$m.

2-10 Write the equation of the ellipse described, in the time domain, by the instantaneous vector of unit amplitude and angular frequency $\omega = 2 \quad 10^8$ rad/s which corresponds to the following Stokes parameters: $S_1 = 0$, $S_2 = 1/\sqrt{2}$, $S_3 = \quad 1/\sqrt{2}$. Find the polar coordinates of this point on the Poincare sphere.

2-11 Write the Jones matrix for a *half-wave retardation plate*, i.e., an element which introduces a phase delay of radians between its principal axes. First, write the matrix in a reference system which coincides with its principal axes, as was shown in the text for the quarter-wave plate. Then, rewrite it in a reference system rotated by an angle ϑ with respect to the previous one.

2-12 Write the Jones matrix for the following combinations of cascaded elements.

a) A quarter-wave plate followed by an ideal polarizer (with an arbitrary rotation between the axes of the two elements).

b) An ideal polarizer followed by a quarter-wave plate (again with an arbitrary rotation).

Show that, for a generic input state of polarization, the output states of the two combinations are not equal. Discuss the reason for this situation to occur.

2-13 A birefringent plate (see Chapter 6) is characterized, as a function of frequency, by the following Jones matrix: $J_{11} = 1$, $J_{22} = \exp(j\omega \quad n\ell/c)$, $J_{12} = J_{21} = 0$, where $c = 3 \quad 10^8$ m/s (speed of light in free space), $n = 10 \quad ^3$ (index di erence, or birefringence), $\ell = 1$ mm (plate thickness).
Find *all* the frequencies (or wavelengths, measured in free space) at which this device acts as a quarter-wave plate.

2-14 Suppose that the device studied in the previous problem can, in practice, be considered as a quarter-wave plate when the phase shift between J_{11} and J_{22} deviates by less than 1% from its nominal value of $/2$.
Calculate the relative bandwidths, ω/ω_i, $(i = 1, 2, ...)$, over which this tolerance is satis ed, around the nominal values ω_i found in the previous problem. Discuss how these bandwidths behave, as ω_i increases.

References

Azzam, R.M.A. and Bashara, N.M. (1987) *Ellipsometry and Polarized Light.* Paperback edition, Elsevier, Amsterdam, Chapters 1–2.

Galtarossa, A. and Menyuk, C.R. (Eds.) (2005) *Polarization Mode Dispersion.* Springer, New York.

Guenther, R.D. (1990) *Modern Optics.* Wiley, New York.

Halmos, P.R. (1958) *Finite-Dimensional Vector Spaces.* Van Nostrand Reinhold, New York.

Kliger, D.S., Lewis, J.W. and Randall, C.E. (1990) *Polarized Light in Optics and Spectroscopy.* San Diego, Academic.

Someda, C.G. and Stegeman, G.I. (Eds.) (1992) *Anisotropic and Nonlinear Optical Waveguides.* Elsevier, Amsterdam.

Yariv, A. and Yeh, P. (1984) *Optical Waves in Crystals: Propagation and Control of Laser Radiation.* Wiley, New York, Sections 1.3 and 5.1.

CHAPTER 3

General theorems

3.1 Introduction

Chapter 1 was devoted to the di erential equations which the e.m. eld satis es, and to some manipulations that allowed us to pass to other equations, either in the same unknowns or in other variables. The small number and the apparent simplicity of those equations could lead one to believe that they could be solved with one procedure, or at most with a few procedures. Still, the number of problems where Maxwell's equations have *exact* closed-form solutions is very small. Furthermore, the parameters in Maxwell's equations can vary by many orders of magnitude. For example, frequencies of technical interest range from zero to over 10^{14} Hertz; conductivity may vary between zero and 10^6 ^{1} m ^{1}, and so forth. Therefore, approximate methods, which are necessary in many instances, are extremely diversi ed.

As a result, *ad hoc* discussions are required in a multitude of cases. Indeed, that is what we will do starting with Chapter 4 and proceeding through until the end of the book. However before we begin this, it is advisable to prove some theorems which follow directly from the equations that were either postulated or derived in Chapter 1. Obviously, these theorems must be satis ed by all the solutions to be found later on. Their purposes are di erent. Some of them clearly aim at "economy". For example, the *uniqueness theorem* of Section 3.3 illustrates the conditions which are necessary in order to obtain one solution to Maxwell's equations. Others (for example, the *equivalence theorem* of Section 3.5) are mainly pointers to approximate methods. In no case, however, does the matter discussed in this chapter yield an algorithm for solving Maxwell's equations. For this, the reader should wait until the following chapters.

In some cases, either the statements, or the proofs, depend on whether we consider a generic behavior in the time domain, or time-harmonic complex vectors. This will be emphasized by splitting those theorems into two subsections. Whenever that is not done, it is understood that the proofs in the two domains are substantially the same.

3.2 Poynting's theorem. Wave impedance

Under this heading we will present, separately for the time and frequency domains, a relation between quantities with the physical dimensions of power. It has an immediate physical interpretation as an energy budget. Furthermore,

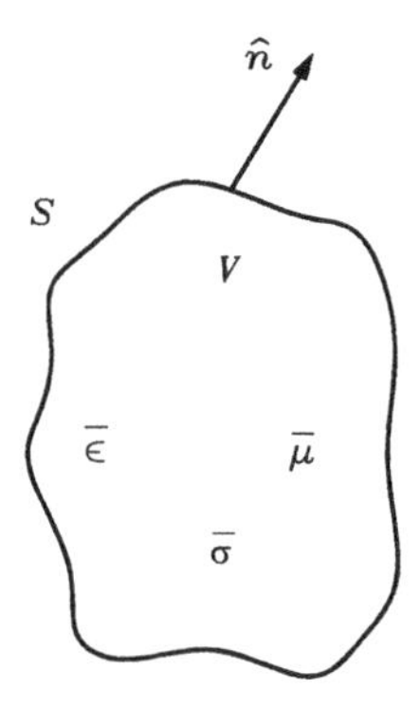

Figure 3.1 *The region of interest in the proof of Poynting's theorem.*

the next section will show that the same relation plays another important role, as the lemma of a subsequent theorem.

The proofs presented here require a *linear anisotropic medium*, whose dyadics $\bar{\ }$, $\bar{\ }$, $\bar{\ }$ are constant with respect to time t and angular frequency ω. It is trivial to reduce them to an isotropic medium, whereas it can be a di cult exercise to extend them to cases in which and depend on ω. This point will be reconsidered in Chapter 5.

3.2.1 *Time domain*

In the *time domain*, the theorem can be stated as follows. Consider a region V, in ordinary 3-D space, bounded by a closed, regular, time-invariant surface, S. Then, the total instantaneous power that the sources in V generate equals the sum of three terms: (i) the rate of change vs. time of the electromagnetic energy stored in V, (ii) the instantaneous dissipated power (i.e., the Joule e ect) in V, and (iii) the instantaneous power transmitted through the surface S.

All those terms should be understood in the algebraic sense (i.e., they may be positive or negative), except for dissipation, which is non-negative by de ni- tion. In particular, the so-called sources can be either *generators*, i.e., imposed currents such that, as a time average, they supply a positive power to the eld, or *utilizers*, i.e., imposed currents which correspond to a time-averaged negative power, like for instance electrical motors.

To prove the time domain version of the theorem, let us write, at any point in the region V (Figure 3.1), Maxwell's equations; in the case of a linear anisotropic medium, they are straightforward generalizations of Eqs. (1.47) and (1.48):

$$\nabla \quad \mathbf{E} = - \frac{\partial \mathbf{H}}{\partial t} , \tag{3.1}$$

$$\nabla \quad \mathbf{H} = \bar{\ } \frac{\partial \mathbf{E}}{\partial t} + \bar{\ } \mathbf{E} + \mathbf{J}_0 . \tag{3.2}$$

We then take the scalar product (from the left) of the first equation by $\mathbf{H}$, and of the second by $\mathbf{E}$. Next we subtract term by term, and use the vector identity (C.4), which reads:

$$\mathbf{a} \cdot \nabla \times \mathbf{b} - \mathbf{b} \cdot \nabla \times \mathbf{a} = -\nabla \cdot (\mathbf{a} \times \mathbf{b}) .$$

We obtain the following relation, holding at any point in V:

$$\nabla \cdot (\mathbf{E} \times \mathbf{H}) = -\frac{\partial}{\partial t}\left(\frac{1}{2}\mu \mathbf{H} \cdot \mathbf{H} + \frac{1}{2}\epsilon \mathbf{E} \cdot \mathbf{E}\right) - \sigma \mathbf{E} \cdot \mathbf{E} - \mathbf{J}_0 \cdot \mathbf{E} . \quad (3.3)$$

Let us then integrate it over the whole region V. Taking account of the time-invariance of V, applying Gauss' theorem (A.8) to the integral of the divergence of $\mathbf{E} \times \mathbf{H}$, and finally reordering the terms, we obtain:

$$\begin{aligned} -\int_V \mathbf{J}_0 \cdot \mathbf{E}\, dV &= \frac{\partial}{\partial t} \int_V \frac{1}{2}\left(\mu \mathbf{H} \cdot \mathbf{H} + \epsilon \mathbf{E} \cdot \mathbf{E}\right) dV \\ &+ \int_V \sigma \mathbf{E} \cdot \mathbf{E}\, dV + \oint_S \mathbf{E} \times \mathbf{H} \cdot \hat{n}\, dS , \end{aligned} \quad (3.4)$$

where $\hat{n}$ denotes the unit vector normal to S, directed outwards from V.

Eq. (3.4) proves the above statement as can be rapidly verified by considering the physical meaning of each term on which we now comment. We first consider the term on the left-hand side. The imposed current density $\mathbf{J}_0$ can be regarded as the motion of a charge distribution of density ρ_0 and velocity $\mathbf{v}_0$, i.e., $\mathbf{J}_0 = \rho_0 \mathbf{v}_0$. Then, if we recall Eq. (1.2), we establish that $\mathbf{E} \cdot \mathbf{J}_0$ is the per-unit-time and per-unit-volume work of the Lorentz force, when a current with density $\mathbf{J}_0$ circulates. Therefore, when we change its sign, i.e., we consider $-\mathbf{E} \cdot \mathbf{J}_0$, this represents the per-unit-volume instantaneous power that is transferred (in the algebraic sense) by the source to the e.m. field $\{\mathbf{E}, \mathbf{H}\}$, at the expense of an energy source that supports the imposed current density $\mathbf{J}_0$, but that does not appear in Maxwell's equations, and that for this reason is often referred to as being of a "non-electrical nature".

Passing to the right-hand side, the reader should already know from the time-independent (or slowly varying) case, that the quadratic forms $(\mu \mathbf{H} \cdot \mathbf{H})/2$ and $(\epsilon \mathbf{E} \cdot \mathbf{E})/2$ are, respectively, the instantaneous magnetic and electric *energy densities* stored in V, measured in joules per cubic meter. The quadratic form $\sigma \mathbf{E} \cdot \mathbf{E}$ is the instantaneous dissipated power density measured in watts per cubic meter. The remaining term, consisting of the surface integral of the vector product $\mathbf{E} \times \mathbf{H} \cdot \hat{n}$, is the *power transmitted through the surface* S, as one can deduce from two observations. The first is that inside the region V there are no other possible uses for electromagnetic power, in addition to those already discussed. The second is that the sign of this surface integral changes its sign (while its absolute value remains the same) if we deal with the region that is complementary to V.

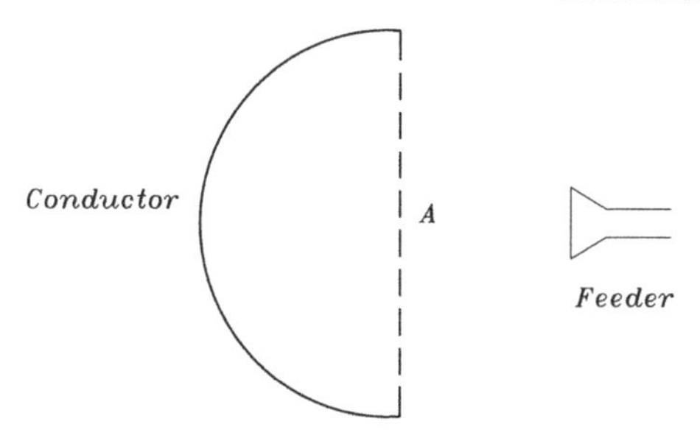

Figure 3.2 *Schematic diagram of a re ector antenna.*

The vector function:

$$\mathbf{P} = \mathbf{E} \quad \mathbf{H} \qquad (\mathrm{W/m^2}) \quad , \tag{3.5}$$

is referred to as the *Poynting vector*. Its physical dimensions are those of a power density, i.e., watts per square meter. However, it is not legitimate to state that it has a physical meaning point by point. Only its ux through a closed surface is physically identi ed as a power. Nonetheless, in the following chapters we will encounter many cases in which we consider the ux of $\mathbf{P}$ through open surfaces, S', and attribute to it the physical meaning of power transmitted through S'. Whenever we do so, this will be justi ed by the fact that there exists another surface, S'', such that (i) $S'' + S'$ is a closed surface, and (ii) the ux of $\mathbf{P}$ through S'' is zero. A typical example in this sense is a so-called *aperture antenna*, similar to a parabolic-re ector antenna (see Figure 3.2). The total power received by the antenna is usually computed as the Poynting vector ux through its "aperture," because the contribution through the paraboloidal surface, consisting of a good conductor, is supposed to be negligible.

3.2.2 Frequency domain

In the complex-vector representation of time-harmonic vectors, Eq. (3.4) is replaced by a relation involving complex quantities. We assume that the reader is familiar with the concept of *complex power* for a lumped-element linear one-port network (Figure 3.3).

Poynting's theorem can then be stated as follows. Consider a region V in ordinary 3-D space, bounded by a regular closed time-invariant surface, S. Then, the total complex power delivered by the sources present in V equals

For a linear one-port electrical device, this quantity (having physical dimensions of power) can be de ned as follows: (i) its real part is the *active power*, i.e., the time-average over a period of the instantaneous power; (ii) its imaginary coe cient is the so-called *reactive power*, i.e., the product of twice the angular frequency, 2ω, times the amplitude of the sinusoidal component (at an angular frequency 2ω) of the energy stored in the device, with sign + or depending on whether magnetic or electric energy dominates in the storage.

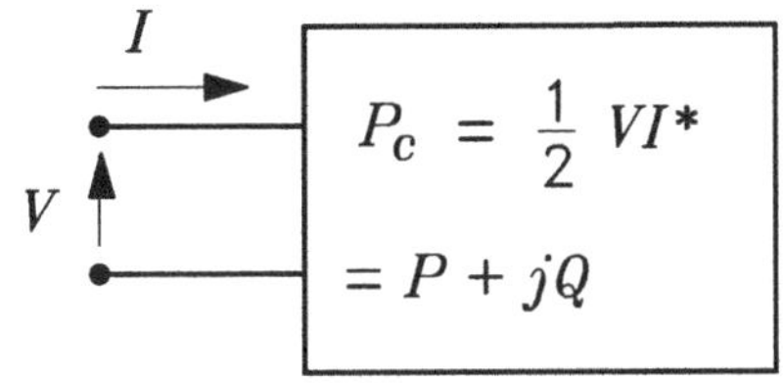

Figure 3.3 *The concept of complex power in linear circuit theory.*

the sum of three terms: (i) the reactive power corresponding to the periodic variation in the energy stored in V, (ii) the active power dissipated in V, and (iii) the complex power transmitted through the surface S.

The proof is quite similar to that of the previous case, but the results should not be mixed; let us stress the main di erences. We start from the complex form of Maxwell's equations, and multiply scalarly from the left the rst equation by $\mathbf{H}$, and the *complex conjugate* of the second equation by $\mathbf{E}$. After subtracting, we use the identity (C.4). Then, we integrate over the region V. Finally, we divide all terms by 2, in order to correctly identify their physical meaning. We obtain:

$$\int_V \frac{\mathbf{J}_0 \ \mathbf{E}}{2} \ = \ \int_V \frac{\mathbf{E} \ \overline{\ } \ \mathbf{E}}{2} \, \mathrm{d}V + 2j\omega \int_V \left[\frac{\mathbf{H} \ \ \overline{\ } \ \mathbf{H}}{4} \qquad \frac{\mathbf{E} \ \overline{\ } \ \mathbf{E}}{4} \right] \mathrm{d}V$$

$$+ \int_S \frac{\mathbf{E} \quad \mathbf{H}}{2} \ \hat{n} \, \mathrm{d}S \quad . \tag{3.6}$$

It is straightforward to check that Eq. (3.6) corresponds to what we stated. Notice that the rst term on the right-hand side is real, while the second is imaginary; the left-hand side and the nal term are, in general, complex. Also note that, as we pointed out previously in Chapter 2, the complex Poynting vector:

$$\mathbf{P} = \frac{\mathbf{E} \quad \mathbf{H}}{2} \quad , \tag{3.7}$$

is *not* related to the time-domain Poynting vector, Eq. (3.5), by the Steinmetz method. Indeed, the one-to-one and onto mapping of complex vectors into time-harmonic vectors commutes only with *linear operations* in either of the two spaces, while Poynting vectors, being de ned via a vector product, are not linear with respect to the e.m. elds.

We conclude this section giving a broad-scope de nition, whose meaning probably looks obscure at this stage; hopefully, however, its application will clarify it in the following chapters. Let $\hat{u}$ be a unit-vector in ordinary 3-D real space; for any eld $\{\mathbf{E}, \mathbf{H}\}$, we de ne as the *wave impedance in the direction of* $\hat{u}$, at any point Q, the dyadic operator $\overline{\ }_{\hat{u}}(Q)$ de ned by the following relationship:

$$\hat{u} \quad \mathbf{E} \quad \hat{u} = \overline{\ }_{\hat{u}}(Q) \ (\mathbf{H} \quad \hat{u}) \quad . \tag{3.8}$$

We will see in the following chapters that the dyadic $\overline{\ }_{\hat{u}}(P)$ reduces to a scalar in most cases of practical interest, where the components of **E** and **H** that are transverse with respect to $\hat{u}$ are orthogonal to each other. Hence, we may think of it as a generalized "ratio" of the components of **E** and **H** that are transverse with respect to $\hat{u}$. Its usefulness consists in allowing us to express the component of the Poynting vector parallel to $\hat{u}$ (and consequently, the ux of **P** through a surface) in terms of only one of the two vectors **E** and **H**. The role is then the same as that of an impedance (or an admittance) in network theory, quantities which allow us to express the complex power owing into a port in terms of either only the voltage, or only the current at that port (Figure 3.3). In particular, if **E** and $\hat{u}$ are orthogonal to each other, we have:

$$\mathbf{P}\ \ \hat{u} = \frac{1}{2}\,(\mathbf{H}\ \ \ \hat{u})\ \ \bar{\ }_{\hat{u}}\ \ (\mathbf{H}\ \ \ \hat{u}) = \frac{1}{2}\,\mathbf{E}\ \ (\bar{\ }_{\hat{u}})^{\ \ 1}\ \ (\hat{u}\ \ \mathbf{E}\ \ \ \hat{u})\ \ , \qquad (3.9)$$

where, to simplify the notation, the dependence on the point Q is implicitly understood. We leave it as an exercise for the reader to nd how this expression should be modi ed when **H** and $\hat{u}$ are orthogonal.

3.3 Uniqueness theorem

Whenever one has to solve di erential equations, it is advisable, before trying to determine the unknowns, to identify conditions that guarantee, when they hold, the uniqueness of the solution. In this section we shall do that with reference to Maxwell's equations. After that, we will be sure that, if we can solve the problem, there are no risks of other solutions, which might contradict experimentally our analytical (or numerical) results. At the same time, we will identify the *minimum* set of conditions that guarantee the uniqueness of the solution. However, we will also see later that there are situations, not at all rare, in which Maxwell's equations can be solved only if one starts with a set of data larger than this minimum. What the *uniqueness theorem* tells us, is that the "extra" data can not be independent of the essential data.

3.3.1 Time domain

In the *time domain*, the theorem can be stated as follows. Consider a region V, in ordinary 3-D space, bounded by a closed, regular, time-invariant surface, S. Maxwell's equations therein have a unique solution $\{\mathbf{E}, \mathbf{H}\}$, if the following quantities are known:

the *eld sources*, i.e., the imposed current density, $\mathbf{J}_0$, at each point of V, and at each time t after an initial time t_0,

the *initial conditions*, i.e., the values of **E** and **H** at each point of V at the initial time t_0, and

the *boundary conditions*, i.e., the components of *either* **E** *or* **H** tangential to the surface S, at each point on S, for all times $t > t_0$.

To prove this, we proceed *ab absurdo*, i.e., we start by assuming that there are two solutions, $\{\mathbf{E}', \mathbf{H}'\}$ and $\{\mathbf{E}'', \mathbf{H}''\}$, and then we show that they must be the same in order to satisfy all the assumptions. Maxwell's equations being linear, the difference between the two solutions:

$$\mathbf{e} = \mathbf{E}' \quad \mathbf{E}'' \;, \qquad \mathbf{h} = \mathbf{H}' \quad \mathbf{H}'' \;, \tag{3.10}$$

is in turn an e.m. field. The term $\mathbf{J}_0$ being identical for the two solutions implies that the field (3.10) satisfies homogeneous Maxwell's equations, without sources, in V. Then, if we apply Poynting's theorem in Eq. (3.4) to $\{\mathbf{e}\,,\mathbf{h}\}$, we find:

$$\begin{aligned} 0 \;=\; & \frac{\partial}{\partial t}\int_V \frac{1}{2}\,(\mathbf{h} \;\bar{}\; \mathbf{h} + \mathbf{e} \;\bar{}\; \mathbf{e}\,)\,\mathrm{d}V \\ & + \int_V \mathbf{e} \;\bar{}\; \mathbf{e}\;\,\mathrm{d}V + \int_S \mathbf{e} \quad \mathbf{h}\;\hat{n}\,\mathrm{d}S\,. \end{aligned} \tag{3.11}$$

It was assumed that we know unambiguously either $\mathbf{E}_{\tan}$ or $\mathbf{H}_{\tan}$ at each point on S; this allows us to determine that the last term of Eq. (3.11) is zero. Integrating the remaining terms with respect to time, from initial time t_0, and exploiting the initial-condition assumption for $t = t_0$ at each point of V, we finally write:

$$\int_V \frac{1}{2}\,(\mathbf{h} \;\bar{}\; \mathbf{h} + \mathbf{e} \;\bar{}\; \mathbf{e}\,)\,\mathrm{d}V + \int_{t_0}^{t} \mathrm{d}t \int_V \mathbf{e} \;\bar{}\; \mathbf{e}\;\,\mathrm{d}V = 0 \;\;. \tag{3.12}$$

The first term on the left-hand side of Eq. (3.12) is an integral of definite positive quadratic forms; the second term is an integral of a semidefinite positive quadratic form. Therefore, this equation can only be satisfied in the case:

$$\mathbf{e} = 0 \;\;, \qquad \mathbf{h} = 0 \;\;, \tag{3.13}$$

as we intended to show.

3.3.2 Frequency domain

In the complex-vector representation of time-harmonic vectors, the uniqueness theorem has to be modified, slightly but in a non-trivial manner. The assumptions, under which we will show that Maxwell's equations have a unique solution in a region V, are the following:

- the *field sources*, i.e., the imposed current densities, $\mathbf{J}_0$, are known at each point in V,
- the *boundary conditions*, i.e., the components of *either* $\mathbf{E}$ *or* $\mathbf{H}$ tangential to the surface S, are known at each point on S, and
- the medium contained in the region V has everywhere a nonzero conductivity, i.e., it is a *lossy medium.*

The proof of this theorem starts off as does the previous one, but it diverges from it at the moment of applying Poynting's theorem to the difference between the two solutions. Indeed, for complex vectors, Poynting's theorem reads:

$$0 = \int_V \frac{\mathbf{e}\ ^{-}\ \mathbf{e}}{2}\, \mathrm{d}V + 2j\omega \int_V \left(\frac{\mathbf{h}\ ^{-}\ \mathbf{h}}{4} \quad \frac{\mathbf{e}\ ^{-}\ \mathbf{e}}{4} \right) \mathrm{d}V \quad . \tag{3.14}$$

We omitted the surface integral, equal to zero, for the same reason as in the previous case. The real and the imaginary parts of the right-hand side of Eq. (3.14) have to be both equal to zero, so we have:

$$\int_V \frac{\mathbf{e}\ ^{-}\ \mathbf{e}}{2}\, \mathrm{d}V = 0 \quad , \tag{3.15}$$

$$\int_V \left(\frac{\mathbf{h}\ ^{-}\ \mathbf{h}}{4} \quad \frac{\mathbf{e}\ ^{-}\ \mathbf{e}}{4} \right) \mathrm{d}V = 0 \quad . \tag{3.16}$$

It is straightforward to establish that *if and only if* we have $^{-} \neq 0$, these two relations imply $\mathbf{e} = \mathbf{h} = 0$, as required. This explains why it was necessary to add the third assumption.

If $^{-} = 0$, i.e., in a lossless medium, we see that elds not sustained by any imposed current can indeed exist in V, simply provided that, as required by Eq. (3.16), the time-averaged energies stored in V in magnetic and electric forms are equal. Such an e.m. eld is referred to as a *resonant eld*, because of its clear analogy with a lumped-element LC circuit at resonance, i.e., at an angular frequency ($\omega = 1/\sqrt{LC}$).

It should be pointed out that assuming a lossless medium often simpli es Maxwell's equations remarkably. For this reason, we will use it quite often in the following chapters, but we must reconcile it with the uniqueness theorem. There is no contradiction, in those cases where the solution for $= 0$ is indeed the limit of the solution for the lossy case, as $\rightarrow 0$. In fact, the third assumption of the uniqueness theorem requires only a non-vanishing loss, without requiring a minimum value for this loss, which can thus be arbitrarily small.

Finally, let us mention that the uniqueness theorem can also be proved for a eld de ned *throughout 3-D space*, provided that the assumption on either $\mathbf{E}_{\mathrm{tan}}$ or $\mathbf{H}_{\mathrm{tan}}$ on a closed surface S is replaced by the following postulate (which is known as *Sommerfeld's radiation condition*):

$$\lim_{r \to \infty} r|\mathbf{E}| = \lim_{r \to \infty} r|\mathbf{H}| = 0 \quad , \tag{3.17}$$

where r is the distance from an arbitrarily chosen origin 0 (not at in nity) to the point where the eld is calculated. The proof can be undertaken by the reader by following the guidelines of the previous proofs, applying Poynting's theorem to a sphere of radius r centered at 0, and then letting r tend to in nity.

Comments on how the last proof should be adapted to problems where there are structures of cylindrical shape and in nite length, such as waveguides, are postponed until Chapter 7.

3.4 Reciprocity theorem

This section is devoted to a theorem that we will only prove for complex vectors representing time-harmonic elds: we leave as an exercise for the reader the version in the time domain. Before the proof, we need to introduce a de nition, whose scope of applicability includes also circuit theory. Let $\mathbf{J}_0$ be an imposed current density, de ned in a region V of ordinary 3-D space. Let $\{\mathbf{E}, \mathbf{H}\}$ be an electromagnetic eld, de ned in the same region V, but not necessarily the eld sustained by the source $\mathbf{J}_0$. We de ne the *reaction* of the eld $\{\mathbf{E}, \mathbf{H}\}$ on the source $\mathbf{J}_0$, in the region V, as the quantity:

$$R(\mathbf{E}, \mathbf{J}_0) = \int_V \frac{\mathbf{E}\ \mathbf{J}_0}{2}\, \mathrm{d}V \quad . \tag{3.18}$$

Eq. (3.18) has the physical dimensions of power. However, even when $\{\mathbf{E}, \mathbf{H}\}$ is the eld sustained by $\mathbf{J}_0$, it is not the complex power supplied by the source $\mathbf{J}_0$; the latter, in fact, does not involve $\mathbf{J}_0$, but $\mathbf{J}_0$ (see Section 3.2).

The main use of this new de nition results in the theorem that we will prove shortly, which establishes under what conditions two di erent current densities, $\mathbf{J}_{0a}$ and $\mathbf{J}_{0b}$, de ned in the same region V of 3-D space, play interchangeable roles. Subscripts a and b will be used also to distinguish the elds sustained by these two di erent sets of sources. For brevity, let us use terms like "source a, source b, eld a, eld b," without fear of ambiguity.

The *Lorentz reciprocity theorem* states that the reaction of the eld a on the source b is equal to that of the eld b on the source a, provided that:

the region V contains either a linear isotropic medium, or a linear anisotropic medium whose constitutive dyadics are all represented by diagonal matrices in a real orthogonal reference frame;

either V is the whole 3-D space, and then the elds a and b satisfy the Sommerfeld's radiation conditions in Eq. (3.17), or, alternatively, V is bounded by a closed surface S, on which any e.m. eld de ned in V satis es a boundary condition of the following type, whose physical meaning will become clear throughout the following chapters:

$$\mathbf{E}_{\tan} = Z\, \mathbf{H}_{\tan}\ \ \hat{n} \quad , \tag{3.19}$$

where $\hat{n}$ is the unit vector normal to S; Z [ohms] may be a function of spatial coordinates, but is independent of the eld $\{\mathbf{E}, \mathbf{H}\}$, and is referred to as the *wall impedance*.

For simplicity, we will write in full detail the proof only for an isotropic medium in a bounded region, leaving as exercises the extensions to the other cases just stated. Maxwell's equations for the source a and the associated eld read:

$$\nabla\ \ \mathbf{E}_a = \ \ j\omega\ \ \mathbf{H}_a \quad , \tag{3.20}$$

$$\nabla\ \ \mathbf{H}_a = j\omega\ {}_c\mathbf{E}_a + \mathbf{J}_{0a} \quad . \tag{3.21}$$

We take the scalar product of the rst equation with $\mathbf{H}_b$, and of the second with $\mathbf{E}_b$; then, we add term by term the two thus obtained relations. After that, we swap the roles of a and b, and repeat the same operations. Finally, we subtract term by term the two equations obtained in the two cases. The nal result is:

$$\mathbf{H}_b \ \nabla \ \ \mathbf{E}_a \ \ \mathbf{E}_a \ \nabla \ \ \mathbf{H}_b + \mathbf{E}_b \ \nabla \ \ \mathbf{H}_a \ \ \mathbf{H}_a \ \nabla \ \ \mathbf{E}_b = \mathbf{J}_{0a} \ \mathbf{E}_b \ \ \mathbf{J}_{0b} \ \mathbf{E}_a \,, \quad (3.22)$$

while all the other terms cancel out.

We now divide Eq. (3.22) by 2 and integrate it over the region V. Then we apply to the two pairs of terms on the left-hand side the vector identity $\mathbf{a} \ \nabla \ \ \mathbf{b} \ \ \mathbf{b} \ \nabla \ \ \mathbf{a} = \nabla \ (\mathbf{b} \ \ \mathbf{a})$. We obtain:

$$\int_S \left(\frac{\mathbf{E}_a \ \ \mathbf{H}_b}{2} \ \ \frac{\mathbf{E}_b \ \ \mathbf{H}_a}{2} \right) \ \hat{n}\, \mathrm{d}S = \int_V \frac{\mathbf{J}_{0a} \ \mathbf{E}_b}{2} \, \mathrm{d}V \ \ \int_V \frac{\mathbf{J}_{0b} \ \mathbf{E}_a}{2} \, \mathrm{d}V \,. \quad (3.23)$$

We can now apply the cross-product cyclic permutation rule to both terms in the integral on the left-hand side of Eq. (3.23). After that, we apply Eq. (3.19) to the eld a in one of the two terms, and to the eld b in the other term. We conclude that the left-hand side of Eq. (3.23) is zero. This nal result is exactly what we wanted to prove.

As mentioned at the beginning of the section, this theorem means that, when its assumptions hold, one can interchange roles between two distinct sets of sources. A source a, taken alone, gives rise to a certain electric eld $\mathbf{E}_a$ at some points, where, in a di erent situation, there is another source b; then the electric eld $\mathbf{E}_b$ generated by source b at the points that are sites of source a can not be independent of the eld $\mathbf{E}_a$. It must in fact satisfy the equality between the two integrals on the right-hand side of Eq. (3.23).

This theorem expresses, in more physical terms, the same concept as reciprocity in an electrical network consisting of linear, passive and reciprocal two-terminal components (resistors, inductors, mutually coupled inductors, and capacitors). In elementary circuit theory, this concept is often expressed in rather simplistic terms as follows: if we interchange the positions of a generator and of a current meter in a network, the reading of the latter remains unchanged.

This familiarity with reciprocity may induce the reader to believe that all the consequences of the theorem match easily with intuition. This is not always true. To show this, let us consider the example of a so-called Y-junction (Figure 3.4). In simple intuitive terms, it is the con uence of three in nitely long waveguide structures, each of which, we will suppose, can propagate only one solution of Maxwell's equations (apart from an arbitrary complex multiplicative constant). What is easy to reconcile with intuition, is that one can design such a structure so that, when a wave is coming in from arm 1, it is split in equal parts between arms 2 and 3. What, on the other hand, is not obvious at rst sight, is that, for the same structure, when a wave is coming in from arm 2, then, *because of reciprocity*, only one half of its power, at most, can be coupled to the output of arm 1. The reader may come back to this example at the end of Section 3.9, and rediscuss it in terms of symmetries.

By far and large, the majority of the subjects that are dealt with in this

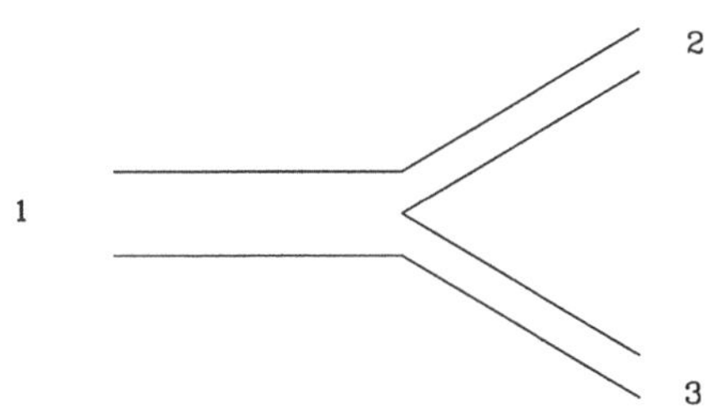

Figure 3.4 *A microwave Y-junction.*

book satisfy the assumptions of the reciprocity theorem. However, this should not induce the reader to draw the erroneous conclusion that *all* electromagnetic phenomena of practical interest are reciprocal. Nonreciprocal behavior is present in nature as well as in man-made devices. The most classical example of the first class is *ionospheric propagation.* As for the second one, we may quote two-port and three-port devices that are widely used at microwave and optical frequencies. *Isolators* are two-port devices which, in their ideal version, transmit a signal without any loss in one sense, and introduce an infinite attenuation in the opposite sense. *Circulators* are three-port devices which, again in their ideal version, pass all the power entering one port to the next one, and nothing to the other one. The usefulness of such devices in communications or radar systems is quite evident.

From the proof that we gave in this section, it should be easy to grasp that, in order to get nonreciprocal behavior, it is strictly necessary to have a *suitable* anisotropic medium, and also to exploit it in a convenient way. In the ionosphere, anisotropy is caused by the Earth's static magnetic field. Isolators and circulators exploit either magnetized ferrites, or some types of crystals. We postpone discussion of more details regarding these subjects until Chapter 6.

3.5 Equivalence theorem

3.5.1 Preliminaries

The contents of this section are more complicated than the previous ones. The path that we will take might initially look unnecessarily cautious and slow, but eventually it will turn out to be beneficial to the reader.

In Section 1.4, a known term was introduced into Maxwell's equations, in order to model the physical sources of a field. This procedure could lead one to believe that, in every problem, one must first identify a unique, well-defined function of space coordinates, $\mathbf{J}_0$, that corresponds to the actual physical sources. The aim of this section is to illustrate that indeed the contrary is true: in order to determine the field in a given region of 3-D space, multiple choices exist for replacing assigned sets of sources with other *equivalent* sets. Equivalent means that, if the boundary conditions are left unchanged,

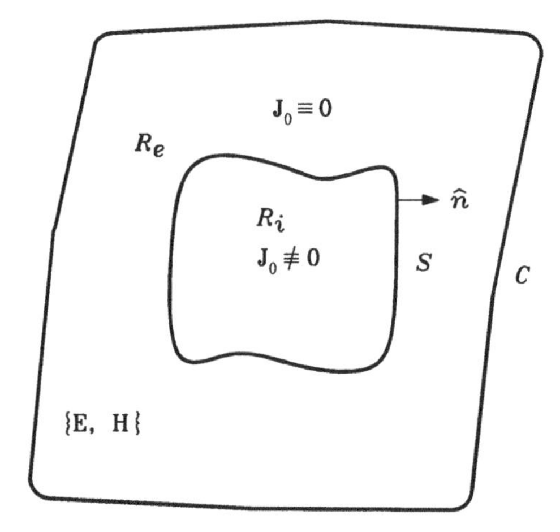

Figure 3.5 *The region of interest in the proof of the equivalence theorem, and its subdivision as an internal and an external region.*

the eld remains the same as that given by the original sources. The reader may correlate these results with others in circuit theory, in particular, with Thevenin's and Norton's theorems on the equivalent generators (Kuo, 1966). All analogies between elds and circuits have deep logical roots, as circuits result always from an abstraction process that starts from the elds.

The contents of this section apply only to media that are strictly *linear*, as all proofs involve superposition, and with *losses*, arbitrarily small but not strictly zero, as required by the uniqueness theorem of Section 3.3. The phrase "turn o the generators in a region of space R" will be used to indicate that, starting from a current density not identically zero in R, we pass to another one, which di ers from the previous one *only because* we impose $\mathbf{J}_0$ 0 in R. We will denote with R that region of 3-D space where the e.m. eld on which we elaborate is de ned, and with C the boundary of R. On C, we assume that the boundary conditions are such that, once we specify all the sources $\mathbf{J}_0$ within R, then the eld is unique everywhere in R.

Furthermore, we will assume that all the sources are localized within an internal subset of R, so that (see Figure 3.5) there are an in nite number of regular closed surfaces completely contained inside R such that if S is any of these taken at random, then we have $\mathbf{J}_0 = 0$ in the region contained between S and C. We use the symbol R_e to denote the portion of R external to S, and R_i to denote the portion internal to S. Of course, R can also be the whole space, in which case C is at in nity and the eld satis es the Sommerfeld's radiation conditions in Eq. (3.17).

Suppose that we aim only at determining the eld *in the region* R_e, which does not contain sources. This case is quite frequent in practice. For example, we might be dealing with the far eld of a transmitting antenna, or with a di raction problem. Then, according to the uniqueness theorem of Section 3.3,

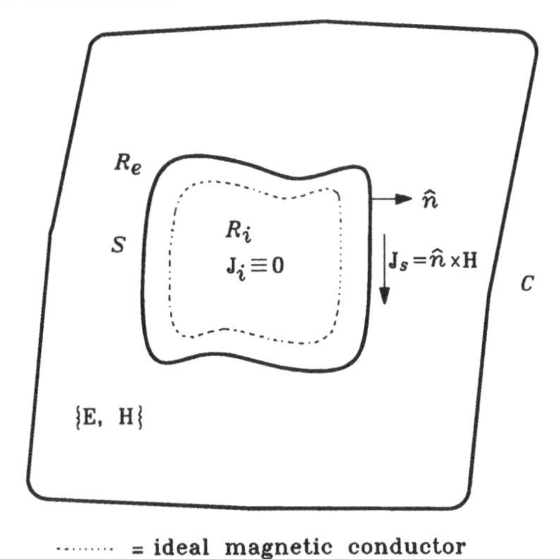

Figure 3.6 *The equivalence theorem, in the version which makes use of a perfect magnetic conductor.*

any set of sources, situated either in R_i or on S, and capable of giving rise to the correct tangential component of $\mathbf{H}$ on the external face of S, would generate in R_e a eld $\{\mathbf{E}, \mathbf{H}\}$, equal to that sustained by the true sources $\mathbf{J}_0$. Recalling what we said in Section 1.6, we note that the problem may be reduced to that of being able to guarantee $\mathbf{H}_{\tan} = 0$ *on the internal face of* S. In fact, once that is done, the correct $\mathbf{H}_{\tan}$ on the external face is supplied by a source consisting of a *surface electric current*, on S, having a (linear) density:

$$\mathbf{J}_S = \hat{n} \quad \mathbf{H} \qquad (\mathrm{A/m}) \quad , \tag{3.24}$$

where $\hat{n}$ is the unit vector normal to S, oriented towards R_e.

The internal face of S does not belong to the region R_e where we are interested in the eld. Hence, in order to ensure $\mathbf{H}_{\tan} = 0$ on this face, we can take the liberty of modifying the medium in R_i. In nature, a material capable of guaranteeing $\mathbf{H}_{\tan} = 0$ on its surface does not exist, not even as a rough approximation. Nevertheless, it is a useful mathematical device to *de ne* such a medium, and call it a *perfect magnetic conductor.*

What we have obtained in this way is schematically illustrated in Figure 3.6. The generators in R_i are turned o , the current density of Eq. (3.24) is imposed at any point on S, and a "perfect magnetic conductor" has been placed all along the internal face of S. The eld $\{\mathbf{E}, \mathbf{H}\}$ coincides in R_e with that of Figure 3.5. In R_i, as the medium has been modi ed, the eld there is no longer de ned.

Our goal has been, to a large extent, reached, because the physical sources have been replaced with equivalent ones. But the price we paid for this, inventing the ideal magnetic conductor, looks high. Another approach may start

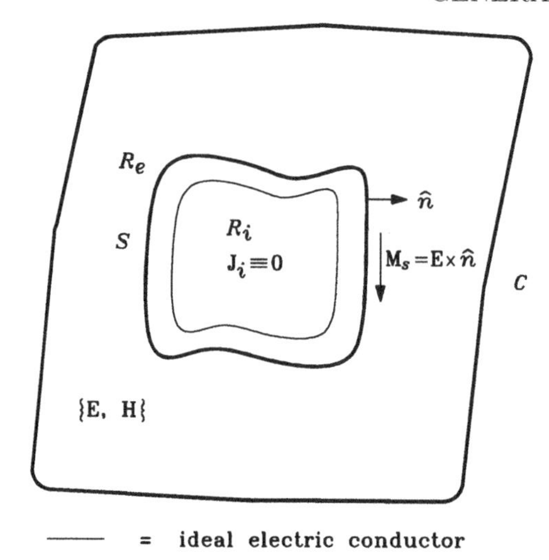

Figure 3.7 *The equivalence theorem, in the version which makes use of a perfect electric conductor.*

from the fact that the tangential magnetic and electric elds play perfectly interchangeable roles in the uniqueness theorem (see Section 3.3). Therefore, there is another recipe in order to get an e.m. eld $\{\mathbf{E}, \mathbf{H}\}$ in R_e which coincides with that of Figure 3.5. Its ingredients are sources which ensure the correct value of $\mathbf{E}_{\tan}$ on the external face of S. To obtain this, one can use another mathematical device, also without experimental correspondence: the de nition of a *magnetic surface current*, whose (linear) density at any point of S is:

$$\mathbf{M}_S = \mathbf{E} \quad \hat{n} \qquad (\text{V/m}) \quad . \tag{3.25}$$

It ensures the correct value of $\mathbf{E}_{\tan}$ on the external face of S, provided we can impose $\mathbf{E}_{\tan} = 0$ on the internal face. This can be done by postulating that in that place there is a *perfect electric conductor*, i.e., a medium having conductivity $= \infty$ (a subject that will be discussed again in Section 4.11).

What we have obtained in this way is schematically illustrated in Figure 3.7. Once again, the eld $\{\mathbf{E}, \mathbf{H}\}$ in R_e coincides with that of Figure 3.5, whereas the eld in R_i is not de ned. Advantages and disadvantages are not the same as in the previous case, but comparable in level.

3.5.2 Theorem

None of the above-mentioned procedures is really attractive, mainly because they require modi cation of the medium in the region R_i, so that the elds therein are no longer de ned. Hence the importance of the following statement, that for the time being will be proved for a source-free region. If we turn o the generators $\mathbf{J}_0$ in R_i, let *both* electric and magnetic surface currents in

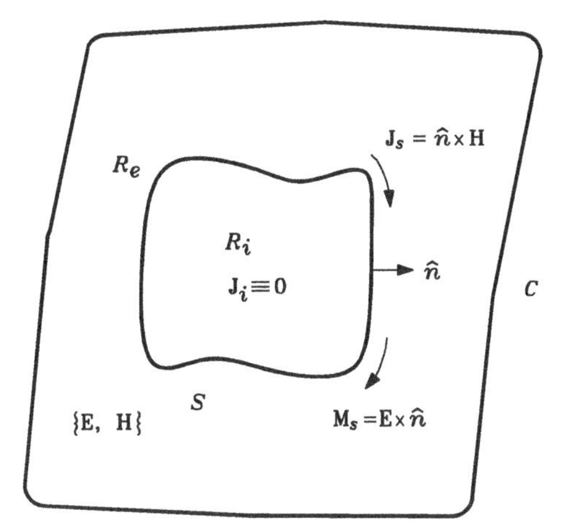

Figure 3.8 *The equivalence theorem, in the version which does not require any perfect conductor.*

Eqs. (3.24) and (3.25) circulate on S, and *leave the medium unchanged over the entire region* R, then:

in the region R_e, we have *the same eld* $\{\mathbf{E}, \mathbf{H}\}$, as that of the situation we started from, which is summarized in Figure 3.5, and

in the region R_i, *the eld is identically zero* ($\mathbf{E} = \mathbf{H} = 0$).

The proof consists just of observing that the vector elds $\mathbf{E}_{eq}$ and $\mathbf{H}_{eq}$, de ned as:

$$\begin{aligned} \mathbf{E}_{eq} = \mathbf{E} \quad , \quad \mathbf{H}_{eq} = \mathbf{H} \quad \text{in} \quad R_e \quad , \\ \mathbf{E}_{eq} = 0 \quad , \quad \mathbf{H}_{eq} = 0 \quad \text{in} \quad R_i \quad , \end{aligned} \tag{3.26}$$

satisfy Maxwell's equations in the whole region R_i, provided their sources are Eqs. (3.24) and (3.25), and also satisfy the boundary conditions on C that were satis ed by $\{\mathbf{E}, \mathbf{H}\}$ in the case shown in Figure 3.5. Applying the uniqueness theorem, we conclude that the Eqs. (3.26) are indeed *the* solution of the problem schematically illustrated in Figure 3.8, as we intended to prove.

It should be self-explanatory why one usually refers to the two surface densities, in Eqs. (3.24) and (3.25), taken together, as *sources equivalent to* $\mathbf{J}_0$ *for the eld in region* R_e.

3.5.3 Absorbing currents

Let us study now what happens if, on the surface S, there are electric and magnetic surface currents with densities $\mathbf{J}_a$ and $\mathbf{M}_a$ *opposite* to the current densities $\mathbf{J}_S$ and $\mathbf{M}_S$ of Eqs. (3.24) and (3.25), i.e.:

$$\mathbf{J}_a = \quad \hat{n} \quad \mathbf{H} \quad , \qquad \mathbf{M}_a = \quad \mathbf{E} \quad \hat{n} \quad . \tag{3.27}$$

If the boundary conditions on C are invariant with respect to a change in sign of the eld (i.e., if the boundary conditions are satis ed by $\{\ \ \mathbf{E}, \ \ \mathbf{H}\}$ when

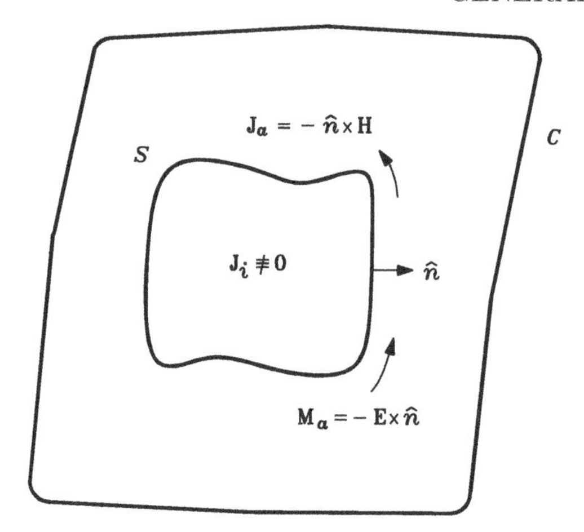

Figure 3.9 *The absorbing currents.*

they are by $\{\mathbf{E}, \mathbf{H}\}$, as is the case, for example, if they are the Sommerfeld's radiation conditions in Eq. (3.17)), then the currents in Eq. (3.27), if alone, generate a eld which is everywhere the opposite of Eq. (3.26). Suppose now that the original sources $\mathbf{J}_0$ in R_i *and* the surface currents in Eq. (3.27) on S are present together. Then, by superposition, we get that the eld in such circumstances is:

$$\begin{aligned} &\mathbf{E}_a = \mathbf{E} \quad \mathbf{E}_{eq} = \mathbf{E} \quad , \quad \mathbf{H}_a = \mathbf{H} \quad \mathbf{H}_{eq} = \mathbf{H} \quad \text{in} \quad R_i \quad , \\ &\mathbf{E}_a = \mathbf{E} \quad \mathbf{E}_{eq} = 0 \quad , \quad \mathbf{H}_a = \mathbf{H} \quad \mathbf{H}_{eq} = 0 \quad \text{in} \quad R_e \quad . \end{aligned} \tag{3.28}$$

So, this eld $\{\mathbf{E}_a, \mathbf{H}_a\}$ coincides then with the original one, $\{\mathbf{E}, \mathbf{H}\}$, in R_i, where the sources are located. It is identically zero in the complementary region R_e. This entails the name of *absorbing currents* (when they are looked at from region R_e) for Eqs. (3.27).

We can also imagine having gone from the original situation (Figure 3.5) to that described by Eqs. (3.28) (Figure 3.9) turning o the generators in R_e (which were zero), and replacing them by the surface current densities in Eq. (3.27). As we saw just now, in doing this, the eld in R_i remained unchanged, while that in R_e became zero. Note that this result proves precisely that *the equivalence theorem is also true for a region* (R_i in our case) *where there are sources.*

Another correct way to read these results is in the following terms. The equivalent sources are always given by Eqs. (3.24) and (3.25), provided that one takes the unit vector $\hat{n}$ oriented always away from the region where the equivalence theorem gives an identically zero eld. Consequently, Eqs. (3.24) and (3.25) can be interpreted also as absorbing currents, for the region R_i. This allows us to conclude by underlining the perfect symmetry in the roles that the two regions R_i and R_e play in the equivalence theorem.

Again making use of superposition, it is straightforward to treat an even more general case, in which one starts from a situation where there are sources both in R_e and R_i. We leave this as an exercise for the reader.

3.5.4 Final remark

It has been shown that equivalent electric and magnetic surface currents can be treated *on a par with* known terms, if Maxwell's equations are to be solved in a part of the region R where the eld is de ned. Unfortunately, however, these equivalent currents are not true known terms. Their expressions contain $\mathbf{E}$ and $\mathbf{H}$ on the surface S, quantities which are, in their turn, unknown. However, to use equivalent currents requires knowledge of $\mathbf{E}$ and $\mathbf{H}$ not in the entire region R, but only on a broadly arbitrary closed surface S. Consequently, the practical usage of the theorem resides in those numerous problems (see in particular Chapter 13), where the elds $\mathbf{E}$ and $\mathbf{H}$ on a suitable surface S can be either measured experimentally, or estimated via a satisfactory approximation. One can then start from the corresponding surface currents as working assumption, and elaborate on them. We will reconsider this at the beginning of Chapter 11.

3.6 Induction theorem

In this section we will prove a corollary of the equivalence theorem. It leads to relations that are *per se* exact. However, their practical use is in general limited to approximate approaches, where the theorem helps one to state a suitable "ansatz".

Let us start from the same assumptions as in Section 3.5 for the medium (linear, with arbitrarily small losses but never lossless). Let R denote the region of real 3-D space wherein the eld $\{\mathbf{E}, \mathbf{H}\}$ is de ned. Suppose that on its contour surface, C, boundary conditions are such that the uniqueness theorem of Section 3.3 holds in R. Consider then a regular closed surface, S; it subdivides R into two domains, R_1 and R_2 (Figure 3.10). There are no imposed currents on it. The subscripts 1 and 2 underline that, as we saw in the previous section, the roles of the regions that were initially called "internal" and "external," are interchangeable.

The *induction theorem* states that the eld $\{\mathbf{E}, \mathbf{H}\}$ in R can be expressed as the sum of two elds:

$$\mathbf{E} = \mathbf{E}_i + \mathbf{E}' \quad , \qquad \mathbf{H} = \mathbf{H}_i + \mathbf{H}' \quad , \tag{3.29}$$

which are characterized by the following properties:

- the eld $\{\mathbf{E}_i, \mathbf{H}_i\}$, known as the *incident eld* (or *inducing*, or *imposed*), is di erent from zero only in one of two regions — to illustrate the idea, in R_1 — and identically zero in the other (which implies that all its sources are in R_1);
- the other eld $\{\mathbf{E}', \mathbf{H}'\}$, referred to as the *induced eld*, is non-zero in all of R. It is sustained by those sources of $\{\mathbf{E}, \mathbf{H}\}$ that are not sources of

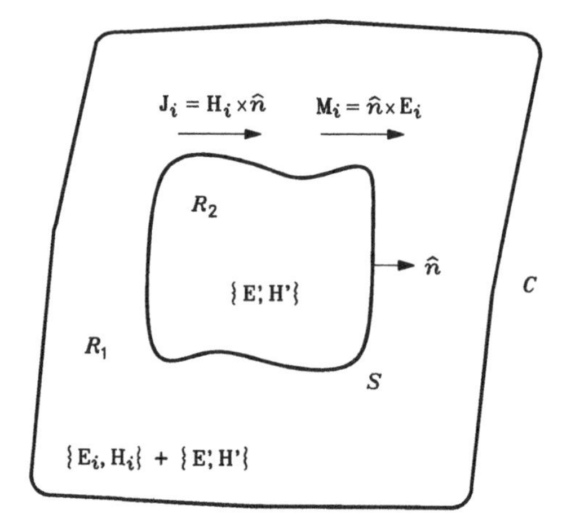

Figure 3.10 *The induction theorem: schematic illustration.*

$\{\mathbf{E}_i, \mathbf{H}_i\}$, and moreover by electric and magnetic surface currents, localized on S, whose densities are, respectively:

$$\mathbf{J}_i = \mathbf{H}_i \quad \hat{n} \quad , \qquad \mathbf{M}_i = \hat{n} \quad \mathbf{E}_i \quad , \tag{3.30}$$

where the unit vector normal to S, $\hat{n}$, is directed towards the region where $\{\mathbf{E}_i, \mathbf{H}_i\}$ is non-zero.

To prove all this, consider, in addition to the actual sources of the total eld $\{\mathbf{E}, \mathbf{H}\}$, four surface current densities, localized on S. Two of them are given by Eq. (3.30), while the other two are the *opposite* of Eq. (3.30). Incidentally, note how arbitrary the eld $\{\mathbf{E}_i, \mathbf{H}_i\}$ in Eq. (3.30) is, in doing this operation.

The superposition of these surface currents is obviously equal to zero and therefore the e.m. eld, in this new situation, is identical to the eld $\{\mathbf{E}, \mathbf{H}\}$. We now subdivide all the currents present into two sets, S_0 and S'. The linearity of the medium allows us to consider the total eld $\{\mathbf{E}, \mathbf{H}\}$ as the sum of the elds due *separately* to S_0 and S'. In the set S_0, we include the sources (all located in R_1) that are able to generate the eld $\{\mathbf{E}_i, \mathbf{H}_i\}$ on the surface S, and furthermore the surface currents opposite to those of Eq. (3.30). In the set S', we put Eq. (3.30), as well as all the sources of $\{\mathbf{E}, \mathbf{H}\}$ that were *not* included in the set S_0.

To reach our conclusion, it is enough to observe that the opposites to Eq. (3.30) are the absorbing currents of the eld $\{\mathbf{E}_i, \mathbf{H}_i\}$ (i.e., they assure $\mathbf{E}_i = \mathbf{H}_i = 0$ in R_2). Thus, S_0 and S' respectively sustain elds $\{\mathbf{E}_i, \mathbf{H}_i\}$ and $\{\mathbf{E}', \mathbf{H}'\}$ that meet all the properties that were listed when we stated the theorem.

A di erent point of view regarding the current densities in Eq. (3.30) can help clarify the practical impact of this theorem. Let us use di erent names and symbols for the fields induced in the two regions R_1 and R_2, for example:

$$\{\mathbf{E}', \mathbf{H}'\} = \{\mathbf{E}_r, \mathbf{H}_r\} \ , \qquad \text{"reflected field" in } R_1 \ ;$$

$$\{\mathbf{E}', \mathbf{H}'\} = \{\mathbf{E}_t, \mathbf{H}_t\} \ , \qquad \text{"transmitted field" in } R_2 \ . \tag{3.31}$$

One of the assumptions was that the initial field $\{\mathbf{E}, \mathbf{H}\}$ does not have sources located on S. Therefore, its tangential components are continuous across that surface. With our new symbols, this continuity can be expressed as:

$$\hat{n} \times \mathbf{E}_i = \hat{n} \times (\mathbf{E}_t - \mathbf{E}_r) \ ,$$

$$\mathbf{H}_i \times \hat{n} = (\mathbf{H}_t - \mathbf{H}_r) \times \hat{n} \ . \tag{3.32}$$

Comparing Eq. (3.32) with Eq. (3.30), we see that the surface currents defined by Eq. (3.30) are those that are required in order to have, in the passage across S, discontinuities in the induced field $\{\mathbf{E}', \mathbf{H}'\}$ equal to the right-hand sides of Eqs. (3.32).

As we said at the beginning, this theorem turns out to be really useful in practice only when one is clever in estimating the inducing field. This must be a good approximation to the exact solution, and at the same time be simple to handle, in order to make the subsequent calculations simple and rapid.

3.7 Duality theorem

The proof of the equivalence theorem of Section 3.5 required the new concept of magnetic currents, which, as already stated, is a useful mathematical tool, with no experimental significance. So far, we have only made use of surface magnetic currents, but nothing prevents us from also considering magnetic currents in a 3-D domain. Their use will be thoroughly justified much later, in Section 11.5.

With this, Maxwell's equation for complex vectors in a linear, time-invariant, anisotropic (or, as a special case, isotropic) medium, become symmetrical (apart from the signs) not only with respect to the unknowns, but also with respect to the imposed terms:

$$\nabla \times \mathbf{E} = -j\omega\bar{\bar{\mu}}\,\mathbf{H} - \mathbf{M}_0 \ , \tag{3.33}$$

$$\nabla \times \mathbf{H} = j\omega\bar{\bar{\epsilon}}_c\,\mathbf{E} + \mathbf{J}_0 \ , \tag{3.34}$$

where the imposed magnetic current density is denoted as $\mathbf{M}_0$.

The so-called *duality theorem* is simply the following observation. If $\{\mathbf{E}, \mathbf{H}\}$ is a solution of Eqs. (3.33) and (3.34) in a region R of real 3-D space, then another e.m. field, also defined in R, but in a different medium and with different sources, is obtained via the following set of formal substitutions (where implicit factors of unitary magnitude are understood, to provide suitable conversion of physical dimensions):

$$\mathbf{E}' = -\mathbf{H} \ , \quad \mathbf{H}' = \mathbf{E} \ , \quad \mathbf{J}_0' = -\mathbf{M}_0 \ , \quad \mathbf{M}_0' = \mathbf{J}_0 \ ,$$
$$\bar{\bar{\epsilon}}_c' = \bar{\bar{\mu}} \ , \quad \bar{\bar{\mu}}' = \bar{\bar{\epsilon}}_c \ . \tag{3.35}$$

The eld $\{\mathbf{E}', \mathbf{H}'\}$ obtained in this way is referred to as the *dual* (or the *conjugate*) of the eld $\{\mathbf{E}, \mathbf{H}\}$. The same word is used also for sources and media. Obviously, the dual of the dual of a eld corresponds to the eld itself.

Readers who have always seen Maxwell's equations in the MKS SI system of units may be very surprised by this theorem, and wonder whether it bears any adherence to physical reality. If they consult a textbook (e.g., Jackson, 1975) where the so-called Gaussian CGS system of units is adopted, and pay attention to the fact that in this system, both and in free space are equal to unity, then it becomes rather intuitive that two dual elds may, at least in some cases, be very similar to each other. For example, there are cases where duality corresponds just to a change from one polarization to the orthogonal polarization.

3.8 TE-TM eld decomposition theorem

Electromagnetic potentials, in Chapter 1, gave us the feeling that there is a large degree of arbitrariness in the ways to express an e.m. eld in a linear medium. In particular, we saw that it may be expressed in many ways as the sum of two or more elds, via "superposition," and that it is possible to impose some pre-selected requirements on each of these elds. The equivalence and induction theorems, of Sections 3.6 and 3.7, are two examples of how these degrees of freedom may be exploited. Another example will be supplied by this section, whose results will be used very often in Chapters 7, 9 and 10.

We will show that in a simply connected region — for the relevance of this restriction, see again the footnote in Section 1.8 — without sources, containing a homogeneous linear isotropic medium, any e.m. eld $\{\mathbf{E}, \mathbf{H}\}$ can be expressed as the sum of two e.m. elds (that will be labeled, respectively, with a subscript TE and a subscript TM), which, with respect to a unit vector $\hat{a}$ chosen in a completely arbitrary direction, are *Transverse Electric* (TE) and *Transverse Magnetic* (TM), namely:

$$\begin{gathered}\{\mathbf{E}, \mathbf{H}\} = \{\mathbf{E}_{\mathrm{TE}}, \mathbf{H}_{\mathrm{TE}}\} + \{\mathbf{E}_{\mathrm{TM}}, \mathbf{H}_{\mathrm{TM}}\} \quad , \\ \mathbf{E}_{\mathrm{TE}}\ \hat{a} = 0 \quad , \quad \mathbf{H}_{\mathrm{TM}}\ \hat{a} = 0 \quad .\end{gathered} \tag{3.36}$$

Furthermore, we will prove that each of these two elds can be obtained, via standard vector-calculus operations, from a scalar function of spatial coordinates that satis es the scalar homogeneous Helmholtz equation.

Before we start showing this, let us warn the reader that the proof of this theorem is enormously simpli ed if one is entitled to make two further assumptions. The rst assumption is that we are entitled to disregard *transverse electromagnetic* (TEM) elds, i.e., solutions of Maxwell's equations having $\mathbf{E}\ \hat{a} = 0$ *and* $\mathbf{H}\ \hat{a} = 0$. The second assumption is that one can guess the functional dependence of the eld on the coordinate along the direction of $\hat{a}$, leaving only a complex parameter to be determined later on in this function. However, at this stage, we believe, the reader is totally unprepared to grasp the physical meaning and the relevance of these simplifying assumptions. For this reason, we will present here a very general, but also very demanding proof. If the use that is going to be made of this theorem, by a reader or by an instructor, is just to introduce guided-wave propagation (Chapters 7 and 10), then the rest of this section may be replaced by the simpli ed proof given in Section 7.2.

The general proof begins observing that, given the assumptions on the sources and on the medium, a scalar electric potential $\varphi = 0$ is a Lorentz gauge. This implies, via Eq. (1.79), that $\nabla \;\; \mathbf{A} = 0$, and consequently, the magnetic vector potential satis es an equation of the type:

$$\mathbf{A} = \boldsymbol{\nabla} \;\; \mathbf{W} \quad . \tag{3.37}$$

So we introduce another vector $\mathbf{W}$ (with arbitrary divergence), which in turn must satisfy the following equation, as one can show by inserting Eq. (3.37) into Eq. (1.80):

$$\boldsymbol{\nabla} \;\; (\boldsymbol{\nabla} \;\; \boldsymbol{\nabla} \;\; \mathbf{W} + \;^2\mathbf{W}) = 0 \quad . \tag{3.38}$$

Incidentally, from the following vector identity (which can be proved by the reader as an exercise):

$$\boldsymbol{\nabla} \;\; (\boldsymbol{\nabla} \;\; \boldsymbol{\nabla} \;\; \mathbf{q}) = \;\; \boldsymbol{\nabla} \;\; (\nabla^2\mathbf{q}) \quad , \tag{3.39}$$

(where $\mathbf{q}$ is any vector), it follows that Eq. (3.38) can also be replaced by:

$$\boldsymbol{\nabla} \;\; (\nabla^2\mathbf{W} \quad \;^2\mathbf{W}) = 0 \quad . \tag{3.40}$$

Now, separating the longitudinal and transverse parts with respect to the direction (say z) of the unit vector $\hat{a}$, we get:

$$\mathbf{W} = \hat{a} \quad \mathbf{W} \quad \hat{a} + (\mathbf{W} \quad \hat{a})\,\hat{a} = \mathbf{W}_{\mathrm{TM}} + L\,\hat{a} \quad . \tag{3.41}$$

The divergence of $\mathbf{W}$ being arbitrary (while its curl is xed) implies that the scalar L is only speci ed to within the addition of an arbitrary function g, which must satisfy only the restriction $\boldsymbol{\nabla} \;\; (g\hat{a}) \quad \nabla g \quad \hat{a} = 0$. At the same time, the vector $\mathbf{W}_{\mathrm{TM}}$ is only speci ed to within the addition of a irrotational vector $\mathbf{W}'$, perpendicular to $\hat{a}$.

Now let us express the total e.m. eld as the sum of a contribution due to $\mathbf{W}_{\mathrm{TM}}$, and another one due to $L\,\hat{a}$. As for the latter, it may be immediately shown that it is transverse electric (TE). In fact, from Eqs. (1.66) and (3.37), and from the vector identity (B.6), applied in the present case in which $\hat{a}$ = constant, we have:

$$\mathbf{E}_{\mathrm{TE}} = \;\; j\omega\mathbf{A}_{\mathrm{TE}} = \;\; i\omega\boldsymbol{\nabla} \;\; (L\,\hat{a}) = \;\; j\omega\nabla L \quad \hat{a} \quad , \tag{3.42}$$

and this relation requires $\mathbf{E}_{\mathrm{TE}} \quad \hat{a} = 0$, as we wanted to show.

It is rather simple to prove now that L satis es the scalar homogeneous Helmholtz equation:

$$\nabla^2 L \quad \;^2 L = 0 \quad . \tag{3.43}$$

Actually, Eq. (3.43) does not follow necessarily from Eq. (3.40), but, as we said before, L contains an arbitrary additive function g, and this is enough for us to *impose* that the function L satisfy Eq. (3.43).

At this point, we are able to prove that the eld contributed by $\mathbf{W}_{\rm TM}$ is transverse magnetic (TM). Namely, if we take into account Eq. (3.43), we see that $\mathbf{W}_{\rm TM}$ must also obey Eq. (3.40) and thus Eq. (3.38). Exploiting then the above mentioned existence of an arbitrary irrotational vector $\mathbf{W}'$, we can *impose* that $\mathbf{W}_{\rm TM}$ obeys the following equation:

$$\nabla\quad \nabla\quad \mathbf{W}_{\rm TM} + \;{}^2\mathbf{W}_{\rm TM} = 0 \quad . \tag{3.44}$$

Incidentally, the steps from Eq. (3.38) to Eq. (3.44) are formally identical to those from Eq. (1.64) to Eq. (1.68), and thus equivalent to imposing $\nabla\ \mathbf{W}_{\rm TM} = 0$. From Eqs. (1.60), (3.37) and (3.44), we get:

$$\mathbf{H}_{\rm TM} = (1/\quad)\nabla\quad \mathbf{A}_{\rm TM} = (1/\quad)\nabla\quad \nabla\quad \mathbf{W}_{\rm TM} = \quad (\;{}^2/\quad)\mathbf{W}_{\rm TM} \quad , \tag{3.45}$$

which yields $\mathbf{H}_{\rm TM}\ \hat{a} = 0$, because $\mathbf{W}_{\rm TM}$ is perpendicular to $\hat{a}$ by construction. This is what we intended to prove.

The last thing that remains to be shown is that the TM eld can also be obtained from a scalar function T satisfying the homogeneous Helmholtz equation:

$$\nabla^2 T \qquad {}^2 T = 0 \quad . \tag{3.46}$$

To do this, it is su cient to invoke the *duality theorem* of Section 3.7, and start again from the very beginning of the present theorem, using the electric vector potential $\mathbf{F}$ instead of the magnetic vector potential $\mathbf{A}$. The TE and TM elds interchange their roles, whence the conclusion.

Summarizing, with the help of a few steps that follow simply from some vector identities of Appendix B, we obtain the following expressions for the TE eld:

$$\begin{aligned} \mathbf{E}_{\rm TE} &= \quad j\omega\nabla L \quad \hat{a} \quad , \\ \mathbf{H}_{\rm TE} &= \frac{\nabla\quad \mathbf{E}_{\rm TE}}{j\omega} = \frac{1}{\ }\left(\nabla^2 L\hat{a} + \frac{\partial}{\partial z}\nabla L\right) = \frac{1}{\ }\left(\quad {}^2 L\hat{a} + \frac{\partial}{\partial z}\nabla L\right) , \end{aligned} \tag{3.47}$$

where L is a solution of Eq. (3.43). Dually, for the TM eld, we have:

$$\begin{aligned} \mathbf{H}_{\rm TM} &= \quad j\omega\nabla T \quad \hat{a} \quad , \\ \mathbf{E}_{\rm TM} &= \frac{\nabla\quad \mathbf{H}_{\rm TM}}{j\omega\ _c} = \frac{1}{\ _c}\left(\nabla^2 T\hat{a} \quad \frac{\partial}{\partial z}\nabla T\right) = \frac{1}{\ _c}\left(\;{}^2 L\hat{a} \quad \frac{\partial}{\partial z}\nabla T\right) , \end{aligned} \tag{3.48}$$

where T is a solution of Eq. (3.46).

Comparison between the rst of Eqs. (3.48) and Eq. (3.45) illustrates how the scalar function T is connected with the vector $\mathbf{W}_{\rm TM}$ that was used in the rst part of the present proof, namely:

$$\mathbf{W}_{\rm TM} = \frac{1}{j\omega\ _c}\nabla T \quad \hat{a} \quad . \tag{3.49}$$

This relation also reveals where the residual liberty was hidden that was exploited when we imposed that T obeys Eq. (3.46). Indeed, Eq. (3.49) shows that the z-dependence of the function T is completely irrelevant for calculating the eld of Eq. (3.45), because the longitudinal component of ∇T disappears, when we take the cross product with the unit vector $\hat{a}$. Hence, there was an arbitrary component of ∇T, up to when T was forced to satisfy Eq. (3.46).

3.9 Spatial symmetries. Re ection operators

Geometrical symmetries of the region where an electromagnetic eld is de- ned, if properly exploited, can make its determination signi cantly simpler. This statement has been known for a long time, but it became particularly attractive in recent times, when interactive numerical methods became widely used. Indeed, symmetries may permit great reductions in memory allocation and in processing time.

The simplest case, as well as the most exploited one, is a *symmetry plane*. Other cases of considerable interest are symmetries with respect to an *axis*, and with respect to a *point*. To approach all these cases in a uni ed way, let us take as a starting point Maxwell's equations written by components in *a rectangular coordinate frame* (x_1, x_2, x_3). Our aim is to nd transformation rules which, when applied to one of their solutions, provide another solution, whenever a symmetry of one of those types is present. We will deal with media that are *isotropic*, but not necessarily homogeneous. Extensions to some classes of anisotropic media can be found in the literature, but the reader is advised to look for them only after studying Chapter 6.

Expanding the vectors as $\mathbf{E} = \ _i E_i \hat{a}_i$, Maxwell's equations, in the generalized form which also includes magnetic sources, are written by components, as:

$$
\begin{aligned}
\frac{\partial E_3}{\partial x_2} \quad \frac{\partial E_2}{\partial x_3} &= \quad j\omega \quad H_1 \quad M_{01} \quad , \\
\frac{\partial E_1}{\partial x_3} \quad \frac{\partial E_3}{\partial x_1} &= \quad j\omega \quad H_2 \quad M_{02} \quad , \\
\frac{\partial E_2}{\partial x_1} \quad \frac{\partial E_1}{\partial x_2} &= \quad j\omega \quad H_3 \quad M_{03} \quad , \\
\frac{\partial H_3}{\partial x_2} \quad \frac{\partial H_2}{\partial x_3} &= j\omega \ {}_c E_1 + J_{01} \quad , \\
\frac{\partial H_1}{\partial x_3} \quad \frac{\partial H_3}{\partial x_1} &= j\omega \ {}_c E_2 + J_{02} \quad , \\
\frac{\partial H_2}{\partial x_1} \quad \frac{\partial H_1}{\partial x_2} &= j\omega \ {}_c E_3 + J_{03} \quad .
\end{aligned}
\qquad (3.50)
$$

Let P denote the generic point in the region R in which Eqs. (3.50) are satis ed. Consider now the following set of substitutions:

$$
\begin{array}{llcll}
P & (x_1, x_2, x_3) & \longrightarrow & P' & (x_1, x_2, \;\; x_3) \quad , \\
\mathbf{E} & (E_1, E_2, E_3) & \longrightarrow & \mathbf{E}' & (E_1, E_2, \;\; E_3) \quad , \\
\mathbf{H} & (H_1, H_2, H_3) & \longrightarrow & \mathbf{H}' & (\;\; H_1, \;\; H_2, H_3) \quad , \\
\mathbf{J}_0 & (J_{01}, J_{02}, J_{03}) & \longrightarrow & \mathbf{J}'_0 & (J_{01}, J_{02}, \;\; J_{03}) \quad , \\
\mathbf{M}_0 & (M_{01}, M_{02}, M_{03}) & \longrightarrow & \mathbf{M}'_0 & (\;\; M_{01}, \;\; M_{02}, M_{03}) \quad .
\end{array}
\tag{3.51}
$$

One can verify immediately that if Eqs. (3.51) are applied all at once, then Eqs. (3.50) are unaltered. Therefore, if the point P' belongs to the region R where Eqs. (3.50) are satis ed, then the vectors $\mathbf{E}'(P'), \mathbf{H}'(P')$ satisfy *at the point* P Maxwell's equations, with sources $\{\mathbf{J}'_0(P'), \mathbf{M}'_0(P')\}$ also located at the point P. Therefore, if the region R is symmetrical with respect to the plane $x_3 = 0$, and if the boundary conditions are themselves symmetrical with respect to the same plane, Eqs. (3.51) de ne a new solution of Maxwell's equations. We say that this new solution is obtained starting from the initial solution *by re ection* in the plane $x_3 = 0$.

It may be convenient to write Eqs. (3.51) in the following compact notation:

$$
\{\mathbf{E}', \mathbf{H}'\} = R_3 \quad \{\mathbf{E}, \mathbf{H}\} \quad , \qquad \{\mathbf{J}'_0, \mathbf{M}'_0\} = R_3 \quad \{\mathbf{J}_0, \mathbf{M}_0\} \quad . \tag{3.52}
$$

The operator represented by the symbol R_3 is referred to as the *re ection operator* in the plane $x_3 = 0$.

Provided that the region R and the boundary conditions satisfy the appropriate symmetry conditions, also the quantities:

$$
\{\mathbf{E},'' \mathbf{H}''\} = \quad R_3 \quad \{\mathbf{E}, \mathbf{H}\} \quad , \qquad \{\mathbf{J}''_0, \mathbf{M}''_0\} = \quad R_3 \quad \{\mathbf{J}_0, \mathbf{M}_0\} \quad , \tag{3.53}
$$

which di er from Eq. (3.51) only by a change in sign of all components, form a solution of Maxwell's equations. Moreover, it can be veri ed immediately that $R_3 \quad R_3 \quad = (\quad R_3 \quad)(\quad R_3 \quad) = I$, where the last symbol stands for the identity operator.

An e.m. eld which is invariant with respect to the application of the operator R_3 , namely, is such that in its entire domain of de nition satis es the following equalities:

$$
\begin{array}{rcl}
E_1(x_1, x_2, x_3) & = & E_1(x_1, x_2, \;\; x_3) \quad , \\
E_2(x_1, x_2, x_3) & = & E_2(x_1, x_2, \;\; x_3) \quad , \\
E_3(x_1, x_2, x_3) & = & \quad E_3(x_1, x_2, \;\; x_3) \quad , \\
H_1(x_1, x_2, x_3) & = & \quad H_1(x_1, x_2, \;\; x_3) \quad , \\
H_2(x_1, x_2, x_3) & = & \quad H_2(x_1, x_2, \;\; x_3) \quad , \\
H_3(x_1, x_2, x_3) & = & H_3(x_1, x_2, \;\; x_3)
\end{array}
\tag{3.54}
$$

is called a eld *of even symmetry* (or, simply, an even eld) with respect to the plane $x_3 = 0$. Analogously, a eld which obeys the relation $\{\mathbf{E}, \mathbf{H}\} =$

$R_3 \quad \{\mathbf{E}, \mathbf{H}\}$, i.e., satis es a set of equations that di er from Eqs. (3.54) only by the sign in each right-hand side, is referred to as a eld *of odd symmetry* (or, simply, an odd eld) with respect to the plane $x_3 = 0$.

If we let $x_3 = 0$ in Eqs. (3.54), we see immediately that they imply $E_3 = 0$ and $H_1 = H_2 = 0$ over the entire symmetry plane. Compare with what was said at the beginning of Section 3.5: the conclusion is that an even eld remains unaltered if an *ideal magnetic conductor* is introduced on the symmetry plane. Alternatively, if we let $x_3 = 0$ in the relations that de ne an odd eld, then we establish in the same way that they imply, on the plane $x_3 = 0$, zero tangential components of $\mathbf{E}$, and zero normal component of $\mathbf{H}$ ($E_1 = E_2 = 0$, $H_3 = 0$). Again comparing with Section 3.5, we conclude that an odd eld remains unaltered if an *ideal electric conductor* is introduced on the symmetry plane.

A eld of odd symmetry, i.e., compatible with an ideal electric conductor wall, requires sources satisfying the relation:

$$\{\mathbf{J}_0, \mathbf{M}_0\} = \quad R_3 \quad \{\mathbf{J}_0, \mathbf{M}_0\} \quad . \tag{3.55}$$

This equation expresses in analytical terms the so-called *method of images*, which is probably known to the reader in the electrostatic case (Plonsey and Collin, 1961). Note that Eq. (3.55) involves a change in sign for some components (J_{01}, J_{02}, M_{03}), passing from the "actual" sources to the "image" sources.

It will be well-known to the reader that any scalar function of a single variable, de ned over an interval symmetrical with respect to the origin, can be expressed as the sum of its even and odd parts. Similarly, any e.m. eld, provided it is de ned in a region symmetrical with respect to the plane $x_3 = 0$, may be split into its (i) *even part*:

$$\{\mathbf{E}_e, \mathbf{H}_e\} = (1/2) \left[\{\mathbf{E}, \mathbf{H}\} + R_3 \quad \{\mathbf{E}, \mathbf{H}\}\right] \quad , \tag{3.56}$$

and its (ii) *odd part*:

$$\{\mathbf{E}_o, \mathbf{H}_o\} = (1/2) \left[\{\mathbf{E}, \mathbf{H}\} \quad R_3 \quad \{\mathbf{E}, \mathbf{H}\}\right] \quad . \tag{3.57}$$

Obviously:

$$\{\mathbf{E}, \mathbf{H}\} = \{\mathbf{E}_e, \mathbf{H}_e\} + \{\mathbf{E}_o, \mathbf{H}_o\} \quad .$$

This property, if suitably exploited, is the key towards saving memory and computation time in numerical methods (see comments at the beginning of this section). However, since di erent components in Eqs. (3.51) are treated di erently in respect of their signs, Eqs. (3.56) and (3.57) are non-trivial extensions of the scalar case, and must be handled with care by a programmer.

Let us go back to the beginning of this subject: it is evident that the coordinate x_3 does not play a privileged role in Eqs. (3.50) with respect to the other two coordinates. Therefore, it is straightforward to de ne re ection operators with respect to the coordinate planes x_1 and x_2. They will be denoted by R_1

Table 3.1 *Re ection operator multiplication table.*

	I	R^1	R^2	R^3	R_1	R_2	R_3	R_0
I	I	R^1	R^2	R^3	R_1	R_2	R_3	R_0
R^1	R^1	I	R^3	R^2	R_0	R_3	R_2	R_1
R^2	R^2	R^3	I	R^1	R_3	R_0	R_1	R_2
R^3	R^3	R^2	R^1	I	R_2	R_1	R_0	R_3
R_1	R_1	R_0	R_3	R_2	I	R^3	R^2	R^1
R_2	R_2	R_3	R_0	R_1	R^3	I	R^1	R^2
R_3	R_3	R_2	R_1	R_0	R^2	R^1	I	R^3
R_0	R_0	R_1	R_2	R_3	R^1	R^2	R^3	I

and R_2 , respectively. We leave it as an exercise for the reader to derive their explicit forms.

All the previous considerations regarding even and odd parts, electric and magnetic walls, image sources, etc., can be repeated for these operators without any substantial change.

Next we can de ne *re ection operators with respect to coordinate axes.* For example, the re ection operator with respect to the axis x_3 is de ned by the following substitutions:

$$
\begin{aligned}
P \quad (x_1, x_2, x_3) \quad &\rightarrow \quad P' \quad (\ x_1, \ \ x_2, x_3) \quad , \\
\mathbf{E} \quad (E_1, E_2, E_3) \quad &\rightarrow \quad \mathbf{E}' \quad (\ E_1, \ \ E_2, E_3) \quad , \\
\mathbf{H} \quad (H_1, H_2, H_3) \quad &\rightarrow \quad \mathbf{H}' \quad (\ H_1, \ \ H_2, H_3) \quad , \\
\mathbf{J}_0 \quad (J_{01}, J_{02}, J_{03}) \quad &\rightarrow \quad \mathbf{J}_0' \quad (\ J_{01}, \ \ J_{02}, J_{03}) \quad , \\
\mathbf{M}_0 \quad (M_{01}, M_{02}, M_{03}) \quad &\rightarrow \quad \mathbf{M}_0' \quad (\ M_{01}, \ \ M_{02}, M_{03}) \quad .
\end{aligned}
\tag{3.58}
$$

Finally, a *re ection operator with respect to the origin* can be de ned, as the following set of substitutions:

$$
\begin{aligned}
P \quad (x_1, x_2, x_3) \quad &\rightarrow \quad P' \quad (\ x_1, \ \ x_2, \ \ x_3) \quad , \\
\mathbf{E} \quad (E_1, E_2, E_3) \quad &\rightarrow \quad \mathbf{E}' \quad (E_1, E_2, E_3) \quad , \\
\mathbf{H} \quad (H_1, H_2, H_3) \quad &\rightarrow \quad \mathbf{H}' \quad (\ H_1, \ \ H_2, \ \ H_3) \quad , \\
\mathbf{J}_0 \quad (J_{01}, J_{02}, J_{03}) \quad &\rightarrow \quad \mathbf{J}' \quad (J_{01}, J_{02}, J_{03}) \quad , \\
\mathbf{M}_0 \quad (M_{01}, M_{02}, M_{03}) \quad &\rightarrow \quad \mathbf{M}' \quad (\ M_{01}, \ \ M_{02}, \ \ M_{03}) \quad .
\end{aligned}
\tag{3.59}
$$

Using R^i and R_0 to denote the re ection operators with respect to the i-axis and to the origin, respectively, it can be shown that the seven so-de ned operators, plus the identity operator, form a set which is an *Abelian group* (e.g., Cracknell, 1968). Its elements are linked by the relationships summarized in Table 3.1.

A simple and illustrative example of how useful symmetries can be is the following approach to a so-called *directional coupler* in waveguides, or striplines, etc., whose operation can be explained intuitively, without calculations. Although more formal de nitions exist, let us consider here as a directional coupler a pair of in nitely long transmission lines or waveguides, with a longitudinal symmetry plane, say $x_1 = 0$ (see Figure 3.11). Because of symmetry, any eld in this structure can be decomposed into its even and odd part. Suppose now that the two lines or waveguides are "single-mode," i.e., are designed in such a way, that each of them, if left alone, may support just one solution of Maxwell's equations, uniquely characterized apart from a complex multiplicative constant. Then, at least as a rst approximation, the total eld in the coupler will be a superposition of the even (or in-phase) and odd (or out-of-phase) combinations of these "modes" of the individual waveguides.

The even and odd combinations, corresponding to a magnetic and an electric wall in the $x_1 = 0$ plane, respectively, are elds which satisfy Maxwell's equations in two di erent structures. Therefore, one should not be surprised if we state that they travel at di erent velocities along the in nitely long direction. Consequently, the total eld amplitude in each waveguide may evolve periodically, although the total power is conserved. The example shown in Figure 3.11 refers in particular to the case where all the power in one section was in the upper waveguide (for example, this was the "launching condition" at the input of the coupler). It shows that, perhaps in contrast with intuition, all the power may be transferred to the lower waveguide at suitable distances.

The reader may try, as a demanding exercise, to explain in terms of symmetries (even and odd eld combinations) the reciprocity results for a Y-junction that we saw in Section 3.4.

3.10 Further applications and suggested reading

As all solutions of Maxwell's equations must satisfy the theorems that form the core of this chapter, all the contents of the following chapters may be regarded as a set of further applications of the material in this chapter. In particular, *Poynting's theorem* and the *uniqueness theorem* are quite pervasive, and therefore they may show up in any of the following chapters.

The *reciprocity theorem*, as we already mentioned, will nd one of its main applications in Chapter 12, when we discuss the relationship between transmitting and receiving properties of a given antenna. Nonreciprocal electromagnetic phenomena, which occur when the assumptions of the reciprocity theorem are violated — namely, in some anisotropic media — will be introduced in Chapter 6. However, a much deeper knowledge of nonreciprocities, in terms of basic physics as well as of applications, is instrumental for those who intend to work in such areas as microwave devices and networks, optical telecommunications and sensors, ionospheric propagation, plasma physics. Many textbooks cover these subjects very well. The basic references can be found at the end of Chapter 6.

The *equivalence theorem* is a fundamental starting point for retarded po-

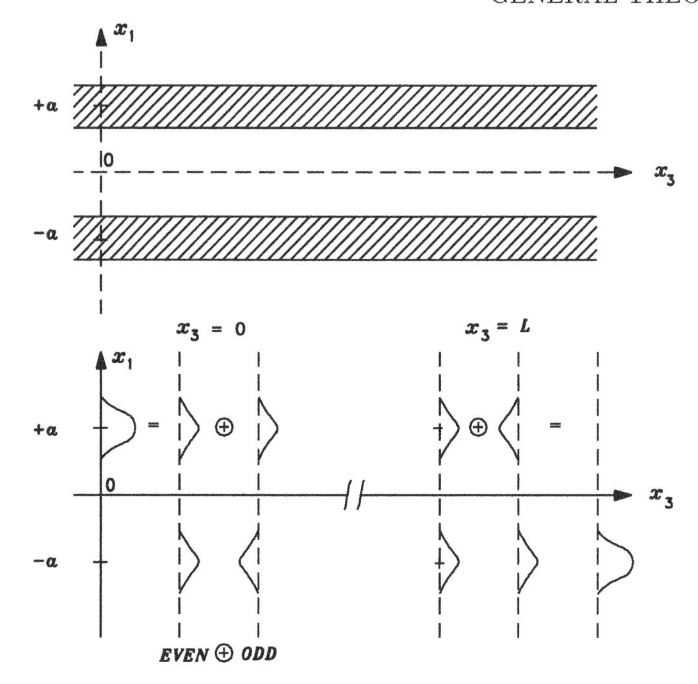

Figure 3.11 *The principle of operation of a directional coupler.*

tentials (Chapter 11) and for di raction (Chapter 13). The *induction theorem* can prove very useful in the same context, but unfortunately space constraints oblige us to make its use minimal throughout this book. Interested readers can nd its applications in advanced books on di raction, scattering, aperture antennas (e.g., Chapter 8 of Stutzman and Thiele, 1981; see also the list of references at the end of Chapter 13). Also, it may be useful to have a thorough understanding of the use of this theorem before getting involved in numerical methods (e.g., Ida, 1995; Press *et al.*, 1992; Stutzman and Thiele, 1981, and references therein).

As for the *duality theorem*, we said at the end of Section 3.7 that it may appear awkward, when presented, as herein, in the MKS system of units, where the orders of magnitudes of dual variables are, in most practical cases, quite di erent. To appreciate why this theorem can be useful in practice, we insist on the point that the reader should deal, at least brie y, with Maxwell's equations written in the so-called *Gaussian CGS system of units*, where electric permittivity and magnetic permeability of free space are both equal to unity. These equations can be found, for example, in Jackson (1975). One important consequence was already pointed out in Section 3.7: there are many problems, in optics and in di raction theory, where two dual elds di er only in *polarization.* Another way to appreciate the potential uses of this theorem is to think, once again, of *numerical methods* for solving Maxwell's equations. All quantities handled by a numerical computer are adimensional, and must be normalized so that their numerical values fall within a suitable range.

The TE–TM decomposition of an e.m. eld is the strategic starting point for waveguide theory, both at microwave and optical frequencies. The basic notions regarding guided-wave propagation, which are given in Chapters 7, 9 and 10, provide a very extensive set of examples of such applications. All of them, however, may be derived from the simpli ed proof to be given in Chapter 7. As we said in Section 3.8, the demanding proof that was given there broadens the theorem's scope of validity. This entitles one to tackle more di cult problems, compared to those of our chapters on waveguides. Advanced problems are dealt with in more specialized books, such as Collin (1991), Lewin *et al.* (1977), and Vassallo (1991). On these occasions, one needs elds that are TE and TM not with respect to a waveguide axis, but rather with respect to other directions.

As for symmetries, several results which will be found in the following chapters are easy to reconcile with physical intuition if they are examined in terms of spatial symmetries of structures and, consequently, of the elds. The very well known example of the directional coupler, has already been mentioned, and will be brie y reexamined in Chapter 8. Symmetries, as we pointed out in the previous section, are related to basic group theory notions. The advantages of a group-theory-based approach are beautifully outlined, for example, in Chapter 12 of Montgomery, Dicke and Purcell (1948), where many fundamental properties of microwave junctions are derived without getting into any detail of eld calculations. Similarly, fundamental properties of optical waveguides can be found in a very elegant way via group theory, as shown by Black and Gagnon (1995). For these reasons, the reader is strongly advised to come back and read again Section 3.9 after reading Chapter 10, particularly Section 10.7.

Finally, we wish to stress that, from a tutorial viewpoint, it is very helpful for the reader to study the basic theorems of *electric circuit theory*, in parallel to these theorems on e.m. elds. Theorems in these two areas *must overlap exactly*, because circuit theory is another axiomatic theory whose postulates were derived from the same basic physical principles as those of our theory. Just to quote a few examples, uniqueness of voltages and/or currents in an electrical network may run into problems for perfectly lossless resonant circuits: compare with Section 3.3. Reciprocity is found in circuit theory to be intimately related with time invariance. *Equivalent generator theorems* (also called *generalized Thevenin's theorems*) are conceptually the same as the *equivalence theorems* of Section 3.5 (Corazza *et al.*, 1969). The classical textbook by Newcomb (1966) remains a highly recommended starting point for those who look for advanced readings in circuit theory.

Problems

3-1 For the eld of the TE_{10} mode of a rectangular waveguide (see Problems 1-23 and 2-4), $\mathbf{E} = E_y\,\hat{a}_y = E_0\,\hat{a}_y \sin\frac{\ x}{a} \exp(\ \ j\ \ z)$, $\mathbf{H} = H_x\,\hat{a}_x + H_z\,\hat{a}_z$, with $H_x = \ \ \frac{\ E_0}{\omega\ } \sin\frac{\ x}{a} \exp(\ \ j\ \ z)$, $H_z = \frac{jE_0}{\omega\ }\frac{\ }{a} \cos\frac{\ x}{a} \exp(\ \ j\ \ z)$, de ned for $0 < x < a$, $0 < y < b$, without sources, show that the ux of the Poynting vector

a) through the planes $x = 0$, $x = a$, $y = 0$, $y = b$, is equal to zero;

b) through any plane $z =$ constant, is real, and independent of z.

Draw your conclusions.

3-2 Show that the results of part a) of the previous problem remain the same when j is replaced by , with real , in the argument of the exponentials and as a factor in the expression of H_x. On the other hand, this change modi es deeply the conclusions of part b). Recalculate and draw suitable conclusions.

3-3 Calculate the wave impedance in the z direction, for the elds of the previous two problems. What happens to these wave impedances if the sign of (or) is changed?

3-4 The following eld represents a *purely standing wave* in free space, to be studied in Chapter 4: $\mathbf{E} = E_y\,\hat{a}_y = E_0 \sin(2\ \ z)\,\hat{a}_y$, $\mathbf{H} = H_x\,\hat{a}_x = j\frac{E_0}{\ _0} \cos(2\ \ z)\,\hat{a}_x$, where $\ _0 = \sqrt{\ _0/\ _0}$.

a) Write the Poynting vector.

b) Find for which values of L the following equation is satis ed:

$$\int_0^L \frac{\ _0}{4}\,|\mathbf{E}|^2\,dz = \int_0^L \frac{\ _0}{4}\,|\mathbf{H}|^2\,dz \quad .$$

c) Does this eld obey the uniqueness theorem?

3-5 In the time domain, two elds, $\{\mathbf{E}', \mathbf{H}'\}$ and $\{\mathbf{E}'', \mathbf{H}''\}$, have identical sources and satisfy identical boundary conditions, but di erent initial conditions. Show that, if the medium is lossy (i.e., for $\ \neq 0$), it must be $\lim_{t\to\infty} (\mathbf{E}'\ \ \mathbf{E}'') = \lim_{t\to\infty} (\mathbf{H}'\ \ \mathbf{H}'') = 0$, where t denotes time.

3-6 Consider a region, limited by two parallel conducting planes and extending to in nity in all directions parallel to these planes. How should the Sommerfeld radiation conditions be modi ed, to ensure uniqueness of a eld de ned in this region?
(*Hint.* Introduce polar cylindrical coordinates.)

3-7 The eld of Problem 3-1 is incident on a source which consists of the following surface electric current density, imposed on the $z = 0$ plane:

$$\mathbf{J}_0 = J_a\,\hat{a}_y \sin k_x x \quad [\mathrm{A/m}] \qquad \text{for } 0\ \ x\ \ a\,,\ 0\ \ y\ \ b\ .$$

a) Calculate the reaction of the eld on this source, as a function of k_x.

b) Discuss, in particular, what occurs when $k_x = n\ /a$, with integer n, and suggest a physical interpretation of these results.

3-8 Two sources, equal to that of Problem 3-7, are separated by a distance L such that $\exp(\ j\ L) = \ 1$. What is the total reaction of the eld of Problem 2-1 on this compound source? Comment on the result.

3-9 In the de nition of reaction, replace the imposed current densities with their complex conjugates. Retain all the other assumptions of the reciprocity theorem. Prove that a "modi ed" reciprocity theorem, involving "reactions" modi ed in this way, holds only if the medium is *strictly lossless*. Is this a meaningful, physically acceptable result?
(*Hint*. Reconsider what was said, about lossless media, after the uniqueness theorem.)

3-10 Consider the eld of Problem 3-1, de ned for $\ \infty < z < +\infty$.

a) Calculate its equivalent electric and magnetic currents on the plane $z = 0$, i.e., the current densities on that plane such that the eld remains unchanged for $z > 0$, and is identically zero for $z < 0$.

b) Calculate its absorbing currents on the plane $z = 0$, i.e., the current densities on that plane such that the eld remains unchanged for $z < 0$, and is identically zero for $z > 0$.

3-11 Repeat Problem 3-10, for a eld which di ers from the previous one in that is replaced by . Comment on the results.

3-12 Consider two e.m. elds. The rst one is identical to that of Problem 3-1 for $0 < x < a$, $0 < y < b$, and identically zero outside that region. The second one has the same analytical expressions, which apply now for $\ \infty < x < +\infty$, $\ \infty < y < +\infty$.
Calculate absorbing currents at $x = 0$, $x = a$, $y = 0$, $y = b$, which transform the second eld into the rst one. Compare the results with those of Problem 1-23.

3-13 Superimposing the eld of Problem 3-1 and a eld, of equal amplitude, which di ers from the former one in that is replaced by , we get (assuming that E_0 is real): $\mathbf{E} = E_y\,\hat{a}_y = 2E_0\,\hat{a}_y \sin\frac{x}{a}\cos(\ z)$, $\mathbf{H} = H_x\,\hat{a}_x + H_z\,\hat{a}_z$, with $H_x = 2j\frac{E_0}{\omega}\sin\frac{x}{a}\sin(\ z)$, $H_z = 2\frac{jE_0}{\omega}\frac{}{a}\cos\frac{x}{a}\cos(\ z)$, for $0 < x < a$, $0 < y < b$.
Calculate the equivalent currents in a generic plane $z =$ constant, and discuss their behavior as functions of z.

3-14 A rectangular waveguide is de ned as an in nitely long region of rectangular cross-section, lled by a homogeneous lossless dielectric and surrounded by a perfectly conducting wall (see Figure 7.4 in Chapter 7). Imagine

a structure which consists of two identical waveguides, one above the other (i.e., $0 \quad x \quad a$, $\infty < z < +\infty$ for both; $0 \quad y \quad b$ for waveguide 1, $b \quad y \quad 0$ for waveguide 2). In the conducting wall between the two waveguides, there is a rectangular aperture S, at $0 \quad x \quad a$, $0 \quad z \quad t$.
Assume that the "incident" eld, $\{\mathbf{E}_0, \mathbf{H}_0\}$, is de ned in waveguide 1, and coincides with that of Problem 3-1. Find the surface current densities of the induction theorem, Eqs. (3.30), on the aperture S.
Next, suppose that between the two waveguides there are two rectangular apertures, each of the same size as the previous one, but with a distance $d = \ /(2 \ )$ between their centers.
Calculate the current densities of the induction theorem on both apertures. What are the consequences which one should expect from this result, for the induced eld in region 2?

3-15 Write the dual of the eld of Problem 3-1, assuming, for simplicity, $= {}_0$, $= {}_0$.

3-16 Consider the following e.m. eld, de ned for $r < a$ in cylindrical coordinates: $E_r = E_z = 0$, $E_\varphi = E_0 J_1(3.83\, r/a) \exp(\ j \ z)$,
$H_r = E_0(\ /\omega \ ) J_1(3.83\, r/a) \exp(\ j \ z)$, $H_\varphi = 0$,
$H_z = j\, 3.83/(a\omega \ ) E_0 J_0(3.83\, r/a) \exp(\ j \ z)$, where E_0 is a constant, and is real (the so-called TE_{01} mode of a circular waveguide). (For the de nition of Bessel functions, see Appendix D).
Write explicit expressions for the components of the dual eld.

3-17 Consider again the eld of the previous problem.

a) Verify that it is TE with respect to the direction of $\hat{a}_z$.
b) Verify that its dual (see again Problem 3-16) is TM with respect to $\hat{a}_z$.
c) Find the corresponding pre-potential functions L and T, and nd the relationships which must be satis ed by .
(*Hint.* Exploit proportionality between longitudinal eld components and pre-potentials.)

3-18 Consider again the eld of Problem 3-1.

a) Decompose it into its TE and TM parts *with respect to* $\hat{a}_x$ *and to* $\hat{a}_y$.
b) Carry out the same procedure for the eld given in Problem 3-16, and for its dual.

3-19 Show that the eld of Problem 3-13 (a mode of a so-called rectangular resonator, see Chapter 9) is:

a) odd with respect to the planes $x = 0$, $y = 0$, $z = 0$;
b) even with respect to re ection in the planes $x = a/2$, $y = b/2$, $z = d/2$, where $d = \ / \ $.

(*Hint.* Change reference frame, $x' = x - a/2$, etc.)

3-20 Consider again a structure like that shown in Problem 3-14, but *without* any opening between the two waveguides. Consider a field which is identical to that of Problem 3-1 in region 1, and identical to zero in region 2. Find its even and odd parts, with respect to the reflection operator in the plane $y = 0$.

3-21 Introduce a rectangular aperture between regions 1 and 2, as in Problem 3-14. Calculate the current densities of the induction theorem, assuming that the incident field $\{\mathbf{E}_0, \mathbf{H}_0\}$, defined now *in the whole structure*, is:

a) the even field as found in Problem 3-20;

b) the odd field as found in Problem 3-20.

Comment on these results.

References

Black, R.J. and Gagnon, L. (1995) *Optical Waveguides Modes: Symmetry, Coupling and Polarization.* Elsevier, Amsterdam.

Collin, R.E. (1991) *Field Theory of Guided Waves.* 2nd ed., IEEE Press, Piscataway, New Jersey.

Corazza, G.C., Longo, G. and Someda, C.G. (1969) "Generalized Thevenin's Theorem for Linear n-Port Networks". *IEEE Trans. on Circuit Theory,* **16** (4), 564–566.

Cracknell, A.P. (1968) *Applied Group Theory.* Pergamon, Oxford.

Ida, N. (1995) *Numerical Modeling for Electromagnetic Non-Destructive Evaluation.* Chapman & Hall, London.

Jackson, J.D. (1975) *Classical Electrodynamics.* 2nd ed., Wiley, New York.

Kuo, F.F. (1966) *Network Analysis and Synthesis.* 2nd ed., Wiley, New York, Section 7.2.

Lewin, L., Chang, D.C. and Kuester, E.F. (1977) *Electromagnetic Waves and Curved Structures.* P. Peregrinus, Stevenage, UK.

Montgomery, C.G., Dicke, R.H. and Purcell, E.M. (1948) *Principles of Microwave Circuits.* McGraw-Hill, New York.

Newcomb, R.W. (1966) *Linear Multiport Synthesis.* McGraw-Hill, New York.

Plonsey, R. and Collin, R.E. (1961) *Principles and Applications of Electromagnetic Fields.* McGraw-Hill, New York.

Press, W.H., Flannery, B.P., Teukolsky, S.A. and Vetterling, W.T. (1992) *Numerical Recipes in C: The Art of Scienti c Computing.* 2nd ed., Cambridge University Press, Cambridge, UK.

Stutzman, W.L. and Thiele, G.A. (1981) *Antenna Theory and Design.* Wiley, New York.

Vassallo, C. (1991) *Optical Waveguide Concepts.* Amsterdam, Elsevier.

CHAPTER 4

Plane waves in isotropic media

4.1 Separability of variables in the homogeneous Helmholtz equation

The rst three chapters have provided a general framework. The next six chapters will be devoted to studying electromagnetic waves in regions which contain no sources, i.e., to the fundamentals of propagation. In this view, this section is devoted to some general remarks regarding the solution of the homogeneous Helmholtz equation. The preferential role that we attribute to this equation is related to the attempt made in this book to give a uni ed approach, based on electromagnetic potentials, to *propagation* (Chapters 4–10) and to *radiation* (Chapters 11–14). Note, though, that propagation phenomena could be studied by solving Maxwell's equations directly, without using potentials. However, this loss of a uni ed approach would not entail any bene t, in terms of less mathematical complexity.

The content of this section applies to many of the partial di erential equations of mathematical physics. For this reason, some of the following statements will look very generic, and possibly too vague, to the reader. To grasp their full meaning, we suggest to the reader to come back to this point after reading the rest of the present chapter.

One of the most commonly used, and most deeply studied, techniques to solve partial di erential equations, is the so-called method of *separation of variables*. Essentially, it consists of breaking a partial di erential equation (PDE) into a set of ordinary di erential equations (ODE's), which can be solved separately from one another, by isolating each independent variable in a separate equation. Examples, presented in this chapter as well as in some of the following ones (Chapters 5, 7, 9, 10), will clarify many details on how this procedure works. However, before we proceed, it is worth brie y discussing the scope of applicability of the technique.

When an equation, such as the Helmholtz equation, contains partial derivatives with respect to ordinary coordinates in real space, then the main limitation stems from the fact that the equation is separable only in a few, well-de ned systems of orthogonal curvilinear coordinates. The phrase "*the equation is separable*" means that there is at least one set, S, of solutions obtained by separation of variables, which is dense enough, to allow one to express *any* possible solution of the PDE as a linear combination of the solutions belonging to S. Another way to express the same concept is to say that S is a *complete set* of solutions.

The type of di culties that may arise in practice from this limitation, can be understood as soon as one considers that solutions of any di erential equation must always satisfy suitable *boundary conditions.* Indeed, it is easy to grasp that formal complications will arise as soon as the boundary is not a coordinate surface (or a set of coordinate surfaces), in one of the reference frames where the equation is separable.

Nevertheless, separation of variables is a very widely used method. To a large extent, this is due to the drawbacks of the only method which o ers a broad-scope alternative, the so-called *Green's function method.* This technique consists essentially of writing the solution of a PDE as a volume (or surface) integral of the eld sources multiplied by suitable functions, whose meaning will be identi ed in Chapter 11. It will be exploited in Chapters 11, 12 and 13, which are mainly devoted to the *inhomogeneous* Helmholtz equation, i.e., to elds with sources. We wish to point out that the Green's function method is applicable also in the homogeneous case, but usually, in this case, it leads to more complication, compared with separation of variables, and this justi es our preference. The relationship between the two methods will be discussed brie y in Chapter 11.

4.2 Solution of the homogeneous Helmholtz equation in Cartesian coordinates

Let us suppose that the whole three-dimensional (3-D) real space is lled by a linear, isotropic and homogeneous medium, with parameters , , , and that imposed currents are zero everywhere. This, which will be assumed throughout this chapter, is, strictly speaking, a physically meaningless system. Nevertheless, it may build a model which applies with an adequate accuracy to systems which contain su ciently wide source-free homogeneous regions. A quantitative estimate of the expression "su ciently wide" will be given soon, through the de nition of a length which characterizes the spatial evolution of e.m. phenomena, the *wavelength.*

We aim at determining the electromagnetic elds which can exist in these conditions, using the potentials $(\mathbf{A}, \varphi)$ that were de ned in Section 1.8. Note that, under the present assumptions, $\varphi = 0$ is a Lorentz gauge. Therefore, the whole problem we want to tackle reduces to solving the homogeneous version of Eq. (1.80), i.e.:

$$\nabla^2 \mathbf{A} \qquad {}^2\mathbf{A} = 0 \quad , \tag{4.1}$$

where:

$$ {}^2 = \quad \omega^2 \quad {}_c = \quad \omega^2 \left(\quad j \frac{}{\omega} \right) \quad . \tag{4.2}$$

Let us introduce a cartesian coordinate system (x_1, x_2, x_3). In such a reference frame, the components of the Laplacian of a vector (see Appendix B) are the scalar Laplacians of the components of the vector. This enables us to write Eq. (4.1) simply by components as:

$$\nabla^2 A_i \quad \frac{\partial^2 A_i}{\partial x_1^2} + \frac{\partial^2 A_i}{\partial x_2^2} + \frac{\partial^2 A_i}{\partial x_3^2} = \ {}^2 A_i \qquad (i = 1, 2, 3) \quad . \tag{4.3}$$

As the three scalar equations (4.3), for the three components of **A**, are independent of each other, and the medium is linear, without loss of generality we can take **A** parallel to one of the axes, for example:

$$\mathbf{A} = A\,\hat{x}_1 \quad . \tag{4.4}$$

Separation of variables, anticipated in the previous section, consists of making the *ansatz* that $A(x_1, x_2, x_3)$ be written as a product of three functions, each of them depending on just one coordinate:

$$A(x_1, x_2, x_3) = f_1(x_1)\, f_2(x_2)\, f_3(x_3) \quad . \tag{4.5}$$

Substituting Eq. (4.5) in Eq. (4.3), and dividing by A ($\neq 0$), we get:

$$\frac{f_1''}{f_1} + \frac{f_2''}{f_2} + \frac{f_3''}{f_3} = \ {}^2 \quad , \tag{4.6}$$

where f_j'' denotes the second derivative of f_j.

The right-hand side of Eq. (4.6) is a constant. The left-hand side consists of three terms, each of them depending on a di erent coordinate. Therefore, Eq. (4.6) can be satis ed if and only if the three terms are *separately* constant:

$$\frac{f_i''}{f_i} = S_i^2 \quad , \tag{4.7}$$

with the condition:

$$S_1^2 + S_2^2 + S_3^2 = \ {}^2 \quad . \tag{4.8}$$

The quantities S_i^2 are the *separation constants* for Eq. (4.6). Eq. (4.8) is referred to as the *separation condition.*

As we anticipated in Section 4.1, Eqs. (4.7) are three ordinary, mutually independent di erential equations. In particular, in this case we found three harmonic equations, whose general integral, as well known, can be written in the form:

$$f_i = F_{ai}\, e^{\ S_i x_i} + F_{bi}\, e^{S_i x_i} \quad , \tag{4.9}$$

where F_{ai}, F_{bi} are arbitrary (and in general, complex) constants. Hence, any solution of the type in Eq. (4.5) could be written as a product of three functions of the type in Eq. (4.9). However we may notice that Eq. (4.8) deals with S_i^2, not with $S_i = \ \sqrt{S_i^2}$. Therefore, there is no loss of generality if we write Eq. (4.5) in the following form:

$$A = A_0\, e^{\ (S_1 x_1 + S_2 x_2 + S_3 x_3)} \quad A_0\, e^{\ \mathbf{S}\ \mathbf{r}} \quad , \tag{4.10}$$

where A_0 is an arbitrary constant. We have introduced the position vector $\mathbf{r} = \ _i x_i \hat{x}_i$, and de ned:

$$\mathbf{S} = \ _i\, S_i\, \hat{x}_i \tag{4.11}$$

as the *propagation vector*, whose components in our cartesian reference frame are S_i $(i = 1, 2, 3)$. Due to the constraint in Eq. (4.8) on the separation constants, this vector in Eq. (4.11) has to satisfy:

$$\mathbf{S} \quad \mathbf{S} = \quad {}^2 = \quad \omega^2 \quad {}_c \quad . \tag{4.12}$$

Let us stress that, since $\mathbf{S}$ is a complex vector, $\mathbf{S} \quad \mathbf{S} \neq |\mathbf{S}|^2 = \mathbf{S} \quad \mathbf{S}$, i.e., Eq. (4.12) *is not* the norm of the vector $\mathbf{S}$.

Let us introduce, for future use, explicit symbols for the real and the imaginary part of the propagation vector:

$$\mathbf{S} = \mathbf{a} + j\,\mathbf{k} \quad , \tag{4.13}$$

where $\mathbf{a}$ is referred to as the *attenuation vector*, and $\mathbf{k}$ as the *phase vector*. This notation permits us to rewrite Eq. (4.12) as follows:

$$\mathbf{a}^2 \quad \mathbf{k}^2 + 2j\,\mathbf{a} \quad \mathbf{k} = \quad {}^2 \quad . \tag{4.14}$$

Implications arising from Eq. (4.14) will be the subject of the next section. We will conclude this section by giving explicit expressions for the elds $\mathbf{E}$ and $\mathbf{H}$ that follow from the vector potential in Eq. (4.10). We emphasize that these elds may be derived directly, solving the Helmholtz equation for $\mathbf{E}$ or $\mathbf{H}$ by the method of separation variables. We leave this to the reader as an exercise.

From Eq. (1.66), with $\varphi = 0$, it follows that:

$$\mathbf{E} = \quad j\omega\,\mathbf{A} = \mathbf{E}_0\, e^{\quad \mathbf{S} \quad \mathbf{r}} \quad , \tag{4.15}$$

with $\mathbf{E}_0 = \quad j\omega\, A_0 \hat{x}_1 = E_0\, \hat{x}_1$, which means that E_0 is essentially an arbitrary constant. From Eq. (1.60), or from Maxwell's equation for $\nabla \quad \mathbf{E}$, using the vectorial identity (C.6), it follows that:

$$\begin{aligned}\mathbf{H} &= \frac{1}{\quad} \nabla \quad (A_0\, e^{\quad \mathbf{S} \quad \mathbf{r}}\, \hat{x}_1) = \frac{A_0}{\quad} \nabla \left(e^{\quad \mathbf{S} \quad \mathbf{r}}\right) \quad \hat{x}_1 \\ &= \frac{\mathbf{S} \quad \mathbf{A}}{\quad} = \frac{\mathbf{S} \quad \mathbf{E}}{j\omega} = \frac{\mathbf{S} \quad \mathbf{E}_0}{j\omega}\, e^{\quad \mathbf{S} \quad \mathbf{r}} \quad .\end{aligned} \tag{4.16}$$

Note that the penultimate term in Eq. (4.16) gives a relation between $\mathbf{E}$ and $\mathbf{H}$ which is insensitive to the direction of the vector $\mathbf{A}$. This means that this relationship between the electric and the magnetic eld vectors is always valid, and is not restricted to the case of a linear polarization, which was the restrictive meaning of Eq. (4.4).

The last comment is a good starting point for a short discussion on how one should superimpose three solutions of the kind we just found, each of which gives one component of the vector $\mathbf{A}$ in the cartesian reference frame that we introduced before. From now on, unless the opposite is explicitly stated, we will restrict ourselves to superpositions where all three components of the vector $\mathbf{A}$ have *the same propagation vector*, $\mathbf{S}$, i.e., to solutions of Eq. (4.1) of the kind:

$$\mathbf{A} = \mathbf{A}_0\, e^{\quad \mathbf{S} \quad \mathbf{r}} \quad , \tag{4.17}$$

where now $\mathbf{A}_0$ is a constant complex vector. The corresponding electromagnetic eld $\{\mathbf{E}, \mathbf{H}\}$ is then:

$$\mathbf{E} = \mathbf{E}_0\, e^{\ \mathbf{S}\ \mathbf{r}} \quad , \qquad \mathbf{H} = \frac{\mathbf{S}\ \ \mathbf{E}_0}{j\omega\ }\, e^{\ \mathbf{S}\ \mathbf{r}} \quad , \tag{4.18}$$

where $\mathbf{E}_0 = \quad j\omega\, \mathbf{A}_0$.

Note that the previous steps provide a nontrivial generalization, compared with Eq. (4.4). Indeed, now we are in a position to change the reference frame in a completely arbitrary way. Should one be obliged to deal with a solution of Eq. (4.1) where the components A_i have di erent propagation vectors (but always satisfying Eq. (4.12)), this eld could be thought of as a superposition of solutions of the type (4.17).

A eld represented by Eq. (4.18) does not satisfy Sommerfeld's radiation conditions. Indeed, for some directions of the position vector $\mathbf{r}$ (for example, when it is antiparallel to the attenuation vector), Eq. (3.17) is not satis ed. Therefore, the waves we just found (to be called *plane waves*) do not satisfy the uniqueness theorem, not even in the case of a lossy medium. On the other hand, this was an expected result, as we were looking for *nontrivial* solutions (i.e., solutions which do not coincide with $\mathbf{E} = \mathbf{H} = 0$) of Maxwell's equations. However, in Section 4.7, where we use plane waves as a basis to expand more general elds, we will show that the solutions we just derived, although individually violating the uniqueness theorem, nevertheless can, as a set, be used in the construction of elds that satisfy rigorous physical constraints, such as Sommerfeld's radiation conditions.

Finally, inserting the second of Eq. (4.18) in Maxwell's equation for $\nabla\ \ \mathbf{H}$, we get:

$$\mathbf{E} = \frac{\mathbf{S}\ \ \mathbf{H}}{j\omega\ _c} = \frac{\mathbf{S}\ \ \mathbf{E}\ \ \mathbf{S}}{\omega^2\ \ _c} \quad , \tag{4.19}$$

and, in the same way:

$$\mathbf{H} = \frac{\mathbf{S}\ \ \mathbf{H}\ \ \mathbf{S}}{\omega^2\ \ _c} \quad . \tag{4.20}$$

By similarities with real vectors, Eqs. (4.19) and (4.20) could make us think that the vectors $\mathbf{E}$, $\mathbf{H}$ and $\mathbf{S}$ are three mutually orthogonal vectors. However the three vectors are complex, so that more attention has to be paid to orthogonality. The subject is postponed to the end of the next section for a more accurate discussion.

4.3 Plane waves: Terminology and classi cation

Inserting Eq. (4.13), where $\mathbf{a}$ and $\mathbf{k}$ are real vectors, into Eq. (4.18), it follows that the loci where $|\mathbf{E}| =$ constant (and therefore $|\mathbf{H}| =$ constant, $|\mathbf{A}| =$ constant, also) are given by:

$$\mathbf{a}\ \ \mathbf{r} = \text{constant} \quad , \tag{4.21}$$

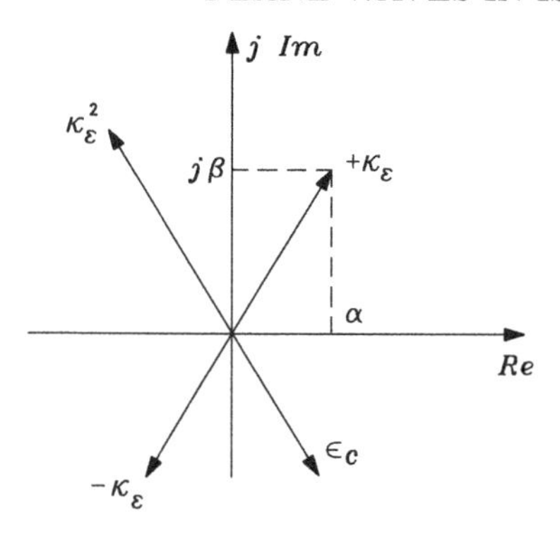

Figure 4.1 *Intrinsic propagation, phase and attenuation constants of a homogeneous medium, represented on the complex plane.*

i.e., are the *planes perpendicular to the attenuation vector* **a**. They are called *constant-amplitude* or *equiamplitude* planes of the wave. We will comment on the term "wave" in the next section.

In the same manner, it can be proved that loci where $\angle E_i$ = constant (the symbol $\angle f$ indicates the phase of the complex number f), E_i being one of the components of the **E** vector in an arbitrary coordinate frame, are given by:

$$\mathbf{k} \quad \mathbf{r} = \text{constant} \quad , \tag{4.22}$$

i.e., are the planes, called *constant-phase* or *equiphase* planes, *perpendicular to the phase vector* **k**. The equiphase surfaces of a wave are also called wavefronts. As its wavefronts are planes, the elds we found in the previous section are called *plane waves.*

Properties of plane waves change drastically, depending on whether **a** and **k** are parallel (i.e., constant-phase and constant-amplitude planes coincide), or not. This subject will now be dealt with in detail. As in Section 1.7, let:

$$\quad = \quad + j \quad , \tag{4.23}$$

be the root of $\ ^2 = \ \omega^2 \ \ _c$ which belongs to the rst quadrant of the complex plane (see Figure 4.1). Then, certainly we have $\ > 0$, $\ \ 0$, with the equality holding only if the medium is lossless. The quantities we denoted with and are called, respectively, the *intrinsic attenuation constant* and *the intrinsic phase constant* of the homogeneous medium we are dealing with. We can then rewrite the real and the imaginary part of Eq. (4.14) as follows:

$$|\mathbf{a}|^2 \quad |\mathbf{k}|^2 \;=\; \mathrm{Re}\{\ ^2\} = \ \omega^2 \quad , \tag{4.24}$$

$$2\,\mathbf{a} \quad \mathbf{k} \;=\; \mathrm{Im}\{\ ^2\} = \omega \quad . \tag{4.25}$$

Comparing signs on the two sides of Eq. (4.24), it follows that $|\mathbf{k}|^2 > |\mathbf{a}|^2$, and therefore $|\mathbf{k}| > 0$. Thus, solutions of Maxwell's equations of the type in Eq. (4.18) *can never have a constant phase* throughout the region where they are de ned.

From Eq. (4.25), we will draw di erent information, depending on whether the medium is lossless ($=0$) or lossy ($\neq 0$). Let us discuss separately the two cases.

A) *Lossless medium.* From Eq. (4.25), it follows:

$$\mathbf{a} \quad \mathbf{k} = 0 \quad , \tag{4.26}$$

which is satis ed either for:

$$\mathbf{a} = 0 \quad , \tag{4.27}$$

or for:

$$\mathbf{a} \neq 0 \quad , \qquad \mathbf{a} \perp \mathbf{k} \quad , \tag{4.28}$$

where $\perp$ means that $\mathbf{a}$ is perpendicular to $\mathbf{k}$.

When Eq. (4.27) holds, then we have $|\mathbf{E}|$ = constant and $|\mathbf{H}|$ = constant over the whole 3-D space, i.e., *any* plane is a constant-amplitude plane, for this kind of wave. Conventionally, we may decide (mainly by analogy with the following case of a lossy medium) to call constant-amplitude planes those which coincide with the constant-phase planes in Eq. (4.22). Such a wave is called a *uniform* plane wave.

When, on the contrary, Eq. (4.28) holds — we notice, incidentally, that in this case, because of Eq. (4.24), $|\mathbf{a}| < |\mathbf{k}|$ — then the *constant-phase planes are orthogonal to the constant-amplitude planes.* Such a wave is called *evanescent.* Many among the following sections and chapters will underline the physical relevance of this kind of waves, which, at rst sight, the reader might consider just as mathematical curiosities.

B) *Lossy medium.* From Eq. (4.25) it follows that:

$$\mathbf{a} \quad \mathbf{k} > 0 \quad . \tag{4.29}$$

This tells us rst, that $|\mathbf{a}| \neq 0$, and second, that the angle between $\mathbf{a}$ and $\mathbf{k}$ is less than $/2$. In the special case where $\mathbf{a}$ and $\mathbf{k}$ are parallel, i.e., the constant-phase planes coincide with the constant-amplitude planes, the wave is still called a uniform plane wave. When the angle between $\mathbf{a}$ and $\mathbf{k}$ is greater than zero, the wave is called *dissociated.*

It is advisable to return to Eqs. (4.19) and (4.20), and discuss them in the light of the results we just found. In order to avoid a substantial amount of needless formal complication, let us restrict ourselves, for the rest of this section, to *linearly polarized* waves, i.e., waves whose eld vectors are proportional to real vectors. As a preliminary, note that, as a consequence of what was shown in Section 2.3, the identity $\mathbf{B} = (\mathbf{B} \ \hat{v})\,\hat{v} + \hat{v} \quad \mathbf{B} \quad \hat{v}$, which, as well known, holds when $\mathbf{B}$ and $\hat{v}$ belong to the real 3-D space, must be modi ed in the following way for the case of complex vectors:

$$\mathbf{B} = (\mathbf{B} \quad \hat{v} \ )\,\hat{v} + (\hat{v} \quad \mathbf{B}) \quad \hat{v} \quad . \tag{4.30}$$

Comparing now Eq. (4.30) with Eqs. (4.19) and (4.20), one sees that $\mathbf{E}$, $\mathbf{H}$ and $\mathbf{S}$ are mutually orthogonal if and only if the wave is linearly polarized and, furthermore, $\mathbf{S}$ is parallel to $\mathbf{S}$. It is straightforward to check that this happens *if and only if* $\mathbf{a}$ *and* $\mathbf{k}$ *are parallel* (including also the particular case where $\mathbf{a} = 0$). Therefore, both in a lossless and in a lossy medium, $\mathbf{E}$, $\mathbf{H}$ and $\mathbf{S}$ are mutually orthogonal *only for linearly polarized uniform plane waves.* On the contrary, this is never possible for evanescent and for dissociated waves. This agrees with geometrical intuition: when $\mathbf{a}$ and $\mathbf{k}$ are not parallel, the vector propagation "covers" by itself two dimensions in space, so that it is impossible for $\mathbf{E}$ and $\mathbf{H}$ to be both mutually orthogonal, and perpendicular to the propagation vector as well.

Finally it is worth noticing that Eqs. (4.19) and (4.20) *always* imply the following relations:

$$\mathbf{E} \quad \mathbf{S} = 0 \quad , \qquad \mathbf{H} \quad \mathbf{S} = 0 \quad , \tag{4.31}$$

and similarly for their complex conjugates:

$$\mathbf{E} \quad \mathbf{S} \quad = 0 \quad , \qquad \mathbf{H} \quad \mathbf{S} \quad = 0 \quad . \tag{4.32}$$

However, note that, because of what was proved in Section 2.3, these relations we just found *are not* orthogonality relations for $\mathbf{E}$, $\mathbf{H}$ and $\mathbf{S}$.

4.4 Traveling waves. Phase velocity

In order to clarify the basic physical features of plane waves, it is useful to transform, according to the usual Steinmetz procedure, complex eld vectors into the corresponding time-domain vectors. It is easy to check that, under this viewpoint, the quantities $\mathbf{A}$, $\mathbf{E}$ and $\mathbf{H}$ given by Eqs. (4.17) and (4.18) have identical properties; therefore, we may concentrate on one of them. For instance, let us consider the electric eld $\mathbf{E}$ and, for the sake of simplicity, let us assume $\mathbf{E}_0$ to be real. The corresponding vectorial function of the time-space coordinates is then:

$$\mathbf{E} = \mathrm{Re}\left[\mathbf{E}_0\, e^{\ \mathbf{S}\ \mathbf{r}}\, e^{j\omega t}\right] = \mathbf{E}_0\, e^{\ \mathbf{a}\ \mathbf{r}} \cos(\mathbf{k} \quad \mathbf{r} \quad \omega t) \quad . \tag{4.33}$$

This shows that $\mathbf{E}$ varies sinusoidally in time, and, if we leave aside the factor exp($\mathbf{a}$ $\mathbf{r}$), in space also. There is a well-de ned relationship between time and space dependencies, as the argument of the cosine function mixes the two coordinates. To underline this character, a wave of the kind in Eq. (4.33) is called a *traveling wave.*

The factor exp($\mathbf{a}$ $\mathbf{r}$) is, as we said before, an attenuation term, which can be sensed in any direction, except for those perpendicular to $\mathbf{a}$, i.e., those that lie in the constant-amplitude planes. The rate of change of this attenuation is maximum in the same direction as the vector $\mathbf{a}$. This factor is not essential, when discussing the physical features of a traveling wave. Therefore, for the sake of simplicity, from now on it will be omitted.

The factor cos($\mathbf{k}$ $\mathbf{r}$ ωt) describes an undulatory motion. This can be veri ed by tracking, as a function of time, those points where the cosine factor

Do not confuse orthogonality between instantaneous time-harmonic vectors, and orthogonality between the corresponding complex vectors. To clarify this point, solve Problem 4.2, at the end of this chapter.

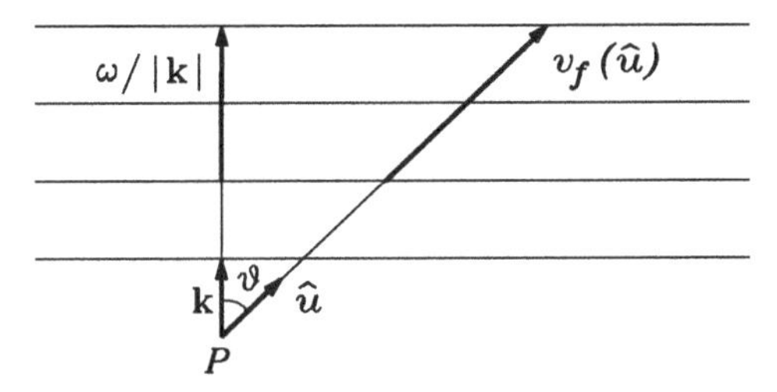

Figure 4.2 *Understanding the behavior of the phase velocity.*

is constant, i.e., points where the argument of the cosine is constant, that correspond to:

$$d(\mathbf{k}\ \ \mathbf{r}\ \ \ \omega t) = 0 \quad , \tag{4.34}$$

where $d(\ )$ denotes the total di erential of the expression in brackets. Since ω and $\mathbf{k}$ are constant, Eq. (4.34) is equivalent to:

$$\mathbf{k}\ \ d\mathbf{r}\ \ \ \omega\, dt = 0 \quad . \tag{4.35}$$

If we let:

$$d\mathbf{r} = \hat{u}\, dr \quad , \tag{4.36}$$

where the unit vector $\hat{u}$ has an arbitrary direction, and originates at the generic point $P = O + \mathbf{r}$, then Eq. (4.35) can be written in terms of the following quantity:

$$v_f(\hat{u}) \quad \frac{dr}{dt} = \frac{\omega}{\mathbf{k}\ \ \hat{u}} \quad , \tag{4.37}$$

whose dimensions are those of a velocity. This quantity is called *phase velocity in the direction of* $\hat{u}$.

It is easy to prove that $v_f(\hat{u})$ is not the same as a velocity of a point motion in kinematics. Indeed, its dependence on the angle ϑ between the direction of $\hat{u}$, which we choose arbitrarily, and that of $\mathbf{k}$, which is xed, is:

$$v_f(\hat{u}) = \frac{\omega}{|\mathbf{k}|\ \cos\vartheta} \quad . \tag{4.38}$$

Accordingly, $v_f(\hat{u})$ is inversely proportional to $\cos\vartheta$, whereas in kinematics the component of a velocity vector $\mathbf{v}$ along the direction de ned by $\hat{u}$ would be directly proportional to $\cos\vartheta$. It can be seen with the help of Figure 4.2 that Eq. (4.38) is the velocity at which, to an observer, located in P and looking in the direction of $\hat{u}$, would appear to move the intersection between the generic constant-phase plane and the direction of $\hat{u}$. In no case is $v_f(\hat{u})$ the speed of a point mass. It will be shown in the following section that it is not even the velocity at which electromagnetic energy moves. So, it should not come as a surprise if it can be larger than the speed of light, eventually *arbitrarily large*, as is the case for $\vartheta \to\ /2$.

Normally, if $\hat{u}$ is not explicitly speci ed, then it is understood that the expression "phase velocity" refers to the value of Eq. (4.37) in the direction of $\mathbf{k}$, i.e., the *minimum* value of Eq. (4.38),

$$v_f = \frac{\omega}{|\mathbf{k}|} \quad . \tag{4.39}$$

Now, following the scheme of the previous section, let us nd what are the values that Eq. (4.39) can take.

A) *Lossless medium.* For a *uniform plane wave*, since $\mathbf{a} = 0$, Eq. (4.24) gives:

$$|\mathbf{k}| = \omega\sqrt{\quad} = \quad = 2\ /\quad , \tag{4.40}$$

where is the *wavelength* in the given medium, at the angular frequency ω. The modulus of the phase vector equals the intrinsic phase constant , so that Eq. (4.39) yields:

$$v_f = \frac{1}{\sqrt{\quad}} = c \quad . \tag{4.41}$$

Therefore, for uniform plane waves in lossless media, the phase velocity (which is, in general, a feature of the wave) equals that quantity which in Section 1.7 was called the *speed of light*, which is, in turn, a feature of the medium.

For an *evanescent plane wave*, since $\mathbf{a} \neq 0$, Eq. (4.24) yields $|\mathbf{k}| > \omega\sqrt{\quad} =$, where denotes, as usual, the intrinsic phase constant of the medium. Hence, the equality (4.41) is replaced by the following inequality:

$$v_f < c \quad . \tag{4.42}$$

Due to Eq. (4.42), an evanescent plane wave in a lossless medium is said to be a *slow wave.*

B) *Lossy medium.* As in this case $\mathbf{a} \neq 0$, the inequality in Eq. (4.42) holds, and therefore only slow waves can exist. Anyway, as $\to 0$, we get $v_f \to c$ for uniform plane waves, so that in media with small losses (ω), a rst-order approximation yields $v_f \simeq c$. A higher-order approximation can be evaluated by the reader as an exercise.

4.5 Standing waves

The previous sections were devoted to the fundamental physical aspects of a solution of the Helmholtz equation of the kind in Eq. (4.10), assuming always that it was the only eld that existed in the region of interest.

From previous experience with di erential equations, and from elementary experimental observations on motion of waves (such as, for example, vibrating strings, or waves on the surface of a liquid), the reader can expect that as soon as *boundary conditions* are imposed on an e.m. eld, they can induce very substantial di erences, with respect to what we saw up to now. At least one of these cases deserves to be investigated immediately.

Consider two uniform plane waves which overlap in a lossless unbounded medium. Each wave is of the kind in Eq. (4.18). Their propagation vectors are

$\mathbf{S} = j\mathbf{k}$ and $\mathbf{S} = \quad j\mathbf{k} = \mathbf{S}$, respectively, so:

$$\mathbf{E} = \mathbf{E}_1\, e^{\quad j\mathbf{k}\ \mathbf{r}} + \mathbf{E}_2\, e^{j\mathbf{k}\ \mathbf{r}} \quad . \tag{4.43}$$

For simplicity, assume that $\mathbf{E}_1$ and $\mathbf{E}_2$ are real. Then, the corresponding time-dependent vector is:

$$\begin{aligned} \mathbf{E} &= \mathrm{Re}(\mathbf{E}\, e^{j\omega t}) \\ &= 2\mathbf{E}_2 \cos(\mathbf{k} \quad \mathbf{r}) \cos \omega t + (\mathbf{E}_1 \quad \mathbf{E}_2) \cos(\mathbf{k} \quad \mathbf{r} \quad \omega t) \quad . \end{aligned} \tag{4.44}$$

Whereas the second term in the nal expression of Eq. (4.44) is of the same type as Eq. (4.33), which has already been investigated in Section 4.4, the rst term is completely di erent, as space and time coordinates appear in two distinct functions, which multiply each other, instead of being combined linearly in the argument of the same function. In particular, the time evolution is in synchronism throughout the whole space: the phase delays from point to point, that characterize Eq. (4.33), have disappeared in Eq. (4.44). What we see now is an oscillation, whose *amplitude* varies in space, according to a $\cos(\mathbf{k} \quad \mathbf{r})$ rule. The points where amplitude reaches its maximum (called *crests*) are given by:

$$\mathbf{k} \quad \mathbf{r} = m \qquad (m = 0, \quad 1, \quad 2, \ldots) \quad , \tag{4.45}$$

whereas the points where the oscillation amplitude is zero (called *nodes*) are given by:

$$\mathbf{k} \quad \mathbf{r} = (2m+1) \quad /2 \qquad (m = 0, \quad 1, \ldots) \quad . \tag{4.46}$$

Eqs. (4.45) and (4.46) de ne two families of planes, perpendicular to $\mathbf{k}$. Comparison with Eq. (4.22) shows that, for each of the two traveling waves that overlap, these planes would have been constant-phase planes. Both families of planes in Eqs. (4.45) and (4.46) are periodic in space, with a period:

$$\mathbf{r} = \frac{\quad}{|\mathbf{k}|}\, \hat{k} = \frac{\quad}{2}\, \hat{k} \quad , \tag{4.47}$$

where $\hat{k}$ is the unit vector of $\mathbf{k}$ and is the wavelength, de ned in Eq. (4.40). Planes belonging to the two families alternate with a constant spacing $\mathbf{r}' = \hat{k} \quad /4$.

The rst term in Eq. (4.44) is referred to as a *pure standing wave*, while the whole Eq. (4.44) is called a *partially standing wave*, given by superimposing a traveling wave with a pure standing wave. In Chapter 8 we will elaborate further on this point. In particular, we will see that the ratio between amplitudes of the traveling wave and of the pure standing one conveys very signi cant information on boundary conditions where partial re ection of the traveling wave occurs.

A partially standing wave, Eq. (4.44), is the most general case of superposition of two uniform plane waves propagating in *opposite* directions. To relax this last constraint, and to consider a superposition of two plane waves (of equal frequency) traveling in any possible direction, the factors $\exp(\quad j\mathbf{k} \quad \mathbf{r})$

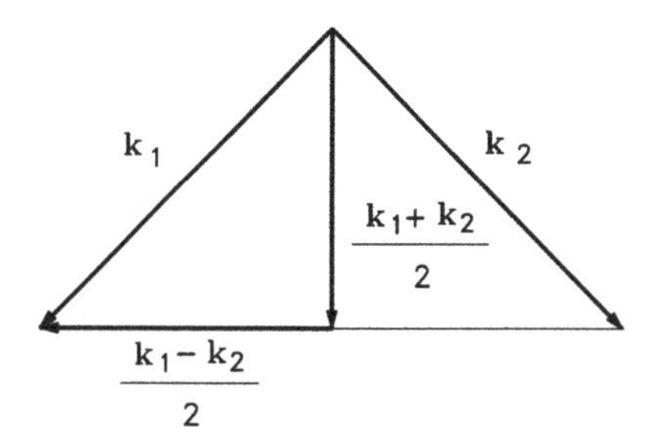

Figure 4.3 *Two uniform plane waves interfering at an angle: decomposition into a standing-wave factor and a traveling-wave factor.*

have to be replaced by exp($j\mathbf{k}_1 \;\; \mathbf{r}$) and exp($j\mathbf{k}_2 \;\; \mathbf{r}$), where $|\mathbf{k}_1| = |\mathbf{k}_2|$. We leave it to the reader to develop the remaining steps, and reach the following conclusions:

the eld, in general, is still a superposition of a standing wave and a traveling wave;

the traveling wave is essentially the same as the one in Eq. (4.44);

the term representing the standing wave consists of a factor which is periodic in space, like:

$$\cos\left[\frac{\mathbf{k}_1 \quad \mathbf{k}_2}{2} \;\; \mathbf{r}\right] \quad , \tag{4.48}$$

and has to be multiplied by another factor, like:

$$e^{\quad j(\mathbf{k}_1+\mathbf{k}_2)\;\mathbf{r}/2} \quad , \tag{4.49}$$

so that the wave is a standing one if we look at it in the direction of the vector $\mathbf{k}_1 \quad \mathbf{k}_2$, as shown by Eq. (4.48); but at the same time, as shown by Eq. (4.49), it travels along the direction of the vector $\mathbf{k}_1 + \mathbf{k}_2$, i.e., along the line that bisects the angle between the phase vectors of the two traveling waves (see Figure 4.3).

Let us now return to the case described by Eq. (4.44), and, taking only the pure standing wave term, nd its instantaneous magnetic eld. It can be calculated by applying Eq. (4.18) to each of the terms that form the electric eld. The nal result reads:

$$\begin{aligned} \mathbf{H} &= \mathrm{Re}\left\{ j\,\frac{\mathbf{k} \quad \mathbf{E}_2}{j\omega}\left(e^{\quad j\mathbf{k}\;\mathbf{r}} \quad e^{j\mathbf{k}\;\mathbf{r}}\right) e^{j\omega t}\right\} \\ &= 2\,\frac{\mathbf{k} \quad \mathbf{E}_2}{\omega}\,\sin(\mathbf{k} \;\; \mathbf{r})\,\sin\omega t \quad . \end{aligned} \tag{4.50}$$

It shows that the nodes of the **H** eld coincide with the crests of the **E** eld, and vice-versa. In the time domain, **E** is in quadrature with respect to **H**.

The reader can prove that, for a pure standing wave, the time-averaged electric and magnetic energies, stored in a cylindrical volume of arbitrary cross-section, and of length /2 along the direction of the phase vectors $\mathbf{k}$ of the superimposed plane waves, are equal to each other.

4.6 Poynting vector and wave impedance

For a *traveling wave* of the kind in Eq. (4.18), which satis es Eq. (4.19), the Poynting vector $\mathbf{P} = \mathbf{E} \quad \mathbf{H} /2$ can be evaluated by means of the vector identity (C.11). As Eq. (4.13) yields $\mathbf{S} + \mathbf{S} \quad = 2\mathbf{a}$, one gets:

$$\begin{aligned} \mathbf{P} &= \frac{\mathbf{E}_0 \quad (\mathbf{S} \quad \mathbf{E}_0)}{2j\omega} \, e^{\;(\mathbf{S}+\mathbf{S}\;)\;\mathbf{r}} \\ &= \left[\frac{|\mathbf{E}_0|^2}{2j\omega} \, \mathbf{S} \; + \frac{\mathbf{E}_0 \quad \mathbf{S}}{2j\omega} \, \mathbf{E}_0 \right] e^{\;2\mathbf{a}\;\mathbf{r}} \\ &= \left[\frac{|\mathbf{H}_0|^2}{2j\omega} \, \mathbf{S} \quad \frac{\mathbf{H}_0 \quad \mathbf{S}}{2j\omega} \, \mathbf{H}_0 \right] e^{\;2\mathbf{a}\;\mathbf{r}} \quad . \end{aligned} \tag{4.51}$$

So, $\mathbf{P}$ is a function of space coordinates only through the factor exp($2\mathbf{a}\;\mathbf{r}$). It is then constant on any constant-amplitude plane in Eq. (4.21). This would imply that the ux of the $\mathbf{P}$ vector through any plane in space is in nite. This is physically untenable. Indeed, Eq. (4.18) is perfectly legitimate as a mathematical solution of Maxwell's equations, but, as it requires an inde nitely extended homogeneous medium, there are limitations to its physical validity. When these constraints are violated, one gets meaningless results, as the previous example shows. This point will be further clari ed in Section 4.7 and in Chapter 5.

As for the direction of $\mathbf{P}$, it is not easy to recognize it from Eq. (4.51), because, as we said in Section 4.3, $\mathbf{E}$ and $\mathbf{S}$, or $\mathbf{H}$ and $\mathbf{S}$, are not always orthogonal. It is advisable to analyze separately the peculiar aspects that Eq. (4.51) can assume for the various types of plane waves, following the scheme given in Section 4.3.

A) *Lossless medium, uniform plane wave.* The propagation vector $\mathbf{S} = j\mathbf{k}$ is orthogonal to both electric and magnetic elds. Moreover, since Eq. (4.40) holds, it is:

$$\mathbf{P} = \sqrt{-} \; \frac{|\mathbf{E}_0|^2}{2} \, \hat{k} \quad . \tag{4.52}$$

Poynting vector is hence real, so the complex power carried by a uniform plane wave consists only of active power. It is simple to verify that under the same circumstances, at any point in space it is:

$$\frac{1}{4} \quad |\mathbf{E}|^2 = \frac{1}{4} \quad |\mathbf{H}|^2 \quad , \tag{4.53}$$

as required by Poynting's theorem (see Section 3.2) whenever $\mathbf{P}$ is real.

Let us now evaluate the *wave impedance*, as de ned in Section 3.2. Along the direction of the phase vector, $\hat{k}$, which is also called the direction of the

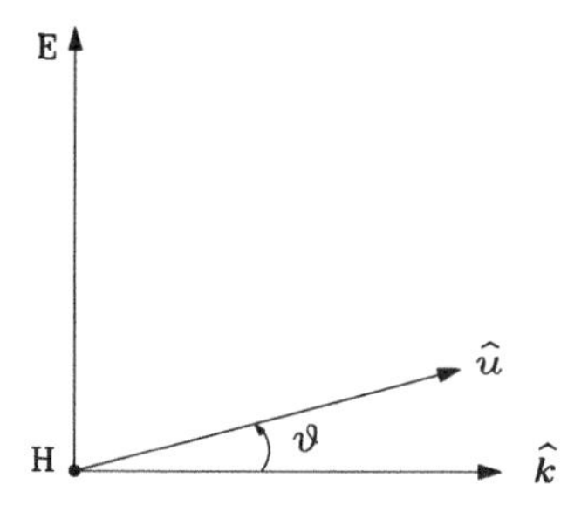

Figure 4.4 *The geometry of a TM plane wave.*

wave normal, and can be seen from Eq. (4.52) to coincide with the direction of **P**, the wave impedance can be evaluated using Eq. (B.11). It reads:

$$Z_{\hat{k}} = \sqrt{-} \qquad . \tag{4.54}$$

Thus, the wave impedance in this case coincides with the quantity $\ = \sqrt{\ /\ }$ which is referred to as the *intrinsic impedance* of the medium, this name coming from the fact that it depends only on the medium parameters, not on the features of a speci c wave propagating in it. It is very useful to memorize that the free-space intrinsic impedance $_0 = \sqrt{\ _0/\ _0}$ equals 120 $\ \simeq 377\ $.

As for wave impedances in directions which di er from that of **k**, it is of primary interest to derive and emphasize two results, to be used in Section 4.8. Both deal with linearly polarized waves, for which the electric and magnetic vectors can be assumed to be real. The rst one refers to a unit vector $\hat{u}$ lying in the plane de ned by **E** and **k**, and is referred to as the *transverse magnetic* or TM case. Application of Eq. (3.8) gives then (see Figure 4.4):

$$Z_{\mathrm{TM}} = \quad \cos\vartheta \quad , \tag{4.55}$$

where $\vartheta = (\hat{u}, \hat{k})$ is the angle between $\hat{u}$ and the wave normal.

The second result refers to the case where $\hat{u}$ lies in the plane de ned by **H** and **k**, and is referred to as the *transverse electric* or TE case. It yields:

$$Z_{\mathrm{TE}} = \quad / \cos\vartheta \quad , \tag{4.56}$$

where, again, $\vartheta = (\hat{u}, \hat{k})$ (see Figure 4.5).

B) For an *evanescent plane wave* in a lossless medium, Eq. (4.51) becomes more complicated. Also for evanescent waves, it will become clear in the following sections that there is a special interest in the case where **E** is orthogonal to **S** (a case that, by analogy with the previous discussion, is called TE), and in the case where **H** is orthogonal to **S** (TM). Under these circumstances, we nd, respectively:

$$\mathbf{P} = \frac{|\mathbf{E}_0|^2}{2\omega} e^{\ 2\mathbf{a}\ \mathbf{r}} (\mathbf{k} + j\mathbf{a}) \qquad (\mathrm{TE}) \quad , \tag{4.57}$$

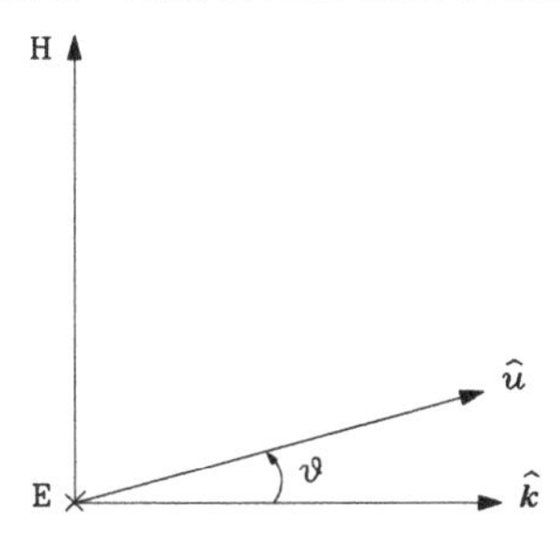

Figure 4.5 *The geometry of a TE plane wave.*

$$\mathbf{P} \;=\; \frac{|\mathbf{H}_0|^2}{2\omega}\, e^{\;\;2\mathbf{a}\;\mathbf{r}} \,(\mathbf{k} \quad j\mathbf{a}) \qquad \text{(TM)} \quad . \tag{4.58}$$

In both these cases, the Poynting vector is complex. Its real part is parallel to **k**, and hence perpendicular to the constant-phase planes. Its imaginary part is parallel to **a**, and hence lies in the constant-phase planes. There is a change in the sign of reactive power, passing from the TE to the TM case.

C) *Lossy medium.* In both the TE and TM cases, Eq. (4.51) still becomes Eqs. (4.57) and (4.58), respectively. However, now **k** is no longer orthogonal to **a**. This implies, on one hand, that along every direction lying in the constant-amplitude planes there is a real component of **P**, but, on the other hand, that the wave normal direction *can never lie* on the constant-amplitude planes. Therefore, in the **k** direction there is always a ow of both active and reactive power.

D) *Standing waves.* Let us stress that the Poynting vector for these waves cannot be obtained by superposition of results calculated separately for the traveling waves which build up the standing wave, since the Poynting vector is derived through a nonlinear operation, with respect to the elds (the outer product of two vectors). We will deal only with the case of a pure standing wave in a lossless medium. We leave it to the reader to consider the other cases as problems, to be solved more conveniently after reading Chapter 8.

Let us start from:

$$\mathbf{E} = \mathbf{E}_0\, e^{\;\;j\mathbf{k}\;\mathbf{r}} + \mathbf{E}_0\, e^{j\mathbf{k}\;\mathbf{r}} \quad , \tag{4.59}$$

where $\mathbf{E}_0$ is orthogonal to $\mathbf{S} = \; j\mathbf{k}$, and **H** can be derived by means of Eq. (4.50). One then gets:

$$\mathbf{P} = \frac{|\mathbf{E}_0|^2}{2\omega}\left(e^{j2\mathbf{k}\;\mathbf{r}} \quad e^{\;\;j2\mathbf{k}\;\mathbf{r}}\right)\mathbf{k} = j\,\frac{|\mathbf{E}_0|^2}{}\;\sin(2\mathbf{k}\quad\mathbf{r})\,\hat{k} \quad . \tag{4.60}$$

Therefore the Poynting vector is parallel, as in the case A), to the **k** vector of a traveling wave, but it is *purely imaginary*, and it is *varying periodically in space*. It goes through zero on the planes:

$$\mathbf{k} \quad \mathbf{r} = \frac{m}{2} \qquad (m = 0, \quad 1, \quad 2, \ldots) \quad , \tag{4.61}$$

i.e. (see Section 4.5), both at the nodes of **E** and at those of **H**. Its modulus reaches its maximum on the planes at half way between two successive nodes. It is suggested that the reader complete these results with a careful analysis of reactive power and of the distribution in space of stored electric and magnetic energy, relating this subject with the discussion at the end of Section 4.5.

The wave impedance in the wave normal direction is purely reactive, and periodic in space:

$$Z_{\hat{k}} = j \quad \mathrm{cotg}\,(\mathbf{k} \quad \mathbf{r}) \quad . \tag{4.62}$$

This is strictly analogous to situations involving transmission lines, to be dealt with in Chapter 8. The reader should reconsider this section after dealing, in particular, with the input impedance of lossless lines terminated on purely reactive loads.

E) *Waves interfering at an angle.* The last case we wish to discuss brie y is that which was schematically shown in Figure 4.3, and discussed in the previous section for what refers to amplitudes and phases of the two elds. Once again, we insist on the fact that the Poynting vector for the overall eld cannot be inferred in an elementary way from those of the two interfering waves. We will brie y summarize the main ndings for the case in which the two interfering waves have *equal amplitudes*, leaving their derivation to the reader as an exercise. In such a case:

neither the Poynting vector, nor the wave impedance (the latter taken along any direction in space) change, as one moves along the direction of the vector $\mathbf{k}_1 + \mathbf{k}_2$, because exponential factors like Eq. (4.49) and its complex conjugate disappear in the calculations;

the Poynting vector and the wave impedance vary periodically along the direction of the vector $\mathbf{k}_1 \quad \mathbf{k}_2$. The laws for their variations in space are easily derived from Eq. (4.60) and from Eq. (4.62), respectively, by replacing $\mathbf{k}$ with $(\mathbf{k}_1 \quad \mathbf{k}_2)/2$;

the component of the Poynting vector along the direction of $\mathbf{k}_1 + \mathbf{k}_2$, and the wave impedance in the same direction, are *real* at any point in space, indicating that active power is owing in that direction;

the component of the Poynting vector along the direction of $\mathbf{k}_1 \quad \mathbf{k}_2$, and the wave impedance in the same direction, are *imaginary* at any point in space, indicating that reactive power is owing in that direction;

in general, the Poynting vector and the wave impedance are polarization dependent.

A more demanding exercise is to repeat all these calculations for the case where the two interfering waves are of di erent amplitudes. At this stage, the reader should know how to do it. However, we suggest to wait until after reading Chapter 8, because analogies with transmission lines will be a source of self-con dence, for this type of exercise.

4.7 Completeness of plane waves

In this section, which is linked to the introductory remarks of Section 4.1, we shall investigate the possibility of representing a generic electromagnetic field as a linear superposition of plane waves. In order to proceed towards this aim, it is necessary first to state explicit requirements that this field must satisfy. This sentence is meant to underline that the question of what fields can be actually expanded into plane waves, does not have a unique answer. The more restrictive the requirements one places on the fields, the easier to prove the completeness of plane waves. Here, our main task is to show merely that *it is possible* to make such an expansion. Thus, we will start from fairly relaxed requirements, and consequently give a rather simple proof. More demanding ones can be found in some of the books that are listed among our references.

Let us consider fields defined in a region R which contains *no sources*, and is filled by an *isotropic and linear medium.* Moreover, let us suppose that this medium is *stepwise homogeneous*, i.e., the parameters $_c$ and , in R, vary in steps, at most, a finite number of times, as functions of the space coordinates. As we saw in Chapter 1, this implies that any field $\{\mathbf{E}, \mathbf{H}\}$ that may exist in R satisfies two requirements:

a) both $\mathbf{E}$ and $\mathbf{H}$ satisfy almost everywhere (i.e., excluding at most a set of measure zero) the homogeneous Helmholtz equation, $\nabla^2 \mathbf{a} = {}^2\mathbf{a}$, where ^{2} is stepwise constant;

b) all the components of $\mathbf{E}$ and $\mathbf{H}$ in a fixed reference frame are piecewise constant, i.e., have at most a finite number of discontinuities of the first kind.

We will now add a third assumption, whose purpose is to restrict our proof to physically meaningful fields, actually to those fields which store a finite energy, and carry a finite power. We will suppose that there is at least one direction in ordinary 3-D space, labeled z, such that the surface integral:

$$\iint_N \frac{1}{4} \left(|\mathbf{H}|^2 + |\mathbf{E}|^2 \right) \mathrm{d}n \quad , \tag{4.63}$$

is *finite* when N is any plane orthogonal to z. This is equivalent to saying that the field $\{\mathbf{E}, \mathbf{H}\}$ is *square integrable*, over any plane orthogonal to z. This assumption is well matched to situations (such as those treated in Chapters 7 and 9) where propagation takes place essentially in one direction. However, it is not difficult to grasp that even for multidimensional propagation phenomena (such as, for example, waves of spherical symmetry), the physical need that the total energy must be finite can be translated into a mathematical requirement of the kind (4.63) on an arbitrary plane.

We will show now that, under these assumptions, the field $\{\mathbf{E}, \mathbf{H}\}$ can be expanded as a linear superposition of an infinite number of plane waves. Let us introduce a rectangular reference frame, (x_1, x_2, x_3), whose x_3 axis is parallel to the direction that we called z. Consider the generic E_i component in this system. Due to the assumptions, it is possible to *Fourier transform* E_i *with*

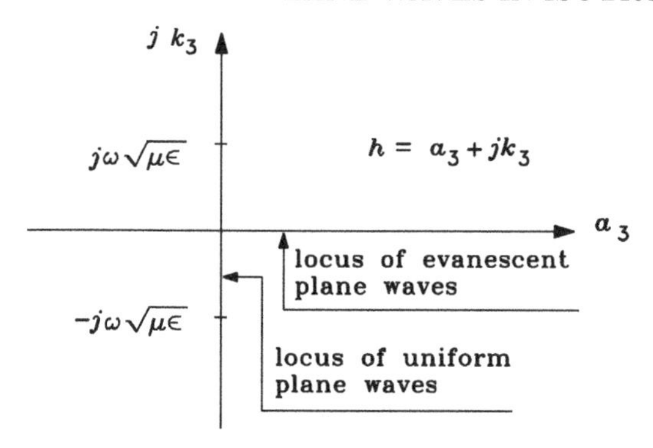

Figure 4.6 *The loci of plane waves on the complex plane of the propagation constant.*

respect to the "transverse" coordinates x_1, x_2:

$$\mathcal{E}_i(k_1, k_2, x_3) = \frac{1}{2} \int_{\infty}^{+\infty} \mathrm{d}x_1 \int_{\infty}^{+\infty} \mathrm{d}x_2 \; E_i \; e^{j(k_1 x_1 + k_2 x_2)} \quad . \tag{4.64}$$

Moreover, since E_i satis es the scalar Helmholtz equation:

$$\nabla^2 E_i \qquad {}_l \, \frac{\partial^2 E_i}{\partial x_l^2} = {}^2 \, E_i \quad , \tag{4.65}$$

exploiting the correspondence, in the Fourier transformation, between the $\partial/\partial x_l$ operator and multiplication times jk_l, the transformed component $\mathcal{E}_i$ satis es:

$$(k_1^2 + k_2^2)\, \mathcal{E}_i \qquad \frac{\partial^2 \mathcal{E}_i}{\partial x_3^2} = \qquad {}^2 \, \mathcal{E}_i \quad . \tag{4.66}$$

Rearranging the terms, and letting:

$$h^2 = \; {}^2 + k_1^2 + k_2^2 \quad , \tag{4.67}$$

we see that Eq. (4.66) can be written as an ordinary di erential equation:

$$\frac{\mathrm{d}^2 \, \mathcal{E}_i}{\mathrm{d}x_3^2} = h^2 \, \mathcal{E}_i \quad , \tag{4.68}$$

whose general integral can be expressed as the linear superposition of $\exp(\;\; hx_3)$, if $h =$ constant, and as follows, if h varies (independently of x_3):

$$\mathcal{E}_i = f_i(h) \; e^{\;\; hx_3} \quad , \tag{4.69}$$

f_i being an arbitrary function of h.

From Eq. (4.67), noting that k_1 and k_2 are real and range from $\;\infty$ to $+\infty$, it follows that, if ${}^2 = \;\; \omega^2$ is real and negative (as is the case in lossless media), then h^2 ranges from $\;\omega^2\;$ to $+\infty$. Consequently, h spans the domain that consists of the real axis of the complex plane, plus the segment of the imaginary axis that lays between $\; j\omega \sqrt{}$ (see Figure 4.6).

The inverse Fourier transformation then gives:

$$E_i = \frac{1}{2} \int_{\infty}^{+\infty} dk_1 \int_{\infty}^{+\infty} dk_2 \, f_i \left(\sqrt{\ ^2 + k_1^2 + k_2^2} \right) e^{\ jk_1x_1 \ \ jk_2x_2 \ \ hx_3} \ , \tag{4.70}$$

which expresses E_i as a linear superposition of plane waves, each of which has a propagation vector $\mathbf{S} = jk_1\hat{x}_1 + jk_2\hat{x}_2 + h\hat{x}_3$.

Making use of Fourier transforms, we have proved plane wave completeness in the sense of the so-called *averaged linear approximation*: the right-hand side of Eq. (4.70) converges towards E_i, when the integration interval goes to in nity, *almost everywhere*, i.e., at most with the exception of a set of discrete points where E_i has step discontinuities.

This ends the proof that it is possible to expand $\mathbf{E}$ as a linear superposition of plane waves. It remains to be proved that the same expansion also yields the magnetic eld. It would not be enough to repeat the same procedure, starting from the Helmholtz equation for $\mathbf{H}$. What has to be proved is that $\mathbf{H}$ can be expanded into plane waves *with the same coe cients as* $\mathbf{E}$. This step is necessary in order to prove that the whole eld $\{\mathbf{E}, \mathbf{H}\}$ can be expressed as a linear combination of plane waves.

Let us substitute $\mathbf{E}$, by components expanded like Eq. (4.70), into Maxwell's equation $\nabla \ \ \mathbf{E} = \ \ j\omega \ \ \mathbf{H}$. To prove that $\mathbf{H}$ is the linear superposition of same plane waves as $\mathbf{E}$, is equivalent to showing that we can interchange the order between integration over k_1 and k_2, and partial derivatives with respect to the coordinates. This is typically a question of *uniform convergence* of the integrals such as Eq. (4.70). It certainly applies where the components of $\mathbf{E}$ are continuous, i.e., within each one of the homogeneous media, which build up the piecewise homogeneous medium. So, this proves that the magnetic eld $\mathbf{H}$ can be expanded into plane waves with the same expansion coe cients as $\mathbf{E}$, except, at most, for what happens at the discrete points where the medium is discontinuous. The problem of what happens at these discontinuities is beyond the scope of this book, and the interested reader is referred to more advanced texts (e.g., Morse and Feshbach, 1953).

The results of this section deserve some further comments.

1) Eq. (4.70), together with Eq. (4.67), shows that, in order to expand a generic eld $\{\mathbf{E}, \mathbf{H}\}$, *uniform plane waves are not su cient*, as they cannot span the whole range of values which are taken by h. In particular, in lossless media ($^2 = \ \ \omega^2 \ \ < 0$, and real) we see that it is necessary to take into account the whole set of the evanescent waves whose attenuation vector is along z, in addition to the set of plane waves traveling in *all directions* in 3-D real space. This point must be stressed, as it di ers signi cantly from what the reader may know about Fourier transforms in the time domain, where the conjugate variable (frequency) spans only the real axis.

2) We have demonstrated that a square-integrable eld (which is a physically meaningful one) can be expanded as a superposition of non-square-integrable waves. In this way, we have reconciled with physical intuition

plane waves which, although physically absurd if taken literally, nevertheless are very powerful and simple mathematical tools.

3) A procedure which makes use of the Fourier transform automatically provides a quantitative criterion, to decide when a single plane wave is acceptable, at least as an approximation. Indeed, the "conjugate variables" relation between x_1, x_2 and k_1, k_2 allows us to write a well-known inequality, which links the minimum ranges where E_i and $\mathcal{E}_i$ have a nonnegligible amplitude:

$$k_i \quad x_i \quad 2 \qquad (i = 1, 2) \quad . \tag{4.71}$$

Therefore, in order to express a eld as a single plane wave (i.e., in order to consider k_i k_i), it is *necessary* for this eld to have its amplitude nonnegligible over a region wide enough to satisfy the following condition:

$$x_i \quad \frac{2}{k_i} \quad . \tag{4.72}$$

The components k_i of the $\mathbf{k}$ vector depend, of course, on the direction of $\mathbf{k}$ itself. Nevertheless, a good term for comparison, which does not depend on the direction of $\mathbf{k}$, can be easily found, as:

$$= \frac{2}{|\mathbf{k}|} \quad . \tag{4.73}$$

This quantity was de ned in Section 4.4, and is called the *wavelength* of the plane wave of angular frequency ω.

4.8 Re ection and refraction of plane waves

Plane waves lend themselves to the description of how e.m. elds behave at a sharpe discontinuity in the parameters of the medium. The validity of the results one gets there in terms of single plane waves has to be checked with the criterion that we outlined in the previous section.

For the sake of extreme simplicity, let us suppose that the whole 3-D space is divided by a *plane surface*, which separates two homogeneous media, whose parameters are labeled with a subscript $m = 1, 2$. Moreover, suppose that the medium 1 is *lossless* ($_1 = 0$). Note, in particular, that the uniqueness theorem and the induction theorem do not hold under these assumptions. However, physical intuition, based on experimental evidence, suggests testing whether continuity of the tangential eld components on the discontinuity plane can be satis ed by superimposing an incident and a re ected wave in region 1, and assuming a transmitted wave in region 2. Let us suppose that the incident eld is a *uniform* plane wave. Let us also introduce a cartesian reference frame (Figure 4.7), whose plane $x_1 = 0$ is placed at the discontinuity between the two media, and whose x_3 axis is chosen so that the phase vector of the incident wave, $\mathbf{k}_i$, lies on the $x_3 = 0$ plane.

All these assumptions translate into the following formulas: for the incoming

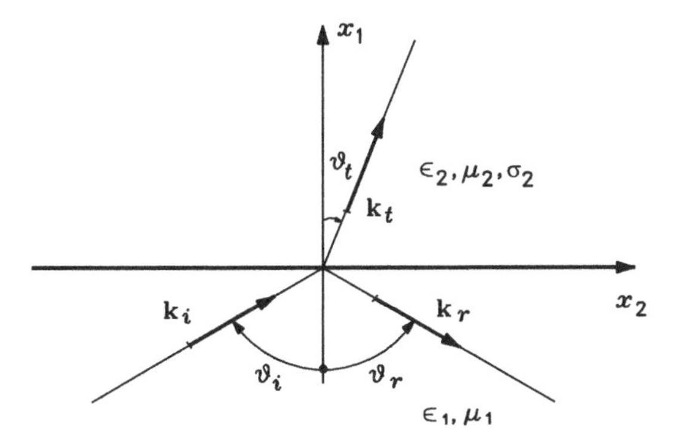

Figure 4.7 *Schematic representation of incident, re ected and transmitted waves, and the reference frame.*

wave:

$$\mathbf{E}_i = \mathbf{E}_{0i}\, e^{\ j\mathbf{k}_i\ \mathbf{r}} \ ,$$

$$\mathbf{H}_i = \frac{\hat{k}_i \quad \mathbf{E}_{0i}}{\ _1} e^{\ j\mathbf{k}_i\ \mathbf{r}} \ . \tag{4.74}$$

For the re ected wave:

$$\mathbf{E}_r = \mathbf{E}_{0r}\, e^{\ \mathbf{S}_r\ \mathbf{r}} \ ,$$

$$\mathbf{H}_r = \frac{\mathbf{S}_r \quad \mathbf{E}_{0r}}{j\omega \ _1} e^{\ \mathbf{S}_r\ \mathbf{r}} \ , \tag{4.75}$$

and, for the transmitted wave:

$$\mathbf{E}_t = \mathbf{E}_{0t}\, e^{\ \mathbf{S}_t\ \mathbf{r}} \ ,$$

$$\mathbf{H}_t = \frac{\mathbf{S}_t \quad \mathbf{E}_{0t}}{j\omega \ _2} e^{\ \mathbf{S}_t\ \mathbf{r}} \ . \tag{4.76}$$

Let us check whether these waves can satisfy, over the entire $x_1 = 0$ plane, the continuity conditions in Eqs. (1.43) and (1.44), which can be written in the following form, usually referred to as *Fresnel's equations*:

$$\hat{x}_1 \quad (\mathbf{E}_i + \mathbf{E}_r) = \hat{x}_1 \quad \mathbf{E}_t \ ,$$

$$\hat{x}_1 \quad (\mathbf{H}_i + \mathbf{H}_r) = \hat{x}_1 \quad \mathbf{H}_t \ . \tag{4.77}$$

In particular, these equations must hold at $\mathbf{r} = 0$, i.e., we must have:

$$\hat{x}_1 \quad (\mathbf{E}_{0i} + \mathbf{E}_{0r}) = \hat{x}_1 \quad \mathbf{E}_{0t} \ ,$$

$$\hat{x}_1 \quad (\mathbf{H}_{0i} + \mathbf{H}_{0r}) = \hat{x}_1 \quad \mathbf{H}_{0t} \ . \tag{4.78}$$

In Eq. (4.77), each term can be written as the product of a factor like E_{0i}, E_{0r}, E_{0t}, and of an exponential function, according to Eqs. (4.74), (4.75) and (4.76). Then, Eqs. (4.77) and (4.78) look like a system of four linear homogeneous equations in the three unknowns E_{0i}, E_{0r}, E_{0t}. For this system to have nontrivial solutions, it is necessary that at most two of these equations are linearly independent. This happens if the three exponential functions have the same argument, for any value of $\mathbf{r} = x_2\hat{x}_2 + x_3\hat{x}_3$, as in this case Eqs. (4.77) coincide with Eqs. (4.78). This provides the following formulas, which describe the *geometrical relationships* of the three waves:

$$j\mathbf{k}_i \quad \mathbf{r} = \mathbf{S}_r \quad \mathbf{r} = \mathbf{S}_t \quad \mathbf{r} \quad . \tag{4.79}$$

For the re ected wave, writing $\mathbf{S}_r = \mathbf{a}_r + j\mathbf{k}_r$, Eq. (4.79) yields:

$$\mathbf{a}_r \quad \mathbf{r} = 0 \quad , \qquad \mathbf{k}_r \quad \mathbf{r} = \mathbf{k}_i \quad \mathbf{r} \quad . \tag{4.80}$$

This pair of equations cannot be satis ed by an evanescent wave. Indeed, $\mathbf{a}_r$ being orthogonal to $\mathbf{k}_r$, from the rst equation it follows that $\mathbf{k}_r$ should lie on the plane $x_1 = 0$; but then (remembering also that $\mathbf{k}_i$ does not belong to this plane) the second equation would imply $|\mathbf{k}_r| < |\mathbf{k}_i| = \ _1 = \omega\sqrt{\ _1\ _1}$, thus violating Eq. (4.24). Therefore, we can conclude that the *re ected wave is a uniform plane wave*, with $|\mathbf{k}_r| = |\mathbf{k}_i| = \ _1$.

The second of Eqs. (4.80), for $\mathbf{r} = \hat{x}_3$, gives $\mathbf{k}_r \quad \hat{x}_3 = 0$, showing that $\mathbf{k}_r$ lies on the plane $x_3 = 0$, de ned by the vector $\mathbf{k}_i$ and by the normal (x_1 axis) to the discontinuity between the two media, and called the *incidence plane.* Moreover, as $|\mathbf{k}_r| = |\mathbf{k}_i|$, we must have:

$$\vartheta_r = \vartheta_i \quad , \tag{4.81}$$

where ϑ_r (ϑ_i) is the angle between $\mathbf{k}_r$ ($\mathbf{k}_i$) and the x_1 axis. They are called the *re ection angle* and the *incidence angle*, respectively.

As for the *transmitted wave*, Eq. (4.79) reads:

$$\mathbf{a}_t \quad \mathbf{r} = 0 \quad , \qquad \mathbf{k}_t \quad \mathbf{r} = \mathbf{k}_i \quad \mathbf{r} \quad . \tag{4.82}$$

A priori, there is no reason why this wave cannot be a dissociated one, or an evanescent one if medium 2 is lossless. We will see soon that, in some circumstances, this is indeed the case. In any case, the second of Eqs. (4.82) shows that also $\mathbf{k}_t$ lies always in the incidence plane. De ning then the *transmission angle* (or refraction angle) ϑ_t as the angle between $\mathbf{k}_t$ and the x_1 axis, the second of Eqs. (4.82) can be rewritten as:

$$\sin\vartheta_t = \frac{\ _1}{|\mathbf{k}_t|} \sin\vartheta_i \quad . \tag{4.83}$$

In order to evaluate $|\mathbf{k}_t|$, which is still unknown, it is necessary to distinguish whether medium 2 is lossless or lossy.

A) *Lossless medium.* In Eqs. (4.82), we can have either $\mathbf{a}_t = 0$, or $\mathbf{a}_t$ perpendicular to the $x_1 = 0$ plane. For $\mathbf{a}_t = 0$, the transmitted eld is a uniform plane wave, and is called the *refracted wave.* It has $|\mathbf{k}_t| = \ _2 = \omega\sqrt{\ _2\ _2}$. Eq. (4.83) can then be rewritten in the following form, which is known as *Snell's law*, or sometimes as *Descartes' law*:

$$\sin\vartheta_t = \frac{1}{2}\ \sin\vartheta_i = \sqrt{\frac{1\ 1}{2\ 2}}\ \sin\vartheta_i \quad , \tag{4.84}$$

where:

$$n_m = \sqrt{\frac{m\ m}{0\ 0}} \qquad (m = 1, 2) \quad , \tag{4.85}$$

is the *refractive index* in the m-th medium. Eq. (4.84) then reads as:

$$n_2\ \sin\vartheta_t = n_1\ \sin\vartheta_i \quad . \tag{4.86}$$

B) Again in a *lossless medium*, but considering now the possibility that:

$$\mathbf{a}_t\ \ \mathbf{r} = 0 \qquad (\mathbf{a}_t \neq 0) \quad , \tag{4.87}$$

we see that this is an acceptable solution only if $\mathbf{k}_t$ (which, because of what was shown in Section 4.3, must be orthogonal to $\mathbf{a}_t$) is parallel to $\hat{x}_2$ (i.e., if $\sin\vartheta_t = 1$). Moreover, we saw in Section 4.3 that $|\mathbf{k}_t| >\ \ _2$, and therefore Eq. (4.83) tells us that an evanescent wave can exist in medium 2 *only if* the inequality:

$$\sin\vartheta_i > \frac{2}{1} = \frac{n_2}{n_1} \quad , \tag{4.88}$$

can be satis ed by a real value of the angle ϑ_i, as required by the initial assumption that the incident eld is a uniform plane wave. It is obvious that this condition can never be satis ed if $n_2 > n_1$. What happens for $n_2 < n_1$ is extremely important, both conceptually and practically. It will be the subject of Section 4.11.

C) *Lossy medium.* The transmitted wave has still to satisfy Eqs. (4.82). Hence, it is a dissociated wave, with constant-amplitude planes parallel to $x_1 = 0$ and constant-phase planes orthogonal to the direction de ned by Eq. (4.83). Note that in this case $|\mathbf{k}_t|$ becomes *a function of the angle* ϑ_i. It can be proved (see Problems) that, instead of Eq. (4.83), one gets:

$$\sin\vartheta_t = \frac{n_1}{n_2}\ \sin\vartheta_i \left\{ \frac{1}{2}\left[1 + \left(1 + \frac{\ \ _2^2}{\omega^2\ \ _2^2\ \cos^2\vartheta_t} \right)^{1/2} \right] \right\}^{1/2} \quad . \tag{4.89}$$

This formula says that, for $\ _2 \neq 0$, the ratio $\sin\vartheta_t / \sin\vartheta_i$ does not depend only on the ratio between the refractive indices, but also on $\ _2$. Snell's law has then to be looked at as a limit, which is approached when $\ _2 \to 0$. We will come back to this point in Section 4.12, when dealing with the case $\quad \omega\ \ _2$.

Before we close this section, let us brie y mention what happens in the case (a very frequent one in practice) where there are more than two media, separated by parallel planes (see Figure 4.8).

In all media but the last one, one wave is no longer su cient: the number of continuity conditions to be satis ed by the eld in that medium doubled: in the previous case we had one plane of discontinuity in the parameters, in the

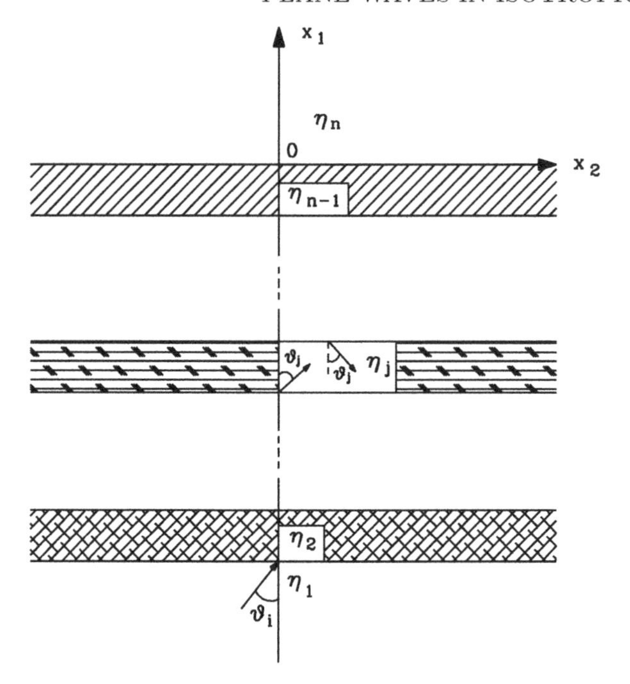

Figure 4.8 *A typical "multilayer" structure.*

present case we have two such planes. Therefore, *two waves* are necessary in each medium, except the last one; they must di er in the sign of the component along $+x_1$ of their propagation vectors. Obviously, in general they will di er in amplitude and phase too.

When all continuity conditions are written explicitly, then one sees that the only way for this system to have nontrivial solutions is that conditions like Eqs. (4.81) and (4.83) be satis ed at each separation plane (the parameters in there must be those of the two media before and after that plane). However, at least as long as all the waves are uniform, to nd the transmission angle in the n-th medium it is not at all necessary to pass through explicit calculations which involve the other media. Indeed, exploiting the transitive law of equality, one may apply directly Snell's law, Eq. (4.84), to the rst and to the n-th medium:

$$\sin\vartheta_n = \frac{1}{n} \sin\vartheta_i = \sqrt{\frac{1\ 1}{n\ n}}\ \sin\vartheta_i \quad . \tag{4.90}$$

The transitive law of equality makes the case where evanescent waves show up in one or more layers even more interesting than the present one. We will brie y deal with it in Section 4.11.

4.9 Fresnel formulas

The previous section, devoted to the geometry of the re ection problem, has led us to conclude that nontrivial solutions of the system in Eq. (4.78) exist. We may then proceed to express the re ected and transmitted elds as function of the incident eld. These relationships depend on *polarization of the incident eld.* To determine them, the simplest way is to exploit some of the results of Chapter 2: we expand the incident wave, whose polarization is completely arbitrary, as the sum of two waves which have two preselected, linearly independent polarizations. Re ection and transmission of these two components can be studied separately, and the results can then be superimposed, to reconstruct what happens to the generic incident wave.

Since $\mathbf{E}_i$ and $\mathbf{H}_i$ are, in any case, both transverse to $\mathbf{k}_i$, a convenient basis for the decomposition consists of two orthogonal linear polarizations, lying in the plane orthogonal to $\mathbf{k}_i$. The simplest results are obtained if we take the following pair: $\mathbf{E}_i$ perpendicular to the incidence plane, and $\mathbf{H}_i$ perpendicular to the incidence plane. They are referred to, respectively, as *TE and TM polarizations.*[†]

We will discuss in full detail the case of *lossless medium 2*, and *uniform plane transmitted wave.* Suggestions will be given later on to the reader in order to extend the results to the other cases.

For *TM polarization* (see Figure 4.9), Eqs. (4.78) read:

$$\begin{aligned} E_{0i}\ \cos\vartheta_i + E_{0r}\ \cos\vartheta_r &= E_{0t}\ \cos\vartheta_t \quad , \\ \frac{E_{0i}}{1} \quad \frac{E_{0r}}{1} &= \frac{E_{0t}}{2} \quad . \end{aligned} \tag{4.91}$$

Eliminating once E_{0t}, and once E_{0r}, we can nd the following expressions for the ratios $E_{0r}/E_{0i} =$ and $E_{0t}/E_{0i} =$, which are called *re ection and transmission coe cients*, respectively:

$$\begin{aligned} {}_{\mathrm{TM}} &= \frac{{}_2 \cos\vartheta_t \quad {}_1 \cos\vartheta_i}{{}_2 \cos\vartheta_t + {}_1 \cos\vartheta_i} \quad , \\ {}_{\mathrm{TM}} &= (1 \quad {}_{\mathrm{TM}})\ {}_2/\ {}_1 = \frac{2\ {}_2 \cos\vartheta_t}{{}_2 \cos\vartheta_t + {}_1 \cos\vartheta_i} \quad . \end{aligned} \tag{4.92}$$

Similarly, for *TE polarization*, replacing the electric eld vectors in Figure 4.9 with the magnetic ones (and keeping track correctly of all the direction

[†] There are other terminologies which are widely used for these two polarizations, like "parallel" and "perpendicular" polarizations, or "horizontal" and "vertical" polarizations. We prefer to avoid these terminologies because there is not a general consensus on them. Radio engineers tend to refer, when using these names, to the direction of the electric eld, while physicists, in particular in optics, tended, in the past, to refer to that of the magnetic eld. Recently (see, e.g., Born and Wolf, 1980), some very authoritative textbooks in optics adopted the convention related to the electric eld.

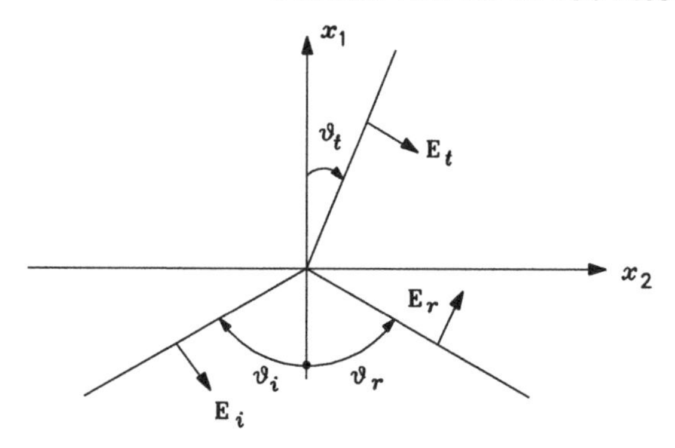

Figure 4.9 *The eld vectors of the incident, re ected and transmitted waves, in the TM case.*

of the vectors that are orthogonal to the plane of the gure), we nd:

$$\text{TE} = \frac{{}_2 \cos\vartheta_i \quad {}_1 \cos\vartheta_t}{{}_2 \cos\vartheta_i + {}_1 \cos\vartheta_t} = \frac{{}_2 \sec\vartheta_t \quad {}_1 \sec\vartheta_i}{{}_2 \sec\vartheta_t + {}_1 \sec\vartheta_i} ,$$

$$\text{TE} = 1 + \text{TE} = \frac{2 \; {}_2 \sec\vartheta_t}{{}_2 \sec\vartheta_t + {}_1 \sec\vartheta_i} . \quad (4.93)$$

Eqs. (4.92) and (4.93) are known as *Fresnel formulas.*

If we recall the *TM and TE wave impedance* formulas, Eqs. (4.55) and (4.56), the re ection coe cients can be given a uni ed expression:

$$= \frac{Z_2(\hat{x}_1) \quad Z_1(\hat{x}_1)}{Z_2(\hat{x}_1) + Z_1(\hat{x}_1)} , \quad (4.94)$$

where $Z_m(\hat{x}_1)$ $(m = 1, 2)$ is the TM or TE wave impedance in the m-th medium, in the direction normal to the plane that separates the two media. It can be proved that Eq. (4.94) holds also in the case of lossy medium 2. The reader who is familiar either with common transmission lines, or with the scattering matrix formalism in circuit theory, will recognize Eq. (4.94) as a very familiar formula. Those who see it for the rst time will understand the meaning of this comment as soon as they get to Chapter 8.

Finally, in next section we will show that, when the transmitted wave in medium 2 is evanescent, it is possible to replace the angle $\vartheta_t = (\hat{k}_t, \hat{x}_1)$ with a generalized transmission angle, in general complex, such that Eqs. (4.92) and (4.93) remain formally valid even in this case, although the physics of the transmitted wave changes completely.

An interesting question is whether there is any value for the angle of incidence, such that the re ection coe cients vanish. To answer it, one has to remember that, under the assumptions which hold in this section, the transmission angle is related to the angle of incidence by Snell's law, Eq. (4.84). Therefore, the larger angle is in the medium where the index is smaller. Before we proceed further, we must specify whether we deal with dielectric or with magnetic media, because permittivity and permeability play equal roles in the refractive index but different roles in the wave impedance. Let us deal with the first case (i.e., $\mu_1 = \mu_2 = \mu_0$). Then, we see that the equation $\rho_{TE} = 0$ does not have any real solution, being impossible to make $\eta_2 \cos\vartheta_i = \eta_1 \cos\vartheta_t$. On the contrary, the equation $\rho_{TM} = 0$ has a solution, which, as the reader may verify by imposing this equation plus Snell's law, Eq. (4.84), reads:

$$\sin\vartheta_i = n_2/(n_1^2 + n_2^2)^{1/2} \quad . \tag{4.95}$$

This angle is referred to as the *Brewster angle* or the *polarizing angle*. The second name points out that when a plane wave in a completely arbitrary state of polarization impinges on a discontinuity at the Brewster angle, the reflected wave is linearly polarized in the TE state. This fact is exploited in practice, for example to build a laser which emits linearly polarized light (see Chapter 9).

4.10 Reflection in multilayer structures

At the end of Section 4.8, we stressed that if there are more than two media, separated by parallel planes (Figure 4.8), then *two waves* are necessary in each medium except the last one, to satisfy all the continuity conditions. We saw then that Eqs. (4.81) and (4.83) must be satisfied at each separation plane. This means that the geometrical part of the problem, i.e., the laws of reflection and refraction in a multilayer structure, are almost trivial extensions of the two semi-infinite media. It is not so concerning the reflection and transmission coefficients that we found in the previous section.

In this section, we will just give a very schematic description of how one deals with this new problem, and then provide two simple examples. The reason why we postpone most of the applications, is because the same calculations, which would be very heavy now in terms of computational effort, will be enormously simplified by the analogies that one can establish after reading Chapter 8.

The procedure goes as follows. The Fresnel reflection coefficients, ρ_{TM} and ρ_{TE} given by Eqs. (4.92) and (4.93), respectively, can still be used at the interface between the $(n-1)$-th and the n-th medium, located (Figure 4.8) at $x_1 = 0$. Then, one tracks the two waves in the $(n-1)$-th medium, back to the plane $x_1 = -\ell_{n-1}$. Note that the up-going and the down-going waves undergo opposite phase shifts, in this operation. Therefore, the total field in the $(n-1)$-th medium at $x_1 = -\ell_{n-1}$ will not be the same as that at $x_1 = 0$. Still, at this stage we may think of it as a quantity which contains *one* undetermined complex multiplicative constant, because the relationship between the two waves is known. This means that we are facing again the same situation as at the beginning, namely, a plane of discontinuity where the field has two degrees of freedom on the upstream side, and one degree of freedom on the downstream side. If we apply again the continuity conditions, we may now calculate the reflection coefficient at the interface between the $(n-2)$-th and the $(n-1)$-th medium. We iterate this procedure as many times

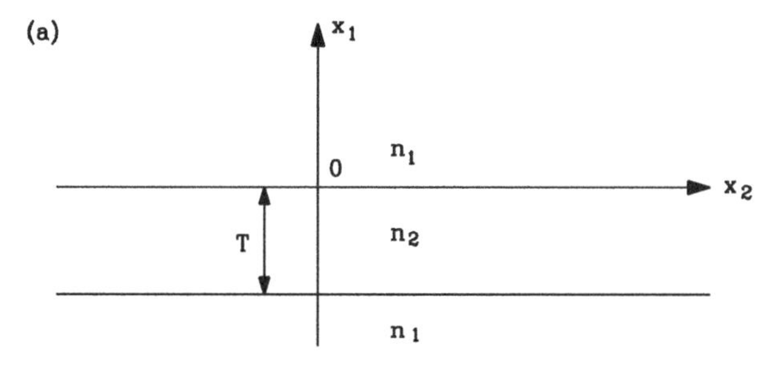

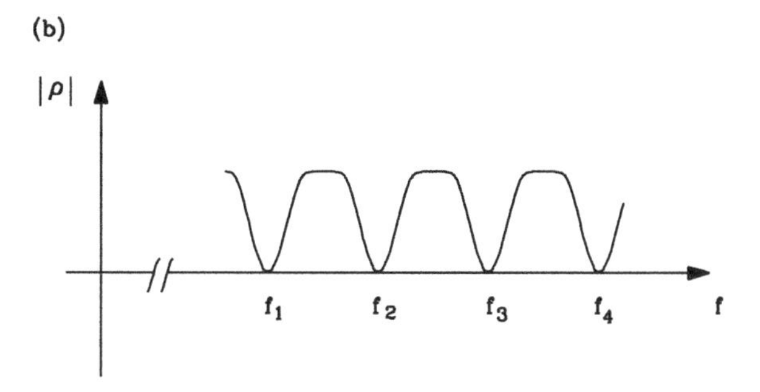

Figure 4.10 *Schematic illustration of a Fabry-Perot interferometer.*

as there are layers, and nally get the re ection coe cients at the input of the structure.

As we said before, the procedure that we outlined does often require very cumbersome calculations, especially at oblique incidence, i.e., for an incidence angle $\vartheta_i \neq 0$, because Snell's law has to used at the same time to nd the various ϑ_j's, $j = 2, \ldots, n$, which in turn a ect the re ection and transmission coe cients. The following examples are quite simple, as they refer to *normal incidence*, $\vartheta_i = 0$, and to three media, i.e., two half-spaces and just one layer in between.

Let us study rst a layer between two semi-in nite media having equal refractive indices, $n_1 = n_3 \neq n_2$. The media are supposed to be non-magnetic dielectrics.

The re ection coe cient at $x_1 = 0$, obviously independent of polarization, is:

$$= \frac{\ _1 \ \ _2}{\ _1 + \ _2} = \frac{n_2 \ \ n_1}{n_2 + n_1} \quad . \tag{4.96}$$

We now track back the two waves in medium 2, from $x_1 = 0$ to $x_1 = \ T$ (Figure 4.10). At those frequencies at which the thickness T of the layer is *an integer multiple of the half-wavelength* in medium 2, i.e., $T = m \ / \ _2$, $m = 1, 2, \ldots$, the relative phase of the two waves in medium 2 at $x_1 = \ T$ is the same as at $x_1 = 0$, exactly as if there were no gap between medium 1 and medium 3. As these are identical, *the re ection coe cient at* $x_1 = \ T$, as seen from medium 1, *is equal to zero.* At these frequencies, all the power

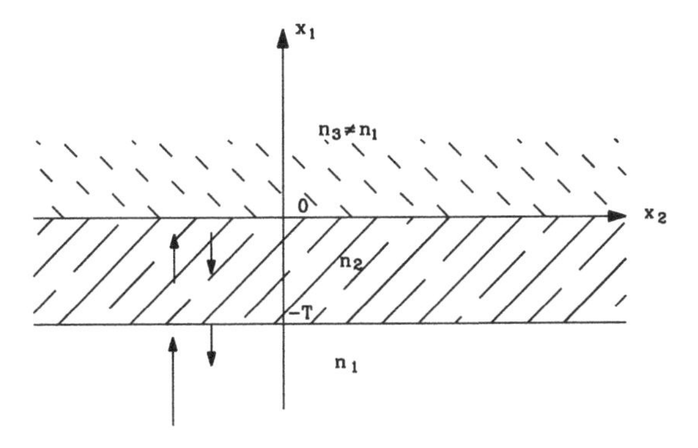

Figure 4.11 *Design of an anti-re ection layer.*

incident on the rst discontinuity is transmitted through layer 2, and emerges in medium 3. At all other frequencies, the re ection coe cient at $x_1 = \quad T$ is not zero. Its modulus reaches its maximum value at those frequencies at which T equals an odd number of quarters of the wavelength in medium 2, a result that the reader may interpret in rather intuitive terms. The details of how the re ection coe cient varies as a function of frequency depend on other details that will be easy to understand after reading Chapters 8 and 9. In any case, however, it is a periodic function of frequency, and its minima are equal to zero.

The device schematically shown in Figure 4.10 is known as a *Fabry-Perot interferometer* or as an *etalon.*

In our second example, Figure 4.11, the two semi-in nite media have different refractive indices, $n_1 \neq n_3$, and the purpose of the layer between them is to *cancel re ection at the front of this layer*, i.e., at $x_1 = \quad T$. The media are again supposed to be non-magnetic dielectrics.

Physical intuition tells us that one way to reach our goal is to design the thickness and the indices in such a way, that two backward waves, due to re ections at $x_1 = 0$ and at $x_1 = \quad T$, cancel each other. Aiming at this, we assume that the thickness of layer 2 is *an odd integer multiple of a quarter of the wavelength* in medium 2, i.e., $T = (2m + 1) \quad /(2 \quad _2)$, $m = 0, 1, \ldots$. This entails that the phase di erence between the forward and backward traveling waves shifts by an odd integer multiple of radians, as we move from the upper side of layer 2 to the lower side.

The re ection coe cient at $x_1 = 0$, again independent of polarization, is:

$$_0 = \frac{_3 \quad _2}{_3 + \ _2} = \frac{n_2 \quad n_3}{n_2 + n_3} \quad , \tag{4.97}$$

and then, from what we just said, it follows that the ratio between the backward and the forward traveling waves in medium 2 at $x_1 = \quad T$ is $' = \quad _0 = (n_3 \quad n_2)/(n_3 + n_2)$. What this implies is that, in the eld continuity conditions, Eqs. (4.78), if written at $x_1 = \quad T$, the transmitted electric eld E_{0t} must be replaced by $E_{0t}(1 + {}')$, and the transmitted magnetic eld H_{0t} must be replaced by $H_{0t}(1 \quad {}')$. The ratio between these two elds is then not equal any more to the intrinsic impedance of medium 2, ${}_2$, but to:

$$_{\mathrm{equiv}} = {}_2 \, \frac{1 + {}'}{1 \quad {}'} = {}_2 \, \frac{n_3}{n_2} \quad . \tag{4.98}$$

If we insert Eq. (4.98) into Eq. (4.94) as the impedance after the discontinuity plane, we nd a re ection coe cient:

$$_1 = \frac{_{\mathrm{equiv}} \quad _1}{_{\mathrm{equiv}} + {}_1} = \frac{n_1 \quad n_2^2/n_3}{n_1 + n_2^2/n_3} \quad , \tag{4.99}$$

which clearly indicates that there is no re ection there (${}_1 = 0$) for:

$$n_2 = (n_1 \, n_3)^{1/2} \quad . \tag{4.100}$$

For reasons that will become clearer after reading Chapter 8, the anti-re ection device which has been illustrated in this example is referred to as a *quarter-wave impedance-matching transformer.*

4.11 Total re ection

Let us now study what happens if, for medium 2 being lossless, we have,

$$n_2 < n_1 \quad , \qquad \vartheta_i > \vartheta_c = \arcsin \frac{n_2}{n_1} \quad , \tag{4.101}$$

where the symbols are the same as in the previous section. The angle ϑ_c de ned by Eq. (4.101) is referred to as the *critical* (or *limiting*) *angle.*

In this case, Snell's law Eq. (4.83) has no real solutions for ϑ_t, since it should require $\sin \vartheta_t > 1$. We may conclude that the transmitted wave cannot be a uniform plane wave. On the other hand, there is a solution which, as we mentioned in Section 4.8, consists of an evanescent wave. In this entire chapter, we are working in such conditions that the uniqueness theorem can never be invoked. Yet, the evanescent wave is *the* solution, as demonstrated experimentally. Its study provides the best possible example of why evanescent waves are important in practice.

In the reference frame of Figure 4.7, as we saw in Section 4.8, the evanescent wave must have its attenuation vector $\mathbf{a}_t$ parallel to the x_1 axis, and its phase vector $\mathbf{k}_t$ parallel to the x_2 axis. The modulus of $\mathbf{k}_t$ is then given by Eq. (4.83) with $\vartheta_t = \ /2$, i.e.:

$$|\mathbf{k}_t| = {}_1 \sin \vartheta_i \qquad (> {}_2) \quad , \tag{4.102}$$

while from Eq. (4.24) we get:

$$|\mathbf{a}_t| = \left(|\mathbf{k}_t|^2 \quad {}_2^2\right)^{1/2} = \left({}_1^2 \sin^2 \vartheta_i \quad {}_2^2\right)^{1/2} \quad . \tag{4.103}$$

Physically, it is evident that $\mathbf{a}_t$ must be oriented in the positive x_1 sense; otherwise, the amplitude would grow exponentially as the eld penetrates into medium 2.

Now let us show that, under these circumstances, the *wave impedances*, both TE and TM, in medium 2, *are purely imaginary*, i.e., pure reactances. Let us deal first with the TE case. Because of the continuity condition in Eq. (1.46) on the normal component of the vector $\mathbf{D}$ across the plane $x_1 = 0$, we see that $\mathbf{E}_t$ has to be parallel to x_3, like $\mathbf{E}_i$ and $\mathbf{E}_r$. Moreover, since the propagation vector $\mathbf{S}_t = \mathbf{a}_t + j\mathbf{k}_t$ has a purely real component along x_1, from the second of Eqs. (4.18) it follows that $\mathbf{H}_{\tan} = (\mathbf{a}_t \quad \mathbf{E}_t)/j\omega$ is in quadrature with $\mathbf{E}_t$, so that $Z_{2\mathrm{TE}}(\hat{x}_1)$ is imaginary, q.e.d.. The TM case can be analyzed in the same way, starting from the continuity condition for the normal component of $\mathbf{B}$, and exploiting now Eq. (4.19). The two procedures yield:

$$Z_{2\mathrm{TE}}(\hat{x}_1) = \frac{j\omega}{|\mathbf{a}_t|} \quad , \qquad Z_{2\mathrm{TM}}(\hat{x}_1) = \frac{|\mathbf{a}_t|}{j\omega} \quad , \tag{4.104}$$

where $|\mathbf{a}_t|$ is given by Eq. (4.103).

Given the physical meaning of the wave impedance, these results show that the Poynting vector component along x_1 is imaginary, which is equivalent to saying that the evanescent wave does not carry any active power in the x_1 direction. This result should be linked with what was said in Section 4.6.

We will show now that, in these cases, both $_{\mathrm{TE}}$ and $_{\mathrm{TM}}$, on the plane $x_1 = 0$, are of unit modulus. Indeed, from Eq. (4.78), recalling that $\mathbf{E}_{0t}$ and $\mathbf{H}_{0t}$ are related by Eq. (4.104), we find that,

$$\begin{aligned} {}_{\mathrm{TM}} &= \frac{(|\mathbf{a}_t|/j\omega \ _2) \quad _1 \cos\vartheta_i}{(|\mathbf{a}_t|/j\omega \ _2) + \ _1 \cos\vartheta_i} \quad , \\ {}_{\mathrm{TE}} &= \frac{(j\omega \ _2/|\mathbf{a}_t|) \quad _1 \sec\vartheta_i}{(j\omega \ _2/|\mathbf{a}_t|) + \ _1 \sec\vartheta_i} \quad , \end{aligned} \tag{4.105}$$

and hence:

$$| \ _{\mathrm{TM}}| \quad 1 \quad , \qquad | \ _{\mathrm{TE}}| \quad 1 \quad . \tag{4.106}$$

This implies that we have, for both the TE and the TM case (we shall therefore omit the subscript):

$$\frac{\mathbf{E}_r \quad \mathbf{H}_r}{2} \quad \hat{x}_1 \qquad \qquad \frac{\mathbf{E}_i \quad \mathbf{H}_i}{2} \quad \hat{x}_1 = \frac{\mathbf{E}_i \quad \mathbf{H}_i}{2} \quad \hat{x}_1 \quad . \tag{4.107}$$

This relation shows that the reflected wave carries, in the direction orthogonal to the separation plane, an active power density equal to that of the incoming wave. This yields the name *total reflection* for this phenomenon, which — let us emphasize it again — can occur only for $n_2 < n_1$, i.e., when passing from a medium of "higher optical density" to a medium of "lower optical density" (such as, for example, from glass to air, but not vice-versa).

The reader can prove that, when total reflection occurs, superimposing the incoming and the reflected wave one gets a field like that which was described at the end of Section 4.5 and in Problem 3.7. This field behaves as a purely standing wave along x_1 (its nodes and crests are $x_1 =$ constant planes, spaced by $/(4\cos\vartheta_i)$), and as a traveling wave along x_2, with phase velocity

$v_f(\hat{x}_2) = c_1/\sin\vartheta_i = \omega/(\ \ _i \sin\vartheta_i)$, i.e., synchronous with the phase velocity in medium 2 where the wave normal is parallel to the x_2 axis.

In the rest of this section, we will show how total re ection can be described with a formalism which is very similar to that of Section 4.8, provided one de nes a suitable complex transmission angle, ϑ_t. Let us start from the *postulate* that Snell's law holds also beyond the critical angle, i.e., for $\vartheta_i > \vartheta_c$. This implies:

$$\sin\vartheta_t = \frac{n_1}{n_2}\sin\vartheta_i > 1 \quad . \tag{4.108}$$

Then, let $\vartheta_t = \varphi + j\ \ $, with real φ and . Exploiting the identity $\sin\vartheta_t =$ $j\sinh(j\vartheta_t)$, and expanding the sinh function into its real and imaginary parts, as:

$$\sin\vartheta_t = \cosh\ \ \sin\varphi\ \ \ j\sinh\ \ \cos\varphi\ \ , \tag{4.109}$$

we see that Eq. (4.108) is satis ed when:

$$\vartheta_t = \frac{\ }{2} + j\ \ \ , \qquad \sin\vartheta_t = \cosh\ \ \ . \tag{4.110}$$

Moreover, also for ϑ_t complex, we have $\cos^2\vartheta_t + \sin^2\vartheta_t = 1$; thus, $\sin\vartheta_t$ larger than 1 implies that $\cos\vartheta_t$ is imaginary if, as recon rmed by an expansion of $\cos\vartheta_t$ analogous to Eq. (4.109), which, using Eq. (4.110), is:

$$\cos\vartheta_t = \ \ j\sinh\ \ \ . \tag{4.111}$$

Now comparing Eq. (4.102) with Eq. (4.107), and using Eq. (4.110), it follows that:

$$|\mathbf{k}_t| = \ _2\sin\vartheta_t = \ _2\cosh\ \ \ . \tag{4.112}$$

From Eq. (4.103), we get:

$$|\mathbf{a}_t| = j\ _2\cos\vartheta_t = \ _2\sinh\ \ \ . \tag{4.113}$$

Eqs. (4.112) and (4.113) allow us to say that ϑ_t is the angle that a complex vector, of modulus $\ _2$, must form with the x_1 axis, in order that its components along x_2 and x_1 equal $\mathbf{k}_t$ and $j\mathbf{a}_t$, respectively. Note that ϑ_t is *not* the angle between x_1 and the propagation vector $\mathbf{S}_t = \mathbf{a}_t + j\mathbf{k}_t$, as, for an evanescent wave, $|\mathbf{S}_t| \neq \ _2$.

Finally, thanks to Eq. (4.113) we see that there is a perfect formal matching between Eq. (4.105) on one side, and Eqs. (4.92), (4.93) on the other side. This means that the complex angle in Eq. (4.110) also provides correct results for the re ection coe cients.

All the contents of this section, up to now, hold provided that medium 2 extends to in nity along the positive x_1 half-axis. Let us now consider a case where the medium with refractive index n_2, smaller than n_1, has a nite thickness. For the sake of simplicity, let us con ne ourselves to the case, shown in Figure 4.12, where $n_3 = n_1$; more complicated examples can be found among the problems.

Continuity conditions like in Eq. (4.77) must be satis ed, now, on two parallel planes, $x_1 = 0$ and $x_1 = d$. It is straightforward to check that the minimum

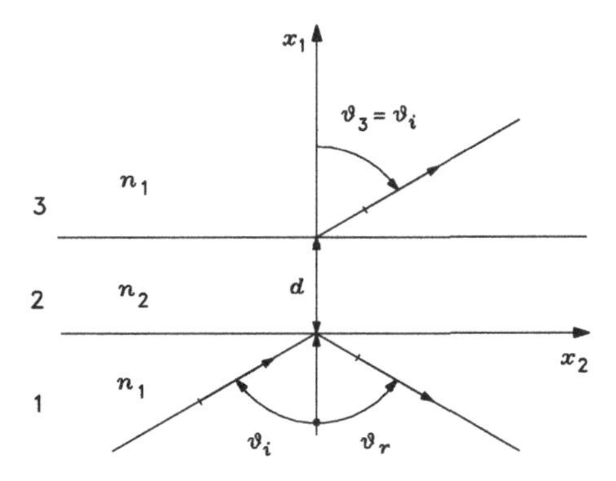

Figure 4.12 *Illustration of partially frustrated total re ection.*

number of waves which are needed to solve this problem for an arbitrary value of the incidence angle is ve: an incoming and a re ected wave in medium 1, *two evanescent waves* in medium 2 (with opposite directions, along $+x_1$ and x_1, for their attenuation vectors), and a transmitted wave in medium 3. For $n_3 = n_1$, it is easy to show, using Snell's law and the transitive law of equality, that:

$$\vartheta_3 = \vartheta_i \quad , \tag{4.114}$$

so the transmitted wave goes in the *same direction* as the incident wave in medium 1.

In medium 2, the eld is the sum of the two evanescent waves:

$$\begin{aligned} \mathbf{E} &= \mathbf{E}_{2+} \, e^{\;\mathbf{S}_2 \;\mathbf{r}} + \mathbf{E}_2 \;\; e^{\mathbf{S}_2 \;\mathbf{r}} \quad , \\ \mathbf{H} &= \frac{\mathbf{S}_2 \quad \mathbf{E}_{2+}}{j\omega \quad 2} \, e^{\;\mathbf{S}_2 \;\mathbf{r}} \quad \frac{\mathbf{S}_2 \quad \mathbf{E}_2}{j\omega \quad 2} \, e^{\mathbf{S}_2 \;\mathbf{r}} \quad . \end{aligned} \tag{4.115}$$

The reader can prove that, for this eld, the wave impedance in the x_1 direction has a nonzero real part and depends on x_1. This is in deep contrast with a single evanescent waves, whereas it has analogies and di erence with stationary waves arising as superposition of two uniform plane waves, discussed in Section 4.5. Indeed, one uniform traveling wave carries zero reactive power, while superposition of two counter-propagating ones results in a stationary wave with nonzero reactive power. For evanescent waves, each carries zero active power, while superposition of two counter-attenuating ones gives rise to a wave that carries an active power in the direction of the attenuation vector. This provides a physical explanation for the presence of a uniform plane wave, carrying active power, in medium 3, which otherwise would appear as an inconsistency under the viewpoint of conservation of energy.

The physical phenomenon of Figure 4.12 is called *partially frustrated total re ection* or, by analogy with quantum mechanics, *electromagnetic tunneling.* The analogy with quantum mechanics is reinforced by the impossibility of explaining this phenomenon in terms of ray optics, which would always predict

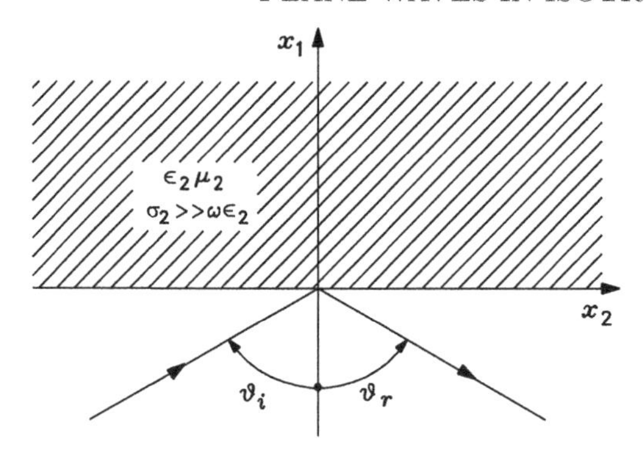

Figure 4.13 *Plane-wave reflection at the surface of a good conductor.*

zero power in medium 2 as soon as the critical angle is exceeded, independently of its thickness.

In Chapter 10, it will be shown that electromagnetic tunneling plays a fundamental role in guided-wave propagation along a dielectric rod, such as an optical fiber. When the geometry is not planar (e.g., when it is cylindrical), there are interesting variations to be added on the same theme, as we will see later.

4.12 Reflection on the surface of a good conductor

In technical application of electromagnetic waves, very commonly one is required to find a field $\{\mathbf{E}, \mathbf{H}\}$ defined in a region filled by a low-loss (or lossless) dielectric medium, bordered by a high-conductivity medium. Usually, knowing the field inside the conducting medium is not of primary interest. Thus, it is unnecessary to solve Maxwell's equations in the conductor, and impose continuity conditions at its borders. Rather, one looks for suitable *boundary conditions*, to be imposed on the field defined in the dielectric medium, at the dielectric-conductor interface. It is our aim, in this section, to find such conditions starting from Maxwell's equations inside the conductor.

Let us consider a plane surface between two semi-infinite media, as in Figure 4.13. As we saw in Section 4.8, in medium 1 a uniform plane wave is reflected with $\vartheta_r = \vartheta_i$, independently of the nature of medium 2. What happens in medium 2 will determine the value of the reflection coefficient.

Suppose that medium 2 is a *good conductor*, i.e.:

$$\sigma_2 \gg \omega \epsilon_2 \,. \tag{4.116}$$

Before we proceed, let us point out that, in some cases, approximations following from Eq. (4.116) may cause errors: we are comparing real and imaginary quantities, and this can result in unacceptable consequences for the phases of complex numbers, although the results are correct for their moduli.

As we saw in Section 4.8, the wave in medium 2 is in general dissociated.

Its attenuation vector $\mathbf{a}_t$ is parallel to x_1. Moreover, from Eq. (4.89) we see that:

$$\lim_{(\ _2/\omega\ _2)\rightarrow\infty} \vartheta_t = 0 \quad . \tag{4.117}$$

Thus, when Eq. (4.116) holds, we can take $\vartheta_t \simeq 0$, i.e., $\mathbf{k}_t$ is almost parallel to $\mathbf{a}_t$. Moreover, from Eqs. (4.116), (4.24) and (4.25), we get:

$$|\mathbf{k}_t|^2 \quad |\mathbf{a}_t|^2 \qquad 2\,\mathbf{a}_t \quad \mathbf{k}_t \quad . \tag{4.118}$$

Eq. (4.118) means that $|\mathbf{a}_t| \simeq |\mathbf{k}_t|$, so that Eq. (4.25) becomes:

$$|\mathbf{a}_t| \simeq |\mathbf{k}_t| \simeq (\omega\ _2\ _2/2)^{1/2} = (\ f\ _2\ _2)^{1/2} \quad , \tag{4.119}$$

where $f = \omega/2$ is the frequency of the incident wave. Therefore, the transmitted wave is attenuated along x_1 as $\exp[\ (\ f\ _2\ _2)^{1/2}\, x_1]$ as it penetrates into the conductor. The quantity:

$$= 1/(\ f\ _2\ _2)^{1/2} \quad , \tag{4.120}$$

whose dimensions are those of a length, gives an estimate of the depth up to which the transmitted wave has a nonnegligible amplitude. It is referred to as the *penetration depth* (or, especially in the past, as the skin depth). In practice, the results that we found for the inde nitely extended medium 2 are reliable if the conductor thickness is equal to a few times the penetration depth.

The wave impedance along x_1 in the conducting medium can be obtained from Eq. (4.18), taking into account that, because of what we just saw:

$$\mathbf{S}_t = \mathbf{a}_t + j\mathbf{k}_t \simeq (1+j)\ (\ f\ _2\ _2)^{1/2}\, \hat{x}_1 \quad , \tag{4.121}$$

and noting that, since $\mathbf{S}$ is parallel to $\hat{x}_1$, both $\mathbf{E}_t$ and $\mathbf{H}_t$ are perpendicular to $\hat{x}_1$, irrespective of the incidence angle ϑ_i. In conclusion, it reads:

$$Z_{\hat{x}_1} = \frac{j\omega\ _2}{\mathbf{S}_t\ \ \hat{x}_1} = (1+j)\left(\frac{\ f\ _2}{\ _2}\right)^{1/2} \quad . \tag{4.122}$$

This impedance depends only on the frequency and on the medium parameters. In contrast to previous results, it *does not depend on the direction of incidence* of the incoming wave. Therefore, it is a characteristic of the conducting medium, not of the wave propagating in it. This justi es its name, *wall impedance*, and entitles us to use Eq. (4.122) as a *boundary condition* on the plane which limits the dielectric medium. The scope of validity to this approach may be limited by *curvature* of the separation surface; this point will be clari ed in Chapter 7.

From Eq. (4.122), it follows that:

$$\lim_{\ _2\rightarrow\infty} Z_{\hat{x}_1} = 0 \quad , \tag{4.123}$$

where no quantity characterizing the incident wave appears.

This reminds us of the concept of an *ideal electric conductor*, found in Chapter 3, which is a useful mathematical tool in many problems, and tells us that it can be regarded as a medium with in nite conductivity, imposing the boundary condition:

$$\mathbf{E}_{\tan} = 0 \tag{4.124}$$

on its surface.

Note that Eq. (4.120) gives $\rightarrow 0$ as $_2 \rightarrow \infty$, so that the thickness of an ideal electric conductor can be arbitrarily small.

4.13 Further applications and suggested reading

In Chapter 1, we presented Maxwell's equations, and other equations which follow from them. In Chapter 3, we derived theorems which are based on Maxwell's equations. In Chapter 4, we began to *solve* Maxwell's equations, and this process will now occupy the remainder of the book. Solutions, subject to boundary conditions and often to other limiting assumptions, never have a scope of validity as general as the fundamental equations and theorems of the rst two chapters. Nevertheless, we may say that, in this chapter, we started o with solutions whose validity and whose usefulness are very broad. Plane waves appear to be much more ubiquitous than any other time-varying electromagnetic eld. Rather than writing a list of applications, which, irrespective of its length, could never aim at being complete, this section will limit itself to a few basic recommendations to the reader, and to the explanation of why some subjects have been omitted in this chapter, although they may be very important.

First, let us remind the reader that electromagnetic plane waves in a homogeneous medium, being solutions of one of the most widely used equations in mathematical physics — the Helmholtz scalar equation — share an extremely large number of properties with waves of completely di erent physical natures (e.g., Elmore and Heald, 1969; Pain, 1993; Young, 1976), such as acoustic waves in uids (e.g., Ford, 1970; Frisk, 1994; Brekhovskikh, 1960; Brekhovskikh and Godin, 1990/1991; Meyer and Neumann, 1972; Pierce, 1981), elastic waves in solids (e.g., Ashcroft and Mermin, 1976; Kittel, 1986), wave functions associated with probability distributions of particles in Quantum Mechanics.

An impressive example in this sense is provided by the Fresnel re ection and transmission formulas (Section 4.9). They were regarded, historically, as the rst irrefutable proof of the wave nature of light, and were published in 1816, i.e., more than 50 years before Maxwell's equations. The *wave* nature of light is by no means a synonym for the *electromagnetic* nature, a concept which was rst proposed by Maxwell many decades later. Incidentally, the history of undulatorial theories in physics is dealt with in an *ad hoc* introductory chapter of Born and Wolf (1980).

It may be very useful for the reader to compare the content of this chapter with that of suitable textbooks on acoustics (e.g., Ford, 1970), classical analytical mechanics (e.g., Goldstein, 1980), and quantum mechanics (e.g., Cohen-Tannoudji *et al.*, 1977; Dicke and Wittke, 1960; Merzbacher, 1970;

Schi , 1968). As we will see later, very strong similarity between electromagnetic waves and other waves exists also when dealing with *guided propagation* phenomena. We recommend that the reader return to these references, after reading Chapters 7, 9 and 10. In particular, consulting the more-than-one-century old *Theory of Sound* (1877) by Lord Rayleigh would, at the same time, satisfy cultural appetites from the viewpoint of physics, and make the reader more curious about history of science.

Plane waves in unbounded isotropic media su er from a number of very severe limitations, which make them inadequate to model some basic phenomena. Just to quote one, at the end of this chapter, the reader does not yet know at which velocity the energy associated with an electromagnetic wave travels. The weak points of plane waves have not been emphasized in this chapter just because it would have been pedagogically wrong and pointless not to insist rst on their strong points. Their disadvantages, and how to correct them, e.g., using "local plane waves" (Snyder and Love, 1983, Chapter 35), will be contents of the next chapter. We postpone more suggestions for further readings until the end of Chapter 5.

Some experienced readers will probably be surprised not to nd in this chapter an analysis (or design) of multilayer dielectrics. Indeed, this is one of the most important problems that can be tackled using plane waves. The reason why it was not dealt with in detail now is, again, a pedagogical choice. We believe that problems of this kind are dealt with in a much simpler fashion, especially for what concerns calculations, by analogy with electrical transmission lines. These will be covered in depth in Chapter 8. So, it will be advisable to come back to multilayers, and solve again some of the problems presented at the end of this chapter, after studying Chapter 8. Of course, talking about multilayers, under the viewpoint of theory and under that of a broad choice of examples of application, a classical reference that it is highly recommended for consultation is Born and Wolf (1980). We also refer to the work of Brekhovskikh (1960) for waves in multilayer media.

Finally, with regard to plane waves in good conductors, in telecommunications they are of primary importance for microwave transmission, to evaluate waveguide loss (see Chapter 5), and in radio propagation, to study penetration into natural conductors, such as salt water at low frequencies, the ionosphere in the MHz range (Budden, 1961), etc. Outside the telecom world, keeping in mind what was said in the nal section of Chapter 1, interaction of electromagnetic elds with living tissues, and applications to material diagnostics, are just two among the most eminent examples. The relevant bibliographical suggestions are those that were already quoted in Chapter 1.

Problems

4-1 Show that the Helmholtz equation for the eld vectors $\mathbf{E}$ and $\mathbf{H}$ can be solved directly by separation of variables. Show that a generic solution of the equation for **E** and one of the equation for **H** do not form a solution of Maxwell's equations.

4-2 A plane wave traveling in air has an electric eld amplitude of 10 mV/m. Derive its magnetic eld in the following cases.

a) The wave is uniform, travels in the x_3 direction, and the electric eld is linearly polarized along the x_1 axis.

b) The wave is uniform, travels in the x_3 direction, and the electric eld is circularly polarized in the (x_1, x_2) plane.

c) The wave is evanescent, with its phase vector in the x_3 direction, its attenuation vector in the x_2 direction, and the electric eld is linearly polarized along the x_1 axis.

Show that in the rst case only the electric eld and the magnetic eld are orthogonal complex vectors. Show that also in the second case the corresponding vectors in the time domain are orthogonal at any instant.

4-3 Find a better approximation than $v_f \simeq c$ for a uniform plane wave in a low-loss medium where $\quad \omega$. Show that the rst correction is proportional to the square of the loss tangent.

4-4 Derive explicitly all the features of the eld resulting from two waves of equal amplitude interfering at an angle, as described in Section 4.6. Repeat the same exercise for two waves of di erent amplitudes.

4-5 Show that the following expressions for *power re ectivity* (i.e., ratio of re ected and incident power densities), for plane waves at the separation between two semi-in nite media, are equivalent to those which can be derived from the Fresnel formulas:

$$R_{\mathrm{TE}} = \frac{\sin^2(\vartheta_i \quad \vartheta_t)}{\sin^2(\vartheta_i + \vartheta_t)}$$

for TE waves, and:

$$R_{\mathrm{TM}} = \frac{\tan^2(\vartheta_i \quad \vartheta_t)}{\tan^2(\vartheta_i + \vartheta_t)}$$

for TM waves, respectively.
Reconsider the Brewster angle problem in the framework of these new results.

4-6 A layer of a lossless medium, 1 m in thickness, with a refractive index $n = 4$, is placed between an air- lled half-space and an ideally conducting plane (located at $x_1 = 0$). A uniform plane wave, at a frequency $f = 6$ GHz, impinges from air, travels through the layer, and is re ected back.
Find the period of the standing wave in the dielectric layer, along the x_1 direction, for the following cases:

a) incidence along the x_1 axis;

b) incidence along a direction such that the refracted wave in the layer travels at 45 degrees with respect to the x_1 axis;

c) incidence at an angle (in air) of 45 degrees with respect to the x_1 axis.

4-7 A uniform plane wave, impinging at normal incidence on the plane ($x_1 = 0$) of separation between two lossless dielectric media, conveys a power density of 1 mW/mm^2. The power density of the transmitted wave is 3 dB below that of the incident wave. Find the modulus of the electric eld of the re ected wave in the following two cases:

a) the medium in which the incident and the re ected waves travel is air;

b) the medium in which the transmitted wave travels is air.

Repeat the problem assuming that the incidence angle is /4 and that the power densities on the $x_1 = 0$ plane remain the same as in the previous cases.

4-8 A TE-polarized uniform plane wave impinges with an incidence angle of /4 radians on the plane $x_1 = 0$, separating air from a lossless dielectric medium. The re ection coe cient equals 0.5.
Find the refractive index of the second medium, the re ection coe cient for a TM-polarized uniform plane wave coming along the same direction, and the re ection coe cients for a TE- and a TM-polarized plane wave coming from the second medium, and giving rise to refracted waves with a transmission angle (in air) equal to /4 radians.

4-9 Two semi-in nite lossless dielectric media have refractive indices equal to 3 and to 1, respectively. At a frequency of 3 10^{14} Hz, experimental observation of an evanescent wave in the second medium shows that it decays in the ratio $1/e$ over a distance of 1 m, measured along the normal to the plane of separation.

a) Find the corresponding angle of incidence, in medium 1.

b) Evaluate the phases of the corresponding TE and TM re ection coe cients in medium 1, on the separation plane.

4-10 Show that when a wave is incident at the Brewster angle, Eq. (4.95), then the transmitted wave travels at an angle $\vartheta_t =$ /2 ϑ_i. Hence, show that it is impossible to nd two media such that the Brewster angle at their interface equals /4.

4-11 A lossless layer, with a refractive index equal to $\sqrt{3}$ and a thickness of 1/8 m, has been sandwiched between the semi-in nite media of Problem 4-9.

a) At what frequencies does it operate as an impedance-matching layer (i.e., re ection is canceled at normal incidence)?

b) Which is, in this structure with an intermediate layer, the critical angle for total internal re ection of a plane wave coming from the medium whose index equals 3 ?

4-12 Calculate the Brewster angle for the two semi-in nite media of Problem 4-9, and the re ection coe cient for a TE polarized wave incident at that angle. Suppose now that a uniform plane wave of unspeci ed polarization is incident at the same angle as in the previous question of this problem, but subsequent to the introduction of the intermediate layer discussed in the previous problem. Is the re ected wave still TE-polarized ?

4-13 A lossless dielectric layer is located between two air- lled semi-in nite half-spaces. A uniform plane wave impinges normally on it. The resulting partially-standing wave pattern has a period of 4 mm in air, of 1 mm in the dielectric layer.

a) Find the frequency of the wave.

b) Find the thicknesses of the layer for which the peak-to-valley ratio of the standing wave, in the half-space where the incident wave comes from, is maximum, and those for which it is minimum.

c) Does the layer thickness a ect the peak-to-valley ratio of the partially standing wave in the layer ?

4-14 A uniform plane wave, coming from a lossless medium, impinges at oblique incidence onto the plane surface which delimits another semi-in nite medium. Compare the following two cases:

a) medium 2 is lossless, and the incident wave undergoes total re ection;

b) medium 2 is a good conductor.

1. Are the re ection coe cients comparable in modulus ?
2. Are they dependent on the polarization of the incident wave ?
3. Are their phases identical, or in general not ?
4. Are their phases polarization-dependent or not ?
5. Are the attenuation vectors in medium 2 in the same direction ?
6. Are the phase vectors in medium 2 in the same direction ?

4-15 A typical value for the conductivity of copper is $5.8 \quad 10^{7} \quad {}^{1}\mathrm{m} \ {}^{1}$. Permittivity and permeability may be taken, as an approximation, equal to those of free space.

a) Evaluate the order of magnitude of the maximum frequencies at which copper can still be considered to be a good conductor. Calculate the frequency at which the penetration depth in copper equals 1 mm.
Find the value of the surface impedance of a at copper wall at that frequency.

b) Repeat these calculations for the case of graphite, considered as a good conductor at industrial frequencies. Its conductivity, strongly temperature-dependent, is of the order of 10^5 ^{1}m ^{1}.
Are the numbers obtained in this case consistent with each other? If not, explain why.

4-16 To establish a radio communication link with an immersed submarine, one requires a wave penetration depth of at least 10 meters. Find the maximum usable frequency. For the conductivity of sea water, a typical value is 5 ^{1}m ^{1}. Magnetic permeability is practically the same as that of free space.

4-17 Evaluate the modulus of the phase vector for a plane wave in sea water, at a frequency of 500 Hz. Compare it with the modulus that the wave vector has, at the same frequency, for a plane wave in a *lossless* medium whose dielectric permittivity is equal to that of sea water (whose relative permittivity is of the order of 80). Explain the origin of this very large di erence.
(*Hint.* Use a graphical representation of ω^2 $_c$ and of its roots on the complex plane.)

References

Ashcroft, N.W. and Mermin, N.D. (1976) *Solid State Physics.* Holt, Rinehart and Winston, New York.

Born, M. and Wolf, E. (1980) *Principles of Optics.* 6th ed., Pergamon, Oxford.

Brekhovskikh, L.M. (1960) *Waves in Layered Media.* Academic, New York.

Brekhovskikh, L.M. and Godin, O.A. (1990) *Acoustics of Layered Media I: Plane and Quasi-Plane Waves.* Springer-Verlag, Berlin.

Brekhovskikh, L.M. and Godin O.A. (1991) *Acoustics of Layered Media II: Point Source and Bounded Beams.* Springer-Verlag, Berlin.

Budden, K.G. (1961) *Radio Waves in the Ionosphere.* Cambridge University Press, Cambridge, UK.

Budden, K.G. (1988) *The Propagation of Radio Waves: The Theory of Radio Waves of Low Power in the Ionosphere and Magnetosphere.* 1st paperback ed. with corrections, Cambridge University Press, Cambridge, UK.

Cohen-Tannoudji, C., Diu, B. and Laloe, F. (1977) *Mecanique Quantique.* Hermann, Paris.

Dicke, R.H. and Wittke, J.P. (1960) *Introduction to Quantum Mechanics.* Addison-Wesley, Reading, MA.

Elmore, W.C. and Heald, M.A. (1969) *Physics of Waves.* McGraw-Hill, New York (Dover, New York, 1985).

Ford, R.D. (1970) *Introduction to Acoustics.* Elsevier, Amsterdam.

Frisk, G.V. (1994) *Ocean and Seabed Acoustics: A Theory of Wave Propagation.* Prentice-Hall, Englewood Cli s, NJ.

Goldstein, H. (1980) *Classical Mechanics.* 2nd ed., Addison-Wesley, Reading, MA.

Kittel, C. (1986) *Introduction to Solid State Physics.* Wiley, New York.

Merzbacher, E. (1970) *Quantum Mechanics.* Wiley, New York.

Meyer, E. and Neumann, E.-G. (1972) *Physical and Applied Acoustics: An Introduction.* Academic, New York.

Morse, P.M. and Feshbach, H. (1953) *Methods of Theoretical Physics.* McGraw-Hill, New York.

Pain, H.J. (1993) *The Physics of Vibrations and Waves.* 4th ed., Wiley, New York.

Pierce, A.D. (1981) *Acoustics: An Introduction to its Physical Principles and Applications.* McGraw-Hill, New York.

Schi , L.I. (1968) *Quantum Mechanics.* McGraw-Hill, International.

Snyder, A.W. and Love, J.D. (1983) *Optical Waveguide Theory.* Chapman & Hall, London.

Young, H.D. (1976) *Fundamentals of Waves Optics and Modern Physics.* 2nd ed., McGraw-Hill, New York.

CHAPTER 5

Plane wave packets and beams

5.1 Modulated waves. Group velocity

Throughout the previous chapter, the reader had to bear in mind two fundamental warnings. One was that a *single* plane wave of finite amplitude does not make physical sense, as it would fill the whole space with an infinite energy. The second was that, on the other hand, plane waves form a *complete set.* Consequently, combinations of plane waves may build up physically meaningful and, more significantly, technically important results. Some of them are still mathematically simple, and fairly intuitive, and this chapter is devoted to them. At first sight, the subjects which form the chapter might look rather heterogeneous, but this is more apparent than real, as we shall see while we proceed.

As we just said, a single plane wave, with an angular frequency ω, invades the whole 3-D space. At any point, its field varies sinusoidally vs. time t, for t going from $-\infty$ to $+\infty$; consequently, it does not convey information from one point to another. It is because of this that we accepted that the phase velocity v_f, defined in Section 4.4, could be arbitrarily large. This is not a violation of the fundamental postulate of the special theory of relativity, since v_f can never be the speed of propagation for a signal that synchronizes two "clocks".

Those readers who have been exposed to the fundamentals of communication theory know that, in order to transmit information, one needs a signal with a finite spectral width, such as a modulated wave. In order to determine the speed at which the information content carried by an e.m. wave travels, we must then start from a non-monochromatic field $\{\mathbf{E}, \mathbf{H}\}$. Actually, in Section 4.7, completeness was proved only for strictly monochromatic plane waves. However, when regularity of functions and derivatives assures that operators in the time domain and in the space domain commute, a field having a finite spectral width in the frequency domain can also be expanded into plane waves, provided it satisfies, as a function of space coordinates, conditions similar to those given in Section 4.7.

A very simple model, which enables us to determine the speed we are looking for, at least in a lossless medium, consists of two uniform plane waves, traveling in the same direction (i.e., having parallel phase vectors $\mathbf{k}$), but different in angular frequencies, $\omega_0 \pm \Delta\omega$, respectively. This notation is reminiscent of an amplitude modulation, with a modulation angular frequency $\Delta\omega$. In the following Section 5.2, however, we will show that the results obtained in this way are quite general. To simplify the calculations, without any loss in the scope of validity of the results, we will suppose that these waves have equal polarizations and equal amplitudes, represented by a vector $\mathbf{E}$ which, again for simplicity, we assume to be real.

Let us introduce rectangular coordinates such that the waves propagate along the x_3 axis. In the time domain, the total electric eld is then expressed by

$$\mathbf{E}\,(\mathbf{r},t) = \mathrm{Re}\left[\mathbf{E}\,e^{\;j(\;\;_{+}x_3\;\;\omega_0 t\;\;\;\;\omega t)} + \mathbf{E}\,e^{\;j(\;\;\;\;x_3\;\;\omega_0 t+\;\;\omega t)}\right] \quad , \tag{5.1}$$

where $_{+}$ and are the phase constants $= \omega\sqrt{\;\;(\omega)\;\;(\omega)}$ at the angular frequencies $\omega_0 +\;\;\omega$ and $\omega_0\;\;\;\;\omega$, respectively.

One may always write

$$_{+} = \;\;_{m} + \qquad , \qquad\qquad = \;\;_{m} \qquad\qquad , \tag{5.2}$$

where $_{m} = (\;\;_{+} + \;\;\;\;)/2$. Quite often, (ω) is a smooth function of ω, between $\omega_0\;\;\;\;\omega$ and $\omega_0 +\;\;\omega$. In that case, there is an angular frequency ω_m, in that interval, such that $_{m} = \;\;(\omega_m)$. In general, ω_m di ers from ω_0, but if ω is small enough, then we can let $_{m} \simeq \;\;(\omega_0)$. Then, inserting Eqs. (5.2) into Eq. (5.1), we nd

$$\mathbf{E} = \mathbf{E}\;\cos(\omega_0 t\;\;\;\;\;_{m}x_3)\;2\cos(\;\;\;\;x_3\;\;\;\;\;\;\omega t) \quad . \tag{5.3}$$

This expression looks like a traveling wave (with angular frequency ω_0 and phase constant $_{m}$), wrapped in an envelope $\cos(\;\;\;x_3\;\;\;\;\;\omega t)$. The loci of constant envelope amplitude are the solutions of $d(\;\;\;x_3\;\;\;\;\;\omega t) = 0$. They are planes perpendicular to the x_3 axis, moving along that axis with a velocity

$$v = \;\;\omega/ \qquad . \tag{5.4}$$

If the function (ω) can be expanded as a Taylor series

$$(\omega) = \;\;(\omega_0) + \frac{d}{d\omega}\,(\omega\;\;\;\;\omega_0) + \frac{1}{2}\,\frac{d^2}{d\omega^2}\,(\omega\;\;\;\;\omega_0)^2 + \ldots \quad , \tag{5.5}$$

then, if we let the spectral width of the signal tend to zero, we see that the velocity (5.4) tends to

$$v_g(\omega_0) = \frac{1}{\left.\dfrac{d}{d\omega}\right|_{\omega_0}} = \left.\frac{d\omega}{d}\right|_{\omega=\omega_0} \quad . \tag{5.6}$$

The quantity de ned by Eq. (5.6) is referred to as the *group velocity* of a uniform plane wave with angular frequency ω_0. The procedure that led us to this result indicates that, for a narrow-band signal ($\omega\;\;\;\;\omega_0$), traveling in a medium where (ω) is a smooth function satisfying all the requirements that we introduced, the group velocity (5.6) equals indeed the velocity at which the signal travels. But let us stress that this is not always true. For example, in a case where $d\;\;/d\omega = 0$, Eq. (5.6) would yield an in nite speed, whereas Eq. (5.4) shows that, for $\;\;\;\;\neq 0$, the envelope propagates at a nite speed. Furthermore, this value must be less than the speed of light, c, to be compatible with the special theory of relativity. In all cases where these requirements are not satis ed by the quantity in Eq. (5.6), we will keep calling it "group velocity," but we will *not* state that, in these cases, the signal envelope propagates at the group velocity.

While we wait until the next section to show what happens to signals whose spectrum is more complicated than just two frequencies, let us show that Eq. (5.6) has also another physical meaning, namely that of the velocity of energy transport along the direction of propagation. For a uniform plane wave in a lossless medium, in Maxwell's equations we may write $\nabla \quad = (\quad j\mathbf{k}) \quad =$ $(\quad j \quad \hat{k}) \quad$. Then, di erentiating all terms in Maxwell's equations with respect to ω, we get the following relations:

$$
\begin{aligned}
j\,\frac{\partial}{\partial\omega}\,(\mathbf{k} \quad \mathbf{E}) &= \quad j\,\frac{\partial(\omega\quad)}{\partial\omega}\,\mathbf{H} \quad j\omega \quad \frac{\partial\mathbf{H}}{\partial\omega} \quad , \\
j\,\frac{\partial}{\partial\omega}\,(\mathbf{k} \quad \mathbf{H}) &= \quad j\,\frac{\partial(\omega\quad)}{\partial\omega}\,\mathbf{E} + j\omega \quad \frac{\partial\mathbf{E}}{\partial\omega} \quad .
\end{aligned}
\tag{5.7}
$$

We take then the scalar product of the rst one by $\mathbf{H}$, and that of the second one by $\mathbf{E}$. Then, we subtract the rst one from the second one, and we take into account also the following relationships, which stem directly from Maxwell's equations and allow us to get rid of several terms:

$$
\begin{aligned}
j\,\mathbf{k} \quad \frac{\partial\mathbf{E}}{\partial\omega} \quad \mathbf{H} &= \quad j\omega \quad \mathbf{E} \quad \frac{\partial\mathbf{E}}{\partial\omega} \quad , \\
j\,\mathbf{k} \quad \frac{\partial\mathbf{H}}{\partial\omega} \quad \mathbf{E} &= \quad j\omega \quad \mathbf{H} \quad \frac{\partial\mathbf{H}}{\partial\omega} \quad .
\end{aligned}
$$

Finally, we obtain:

$$
(\mathbf{E} \quad \mathbf{H} \quad \hat{k} + \mathbf{E} \quad \mathbf{H} \quad \hat{k})\,\frac{\partial}{\partial\omega} = \frac{\partial(\omega\quad)}{\partial\omega}\,\mathbf{H} \quad \mathbf{H} \quad + \frac{\partial(\omega\quad)}{\partial\omega}\,\mathbf{E} \quad \mathbf{E} \quad . \tag{5.8}
$$

This equation can be rewritten as:

$$
\mathrm{Re}\{\mathbf{P} \quad \hat{k}\} = v_g\,\frac{1}{4}\left\{\frac{\partial(\omega\quad)}{\partial\omega}\,\mathbf{H} \quad \mathbf{H} \quad + \frac{\partial(\omega\quad)}{\partial\omega}\,\mathbf{E} \quad \mathbf{E} \quad \right\} \quad . \tag{5.9}
$$

It shows that the real part of the Poynting vector ux through a unit-area surface, orthogonal to the direction of propagation, equals the group velocity times a term whose dimensions are those of an energy density. As in uid dynamics, this tells us that the group velocity is the *speed at which energy is carried* by this e.m. eld. We note, though, that the expression for the energy density in general is not the same as the one which appears in Poynting's theorem, Section 3.2, which was:

$$
\frac{1}{4}\,(\quad \mathbf{H} \quad \mathbf{H} \quad + \quad \mathbf{E} \quad \mathbf{E} \quad) \quad . \tag{5.10}
$$

The two expressions coincide if and only if:

$$\frac{\partial}{\partial\omega} = 0 \quad , \qquad \frac{\partial}{\partial\omega} = 0 \quad , \tag{5.11}$$

that is, in what is called a *nondispersive medium.* The apparent discrepancy between the results can be explained if we remember that in Section 3.2 we dealt with a purely sinusoidal time-harmonic eld, while now we are studying what happens when we superimpose two plane at di erent frequencies. As powers and energies result from nonlinear operations, even a very simple superposition can deeply modify previous results. In Chapter 4, we saw that, for a monochromatic uniform plane wave in a lossless medium, energy is uniformly distributed throughout the whole space. We could not track an energy "carried" by the wave, and indeed, for such a wave, the divergence of Poynting vector, $\mathbf{P} = \mathbf{E} \quad \mathbf{H} \ /2$, is zero everywhere. On the contrary, the modulated wave that we saw in this section does carry an energy. Its Poynting vector has a nonvanishing divergence, as we can prove in the following way. Consider, as an extension of the previous argument, the following e.m. eld:

$$\begin{aligned}
\mathbf{E} &= \frac{1}{2}\Big\{\mathbf{E}(\omega_0 + \quad \omega)\, e^{j(\omega_0 + \quad \omega)t} + \text{c.c.}\Big\} \\
&+ \frac{1}{2}\Big\{\mathbf{E}(\omega_0 \quad \omega)\, e^{j(\omega_0 \quad \omega)t} + \text{c.c.}\Big\} \\
&\simeq \frac{1}{2}\left\{\left[\mathbf{E}(\omega_0) + \frac{\partial\mathbf{E}}{\partial\omega} \quad \omega\right] e^{j(\omega_0 + \quad \omega)t} + \text{c.c.}\right\} \\
&+ \frac{1}{2}\left\{\left[\mathbf{E}(\omega_0) \quad \frac{\partial\mathbf{E}}{\partial\omega} \quad \omega\right] e^{j(\omega_0 \quad \omega)t} + \text{c.c.}\right\}
\end{aligned} \tag{5.12}$$

$$\begin{aligned}
\mathbf{H} &= \frac{1}{2}\left\{\left[\mathbf{H}(\omega_0) + \frac{\partial\mathbf{H}}{\partial\omega} \quad \omega\right] e^{j(\omega_0 + \quad \omega)t} + \text{c.c.}\right\} \\
&+ \frac{1}{2}\left\{\left[\mathbf{H}(\omega_0) \quad \frac{\partial\mathbf{H}}{\partial\omega} \quad \omega\right] e^{j(\omega_0 \quad \omega)t} + \text{c.c.}\right\} \quad ,
\end{aligned} \tag{5.13}$$

where c.c. means the complex conjugate of the preceding term. Suppose that $\{\mathbf{E}(\omega_0), \mathbf{H}(\omega_0)\}$ is a uniform plane wave, then, $\partial\mathbf{E}/\partial\omega$ and $\partial\mathbf{H}/\partial\omega$ are related to each other by two equations that can be obtained by di erentiating Maxwell's equations. Now, let us calculate the instantaneous Poynting vector which corresponds to the eld in Eqs. (5.12), (5.13), including also the products between elds at di erent frequencies, because, as one lets $\omega \to 0$, the time-averaged values of these products tend to become quantities that are not zero. Let us drop the terms of order $(\quad \omega)^2$.

The assumption that $\{\mathbf{E}(\omega_0), \mathbf{H}(\omega_0)\}$ is a uniform plane wave implies that the contribution $\mathbf{E}(\omega_0) \quad \mathbf{H}\ (\omega_0)$ and alike have zero divergence. Then, from Maxwell's equations and from their derivatives with respect to ω, taking into account the relation (C.4) $\nabla \ (\mathbf{a} \quad \mathbf{b}) = (\nabla \quad \mathbf{a}) \ \mathbf{b} \quad \mathbf{a} \ (\nabla \quad \mathbf{b})$, it follows that:

$$\nabla \left(\mathbf{E} \quad \frac{\partial \mathbf{H}}{\partial \omega} + \frac{\partial \mathbf{E}}{\partial \omega} \quad \mathbf{H} \right) = j \left\{ \frac{\partial (\omega \ )}{\partial \omega} \mathbf{H} \ \mathbf{H} \ + \frac{\partial (\omega \ )}{\partial \omega} \mathbf{E} \ \mathbf{E} \ \right\} . \tag{5.14}$$

The Poynting vector computed from Eqs. (5.12) and (5.13) contains the terms on the left-hand side of Eq. (5.14), minus their complex conjugates. If we subtract Eq. (5.14) from its complex conjugate, we nd twice the quantity on the right-hand side. So, we see that the energy density which appears on the right-hand side of Eq. (5.9) is indeed related, in the sense of Poynting's theorem, to that part of the Poynting vector which originates from products of elds at two di erent frequencies. As we said, only this part of the e.m. energy travels, so these ndings recon rm that the interpretation we gave, for the physical meaning of group velocity, is correct.

5.2 Dispersion

In the previous section, group velocity, on one hand, was calculated at a single frequency, according to Eq. (5.6). On the other hand, its physical meaning can be grasped only if we consider non-monochromatic waves. The whole picture is clear and unambiguous if Eq. (5.6) is independent of ω, over a nite frequency interval around ω_0. But if v_g varies within the signal spectral range, then further developments become necessary.

The phenomena which occur when $dv_g/d\omega \neq 0$ go under the generic name of *dispersion*. The historical origin of this term comes from *chromatic dispersion* of prisms, which, as well known, separate spectral components of white light because glass has a wavelength dependent refractive index. Nowadays, the word dispersion, if used in a transmission context, is essentially a synonym to *delay distortion*, and means that di erent spectral components travel at di erent speed.

Other phrases that are also of general use, regardless of their historical origins, are:

normal dispersion: $dv_g/d\omega < 0$ $(dv_g/d \ \ > 0)$;

anomalous dispersion: $dv_g/d\omega > 0$ $(dv_g/d \ \ < 0)$.

In the following chapters we will see that these concepts and these names apply not only to plane waves, but to electromagnetic propagation in general. This section is devoted to an example which is based on plane waves, but the concepts we develop here have a much broader scope of validity.

Let us consider a packet of uniform plane waves that travel along one direction, say z. Let their spectrum be centered on a "carrier" angular frequency ω_0, and their time-domain envelope, at an initial abscissa $z = 0$, be a Gaussian function, which simpli es calculations and so allows us to stress physical results. We choose the time reference system such that at $t = 0$ the center of the Gaussian pulse is at $z = 0$.

Any component of the e.m. eld can be written as:

$$(z = 0, t) = A \exp(\ \ t^2 / \ \ ^2) \ \cos \omega_0 t \quad . \tag{5.15}$$

The symbol is intended to underline that the following results are independent of the physical dimensions of the quantity one explicitly deals with. To follow the temporal and spatial evolution of the wave packet, it is convenient to use the (time-domain) Fourier transform of Eq. (5.15), (ω), which is easily calculated, as the Fourier transform of a Gaussian function is a Gaussian function also:

$$\int_{\infty}^{+\infty} e^{\ \ {}^2 x^2} \, e^{j \ x} \, dx = \frac{\sqrt{\ }}{\ } \, e^{\ \ {}^2/(2\ )^2} \quad . \tag{5.16}$$

We get therefore:

$$\begin{aligned} (\omega) &= \frac{1}{\sqrt{2\ }} \int_{\infty}^{+\infty} \ e^{\ j\omega t} \, dt \\ &= \frac{A}{2\sqrt{2}} \left[e^{\ (\omega \ \ \omega_0)^2 \ {}^2/4} + e^{\ (\omega+\omega_0)^2 \ {}^2/4} \right] \quad . \end{aligned} \tag{5.17}$$

At any time t 0 and at any abscissa z 0, (z,t) can be calculated as an inverse Fourier transform:

$$(z,t) = \frac{1}{\sqrt{2\ }} \int_{\infty}^{+\infty} (\omega) \, e^{\ j[k(\omega)\, z \ \ \omega t]} \, d\omega \quad . \tag{5.18}$$

Let us assume that the phase constant k, as a function of ω, can be expanded as a Taylor series around ω_0:

$$k(\omega) = k(\omega_0) + \frac{dk}{d\omega}(\omega \ \ \omega_0) + \frac{d^2k}{d\omega^2}\frac{(\omega \ \ \omega_0)^2}{2} + \ldots \quad . \tag{5.19}$$

Suppose now that, throughout the spectral range where (ω) has a nonnegligible amplitude compared with (ω_0), k varies slowly enough to neglect, in Eq. (5.19), all the terms of order larger than 2 in $(\omega \ \ \omega_0)$. Moreover, take into account that, by de nition, it is $dk/d\omega = 1/v_g$, and introduce the symbol $k'' = d^2k/d\omega^2 = \ (1/v_g^2)\, dv_g/d\omega$. Substituting Eq. (5.19), within the approximation that we just made, into Eq. (5.18), yields:

$$\begin{aligned} (z,t) &= \frac{A}{4\sqrt{\ }} e^{j[k(\omega_0)\, z \ \ \omega_0 t]} \int_{\infty}^{+\infty} e^{\ \ {}^2\ {}^2/4} \, e^{\ j\left[\frac{z}{v_g}\ \ t\right]} \, e^{jk'' z\ {}^2/2} \, d \\ &+ \frac{A}{4\sqrt{\ }} e^{\ j[k(\omega_0)\, z \ \ \omega_0 t]} \int_{\infty}^{+\infty} e^{\ \ {}^2\ {}^2/4} \, e^{\ j\left[\frac{z}{v_g}\ \ t\right]} \, e^{\ jk'' z\ {}^2/2} \, d \quad . \end{aligned} \tag{5.20}$$

For $k'' = 0$, i.e., in a *dispersionless medium*, Eq. (5.20) becomes much simpler, and, using again Eq. (5.16), it gives:

$$(z,t) = A \cos\left(\omega_0 t \ \ k(\omega_0)\, z\right) e^{\ (t \ \ z/v_g)^2/\ {}^2} \quad . \tag{5.21}$$

Hence, if there is no dispersion, at any z the wave packet is still modulated

As for the part of the spectrum which is centered on $\ \omega_0$, it is enough to recall that, for uniform plane waves, $k(\ \omega) = \ k(\omega)$.

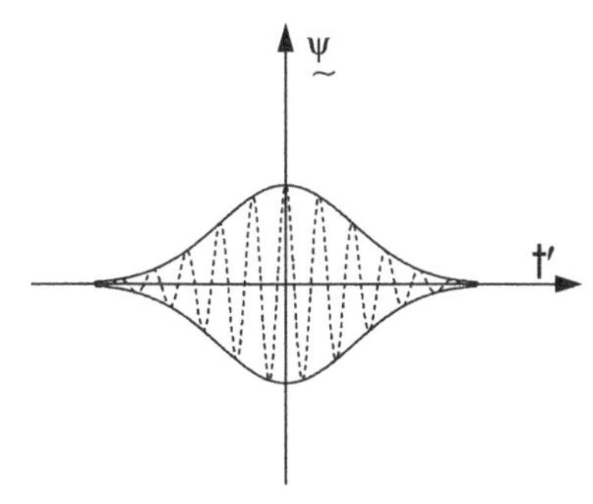

Figure 5.1 *An amplitude-modulated Gaussian pulse.*

by a Gaussian envelope whose temporal width, measured at $1/e$ amplitude, is still 2 (see Figure 5.1). Accordingly, with no dispersion the envelope travels undistorted, and its speed of propagation coincides with the group velocity, v_g.

In the presence of dispersion (i.e., for $k'' \neq 0$), the pulse envelope can no longer travel unchanged, as t grows. To show this, let us collect the rst and the third factor, in each integral in Eq. (5.20), into one complex exponential, and simplify the notation letting:

$$t' = t \quad \frac{z}{v_g} \quad , \quad D = \frac{2k''z}{\ ^2} \quad , \quad T^4 = \ ^4 + 4k''^2 z^2 = \ ^4(1 + D^2) \quad . \qquad (5.22)$$

Then, use again Eq. (5.16), which also holds for complex ^{2}. We get:

$$\begin{aligned} (z,t) &= \frac{A}{2\sqrt{1+D^2}} \left\{ \sqrt{1 + jD}\, e^{j[k(\omega_0)\, z \quad \omega_0 t \quad 2k'' z t'^2/T^4]} \right. \\ &+ \left. \sqrt{1 \quad jD}\, e^{\quad j[k(\omega_0)\, z \quad \omega_0 t + 2k'' z t'^2/T^4]} \right\} e^{\quad t'^2 \quad ^2/T^4} . \qquad (5.23) \end{aligned}$$

This new expression di ers signi cantly from Eq. (5.21), in the traveling wave factor at angular frequency ω_0 as well as in the envelope. Changes in the traveling wave (terms in parenthesis) a ect mainly its phase, as the factor $\sqrt{1 \quad jD}$ is trivial, being constant in time. On the contrary, the change represented by the last term in the argument of the exponentials is signi cant, because for $z =$ constant it introduces a t'^2 time dependence, i.e., a distortion that can be detected with a phase-sensitive receiver.

As for the envelope, changes due to dispersion do not a ect the motion of the pulse center, which travels again at the velocity v_g, nor, in a qualitative sense, the pulse shape, which remains Gaussian. They a ect the *pulse width*, at $1/e$ amplitude, which is now given by:

$$2 \quad t' = 2 \frac{T^2}{\ } = 2\sqrt{\ ^2 + 4k''^2 z^2 / \ ^2} = 2\sqrt{\ ^2 + 4k''^2 v_g^2 t_M^2 / \ ^2} \quad , \qquad (5.24)$$

where $t_M = z/v_g$ is the time at which, at a given distance z, the pulse amplitude reaches its maximum. So, as it propagates, the wave packet spreads, and its width tends asymptotically to be proportional to the distance it covered: 2 $t' \rightarrow 4\,|k''|\,v_g t_M/$. This occurs independently of the sign of k'', i.e., both in the normal and in the anomalous dispersion regime. It is important to stress that this last point is a consequence of our choice for the wave packet model. There are other cases (typical ones are frequency-modulated pulses, or "chirped" ones, i.e., pulses that are both frequency and amplitude modulated), where a suitable choice in the dispersion sign may lead to pulse *compression* as propagation distance grows, until a minimum width is reached. After that, the pulse width starts growing again. Readers who are interested in this subject can nd further hints at the end of this chapter.

The initial width appears *in the denominator* of the parameter which determines the rate at which the Gaussian pulse spreads vs. z. This is because, as one sees from Eq. (5.17), ^{1} is a measure of the *spectral width* of the pulse. The broader the pulse spectrum in the frequency domain, the higher its sensitivity to a given dispersion in the medium. The last comment should also be read as a warning, when one is dealing with pulses whose is very small. In such cases, the spectrum may be so broad, that one cannot neglect the higher-order terms in Eq. (5.19), and the whole content of this section may become inaccurate. Another warning is for the cases where k is not a smooth function of frequency, as happens often around resonant molecular or atomic transitions in material media. Also in these cases the results of this section may become invalid.

5.3 The scalar approximation

In the two previous sections, as well as in most of Chapter 4, we elaborated on one or more components of the e.m. eld, in a cartesian reference frame, in ways that were not signi cantly a ected by the vector nature of these elds. It is self-explanatory why changing a vector problem into a scalar one is always a very attractive target, but in the case of e.m. waves, it can be reached in an exact way only in rare cases, like those of Chapter 4. Much more often, it can be reached through approximations, as we shall see, for instance, in Chapter 13.

Most of the problems that we will deal with in the rest of this chapter belong to the latter class. Therefore, it may be useful, before considering individual cases, to discuss brie y the reasons why a scalar treatment involves approximations, as well as the reasons why, notwithstanding the approximate nature of the approach, we use it; and nally, to nd criteria to assess the order of magnitude of the errors that this approach entails.

The rst reason why, in most sections of this chapter, a scalar treatment cannot be rigorous, is because we deal often with waves in *non-homogeneous media*. In Sections 4.8 and 4.9, it was shown that, at a sharp discontinuity in either or , *polarization* of the incident eld plays a fundamental role. In a smoothly varying medium (i.e., where and are continuous and di erentiable functions of space coordinates, as we shall suppose in the following sections), we must rediscuss the steps which in Chapter 1 led us from Maxwell's equations to the Helmholtz equation. In general, $\nabla \quad \mathbf{E} \neq 0$ and $\nabla \quad \mathbf{H} \neq 0$, so we must take into account the terms $\nabla(\nabla \quad \mathbf{E})$ and $\nabla(\nabla \quad \mathbf{H})$ when we want to pass from $\nabla \quad \nabla$ to ∇^2. On the other hand, we know that $\nabla \quad (\quad \mathbf{E}) = 0$ and $\nabla \quad (\quad \mathbf{H}) = 0$, always. These relations can be rewritten in the following form, thanks to the vector identity (C.3):

$$\nabla \quad \mathbf{E} + \quad \nabla \quad \mathbf{E} = 0 \quad ,$$

$$\nabla \quad \mathbf{H} + \quad \nabla \quad \mathbf{H} = 0 \quad . \tag{5.25}$$

From this we nd:

$$\nabla \quad \nabla \quad \mathbf{E} \quad = \quad \nabla^2\mathbf{E} + \nabla(\nabla \quad \mathbf{E}) = \quad \nabla^2\mathbf{E} \quad \nabla\left(\frac{\nabla \quad \mathbf{E}}{\quad}\right) ,$$

$$\nabla \quad \nabla \quad \mathbf{H} \quad = \quad \nabla^2\mathbf{H} + \nabla(\nabla \quad \mathbf{H}) = \quad \nabla^2\mathbf{H} \quad \nabla\left(\frac{\nabla \quad \mathbf{H}}{\quad}\right) , \tag{5.26}$$

so in an inhomogeneous medium the Helmholtz equation must encompass one more term, which is sensitive to the direction of the vector $\mathbf{E}$ (or $\mathbf{H}$) through the scalar product times ∇ (or ∇). This is an equation which is not easy to break down into scalar problems in an exact way.

The reasons why, nonetheless, even in inhomogeneous media the scalar approximation is used very often, come on one hand from mathematical di culties due to the additional terms in Eqs. (5.26); on the other hand, from the richness of the scienti c literature on scalar waves (acoustic, elastic, quantum-mechanical waves, etc.), which obey the scalar Helmholtz equation also in inhomogeneous media. Results from these neighboring elds are often acceptable approximations for e.m. wave propagation also, especially when they lead to numerical methods.

One consequence of the scalar approximation is that physical dimensions of the wave lose importance: all the cartesian components of elds and of the potentials satisfy the same equation, namely:

$$\nabla^2 \quad\quad ^2 \quad = 0 \quad . \tag{5.27}$$

So, the *wave function* can have arbitrary physical dimensions. It is often convenient to assume that it is adimensional.

Finally, let us look for conditions that must be satis ed in order to insure that errors due to the scalar approximation are acceptable. From Eqs. (5.25) and (5.26), we can extract a *necessary condition*: the medium has to be *slowly varying*, this phrase meaning that the moduli of the following vectors (whose dimensions are $m \quad ^1$):

$$\frac{\nabla}{\quad} \quad , \quad \frac{\nabla}{\quad} \quad , \tag{5.28}$$

must be small, compared with the scale length over which $\mathbf{E}$ and $\mathbf{H}$ evolve in space. In some cases (as we will see in Section 5.6), one may nd additional, more restrictive constraints, but whenever this is not the case, then the typical eld evolution scale is the *wavelength*, . Let us stress once more that this condition is only a *necessary* one. There are no su cient conditions with a broad scope of validity; they must be assessed individually for each problem.

5.4 The equations of geometrical optics

Propagation of light, meaning electromagnetic radiation whose wavelength is small compared to the size of all the objects which are present along its path, is often approached, at the most elementary level, in a purely geometrical way. Let us see how we may connect rigorous e.m. theory, i.e., Maxwell's equations, with simpli ed models in terms of "rays," starting from the axiomatic theory of Chapter 1 and introducing suitable approximations.

Taking completeness of plane waves as a generic hint, let us suppose that, *in a lossless inhomogeneous medium*, any solution of Eq. (5.27) can be written in the following form:

$$(\mathbf{r}) = C(\mathbf{r})\, e^{\ \ jk_0 S(\mathbf{r})} \quad , \tag{5.29}$$

where $k_0 = \omega\sqrt{\ _0\ _0}$ is the intrinsic phase constant of free space, while $C(\mathbf{r})$ and $S(\mathbf{r})$ are two *real* functions of spatial coordinates, which we will call *amplitude* and *phase functions*, respectively.

At a generic (but xed) point $P = O + \mathbf{r}_0$, Eq. (5.29) looks like a uniform plane wave. For this reason, the assumption we have introduced now is usually referred to as a " eld expansion into *local* plane waves".†

The ansatz Eq. (5.29) enables us to introduce easily some approximations. Indeed, if we substitute Eq. (5.29) into Eq. (5.27), make use of some vector identities (see Appendix C), and divide by $\exp(\ \ jk_0 S(r)) \neq 0$, we get:

$$\nabla^2 C \quad jk_0(2\nabla C \quad \nabla S + C\nabla^2 S) \quad \left[(\nabla S)^2\, k_0^2 + \ \ ^2\right] C = 0 \quad . \tag{5.30}$$

Having assumed that C and S are real, we may equate to zero *separately* the second term, which is imaginary, and the sum of the rst and the third ones, which are real. Considering these, we have:

$$(\nabla S)^2 = \frac{\ \ ^2}{k_0^2} + \frac{1}{k_0^2}\frac{\nabla^2 C}{C} \quad . \tag{5.31}$$

The rst term on the right-hand side in this equation, equal to $\omega^2 \ \ /\omega^2 \ _0\ _0$, is always 1. The second term would be equal to 1, if C were a solution of the Helmholtz equation in free space. In such a case, C would vary in space with a period equal to $\ _0 = 2\ /k_0$. But in our case, remember that C is just the *amplitude* of the wave Eq. (5.29). If it varies appreciably on a length scale comparable to $\ _0$, then the applicability of the idea of local plane waves is not

† If we allow the functions $C(\mathbf{r})$ and $S(\mathbf{r})$ to take complex values, then we encompass *evanescent* local plane waves in the expansion. This entails a remarkable broadening in the scope of validity, but the subject is now premature. We will come back to it in Section 5.13.

justi ed. Hence, we may assume:

$$|\nabla^2 C| \quad k_0^2 \, |C| \quad . \tag{5.32}$$

Some constraints on the results that follow from this assumption will be outlined in Section 5.13.

Eq. (5.31) is then approximated as follows:

$$(\nabla S)^2 = n^2 \quad , \tag{5.33}$$

where $n = (\quad / \, _0 \, _0)^{1/2}$ is the *refractive index*, which was already de ned in Section 4.9, but now becomes, in general, a function of spatial coordinates. If one is able to solve Eq. (5.33), then one can determine the surfaces over which $S(\mathbf{r})$ = constant, i.e., the *constant-phase* surfaces of the wave Eq. (5.29). As the reader may know from elementary optics, the shape of such surfaces is strictly related to the capability of a wave of creating "images" of its sources. For this reason, Eq. (5.33) is often referred to as the *eikonal equation* (or the equation of images, from Greek $\tilde{\omega}$ = image).

Within the approximations that we have accepted when we wrote Eq. (5.32), geometry of the wave propagation is fully contained in Eq. (5.33). Indeed, if the eld can be expanded, at any point $R = O + \mathbf{r}$, into local plane waves, then each of these waves, utilizing what we saw in Section 4.3, is characterized by a phase vector perpendicular to the constant-phase surface which passes through R. The lines which envelope the local phase vectors, i.e., the lines orthogonal to the surfaces $S(\mathbf{r})$ = constant, are called *ray paths.*

If what one aims at is to nd the ray paths, the procedure can be simpli ed, with respect to solving Eq. (5.33) and then calculating the orthogonal trajectories. We can nd from Eq. (5.33) another di erential equation where the unknown is the position along the generic ray. To this purpose, we introduce (see Figure 5.2) a curvilinear coordinate u which varies along the ray with a metric coe cient equal to unity (i.e., the ratio between any path length along the ray, and the corresponding change in the coordinate u, equals one). Then, $\hat{u} = d\mathbf{r}/du$ equals the unit vector tangent to the ray, and $d/du = \hat{u} \;\; \nabla$ is the direction-derivative operator along the direction of the ray, orthogonal everywhere to a constant-phase surface. Thus we may write:

$$\nabla S = n \, \frac{d\mathbf{r}}{du} = n\hat{u} \quad . \tag{5.34}$$

To reach our goal, we must now get from Eq. (5.34) an equation where the function S does not appear explicitly. Let us consider the quantity $\nabla(n^2) = 2n \, \nabla n$, which is also equal to $\nabla(n\hat{u} \;\; n\hat{u})$ and therefore can be expanded (see Eq. (C.2)) as follows:

$$\nabla(\mathbf{A} \;\; \mathbf{B}) = (\mathbf{A} \;\; \nabla)\,\mathbf{B} + (\mathbf{B} \;\; \nabla)\,\mathbf{A} + \mathbf{A} \quad (\nabla \quad \mathbf{B}) + \mathbf{B} \quad (\nabla \quad \mathbf{A}) \quad .$$

Remembering that:

$$\nabla \quad (n\hat{u}) = \nabla \quad \nabla S = 0 \quad , \tag{5.35}$$

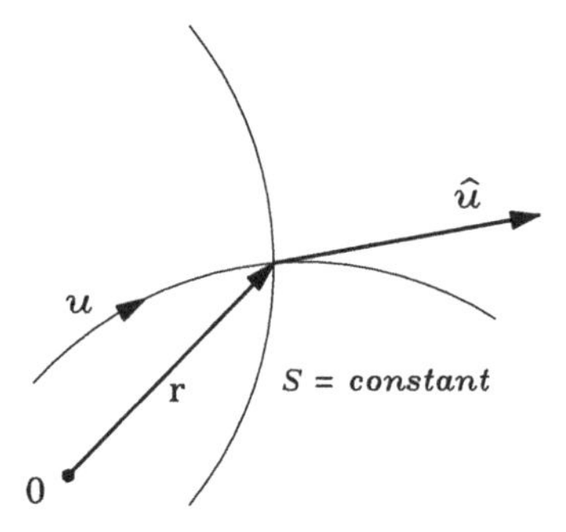

Figure 5.2 *The curvilinear coordinate u, along the trajectory of a generic ray.*

we get:

$$2n\,\nabla n = 2n(\hat{u} \quad \nabla)\,(n\hat{u}) \quad . \tag{5.36}$$

Dividing by $2n$ (di erent from zero everywhere) and rewriting the direction derivative $\hat{u} \quad \nabla$ as d/du, we have:

$$\frac{d}{du}\left(n\,\frac{d\mathbf{r}}{du}\right) = \nabla n \quad . \tag{5.37}$$

Eq. (5.37) is called the *ray equation.* The procedures for its direct integration, frequently used for propagation in non-homogeneous media (like the atmosphere and the ionosphere at radio frequencies, graded-index lenses and optical bers at optical frequencies), are called *ray tracing.* Nowadays, in most cases they are implemented numerically, and many packages for computer-aided design based on ray tracing are commercially available.

In several cases of practical interest, Eq. (5.37) is still too complicated to be integrated directly. Then, one makes use of a further simplifying approximation, based on the assumption that all the rays that matter in practice are nearly parallel to a direction, z, called the *optical axis* of the system. Examples in this sense can be optical instruments with several lenses, whose spacing is not small compared to their transverse dimensions, but they extend also to multiple-re ector antennas, together with their feeders. In all these cases we may introduce the so-called *paraxial approximation*:

$$du \simeq dz \quad , \tag{5.38}$$

where z indicates now a cartesian coordinate, measured along the optical axis. So, Eq. (5.37) becomes:

$$\frac{d}{dz}\left(n\,\frac{d\mathbf{r}}{dz}\right) = \nabla n \quad . \tag{5.39}$$

This equation is much simpler than Eq. (5.37), as the derivative along the unknown direction of a ray has been replaced by an ordinary derivative. However, one has to memorize that Eq. (5.39) is a further approximation, compared with the eikonal equation. This can be proved, for instance, by computing the curl

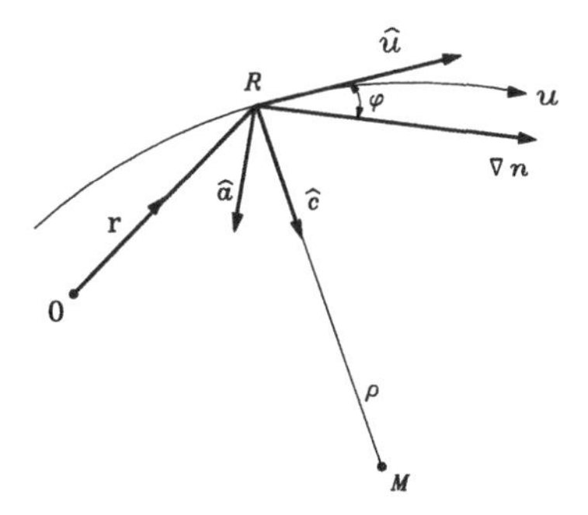

Figure 5.3 *The ray equation as generalized Snell's law: geometrical scheme.* M = *center of curvature.*

of both sides of Eq. (5.39). Provided that all the continuity requirements be satis ed, so that the operators $\nabla \quad$ and d/dz commute, one obtains:

$$\frac{d}{dz}\Big[\nabla \quad (n\hat{u})\Big] = 0 \quad , \tag{5.40}$$

i.e., $\nabla \quad (n\hat{u})$ = constant *along the z-axis*; on the contrary, as we have shown shortly ago, the eikonal equation yields $\nabla \quad (n\hat{u}) = 0$ *along a ray.* These two results coincide exactly only for a ray strictly parallel to the z axis, whence the conclusion.

Let us show now that Eq. (5.37) can be thought of as a generalized Snell's law (Eq. (4.84)). Take the scalar product of Eq. (5.37) by any unit-vector $\hat{v}$ lying in the plane orthogonal to ∇n, i.e., tangent to the n = constant surface that passes through the point $R = O + \mathbf{r}$, which belongs to the ray we are dealing with (see Figure 5.3). We get:

$$\hat{v} \quad \frac{d}{du}(n\hat{u}) = \hat{v} \quad \hat{u}\ \frac{dn}{du} + n\ \frac{d\hat{u}}{du} \quad \hat{v} = \nabla n \quad \hat{v} = 0 \quad . \tag{5.41}$$

In particular, if we let $\hat{v} = \hat{b}$ (the unit vector orthogonal to the plane de ned by the vectors ∇n and $\hat{u}$), then the scalar product $\hat{u} \quad \hat{b}$ vanishes. So (being $n \neq 0$), we obtain:

$$\frac{d\hat{u}}{du} \quad \hat{b} = 0 \quad . \tag{5.42}$$

This equation shows that, in the proximity of the point R, variations of the vector $\hat{u}$ along the ray must also be orthogonal to the unit vector $\hat{b}$, and consequently, co-planar with $\hat{u}$ and with ∇n. Hence, in the proximity of any point R, a ray is well tted by a planar curve. The restrictions which are expressed by the phrases "in the proximity of R" and " tted" can be relaxed, if, in the whole medium we are dealing with, the index distribution is such that the vector ∇n remains everywhere parallel to a given plane (see an example in Section 5.6).

From what we said, we may conclude that $d\hat{u}/du$ is the *curvature vector* of the ray trajectory:

$$\frac{d\hat{u}}{du} = \mathbf{c} = \frac{\hat{c}}{\ } \quad , \tag{5.43}$$

where is the *radius of curvature* at the point R (see again Figure 5.3). Let φ now be the angle between $\hat{u}$ and ∇n, whose physical signi cance is the same as that of the *incidence angle* of Section 4.8. Set now $\hat{v} = \hat{a}$, the unit vector orthogonal to ∇n and belonging to the plane $(\hat{u}, \nabla n)$. Then, from Eq. (5.43) we nally get:

$$\frac{dn}{du} \sin\varphi + n \frac{1}{\ } \cos\varphi = 0 \quad . \tag{5.44}$$

Note that, if the direction of ∇n remains constant in the neighborhood of R, then it is $du = \ \ d\varphi$. Therefore, in such a case Eq. (5.44) can be rewritten in the form:

$$\frac{d}{du} (n \sin\varphi) = 0 \quad , \tag{5.45}$$

which is, as we anticipated, a generalized Snell's law which holds, as we just said, when the direction of ∇n remains constant. This happens, for example, in the media where the refractive index depends only on one cartesian coordinate. In cylindrically symmetrical media, where ∇n is everywhere in the radial direction (see an example in the next section), this statement applies only to rays whose trajectories belong to planes passing through the symmetry axis, usually referred to as *meridional rays*. For *skew rays*, the reader can show as an exercise that what follows from the ray equation, Eq. (5.37), is $n \cos\vartheta_z =$ constant, where ϑ_z is the angle between the symmetry axis of the medium and the tangent to the ray trajectory. This also can be though of as a generalized Snell's law. The case of spherical symmetry is also discussed in some of the problems at the end of this chapter. In the paraxial approximation, direction derivatives are replaced by the ordinary derivative with respect to z. Restrictions on the consequent results have already been pointed out a shortwhile ago.

Let us now go back to the equation that we get if we set the imaginary part of Eq. (5.30) equal to zero, namely:

$$2\nabla C \ \ \nabla S + C\,\nabla^2 S = 0 \quad . \tag{5.46}$$

Multiply it by C, and make use of the identity (C.3) $\nabla \ (\varphi\mathbf{a}) = \nabla\varphi \ \mathbf{a} + \varphi\nabla \ \mathbf{a}$, as well as of Eq. (5.34). The result is:

$$\nabla \ \ (C^2 \nabla S) = \nabla \ \ (C^2 n \hat{u}) = 0 \quad . \tag{5.47}$$

In order to grasp the meaning of this relation, let us imagine a surface, T, which consists entirely of ray trajectories (see Figure 5.4). By de nition, $\hat{u}$ is parallel to a ray everywhere. Hence, Eq. (5.47) indicates that this surface T is a " ux tube" of the quantity $C^2 n$. In other words, if A_S denotes that portion of a generic constant-phase surface (S = constant) which is inside T, then

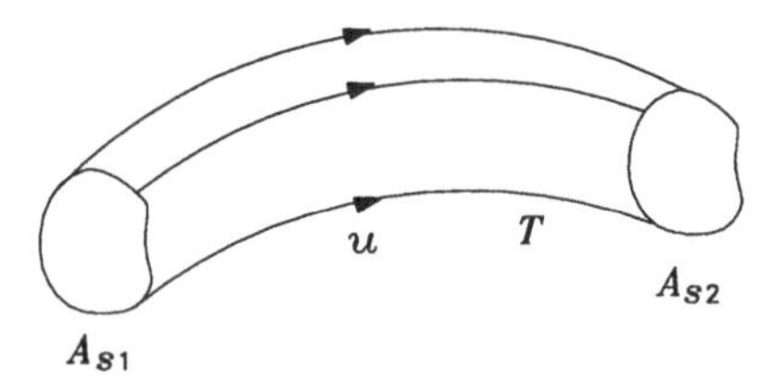

Figure 5.4 *A family of rays seen as a " ow tube".*

Eq. (5.47) is equivalent to writing:

$$\int_{A_S} C^2 \, n \, dA_S \quad \int_{A_S} | \quad |^2 \, n \, dA_S = \text{constant} \quad , \tag{5.48}$$

(constant meaning independent of the curvilinear coordinate u measured along one of the rays lying on T).

Eq. (5.47), or equivalently Eq. (5.48), express the *intensity conservation law* of geometrical optics. Its electromagnetic implications will be discussed in the next section. These equations provide a " uid-dynamics" model of wave propagation. They are also a good starting point if one wants to point out the limitations of geometrical optics. Indeed, let us consider a *focus*, that is a point where, according to Eqs. (5.37), an in nite number of rays converge into an in nitesimal cross-section A_s. Then, Eq. (5.48) implies $C \to \infty$. This result is not only physically unrealistic, it is also in contrast with the approximations that we introduced at the beginning of this section, based on Eq. (5.32). This point will be re-examined in the following sections.

Going back to Eq. (5.46), we can deduce from it another relationship: by simply taking into account Eq. (5.34), we get:

$$\frac{2}{C}\frac{dC}{du} = \quad \frac{1}{n} \nabla^2 S = \quad \frac{1}{n} \nabla \quad (n\hat{u}) \quad . \tag{5.49}$$

This equation is easy to integrate along a generic ray path, yielding:

$$| \quad (u)|^2 \quad C^2(u) = C^2(0) \, e^{\quad \int_0^u \frac{\nabla^2 S}{n} du} = C^2(0) \, e^{\quad \int_0^u \frac{\nabla \quad (n\hat{u})}{n} du} \quad . \tag{5.50}$$

The last formula can be used while tracing the rays, in order to track, from one point to the next, along a ray, how the intensity $C^2(u)$ evolves.

5.5 Geometrical optics: Electromagnetic implications

The basic equations of geometrical optics, Eqs. (5.33), (5.37) and (5.48), were obtained after the scalar approximation. In fact, they can be obtained also from the e.m. eld vectors, provided one makes an ansatz similar to Eq. (5.29), namely:

$$\mathbf{E}(\mathbf{r}) = \mathbf{e}(\mathbf{r}) \, e^{\quad jk_0 S(\mathbf{r})} \quad , \qquad \mathbf{H}(\mathbf{r}) = \mathbf{h}(\mathbf{r}) \, e^{\quad jk_0 S(\mathbf{r})} \quad , \tag{5.51}$$

where the vectors **e** and **h** have to be supposed as *real.* If we insert Eqs. (5.51) into Maxwell's equations (without losses and without sources), the conclusions we reach are the same as in the previous section. The main reason why we repeat this procedure is simply because we want to avoid a misunderstanding. We do not want the reader to believe that a "vectorial" theory of geometrical optics could be more accurate that the "scalar" one. Indeed, let us show that, when one arrives at the equations of geometrical optics, then the problem has already become a scalar one, perhaps without us noticing it, even if one starts from the ansatz in Eq. (5.51) and makes use of Maxwell's equations.

To show that, we introduce Eqs. (5.51) into Maxwell's equations. We get:

$$\nabla \quad \mathbf{E} \quad (\nabla \quad \mathbf{e})\, e^{\ jk_0 S} + (\ jk_0 \nabla S) \quad \mathbf{e}\, e^{\ jk_0 S} = \quad j\omega \quad \mathbf{h}\, e^{\ jk_0 S} , \quad (5.52)$$

and likewise for $\nabla \quad \mathbf{H}$. Divide by $k_0 \exp(\ jk_0 S)$ and neglect, for the same reasons that were outlined at the beginning of the previous section, the terms in $k_0 \ ^1$. We obtain:

$$\nabla S \quad \mathbf{e} \simeq n \ \mathbf{h} \ ,$$

$$\nabla S \quad \mathbf{h} \simeq \quad n \frac{\mathbf{e}}{} \ , \quad (5.53)$$

where the refractive index $n = (\ / \ _0 \ _0)^{1/2}$ and the wave impedance $=$ $(\ / \ )^{1/2}$ are now functions of spatial coordinates, since in general the medium is not homogeneous. Taking into account Eq. (5.34), we get:

$$\hat{u} \quad \mathbf{e} \simeq \ \mathbf{h} \ , \qquad \hat{u} \quad \mathbf{h} \simeq \frac{\mathbf{e}}{} \ . \quad (5.54)$$

From Eqs. (5.54) we may deduce some interesting results, that were implicit in the scalar theory of the previous section. In particular, note that these results depend critically on $S(\mathbf{r})$ being a *real-valued function.*

A) The vectors **e**, **h** and ∇S (or $\hat{u}$) are *mutually orthogonal.* Therefore, **E** and **H** are transverse, at any point R, with respect to the ray path passing through R.

B) As is real, **e** and **h**, and consequently **E** and **H**, are *in phase*, at any point R.

C) As a consequence, the Poynting vector $\mathbf{P} = \mathbf{E} \quad \mathbf{H} \ /2$ is everywhere *real and parallel to the ray direction.* This is the reason why it became common, even in other contexts, to call "direction of the electromagnetic ray" the direction of the real part of the Poynting vector. We will speak again about this terminology in Chapter 6.

D) The wave impedance in the ray direction is, at any point R, equal to (R), the intrinsic impedance of the medium (or rather, its local value at the point R).

E) From the previous points A) and D) we can infer that, at any point R, the time-averaged values of the electric and magnetic energy density equal each other:

$$\frac{1}{4}\ (R)\,\mathbf{E}(R)\ \ \mathbf{E}\ (R) = \frac{1}{4}\ (R)\,\mathbf{H}(R)\ \ \mathbf{H}\ (R)\quad . \tag{5.55}$$

F) From Eq. (5.55) and from Poynting's theorem, as there are neither losses nor sources, it follows that the divergence of the Poynting vector equals zero everywhere. Based on this and on the previous point C), Eq. (5.48) indicates that the conserved quantity, referred to as the *intensity*, in the scalar approximation $|\ \ |^2 n$, is *proportional to the modulus of the Poynting vector*:

$$|\ \ |^2 n \propto |\mathbf{P}|\quad \frac{|\mathbf{E}|^2}{2}\qquad \frac{|\mathbf{H}|^2}{2}\quad . \tag{5.56}$$

The physical dimensions of the proportionality constant in the previous relationship are not yet speci ed, since the physical dimensions of the scalar wave function are unspeci ed as well. However, we may notice that Eq. (5.56) makes it necessary that the refractive index n shows up explicitly in the relationship between and the e.m. vectors $\mathbf{E}, \mathbf{H}$. Furthermore, n is involved implicitly, through the local wave impedance . This point will be raised again and clari ed in Section 5.7.

In conclusion, we have shown in this section that the approximations of geometrical optics imply that, locally at any point, the e.m. eld has all the attributes of a *uniform plane wave*. Therefore, referring to such a eld as a set of *local plane waves*, as we did before (after Eq. (5.29)), appears to be a legitimate procedure.

Clearly, this approach is intrinsically an approximation. Indeed, we have shown in Chapter 4 that a wave, when impinging on a steep change in parameters of the medium, *must* be re ected in part. The theory we developed here loses any track of re ection, as shown by the invariance of the ux of Poynting vector along a ux tube. This is one of its basic drawbacks, or at least limitations.

5.6 Examples of ray tracing in radio propagation and in optics

An isotropic *spherically symmetric* medium, whose refractive index decreases monotonically as the radial coordinate increases, is a reasonable model for studying radio wave propagation in the earth atmosphere, at heights below the ionosphere.

Assume that above the ground (i.e., for $r > r_0$) the refractive index varies as:

$$n(r) = n_0\Big[1 \quad a(r \quad r_0)\Big] = n_0[1 \quad ah]\quad , \tag{5.57}$$

where $n_0 = n(r_0)$ is the refractive index at the ground level, and $h = r \quad r_0$ is the height above the ground. For $r < r_0$, $n(r)$ is not de ned. If we restrict ourselves to heights such that $|a|\,h \quad 1$, then Eq. (5.57) can be looked at as a truncated Taylor series, for a widely arbitrary function $n(r)$, being $a =$

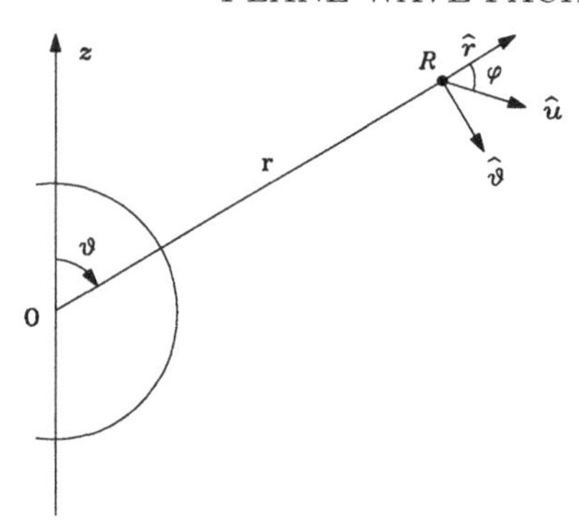

Figure 5.5 *A spherical coordinate reference frame, to solve the ray equation in a spherically symmetrical medium.*

$(1/n_0)\,(dn/dr)_{r=r_0}$. Therefore, the model is more general than one may think at rst sight.

In spherical coordinates‡ (Figure 5.5), it is convenient to re-write the ray equation (5.37) recalling Eq. (5.43) and using the following identity:

$$\frac{d}{du} = \hat{u} \quad \nabla = \cos\varphi\,\frac{\partial}{\partial r} + \sin\varphi\,\frac{1}{r}\,\frac{\partial}{\partial\vartheta} \quad . \tag{5.58}$$

We obtain then:

$$\frac{d}{du}\,(n\hat{u}) \quad \cos\varphi\,\frac{\partial n}{\partial r}\,\hat{u} + \sin\varphi\,\frac{1}{r}\,\frac{\partial n}{\partial\vartheta}\,\hat{u} + n\,\frac{1}{\ }\,\hat{c} = \nabla n \qquad n_0\,a\hat{r}\,. \tag{5.59}$$

The term $\partial n/\partial\vartheta$ is equal to zero, because of the spherical symmetry. Since $\hat{c}$ and $\hat{u}$ are orthogonal, we see that $\hat{u}$, $\hat{c}$ and $\hat{r}$ lay on the same plane and, consequently, the trajectory of any ray is on a plane passing through the origin.

Integration of Eq. (5.59) becomes very simple in the cases (quite relevant for terrestrial radio links) where the ray trajectories have little inclination with respect to the horizon. Then, we may assume $\varphi \simeq \ /2$, $|\cos\varphi| \quad 1$ everywhere, while $\partial n/\partial r$ and $n/\ $ are of the same order of magnitude. Therefore we get the following equation:

$$n\,\frac{1}{\ }\,\hat{c} = \quad n_0\,a\hat{r} \quad , \tag{5.60}$$

which shows that, for $a > 0$, the unit vector $\hat{c}$ is oriented towards the center of the sphere, indicating that rays bend towards the ground as they propagate. Furthermore, if we assume $n \simeq n_0$ on the left-hand side, Eq. (5.60) gives:

$$= \frac{1}{a} = \text{constant} \quad . \tag{5.61}$$

‡ Note that, to be consistent with symbols used in the previous section, φ is not a coordinate angle in the spherical reference frame, but the angle between the ray and the index gradient. Anyhow, thanks to symmetries in the problem, two spherical coordinates, r and ϑ, will be enough in this section.

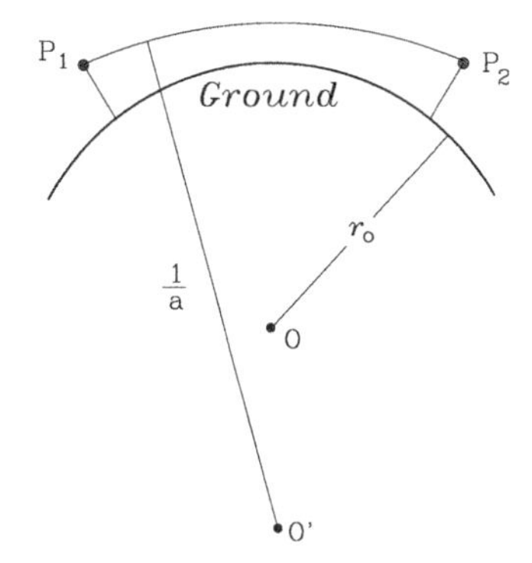

Figure 5.6 *Radio transmission beyond the geometrical horizon.*

This means that, within this approximation, all the ray trajectories are circles of radius $(1/a)$. If $(1/a) > r_0$ (which is the normal situation in the atmosphere), then it is possible to draw circles of this family, which pass through two points above the ground, without intersecting the earth surface (see Figure 5.6). This result implies that a radio link can be established between two points each of which is *beyond the geometrical horizon* with respect to the other.

As a second example, let us deal with *cylindrical symmetry*, i.e., a medium whose refractive index depends only on the distance r from an axis, which we take as the z axis of a cylindrical reference frame (Figure 5.7). This can be a model for some types of graded-index optical waveguides (or lenses), to be studied in more detail in Section 10.9 and in the following ones. In particular, we concentrate on the so-called parabolic index pro le:

$$n^2 = n_0^2\left[1 \quad (r/a)^2\right] \quad , \tag{5.62}$$

where n_0 is the maximum index, on axis, and a is a suitable normalizing radius, which sets the scale on which the index decreases as we get away from the axis. The region of interest for ray propagation is where one can assume:

$$r/a \quad 1 \quad . \tag{5.63}$$

The index gradient is found to be $\quad (r/a^2)\,(n_0^2/n)\,\hat{r}$, in the radial direction and always inwards: the cylindrical medium is a focusing one. In the paraxial approximation $du \simeq dz$, no derivative of n appears on the left-hand side of the ray equation, Eq. (5.37), which then reads:

$$\frac{d\hat{u}}{dz} = \quad \hat{r}(r/a^2)\,(n_0/n)^2 \simeq \quad \hat{r}(r/a^2) \quad . \tag{5.64}$$

This shows that the ray is subject to a radial "pull-back" force, proportional to the elongation away from the z axis. Therefore, by analogy with mechanics of a simple pendulum, the reader may conclude that those solutions of Eq. (5.64) that are contained in a plane passing through the z axis are *sinusoidal trajec-*

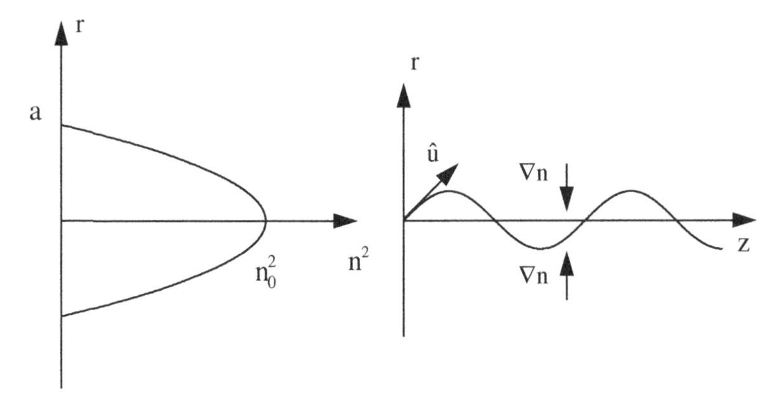

Figure 5.7 *Ray trajectories in a parabolic-index slab.*

tories (see again Figure 5.7). They are called *meridional rays.* Combinations of sinusoidal motions with equal periods in two orthogonal planes give rise to *skew rays*, which are, in the most general case, helical trajectories. The reader should, as an exercise, explain why the assumption that the two periods are equal is justi ed by Eq. (5.64).

For a signi cant number of additional examples of ray tracing in structures which are of interest to optical communications, the reader may consult, for instance, van Etten and van der Plaats (1991).

5.7 The WKBJ method

The subject of this section, although closely related to the previous one, is not easy to link with intuition. It looks advisable to postpone the discussion of its physical meaning and of its relationship with the previous topics, and to outline its mathematical treatment rst.

The technique which is currently referred to as the WKBJ method (from the initials of the four mathematicians, Wentzel, Kramers, Brillouin, and Je reys, who initiated it, independently of each other) will be outlined in this section in its version for uniform plane waves. The same method can be applied to many other circumstances; examples in this sense will be discussed in the next section and again in Chapter 10.

Similarly to Section 5.4, let us suppose that the equation to be solved is (5.27). So, we assume that the scalar approximation holds. Consider a lossless medium, and suppose it is inhomogeneous just because and vary *only in one direction* which we take as the z axis of an orthogonal reference frame:

$$= (z) \quad , \quad = (z) \quad , \quad {}^2 = \omega^2 (z) (z) = k^2(z) \quad . \qquad (5.65)$$

Another assumption, which links the WKBJ method to plane waves, but makes an important di erence with respect to the previous section, is that we want *the wave function be separable*, namely:

$$(x, y, z) = \quad (x, y)\, f(z) \quad . \tag{5.66}$$

Consequently, to solve the scalar Helmholtz equation:

$$\nabla^2 \quad\quad ^2 \quad = 0 \quad , \tag{5.67}$$

we have merely to solve the following ordinary di erential equation:

$$\frac{d^2 f}{dz^2} = \left[\; ^2(z) + k_t^2 \right] f \qquad ^2(z)\, f \quad , \tag{5.68}$$

where k_t^2 is a separation constant. We will restrict ourselves to the case $k_t^2 < \omega^2$, i.e., $^2(z) > 0$. The case $^2(z) < 0$ is thoroughly treated in the literature (e.g., Budden, 1961).

For ^{2} an arbitrary function of z, Eq. (5.68) does not admit an *exact* solution. The WKBJ *approximate* solution requires that $^2(z)$ be slowly varying on the scale of a *local characteristic length*, $(z) = 2 \;/|\; (z)|$:

$$\left| \frac{d}{dz} \right| \qquad \frac{|\;|}{} = \frac{^2}{2} \quad . \tag{5.69}$$

To clarify the meaning of this characteristic length, let us refer to the case where the unknown function is a plane wave; then, we have $k_t^2 = 0$, $^2 = \quad ^2$, and the condition in Eq. (5.69) is equivalent to assuming that the parameters of the medium vary slowly with respect to the local wavelength $= 2 \; /k(z)$:

$$\left| \frac{d(\quad)}{dz} \right| \qquad \overline{} \quad . \tag{5.70}$$

Note that Eq. (5.69) is more general than Eq. (5.70). This will be very useful in the following, mainly in Chapter 10.

Let us now set:

$$f(z) = e^{g(z)} \quad . \tag{5.71}$$

Di erentiating with respect to z, we obtain:

$$f'' = (g'^2 + g'')\, f \quad . \tag{5.72}$$

Then, Eq. (5.68) becomes:

$$g'^2 + g'' = \quad ^2(z) \quad . \tag{5.73}$$

Now it is time to exploit the assumption in Eq. (5.69). Remember that, in a homogeneous medium ($=$ constant), Eq. (5.27) has uniform plane waves as solutions, for which $g(z) = \quad j \; z$, and then $g' = \quad j \;$. Therefore, let us presume now, and verify later, that, when the condition in Eq. (5.69) holds, we are entitled to say that:

$$|g'^2| \qquad |g''| \quad . \tag{5.74}$$

Then, a rst approximation to Eq. (5.73) reads:

$$g'^2 \simeq \quad ^2(z) \quad . \tag{5.75}$$

Anyway, it is easy to check that this approximation is not satisfactory. Eq. (5.75) would give an imaginary g', whence also an imaginary $g(z)$, and so, through Eq. (5.71), $|f(z)| =$ constant. The corresponding wave function $= \ f$ would not even satisfy energy conservation, which is respected by geometrical optics. In conclusion, this approximation is actually worse than geometrical optics.

In order to improve on Eq. (5.75), we can use a result which is a straightforward consequence of Eq. (5.75) itself, namely:

$$g'' \simeq \ j \frac{d}{dz} \ , \tag{5.76}$$

and substitute it into the exact equation (5.73). The binomial expansion of the square root, cut o at the second term (as allowed by Eq. (5.74)), gives then:

$$g' \simeq \ j \ (z) \left\{ 1 \ \ j \frac{1}{2 \ ^2(z)} \frac{d}{dz} \right\} = \ j \ (z) \ \ \frac{d}{dz} \ln(\ ^{1/2}) \ . \tag{5.77}$$

Integrating Eq. (5.77), and substituting the result into Eq. (5.71), we get:

$$f(z) = \frac{A}{\ ^{1/2}} e^{\ j \int_0^z \ (z')\, dz'} \ , \tag{5.78}$$

($A =$ arbitrary constant), which is the rst-order WKBJ solution, one of the key results of the method.

Eq. (5.78) expresses the phase delay of the wave as the result of an in nite number of in nitesimal phase shifts. This is the reason why the integral in the previous exponential function is sometimes referred to as *phase memory*. The factor $\ ^{1/2}$ in front of the exponential means intensity conservation: in the one-dimensional case (i.e., for $\ (x, y) =$ constant), this factor plays the same role as $n \ ^{1/2}$ in the intensity conservation law of geometrical optics $| \ |^2 n$ area $=$ constant. Hence, we have many indications that geometrical optics and the WKBJ method are strictly interconnected. Indeed, it can be shown that the two methods give results of *equal accuracy*, measured as the number of powers of that are accounted for in a series expansion of the unknown wave functions and of the equations they must satisfy.

Nevertheless, the two theories di er in their scope of application. In principle, the assumption of *separation of variables* makes the WKBJ method less general than geometrical optics. On the other hand, in practice that assumption makes the WKBJ method more versatile than geometrical optics. For example, one can apply WKBJ to study the dependence of the wave function on a coordinate which, in the paraxial approximation (see Section 5.4), is orthogonal to the z direction. This point will be dealt with in detail in Section 10.10.

It is easy to show that also the WKBJ method collapses under some critical conditions, which make at least its rst-order version, which we saw so far, not applicable. These conditions coincide with those that are critical for geometrical optics, that is to say, they occur near the focal points. Indeed, if we insert the approximate solution in Eq. (5.78) into the di erential equation (5.73), we nd that the validity of that solution is subject to the following *necessary condition*:

$$\left| \frac{3}{4}\left(\frac{1}{\ }\ \frac{d}{dz}\right)^2 \quad \frac{1}{2}\ \frac{1}{\ }\ \frac{d^2}{dz^2}\right| \qquad {}^2 \ . \tag{5.79}$$

This condition is not satis ed where $(z) = 0$, i.e., where $k_t^2 = \ {}^2$. The surfaces where $(z) = 0$ are called *caustics*. It can be shown, with a rather lengthy procedure, that the same surfaces are loci of focal points when the same problem is examined in terms of ray tracing. In Section 5.13 we will quote brie y some procedures to overcome these restrictions.

Eq. (5.79) is more restrictive than Eq. (5.69), which was our starting point. Indeed, Eq. (5.69) is veri ed also in the proximity of the extrema of (z), but in general near such points there are no restrictions on the second derivative of (z), so Eq. (5.79) is not necessarily satis ed. One must then adopt additional care when applying the WKBJ solution near a maximum or minimum of (z).

Finally, let us stress that the WKBJ method is intrinsically a *scalar method*, like geometrical optics. It is not possible to gain in accuracy by going to a vectorial formulation of the same method. In fact, if one looks for a WKBJ solution for an e.m. eld, i.e., for a behavior like Eq. (5.78) for all components of the vectors **E** and **H**, then one *must suppose* that every component *separately* satis es an equation like Eq. (5.73). This entails necessarily a scalar approximation, i.e., an implicit assumption that the double curl operator is replaced by a Laplacian operator even in non-homogeneous media, where this step is not rigorous.

5.8 Further comments on the WKBJ method

Several propagation experiments (for example, radio wave sensing of the ionosphere, or testing of graded-index optical waveguides) deal with slowly varying media of unknown characteristic. One tries to exploit the results of the WKBJ method to evaluate the parameters of the medium from measurements on waves that propagate through it ("inverse WKBJ").

If what is measured is a wave *transmitted* through the unknown medium, then the equations to be used are those of the previous section. In particular we may use Eq. (5.78), where the value of $f(z)$ at the receiver ($z = z_0$) is now a known quantity, and the quantity to be determined is (z') for $0 \quad z' \quad z_0$. It is evident that the known quantities are not enough to evaluate (z'). One needs additional information. Usually, one starts from a model of (z') which contains a number of unknown parameters equal to the number that can be evaluated from Eq. (5.78). These unknown parameters are then tted to the experimental data.

Often, however, transmitted waves are much more di cult to measure than *re ected* waves. For example, ionospheric soundings become much simpler to perform if the receiver is placed near the transmitter. This requires the test waves to propagate vertically. However, the WKBJ method is not the most suitable one in order to evaluate *distributed* re ections, due to slow variations in (z). Indeed, the two solutions of Eq. (5.78), corresponding to waves traveling in opposite directions, look as if they were completely decoupled; but physical intuition says that a strong localized re ection should be expected to occur where the incident wave (e.g., Eq. (5.78) with a sign) reaches a caustic, i.e., points where $= 0$. This intuitive idea relies on the fact that beyond the caustic (z) becomes imaginary, meaning that there our wave becomes evanescent, a situation which is reminiscent of total re ection (Section 4.11).

To solve this problem with the inverse WKBJ method, one is obliged to work in a region where the approximations of the previous section fail. Anyway, it can be shown (although the proof is too time-consuming to be reported here) that the corrections required because Eq. (5.69) does not hold are small, and physically intuitive. We may then begin to discuss re ections starting from the results of the previous section, in particular from Eq. (5.78).

Let us set $z = 0$ at the point of observation of the re ected wave, and call z_0 the height of the caustic, i.e., $(z_0) = 0$. Because of what we said above, we suppose that Eq. (5.78) applies over the whole range $0, z_0$. If f_i and f_r are the incident and the re ected wave, respectively, then we have:

$$f_i(z) = f_i(0)\sqrt{\frac{(0)}{(z)}}\; e^{\;\; j\int_0^z \;\; (z')\,dz'} \quad , \tag{5.80}$$

$$\begin{aligned} f_r(0) &= f_r(z_0)\sqrt{\frac{(z_0)}{(0)}}\; e^{\;\; j\int_{z_0}^{0} \;\; (z')\,dz'} \\ &= f_r(z_0)\sqrt{\frac{(z_0)}{(0)}}\; e^{j\int_0^{z_0} \;\; (z')\,dz'} \quad . \end{aligned} \tag{5.81}$$

Our goal is a relationship between the re ection coe cient at $z = 0$, $=$ $f_r(0)/f_i(0)$, and that at z_0. We need rst to nd a relationship between $f_i(z_0)$ and $f_r(z_0)$. This cannot be derived rigorously from Eq. (5.78), because this equation gives $f_i \to \infty$ and $f_r \to \infty$ as $z \to z_0$. One way to overcome the problem could be to *suppose* that these two in nites are equal: $f_i(z_0) = f_r(z_0)$. Then we would obtain:

$$= e^{\;\; j2\int_0^{z_0} \;\; (z')\,dz'} \quad . \tag{5.82}$$

In reality, this is an arbitrary statement. Eq. (5.78) says only that both wave amplitudes diverge when $(z) \to 0$. To be more accurate, one can show (e.g., Budden, 1961), through a more careful approximation, that the solutions of Eq. (5.73) near a caustic can be expressed as combinations of Airy functions (Abramowitz and Stegun, Eds., 1965). The only change that this entails with respect to Eq. (5.82) is simply a phase shift of $/2$ radians. The nal result is then:

$$= j\,e^{\;\; j2\int_0^{z_0} \;\; (z')\,dz'} = e^{\;\; j\left[2\int_0^{z_0} \;\; (z')\,dz' + \frac{}{2}\right]} \quad . \tag{5.83}$$

As we anticipated, this phase correction is easy to link to physical intuition.

Indeed, it expresses, within the limits of the scalar approximation, the phase shift that we found in Section 4.11 for total re ection, which we called then the *Goos-Hanchen shift*. Here as well as there, the shift is a consequence of passing from a real wave impedance to an imaginary one, at $z = z_0$.

5.9 Gaussian beams

In the previous sections, we gained some experience on how to nd approximate solutions of Maxwell's equations in slowly varying media. We can now exploit that experience, to nd some other approximate solutions, rst in a *homogeneous medium*, although they can be extended to some classes of nonhomogeneous media as well. They are physically more realistic than a single plane wave, particularly because they carry a nite power. At the same time they are still reasonably simple.

In a homogeneous lossless medium, the vector Helmholtz equation is *strictly* valid. It can be written by components in any cartesian reference frame, like for plane waves. So, it is easy to tackle in scalar terms. Therefore, let us consider just one component of the electric eld, with an angular frequency ω. Let us call (z) the direction, for the time being chosen with total freedom, of a phase vector $\mathbf{k}$ ($|\mathbf{k}| = \omega\sqrt{\quad}$). We may always write, without any physical implication:

$$E(\mathbf{r}) = \quad (\mathbf{r})\, e^{\quad j\mathbf{k}\ \mathbf{r}} \quad . \tag{5.84}$$

This is simply a change of variables, but soon it will allow us to give a physical meaning to some approximations that we shall make. For a uniform plane wave traveling in the direction of $\mathbf{k}$, we know that $(\mathbf{r})$ is a constant. Within the approximation of geometrical optics, the amplitude function C is slowly varying. Therefore, it appears legitimate to see what happens to Eq. (5.84) if the variations of vs. z are slow, on the length scale of $= 2\ /k$. If we recall the eld expansion as an integral of plane waves (Section 4.7), then we understand why a eld like Eq. (5.84) with slowly varying is called a *beam of plane waves* propagating in the z direction. Sometimes it is also called a *ray beam.*

Quantitatively, the approximation we need to make reads:

$$\left|\frac{\partial^2}{\partial z^2}\right| \qquad 2k\left|\frac{\partial}{\partial z}\right| \quad . \tag{5.85}$$

If we insert Eq. (5.84) into the scalar Helmholtz equation, and neglect the term $\partial^2\ /\partial z^2$ because of Eq. (5.85), we get:

$$\frac{\partial^2}{\partial x^2} + \frac{\partial^2}{\partial y^2} = 2jk\,\frac{\partial}{\partial z} \quad . \tag{5.86}$$

This equation is rather unusual in classical electrodynamics, but, on the contrary, is very well known in Quantum Mechanics, where it is referred to as the (two-dimensional) *Schrodinger equation* for a free particle, and is usually written as:

$$\nabla^2 \quad = \quad j\,\frac{2m}{h}\,\frac{\partial}{\partial t} \quad ,$$

where   is the wave function associated with the particle, m is the particle mass, t is time, $h = h/2$, h being Planck's constant. This equation has been studied so widely, that we can adapt to our needs solutions that can be found in the literature. In particular, we will now study in depth the simplest element in a complete set of discrete, linearly independent solutions. The other elements of this set will be examined in Section 5.10, and shown to share many basic features of the rst solution.

Let then   be a complex function, whose amplitude, in the generic transverse plane z = constant, varies like a circularly symmetrical Gaussian function:

$$= A\,\exp\left\{ \quad j\left[P(z)+\frac{k}{2q(z)}\,r^2\right]\right\} \quad , \tag{5.87}$$

where $r^2 = x^2 + y^2$. The *complex* quantities $P(z)$ and $q(z)$ are still unknown. Inserting the ansatz (5.87) into Eq. (5.86), we nd:

$$2k\left(P' + \frac{j}{q}\right) + r^2\,\frac{k^2}{q^2}\,(1 \quad q') = 0 \quad , \tag{5.88}$$

where the primes denote derivatives with respect to z. The quantities in brackets are functions only of z, the rst bracket is multiplied by a constant, while the second one is multiplied by a function of x and y. Therefore, Eq. (5.88) can be satis ed everywhere only if the two quantities in brackets are *separately* equal to zero:

$$q' \quad = \quad 1 \quad , \tag{5.89}$$

$$P' \quad = \quad \frac{j}{q} \quad . \tag{5.90}$$

The general integrals of these two equations are simply found to be respectively:

$$q \quad = \quad z + q_0 \quad , \tag{5.91}$$

$$P \quad = \quad j\,\ln\left[1+\frac{z}{q_0}\right] \quad . \tag{5.92}$$

In the second integral, an arbitrary integration constant B has been omitted, since in Eq. (5.87) P appears in the argument of an exponential function, so the factor $\exp(B)$ can be incorporated into the arbitrary constant A.

The function $q(z)$, being complex, plays two roles, through Eq. (5.87). It a ects the length scale of the Gaussian dependence on x and y in the transverse plane z = constant, and also the shape of the constant-phase surfaces of the wave. As for $P(z)$, it characterizes the slow variations in amplitude and phase of   as a function of z. However, Eq. (5.87) becomes simpler to analyze if we replace $P(z)$ and $q(z)$, which were useful in making the di erential equations

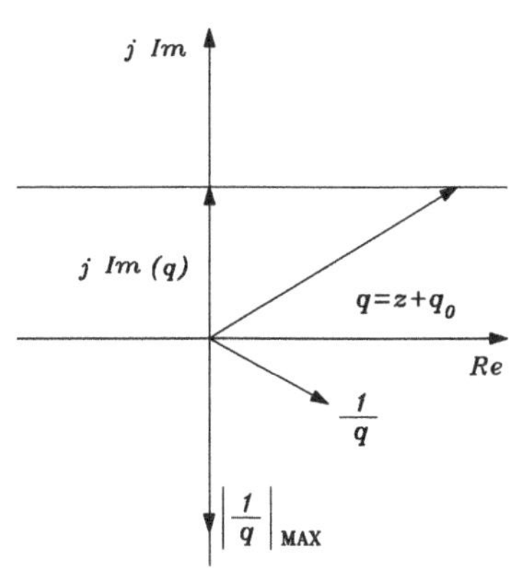

Figure 5.8 *Behavior of the complex function q vs. the longitudinal coordinate z.*

easy, with new variables, whose physical meaning is more direct. First, let us separate real and imaginary parts of $1/q$:

$$\frac{1}{q} = \frac{1}{R} \quad 2j \, \frac{1}{kw^2} \quad , \tag{5.93}$$

where R and w are real function of z. Eq. (5.87) then becomes:

$$= A \exp\left\{ \quad j\left(P(z) + \frac{kr^2}{2R}\right) \quad \frac{r^2}{w^2} \right\} \quad . \tag{5.94}$$

From this expression we can easily grasp the signi cance of $R(z)$ and $w(z)$. Recalling Eq. (5.84), we see that the eld is given by multiplied by exp(jkz); variations in the phase of due to $P(z)$ are slow, with respect to (see Eq. (5.92)). Consequently, the constant-phase surface passing through the generic point of coordinates (x, y, z) satis es the equation $k(z+r^2/2R)$ = constant. This says that the surface is an ellipsoid, but near that point it is well approximated by a spherical surface, centered on the z axis, and of radius R. So, physically $R(z)$ is the *radius of curvature of the wavefront* at the point $(0, 0, z)$.

As for $w(z)$, from Eq. (5.94) we see that it equals the radius of a circle, on the transverse plane z = constant, where the modulus of equals $1/e$ times its value on the z axis. It is clear why $2w$ is referred to as the Gaussian beam *diameter*, or the *spot size*.

From Eq. (5.91) it is clear that there is one (and only one) value of z, for which the real part of q vanishes (while the imaginary part is independent of z). This is in fact a peculiar z = constant plane, because Eq. (5.93) shows that, when $\mathrm{Re}(q) = 0$, then $1/R = 0$. Furthermore (see Figure 5.8), w takes its minimum value on the same plane. Hence, this plane is referred to as the *beam waist* plane. From now on, it is convenient to choose it as the $z = 0$ plane.

Let w_0 denote the minimum of w. In the new reference frame, we have

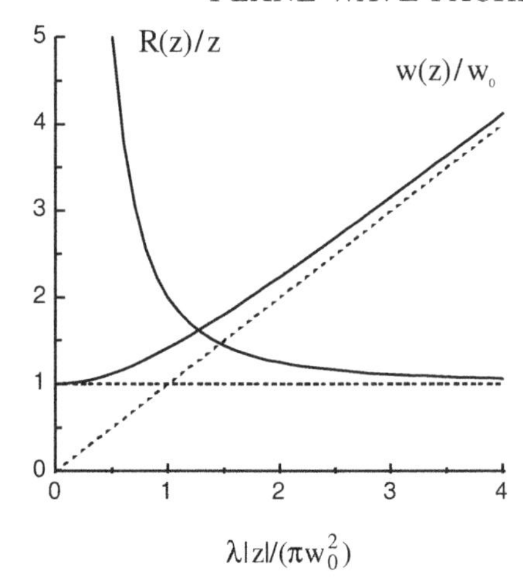

Figure 5.9 *Beam width and radius of curvature, vs. normalized distance from the beam waist.*

$q_0 = q(z = 0)$, so that:

$$q_0 = jk\,\frac{w_0^2}{2} \quad . \tag{5.95}$$

This simpli es very much the z-dependencies of w and R. Indeed, from Eqs. (5.91), (5.93) and (5.95), letting $k = 2\ /\ $, we get:

$$w^2(z) = w_0^2\left[1 + \left(\frac{z}{w_0^2}\right)^2\right] \quad , \tag{5.96}$$

$$R(z) = z\left[1 + \left(\frac{w_0^2}{z}\right)^2\right] \quad . \tag{5.97}$$

We see that the wavefront curvature and the relative width of the beam depend on only one variable, namely the ratio between the area of the beam waist, w_0^2, and the product z. They are plotted in Figure 5.9. As we see, the narrower the spot on the beam waist, with respect to , the faster the beam diverges as we move away from the waist, and the larger the curvature of the wavefront at a given distance $|z|$. This is in perfect agreement with di raction theory (see Chapter 13), but in sharp contrast with geometrical optics. This is the rst important indication that Gaussian beams are indeed an improved approximation, compared to ray optics, especially in the vicinity of focal planes.

Eq. (5.96) is, on the plane (z, w), the equation of a hyperbola (Figure 5.10) whose asymptotes are given by:

$$w = \frac{}{w_0}\,z \quad . \tag{5.98}$$

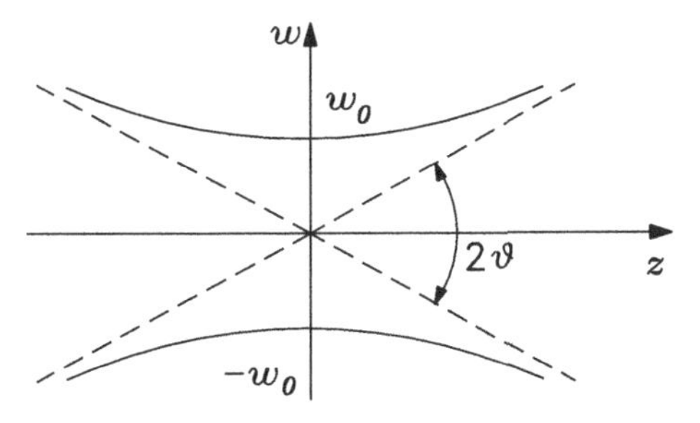

Figure 5.10 *The 1/e hyperbola and its asymptotes.*

Hence, far from the beam waist, i.e., for $z \quad w_0^2/$, the beam diverges linearly vs. z, with an angular aperture:

$$2\vartheta = 2\arctan \frac{}{w_0} = 2\arctan \frac{2}{kw_0} \quad . \tag{5.99}$$

This relation con rms that a tightly focused Gaussian beam (small spot size) diverges rapidly, away from the waist plane, while if one wants a well collimated beam, i.e., a beam consisting of almost parallel rays, it is necessary that the beam cross-section be not too small. This comment is fundamental not only in optics, but also in the case of aperture antennas. It expresses the so-called "uncertainty principle" between "conjugate variables," i.e., between two quantities which span the domains of two functions related to each other by a Fourier transform. It is, in this sense, the same as the relationship between widths of a signal in the time domain and in the frequency domain, which we recalled in Section 4.7. In the present case, the Fourier transform is the plane-wave expansion of the eld on the beam waist plane.

Let us add a short comment, which stems from terminology, but whose real purpose is to stress the underlying links with other subjects, to be dealt with in some of the following chapters. The quantity:

$$z_0 = \quad jq_0 = k\,\frac{w_0^2}{2} = \frac{w_0^2}{} \tag{5.100}$$

is called the *Rayleigh range*. This name is deliberately reminiscent of a result obtained by Lord Rayleigh, at the end of the 19th Century: in a focused pencil of rays, when the distance z from the focal plane is less than z_0, results based on geometrical optics are severely contradicted by experiment. The physical phenomenon that governs proagation in that range is di raction. This is a further con rmation that Gaussian beams are indeed the big improvement we were looking for, with respect to ray optics.

To give Eq. (5.94) a nal shape, let us substitute Eq. (5.95) into Eq. (5.92). We get:

$$j\,P(z) = \ln\left[1 \quad j\,\frac{2z}{kw_0^2}\right] = \ln\left|1 \quad j\,\frac{2z}{kw_0^2}\right| + j\,\arctan\left(\quad \frac{2z}{kw_0^2}\right) \quad . \tag{5.101}$$

Note that:

$$\left|1 \quad j\frac{2z}{kw_0^2}\right| = \left[1+\left(\quad\frac{z}{w_0^2}\right)^2\right]^{1/2} = \frac{w(z)}{w_0} \quad , \tag{5.102}$$

and let:

$$= \arctan\left(\quad\frac{z}{w_0^2}\right) \quad . \tag{5.103}$$

Incidentally, we point out that for any z this quantity is between $/2$. This recon rms our previous statements, that contributions to the phase coming from the factor encompassing $P(z)$ were much smaller than the others. We then get:

$$e^{\quad jP(z)} = \frac{w_0}{w(z)}\, e^{\quad j} \quad , \tag{5.104}$$

and the whole solution of Eq. (5.86) nally reads:

$$E = A\frac{w_0}{w(z)}\exp\left\{\quad j(kz+\quad)\quad j\frac{kr^2}{2R(z)}\right\}\exp\{\quad r^2/w^2(z)\} \quad . \tag{5.105}$$

Note that the factor $w_0/w(z)$ is necessary in order to maintain the power ux through the $z =$ constant planes independent of z, as required by the medium being lossless and homogeneous.

In order to strengthen the ties between the contents of this section and their technical fallouts, Table 5.1 shows a set of basic data regarding some of the most popular state-of-the-art laser sources.

5.10 Hermite-Gauss and Laguerre-Gauss modes

Before making the ansatz (5.87), we justi ed why the previous section was devoted entirely to one solution of Eq. (5.86). A countable in nity of solutions, forming a complete set, share all the basic features of Eq. (5.105). The completeness of the set will be shortly discussed in Section 5.12. In this section, we will look for the other solutions of Eq. (5.86), starting with a less restrictive ansatz than Eq. (5.87), with more freedom for the dependence of (x, y, z) on the transverse coordinates x, y.

As we aim at separating variables, we are tempted to set:

$$= f_1(x)\,f_2(y)\,\exp\left\{\quad j\left[P+\frac{k}{2q}r^2+\quad(z)\right]\right\} \quad , \tag{5.106}$$

where $f_1(x)$, $f_2(y)$ and (z) are unknown functions of one coordinate each.

Suppose that P and q remain the same functions of z as in the previous section. Then, inserting Eq. (5.106) into Eq. (5.86), exploiting Eqs. (5.89) and (5.90), and dividing by $f_1 f_2$, we nd:

$$\left(\frac{f_1''}{f_1}\quad j\frac{2kx}{q}\frac{f_1'}{f_1}\right)+\left(\frac{f_2''}{f_2}\quad j\frac{2ky}{q}\frac{f_2'}{f_2}\right) = 2k\ ' \quad , \tag{5.107}$$

where the primes denote ordinary derivatives, without any ambiguity.

Contrary to what may appear at rst sight, Eq. (5.107) does not allow

Table 5.1 *State-of-the-art lasers.*

Name	*Active medium*	*Wavelength* (m)	*Typical angular divergence (radians)*	*Typical power range*
Ruby	Crystal (Cr^{3+} in Al_2O_3)	0.6943		2.5–25 kW (pulsed)
Neodymium: Yag	Crystal (Nd^{3+} in $Y_3Al_5O_{12}$)	1.0641	0.5–2 10 ^{3}	1–100 W (cw)
Helium-Neon	Gas mixture	0.6328 (1.15, 3.39)	10 ^{3}	0.5–2.5 mW (cw)
Argon Ion	Ionized gas (Ar)	0.35–0.52 (typ. 0.488)	$<$ 10 ^{3}	0.2–20 W
Carbon Dioxide	Gas mixture (CO_2)	10.4	1.5–2.5 10 ^{3}	1 W–1.5 kW (cw)
Gallium Arsenide	Semiconductor (GaAs)	0.85	0.3 (in the juction plane) 1 (normal to the juction)	$<$ 100 mW (cw)
All- ber	Er-doped glass ber	1.55	10 ^{1}	$<$ 100 mW (cw)

separation of variables. In fact, the two terms on the left-hand side contain q, which is a function of z. To reach our goal, rather than writing Eq. (5.106) we have to let:

$$= g_1(t_1)\ g_2(t_2)\ \exp\left\{\ \ j\left[P + \frac{k}{2q}r^2 + \ \ (z)\right]\right\} \quad , \tag{5.108}$$

where:

$$t_1 = \sqrt{2}\,\frac{x}{w} \quad , \qquad t_2 = \sqrt{2}\,\frac{y}{w} \quad , \tag{5.109}$$

are transverse coordinates *normalized* to the width (5.96) of the Gaussian function, which depends on z. If we substitute Eq. (5.108) into Eq. (5.86), take correctly into account all the terms like $(dg_i/dt_i)\,(dt_i/dz)$, and make use of the following relationship, which comes from Eqs. (5.96) and (5.97):

$$2\,\frac{w^2}{R} = \frac{d(w^2)}{dz} \quad , \tag{5.110}$$

then we obtain:

$$\left(\frac{g_1''}{g_1} \quad 2t_1\,\frac{g_1'}{g_1}\right) + \left(\frac{g_2''}{g_2} \quad 2t_2\,\frac{g_2'}{g_2}\right) = kw^2\ ' \quad , \tag{5.111}$$

where now primes denote derivatives with respect to t_i on the left-hand side,

Table 5.2 *Hermite polynomials.*

De nition
$H_m(t) = m! \sum_{n=0}^{[m/2]} \frac{(\;\;1)^n\,(2t)^{m\;\;2n}}{n!\,(m\;\;\;2n)!} = (\;\;1)^m\, e^{t^2}\, \frac{d^m}{dt^m}\, e^{\;\;t^2}$
where $[m/2]$ is the integer part of $m/2$.
Examples
$H_0(t) = 1$
$H_1(t) = 2t$
$H_2(t) = 4t^2 \quad 2$
$H_3(t) = 8t^3 \quad 12t$

and z on the right-hand side. The three independent variables are now t_1, t_2, z, and we can separate them. The equations for g_1 and g_2 are identical, namely:

$$\frac{d^2 g_i}{dt_i^2} \quad 2t_i\,\frac{dg_i}{dt_i} + 2m_i g_i = 0 \qquad (i = 1, 2) \quad . \tag{5.112}$$

When the separation constant m_i equals a *non-negative integer*, then the previous equation has a polynomial solution, known as the *Hermite polynomial* of degree m_i. De nition and examples of Hermite polynomials are shown in Table 5.2. When m_i is not a non-negative integer, then Eq. (5.112) still has power-series solutions, but they diverge at in nity faster than $\exp(t_i^2)$. Hence, when multiplied by the Gaussian function, they would yield elds which are not modulus-square integrable. So they must be disregarded, as non-physical solutions for our problem.

The corresponding equation for (z) reads:

$$' = \quad 2(m_1 + m_2)\,\frac{1}{kw^2} \quad . \tag{5.113}$$

Inserting Eq. (5.96) into it, its general integral is found to be:

$$= \quad (m_1 + m_2)\,\arctan\left(\frac{z}{w_0^2}\right) + \text{constant} \quad . \tag{5.114}$$

Therefore, Eq. (5.86) has the following discrete set of solutions:

$$\begin{aligned} E \;=\; & A\,\frac{w_0}{w}\, H_{m_1}\!\left(\sqrt{2}\,\frac{x}{w}\right) H_{m_2}\!\left(\sqrt{2}\,\frac{y}{w}\right) \exp\{\;\; r^2/w^2\} \\ & \exp\left\{\;\; j\left[kz + \frac{r^2}{R} \quad (m_1 + m_2 + 1)\,\arctan\left(\frac{z}{w_0^2}\right)\right]\right\} . \end{aligned} \tag{5.115}$$

One of their most interesting properties is that they all share the same behavior as functions of z, for what concerns the *beam diameter*, $2w(z)$, and the *wavefront radius of curvature*, $R(z)$. This recon rms that eventually it has been sensible to devote all the previous section to one particular solution, which belongs of course to this set and corresponds to $m_1 = m_2 = 0$ in the equations of this section. The solutions in Eq. (5.115) are called *Hermite-Gauss modes.*

An interesting property that can be proven for the Hermite polynomial of any degree m is that its m zeros are *all real.* Therefore, the orders m_1 and m_2 in Eq. (5.115) have a simple physical meaning: they are the numbers of zeroes of the function E, Eq. (5.84), along the x and y axes, respectively. They are called the *transverse wavenumbers* of the Hermite-Gauss modes, which in turn (for reasons similar to those outlined in Section 5.5) are usually labelled TEM_{m_1,m_2} modes.

The Hermite-Gauss modes are *linearly independent* solutions of Eq. (5.86). This stems from the following *orthogonality relation* between Hermite polynomials (Abramowitz and Stegun, Eds., 1965):

$$\int_{\infty}^{+\infty} e^{\;\; t^2} H_m(t)\, H_\ell(t)\, dt = 0 \qquad \text{for } \ell \neq m \quad . \tag{5.116}$$

At this stage, let us recall that a Gaussian beam is an *approximate* solution of the Helmholtz equation that must satisfy the condition in Eq. (5.85). Indeed, not all the solutions of the paraxial equation (5.86) that we have found now, satisfy the restriction in Eq. (5.85). As m_1 and m_2 grow, the z-derivative of the exponent in Eq. (5.115) grows inde nitely as well. It is left to the reader as an exercise to show that, taking into account Eqs. (5.113) and (5.114) (and assuming, for simplicity, $z \qquad w_0^2$), the inequality in Eq. (5.85) is satis ed only by those modes for which:

$$m_1 + m_2 \quad 4 \;\; {}^2 \frac{w_0^2}{\;\;{}^2} \quad . \tag{5.117}$$

Fortunately, in all cases of practical interest (like laser beams, open resonators — see Chapter 9 — and aperture antennas), most of the eld power is in the lowest-order Hermite-Gauss modes. Thus, an expansion of the total eld as a sum of Hermite-Gauss modes does not contradict the initial assumption of slow variations in vs. z.

For conceptual reasons, to be discussed in the next section, as well as for practical ones, we are driven to nd solutions of Eq. (5.86) in *cylindrical coordinates* (r, φ, z) as well. Passing from a cartesian system to a cylindrical one, the Gaussian function, circularly symmetrical around the z axis, remains unchanged. The other amplitude factors can be found following essentially the same way as in the previous case. The nal results are:

$$\begin{aligned} E \;\; = \;\; & A \, \frac{w_0}{w(z)} \, h_\ell\left(\frac{r}{w}\right) \, \exp\{\;\; r^2/w^2\} \\ & \exp\left\{\;\; j\left[kz \quad \ell\varphi + \frac{\;\; r^2}{\;\; R} \quad (2p + \ell + 1)\arctan\left(\frac{\;\; z}{\;\; w_0^2}\right)\right]\right\} , \end{aligned} \tag{5.118}$$

Table 5.3 *Associated Laguerre polynomials.*

De nitions

$$L_p^\ell(t) = \sum_{m=0}^{p} \frac{(\ 1)^m}{m!} \binom{p+\ell}{p\ \ m} t^m = (\ \ 1)^\ell \frac{d^\ell}{dt^\ell} L_{p+\ell}(t)$$

where

$$L_s(t) = \sum_{m=0}^{s} \frac{(\ 1)^m}{m!} \binom{s}{m} t^m = e^t \frac{d^s}{dt^s} (e^{\ \ t} t^s)$$

is the ordinary Laguerre polynomial of degree s.

Examples

$$L_0^\ell(t) = 1$$
$$L_1^\ell(t) = \ell + 1 \quad t$$
$$L_2^\ell(t) = \frac{1}{2} (\ell+1)(\ell+2) \quad (\ell+2)\, t + \frac{1}{2} t^2 .$$

where ℓ and p are two non-negative integers. We set:

$$h_\ell = \left(\sqrt{2}\,\frac{r}{w}\right)^\ell L_p^\ell\left(\frac{2r^2}{w^2}\right) \quad , \tag{5.119}$$

where $L_p^\ell(t)$ is the so-called *associated Laguerre polynomial* of orders ℓ, p (see Table 5.3), which are polynomial solutions of the following equation:

$$t\,\frac{d^2L}{dt^2} + (\ell + 1 \quad t)\,\frac{dL}{dt} + pL = 0 \quad . \tag{5.120}$$

From Eq. (5.118) we see that, when φ varies from 0 to 2 , then E, Eq. (5.84), repeats periodically ℓ times. For this reason ℓ is called the *azimuthal wavenumber*, and must be *an integer*, otherwise we would get a multivalued function, which is physically unacceptable. From general properties of Laguerre polynomials (see Abramowitz and Stegun, Eds., 1965), it follows that p equals the number of zeroes of the polynomial L_p^ℓ, which are all real and positive. For this reason p is called the *radial wavenumber.*

Linear independence among the *Laguerre-Gauss modes* (also labeled $\mathrm{TEM}_{\ell p}$) is guaranteed by the following *orthogonality relation* between generalized Laguerre polynomials of equal azimuthal order:

$$\int_0^\infty e^{\ \ t}\, t^\ell\, L_p^\ell(t)\, L_q^\ell(t)\, dt = 0 \qquad \text{for } p \neq q \quad . \tag{5.121}$$

Well known orthogonality relations apply among the functions $e^{\ \ j\ell\varphi}$ over the interval $(0, 2\ \ )$, and so entail modes of di erent azimuthal order to be always orthogonal.

Figure 5.11 shows some very famous experimental results of intensities (proportional to $|E|^2$) on a z = constant plane, for some Hermite-Gauss modes. Figures 5.12 and 5.13 show the behavior of a few Hermite and Laguerre polynomials. They also indicate that there is no one-to-one matching between these

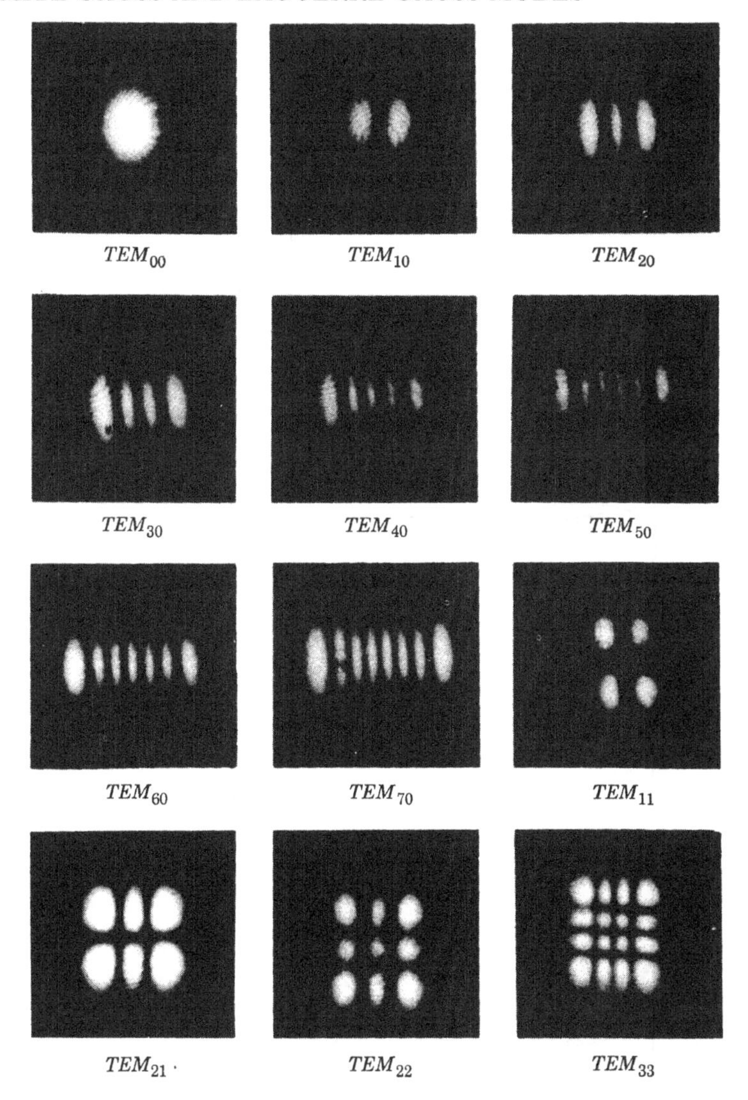

Figure 5.11 *Very famous early experimental results: a laser oscillating in various Hermite-Gauss modes (from Boyd, G.D. and Gordon, J.P. (1961) Bell System Technical Journal,* **40**, *489–508).*

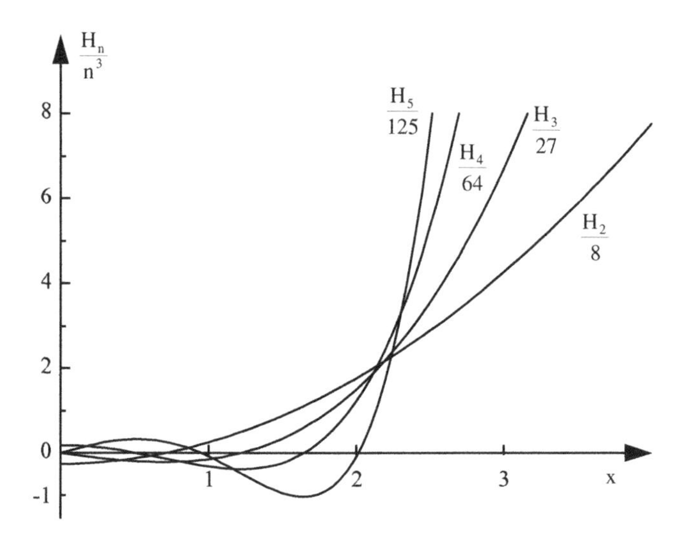

Figure 5.12 *Examples of Hermite polynomials.*

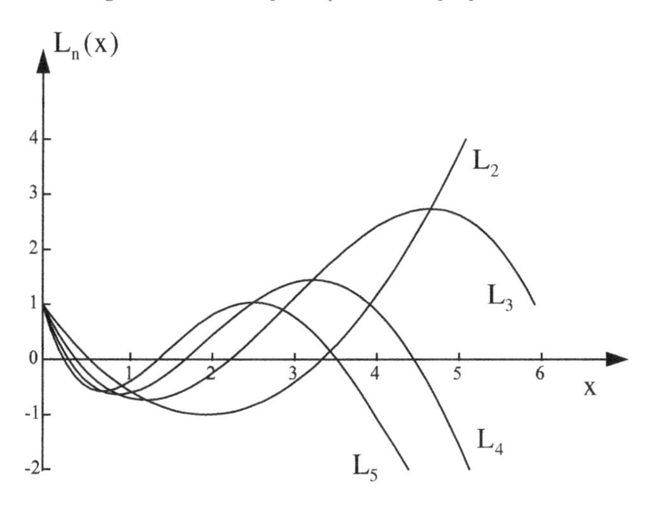

Figure 5.13 *Examples of Laguerre polynomials.*

two families of polynomials. Their only relationship is that any Hermite-Gauss mode can be expanded as a series of Laguerre-Gauss modes, and vice-versa.

5.11 Re ection and refraction of Gaussian beams

At the beginning of Section 5.9, we introduced an important assumption, namely, that the medium in all the region of 3-D space where a Gaussian

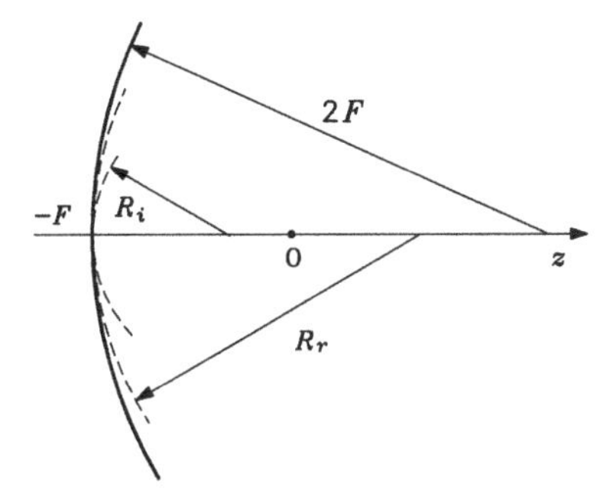

Figure 5.14 *Re ection of a wavefront on a spherical mirror.*

beam is de ned, be homogeneous. It occurs often that one wants to know how a Gaussian beam behaves at a sharp discontinuity in the medium. This is the case when one analyzes the re ection of the beam on a mirror, for example in an optical resonator (see Chapter 9), or the transmission of a beam through a lens.

To be rigorous, one could expand the incident beam into plane waves, and apply the results of Chapter 4 to each individual wave. This procedure is quite cumbersome. Frequently, the problem can be simpli ed through approximations based on ray optics. We shall discuss them now.

Let us start from a perfectly re ecting spherical mirror (Figure 5.14) of focal length F, meaning that its curvature radius equals $2F$. We take the focus of the mirror as the origin of coordinates. Let us deal explicitly only with the TEM_{00} mode. The results can be easily extended to higher-order modes, as well as to a thin lens.

Suppose (this is the source of approximation) that the distance R_i between the waist of the incident Gaussian beam and the point where the mirror intersects the z axis is so large, that we may approximate the eld impinging on the mirror with a non-uniform (i.e., amplitude not constant on a constant-phase surface) spherical wave, of radius R_i. In this case, the elementary laws of ray optics say that the re ected eld is also a spherical wave, whose curvature radius is related to R_i and to F as:

$$\frac{1}{R_r} = \quad \frac{1}{R_i} + \frac{1}{F} \quad . \tag{5.122}$$

The re ected spherical wave is also non-uniform. We assumed a perfect re ector, so the re ected amplitude distribution coincides with that of the incident wave, whence is still Gaussian and its width equals that of the incident wave: $w_r(z = \quad F) = w_i(z = \quad F)$. If R_r, given by Eq. (5.122), is also large enough, so that the re ected wave is in turn a good approximation to a Gaussian beam, then the complex parameter $q_r(z)$ of the re ected beam de ned by Eq. (5.93):

$$\frac{1}{q_r} = \frac{1}{R_r} \quad 2j\,\frac{1}{kw^2} \quad , \tag{5.123}$$

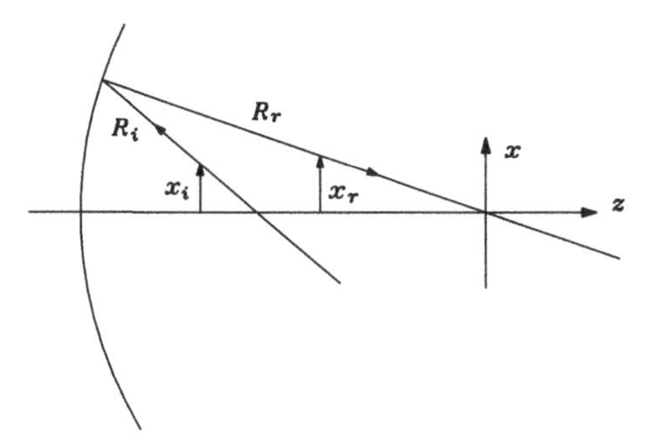

Figure 5.15 *Re ection on a spherical mirror: the ray picture.*

is related *at the mirror* ($z = \;\; F$) to that of the incident beam, q_i, by the following equation:

$$\frac{1}{q_r(\;\; F)} = \frac{1}{q_i(\;\; F)} + \frac{1}{F} \quad . \tag{5.124}$$

The di erence in sign between Eq. (5.122) and Eq. (5.124) is simply a consequence of Eq. (5.97), which gives $R < 0$ for $z < 0$.

Once the value (5.124) of the complex parameter $q_r(\;\; F)$ is known, then q can be calculated at any plane $z =$ constant by means of Eq. (5.91), which can be rewritten as follows:

$$q(z) = q(\;\; F) + (z + F) \quad . \tag{5.125}$$

It is of interest to see whether we can write how q_r at the generic plane $z = z_2$ is related to q_i at another generic plane, $z = z_1$, skipping Eqs. (5.124) and (5.125). Looking at Figure 5.15, we see that $R = (x/x')$, where $x' = dx/dz$ (for paraxial rays). While traveling in the region between mirrors (or lenses), the transverse coordinate $x(z)$ of the generic ray varies linearly vs. z, in proportion to Eq. (5.125), while $x' = dx/dz$ remains constant. At each re ection, $x(z)$ remains constant while x/x' follows Eq. (5.124). Hence, during the whole path from $z = z_1$ to $z = z_2$, the parameter q undergoes exactly the same transformations as the ratio x/x'. Now, it is well known that a spherical mirror maps, in a "one-to-one and onto" way, a family of rays leaving a source point, into another family of rays, passing through an image point. This operation performed by the mirror is fully characterized, in mathematical terms, by a $2 \;\; 2$ matrix:

$$\begin{pmatrix} x_r \\ x'_r \end{pmatrix} = \begin{pmatrix} A & B \\ C & D \end{pmatrix} \begin{pmatrix} x_i \\ x'_i \end{pmatrix} \quad . \tag{5.126}$$

From what we said, it follows that Gaussian beam re ections obey the following relationship:

$$q_r(z = z_2) = \frac{A\, q_i(z = z_1) + B}{C\, q_i(z = z_1) + D} \quad , \tag{5.127}$$

where the coefficients A, B, C and D can be determined by elementary ray optics. Eq. (5.127) is usually referred to as the *ABCD law.*

Also a thin lens maps a family of straight lines into another family of straight lines. So, refraction of Gaussian beams at thin lenses can also be described by Eq. (5.127). When using Eq. (5.124), to distinguish between a lens and a mirror it is important to pay attention to the sign of the focal length F.

Many contents of this section will be used again in Section 9.10, which deals with stability of open resonators.

5.12 On the completeness of a series

In Section 4.7 we discussed completeness of plane waves and we showed that, in general, a monochromatic field, which carries a finite power through a plane, can be expressed as a plane wave integral. In other words, the basis for expanding such a field is a *continuous infinity* of elements. On the other hand, in Section 5.10 we mentioned on several occasions that we were dealing with a *countable set* of functions, and yet we stated that it was a complete set. This means that the most general field we can have under these circumstances is not an integral, but a *series.* There is an apparent decrease in generality, so we must discuss, at least qualitatively, the conditions the field must satisfy in order to allow such an expansion.

Completeness problems, related to a partial differential equation, are well defined only when boundary conditions are specified. In general, the weaker the boundary conditions, the wider the set, in order to be complete. So, the proof given in Section 4.7 referred to rather weak boundary conditions, as the only requirement was that the field be modulus-square integrable. In such a case, a set could be complete only if it was a continuum. One may ask an interesting question: how stringent must the boundary conditions become, in order to reach completeness with a countable set of basis elements?

If we recall how a Fourier integral is related to a Fourier series, it is easy to conclude that a *sufficient* condition for the spatial spectrum of a field to become discrete, is that the field should satisfy *periodic* boundary conditions; or, equivalently, that the field be non-zero only over a *finite spatial interval,* I. In the latter case, as well known from signal analysis, the field can be repeated periodically, all the way to infinity.[§]

Sometimes, the series convergence could be nonuniform at the extremes of the interval I. When this happens, it is not possible to differentiate the series term by term. In this framework, we will speak of completeness of a set only

[§] Maybe the terminology used in this section biases the reader towards rectangular coordinates. However, the same concepts, with minor changes, apply in other types of reference frames. For example, in cylindrical coordinates, we shall consider expansions in series of trigonometric functions of the azimuthal coordinate, and Bessel functions of the radial coordinate.

in the sense that:

$$\lim_{n\to\infty} \int_I \left| \quad \sum_{m=1}^{n} c_m f_m \right|^2 dI = 0 \quad , \tag{5.128}$$

where can be any solution of the partial di erential equation, $\{f_m\}$ is the complete set, the $\{c_m\}$'s are the expansion coe cients, and I, as we said, is the interval where the eld is de ned.

All the subjects of Chapters 7 and 9 (waveguides with conducting walls and resonant cavities) will enable us to use periodic extension of elds de ned over nite intervals. It is easy, in those cases, to expect complete sets to be discrete; but Gaussian beams do not belong to the same class. We infer from this that periodicity is a su cient condition for passing from a continuum to a discrete basis, but not a necessary one. There must be another explanation which applies to Gaussian beams.

The answer can be identi ed as a boundary condition which, although set at in nity, is much more restrictive than simply to require that the eld be square-integrable. To simplify the proof of this statement, let us use a one-dimensional notation, which applies to both Laguerre-Gauss and Hermite-Gauss modes. Assume that F is a function that can be expanded as a series of Gaussian beams:

$$F(u) = \sum_{i=0}^{\infty} P_i(u)\, e^{\quad u^2} = e^{\quad u^2} \sum_{i=0}^{\infty} P_i(u) \quad , \tag{5.129}$$

where $P_i(u)$ is the i-th polynomial of the suitable type. From the last equation it follows that $F(u)$ can be written as the product of a Gaussian function times *a power series* in u. $F(u)$ is de ned in a region which extends to in nity, and Eq. (5.129) must hold in all that region. Then, as u grows, $F(u)$ has to decrease fast enough, so that the Taylor series for $F(u)\, e^{u^2}$ *converges all the way to in nity.* In other words, it must be:

$$\lim_{n\to\infty} \left| F(u)\, e^{u^2} \quad \sum_{i=1}^{n} P_i(u) \right| = 0 \quad , \tag{5.130}$$

for *any* $u \in (0, +\infty)$. As we said above, this constraint is stronger than square-integrability.

In practical problems where Gaussian beams turn out to be convenient (typically, laser beams and aperture antennas), Eq. (5.130) is easily satis ed for very simple physical reasons. For example, in an open resonator (Chapter 9), i.e., in a system consisting of two mirrors facing each other, the nite size of the mirrors in the transverse directions ensures that the eld is con ned, in practice, within a nite region. For u greater than a certain nite value, it is $F(u) \simeq 0$. Similarly, for a laser beam, or for the eld emitted by an aperture antenna, physical intuition leads to the same conclusion.

With these limitations and margins of caution, one can speak of completeness of Gaussian beams in all cases where we need to use them.

5.13 Further comments on rays and beams

There are many links between geometrical optics and Gaussian beams. The rst one is in the terminology. Indeed, at the beginning of Section 5.9 we said that a tentative solution of the type $(x, y, z)\exp(\ jkz)$, where is a slowly varying function of z, could be called a beam of rays. In contrast to a single ray, however, such a beam satis es a basic requirement, the "minimum uncertainty relationship," $x \ k_x \ \ 2$, between its location in real space and its location in the spatial frequency space. As the reader may know from previous studies, and which will be recon rmed in Chapter 13, this relationship is also fundamental for di raction. Geometrical optics misses completely this point because it deals with each ray as an entity that is totally independent of what surrounds it. It comes as a consequence that energy ows in ux tubes, according to a uid-dynamics pattern. A critical analysis would show that WKBJ results are also a ected by the same limitations since they neglect di raction. This shows that Gaussian beams are more accurate solutions of Maxwell's equations than WKBJ solutions.

Consequently, while geometrical optics and the WKBJ method fail near foci and caustics, where ray densities grow inde nitely and ux-tube cross-sections collapse, Gaussian beams do not have critical points, and are acceptable — although approximate — solutions in the whole 3-D space. It takes a more accurate analysis (e.g., Marcatili and Someda, 1987) to point out some inconsistencies in tightly focused Gaussian beams, in the neighborhood of the waist.

As for geometrical optics and the WKBJ method, we wish to point out that suitable "correction procedures" have been developed, which allow one to overcome part of their drawbacks. The results are interesting, but too specialized to be discussed here in detail. Let us just say that there are rules that enable one to track rays *beyond a focus*. What happens *near the focus* can never be described in terms of rays. Similarly, there are connection formulas which link to each other two WKBJ solutions de ned on each side of the return point, or caustic. They go under the generic name "enlarged WKBJ method," and yield, among other things, the result on phase shift that was mentioned in Section 5.8. However, these connection formulas are still rather poor approximations. In particular, they fail when there two or more caustics are close to each other.

However, it would be naive to think that beams are superior to rays, as solutions of Maxwell's equations, only because a larger number of powers of are kept in the approximation. If this were the case, then one might improve very signi cantly on geometrical optics just by keeping more powers of . So, there must be other reasons.

In fact, the true cause is that, in geometrical optics, *the amplitude and phase functions are assumed to be real.* Following this assumption, we divided the Helmholtz equation into two equations, where no term of order $k_0^{\ 1}$ appears. At this stage, we have diminished the initial level of accuracy. Removing the assumption of real amplitude and phase functions, one nds Gaussian beams. In this sense, beams can be looked at as *complex rays*. Physically, what makes the di erence are *evanescent waves*, which are left out of geometrical optics on one side, but on the other hand play a crucial, although partly hidden, role in Gaussian beams, near the beam waist, where the Poynting vector changes gradually its direction.

In the literature, the reader can nd very important examples (e.g., Felsen and Marcuvitz, 1973) that show how other methods can be pushed to any level of accuracy in terms of powers of $1/k_0$, at the price of a rapid growth in computation complexity.

5.14 Further applications and suggested reading

The logical thread of this chapter can be summarized as follows. A monochromatic plane wave is a simple, exact solution (in an unbounded homogeneous medium) of Maxwell's equations, but it can not be taken literally, because it is unphysical under several viewpoints. In order to nd similar but physically acceptable solutions, one has to introduce approximations. Di erent approximations are required under di erent circumstances.

It is rather unusual for a textbook to present several approximations side by side in one chapter. Our purpose was to help the reader to make comparison, stressing the relationships that exist often among di erent approximations. It happens much more frequently to nd books which deal in depth with just one, or with a few, of these approximate solutions. Therefore, it is quite easy, for the interested reader, to proceed, on each individual subject, well beyond the introductory level of this chapter.

Readers who have been exposed to fundamentals of telecommunications know well that the contents of the rst two sections (group velocity and dispersion) are extremely important in that context. What happens to signals which travel through non-ideal transmission media must be studied in depth by telecommunication engineers, regardless of the physical mechanism which is responsible for non-ideal propagation. Just to quote one among many classical reference books in this area, in Sunde (1969) the subject of dispersion-related signal distortion is covered in detail, for both analog and digital signals, and well linked to the other crucial issues in transmission and detection.

In Section 5.2 we brie y mentioned that pulses may be compressed in the time domain by making a proper choice of the dispersion sign in the medium where they propagate. This property has been exploited for decades at microwave frequencies, where shortening radar pulses is intended to increase spatial resolution (Skolnik, 1980; Baden Fuller, 1990). In the last decades, it has been transferred, very successfully, to the optical frequency range (Rudolph and Wilhelmi, 1989). Grating-assisted compression of laser pulses has been the way to generate the shortest man-made events that are known today, i.e., optical pulses in the range of a few femtoseconds (1 fs = 10 ^{15} s). In optical telecommunications, ingenious techniques have been demonstrated for pulse compression such as reversing signal spectrum midway between the transmitter and the receiver with a chirped grating (Hill *et al.*, 1994; Kashyap *et al.*, 1996) or reversing the sign of the dispersion using a DISpersion COmpensating ("DISCO") optical ber.

Section 5.3 was mainly intended to set the stage for the following ones, but it could also be looked at as a hint for those who are interested in understanding more in depth the relation between scalar waves (encountered in many other branches of Physics) and vector waves. Some chapters in the classical treatise by Morse and Feshbach (1953) are quite relevant in this respect. They should be at least sampled by our readers. We also refer to Black and Ankiewicz (1985) for a discussion of analogies between optics and other areas of physics.

The following three sections formed a package devoted to a subject — geometrical optics — which has many applications, in basic physics, in optical engineering, in communication engineering, etc. Chapter III of Born and Wolf (1980) is an in-depth view of basic theoretical arguments, those in favor of this approximation as well as those that show its limits. Chapters IV, V and VI in the same book describe exhaustively applications to imaging systems, and to the theory of aberrations (Buchdahl, 1968/1970), a subject which is also of potential interest to engineers involved in aperture antenna design.

Readers who want to become more familiar with recent developments in imaging systems should invest some of their time studying graded-index (GRIN) components, and their computer-aided design, based on numerical solutions of the ray equation. They may start from the basic concepts outlined in Part I of Snyder and Love (1983), and then proceed to consult more specialized books, like Marchand (1978), for detailed, sophisticated theories, or Iga *et al.* (1984) for descriptions of GRIN engineering applications.

We also said that geometrical optics has been for a long time, and still is, an important background item also for communication engineers. Radio wave propagation in the slowly-varying (and also random-varying) troposphere is instrumental in designing radio-relay and radar systems. Just to mention a few basic references on this subject, Chapters 33 and 34 of the handbook edited by Jordan (1986) focus on radio transmission problems. On the other hand, Sunde (1969) deals mainly with consequences on transmitted signals.

Passing from the troposphere to the ionosphere, in addition to being slowly varying this medium is anisotropic as well. Should one wait for Chapter 6 ? In part, the answer is yes, but we can say now that even in the ionosphere ray tracing remains important for radio engineers. Highly recommended reading is Budden (1961).

Ray tracing application to telecom problems enjoyed a second youth in the 1970's, when optical bers became practical, and most of the interest was concentrated, for about one decade, on multimode bers. Snyder and Love (1983) is one of the most complete — and most popular — among the books which distilled that knowledge. Even at times, like the present ones, when single-mode bers are much more widely used than multimode ones for long-haul systems, it is still mandatory for a telecommunication engineer to be exposed to the geometrical-optics treatment of graded-index multimode bers. This subject will be dealt with, to some extent, in Chapter 10.

Not surprisingly, ionospheric propagation and graded-index optical guides are also among the main examples of application of the WKBJ method to e.m. wave propagation. References given in the previous paragraph, including our own Chapter 10, are still valid for the contents of Sections 5.7 and 5.8. We should also point out that the WKBJ method is applied, very widely, in other branches of pure and applied physics, ranging from underwater propagation (Frisk, 1994) to solid-state electronics (e.g., Kittel, 1996).

Gaussian beams, the subject of Sections 5.9 through 5.12, were invented in the early '60s, to model laser beams and resonators accurately enough to account for their basic di raction limits. It is strongly recommended to come back to this subject after reading Chapter 13. Gaussian beams were successfully used, later on, also for designing aperture antennas (see Chapter 12), especially those with multiple re ectors, each of which is in the near eld of at least another one. Still, their main applications remain in the laser area. Classical textbooks on optical oscillators (see, e.g., Svelto, 1989) devote several chapters to Gaussian beams. All the students who include Quantum Electronics in their curricula must master this subject, and be able to solve problems on Gaussian beams.

A special problem, obviously related to the previous ones, is propagation of Gaussian beams in lens-like media, i.e., in media where a slight radial gradient in refractive index may counteract, at least in part, beam divergence caused by di raction. Gaussian beams in media with gain or loss are also required in some very practical problems. Straightforward introductions to these subjects can be found in Yariv (1991) and in Adams (1981). Other important practical problems that have been modeled successfully in terms of Gaussian beams are those related to source-to- ber and ber-to- ber coupling in optical communication systems (Senior, 1992; Tosco, Ed., 1990). In particular, in the case of multimode bers the latter problem was an interesting example where the application of the whole family of Hermite-Gauss modes (Rizzoli and Someda, 1981) yields fundamental improvements with respect to ray-based approaches.

Finally, the content of Section 5.13 should be taken just as a hint. It suggests that the links between Gaussian beams and geometrical optics are deep and fundamental. Clarifying them was a task that was successfully accomplished by theoreticians who investigated e.m. elds in very abstract terms, without direct interest in designing devices. A very suitable reference, in order to appreciate this, although very demanding reading, is the previously quoted book by Felsen and Marcuvitz (1973).

Problems

5-1 Evaluate the group velocity, v_g, and the dispersion, k'', as a function of the angular frequency ω, in a medium where the phase constant for a plane wave (see Section 6.9) is de ned as $_+ = (\omega/c)\sqrt{1 + \omega_M/(\omega_0 + \omega)}$.

5-2 Repeat the previous exercise with the following expression for the phase constant (see again Section 6.9): $= (\omega/c)\sqrt{1 + \omega_M/(\omega_0 \quad \omega)}$. Note that in this case there is a frequency band where the phase constant is imaginary, i.e., the plane wave is evanescent. The group velocity is not de ned over that band.

5-3 Evaluate the group velocity, v_y, and the dispersion, k'', as a function of frequency, in a dielectric medium, near an atomic or molecular resonant absorption peak at $\omega = \omega_0$, where the permittivity exhibits a "reactive tail" of the Lorentzian absorption line, i.e.:

$$= {}_0 + \frac{2 \quad \omega}{\omega_0^2 + 4Q^2 \quad \omega^2},$$

where $\omega = \omega \quad \omega_0$, and , ω_0 and Q are real and positive, while $= {}_0$. Find the frequency intervals where dispersion is normal and those where it is anomalous.
(*Hint.* $d/d\omega = d/(d \quad \omega)$.)

5-4 A wave is characterized by a phase constant $= (\omega/c)\,[1 \quad (\omega_c/\omega)^2]^{1/2}$ (de ned for $\omega > \omega_c > 0$, with $c =$ constant).
Find its group velocity vs. ω. Show that $v_f v_g = c^2$. Show that $'' \to 0$ for $\omega \to \infty$. Discuss the physical interpretation of these results.

5-5 A wave is characterized by a phase constant $= \omega\sqrt{\quad}\,[1 \quad (\omega_c/\omega)^2]^{1/2}$, but now, in contrast to the previous problem, depends on frequency as in Problem 5-3. Assume $\omega_0 \quad \omega_c$.
Show that the contributions to dispersion due to $d \;/d\omega \neq 0$, and those due to the factor $[1 \quad (\omega_c/\omega)^2]^{1/2}$, have equal signs in a frequency range close to ω_0, where both yield anomalous dispersion; opposite signs far from ω_0, where the "material dispersion" (Problem 5-3) is normal while the "waveguide dispersion" (Problem 5-4) remains anomalous.

5-6 A transmission medium is characterized by a dispersion coe cient $|k''| = 0.05\,\mathrm{ns}^2/\mathrm{km}$, and is 20 km in length. What is the required initial half-width of a Gaussian pulse, for its nal width to be 3 ns? Why are there two solutions to this problem? Which one is more convenient in practice?

5-7 A simple model for a " at" atmosphere is a dielectric medium ($= {}_0$) where the permittivity varies in the vertical direction according to $= {}_1(1 \quad az)$, where ${}_1 = 1.05\;{}_0$, $a = 10 \;\, ^5\,\mathrm{m} \;\, ^1$.
Estimate the minimum frequency at which this can be looked at as a slowly varying medium. Discuss whether polarization (horizontal or vertical electric eld) has an e ect on the validity of the slowly-varying approximation.

5-8 Explain why Eq. (5.64) enables us to assume that the two periods, in two di erent meridional planes over which a meridional ray can be projected, are equal.

5-9 Show that, in a spherically symmetrical medium, i.e., for a refractive index $n = n(r)$, the generalized Snell's law reads:

$$\frac{d}{du}\left(n(r)\, r\, \sin\varphi\right) = 0 \quad ,$$

where φ is the angle between the tangent to the ray and the position vector $\mathbf{r}$, which, due to the spherical symmetry, is parallel to ∇n.
Show that any ray propagating through such a medium is meridional, i.e., belongs entirely to a plane through the origin.
(*Hint.* Show rst that $d(\mathbf{r} \quad n\hat{u})/du = 0$ and make use of the ray equation.)

5-10 The *Luneburg lens* is a sphere where the refractive index varies as $n(r) = [2 \quad (r/a)^2]^{1/2}$ for $r \quad a$, while $n = 1$ outside the sphere. Using the result of the previous problem, show that a beam of parallel rays, incident on this lens, is focused "on the back" of the sphere, i.e., on the point where the surface $r = a$ intercepts the diameter parallel to the beam.

5-11 The so-called *sh-eye lens*, rst investigated by Maxwell, is a spherically symmetrical medium where the refractive index varies, for any r, as $n(r) = n_0[1 + (r/a)^2]^{\quad 1}$ (n_0 is a real constant).

a) Show that all the rays emitted by a point source located on the $r = a$ spherical surface pass through the symmetrical point with respect to the center of the lens.

b) Show that all the rays emitted by a source located at a point where $r > a$ pass through a point, located on the same diameter, on the opposite side with respect to the origin, and having a radial coordinate $r' = a^2/r < a$ (and vice-versa).

5-12 The *numerical aperture* of an optical system, exhibiting rotational symmetry around an "optic axis" z, is de ned as $NA = \sin\vartheta_M$, where ϑ_M is the maximum angle (measured in air) which a ray propagating through the system can form with the z axis.

a) Show that if the refractive index varies along z, either in steps or gradually, then $NA = n(z)\,\sin\vartheta_M(z)$.

b) For a rotationally symmetrical medium where n decreases parabolically vs. r (see Eq. (5.62)), show that the numerical aperture (de ned with reference to meridional rays) is maximum on the axis and decreases gradually for growing r.

5-13 Consider a parabolic graded-index slab, i.e., a medium where the refractive index is independent of one of the transverse coordinates, y, and varies with the other one according to $n^2 = n_0^2\,[1 \quad 2 \quad (x/a)^2]$ for $|x| < a$, and then remains constant for $x > a$. Rays may reach the borderline $r = a$, but are not supposed to penetrate further.

a) Discuss the relationship between the numerical aperture (de ned as in the previous problem, with reference to rays in the (x, z) plane) and the relative index di erence, (which is always supposed to be much less than unity).

b) For $n_0 = 1.5$, $NA = 0.1$, and a maximum index gradient 10^4 m ^{1}, nd the corresponding slab width, $2a$.

5-14 Consider again the parabolic graded-index slab of the previous problem. Suppose that the scalar wave equation, Eq. (5.67), can be approached by separation of the variables x and z, and furthermore, that $\partial^2/\partial z^2 = k_z^2 =$ constant.
Write explicitly the restriction which plays the role of Eq. (5.69), in order to use the WKBJ method, for the equation which governs the x-dependence. Discuss in detail whether it is satis ed in the region where $|x| \simeq a$.

5-15 A lossless slab whose refractive index $n(z)$ varies slowly along z (normal to the slab faces), and is independent of x and y, is sandwiched between two semi-in nite lossless media of refractive indices n_1 and n_2, at $z = 0$ and $z = L$, respectively.
Show that the WKBJ method yields the following expression for the re ection coe cient at $z = 0$:

$$= \frac{n_2\,n(0) \quad n_1\,n(L)}{n_2\,n(0) + n_1\,n(L)} \quad .$$

5-16 To use a graded-index layer as an antire ection coating, one sets $n(L) = n_2$. Find how the expression of the previous problem changes in this case. Compare this device to the quarter-wavelength antire ection coating discussed in Chapter 4, in terms of practical parameters, such as thickness, fabrication tolerances, bandwidth.

5-17 Taking into account Eqs. (5.113) and (5.114) — and assuming, for simplicity, $|z| \qquad w_0^2$ — show that only those TEM modes which satisfy the restriction (5.117) do not violate the inequality (5.85).

5-18 Draw a comparison between $|f(z)|^2$ for the rst-order WKBJ solution, and $| \quad (z)|^2$, where is a $TEM_{0,0}$ Gaussian beam expressed by Eq. (5.87). Discuss the relationship between these results and the conservation law of geometrical optics, Eq. (5.48).

5-19 The emitting surface of a heterojunction semiconductor laser is, roughly, a rectangular strip. Its width in the junction plane is $x = 100$ m, and its thickness in the orthogonal direction is $y = 10$ m. The operating wavelength is 1.3 m. Suppose that the emitted beam can be modeled as a $TEM_{0,0}$

astigmatic (i.e., non-circularly-symmetrical) Gaussian beam. Find the beam angular widths in the (x, z) and (y, z) planes, assuming that the beam waist is located on the output facet of the device. After reading Chapter 13, compare these results with those based on di raction theory.

5-20 In cylindrical coordinates, a CO_2 laser emits a $TEM_{0,2}$ beam at a wavelength $= 10$ m. Far from the waist, the beam full angular divergence is 2 $10^{\ 2}$ rad. The electric eld, measured on the beam axis at $z = 1$ m (distance from the beam waist) equals 100 V/m. Find:

a) the spot size on the plane of the beam waist;

b) the maximum electric eld on the beam waist;

c) the total power of the beam;

d) the coordinates of the points, on the waist plane, where the eld vanishes.

5-21 A Gaussian beam, emitted by a Nd:YAG laser source ($= 1.06$ m) whose waist half-width equals 2 mm and whose angular divergence equals $10^{\ 4}$ radians, has to be focused through a circular aperture, 30 mm in diameter, at a distance of 1.4 m from the waist plane.
Design a suitable lens (i.e., nd its distance from the beam waist and its focal length). As a design criterion, note that about 99% of the power of a Gaussian beam is transmitted through an aperture whose radius is 3 times that of the beam at $1/e$ its maximum value.

5-22 A circular aperture of radius a is illuminated by a uniform eld, $E = E_0 =$ constant (while $E = 0$ for $r > a$). Expand this eld into a set of Laguerre-Gauss modes.
(*Hint.* Use the orthogonality relation Eq. (5.121), and the following identity (Abramowitz and Stegun, 1965, p. 786):

$$\int_x^{\infty} e^{\ t}\, L_p^{\ell}(t)\, dt = e^{\ x}\left[L_p^{\ell(x)} \quad L_{p\ 1}^{\ell(x)}\right] \quad .)$$

5-23 Repeat the previous problem, assuming now that the illuminated surface is annular (i.e., $b < r < a$, $b > 0$) and that the eld on it is $E = E_0\, r^{\ 1} \cos\varphi$ (again $E = 0$ out of this region).

5-24 A successful way to model an o set joint between two multimode optical bers (Rizzoli and Someda, 1981) was to express the eld in each ber as a series of Hermite-Gauss modes. Show that the so-called overlap integral (i.e., the coe cient in the expansion) between the $TEM_{p,q}$ and the $TEM_{m,n}$ Hermite-Gauss modes, of equal widths w, for an o set d between the centers of the two Gaussian functions along the x axis, and perfect alignment along the y axis, is expressed by

$$t \ _{v} = \frac{{}^{n}_{q} \exp\left(\ \ \frac{d^2}{2w^2}\right)}{(\ \ 2^{p+m} p! m!)^{1/2}} \sum_{i=1}^{n} w_i \, H_p \left(x_i + \frac{d}{\sqrt{(2)}w}\right) H_m \left(x_i \qquad \frac{d}{\sqrt{(2)}w}\right) \quad ,$$

where ${}^{n}_{q}$ = Kronecker's symbol, a = core radius, $s = p + q$, $[p, q]$ $\quad v =$ $[s(s + 1)/2] + p + 1$ is the so-called principal mode number, and $[m, n]$ $\quad$. For w_i, x_i see Abramowitz and Stegun (1965, pp. 890–924).

References

Abramowitz, M. and Stegun, I.A. (Eds.) (1965) *Handbook of Mathematical Functions.* Dover, New York.

Adams, M.J. (1981) *An Introduction to Optical Waveguides.* Wiley, Chichester.

Baden Fuller, A.J. (1990) *Microwaves: An Introduction to Microwave Theory and Techniques.* 3rd ed., Pergamon Press, Oxford.

Black, R.J. and Ankiewicz, A. (1985) "Fiber-optic analogies with mechanics". *American Journal of Physics*, **53**, 554–563.

Born, M. and Wolf, E. (1980) *Principles of Optics: Electromagnetic Theory of Propagation, Interference and Di raction of Light.* 6th (corrected) ed., reprinted 1986, Pergamon Press, Oxford.

Buchdahl, H.A. (1968) *Optical Aberration Coe cients.* Dover, New York.

Buchdahl, H.A. (1970) *An Introduction to Hamiltonian Optics.* Cambridge University Press, UK.

Budden, K.G. (1961) *Radio Waves in the Ionosphere.* Cambridge University Press, Cambridge, UK.

Etten, W., van der and Plaats, J., van der (1991) *Fundamentals of Optical Fibers Communications.* Prentice-Hall, New York.

Felsen, L.B. and Marcuvitz, N. (1973) *Radiation and Scattering of Waves.* Prentice-Hall, Englewood Cli s, NJ.

Frisk, G.V. (1994) *Ocean and Seabed Acoustics: A Theory of Wave Propagation.* Prentice-Hall, Englewood Cli s, NJ.

Hill, K.O. *et al.* (1994) "Chirped in- ber Bragg gratings for compensation of optical- ber dispersion". *Optics Letters*, **19** (17), 1314–1316.

Iga, K., Kokubun, Y. and Oikawa, M. (1984) *Fundamentals of Microoptics: Distributed-Index, Microlens, and Stacked Planar Optics.* Academic, Tokyo and Orlando, FL.

Jordan, E.C., (Ed.) (1986) *Reference Data for Engineers: Radio, Electronics, Computer and Communications.* 7th ed., Howard W. Sams & Co., Indianapolis.

Kashyap, R. *et al.* (1996) "1.3 m long super-step-chirped bre Bragg grating with a continuous delay of 13.5 ns and bandwidth 10 nm for broadband dispersion compensation". *Electronics Letters*, **32** (19), 1807–1809.

Kittel, C. (1996) *Introduction to Solid State Physics.* 7th ed., Wiley, New York.

Marcatili, E.A.J. and Someda, C.G. (1987) "Gaussian beam are fundamentally different from free-space modes". *IEEE J. of Quantum Electronics*, **23** (2), 164–167.

Marchand, E.W. (1978) *Gradient Index Optics.* Academic, New York.

Morse, P.M. and Feshbach, H. (1953) *Methods of Theoretical Physics.* Vols. I and II, McGraw-Hill, New York.

Rizzoli, V. and Someda, C.G. (1981) "Mutual in uence among imperfect joints in multimode- bre link". *Electronics Letters*, **24**, 906–907.

Rudolph, W. and Wilhelmi, B. (1989) *Light Pulse Compression.* Harwood, New

York.
Saleh, B.E.A. and Teich, M.C. (1991) *Fundamentals of Photonics.* Wiley, New York.
Senior, J.M. (1992) *Optical Fiber Communications: Principles and Practice.* 2nd ed., Prentice-Hall, New York.
Skolnik, M.I. (1980) *Introduction to Radar Systems.* 2nd ed., McGraw-Hill, New York.
Snyder, A.W. and Love, J.D. (1983) *Optical Waveguide Theory.* Chapman & Hall, London.
Sunde, E.D. (1969) *Communication System Engineering Theory.* Wiley, New York.
Svelto, O. (1989) *Principles of Lasers.* 3rd ed., translated from Italian and edited by D.C. Hanna, Plenum, New York.
Tosco, F. (Ed.) (1990) *Fiber Optic Communications Handbook.* TAB Professional and Reference Books, Blue Ridge Summit, PA.
Yariv, A. (1991) *Optical Electronics.* 4th ed., Saunders College Publishing, Philadelphia.

CHAPTER 6

Plane waves in anisotropic media

6.1 General properties of anisotropic media

We brie y mentioned in Section 1.3 that a medium is said to be *anisotropic*, from an electromagnetic point of view, when it has the following property: at any point, and considering generic directions for the vectors $\mathbf{E}, \mathbf{H}$, at least one of the following statements is true:

- the total current density vector $\overline{\mathbf{J}}_{\mathrm{t}} = \mathbf{J} + j\omega\mathbf{D}$ is not parallel to the electric eld vector $\mathbf{E}$;

- the magnetic induction vector $\mathbf{B}$ is not parallel to the magnetic eld intensity vector $\mathbf{H}$.

Tackling anisotropic media analytically is not too complicated when the above mentioned vectors are linked to one another by *linear relationships.* As is well known, it is absolutely general to express a linear relationship between two vectors in terms of a *dyadic* operator, i.e., in terms of a second-rank tensor accompanied by the rule of taking its inner product with the vector on its right.

If $\mathbf{B}$ and $\overline{\mathbf{J}}_{\mathrm{t}}$ are linked by a linear relationship to *both* $\mathbf{E}$ and $\mathbf{H}$, i.e., if we can write:

$$\begin{aligned} \overline{\mathbf{J}}_{\mathrm{t}} &= \overline{\mathrm{g}}\ \mathbf{E} + \overline{\ }\ \mathbf{H} \ , \\ \mathbf{B} &= \overline{\ }\ \mathbf{E} + \overline{\ }\ \mathbf{H} \ , \end{aligned} \tag{6.1}$$

then one refers to this as a *linear bianisotropic* medium. The special case $\overline{\ } = \overline{\ } = 0$ is referred to as a *linear anisotropic* medium. In the latter case, it is a generalized habit to use the notation of Eqs. (1.23), (1.24) and (1.25), which we rewrite here for future convenience:

$$\mathbf{J} = \overline{\ }\ \mathbf{E} \ , \tag{6.2}$$

$$\mathbf{D} = \overline{\ }\ \mathbf{E} \ , \tag{6.3}$$

$$\mathbf{B} = \overline{\ }\ \mathbf{H} \ . \tag{6.4}$$

In a linear *isotropic* medium, one may always think that Eqs. (6.2), (6.3), and (6.4) still hold, with $\overline{\ } = \ \mathbf{i}$, $\overline{\ } = \ \mathbf{i}$, $\overline{\ } = \ \mathbf{i}$, where $\mathbf{i}$ is the identity dyadic. In the following, whenever we speak about anisotropic media, we will implicitly exclude this trivial case. An *anisotropic conductor* will be, in our terminology, a medium with $\overline{\ } \neq \ \mathbf{i}$, but with $\overline{\ } = \ \mathbf{i}$ and $\overline{\ } = \ \mathbf{i}$. The de nitions of *anisotropic dielectric* and *anisotropic magnetic* media are now obvious.

Before studying how plane waves propagate in anisotropic media, let us see how some fundamental results, which we obtained in the previous chapters for isotropic media, can either be extended, or modi ed.

A) *Lossless media.* Like in the isotropic case, in Maxwell's equations:

$$\nabla \quad \mathbf{H} = j\omega\bar{\ } \ \mathbf{E} + \bar{\ } \ \mathbf{E} + \mathbf{J}_0 \quad , \tag{6.5}$$

$$\nabla \quad \mathbf{E} = \quad j\omega\bar{\ } \ \mathbf{H} \quad , \tag{6.6}$$

the right-hand side of Eq. (6.5) can be written in a more compact form if we de ne the following dyadic:

$$\bar{\ }_c = \bar{\ } + \frac{1}{j\omega}\bar{\ } \quad . \tag{6.7}$$

By similarity with the isotropic case, we are inclined to refer to this quantity as the *complex permittivity dyadic.* However, there is a fundamental di erence, with respect to the isotropic case. Here, in fact, for the medium to be lossless it is not required that all the elements of the dyadic $\bar{\ }_c$ should be real. This conclusion can be drawn directly from Poynting's theorem. If we introduce the dyadic in Eq. (6.7) into Maxwell's equations, then the proof of the theorem (see Section 3.2) yields an expression which no longer contains an explicit dissipation term (i.e., a term proportional to the conductivity $\bar{\ }$), but instead we nd that the other volume integral becomes now:

$$2j\omega \int_V \left(\frac{\mathbf{E} \quad \bar{\ }_c \quad \mathbf{E}}{4} \quad \frac{\mathbf{H} \quad \bar{\ } \quad \mathbf{H}}{4} \right) dV \quad . \tag{6.8}$$

For this quantity to be imaginary (i.e., to represent a reactive power), it is not necessary that the dyadics $\bar{\ }_c$ and $\bar{\ }$ be real. In fact, the real part of Eq. (6.8) is zero, regardless of the particular integration volume V (i.e., the medium is lossless), if and only if:

$$\mathbf{E} \quad \bar{\ }_c \quad \mathbf{E} \quad \mathbf{E} \quad \bar{\ }_c \quad \mathbf{E} + \mathbf{H} \quad \bar{\ } \quad \mathbf{H} \quad \mathbf{H} \quad \bar{\ } \quad \mathbf{H} = 0 \quad , \tag{6.9}$$

where $\mathbf{E}, \mathbf{H}$ are arbitrary. This occurs if and only if both dyadics $\bar{\ }_c$ and $\bar{\ }$ are *Hermitian*, i.e.:

$$\bar{\ }_c = \bar{\ }_c^{\dagger} \quad , \qquad \bar{\ } = \bar{\ }^{\dagger} \quad , \tag{6.10}$$

where the symbol † means transposed and conjugate.

A medium which can be lossless even with $\bar{\ } \neq 0$, i.e., a "lossless conductor," sounds puzzling. In fact, we will see later in this chapter that, under some circumstances, charged particles can move, under the in uence of an e.m. wave, in such a way that their convection current is *in quadrature* with the electric eld of the wave. The eld-current product is imaginary, giving rise to a purely reactive power, although the current is not a displacement one. It is evident, however, that this conduction without dissipation results from a model of particle motion where any friction is completely neglected, hence necessarily an approximate model.

After learning that complex elements in $\bar{\epsilon}_c$ do not necessarily imply that the medium is lossy, we could drop the distinction between the symbols $\bar{\epsilon}$ and $\bar{\epsilon}_c$. However, using two different symbols may help to prevent the mistake of thinking that $\mathbf{E}^* \cdot \bar{\epsilon}_c \cdot \mathbf{E}/4$ is always the (time-averaged) electric energy density. Indeed, it can be shown that, in some lossless media, and for some values of the frequency, this quantity can even become negative, which is clearly unacceptable for an energy density. This apparent nonsense can be explained, taking into account that, in such a medium, there is also a storage of distributed *kinetic* energy, because of the motion of the charged particles. Without getting now into details that are difficult to grasp at this stage, let us repeat that it is because of these problems that, in the following, we will preserve the distinction between $\bar{\epsilon}$, which refers to a zero-conductivity anisotropic dielectric medium, and $\bar{\epsilon}_c$, for an anisotropic medium with a finite conductivity.

Another surprise for the reader may be the fact that, similarly, some media can be characterized by a *complex permeability* dyadic $\bar{\mu}$. This will also be clarified by means of an example, later on in this chapter.

It is well known that, in a complex space, *any* Hermitian operator is described by a diagonal matrix in a suitable orthogonal reference frame (Smirnov, 1961). Therefore, we can state that for *any* lossless anisotropic dielectric or magnetic medium, there is a basis — in general, a complex one — where $\bar{\epsilon}_c$ or $\bar{\mu}$ is represented by a diagonal matrix. However, in a medium characterized by $\bar{\epsilon}_c$ and $\bar{\mu}$ that are *both* Hermitian, we cannot state for sure that there exists an orthogonal basis where both dyadics are represented by diagonal matrices.

B) *Reciprocity.* Let us repeat the proof of the reciprocity theorem (Section 3.4), for a linear anisotropic medium, using a real rectangular reference frame, and writing all the vectors by components. Then, we see that the theorem holds *if and only if* $\bar{\epsilon}_c$ and $\bar{\mu}$ are both represented by *symmetric* matrices:

$$(\bar{\epsilon}_c) = \left(\tilde{\bar{\epsilon}}_c\right) \quad , \qquad (\bar{\mu}) = \left(\tilde{\bar{\mu}}\right) \quad , \tag{6.11}$$

where the symbol $\tilde{}$ means transposed. As known from linear algebra (Smirnov, 1961), symmetry of a matrix is a property which is invariant under an *orthogonal* transformation, i.e., a change of basis in real 3-D space. In general, it is not invariant under a *unitary* transformation, i.e., the most general change of basis in a complex space. Incidentally, this is in contrast with the Hermitian property, which is preserved by unitary transformations. This implies that, while we could speak of lossless media without any need to specify what kind of reference frame we were taking, on the contrary, when the subject is reciprocity in an anisotropic medium, statements are unambiguous only as long as they are restricted to *linearly polarized* fields. In engineering contexts, this point is often taken for granted. However, let us stress that it may entail important consequences, that will be outlined in detail in Section 6.6. For the time being, in order to grasp the first meaning of this discussion, the reader may verify that the *reaction*, defined in Section 3.4 and intended to be a measure of reciprocity, is not an invariant with respect to an arbitrary change of basis in the complex field. Indeed, even a very trivial change — multiply one of the basis vectors by a phase factor — modifies the reaction. Mathematically, this surprising behavior is related to the fact that Steinmetz's representation of time-harmonic quantities is not an isomorphism between the space of real time-harmonic vectors and the 3-D complex vector space, and does not preserve the inner product.

6.2 Wave equations and potentials in anisotropic media

Dyadic operators do not always commute with the curl operator, neither with the divergence one. Accordingly, Maxwell's equations in anisotropic media cannot be handled in the same way as in the first chapter. In general, it is not possible to obtain from Eqs. (6.5) and (6.6) a simple equation, like the Helmholtz equation, containing either **E** or **H**, but not both.

For a *dielectric* anisotropic medium (scalar μ), it is possible to obtain an equation where only **E** appears. To do that, take the curl of Eq. (6.6), replace $\nabla \times \mathbf{H}$ with Eq. (6.5), and make use of the definition of the Laplacian of a vector. This yields:

$$\nabla^2 \mathbf{E} - \nabla(\nabla \cdot \mathbf{E}) + \omega^2 \mu \,\bar{\bar{\epsilon}}_c \cdot \mathbf{E} = j\omega\mu \,\mathbf{J}_0 \quad . \tag{6.12}$$

On the other hand, in such a medium it is impossible to derive an equation where the only unknown is **H**, at least as long as further information on the dyadic $\bar{\bar{\epsilon}}_c$ is not provided.

Eq. (6.12) in general is not the same as the Helmholtz equation, because the identity (C.5) $\nabla \cdot \nabla \times \mathbf{a} = 0$, applied to Eq. (6.5), gives:

$$\nabla \cdot (\bar{\bar{\epsilon}}_c \cdot \mathbf{E}) = -\frac{1}{j\omega} \nabla \cdot \mathbf{J}_0 \quad . \tag{6.13}$$

If there are no sources, or if the sources have zero divergence, this yields:

$$\nabla \cdot (\bar{\bar{\epsilon}}_c \cdot \mathbf{E}) = 0 \quad . \tag{6.14}$$

In general the last equation does not entail $\nabla \cdot \mathbf{E} = 0$. Therefore, the second term on the left-hand side in Eq. (6.12) does not disappear. In the following sections, we will see that only in a few special cases Eq. (6.12) can undergo significant simplification.

Duality tells us that in an anisotropic magnetic medium the governing equation is:

$$\nabla^2 \mathbf{H} - \nabla(\nabla \cdot \mathbf{H}) + \omega^2 \epsilon_c \,\bar{\bar{\mu}} \cdot \mathbf{H} = 0 \quad , \tag{6.15}$$

where the second term cannot be eliminated, since Eq. (6.6) and the identity $\nabla \cdot \nabla \times \mathbf{a} = 0$ entail:

$$\nabla \cdot (\bar{\bar{\mu}} \cdot \mathbf{H}) = 0 \quad , \tag{6.16}$$

which does not necessarily lead to $\nabla \cdot \mathbf{H} = 0$. In this type of medium, we are unable to write a simple and general equation for the vector **E**.

These problems in writing a differential equation, either for **E** or for **H**, are accompanied by parallel problems if we try to develop a theory of e.m. potentials, similar to the one we built in Chapter 1 for isotropic media. Indeed, let us omit the very complicated case where $\bar{\ }_c$ and $\bar{\ }$ are both dyadics (fortunately, not a case of primary importance), and take the case of a *homogeneous anisotropic dielectric*. We try to repeat the initial steps of Section 1.8. So, we introduce a *magnetic vector potential* **A**, whose scope of de nition can be very broad. The eld vectors can be derived from it, as follows:

$$\mathbf{H} = \frac{1}{\ }\nabla \quad \mathbf{A} \quad , \qquad \mathbf{E} = \frac{1}{j\omega}\,\bar{\ }_c^{\ \ 1}\,(\nabla \quad \nabla \quad \mathbf{A} \qquad \mathbf{J}_0) \quad , \tag{6.17}$$

where $\bar{\ }_c^{\ \ 1}$ is the inverse of $\bar{\ }_c$. The vector **A** is easily shown to obey the equation:

$$\nabla \quad \left(\bar{\ }_c^{\ \ 1}\ \nabla \quad \nabla \quad \mathbf{A} \quad \omega^2\ \mathbf{A} \qquad \bar{\ }_c^{\ \ 1}\ \mathbf{J}_0\right) = 0 \quad . \tag{6.18}$$

In a simply connected domain, we may try to solve Eq. (6.18) introducing an arbitrary scalar function φ, as in isotropic media. This yields an equation which di ers formally from Eq. (1.65) only because the dyadic $\bar{\ }_c$ replaces the scalar $\ _c$:

$$\nabla \quad \nabla \quad \mathbf{A} \quad \omega^2\ \bar{\ }_c\ \mathbf{A} \qquad \mathbf{J}_0 = \quad j\omega\ \bar{\ }_c\ \nabla\varphi \quad . \tag{6.19}$$

In reality, we run into problems as soon as we try to simplify Eq. (6.19) through the choice of φ. Indeed, Eq. (6.19) and the usual identity $\nabla \quad \nabla \quad \mathbf{a} = 0$ give us an equation which contains $\nabla \quad (\bar{\ }_c \quad \mathbf{A})$, not $\nabla \quad \mathbf{A}$. As a consequence, even in the simplest case where $\nabla \quad \mathbf{J}_0 = 0$ (let us recall that we showed that under this assumption, in an isotropic medium, $\varphi = 0$ is a Lorentz gauge), now we are unable to get the Helmholtz equation from Eq. (6.19), because the term $\nabla(\nabla \quad \mathbf{A})$ in general is not zero. The only result that we can reach is that, in those special cases where Eq. (6.12) becomes the Helmholtz equation, then the same occurs to Eq. (6.19), provided the $\varphi = 0$ gauge is adopted.

The duality theorem of Section 3.7 indicates how to cover the case of a magnetic anisotropic medium, in terms of an electric vector potential **F** of the same type as the one we saw in Section 1.9. The details are left to the reader as an exercise.

In conclusion, we saw in this section that anisotropic media are impossible to deal with at the same level of generality as we did in previous chapters. Consequently, we will be often forced to introduce some "ansatz" on the elds. The most widely used assumption is to deal only with plane waves of a given polarization. Another consequence is that it appears convenient to classify anisotropic media into classes, which di er from each other in the manner of simplifying the equations that the elds must obey.

6.3 Birefringent media

In this section we will deal with one of the most important classes of anisotropic media. We will assume that they are *lossless dielectrics*. Including loss would make the mathematics much heavier, without any major gain in understanding the physics. The dual case of magnetic media with identical formal properties is left to the reader as an exercise.

A lossless anisotropic dielectric medium is referred to as *birefringent* if its dyadic $\bar{}$ is *real*, i.e., if all the elements of the matrix ( ) which represent the dyadic in a real reference frame, are real. As we said, we maintain a distinction between $\bar{}$ and $\bar{}_c$; otherwise, some consequences that we are going to draw from Eq. (6.22) would no longer be valid.

This de nition, combined with the assumption that the medium is lossless, Eq. (6.10), implies that, in any real rectangular coordinate system (x_1, x_2, x_3), the matrix ( ) *is symmetric*:

$$_{ij} = \ _{ji} \qquad (i, j = 1, 2, 3) \quad . \tag{6.20}$$

Any (3 3) real symmetric matrix has three, mutually orthogonal, real eigenvectors (Smirnov, 1961). Therefore, in any birefringent dielectric, there is a particular rectangular reference frame — from now on, referred to as the (x, y, z) system — where its matrix ( ) is diagonal:

$$\begin{bmatrix} D_x \\ D_y \\ D_z \end{bmatrix} = \ _0 \begin{bmatrix} _x & 0 & 0 \\ 0 & _y & 0 \\ 0 & 0 & _z \end{bmatrix} \begin{bmatrix} E_x \\ E_y \\ E_z \end{bmatrix} \quad . \tag{6.21}$$

If the medium is homogeneous, this reference frame is constant in space throughout the region occupied by the medium. In this case, it is called the *system of principal axes* of the dielectric. The matrix elements $_x$, $_y$, $_z$ in Eq. (6.21) are referred to as the *principal relative permittivities.*

The steps one has to go through, to pass from permittivity in a generic frame to Eq. (6.21), lend themselves to a geometrical interpretation, which might help us later on in understanding e.m. wave propagation in such a medium. Let us start from a generic frame (x_1, x_2, x_3), where $\bar{}$ is described by the matrix ($_{ij}$), and de ne the following quadratic form:

$$\sum_{i,j=1}^{3} \ _{ij}\, x_i\, x_j = \sum_{i,j=1}^{3} \ _{ij}\, x_i\, x_j \quad , \tag{6.22}$$

where $x_j = x_j$, since the coordinates are real. This form is easily seen to be *de nite positive.* In fact, if we replace x_i with the component E_i of the electric eld, Eq. (6.22) becomes four times the time-averaged electric energy density, which must be positive by de nition for any nonvanishing eld. Writing that a de nite positive form like Eq. (6.22) is equal to a constant, one gets an equation which, in analytical geometry, is the equation of an *ellipsoid.* Passing from a generic reference frame to the coordinate system where Eq. (6.21) holds, the ellipsoid equation reads:

$$_x\, x^2 + \ _y\, y^2 + \ _z\, z^2 = \text{constant} \quad . \tag{6.23}$$

Using the language of geometry, we call this the ellipsoid equation in the frame of its *principal axes.* This explains the names that were introduced earlier.

Eq. (6.21) recon rms that, in general, in an anisotropic dielectric medium the vectors **D** and **E** are not parallel. Further information on how these two vectors lay can be obtained if we make the assumption (to be confirmed later on), that a *linearly polarized plane wave* is allowed to propagate in this medium. Assume then, on the basis of what we saw in Chapter 4, that all the field vectors vary in space according to $\exp(-j\mathbf{k} \cdot \mathbf{r})$, where $\mathbf{k}$ is a constant real vector. Then, the curl operator may be replaced by $(-j\mathbf{k}) \times$. Maxwell's Eqs. (6.5) and (6.6), in a region without sources, read:

$$-j\mathbf{k} \times \mathbf{H} = j\omega \mathbf{D} \ , \tag{6.24}$$

$$-j\mathbf{k} \times \mathbf{E} = -j\omega \mu \mathbf{H} \ . \tag{6.25}$$

From these we see that **H** is perpendicular to **D**, to **E**, and to **k**. Consequently, these three vectors must all lay on one plane. Eq. (6.24) also shows that **D** is perpendicular to **k**. On the other hand, Eqs. (6.24) and (6.25) do not yield any information on the direction of **E**, except for the previous point that it belongs to the plane defined by **k** and **D**. This entails that, in general, the vector **E** has a nonvanishing component along the direction of propagation.

Consistently with the previous assumption of linear polarization, let us assume $\mathbf{H} = \mathbf{H}_0 \exp(-j\mathbf{k} \cdot \mathbf{r})$, with a real $\mathbf{H}_0$. Then, the vectors $\mathbf{E}_0 = \mathbf{E}/\exp(-j\mathbf{k} \cdot \mathbf{r})$ and $\mathbf{D}_0 = \mathbf{D}/\exp(-j\mathbf{k} \cdot \mathbf{r})$ are also real, and can be depicted as in Figure 6.1. The Poynting vector $\mathbf{P} = \mathbf{E} \times \mathbf{H}^*/2$ is also real, and furthermore, being perpendicular to $\mathbf{H}_0$, it belongs to the same plane as $\mathbf{D}_0$, $\mathbf{E}_0$ and **k**. By construction, it is also orthogonal to $\mathbf{E}_0$, and therefore in general it is not parallel to the phase vector **k**. The angle that the Poynting vector makes with the phase vector equals the angle between $\mathbf{E}_0$ and $\mathbf{D}_0$. Accordingly, for a plane wave in a birefringent medium power flows, generally speaking, in a direction (called the *ray direction*) which is not the same as the normal to the constant-phase planes (the *wave normal direction*). In Section 4.11, we saw that superpositions of evanescent waves in isotropic media do convey power in a direction perpendicular to their phase vector, but let us stress, to avoid any ambiguity, that the phenomenon that we are discussing here is completely different, from a physical point of view.

Let us aim now at finding a link between the vectors **D** and **E**, for any direction of the phase vector. Let us define a *refractive index* n as a function of the direction of **k**, as follows:

$$\mathbf{k} = n k_0 \hat{k} \ , \tag{6.26}$$

where $k_0 = \omega(\epsilon_0 \mu_0)^{1/2}$, while $\hat{k}$ is the unit vector in the direction of **k**. Such a definition stems from one of the physical meanings of the refractive index in an isotropic medium, namely that of ratio between phase velocity in free space, and phase velocity in the medium. On the other hand, it does not entail anything about the relationship between **D** and **E**. Insert then Eq. (6.26) into Eqs. (6.24) and (6.25), assume for simplicity $\mu = \mu_0$, and eliminate **H**. The result is:

$$\mathbf{D} = n^2 \epsilon_0 \hat{k} \times \mathbf{E} \times \hat{k} = n^2 \epsilon_0 \mathbf{E}_t \ , \tag{6.27}$$

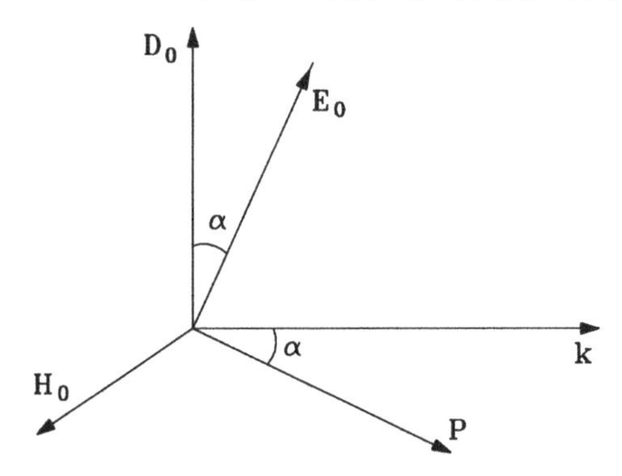

Figure 6.1 *A plane wave in an anisotropic medium: relationships between its electromagnetic vectors.*

where $\mathbf{E}_t$ is the transverse component of $\mathbf{E}$, with respect to the direction of propagation. To determine n as a function of the parameters of the medium, it is convenient to exploit the *inverse permittivity* (also called *impermittivity*) dyadic $\overline{\mathbf{b}}$, de ned as:

$$\overline{\mathbf{b}} = {}_0 {}^{-1} \quad , \tag{6.28}$$

such that:

$${}_0 \mathbf{E} = \overline{\mathbf{b}} \quad \mathbf{D} \quad . \tag{6.29}$$

Let us project Eq. (6.29) on a constant-phase plane (perpendicular to $\hat{k}$), and take into account that $\mathbf{D}$ belongs to this plane. Thus, we obtain:

$${}_0 \mathbf{E}_t = \overline{\mathbf{b}}_t \quad \mathbf{D} \quad , \tag{6.30}$$

where the dyadic $\overline{\mathbf{b}}_t$ is the projection of $\overline{\mathbf{b}}$ onto a plane orthogonal to $\hat{k}$, therefore it is a dyadic in a 2-D subspace and can be described by a 2 2 matrix (Smirnov, 1961).

The system consisting of Eqs. (6.30) and (6.27) yields an eigenvalue equation:

$$\overline{\mathbf{b}}_t \quad \mathbf{D} = \frac{1}{n^2} \mathbf{D} \quad . \tag{6.31}$$

Therefore, n^2 can be identi ed as the inverse of one of the eigenvalues of the operator $\overline{\mathbf{b}}_t$. This enables us, rst of all, to *count* how many values the refractive index can take for a given direction of propagation. Let us take a rectangular coordinate frame with an axis parallel to the direction of propagation (i.e., to $\hat{k}$). Then, thanks to our initial assumptions on $^{-}$, the 2 2 matrix (b_{ij}) representing $\overline{\mathbf{b}}_t$ is real and symmetric. Consequently, it has *two real eigenvalues*, in general di erent from each other. In conclusion, for waves traveling along a generic direction, the refractive index may take two di erent values. It is self-explanatory, now, why these media are called *birefringent.*

Eq. (6.22), as already said, is a positive de nite quadratic form (in lossless media). Hence, both eigenvalues are *positive.* Each eigenvalue gives rise to two opposite values of n, which correspond, obviously, to two waves traveling in opposite directions, $\hat{k}$.

Under the assumptions that were made at the beginning of this section, the operator $\overline{\mathbf{b}}_t$ has two real orthogonal eigenvectors. Each of them belongs to one of the above mentioned eigenvalues. This means that there are *two possible, mutually orthogonal linear states of polarization* for the D vector of plane waves traveling in a birefringent medium. Any other state of polarization is a field whose dependence on spatial coordinates cannot be written in the simple form $\exp(\; j\mathbf{k} \;\mathbf{r})$. So, to see how a field, known at a given point, in an unspecified state of polarization, evolves in space, first we must decompose it as the sum of the two linear polarizations corresponding to the eigenstates that we have just encountered. These are then two waves which travel independently of each other. The total field at any distance is to be written as the sum of those two linearly polarized waves, after they propagate over that distance. As these two waves have two different phase constants, the net result is that the state of polarization of the total field, in general, is a function of distance, along the direction of propagation. The reader should be able to explain, as an exercise, why this function is periodic in space.

6.4 Fresnel's equation of wave normals

In the previous section we studied the fundamentals of wave propagation in birefringent media, but we did not arrive at operative rules. We will show now how to calculate the main features of plane waves which travel in these media.

Solving the eigenvalue equation (6.31) requires two cumbersome operations, to invert the matrix which represents $\bar{\ }$, and to change the coordinate system. There is a faster way to compute the refractive indices, for a given direction of propagation, starting from Eqs. (6.3) and (6.27), which ($\bar{\mathbf{i}}$ being the identity dyadic) entail:

$$(\bar{\ } \quad n^2 \;{}_0\bar{\mathbf{i}}) \quad \mathbf{E} = \quad n^2 \;{}_0(\mathbf{E} \quad \hat{k})\,\hat{k} \quad . \tag{6.32}$$

In the principal-axes reference frame (where $\bar{\ }$ is diagonal), it becomes very simple to write Eq. (6.32) by components. Let $\hat{k} \quad (v_x, v_y, v_z)$. We get:

$$\begin{aligned}
(n^2 \quad {}_x)\,E_x &= \quad n^2\, v_x(\hat{k} \quad \mathbf{E}) \quad , \\
(n^2 \quad {}_y)\,E_y &= \quad n^2\, v_y(\hat{k} \quad \mathbf{E}) \quad , \\
(n^2 \quad {}_z)\,E_z &= \quad n^2\, v_z(\hat{k} \quad \mathbf{E}) \quad .
\end{aligned} \tag{6.33}$$

Many results of this chapter can be extended to the case where $\overline{\boldsymbol{\epsilon}}_c = \overline{\boldsymbol{\epsilon}} + (1/j\omega)\overline{\boldsymbol{\sigma}}$ is real. In that case, however, as we saw in Section 6.2, the quadratic form $\mathbf{E} \quad \overline{\boldsymbol{\epsilon}}_c \quad \mathbf{E}$ does not have a physical meaning which assures that n^2 is always positive. When a negative value is found for n^2, thus an imaginary value is found for n, the corresponding plane wave becomes extremely similar, physically, to the evanescent waves of Chapter 4. Such a formalism can be very useful in modeling ionospheric reflections (see, e.g., Budden, 1961).

Multiply the rst of these equations by v_x, the second by v_y, the third by v_z, sum the three equations, and divide both sides by $\hat{k} \quad \mathbf{E}$ (*leaving out isotropic media*, where $\hat{k} \quad \mathbf{E}$ vanishes). We get the following equation:

$$\frac{v_x^2}{n^2 \quad _x} + \frac{v_y^2}{n^2 \quad _y} + \frac{v_z^2}{n^2 \quad _z} = \frac{1}{n^2} \quad , \tag{6.34}$$

currently referred to as *Fresnel's equation of the wave normals.* It allows an easy computation of n^2, once the principal permittivities and the propagation direction (v_x, v_y, v_z) are given. Eq. (6.34) is biquadratic in n, because of the identity $v_x^2 + v_y^2 + v_z^2 = 1$. Hence, it always gives two solutions for n^2, in agreement with what we saw in Section 6.3. When written in a rational form, it holds also in isotropic media ($_x =$ $_y =$ $_z =$ $_r$), and in this case it gives, as it must give, solutions $n = \quad \sqrt{\ _r}$ that are independent of the direction of propagation.

The directions of the two linear polarizations allowed for $\mathbf{D}$ are now simple to determine by means of geometry. In the principal-axes reference frame (x, y, z), consider the ellipsoid, centered in the origin, de ned by the following equation (note that it is not the same as Eq. (6.23)):

$$\frac{x^2}{\ _x} + \frac{y^2}{\ _y} + \frac{z^2}{\ _z} = 1 \quad . \tag{6.35}$$

This surface is called *index ellipsoid*, or the *optical indicatrix.* We will show now that, for a wave traveling parallel to a unit vector $\hat{k} = (v_x, v_y, v_z)$, the two directions that are allowed for linearly polarized $\mathbf{D}$ vectors are the major and minor axes of the ellipse which is the line where the ellipsoid Eq. (6.35) intersects the plane orthogonal to $\hat{k}$ passing through the origin whose equation reads:

$$v_x x + v_y y + v_z z = 0 \quad . \tag{6.36}$$

We will also prove that, with a suitable normalization, the lengths of these two axes are equal to the indices n given by Eq. (6.34).

To do this, let us rst point out that the axes we are looking for are the minimum and maximum diameter lengths for the ellipse. Therefore, they may be found as conditional extrema of the distance from the origin, r, or of a convenient power of r. The simplest choice is to deal with:

$$r^2 = x^2 + y^2 + z^2 \quad . \tag{6.37}$$

The constraints we must impose on r^2 are Eqs. (6.35) and (6.36), meaning that the ellipse axes terminate at points that belong to the ellipsoid and to the plane orthogonal to $\hat{k}$. The problem can be tackled with the standard method of Lagrange multipliers. Namely, we write a linear combination of Eqs. (6.37), (6.36) and (6.35), with auxiliary parameters $2 \ _1$, $\ _2$, i.e.:

$$F(x, y, z) = x^2 + y^2 + z^2 + 2 \ _1(v_x x + v_y y + v_z z) + \ _2\left(\frac{x^2}{\ _x} + \frac{y^2}{\ _y} + \frac{z^2}{\ _z}\right) \ ; \tag{6.38}$$

next, we take the partial derivatives of Eq. (6.38) with respect to x, y, z, and

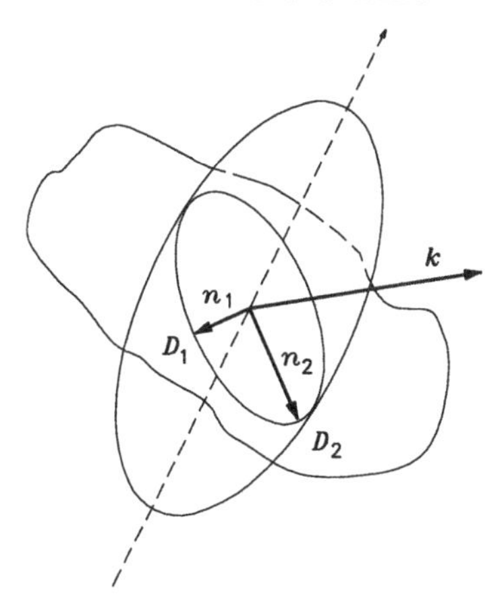

Figure 6.2 *Geometrical construction of the refractive indices and the allowed states of polarization, using the optical indicatrix.*

set them equal to zero. We get the following system of equations:

$$m + {}_1 v_m + {}_2 \frac{m}{{}_m} = 0 \qquad (m = x, y, z) \quad . \qquad (6.39)$$

The parameters ${}_1$ and ${}_2$ can be eliminated combining Eqs. (6.35), (6.36) and (6.39). So, we get three equations, the rst of which reads:

$$\left(\frac{r^2}{{}_x} \quad 1\right) x = r^2 \, v_x \left(\frac{v_x x}{{}_x} + \frac{v_y y}{{}_y} + \frac{v_z z}{{}_z}\right) \quad . \qquad (6.40)$$

The other two can be obtained simply by permuting x with y and with z. These equations de ne, by construction, the extrema of the ellipse axes. Now, we observe that the rst of Eqs. (6.33) becomes Eq. (6.40) if we set $n^2 = r^2$, $x = D_x = {}_x E_x$. The same occurs with the other two components and the other two coordinates. This proves that n is proportional to the axis length, and, at the same time, it proves that the directions of the axes coincide with those allowed for **D** (see Figure 6.2). In other words, this completes our proof.

We run into a case of special interest when two out of the three principal permittivities are identical, i.e., for ${}_x = {}_y \neq {}_z$. The ellipsoid described by Eq. (6.35) has then a rotational symmetry around the z axis. In this case, for *any* direction of $\hat{k}$, one of the axes of the ellipse where the ellipsoid intersects the plane normal to $\hat{k}$ is, at the same time, a diameter of the circle where the ellipsoid intersects the plane $z = 0$ (which, in turn, is either a minimum or a maximum in length among all the intersection ellipses, depending on whether ${}_x$ is smaller or larger than ${}_z$). Consequently, in these media one of the two values for n (usually referred to as the *ordinary index*, n_0) is *independent of the direction of propagation*, while the other one (referred to as the *extraordinary index*, n_e) does depend on the direction of $\hat{k}$. The names given to the indices re ect the habit of calling ordinary and extraordinary waves, respectively, the plane waves associated with n_0 and with n_e. The term *birefringence*, in a given direction, is in current use to denote the absolute value of the di erence, $|n_0 \quad n_e|$, for a pair of plane waves traveling in that direction. Note, however, that sometimes it is used for the *relative* di erence, $|n_0 \quad n_e|/n_0$, and on other occasions it means the absolute di erence between the two phase constants, $|n_0 \quad n_e| \ _0$. The quantity $L_B = 2 \ /|n_0 \quad n_e| \ _0$ is called the *beat length.* It is simple to show, as an exercise, that, if an ordinary wave and an extraordinary one travel in the same direction, then the state of *polarization* of the total eld, resulting from the overlap of the two waves, evolves *periodically* as a funtion of distance, and L_B is its spatial period. The values of n are both independent of $\hat{k}$ only in a medium where $\ _x = \ _y = \ _z$, i.e., in an isotropic medium. In this case, the ellipsoid given by Eq. (6.35) becomes a sphere, and all directions orthogonal to $\hat{k}$ are legitimate for **D**. If, on the contrary, $\ _z \neq \ _x = \ _y$, then there is only one direction for $\hat{k}$ — the z-direction — for which we nd $n_e = n_0$. For waves traveling along this axis, the directions allowed for **D** are all those in the plane orthogonal to z. A medium of this kind is called a *uniaxial medium.* Trigonal, tetragonal and hexagonal crystals fall within this category (Nye, 1985).

When the three principal permittivities di er from each other, it is simple to prove that there are two (and only two) planes through the origin that intersect the ellipsoid in circles. Hence, there are two directions of propagation (normal to those planes), for each of which the two values of n^2 are equal, and any direction normal to propagation is allowed for **D**. These media are referred to as *biaxial.*

Refraction at the separation between an isotropic and an anisotropic medium can be now studied along the guidelines of Section 4.8. For the sake of brevity, let us omit evanescent waves, which can be dealt with by the reader as an exercise. Then, in general (i.e., except for the special cases where the refracted waves travel along one of the optic axes) there are *two transmitted waves*, traveling in two di erent directions, and orthogonally polarized. This phenomenon is referred to as *double refraction.* Each refracted wave obeys Snell's law, Eq. (4.84), with the value of n which is relevant to that wave (either the ordinary, or the extraordinary index). Re ection and transmission coe cients can be evaluated by decomposing any incident wave into the two polarizations that correspond to the ordinary and extraordinary waves in the birefringent medium, respectively, and then applying Fresnel's formulas of Section 4.9 to each of them.

Another refraction phenomenon, di erent from those we saw so far, is peculiar of biaxial media. It is referred to as *conical refraction.* To outline it, let us consider a biaxial slab, de ned by two planes normal to one of the optic axes, and illuminated by a uniform plane wave at normal incidence. By definition of optic axis, any direction in the normal plane is permissible for **D**, and the refractive index n is independent of the direction of **D**. Snell's law in Eq. (4.84) shows that, for normal incidence, the direction of the phase vector does not change as the wave passes from one medium to the other. However,

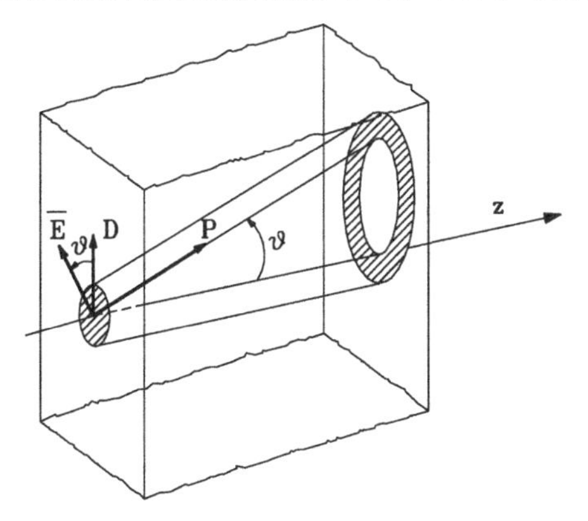

Figure 6.3 *Conical refraction.*

as we saw in Section 6.2, the vector **E** in general is not in the same direction as the vector **D**, having a longitudinal component (see Figure 6.1). Consequently, the Poynting vector is not parallel to the optic axis, and furthermore, its direction, belonging always to the plane orthogonal to the magnetic eld, depends on that of the vector **D**. Suppose now that the incident wave is *depolarized*, i.e., is a random superposition of waves whose **D** vector is linearly polarized in all directions orthogonal to the propagation direction, with a at probability distribution. The corresponding directions for the vector **E** form a surface, which can be shown (see Born and Wolf, 1980, Chapter 14) to be a cone. Note, however, that the axis of this cone cannot be parallel to $\hat{k}$: by de nition of biaxial medium (the three principal permittivities are di erent), an optic axis can never coincide with one of the principal axes, and therefore the length of the longitudinal component of **E** depends on the direction of **D**.

Consequently, the directions of the Poynting vector also form a cone, whose axis is not parallel to $\hat{k}$. Hence, its intersection with the slab output plane is an ellipse. So, if the slab is illuminated over a nite round spot, and if the birefringent slab is thick enough, compared to the spot radius, then light (i.e., active power) emerges at the output face through an elliptical-ring-shaped area (see Figure 6.3). It is self-explanatory why this phenomenon is called, as we said before, conical refraction. Note that, as the wave normals remain always orthogonal to the input and output planes, there is no further refraction when the light leaves the slab.

A few numerical data on typical uniaxial and biaxial crystals will be given at the end of the next section.

6.5 An application: Phase matching of two waves

Quite often, at least at optical frequencies, the nonlinear phenomena of highest practical interest (e.g., harmonic generation, frequency mixing, parametric ampli cation, etc.) take place in birefringent media. Physics of nonlinear phenomena in crystals, and nonlinear e.m. wave propagation, are outside the scope of this book. Those readers who are interested in such topics are strongly encouraged to consult several specialized books (Bloembergen, 1996; Zernike and Midwinter, 1973; Agrawal, 1995; Someda and Stegeman, Eds., 1992; etc.). However, most nonlinear wave interactions share an important problem that lends itself to be tackled now, as an application of what we developed in the previous section.

For two waves to reach a signi cant e ciency in a nonlinear interaction, they must travel with *the same phase velocity.* Otherwise, local contributions might still be strong, but would not add up coherently, and the average e ect over long distances would be essentially zero. Similar considerations apply to nonlinear processes involving three or more waves. Satisfying such constraints among interacting waves is referred to as *phase matching.* We will provide an example, in a birefringent medium, showing that phase matching is achieved only if the waves travel along selected directions.

For simplicity, let us deal with a uniaxial crystal, and leave to the reader, as an exercise, biaxial ones. So, we let† $n_0^2 = {}_x = {}_y$; $n_z^2 = {}_z$. Suppose that, over the spectral range of interest, n_0 and n_z are both monotonic functions of frequency. Then, if f_1 and f_2 are the frequencies of two waves involved in a given nonlinear process, we have, inevitably, $n_0(f_1) \neq n_0(f_2)$, and $n_z(f_1) \neq n_z(f_2)$. Phase matching between two ordinary waves is not possible, regardless of their direction of propagation, neither it is possible between two extraordinary waves. One must try to phase match an ordinary wave and an extraordinary one.

Let ϑ be the angle between the direction along which these two waves propagate, z', and the optic axis, z. We are looking for a value of ϑ such that the following equation:

$$n_0(f_1) = n_e(f_2) \quad , \tag{6.41}$$

is satis ed, n_e being the extraordinary index along z'. To that purpose, let us express n_e as a function of n_0, n_z and ϑ, starting from Fresnel's equation of the wave normal. To avoid a 0/0 form, let us rst rationalize Eq. (6.34), and then make use of the following symbols (see Figure 6.4):

$$v_x = \sin\vartheta \quad , \quad v_y = 0 \quad , \quad v_z = \cos\vartheta \quad . \tag{6.42}$$

After simple algebra, we get:

$$\frac{1}{n_e} = \left(\frac{\cos^2\vartheta}{n_0^2} + \frac{\sin^2\vartheta}{n_z^2}\right)^{1/2} \quad . \tag{6.43}$$

† While n_0 is called ordinary index without any ambiguity, the term "extraordinary index" is used by some authors to denote what we labelled here as n_z, by other authors to denote the quantity (6.43). In this book, we will adhere consistently to the latter rule.

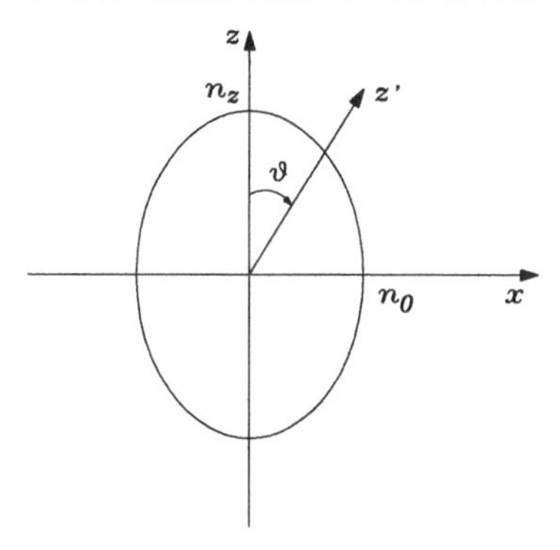

Figure 6.4 *Relationship between the propagation direction, z', and the optic axis, z.*

So, the condition in Eq. (6.41) becomes an equation in ϑ which reads:

$$\sin^2\vartheta\left[\frac{1}{n_0^2(f_2)} \quad \frac{1}{n_z^2(f_2)}\right]=\left[\frac{1}{n_0^2(f_2)} \quad \frac{1}{n_0^2(f_1)}\right] \quad . \tag{6.44}$$

A real solution for ϑ, i.e., a direction of propagation satisfying Eq. (6.41), exists if the two quantities in brackets have the same sign, and if the one on the left-hand side is larger, in absolute value, then the one on the right-hand side. Hence, there are solutions to the problem either for:

$$n_0(f_2) > n_0(f_1) > n_z(f_2) \quad , \tag{6.45}$$

or for:

$$n_0(f_2) < n_0(f_1) < n_z(f_2) \quad . \tag{6.46}$$

In fact, in our formalism the two symbols f_1 and f_2 are completely interchangeable. (They are no longer interchangeable only if at a certain frequency, between f_1 and f_2, the inequality between n_0 and n_z is reversed). So, the actual meaning of Eqs. (6.45) and (6.46) is that, given n_0 and n_z at one frequency (say f_1), phase matching is possible at all those frequencies f_2 at which $n_0(f_2)$ is within the range between $n_0(f_1)$ and $n_z(f_1)$.

The *polarizations* of the two waves, on the other hand, cannot be interchanged. It is easy to prove, as an exercise, that the extraordinary wave must be the higher frequency, if n_0 is an increasing function of frequency f, and Eq. (6.45) applies; the lower frequency, if n_0 is still an increasing function of frequency f, but Eq. (6.46) applies. The reverse is true if the ordinary index is a decreasing function of frequency.

A numerical example may help clarifying how one proceeds. Let us deal with second-harmonic generation in KDP (Deuterated Potassium Phosphate), one of the most widely used materials in bulk integrated optics. It is a uniaxial crystal. Suppose that the light source is a ruby laser, emitting in the red, at $_1 = 694\,\text{nm}$. The second-harmonic wavelength is then $_2 = {}_1/2 = 347\,\text{nm}$, in the green. The measured values for the refractive indices can be found in

Table 6.1
Typical values of KDP refractive indices
at various wavelengths ($T = 25$ C).

[nm]	n_O	n_z
266	1.5593	1.5099
355	1.5311	1.4858
532	1.5123	1.4705
694	1.5050	1.4653
1064	1.4938	1.4599

Typical values of KDP refractive indices
at $T =$ 100 C.

[nm]	n_O	n_z
266	1.5651	1.5133
355	1.5363	1.4890
532	1.5172	1.4736
694	1.5097	1.4685
1064	1.4985	1.4630

the literature, and are:

$$n_O(f_1) = 1.506 \quad , \quad n_z(f_1) = 1.466 \quad , \quad n_O(f_2) = 1.534 \quad , \quad n_z(f_2) = 1.487 \, .$$

We see that $n_O(f_1)$ falls in the range between $n_O(f_2)$ and $n_z(f_2)$. Hence, an ordinary wave at the fundamental frequency can be phase matched to an extraordinary second-harmonic wave. To nd the direction in which the two waves have to travel, we take Eq. (6.43) and rewrite it as:

$$1.506 = \left(\frac{\cos^2 \vartheta}{(1.534)^2} + \frac{1 \quad \cos^2 \vartheta}{(1.487)^2} \right)^{1/2} \quad .$$

The solution is easily found to be $\vartheta \simeq 50$.

One of the features of birefringent media that the previous example is intended to point out is that, usually, the refractive indices and the birefringence (i.e, the di erence between indices) depend signi cantly on wavelength. They also depend strongly on temperature. Table 6.1 stresses this message, showing, as an example, ordinary and extraordinary indices in KDP at a few selected wavelengths and at two temperatures.

On the other hand, Table 6.2 shows a list of birefringent crystals of practical interest. Because of what was pointed out just now, the gures in the table are indicative, and are meant to give the reader just a feeling of the orders of magnitude of the indices and of the birefringence. To solve actual technical problems, the reader should look for more detailed data vs. wavelength and vs. temperature, which can be found in many books and handbooks (e.g., Yariv and Yeh, 1984; Dmitriev *et al.*, 1991; Tosco, Ed., 1990).

Table 6.2 *Some typical birefringent crystals.*

Uniaxial: positive birefringence			
		n_O	n_z
Quartz		1.54	1.55
Zircon		1.92	1.97
Rutile		2.62	2.90
ZnS		2.35	2.36
negative birefringence			
(ADP) $(NH_4)H_2PO_4$		1.52	1.48
(KDP) KH_2PO_4		1.51	1.47
Calcite		1.66	1.49
$LiTaO_3$		2.19	2.18
$LiNbO_3$		2.29	2.20
$BaTiO_3$		2.42	2.37
Biaxial			
	n_x	n_y	n_z
Feldspar	1.522	1.526	1.530
Mica	1.552	1.582	1.588
Topaz	1.619	1.620	1.627
$YAlO_3$	1.923	1.938	1.947

6.6 Gyrotropic media

An anisotropic dielectric is referred to as *gyrotropic*, if there is a real rectangular reference frame, (x_1, x_2, x_3), where the dyadic $\bar{\ }_c$ is represented by a matrix of the form:

$$(\ _c) = \begin{pmatrix} _{11} & _{12} & 0 \\ \ _{12} & _{11} & 0 \\ 0 & 0 & _{33} \end{pmatrix} . \tag{6.47}$$

This de nition makes use of a particular reference frame, and this may appear as a weak point. Indeed, we might replace it with other de nitions based on intrinsic properties. The reason why this one was preferred, is because, in the cases of practical interest, physical considerations make it quite evident that one direction is "special" for that medium. Then, we choose it as the x_3 axis of the reference frame. On the other hand, all the directions orthogonal to the special one are interchangeable. In fact, let us show that the matrix in Eq. (6.47) is invariant under any rotation of the reference frame around the x_3 axis. Such a rotation does not a ect, obviously, the x_3 components, while its e ects in the (x_1, x_2) plane are described by the sub-matrices:

$$(M) = \begin{pmatrix} \cos\vartheta & \sin\vartheta \\ \ \sin\vartheta & \cos\vartheta \end{pmatrix} , \qquad (M)^{\ 1} = \begin{pmatrix} \cos\vartheta & \ \sin\vartheta \\ \sin\vartheta & \cos\vartheta \end{pmatrix} , \tag{6.48}$$

where ϑ is the angle of rotation around x_3, in the sense that in the new frame the upper-left (2 2) submatrix of Eq. (6.47) is replaced by:

$$(M)^{\ 1}\begin{pmatrix} {}_{11} & {}_{12} \\ {}_{12} & {}_{11} \end{pmatrix}(M) = \begin{pmatrix} {}_{11} & {}_{12} \\ {}_{12} & {}_{11} \end{pmatrix} \quad , \tag{6.49}$$

whence we draw the conclusion that Eq. (6.47) does not change. In short, the choice of x_1 and x_2, on the plane normal to x_3, is completely arbitrary.

The examples given in the following sections will enlighten all these statements, and provide them with a stronger physical meaning.

An anisotropic magnetic medium such that, in a real rectangular reference frame, its dyadic $\bar{\ }$ is represented by a matrix with the same features as Eq. (6.47), is also called gyrotropic, or rather *gyromagnetic*. Most of what we say in this section on dielectric gyrotropic media applies in a straightforward way to gyromagnetic media as well. In Section 6.9 we will discuss an example of such a medium.

Combining Eq. (6.47) with Eq. (6.10), i.e., the condition $(\ _c) = (\ _c)^{\dagger}$ for an anisotropic medium to be lossless, we see that in a lossless gyrotropic medium all the diagonal elements of the matrix in Eq. (6.47) are real, whereas all the o -diagonal ones are imaginary:

$$_{11} = \ _{11} \quad , \qquad _{33} = \ _{33} \quad , \qquad _{12} = \ \ _{12} \quad . \tag{6.50}$$

In this section, similarly to what we did in the previous ones, we will focus on lossless media, gaining in simplicity without losing too much in generality. We will study plane waves which travel along one of the axes of the reference frame where Eq. (6.47) applies. We will postpone propagation in an arbitrary direction to the next section.

First, let us see whether uniform plane waves can propagate in a direction orthogonal to x_3, for example along x_1. These results will apply, of course, to any direction in the (x_1, x_2) plane, as a consequence of Eq. (6.49). By de nition of uniform plane waves, it must be $\nabla^2\mathbf{E} = \ \ k^2\mathbf{E}$ and $\nabla(\nabla\ \mathbf{E}) = \ \ k^2(\mathbf{E}\ \hat{a}_1)\,\hat{a}_1$, where $\hat{a}_1$ is the unit vector of the x_1 axis. Thus, Eq. (6.12) reads now:

$$\omega^2\ \ \bar{\ }\ \ \mathbf{E} = k^2\Big[\mathbf{E}\quad(\mathbf{E}\ \ \hat{a}_1)\,\hat{a}_1\Big] \quad . \tag{6.51}$$

Note that this equation is reminiscent of Eq. (6.30); this is not a pure coincidence. Writing now Eq. (6.51) by components, we reach easily the conclusion that it has two solutions, namely:

$$\begin{aligned} \mathbf{E} &= E_3\,\hat{a}_3 \qquad\qquad\qquad\qquad , & k^2 &= \omega^2\ \ \ _{33} \quad , \\ \mathbf{E} &= E_2\left(\ \ \frac{_{12}}{_{11}}\,\hat{a}_1 + \hat{a}_2\right) \quad , & k^2 &= \omega^2\ \left(\ _{11} + \frac{^{2}_{12}}{_{11}}\right) \quad . \end{aligned} \tag{6.52}$$

This shows that there are two possible states of polarization, and that their phase velocities are di erent. At rst sight, this is similar to birefringent media (see Section 6.3), but there is a subtle di erence, as one of the allowed states of polarization (the one in the (x_1, x_2) plane) is elliptical, since $\ _{12}$ is imaginary.

Furthermore, note that this wave can be evanescent, and actually is evanescent for $|\ _{11}| < |\ _{12}|$.

When the phase vector **k** is *parallel to x_3 axis*, the picture is completely di erent. The matrix representation of Eq. (6.51) reads then:

$$\omega^2 \begin{pmatrix} _{11} & _{12} \\ _{12} & _{11} \end{pmatrix} \begin{pmatrix} E_1 \\ E_2 \end{pmatrix} = k^2 \begin{pmatrix} E_1 \\ E_2 \end{pmatrix} \quad . \tag{6.53}$$

The characteristic equation of the matrix in Eq. (6.53) is:

$$(\ _{11} \quad)^2 = \ ^2_{12} \quad , \tag{6.54}$$

whose solutions are:

$$_{L,R} = \ _{11} \quad j\ _{12} \quad . \tag{6.55}$$

Because of the conditions in Eq. (6.50), or, equivalently, since the matrix ( ) is Hermitian, both eigenvalues of ($_t$), in Eq. (6.55), are *real*. The corresponding normalized eigenvectors:

$$(\ _L) = \frac{1}{\sqrt{2}} \begin{pmatrix} 1 \\ j \end{pmatrix} \quad , \qquad (\ _R) = \frac{1}{\sqrt{2}} \begin{pmatrix} 1 \\ \ j \end{pmatrix} \tag{6.56}$$

represent the clockwise and counterclockwise rotating unit-vectors, respectively, well known from Chapter 2, namely $\hat{L} = 1/\sqrt{2}\,(\hat{a}_1 + j\hat{a}_2)$ and $\hat{R} = 1/\sqrt{2}\,(\hat{a}_1 \quad j\hat{a}_2)$. The net result is that uniform plane waves may travel along the x_3 axis, provided they are *circularly polarized*. Their propagation constants are given by:

$$k_L^2 = \omega^2 \ (\ _{11} + j\ _{12}) \quad , \tag{6.57}$$

$$k_R^2 = \omega^2 \ (\ _{11} \quad j\ _{12}) \quad . \tag{6.58}$$

These results entail very important consequences, to be discussed in Section 6.10.

6.7 The Appleton-Hartree formula

Continuing along the guidelines of the previous section, let us deal now with waves traveling in an arbitrary direction, neither x_3 nor orthogonal to it. It may be helpful to change reference frame. We take new rectangular coordinates, y_1, y_2, y_3, with the y_3 axis parallel to the wave vector, and with the medium preferential direction x_3 lying in the (y_1, y_3) plane. Let $\ell_1, \ell_2\ (= 0)$, and $\ell_3 = (1 \quad \ell_1^2)^{1/2}$ be the direction cosines of the "old" x_3 axis with respect to the new axes y_i. Then, the reader may check, as an exercise, that in the new frame the dyadic $\bar{\ }$ is represented by the following matrix:

$$\begin{pmatrix} \ell_3^2\ _{11} + \ell_1^2\ _{33} & \ell_3\ _{12} & \ell_1\ell_3(\ _{33} \quad _{11}) \\ \ell_3\ _{12} & _{11} & \ell_1\ _{12} \\ \ell_1\ell_3(\ _{33} \quad _{11}) & \ell_1\ _{12} & \ell_1^2\ _{11} + \ell_3^2\ _{33} \end{pmatrix} \quad . \tag{6.59}$$

We saw in Section 6.3 that for a uniform plane wave in a birefringent medium the vector **D** is orthogonal to y_3, and belongs to the plane de ned

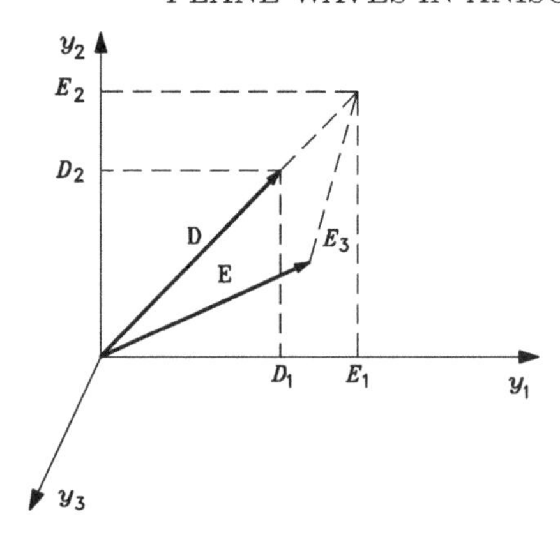

Figure 6.5 *The electric eld vector and the electric displacement vector in the* (y_1, y_2, y_3) *reference system.*

by y_3 and by the vector **E**. This result still holds in the case at hand, as one can prove as an exercise. In the reference frame $\{y_i\}$, we can express these statements through the following formulas (see Figure 6.5):

$$D_3 = 0 \quad ,$$

$$\frac{D_2}{D_1} = \frac{E_2}{E_1} = \quad jp \quad , \tag{6.60}$$

where p is the linear polarization ratio of the D eld, as de ned in Chapter 2. Combining Eq. (6.60) with the constitutive relation $\mathbf{D} = \bar{\ } \ \mathbf{E}$ (written by components using the matrix in Eq. (6.59)), we get the following second-degree equation in p:

$$\ell_3 \ \ _{12} \ \ _{33} p^2 + j\ell_1^2(\ _{11} \ \ _{33} \quad ^2_{11} \quad ^2_{12})p \quad \ell_3 \ \ _{12} \ \ _{33} = 0 \quad . \tag{6.61}$$

Solving Eq. (6.61) we get the *polarizations* of the two uniform plane waves which can propagate in the y_3 direction. Consider a *lossless* gyrotropic medium (i.e., $_{12}$ imaginary and $_{11}$, $_{33}$ real). Then, when both Eqs. (6.57) and (6.58) turn out to be positive (i.e., if $| \ _{12}| < \ _{11}$), Eq. (6.61) divided by the imaginary unit has real coe cients, and a real and positive discriminant, which automatically entails two real roots. From the properties of the polarization ratio p discussed in Chapter 2, we conclude then that *the axes of the polarization ellipse coincide with* y_1 *and* y_2.

When the medium is lossy, the last points are no longer true. However, an interesting property can be derived in any case, as follows: the rst coe cient and the last one in Eq. (6.61) are equal and opposite; hence, the product of the two roots of the equation has to be 1:

$$p_1 \ \ p_2 = \quad 1 \quad . \tag{6.62}$$

Again because of the properties of p seen in Chapter 2, Eq. (6.62) entails that the two corresponding time-dependent vectors **D** describe ellipses which

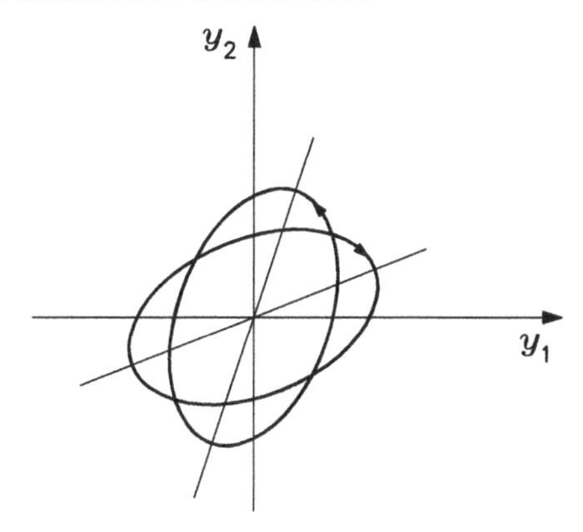

Figure 6.6 *Schematic illustration of the two allowed states of elliptical polarization.*

are *mirror images* of each other, with respect to the line $y_1 = y_2$, i.e., the straight line at 45 degrees between the y_1 and y_2 axes (see Figure 6.6).

We can derive now an expression for the refractive index, de ned always through Eq. (6.26). To this purpose, rewrite Eq. (6.27) as

$$\begin{pmatrix} D_1 \\ D_2 \end{pmatrix} = n^2 \ {}_{0} \begin{pmatrix} E_1 \\ E_2 \end{pmatrix} \quad , \qquad D_3 = 0 \quad . \tag{6.63}$$

Next, write D_1, D_2 and D_3 $(= 0)$ as functions of E_1, E_2 and E_3, using the matrix in Eq. (6.59). The third equation obtained in this way allows us to express E_3 as a function of E_2 and E_1, namely:

$$E_3 = \frac{1}{\ell_1^2 \ {}_{11} + \ell_3^2 \ {}_{33}} \left[(\ {}_{11} \quad {}_{33}) \ell_1 \ell_3 E_1 + \ell_1 \ {}_{12} E_2 \right] \quad . \tag{6.64}$$

If we insert this into the expressions for D_1, and eliminate E_2 by means of Eq. (6.60), we end up with the following general expression for n^2:

$$n^2 = \frac{{}_{33}}{{}_{0}} \frac{{}_{11} + j\ell_3 \ {}_{12} p}{\ell_1^2 \ {}_{11} + \ell_3^2 \ {}_{33}} \quad . \tag{6.65}$$

Replacing p with $\quad 1/p$ thanks to Eq. (6.62), we get

$$n^2 = \frac{{}_{33}}{{}_{0}} \frac{\ell_3 \ {}_{12} + j \ {}_{11} p}{jp(\ell_1^2 \ {}_{11} + \ell_3^2 \ {}_{33})} \quad . \tag{6.66}$$

If p is real (lossless medium, ${}_{12}$ imaginary), then n^2 is real. If we suppose $j\ell_3 \ {}_{12} p < 0$, then, Eq. (6.66) is always positive, while Eq. (6.65) is either positive or negative, depending on whether ${}_{11} > \ell_3 | \ {}_{12} p|$, or ${}_{11} < \ell_3 | \ {}_{12} p|$. However, the complete opposite occurs if we suppose $j\ell_3 \ {}_{12} p > 0$. When n^2 is negative, we have an *evanescent wave*, whose physical meaning will be outlined in Section 6.10.

In a lossy medium, n^2 is complex. Hence, *any* wave traveling in such a medium undergoes an attenuation.

What we said concerning waves traveling in an arbitrary direction in a gyrotropic medium is of practical interest to radio engineers, as it applies to *ionospheric propagation*, where the preferential direction x_3 is that of the terrestrial magnetic eld. The direction of propagation, y_3, and the direction x_3 de ne a plane (in our notation, it is the $y_2 = 0$ plane) which is usually referred to as the "magnetic meridian plane". The form taken by Eq. (6.65) or by Eq. (6.66), when one inserts values for $_{ij}$ which are typical of the ionosphere, is known in the literature as the *Appleton-Hartree formula* (Budden, 1961).

6.8 An example of permittivity dyadic

An e.m. wave traveling in an ionized gas is always accompanied by a convection current, because of the motion of electrons and, to a smaller extent, of positive ions, although they are much heavier than electrons. In this section we will see how, under suitable assumptions, this convection current can be accounted for, in a complex permittivity dyadic.

The equation of motion for an individual electron in an e.m. eld, driven by the Lorentz force in Eq. (1.2), neglecting collisions, is:

$$m\,\frac{d\,\mathbf{v}}{dt} = \quad e(\mathbf{E} + \mathbf{v} \quad \mathbf{B}) \quad , \tag{6.67}$$

where m = electron rest mass (= $9.1091\ 10^{\ \ 31}$ kg), e = electron charge magnitude (= $1.6021\ 10^{\ \ 19}$ C), $\mathbf{v}$ = instantaneous velocity. For a uniform plane wave in a medium whose intrinsic impedance does not di er too much from that of free space, the ratio $|\mathbf{B}|/|\mathbf{E}|$ is so small that, for non-relativistic particles (i.e., for $|\ \mathbf{v}\ | \quad c$), the second term on the right-hand side of Eq. (6.67) is negligible, compared to the rst one. From now on, this term will be omitted. However, it happens often, in plasma physics laboratories as well as in nature in the ionosphere, that an ionized gas is placed in a strong *static* magnetic eld, whose induction vector will be denoted as $\mathbf{B}_0$. In the presence of this $\mathbf{B}_0$ and in the absence of a time-harmonic eld (i.e., for $\mathbf{E} = 0$), Eq. (6.67) becomes the equation of motion along a helix, whose axis is parallel to the ow lines of $\mathbf{B}_0$. It is rather intuitive that propagation of a wave in such a medium is strongly anisotropic, with signi cant di erences between the direction of $\mathbf{B}_0$ and all the orthogonal ones.

Let us assume, for simplicity, that $\mathbf{B}_0$ is constant throughout the region of interest, and introduce rectangular coordinates with the x_3 axis parallel to $\mathbf{B}_0$. From Eq. (6.67) one sees that for $B_0 \to \infty$ the radius of curvature of the helical electron trajectories tends to zero. So, let us rst make a simpli ed analysis under the assumption $B_0 = \infty$, which forbids the electrons to move in the plane transverse to x_3 so that all quantities of interest depend only on x_3 and time.

In the time domain, the convection current density can then be written as:

$$J_3 = \quad v_3 \quad , \tag{6.68}$$

where is the instantaneous electron charge density.[‡] The electrons may move with a nonvanishing average velocity, v_0, referred to as the *drift velocity*. In the rest of this section, we will use a subscript 0 for the dc components, and a subscript a for the ac components. Then, Eq. (6.68) reads:

$$J_0 + J_a = (\ _0 + \ _a)\,(v_0 + v_a) \quad . \tag{6.69}$$

Clearly, $J_0 = \ _0 v_0$. Assuming that $_a$ and v_a are both time-harmonic, with an angular frequency ω, then the time-dependent part contains terms at angular frequencies ω and 2ω. Assuming that the latter are negligibly small, we may write:

$$J_a = \ _0\, v_a + \ _a\, v_0 \quad . \tag{6.70}$$

Our next, time-consuming task is to derive from Eq. (6.70) how the ac electric eld is related to the ac current density J_a. In fact, to link v_a with the electric eld, we note rst that, if we adopt a uid-dynamics viewpoint, we may write, in general:

$$\frac{d\,v_3}{dt} = \frac{\partial\,v_3}{\partial t} + \frac{\partial\,v_3}{\partial x_3}\frac{\partial x_3}{\partial t} = \frac{\partial\,v_3}{\partial t} + v_3\,\frac{\partial\,v_3}{\partial x_3} \quad . \tag{6.71}$$

The last term, a product of two quantities at ω, can be neglected as we did a moment ago with $_a v_a$. Introducing then new symbols (capital letters) such that $v_a = \mathrm{Re}(V_a\, e^{j\omega t})$, etc., the left-hand side of Eq. (6.67) can be replaced by:

$$m\left(j\omega\, V_a + v_0\,\frac{\partial V_a}{\partial x_3}\right) \quad . \tag{6.72}$$

Assume now that all ac quantities depend on x_3 as exp(x_3), where is still unknown. Using calligraphic capital letters to indicate that $V_a = \mathcal{V}_a\, e^{\ \ x_3}$, etc., Eq. (6.67) becomes:

$$(j\omega \quad v_0)\,\mathcal{V}_a = \ \frac{e}{m}\,\mathcal{E}_3 \quad . \tag{6.73}$$

To reach our nal goal, we need now a suitable expression for $_a$, to be used in Eq. (6.70). We get it from the continuity Eq. (1.29), namely:

$$\nabla \ \mathbf{J} = \ \frac{\partial}{\partial t} \quad , \tag{6.74}$$

which, under our assumptions, yields:

$$J_a = \ j\omega\, \mathcal{P}_a \quad , \tag{6.75}$$

where $_a = \mathrm{Re}\{\mathcal{P}_a\, e^{j\omega t}\, e^{\ \ x_3}\}$. Re-writing now Eq. (6.70) for the complex

[‡] This analysis neglects electrostatic forces. This implies an underlying assumption, namely that the ionized gas, on average, is neutral, i.e., the average ion charge density is equal, in absolute value, to the electron charge density. When this assumption is not veri ed, as in a particle beam, then the results that we will nd here hold only as a rst approximation.

quantities represented by capital letters, and using for $\mathcal{V}_a$ and $\mathcal{P}_a$ the expressions which follow from Eqs. (6.73) and (6.75), we get:

$$J_a = \quad j\omega \frac{\quad_a e}{m(j\omega \quad v_0)^2} \mathcal{E}_3 \quad . \tag{6.76}$$

De ning the quantity:

$$\omega_p = \left(\frac{N e^2}{_0 m}\right)^{1/2} \quad , \tag{6.77}$$

which is referred to as the *plasma angular frequency*, Eq. (6.76) becomes:

$$J_a = \quad j\omega \;_0 \frac{\omega_p^2}{(j\omega \quad v_0)^2} \mathcal{E}_3 \quad . \tag{6.78}$$

From our initial assumption that any electron motion perpendicular to x_3 is forbidden, it follows that in the planes orthogonal to x_3 the medium behaves like free space. So, we may write:

$$(\;_c) = \begin{pmatrix} _0 & 0 & 0 \\ 0 & _0 & 0 \\ 0 & 0 & _{33} \end{pmatrix} \quad , \tag{6.79}$$

with:

$$_{33} = \;_0 \left[1 \quad \frac{\omega_p^2}{(\omega + j \quad v_0)^2}\right] \quad . \tag{6.80}$$

In these expressions, $\;_a v_a$ has been neglected with respect to ($_0\, v_a +\; _a$ v_0), and $V_a\, \partial V_a / \partial z$ has been neglected with respect to $(j\omega V_a + v_0\, \partial V_a/\partial z)$. Consequently, when $v_0 = 0$, i.e., for a plasma at rest, these formulas are correct only if $_0 \quad _a$ and $|\omega/ \quad| \quad V_a$, i.e., if the wave phase velocity is large, compared with the electron speed.

Now let us remove the assumption $B_0 = \infty$, and take into account, in the electron motion, the components orthogonal to $\mathbf{B}_0$. The starting point is again Eq. (6.67), where we still neglect the contribution $\mathbf{v} \quad \mathbf{B}$, due to the ac component of $\mathbf{B}$. Eq. (6.68) is replaced by the complete vector equation:

$$\mathbf{J} = \quad \mathbf{v} \quad . \tag{6.81}$$

Assume, for simplicity, that the average velocity is equal to zero ($\mathbf{v}_0 = 0$). Then, the ac component of Eq. (6.81) can be written (neglecting, as usual, products of ac components on the right-hand side), as:

$$\mathbf{J}_a = \;_0\, \mathbf{v}_a \quad . \tag{6.82}$$

Suppose now that a uniform plane wave propagates in this medium, in an arbitrary direction. Then, all complex quantities representing time-harmonic functions are products of constants times a function of spatial coordinates of the type:

$$\exp(\quad \mathbf{r}) = \exp(\quad _1 x_1 \quad _2 x_2 \quad _3 x_3) \quad . \tag{6.83}$$

Under these circumstances, the left-hand side of Eq. (6.67) can be written as:

$$m\,(j\omega \qquad {}_1\mathcal{V}_{a_1} \qquad {}_2\mathcal{V}_{a_2} \qquad {}_3\mathcal{V}_{a_3})\,\overline{\mathcal{V}}_a \quad . \tag{6.84}$$

If the particle velocities are small compared to the wave phase velocity, then Eq. (6.84) can be approximated as:

$$j\omega\, m\,\overline{\mathcal{V}}_a \quad . \tag{6.85}$$

In the same reference frame as before (x_3 parallel to $\mathbf{B}_0$), Eq. (6.67) may be written by components in a matrix form as:

$$\begin{pmatrix} j\omega & \dfrac{eB_0}{m} & 0 \\ \dfrac{eB_0}{m} & j\omega & 0 \\ 0 & 0 & j\omega \end{pmatrix} \begin{pmatrix} \mathcal{V}_{a_1} \\ \mathcal{V}_{a_2} \\ \mathcal{V}_{a_3} \end{pmatrix} = \quad \frac{e}{m} \begin{pmatrix} \mathcal{E}_1 \\ \mathcal{E}_2 \\ \mathcal{E}_3 \end{pmatrix} \quad . \tag{6.86}$$

Inverting the matrix in Eq. (6.86), and setting:

$$\omega_c = \frac{e}{m}\, B_0 \quad , \tag{6.87}$$

(this quantity is referred to as the *cyclotron angular frequency*), we get:

$$\begin{pmatrix} \mathcal{V}_{a_1} \\ \mathcal{V}_{a_2} \\ \mathcal{V}_{a_3} \end{pmatrix} = \quad j\,\frac{e}{m} \begin{pmatrix} \dfrac{\omega}{\omega_c^2 \quad \omega^2} & \dfrac{j\omega_c}{\omega_c^2 \quad \omega^2} & 0 \\ \dfrac{j\omega_c}{\omega_c^2 \quad \omega^2} & \dfrac{\omega}{\omega_c^2 \quad \omega^2} & 0 \\ 0 & 0 & \dfrac{1}{\omega} \end{pmatrix} \begin{pmatrix} \mathcal{E}_1 \\ \mathcal{E}_2 \\ \mathcal{E}_3 \end{pmatrix} \quad . \tag{6.88}$$

Going back to Eq. (6.82), and from this to Maxwell's equations, we nally nd the dyadic $\overline{}_c$ we were looking for. In our reference frame, it is represented by a matrix of the type in Eq. (6.47) with the following elements:

$$\begin{aligned} {}_{11} &= \quad {}_0\left[1 + \frac{\omega_p^2}{\omega_c^2 \quad \omega^2}\right] \quad , \\ {}_{12} &= \quad {}_{21} = j\,\frac{{}_0\omega_c}{\omega}\,\frac{\omega_p^2}{\omega_c^2 \quad \omega^2} \quad , \\ {}_{33} &= \quad {}_0\left[1 \quad \frac{\omega_p^2}{\omega^2}\right] \quad . \end{aligned} \tag{6.89}$$

Note that the denominators in ${}_{11}$, ${}_{12}$ and ${}_{21}$, all vanish if $\omega = \omega_c$. The medium has a resonant behavior when the angular frequency of the e.m. wave equals the angular velocity of the helical electron motion around the ow lines of the vector $\mathbf{B}_0$.

It is left to the reader as an exercise to show that, under suitable conditions, the medium exhibits an evanescence, as brie y mentioned in Section 6.7. The reader should also elaborate on the physical relationship between this evanescence and the frictionless motion of charged particles.

6.9 Second example of permeability dyadic

In this section, we will be obliged to use concepts, terminology and notation taken from magnetostatics, a discipline which is not dealt with in this book. We take it for granted that the reader had been exposed to these notions in basic physics courses. An interest in refreshing them can easily be satis ed by consulting classical textbooks on static elds (e.g., Plonsey and Collin, 1961).

In the so-called semiclassical treatment of ferromagnetic media (i.e., if a minimal amount of Quantum Mechanics is invoked), the magnetic dipole moment due to the individual electron spin, $\mathbf{m}$, is looked at as the result of a rotation of the electron around its axis, with an angular momentum, $\mathbf{Q}$, proportional to $\mathbf{m}$. In formulas:

$$\mathbf{m} = g\,\mathbf{Q} \quad , \tag{6.90}$$

where g is called the *gyromagnetic constant* of the electron ($g = \;\; 1.76 \;\; 10^{11}\,\mathrm{T/s}$). When immersed in a magnetic induction eld $\mathbf{B}$, the dipole moment $\mathbf{m}$ experiences a torque:

$$= \mathbf{m} \quad \mathbf{B} \quad . \tag{6.91}$$

Hence, the dynamics of the vector $\mathbf{m}$ obeys the equation:

$$\frac{d\mathbf{m}}{dt} = g\,\mathbf{m} \quad \mathbf{B} \quad . \tag{6.92}$$

In a saturated magnetic material, if we neglect, as a rst approximation, demagnetizing factors related to the shape of the material itself, we may write:

$$\mathbf{B} = \;_0(\mathbf{H} \quad \mathbf{M}) = \;_0(\mathbf{H} + N_0\,\mathbf{m}) \quad , \tag{6.93}$$

where N_0 is the spin moment density (number per unit volume). Multiplying both sides of Eq. (6.92) times N_0, and using Eq. (6.93), we get:

$$\frac{d\,\mathbf{M}}{dt} = \quad g \;_0\,\mathbf{H} \quad \mathbf{M} \quad . \tag{6.94}$$

This is easily recognized as the equation of a precession around the vector $\mathbf{H}$.

If the $\mathbf{H}$ eld has a dc component, called *orienting eld*, and if this component, $H_0\hat{a}_3$, is parallel to the x_3 axis, then, in the saturation regime, the static component of $\mathbf{M}$ is also parallel to $\hat{a}_3$. We will call it $M_0\hat{a}_3$, and use a subscript a to denote complex quantities which represent ac components.

From Eq. (6.94), we then get:

$$\begin{pmatrix} j\omega & g \ {}_0 H_0 & 0 \\ g \ {}_0 H_0 & j\omega & 0 \\ 0 & 0 & j\omega \end{pmatrix} \begin{pmatrix} M_{a_1} \\ M_{a_2} \\ M_{a_3} \end{pmatrix} = g \ {}_0 H_0 \begin{pmatrix} H_{a_2} \\ H_{a_1} \\ 0 \end{pmatrix} \quad . \tag{6.95}$$

Solving this for the vector $\mathbf{M}_a$, and replacing it into:

$$\mathbf{B}_a = {}_0(\mathbf{H}_a + \mathbf{M}_a) \quad \bar{} \ \mathbf{H}_a \quad , \tag{6.96}$$

we get, in our reference frame, a matrix representation for the dyadic $\bar{}$, whose elements are:

$${}_{11} = {}_{22} = {}_0 \left[1 + \frac{\omega_0 \omega_M}{\omega_0^2 \quad \omega^2} \right] \quad ,$$

$${}_{12} = {}_{21} = j \frac{{}_0 \omega \omega_M}{\omega_0^2 \quad \omega^2} \quad ,$$

$${}_{33} = {}_0 \quad , \quad {}_{13} = {}_{23} = {}_{31} = {}_{32} = 0 \quad , \tag{6.97}$$

where we set:

$$\omega_0 = g \ {}_0 H_0 \quad ,$$

$$\omega_M = g \ {}_0 M_0 \quad . \tag{6.98}$$

The minus signs are due to the fact that $g < 0$ for an electron.

Eqs. (6.97) show that, according to the de nition given in Section 6.6, a magnetized ferrite is indeed a gyrotropic medium. The angular frequency ω_0 plays a role which is very similar to that of the cyclotron angular frequency ω_c in a gyrotropic plasma. Physically, in fact, both these quantities are related to angular velocities (of rotation around $\mathbf{B}_0$ for ω_c, of precession around $\mathbf{H}_0$ for ω_0). Eqs. (6.97) exhibit a resonant behavior centered at $\omega = \omega_0$, like Eqs. (6.89).

Examples of actual microwave devices which exploit gyromagnetic media, as well as of very similar optical devices, will be given at the end of the next section.

6.10 Faraday rotation

Some features of plane waves traveling along the preferential direction, x_3, in a gyrotropic medium, were derived in Section 6.6. However, the subject was left open for further discussion. We will see in this section that those waves provide a very interesting example — probably, the example which is most widely used in practice — of nonreciprocal electromagnetic behavior. Let us start o with two preliminary remarks, restricting ourselves, once again, to lossless media.

A) As a matter of argument, assume that ${}_{12}$ is imaginary and positive. Then, the right-hand side of Eq. (6.58) is always positive. The right-hand side of Eq. (6.57) is positive or negative, depending on whether $_{11} > | \ _{12}|$, or $_{11} < | \ _{12}|$. In the latter case, the root k_L is imaginary: the corresponding wave undergoes an attenuation along x_3, while its phase does not change in this direction. The medium is assumed to be lossless, so an attenuation cannot be attributed to power dissipation. It must originate from an *evanescence*, similarly to what we saw in Chapter 4.

In general, $_{11}$ and $_{12}$ are frequency dependent. So, if ω is changed, one can pass from a situation where both circularly polarized waves are uniform plane waves, to a situation where one of them is evanescent, while the other one is still uniform.

B) The terminology "clockwise" or "counterclockwise," that we adopted in Section 6.6 for the rotation of the instantaneous vectors represented either by $\hat{R}$ or by $\hat{L}$, Eqs. (6.56), refers to *an observer looking at the positive face of the* $x_3 = 0$ *plane*, regardless of whether the wave is traveling in the positive or negative sense of the x_3 axis. This entails that:

> as long as we deal with waves traveling in the sense of increasing x_3, the wave which rotates like a left-handed screw has a propagation constant equal to the positive square root of Eq. (6.57), the wave which rotates like a right-handed screw has a propagation constant equal to the positive square root of Eq. (6.58);
>
> when we deal with waves traveling in the sense of decreasing x_3, *the two propagation constants are interchanged*, i.e., the negative square root of Eq. (6.58) for the left-handed screw, the negative square root of Eq. (6.57) for the right-handed screw.

It is precisely this exchange between propagation constants when the sense in which the waves travel is reversed, that is responsible for nonreciprocity, as we will see in a shortwhile.

Suppose now that, at the frequency of interest, k_L and k_R are both real. Let a uniform plane wave impinge on a plane orthogonal to x_3, for example on the $x_3 = 0$ plane, traveling in the direction of increasing x_3, and assume that its electric eld is linearly polarized along a direction lying in the $x_3 = 0$ plane, say x_1 (Figure 6.7). This vector **E** can be written by components in the basis Eq. (6.56), which consists of the eigenvectors of the matrix ( ) and therefore is easy to track as we move through the gyrotropic medium. In fact, the electric eld at any value of x_3 can be simply written as:

$$\mathbf{E} = \frac{E_0}{\sqrt{2}} \hat{L}\, e^{\ jk_L x_3} + \frac{E_0}{\sqrt{2}} \hat{R}\, e^{\ jk_R x_3} \quad . \tag{6.99}$$

The vector in Eq. (6.99), sum of two circularly polarized vectors which are equal in modulus (for any x_3), remains (see Chapter 2) linearly polarized everywhere. However, at a generic x_3, its direction no longer coincides with the input one. It is at an angle ϑ, with respect to the x_1 axis, that can be

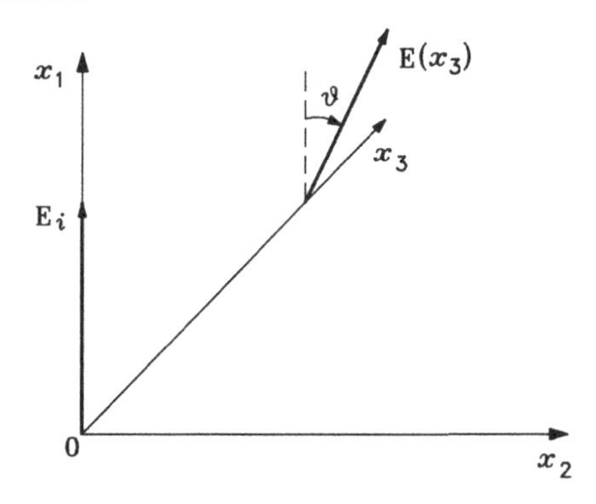

Figure 6.7 *Schematic illustration of Faraday rotation.*

found if we rewrite Eq. (6.99) in the form:

$$\mathbf{E} = E_0 \left[\hat{a}_1 \cos\left(\frac{k_L \quad k_R}{2}\, x_3\right) + \hat{a}_2 \sin\left(\frac{k_L \quad k_R}{2}\, x_3\right)\right] e^{\;j\frac{k_L+k_R}{2}x_3} . \tag{6.100}$$

From this, we nd:

$$\vartheta = \arctan\frac{E_2}{E_1} = \frac{k_L \quad k_R}{2}\, x_3 \quad . \tag{6.101}$$

So, we see that the direction of polarization is rotated by an angle proportional to x_3, the proportionality factor (sometimes referred to as *circular birefringence*) being $(k_L \quad k_R)/2 \neq 0$ for $_{12} \neq 0$, i.e., for a gyrotropic medium. This phenomenon is known as the *Faraday rotation.*

As k_L and k_R are interchanged when propagation is reversed (see the previous comment B) this rotation is peculiar in that the *sign of the rotation angle* (with respect to a xed reference frame) *does not change when the direction of propagation is reversed.* Consequently, let us write, rather than Eq. (6.101):

$$\vartheta = \frac{k_L \quad k_R}{2}\, |x_3| \quad . \tag{6.102}$$

This implies that the reciprocity theorem (Section 3.4) does not apply to the medium at hand. To prove this statement, let us deal with two sets of sources, satisfying the following requirements. The rst set generates a wave which travels in the positive x_3 direction, and whose electric eld equals Eq. (6.99) on the $x_3 = 0$ plane. The second set generates a wave which travels in the negative x_3 direction, and whose electric eld on the plane $x_3 = \ell$ equals Eq. (6.99) in amplitude, but is polarized at an angle $\vartheta_0 = (k_L \quad k_R)\ell/2$ with respect to the x_1 axis. Let us compute the *reactions* of these two elds on the sources, according to the de nition given in Section 3.4. It is easily seen that, except for the very special cases where ϑ_0 is an integer multiple of 2 , the two reactions *are not equal.* This proves, as we intended to do, that we are facing a nonreciprocal phenomenon.

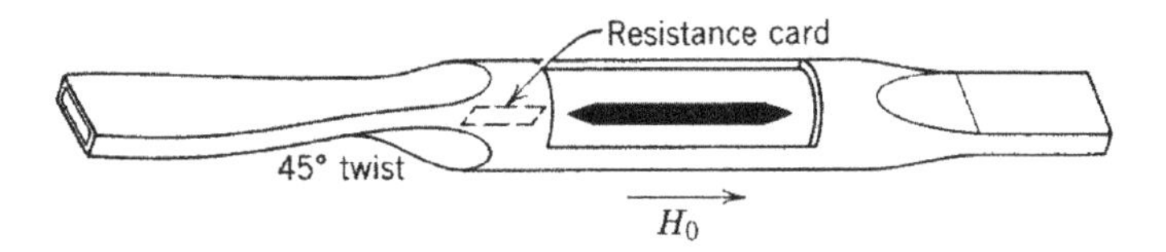

Figure 6.8 *Illustration of a microwave isolator.*

Faraday rotation is widely exploited in practice, especially at microwave and optical frequencies. One of the simplest nonreciprocal devices, and at the same time one of the most useful ones, is the so-called *isolator*. Its function is to introduce a strongly nonreciprocal attenuation — ideally, no attenuation in one direction, an in nite attenuation in the opposite one — aiming at making an oscillator insensitive to any change in its load. The most common types of isolators (although not the only ones) are based on Faraday rotation. Figure 6.8 provides a schematic illustration of its waveguide version at microwave frequencies, Figure 6.9 shows an optical implementation. The two are much more similar in their essence than at rst sight. Although the rst one was developed several decades before the second, we will brie y comment only on the latter one, because understanding some features of the rst requires concepts to be developed in Chapter 7. The reader should come back to it after the chapter on conducting-wall waveguides.

The optical isolator is a Faraday rotator (whose gyromagnetic medium is, usually, a garnet crystal), designed in such a way that the rotation angle ϑ equals /4, and placed between two linear polarizers that are rotated with respect to each other also by /4. As illustrated schematically in the gure, a backward wave is strongly attenuated by polarizer A, while a forward wave passes with very low attenuation through polarizer B.

Another Faraday-rotation-based nonreciprocal device which is widely used at microwave frequencies is the so-called *circulator*. It is an n-port junction, with n 3, de ned, in its ideal version, as follows. When a wave impinges on the i-th port, all the incident power comes out from the $i + 1$-th port (understanding that port $n + 1$ is the same as port 1). Nothing is re ected, or transmitted to the remaining "decoupled" n 2 ports. Figure 6.10 provides a schematic illustration of a 3-port circulator in a waveguide con guration (part a), and the symbol used for it in circuit diagrams (part b).

A qualitative explanation of how a Y-junction 3-port circulator operates may proceed as follows. Because of the conducting walls in the planes orthogonal to the static induction eld, the clockwise and counterclockwise waves of the previous analysis are no longer traveling, but standing waves. Their propagation constants along the vertical axis remain di erent. As they are not in an unbounded medium, but in a closed environment, this di erence in longitudinal propagation constant entails, through Maxwell's equations, that their *azimuthal* phase velocities (i.e., the velocities at which they rotate around the static eld) are di erent too. Then, a suitable design can make the two

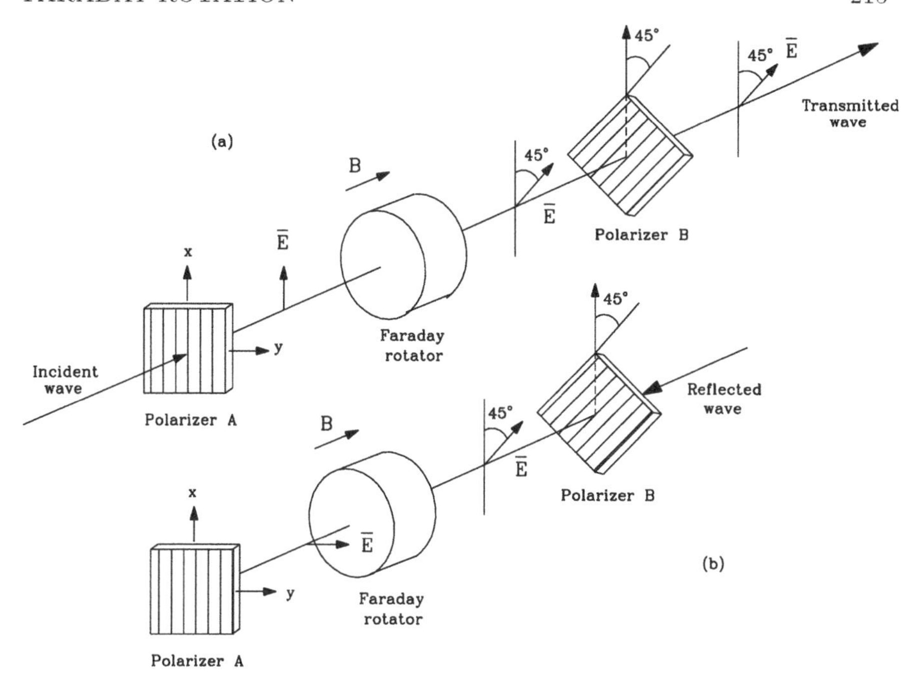

Figure 6.9 *Illustration of an optical isolator.*

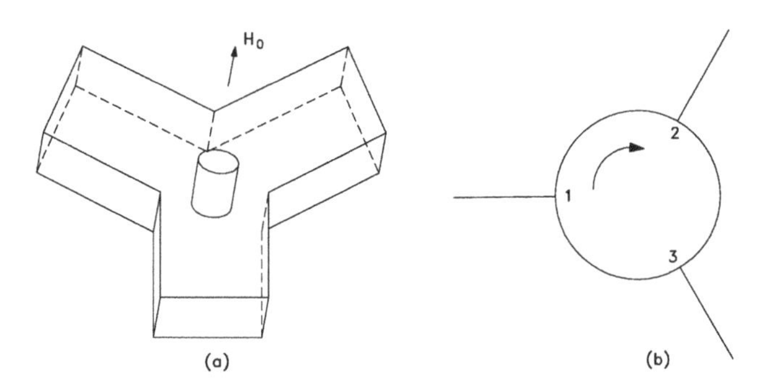

Figure 6.10 *A 3-port microwave circulator. (a) View of a waveguide con guration. (b) Symbol in use for circuit diagrams.*

waves excited by a eld entering through port 1 meet in phase at port 2, in counterphase at port 3, whence the conclusion.

A typical use of the 3-port circulator is in radars, where it allows the transmitter (connected to port 1) and the receiver (connected to port 3) to share the same antenna (connected to port 2) without "seeing" each other, which would be extremely detrimental since the transmitted power is many orders of magnitude larger than the received one.

In addition to its technical relevance, which is illustrated also by some of the problems at the end of the chapter, Faraday rotation is very important as a concept. It stresses that, if we want to avoid any ambiguity in discussing reciprocity in an anisotropic medium, it is strictly mandatory to use real reference frames, and to deal only with linearly polarized elds. To clarify the meaning of this statement, let us recall that, as pointed out in Section 6.6, the matrix ( ) of a gyrotropic medium becomes diagonal (and therefore, symmetric) in a complex reference frame where two of the basis unit vectors are $\hat{R}$ and $\hat{L}$. Consequently, circularly polarized plane waves traveling in a gyrotropic medium obey equations that are *formally* identical to those governing uniform plane waves in isotropic media. This might induce one to claim that circularly polarized waves in a gyrotropic medium behave reciprocally, but such a statement does not agree with the de nition of reciprocity, based on the reaction, that we gave in Section 3.4. The origin of this apparent ambiguity is that the *reaction*, de ned through a scalar product between two complex vectors, without taking the complex conjugate of one of its factors, *is not an invariant with respect to any change of basis in a complex space.* In particular, the reaction does change, in general, when we pass from linearly polarized basis vectors to rotating basis vectors.

6.11 Further applications and suggested reading

This chapter contains only those fundamental notions about anisotropic media, that may be considered essential ingredients, nowadays, in the education of an advanced engineer who intends to work in the area of information technology. The level of knowledge which is necessary in order to become a specialist in propagation in anisotropic media is orders of magnitude larger than the contents of this short chapter. The list of further suggested readings is potentially unlimited.

Similarly to what we said at the end of Chapter 4, we rst stress that linear anisotropic media are very pervasive in physics and in engineering. This statement applies, of course, also to the mathematical formalism which is normally used for their analysis, i.e., tensors and their matrix representations, that is shared by many di erent elds of application (e.g., Babuska and Cara, 1991; Hijden, 1987; Nayfeh, 1995 and references that treat the propagation of mechanical, e.g., seismic waves in anisotropic media). To broaden one's own general background in physical sciences, one should at least examine a broad-scope, comprehensive book, like, for instance, Nye (1985), where the mathematical properties of anisotropic media are presented in detail, and then followed by applications, ranging from magnetic and electric properties, to theory of elasticity, from transport phenomena to crystal optics. An extremely important concept that the reader can grasp from such a book is the physical relevance, in some problems, of tensors whose rank is higher than 2 (2 is, of course, the rank of dyadics, that we used in this chapter). Other treatments of crystal optics include Juretschke (1974).

For what refers, more speci cally, to further reading on electromagnetic waves in anisotropic media, let us mention the following examples. *Optics of crystals* is the title of a famous chapter in Born and Wolf (1980), where the beauty of the presentation matches the richness of the scienti c contents. Our emphasis is on electrical anisotropy. Materials with magnetic anisotropy are treated in, e.g., Tarling and Hrouda (1993) — see also Negi and Saraf (1989). For a treatment of electromagnetic wave propagation in bianisotropic media, we refer to Kong (1990) and Staelin *et al.* (1993). Propagation in magneto-plasmas, with emphasis on ionospheric phenomena, is thoroughly dealt with in Budden (1961). Those who are interested in the fundamentals of plasma physics and magnetohydrodynamics can nd them, e.g., in Jackson (1975), Chandrasekhar (1989), Sitenko and Malnev (1995), and Dendy, Ed., (1993), while Miyamoto (1989) focuses on applications of plasmas in thermonuclear fusion for energy production, and Sturrock (1994) includes material regarding astrophysical and geophysical as well as laboratory plasmas.

As we already mentioned in Section 6.5, many among the experiments that opened, in the '60s, the new eld of nonlinear optics — in bulk media, at that time — were performed in anisotropic media, birefringence being a crucial ingredient for obtaining phase matching. The fundamentals of nonlinear optics, as well as their links with propagation in anisotropic media, are well covered, e.g., in Bloembergen (1996) — see also Boyd (1992), Butcher and Cotter (1990), Newell and Moloney (1992), Shen (1984), and Yariv and Yeh (1984).

Much more recently, thanks to the availability of very-low-loss optical bers, where interaction lengths could become as large as several kilometers, *guided-wave* nonlinear optics became much more attractive than its bulk precursor. Fundamentals on nonlinear phenomena in waveguides can be found, e.g., in Agrawal (1995). More speci cally, Someda and Stegeman, Eds., (1992) covers the area where guided-wave propagation overlaps with anisotropy (for example, random birefringence in optical bers) and with nonlinearities. Propagation problems in anisotropic waveguides are an area where we may state — anticipating something to be seen in Chapter 10 — very few problems can be solved analytically in closed form. Numerical methods become then extremely important; a relevant example is provided by Zoboli and Bassi in Someda and Stegeman, Eds., (1992). The reader should remember what we said in Section 3.9 about symmetries: careful exploitation of spatial symmetries in numerical methods can yield enormous savings in processing time and in memory occupation. Notice, however, that in anisotropic media symmetry operations are legitimate only if the dyadic operators that characterize materials commute with the symmetry operators that characterize the geometry of the region where the eld is de ned.

Problems

6-1 When the dyadics $\bar{}_c$ and $\bar{}$ are non-singular, i.e., when both have an inverse, then Eqs. (6.3) and (6.4) can be inverted, expressing the electric and magnetic eld vectors as functions of the displacement and induction vectors. Write the corresponding expression of the Poynting vector. Derive conditions under which the factors contained in this expression commute.

6-2 Show that if at least one of the dyadics $\bar{\ }_c$ and $\bar{\ }$ is not Hermitian, then the quantity in Eq. (6.8) may have a nonvanishing real part. Find a way to distinguish media where this real part is positive (lossy media) and those where it is negative (media with gain).

6-3 Write Eq. (6.12) by components, in a rectangular reference frame, in a region without imposed currents.

6-4 Show that, in those cases where Eq. (6.12) becomes a Helmholtz equation, the same occurs to Eq. (6.19), provided the $\varphi = 0$ gauge is adopted.

6-5 Starting from the contents of Section 6.2, and applying the duality theorem of Section 3.7, discuss the case of a magnetic anisotropic medium in terms of an electric vector potential **F**, with the same physical dimensions as in Section 1.9.

6-6 Using duality, analyze a magnetic medium whose constitutive relations are the dual of those of the birefringent dielectric medium studied in Section 6.3.

6-7 Show that, if a plane wave is launched into a lossless birefringent medium in a generic direction and with a generic state of polarization, then its state of polarization evolves *periodically* as a function of the distance traveled by the wave. Hence, express the period as a function of the ordinary and extraordinary indices for that direction of propagation.
Does the state of polarization still evolve periodically in the presence of loss?

6-8 In a rutile crystal (a uniaxial dielectric medium whose refractive indices are on the list in Table 6.2), plane waves propagate along the straight line at 45 degrees between the x and z axes, z being the optic axis. Evaluate the values of the ordinary and extraordinary refractive indices. Use both approaches explained in the text: rst, nd the eigenvalues of the reduced inverse permittivity dyadic $\bar{\mathbf{b}}_t$. Next, solve Fresnel's equation of the wave normals as shown in Section 6.5.

6-9 A rutile slab is cut with planar parallel faces, spaced by 10 mm, orthogonal to the y axis. The surrounding medium is air. The incident wave impinges with an incidence angle $\vartheta_i = \ /4$. The plane of incidence is the (x, y)-plane. Applying Snell's law and the Fresnel formulas of Chapter 4, nd the directions of propagation of the TE and TM refracted waves (note the so-called *double refraction*), and the transmission coe cients. What is the angle between the directions of propagation of ordinary and extraordinary waves when they emerge in air after the slab? Is this result valid in general?

6-10 Find the angles between the phase vectors and the Poynting vectors (ray directions) for the refracted waves within the slab of the previous problem.

6-11 Prove that, in a uniaxial medium where n_0 is an increasing function of frequency f, phase matching between fundamental frequency and second harmonic can be achieved only if the extraordinary wave is the second harmonic when Eq. (6.45) applies; only if it is the fundamental frequency when Eq. (6.46) applies. Prove that the reverse is true, if the ordinary index is a decreasing function of frequency.

6-12 One method of fabricating an optical polarizer is to cut a prism of uniaxial medium in such a way that a plane wave, emerging from it towards air, undergoes total internal re ection for one polarization (either ordinary, or extraordinary), but not for the other one. Find the range of values for the angle of incidence for which this occurs, both for calcite and for zircon (see Table 6.2). Specify the totally re ected wave, in both cases.

6-13 As mentioned brie y in Chapter 2, a *quarter-wave plate* is a slab of birefringent material which converts a linearly polarized input wave into a circularly polarized output wave. Consider three slabs of mica (see Table 6.2), each of them cut normally to one of the axes x, y, z. The incident wave is linearly polarized at 45 degrees with respect to the axes laying in the input plane of the slab.

a) Calculate the (minimum) /4-slab thicknesses in the three con gurations.
b) In reality, if the *incident* wave (in air) is linearly polarized, the output polarization will be slightly elliptical because the re ection coe cients along the two principal axes will be slightly di erent. Evaluate the deviation from circular polarization, in terms of percentage.

6-14 Show that, if a plane wave is launched into a lossless gyrotropic medium in the x_3 direction, with a generic state of polarization, then its state of polarization evolves periodically as a function of the distance traveled by the wave (provided both circularly polarized “eigenwaves” propagate at the frequency of interest). Express the period as a function of the propagation constants of the two circularly polarized eigenwaves. Does the state of polarization still evolve periodically with distance in the presence of loss?

6-15 Prove that a lossless anisotropic medium, where the only plane waves that propagate in one direction (x_3) preserving their state of polarization, are circularly polarized, is gyrotropic.

6-16 Prove that the change in reference frame described at the beginning of Section 6.7 yields Eq. (6.59).

6-17 Show that, analogous to what was shown in Section 6.3, for a uniform plane wave in a gyrotropic medium where the reference frame of Section 6.7 is used, the vector **D** is orthogonal to y_3 and, if the wave is linearly polarized, it belongs to the plane de ned by y_3 and by the electric eld vector **E**.

6-18 Study under which conditions one of the plane waves in the medium studied in Section 6.8 is evanescent. Can both waves be evanescent ?

6-19 In the highly ionized layers of the Earth's ionosphere, electron densities are of the order of 10^{12} m ^{3}. A typical value for the terrestrial static magnetic induction eld is 3 10 ^{5} T.

a) Calculate the corresponding plasma frequency and cyclotron frequency.

b) Repeat the calculation for a laboratory plasma with 10^{16} electrons per cubic meter and a 10 T induction eld.

6-20 Sketch the plots of the dispersion diagrams, $k(\omega)$ vs. ω, for the two circularly polarized waves whose phase constants are expressed by Eqs. (6.57) and (6.58), when the elements of the permittivity matrix are expressed by Eqs. (6.89). Consider separately the cases $\omega_p > \omega_c$, $\omega_p < \omega_c$.

6-21 Evaluate the intensity of the static magnetic eld H_0 which is required in order to have a resonant frequency $f_0 = 1\,\text{GHz}$, in a saturated ferrite.

6-22 To build a microwave isolator (see Section 3.4) based on Faraday rotation, a saturated ferrite bar is used, with a static magnetic eld, $\mathbf{H}_0$, applied in the direction along which the wave propagates.

a) Approaching the problem in the dual way with respect to a gyrotropic dielectric medium, express the phase constants of the circularly polarized waves, and then the per-unit-length Faraday rotation, for a gyromagnetic medium with $= {}_0$, and the permeability dyadic found in Section 6.8. *Note*: assume that the bar cross-section is wide enough to approximate the propagation constants in it with those of plane waves in an unbounded medium.

b) Evaluate the intensity of the static magnetic eld required in order to have a Faraday rotation of /4 radians over a length $\ell = 30\,\text{mm}$ at a frequency $f = 5\,\text{GHz}$. The saturation magnetization M_0 equals 10^6 A/m.

References

Agrawal, G.P. (1995) *Nonlinear Fiber Optics.* 2nd ed., Academic, San Diego.

Babuska, V. and Cara, M. (1991) *Seismic Anisotropy in the Earth.* Kluwer Academic, Boston.

Bloembergen, N. (1996) *Nonlinear optics.* 4th ed., World Scienti c, River Edge, NJ.

Born, M. and Wolf, E. (1980) *Principles of Optics: Electromagnetic Theory of Propagation, Interference and Di raction of Light.* 6th (corrected) ed. (reprinted 1986), Pergamon, Oxford.

Boyd, R.W. (1992) *Nonlinear Optics.* Academic Press, Boston.

Budden, K.G. (1961) *Radio Waves in the Ionosphere.* Cambridge University Press, Cambridge, UK.

Butcher, P.N. and Cotter, D. (1990) *The Elements of Nonlinear Optics.* Cambridge University Press, Cambridge, UK.

Chandrasekhar, S. (1989) *Plasma Physics, Hydrodynamic and Hydromagnetic Stability, and Applications of the Tensor-Virial Theorem.* University of Chicago Press, Chicago.

Dendy, R. (Ed.) (1993) *Plasma Physics: An Introductory Course.* Cambridge University Press, UK.

Dmitriev, V.G., Gurzadyan, G.G. and Nikogosian, D.N. (1991) *Handbook of Nonlinear Optical Crystals.* Springer-Verlag, Berlin.

Hijden, J.H.M.T. van der (1987) *Propagation of Transient Elastic Waves in Strati ed Anisotropic Media.* North Holland, Amsterdam.

Jackson, J.D. (1975) *Classical Electrodynamics.* Wiley, New York.

Juretschke, H.J. (1974) *Crystal Physics: Macroscopic Physics of Anisotropic Solids.* Benjamin, Advanced Book Program, Reading, MA.

Kong, J.A. (1990) *Electromagnetic Wave Theory.* 2nd ed., Wiley, New York.

Miyamoto, K. (1989) *Plasma Physics for Nuclear Fusion.* Rev. ed., MIT Press, Cambridge, MA.

Nayfeh, A.H. (1995) *Wave Propagation in Layered Anisotropic Media, with Applications to Composites.* Elsevier, Amsterdam.

Negi, J.G. and Saraf, P.D. (1989) *Anisotropy in Geoelectromagnetism.* Elsevier, Amsterdam.

Newell, A.C. and Moloney, J.V. (1992) *Nonlinear Optics.* Advanced topics in the interdisciplinary mathematical science, Addison-Wesley, Redwood City, CA.

Nye, J.F. (1985) *Physical Properties of Crystals: Their Representation by Tensors and Matrices.* 1st paperback edition with corrections and new material, Clarendon Press, Oxford, UK.

Plonsey, R. and Collin, R.E. (1961) *Principles and Applications of Electromagnetic Fields.* McGraw-Hill, New York.

Shen, Y.R. (1984) *The Principles of Nonlinear Optics.* Wiley, New York.

Sitenko, A.G. and Malnev, V. (1995) *Plasma Physics Theory.* Chapman & Hall,

London.

Smirnov, V.I. (1961) *Linear Algebra and Group Theory.* McGraw-Hill, New York.

Someda, C.G. and Stegeman, G.I. (Eds.) (1992) *Anisotropic and Nonlinear Waveguides.* Elsevier, Amsterdam.

Staelin, D.H., Morgenthaler, A.W. and Kong, J.A. (1993) *Electromagnetic Waves.* Prentice-Hall, Englewood Cli s, NJ.

Sturrock, P.A. (1994) *Plasma Physics: An Introduction to the Theory of Astrophysical, Geophysical, and Laboratory Plasmas.* Cambridge University Press, UK.

Tarling, D.H. and Hrouda, F. (1993) *The Magnetic Anisotropy of Rocks.* Chapman & Hall, London.

Tosco, F. (Ed.) (1990) *Fiber Optic Communications Handbook.* TAB Professional and Reference Books, Blue Ridge Summit, PA.

Yariv, A. and Yeh, P. (1984) *Optical Waves in Crystals: Propagation and Control of Laser Radiation.* Wiley, New York.

Zernike, F. and Midwinter, J.E. (1973) *Applied Nonlinear Optics.* Wiley, New York.

CHAPTER 7

Waveguides with conducting walls

7.1 Introduction

What we saw in Chapters 4, 5 and 6 indicates that there are no physical reasons why in a homogeneous medium the energy carried by an electromagnetic wave should travel along a given direction. Indeed, Chapters 12 and 13, devoted to the fundamentals of antenna theory, will recon rm that sources in a homogeneous medium cannot radiate just along one linear path, but necessarily in all directions, at least within a given cone. Guided propagation, which means sending almost all the energy conveyed by an electromagnetic wave along a chosen path, inevitably requires an inhomogeneous medium. However, depending on the applications and especially on the frequency in use, nonhomogeneities can either consist of combining dielectrics and conductors, or merely using di erent dielectric media.

In principle, electromagnetic elds penetrate in all media whose conductivity is nite, and satisfy continuity conditions on the surfaces between di erent media. In practice, we may encounter two drastically di erent situations. In a structure which consists only of dielectrics, in none of them is the eld negligible, even as a rst approximation. On the other hand, the step from a good conductor (de ned as in Section 4.12) to an *ideal conductor*, in which the eld is identically zero, is, at least in the rst approach to the problem, an excellent trade-o between being realistic and simplifying the mathematics. In fact, it allows replacement of the continuity conditions at the interface between dielectric and metal with *boundary conditions* for the eld, de ned only in the dielectric.

For historical reasons, but above all for their mathematical simplicity which follows from what we said so far, conducting-wall waveguides will be dealt with rst, in this chapter. Dielectric waveguides will be studied later, in Chapter 10. The subject will not be tackled in terms of plane waves, although their completeness, proved in Section 4.7, would enable us to do that. In fact, a di erent approach will yield more compact results, practically useful and formally elegant at the same time. We will invoke a plane wave approach only at some points where it provides an intuitive explanation for some of the results obtained in the other way.

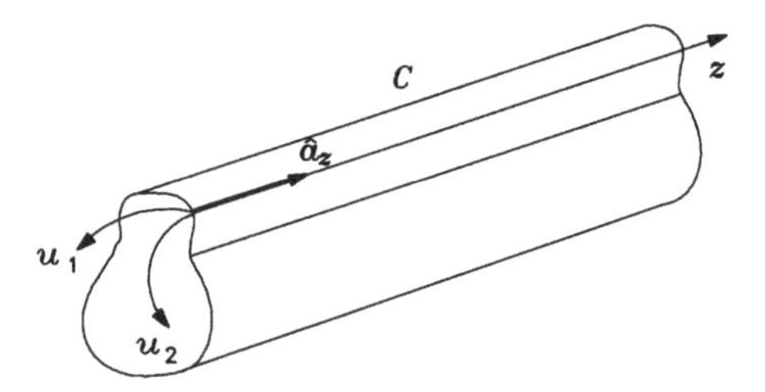

Figure 7.1 *A waveguide of unspeci ed geometry, and in nite length.*

7.2 Homogeneously lled cylindrical structures: Simpli ed proof of the TE-TM decomposition theorem

Let us consider, in real 3-D space, a region lled with a homogeneous medium and surrounded by walls, whose physical nature does not have to be speci ed now, which extend to in nity along one direction, z, and whose cross-section is constant vs. z. We will summarize these assumptions calling this region a "*cylindrical structure* in the z direction". Let $\hat{a}_z$ be the unit vector in the z direction, u_1, u_2 be two generic orthogonal curvilinear coordinates in the plane orthogonal to z (see Figure 7.1). We will show that, under these circumstances, the TE-TM decomposition theorem can be proved in a greatly simpli ed way, compared with Section 3.8, while at the same time a physical identi cation is possible for the rather mysterious pre-potentials that were introduced at that stage. As we anticipated in Section 3.8, there is some price to pay for this simpli cation. One is the need to make an ansatz (easy to justify) for the z-dependence of the elds, the other one is a loss in generality, to be discussed and by-passed later in this section.

As for the ansatz, it consists of the following assumptions, which can nd their justi cation either in what we saw in Section 4.7 or, *a posteriori*, in the results that will follow from them, in particular for what concerns completeness of the solutions:

$$
\begin{aligned}
\mathbf{E}(u_1, u_2, z) &= \mathcal{E}(u_1, u_2)\ \exp(\quad z) \quad , \\
\mathbf{H}(u_1, u_2, z) &= \mathcal{H}(u_1, u_2)\ \exp(\quad z) \quad .
\end{aligned}
\tag{7.1}
$$

In Section 3.8 it was shown that any eld in such a region can be expressed as the sum of two elds which are transverse electric (TE) and transverse magnetic (TM) with respect to an arbitrary real unit vector $\hat{a}$. From now on, unless the opposite is explicitly stated, TE and TM will be understood *with respect to* $\hat{a}_z$. To simplify the proof as much as possible, let us deal, for the time being, with rectangular coordinates ($u_1 = x$, $u_2 = y$), and write Maxwell's equations by components using the symbolic determinant notation (see Appendix B) for the curl operator. We get:

$$
\frac{\partial \mathcal{E}_z}{\partial y} + \quad \mathcal{E}_y = \quad j\omega \quad \mathcal{H}_x \quad ,
$$

$$\begin{aligned}
\mathcal{E}_x \quad \frac{\partial \mathcal{E}_z}{\partial x} &= j\omega \quad \mathcal{H}_y \quad , \\
\frac{\partial \mathcal{H}_z}{\partial y} + \quad \mathcal{H}_y &= j\omega \;_c\, \mathcal{E}_x \quad , \\
\mathcal{H}_x \quad \frac{\partial \mathcal{H}_z}{\partial x} &= j\omega \;_c\, \mathcal{E}_y \quad , \qquad (7.2)
\end{aligned}$$

plus the two components along z, which we omit here as they are not relevant to the following. These equations can be looked at as an algebraic system of four linear equations in the four unknowns $\mathcal{E}_x, \mathcal{E}_y, \mathcal{H}_x, \mathcal{H}_y$, whose solution is simply found to be:

$$\begin{aligned}
\mathcal{E}_x &= \frac{1}{\;^2 \quad ^2}\left(\quad \frac{\partial \mathcal{E}_z}{\partial x} + j\omega \quad \frac{\partial \mathcal{H}_z}{\partial y}\right) \quad , \\
\mathcal{E}_y &= \frac{1}{\;^2 \quad ^2}\left(\quad \frac{\partial \mathcal{E}_z}{\partial y} \quad j\omega \quad \frac{\partial \mathcal{H}_z}{\partial x}\right) \quad , \\
\mathcal{H}_x &= \frac{1}{\;^2 \quad ^2}\left(\quad j\omega \;_c \frac{\partial \mathcal{E}_z}{\partial y} + \quad \frac{\partial \mathcal{H}_z}{\partial x}\right) \quad , \\
\mathcal{H}_y &= \frac{1}{\;^2 \quad ^2}\left(j\omega \;_c \frac{\partial \mathcal{E}_z}{\partial x} + \quad \frac{\partial \mathcal{H}_z}{\partial y}\right) \quad , \qquad (7.3)
\end{aligned}$$

where, as usual, $\;^2 = \quad \omega^2 \quad _c$.

These equations prove the rst part of the decomposition theorem, namely that any eld in the region of interest can be expressed as the sum of a TE and a TM eld. The transverse electric eld consists of $\mathcal{H}_z$ and of four transverse components expressed by the terms in the right column, on the right-hand side of Eqs. (7.3). The transverse magnetic eld consists of $\mathcal{E}_z$ and of four transverse components expressed by the terms in the left column, on the right-hand side of Eqs. (7.3). However, it is extremely important to point out that this proof holds *only for* $\;^2 \neq \;^2$. Otherwise, the quantity in the denominators on the right-hand side goes to zero. Implications of this restriction will be discussed shortly.

Before we proceed to the second part of the decomposition theorem, we note, incidentally, that Eqs. (7.3) can be rewritten in an intrinsic form, independent of the (x, y) reference frame, as follows:

$$\begin{aligned}
\mathcal{E}_t &= \frac{1}{\;^2 \quad ^2}\left(\quad \nabla_t \mathcal{E}_z + j\omega \quad \hat{a}_z \quad \nabla_t \mathcal{H}_z\right) \quad , \\
\mathcal{H}_t &= \frac{1}{\;^2 \quad ^2}\left(\quad \nabla_t \mathcal{H}_z \quad j\omega \;_c\, \hat{a}_z \quad \nabla_t \mathcal{E}_z\right) \quad , \qquad (7.4)
\end{aligned}$$

where the transverse gradient ∇_t is de ned through the position:

$$\nabla = \nabla_t + \hat{a}_z \frac{\partial}{\partial z} = \nabla_t \quad \hat{a}_z \quad , \qquad (7.5)$$

and can be expressed in a generic system of orthogonal curvilinear coordinates u_1, u_2 as:

$$\nabla_t = \left(\frac{\hat{t}_1}{h_1}\right) \left(\frac{\partial}{\partial u_1}\right) + \left(\frac{\hat{t}_2}{h_2}\right) \left(\frac{\partial}{\partial u_2}\right) \quad , \tag{7.6}$$

where h_1, h_2 are the metric coe cients of the (u_1, u_2) reference frame.

The second part of the decomposition theorem says that the TE part of the eld can be derived from a scalar function which satis es the scalar Helmholtz equation, and the same is true for the TM part. In the case at hand, it is straightforward to show that these two scalar functions are (apart from proportionality constants, which are needed to reconcile them with Section 3.8) *the longitudinal eld components* H_z and E_z, respectively. In fact, Eqs. (7.3) show, on one hand, that when these two components are known, the entire eld can be calculated by means of straightforward operations. On the other hand, the elds **H** and **E**, in a homogeneous medium without sources, satisfy the vector Helmholtz equation, as we saw in Chapter 1, and therefore their components along the straight axis z satisfy the scalar Helmholtz equation, as we had to prove.

In more detail, if we de ne a transverse Laplacian ∇_t^2 through the position:

$$\nabla^2 = \nabla_t^2 + \frac{\partial^2}{\partial z^2} = \nabla_t^2 + \quad {}^2 \quad , \tag{7.7}$$

we see that the two functions of the two transverse coordinates u_1, u_2 satisfy an equation of the type:

$$\nabla_t^2 \mathcal{F} = \quad {}^2 \mathcal{F} \quad , \tag{7.8}$$

where $\mathcal{F}$ is either $\mathcal{H}_z$ or $\mathcal{E}_z$, and:

$${}^2 = \quad (\ {}^2 \quad {}^2) \neq 0 \quad . \tag{7.9}$$

This is an eigenvalue equation, similar to the Helmholtz equation but different from it because in the 3-D case the eigenvalue $\ {}^2 = \quad {}^2$ is unique, and is known when the frequency and the medium are speci ed, whereas in Eq. (7.8) the eigenvalue is unknown. We will see shortly that indeed Eq. (7.8), at a given frequency, has an in nity of discrete solutions, for suitable boundary conditions. Each eigenvalue corresponds to a value of , whose physical meaning is, according to Eq. (7.1), that of a propagation constant in the z direction. We understand then that di erent solutions of Maxwell's equations of the form in Eq. (7.1), corresponding to di erent eigenvalues in Eq. (7.8), will have di erent propagation properties. The "spatial spectrum" of the total eld which can exist in a cylindrical structure will be radically di erent from a plane-wave spectrum.

The link between the quantities and the symbols used in this section and those of Section 3.8 are simply found, by comparison, to be:

$$\mathcal{L} = \quad \mathcal{H}_z / \ {}^2 \quad , \qquad T = \quad {}_c\mathcal{E}_z / \ {}^2 \quad . \tag{7.10}$$

Hence, if one follows the approach of Section 3.8, the expressions in Eq. (3.47) and (3.48) for the TE and TM elds, respectively, become the following ones, where symbols like $\mathcal{L}$ and $\mathcal{T}$ indicate, as in the rest of this section, that the exponential dependence on z has been factored out:

$$\begin{aligned} \mathcal{E}_{\mathrm{TE}} &= \quad j\omega \, \nabla_t \mathcal{L} \quad \hat{a}_z \quad , \\ \mathcal{H}_{\mathrm{TE}} &= \quad \frac{1}{\;}\left(\;^2 \quad \;^2 \right) \mathcal{L} \, \hat{a}_z \quad - \nabla_t \mathcal{L} \quad , \end{aligned} \tag{7.11}$$

$$\begin{aligned} \mathcal{E}_{\mathrm{TM}} &= \quad \frac{1}{\;_c}\left(\;^2 \quad \;^2 \right) \mathcal{T} \, \hat{a}_z + \frac{\;}{\;_c} \nabla_t \mathcal{T} \quad , \\ \mathcal{H}_{\mathrm{TM}} &= \quad j\omega \, \nabla_t \mathcal{T} \quad \hat{a}_z \quad . \end{aligned} \tag{7.12}$$

Comparing Eqs. (7.12) (which, being based on the general proof of Section 3.8, are valid also for $\;^2 = 0$) and Eqs. (7.4), we see that a eld corresponding to the eigenvalue $\;^2 = 0$ can be non-identically zero only if:

$$\mathcal{E}_z \quad 0 \quad , \qquad \mathcal{H}_z \quad 0 \quad , \tag{7.13}$$

i.e., for a *transverse electro-magnetic* (TEM) eld. We will see shortly that the loss in generality caused by the restriction $\;^2 \neq 0$ is not harmful, because the guided propagation of a TEM eld is quite peculiar. Indeed, its rst peculiarity, which can be derived starting either from Eqs.(7.11) or from Eqs. (7.12), is that:

$$\mathcal{E} = \quad \mathcal{H} \quad \hat{a}_z \quad , \tag{7.14}$$

where $\; = (\; / \;_c)^{1/2}$ is, as usual, the intrinsic impedance of the homogeneous medium in the cylindrical structure. The sign is either + or depending on whether $\; = + \;$ or $\; = \;$. Hence, for a TEM eld the propagation constant and the wave impedance coincide with those of a uniform plane wave traveling in the z direction, in an unbounded medium with the same characteristics as those of the medium which lls the cylindrical structure. We will see in Section 7.4 that none of the other solutions of Maxwell's equations in a waveguide can have these properties. In the same section we will also see the conditions under which a cylindrical structure can propagate a TEM eld. For the time being, let us point out that for a TEM wave, the elds, the potentials and the pre-potentials, as functions of the transverse coordinates, all obey an equation of the following type:

$$\nabla_t^2 \mathcal{F} = 0 \quad , \tag{7.15}$$

which is the 2-D Laplace equation. Then, we will be entitled to invoke all the uniqueness theorems which hold for this equation, with which the reader should be familiar from electrostatics.

To close this section, let us emphasize that we proved that any eld in a homogeneous cylindrical structure can be expressed as the *sum* of a TE eld, a TM eld and, in some cases to be clari ed, a TEM eld. Whether one of these elds may exist in a given structure *independently* of the others, this can be said only when one knows what *boundary conditions* have to be satis ed in that structure. This will be the subject of Section 7.3 and again of Section 7.9.

7.3 Waveguides with ideal conducting walls

To get involved in the details of guided propagation, let us start from structures where metallic walls shield completely the region where the guided eld is con ned. This approach is convenient, for the reasons that were mentioned in the introduction, both in terms of concepts and of technicalities. The problem is further simpli ed in a drastic manner if we assume that the walls consist of an *ideal* electric conductor (see Section 4.12). Although, obviously, this is not realistic, nevertheless all the results of practical interest can be obtained from this model, at most by introducing further re nements which are usually small and simple. Examples will be given in Section 7.9 and later on.

It is well known that an ideal electric conductor imposes $\mathbf{E}_{\tan} = 0$ on its surface. Then, through Maxwell's equations, also the normal component of $\mathbf{H}$, H_n, vanishes on that surface. Given the ansatz Eq. (7.1), these boundary conditions (not independent of each other) on the contour C of the cylindrical region where the guided eld is de ned (Figure 7.1) read:

$$\mathcal{E}_{\tan} = 0 \quad , \qquad \mathcal{H}_n = 0 \quad . \tag{7.16}$$

We will show now that, if we leave aside the TEM case assuming $^2 \neq 0$, then Eqs. (7.16) are satis ed on C if the following conditions are satis ed on the same surface:

$$\mathcal{E}_z = 0 \qquad \left(\Longleftrightarrow \quad \mathcal{T} = 0 \right) \quad , \tag{7.17}$$

$$\frac{d\mathcal{H}_z}{dn} = 0 \qquad \left(\Longleftrightarrow \quad \frac{d\mathcal{L}}{dn} = 0 \right) \quad , \tag{7.18}$$

where d/dn is, as usual, the normal derivative on C.

To prove this statement, we note rst that Eq. (7.17) implies $\mathcal{E}_z =$ constant on the surface C. Then $\nabla_t \mathcal{E}_z$ is orthogonal to C at any point. Conversely, Eq. (7.18) implies that the normal component of $\nabla_t \mathcal{H}_z$ vanishes on C, so that this vector is tangent to C at any point. Consequently, the general expressions in Eqs. (7.4) for the transverse components show that $\mathcal{E}_t$ is normal to C and $\mathcal{H}_t$ is tangent to C everywhere. If, furthermore, we recall that $\mathcal{E}_z = 0$ on C by assumption, then we conclude that Eqs. (7.16) are satis ed everywhere on C.

Note that Eqs. (7.17) and (7.18) are independent of each other, unlike the two Eqs. (7.16). In particular, Eq. (7.17) is compatible with taking $\mathcal{H}_z$ 0 ($\mathcal{L}$ 0), which means it is applicable to a TM eld. Eq. (7.18) is compatible with taking $\mathcal{E}_z$ 0 ($\mathcal{T}$ 0), meaning it is applicable to a TE eld. A very important consequence that we may draw is that *TE transmission modes* and *TM transmission modes* exist in a cylindrical structure with ideal conducting walls. In Section 7.5 we will see that they form a complete set, in some cases provided that the TEM mode is included also. As we brie y said at the end of Section 7.2, existence and completeness of TE and TM modes are properties which are not shared by all cylindrical structures. They follow from the speci c assumptions in Eq. (7.16).

Finally, note that Eqs. (7.17) and (7.18) are two scalar boundary conditions.

We saw in Section 7.2 that the complete expressions for TE and TM elds may be derived from the solutions of two scalar partial di erential equations. We may then conclude that guided propagation in cylindrical structures with ideal conducting walls can be thought of, under a mathematical point of view, as a *scalar problem*, in spite of the vector physical nature of the guided eld.

7.4 Transmission modes of lossless cylindrical structures

From now on, unless the opposite is explicitly stated, we will assume that the dielectric in the cylindrical structure is *lossless*. The walls were supposed in the previous section to be ideal conductors, so the whole structure is strictly lossless. The quantity $\;^2 = \; \omega^2 \;$ (constant, because the dielectric was supposed to be homogeneous at the beginning of Section 7.2) is real and negative. This property will be exploited several times in this section.

7.4.1 Properties of all transmission modes

A1. Real, positive eigenvalues

All the eigenvalues in Eq. (7.9) of Eq. (7.8) are *real and positive* (except for the zero eigenvalue of the TEM mode). This follows from the 2-D form of the rst Green identity (A.10), namely:

$$\int_S (\varphi \nabla_t^2 \; + \nabla_t \varphi \; \nabla_t \;) \, dS = \oint_\ell \varphi \frac{\partial}{\partial n} \, d\ell \quad , \tag{7.19}$$

which we apply to the cross-section of the structure, S, and to its contour, ℓ, setting $\varphi = \mathcal{F}$, $\; = \mathcal{F}$, $\mathcal{F}$ being any solution of Eq. (7.8). For all modes but the TEM, the boundary conditions in Eqs. (7.17) and (7.18) make the right-hand side of Eq. (7.19) equal to zero. Then, through Eq. (7.8) we get:

$$(\;^2) \int_S \mathcal{F}\mathcal{F} \; dS = \int_S \nabla_t \mathcal{F} \; \nabla_t \mathcal{F} \; dS \quad . \tag{7.20}$$

Both surface integrals in Eq. (7.20) are real and positive, whence the conclusion on $\;^2$.

A2. Critical frequency

For each transmission mode (labeled with an index n, shown later in Section 7.5 to be an integer) there is one positive value (zero for the TEM mode) of the angular frequency ω, at which the longitudinal propagation constant vanishes. Indeed, if we rewrite Eq. (7.9) as:

$$\;_n^2 = \;_n^2 + \;^2 = \;_n^2 \; \omega^2 \quad , \tag{7.21}$$

and set it equal to zero, then it is satis ed for:

$$\omega = \frac{\;_n}{\sqrt{\;}} \quad \omega_{cn} \quad , \tag{7.22}$$

which is a real and positive solution because of the previous property *A1*. The symbol ω_{cn} underlines that this is called the *critical angular frequency* of the n-th mode. It is also called the *cut-o* angular frequency, for reasons to be explained in the next paragraph.

A3. Cut-o of a transmission mode

For the n-th transmission mode, the longitudinal propagation constant $_n$ is *real* at all frequencies smaller than its critical frequency, *imaginary* at all frequencies higher than its critical frequency. To prove it, we start from Eq. (7.21), which indicates that $^2_n > 0$ for $\omega < \omega_{cn}$, $^2_n < 0$ for $\omega > \omega_{cn}$. Hence, we may let:

$$_n \quad = \quad _n = \omega\sqrt{\quad}\ \sqrt{(\omega_{cn}/\omega)^2 \quad 1} \in \mathrm{Re} \qquad \text{for } \omega < \omega_{cn} \quad , \tag{7.23}$$

$$_n \quad = \quad j\ _n = j\omega\sqrt{\quad}\ \sqrt{1 \quad (\omega_{cn}/\omega)^2} \in \mathrm{Im} \qquad \text{for } \omega > \omega_{cn} \quad . \tag{7.24}$$

This tells us that any transmission mode in a waveguide does not propagate (i.e., its phase does not change vs. z) at frequencies below its critical frequency (whence the name "cut-o frequency"). As the structure is strictly lossless, the attenuation described by Eq. (7.23) cannot be caused by energy dissipation. Its physical nature will be clari ed by looking also at the following properties.

A4. Wave impedance

The wave impedance in the z direction is imaginary for $\omega < \omega_{cn}$, and real for $\omega > \omega_{cn}$. To prove this, note that for a TE mode ($\mathcal{E}_z \quad 0$), the electric and magnetic nonvanishing terms in Eqs. (7.4) are in quadrature when Eq. (7.23) holds, are in phase when Eq. (7.24) holds. The same is true for the nonvanishing terms in the case of a TM mode ($\mathcal{H}_z \quad 0$). Further details will be added in the following subsections.

Comparing with what we saw in Chapter 4 concerning propagation constants and wave impedances of plane waves, we may conclude that the physical phenomenon why a mode is attenuated vs. z below its cut-o frequency, is an *evanescence*. Then, the mode cannot convey any active power in the longitudinal direction of the cylindrical structure. However, in strict correspondence to what we saw in Chapter 4, this statement is correct only in the presence of one wave which attenuates in one sense along the z axis, and loses any validity for two superimposed waves which decay in the two opposite senses.

A5. Phase and group velocities

For any mode that can propagate in a cylindrical structure (i.e., at $\omega > \omega_{cn}$), the longitudinal *phase velocity* as a function of frequency can be obtained from Eq. (7.24) and reads:

$$v_{pn} = \frac{\omega}{_n} = c\,\frac{1}{\sqrt{1 \quad \left(\frac{\omega_{cn}}{\omega}\right)^2}} \quad , \tag{7.25}$$

where $c = 1/\sqrt{\ }$ is the speed of light in the dielectric of the cylindrical structure. Eq. (7.25) indicates that v_{pn} tends to in nity for $\omega \to \omega_{cn}$, and decreases monotonically as ω increases, approaching c (from above) as $\omega \to \infty$. From what we saw in Chapter 4 on phase velocity of plane waves in any direction, this result suggests that at its cut-o frequency a guided mode consists of plane waves which travel along directions in the plane orthogonal to $\hat{a}_z$. As ω is increased, i.e., as the wavelength becomes smaller with respect to the transverse geometrical dimensions of the structure, the plane waves the mode consists of, travel along directions which make smaller and smaller angles with that of $\hat{a}_z$ direction, gradually tending to a situation where they ignore the presence of walls.

The *group velocity*:

$$v_{gn} = \frac{1}{\dfrac{\partial \ _n}{\partial \omega}} = c \sqrt{1 \ \left(\frac{\omega_{cn}}{\omega}\right)^2} \quad , \tag{7.26}$$

behaves, vs. frequency, in a way which agrees perfectly with the previous comment. Indeed, it equals zero at cut-o ($\omega = \omega_{cn}$), and increases as the frequency increases, tending asymptotically to c (from below) for $\omega \to \infty$. From Eqs. (7.25) and (7.26) it follows that:

$$v_{gn}\, v_{pn} = c^2 \quad . \tag{7.27}$$

This relationship holds only in the structures that we are studying here. It cannot be extrapolated, unfortunately, to the dielectric waveguides of Chapter 10, where the behavior of group velocity vs. frequency is of paramount importance, and turns out to be much more complicated.

The functions in Eqs. (7.23), (7.24), (7.25), (7.26) are sketched in Figure 7.2. According to the de nitions given in Section 5.2, Eq. (7.26) shows that propagation of a guided mode in a metallic waveguide is a ected by *anomalous dispersion* ($dv_{gn}/d\omega > 0$). Eq. (7.26) also shows that two guided modes having di erent critical frequencies (i.e., di erent eigenvalues) have, at the same frequency, di erent group velocities. Eq. (7.25) shows that they have di erent phase velocities also. Consequently, if several modes propagate together, a modulating signal is distorted because the group delay from the waveguide input to its output has as many values as the propagating modes. This cause of distortion, often referred to as "multipath," in general will be superimposed onto the classical causes of signal distortion, which, as well known, for a two-port linear system read $|H(j\omega)| \neq$ constant, $\angle H(j\omega) \neq a\omega + b$, where $H(j\omega)$ is the system transfer function. By a nity with what we described in Section 5.2, phenomena related to multiple values of the group delay are also referred to as *multimode* (or inter-modal) *dispersion*.

Eq. (7.27) shows that in a waveguide where the dielectric is homogeneous it is never possible, not even at a single value of ω, to have equal phase or group velocities for modes whose cut-o frequencies are di erent. In Chapter 10 we will see that, on the contrary, curves of phase or group velocities vs. ω of

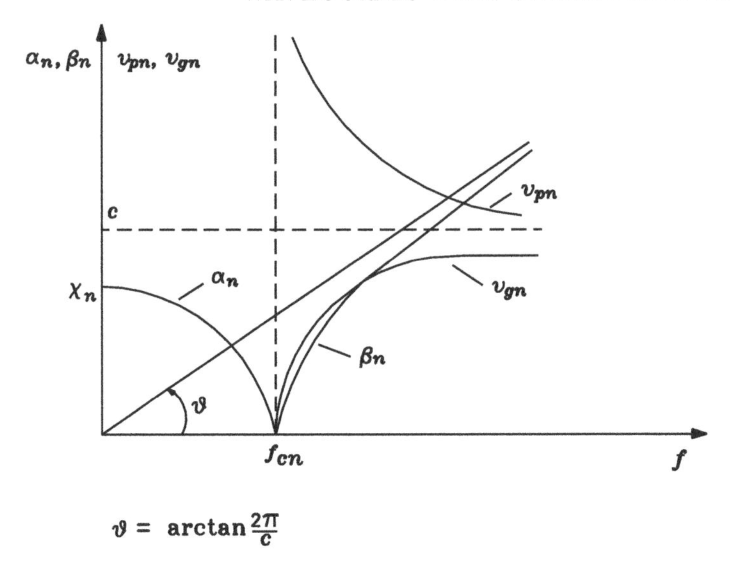

Figure 7.2 *Attenuation constant, phase constant, phase velocity, group velocity vs. frequency, for a generic mode of a lossless metallic waveguide.*

di erent modes may intersect each other in waveguide where the dielectric is not the same in the entire cross-section.

On the other hand, the previous formulas show that two modes having equal eigenvalues have the same phase and group velocities *at any frequency.* Modes which are physically distinct but have equal eigenvalues are called *degenerate.* Their role is quite particular, and becomes extremely important in problems such as power exchange among modes of the same structure or of coupled structures, as we shall clarify later in this chapter and in the two following ones.

A6. The fundamental mode

The mode corresponding to the lowest eigenvalue compatible with the boundary conditions imposed by a given structure, $_0^2$, is called the *fundamental mode* of the structure. If it is not degenerate, it is the only mode which can propagate alone in that structure. This occurs over the frequency range between $f_0 = {}_0c/(2\ )$ and the smallest among the cut-o frequencies of all the other modes. This is the only frequency range where the problems due to intermodal dispersion may be avoided.

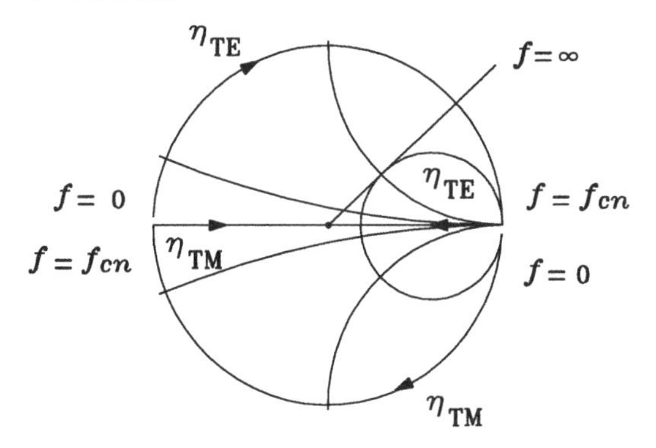

Figure 7.3 *Behavior of the wave impedance of TE and TM modes as functions of frequency, plotted on the Smith chart.*

7.4.2 Properties of TE modes

B1. Wave impedance

The wave impedance in the direction of the unit vector $\hat{a}_z$, which can be calculated letting $\mathcal{E}_z = 0$ in Eqs. (7.4), is:

$$ {}_{\mathrm{TE}} = \frac{j\omega}{{}_n} \quad . \tag{7.28} $$

It is constant over the cross-section of the structure. As for its frequency dependence, Figure 7.3 illustrates it on the Smith chart, to be de ned in the next chapter, but it can be easily described even without this graphical tool. One must distinguish, as we said in the previous subsection, whether the frequency is smaller or larger than $f_{cn} = \omega_{cn}/(2\ )$. For $f < f_{cn}$, Eq. (7.28) is an inductive reactance, equal to zero for $f = 0$, then increasing as f increases, becoming in nite for $f = f_{cn}$. Passing through this value, which corresponds to an ideal open circuit, Eq. (7.28) becomes a resistance for $f > f_{cn}$, and then tends asymptotically to $\ = \sqrt{\ /\ }$ for $f \to \infty$, while ${}_n \to j\omega\sqrt{\ }$, i.e., while the generic TE mode tends to become a plane wave traveling in the z direction. We nd an in nite ${}_{\mathrm{TE}}$ for $f = f_{cn}$ because at this frequency the transverse component of $\mathcal{H}$ vanishes, so that nothing changes if a cross-section of the structure is "metallized" by means of an *ideal magnetic conductor* (see Chapter 3).

B2. Poynting vector

The Poynting vector of the generic TE mode is given by:

$$ \mathbf{P}_n = \frac{1}{2}\mathbf{E}\quad \mathbf{H}_n = e^{\ 2\ {}_n z}\left\{\frac{1}{2\ {}_n^4}\, j\omega\ \ {}_n\, |\nabla_t\mathcal{H}_z|^2\, \hat{a}_z\quad \frac{1}{2\ {}_n^2}\, j\omega\ \ \mathcal{H}_z\, \nabla_t\mathcal{H}_z\right\}, \tag{7.29} $$

where ${}_n = 0$ for $f\quad f_{cn}$. Its longitudinal part (the rst term) is imaginary for $f < f_{cn}$, real for $f > f_{cn}$, and is oriented in the same sense as $\hat{a}_z$. Its transverse component, parallel to $\nabla_t \mathcal{H}_z$, can change in phase from one point to another in the cross-section only if the phase of $\mathcal{H}_z$ depends on the transverse coordinates. If this is not the case, then the phases of $\mathcal{H}_z$ and $\nabla_t \mathcal{H}_z$ cancel each other, and we may state that the transverse component of the Poynting vector is imaginary at any frequency. In fact, we will see in the next section that $\angle \mathcal{H}_z$ = constant is always a legitimate assumption. Therefore, we may indeed state that the transverse Poynting vector is indeed purely imaginary.

7.4.3 Properties of TM modes

These properties are "dual" of those of the TE modes, so they will be stated without proofs.

C1. Wave impedance

The wave impedance in the direction of $\hat{a}_z$ is:

$$_{\mathrm{TM}} = \frac{_n}{j\omega} \quad . \tag{7.30}$$

Its behavior vs. ω is also plotted schematically in Figure 7.3, and can be described as follows. For $f = f_{cn}$ we have $_{\mathrm{TM}} = 0$: at this frequency, $\mathcal{E}_t = 0$, so that nothing changes if the cross-section is short-circuited by means of an ideal electric conductor.

C2. Poynting vector

The Poynting vector of the generic TM mode is given by:

$$\mathbf{P}_n = e^{\ 2 \ _n z} \left\{ \ \frac{1}{2 \ ^4_n} \ j\omega \ _n \left|\nabla_t \mathcal{E}_z\right|^2 \hat{a}_z \quad \frac{1}{2 \ ^2_n} \ j\omega \ \mathcal{E}_z \ \nabla_t \mathcal{E}_z \right\} , \tag{7.31}$$

where again $_n = 0$ for $f \quad f_{cn}$. For its longitudinal and transverse components, refer to Subsection B2.

7.4.4 Properties of the TEM mode

D1. Uniqueness of the TEM mode

We saw in Section 7.2 that, when $^2 = 0$, the eld must be TEM in order not to give inconsistent relationships. Now let us show also that the inverse is true, namely that a eld can be TEM *only* for $^2 = 0$. In fact, for $^2 \neq 0$, Eqs. (7.4) show that if we set $\mathcal{E}_z = 0$, $\mathcal{H}_z = 0$, then we necessarily have $\mathcal{E}_t = 0$, $\mathcal{H}_t = 0$, i.e., the trivial solution of Maxwell's equations in a region without sources. Therefore *the eigenvalue corresponding to a TEM eld is unique.* For this reason we speak of *the* TEM mode.

D2. Existence of the TEM mode

The TEM mode may exist *only* in cylindrical structures where the *cross-section* of the dielectric *is not a simply connected domain.* Indeed, we saw that the TEM field satisfies Eq. (7.15):

$$\nabla_t^2 \mathcal{F} = 0 \qquad (\mathcal{F} = \mathcal{L}, \mathcal{T}) \quad , \tag{7.32}$$

which is the Laplace equation. A well-known theorem ensures the uniqueness of its solution, when the values of either $\mathcal{F}$ or $\partial \mathcal{F}/\partial n$ are known over the whole contour. Suppose that the border, ℓ, consists of only one ideal conductor, where $\mathbf{E}_{\mathrm{tan}} = 0$. Then, elaborating on Eqs. (7.12) one may show that this assumption implies $\partial \mathcal{L}/\partial n = 0$ on ℓ and $\mathcal{T}$ = constant on ℓ. These conditions are compatible with a field identically zero everywhere. Because of the uniqueness theorem that we have just quoted, the vanishing field is *the* solution of the problem. In conclusion, the TEM field cannot exist if the cross-section is a simply connected domain. Examples of non-simply connected regions of practical interest are two-wire lines and coaxial cables. The latter will be studied in this chapter. Most of the contents of the following chapter apply to both. One section in the next chapter will be devoted to *striplines*, which are also very important examples of structures with non-simply connected cross-sections. However, in striplines the dielectric is inevitably not homogeneous throughout the cross-section, and consequently they do not fit rigorously into the patterns of this chapter. Their fundamental mode, as we shall see, is called "quasi-TEM," and can be dealt with only by means of approximations.

D3. TEM cut-off frequency

In non-simply connected structures, *the TEM mode is the fundamental mode.* This is a corollary of the property D1 (the eigenvalue of the TEM mode is zero): through Eq. (7.22) we see that the cut-off frequency of the TEM mode is also equal to zero:

$$f_{c_{\mathrm{TEM}}} = 0 \quad . \tag{7.33}$$

By definition of the fundamental mode, the conclusion is obvious.

Eq. (7.33) tells us that the frequency range where only one mode propagates begins at zero frequency for those structures whose cross-sections are not simply connected. From direct current, it extends to the smallest among the critical frequencies of the higher-order modes. A typical example of higher-order modes in coaxial cable will be dealt with in Section 7.8.

7.5 Mode orthogonality

Propagation in waveguides without sources is described by homogeneous equations. In any mathematical problem of this type, it is important to discuss *linear independence* of solutions. In this section, we will prove some relationships which assess linear independence of different solutions in its strongest possible form, namely orthogonality relationships. They can be looked at as extensions to an infinite-dimensional vector space (the space of the solutions of Eq. (7.8)) of a property which is well known for eigenvalue problems of finite dimension, namely that eigenvectors belonging to different eigenvalues are orthogonal.

7.5.1 Orthogonality between two TE modes or two TM modes

As usual, let S be the cross-section, normal to the unit vector $\hat{a}_z$, of a cylindrical structure surrounded by ideal conductors. Let ℓ be the border of S ($\ell = S \cap C$, where C is the structure wall).

Let us write on S the 2-D form of the Green theorem (A.11), where φ and are two functions of the two transverse coordinates, regular over the domain S:

$$\int_S (\varphi \nabla_t^2 \qquad \nabla_t^2 \varphi)\, dS = \oint_\ell \left(\varphi \frac{\partial}{\partial n} \qquad \frac{\partial \varphi}{\partial n} \right) d\ell \quad . \tag{7.34}$$

Let now φ and be two functions satisfying Eq. (7.8). To be speci c, let $\varphi = \mathcal{H}_{zj}$, $= \mathcal{H}_{zk}$ (i.e., we are dealing with the TE_j and TE_k modes). Both functions satisfy the boundary condition of Eq. (7.18), so the right-hand side of Eq. (7.34) is zero. The left-hand side can be modi ed exploiting Eq. (7.8), so we write:

$$(\ {}_k^2 \quad {}_j^2) \int_S \mathcal{H}_{zj}\, \mathcal{H}_{zk}\, dS = 0 \quad . \tag{7.35}$$

We get immediately the following orthogonality relationship:

$$\int_S \mathcal{H}_{zj}\, \mathcal{H}_{zk}\, dS = 0 \qquad \text{for} \quad {}_j^2 \neq {}_k^2 \quad . \tag{7.36}$$

If we let $\varphi = \mathcal{E}_{zj}$, $= \mathcal{E}_{zk}$ (i.e., if we deal with two TM modes), satisfying the boundary condition of Eq. (7.17), then the right-hand side of Eq. (7.34) is again zero, and we reach the following result:

$$\int_S \mathcal{E}_{zj}\, \mathcal{E}_{zk}\, dS = 0 \qquad \text{for} \quad {}_j^2 \neq {}_k^2 \quad . \tag{7.37}$$

Two additional orthogonality relationships can be proved starting from Eqs. (7.36) and (7.37), and using the rst Green identity written in the form in Eq. (7.19). In fact, a proportionality relationship similar to Eq. (7.20) yields then:

$$\int_S \mathcal{E}_{tj} \quad \mathcal{E}_{tk}\, dS = 0 \qquad \text{for} \quad {}_j^2 \neq {}_k^2 \quad , \tag{7.38}$$

$$\int_S \mathcal{H}_{tj} \quad \mathcal{H}_{tk}\, dS = 0 \qquad \text{for} \quad {}_j^2 \neq {}_k^2 \quad . \tag{7.39}$$

These last relationships hold both *for TE and for TM modes.*

In the case of two (or more) *degenerate modes* (as we said, modes which are physically di erent but belong to equal eigenvalues, ${}_j^2 = {}_k^2$, $j \neq k$), what we may say for certainty is that there exist always two (or more) linear combinations of such modes, that satisfy the same orthogonality relationships that we proved now. These linear combinations are still propagation modes, as we will see in Section 7.7. The algorithm by means of which orthogonal modes can be obtained from non-orthogonal modes is, in abstract terms, the same as that which is known in linear algebra as the *Schmidt orthogonalization procedure* (Dicke and Wittke, 1960). It can be sketched as follows: one chooses, arbitrarily, one among the degenerate modes, and leaves it as it is. Then, one takes the second mode (the order is still arbitrary if the modes are more than two) and subtracts from it its projection onto the rst mode. From the third mode, one subtracts the projection onto the rst one and the projection onto the second one; one proceeds in the same manner with the following modes.

7.5.2 Orthogonality between a TE mode and a TM mode

To prove that the transverse elds of a TE and a TM mode also satisfy Eqs. (7.38) and (7.39), even when their eigenvalues are equal, let us apply Stokes' theorem in Eq. (A.9) to the vector $\nabla\varphi$, where and φ are two regular scalar functions of space coordinates. Exploiting the identity (C.6), so that $\nabla \quad (\quad \nabla\varphi) = \nabla \quad \nabla\varphi$, we have:

$$\int_S (\nabla_t \quad \nabla_t\varphi) \quad \hat{a}_z \, dS = \oint_\ell \quad \nabla\varphi \quad \mathbf{d\ell} \quad . \tag{7.40}$$

Let $= \mathcal{E}_{zk}$ (a TM mode) and $\varphi = \mathcal{H}_{zj}$ (a TE mode). Then, it is 0 on ℓ, so the right-hand side of Eq. (7.40) vanishes. The result is:

$$\int_S \nabla_t \mathcal{E}_{zk} \quad \nabla_t \mathcal{H}_{zj} \quad \hat{a}_z \, dS = 0 \quad . \tag{7.41}$$

Thanks to the cyclic permutation rule for the mixed product of three vectors and to Eq. (7.5), from Eq. (7.41) it follows that Eqs. (7.38) and (7.39) still hold, without any restriction on the eigenvalues.

7.5.3 Consequences of mode orthogonality

The total power owing through the cross-section of a waveguide with ideal conducting walls is the *sum of the powers carried by the individual modes*, since all "cross terms":

$$\frac{1}{2} \int_S \mathbf{E}_{tj} \quad \mathbf{H}_{tk} \quad \hat{a}_z \, dS \qquad (j \neq k) \quad , \tag{7.42}$$

vanish. Strictly speaking, this comment follows not from Eqs. (7.38) and (7.39), but from two similar equations, where $\mathbf{E}_{tk}$ and $\mathbf{H}_{tk}$ are replaced by $\mathbf{E}_{tk}$ and $\mathbf{H}_{tk}$. These relations are easily proved following the previous lines, and can be veri ed as an exercise.

If a eld $\{\mathbf{E}, \mathbf{H}\}$ in a waveguide is expressed as a series of propagation modes, then the expansion coe cients can be simply calculated by an "orthogonal projection," i.e., evaluating integrals of the type:

$$\int_S \mathbf{E} \quad \mathbf{E}_k \, dS \quad , \qquad \int_S \mathbf{H} \quad \mathbf{H}_k \, dS \quad . \tag{7.43}$$

A third fundamental consequence of mode orthogonality will be outlined in the next section.

7.6 Some remarks on completeness

In Section 7.2, we made use (and gave a simplified proof) of the theorem shown in Section 3.8. It says that in a homogeneous cylindrical structure any e.m. field can be split as the sum of a TE and a TM field (leaving aside, for the time being, the peculiar TEM case). In Section 7.3, we proved that, in a structure where the boundary conditions of Eq. (7.16) hold, there are TE and TM fields such that each of them, individually, satisfies these boundary conditions. The scope of validity of this second result is much narrower than that of the first one, which is insensitive to boundary conditions: in fact, Section 7.11 will provide an important counter-example, where the first result holds but the second does not.

There are many problems in which an important prerequisite is to be able to "count" the modes of a structure, i.e., to be able to assess whether a given set of solutions is *complete* or not. In the theory of partial differential equations, boundary conditions and completeness are always linked to each other. This statement matches, in our case, with what we did in Section 4.7, where the assumption of a square-integrable field was instrumental to prove completeness of plane waves, and in Section 5.12, when we discussed completeness of Gaussian beams.

Let us now take the case of *closed* cylindrical structures, where boundary conditions hold on a closed surface which is all at a finite distance from any point where the field is defined. Then, we may proceed by analogy with what happens to signals in the time domain, in particular with the well-known links between the Fourier integral (for boundary conditions at infinity) and the Fourier series (for boundary conditions at the two ends of an interval). This analogy suggests that a complete set may consist of an *infinity of discrete elements*. However, the waveguide problem is 2-D, and in two dimensions a discrete infinity can be built up in an infinite number of possible ways. There is no reason to be sure, *a priori*, that any of these procedures will eventually lead to a complete set. To give an example of a set which is obviously not complete, let us take an infinity of solutions which differ from each other only in their dependence on one coordinate. In summary, cardinality of a set (i.e., the number of its elements) is not enough to assure completeness.

In any case, completeness of a countable set of solutions of a partial differential equation (PDE) is a subject that can be kept within tolerable limits of size and complexity only if the PDE is *separable*, that is to say, it can be reduced to ordinary differential equations (ODE's), each of these in one unknown quantity, a function of only one coordinate. In a 2-D case, like the one at hand, a complete set of solutions of the PDE consists then of all the products of a countable infinity of linearly independent solutions of the ODE with respect to the first coordinate, multiplied by a countable infinity of linearly independent solutions of the ODE with respect to the second coordinate.

As for Eq. (7.8), it can be shown that, for positive real eigenvalues, it is separable *only* in four types of orthogonal coordinates: *rectangular cartesian*;

circular polar (where the coordinate lines are straight half-lines in radial directions, and circles); *parabolic* (whose coordinate lines are two families of confocal, mutually orthogonal parabolas); *elliptical* (whose coordinate lines are confocal ellipses and confocal hyperbolas, mutually orthogonal).

When dealing with a cylindrical structure of any shape, one can always adopt a coordinate frame which belongs to this set, but clearly, such a reference frame is bene cial *only when the boundary conditions are imposed on coordinate lines*, or at least on lines that can be broken into pieces of coordinate lines. When this is not the case, then Eq. (7.8) can still be separated into two ODE's, but their solutions cannot satisfy the boundary conditions on an individual basis, and the nal result is that each mode is not expressed as the product of two functions, but as the product of two series of functions. Nowadays, many numerical methods have been developed on these premises, and can in fact study waveguides of any shape within reasonable processing times and memory occupation, but it remains extremely hard to assess the completeness of a set of solutions obtained in this way.

All the structures dealt with in this book have cross-sections which not only allow the separation of variables, but also enable us to satisfy boundary conditions on an individual basis, i.e., with any single element in a countable in nity of solutions of two ODE's. Modes will be expressed as products of individual functions, not of series. No formal mathematical proof will be given for the completeness of these sets. However, the reader's background on periodic functions will be enough to make him/her con dent that we do not neglect any physically meaningful solution. However, one should not extrapolate from these examples, erroneously, that, given a separable 2-D PDE, the products of *any* two countable in nities of linearly independent solutions of the two ODE's form a complete set. This is not true in general even if all these functions satisfy the boundary conditions. A very simple counter-example goes as follows: given a complete set of linearly independent solutions, take away one of its elements. The set which is left is still a countable in nity, and its elements are still linearly independent, but clearly it cannot be complete, because none of the linear combinations of the elements which are left can express the term which was taken away.

7.7 Rectangular waveguides

Waveguides of rectangular cross-section provide an excellent example of application for the previous results, being at the same time technically signi cant and mathematically simple.

Let us introduce rectangular coordinates (x, y, z), and place the origin in a corner of the cross-section (Figure 7.4), the x-axis along the longer side, of length a, b being the length of the shorter side. In this reference frame, the dif-

The proof of this statement is too long to be developed here. Let us say just that it follows from the theory of analytical functions of a complex variable. It can be found rather easily in the literature (e.g., Morse and Feshbach, 1953).

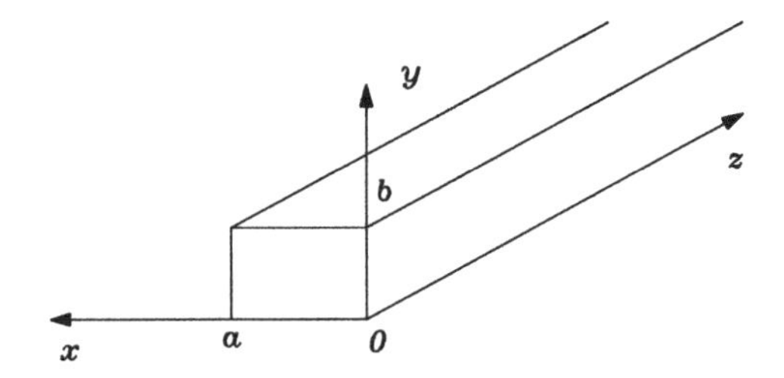

Figure 7.4 *A rectangular waveguide, and its reference frame.*

ferential operator which appears in Eq. (7.8) reads $\nabla_t^2 = (\partial^2/\partial x^2) + (\partial^2/\partial y^2)$, and separation of variables is simply the 2-D version of the procedure we saw in Section 4.2. Let:

$$\mathcal{F}(x,y) = X(x)\,Y(y) \quad . \tag{7.44}$$

Then, the same steps as in Section 4.2 yield a pair of harmonic equations:

$$X'' = \quad k_x^2 X \quad , \qquad Y'' = \quad k_y^2 Y \quad , \tag{7.45}$$

whose separation constants are linked to the eigenvalue $\ _n^2 = (\ ^2 \quad _n^2)$ as follows:

$$k_x^2 + k_y^2 = \ _n^2 \quad . \tag{7.46}$$

We know from Section 7.4 that $\ _n^2$ is real and positive, and from Section 7.3 that $\mathcal{F}$ must satisfy boundary conditions at nite distance. This suggests that it may be convenient to write the general integrals of Eq. (7.45) in terms of sine and cosine functions, i.e.:

$$\begin{aligned} X &= A \sin k_x x + B \cos k_x x \quad , \\ Y &= C \sin k_y y + D \cos k_y y \quad , \end{aligned} \tag{7.47}$$

where A, B, C and D are arbitrary constants.†

The boundary conditions on the walls are di erent, as we saw before, for TE and for TM modes. For TE modes, Eq. (7.18) holds, and once the variables are separated it reads:

$$\begin{aligned} \frac{dX}{dx} &= 0 \qquad \text{for} \quad x = 0\ ,\ x = a \quad , \\ \frac{dY}{dy} &= 0 \qquad \text{for} \quad y = 0\ ,\ y = b \quad . \end{aligned} \tag{7.48}$$

Di erentiating Eqs. (7.47) and imposing the conditions of Eqs. (7.48), we

† So far, we cannot exclude that either k_x or k_y could be imaginary, so that Eqs. (7.47) contain one pair of hyperbolic functions with real argument. However, we will see soon that this is incompatible with the boundary conditions on the ideal conducting wall. Hyperbolic functions, on the other hand, cannot to be disregarded in open structures, like a pair of parallel conducting planes, in which case they represent elds which radiate in the direction along which the structure is open.

Table 7.1 *Field components of TE and TM modes in a rectangular waveguide.*

TE MODES	
$\mathcal{E}_x = E_o \frac{n}{b} \cos \frac{m\ x}{a} \sin \frac{n\ y}{b}$	$\mathcal{H}_x = \frac{\mathcal{E}_y}{\mathrm{TE}}$
$\mathcal{E}_y = \quad E_o \frac{m}{a} \sin \frac{m\ x}{a} \cos \frac{n\ y}{b}$	$\mathcal{H}_y = \frac{\mathcal{E}_x}{\mathrm{TE}}$
$\mathcal{E}_z = 0$	$\mathcal{H}_z = \frac{2}{j\omega} E_o \cos \frac{m\ x}{a} \cos \frac{n\ y}{b}$
TM MODES	
$\mathcal{E}_x = \ {}_{\mathrm{TM}}\,\mathcal{H}_y$	$\mathcal{H}_x = \quad H_o \frac{n}{b} \sin \frac{m\ x}{a} \cos \frac{n\ y}{b}$
$\mathcal{E}_y = \quad {}_{\mathrm{TM}}\,\mathcal{H}_x$	$\mathcal{H}_y = H_o \frac{m}{a} \cos \frac{m\ x}{a} \sin \frac{n\ y}{b}$
$\mathcal{E}_z = \quad \frac{2}{j\omega} H_o \sin \frac{m\ x}{a} \sin \frac{n\ y}{b}$	$\mathcal{H}_z = 0$

get immediately $A = C = 0$, and furthermore:

$$\sin k_x a = 0 \quad \Longrightarrow \quad k_x = \frac{m}{a} \quad ,$$

$$\sin k_y b = 0 \quad \Longrightarrow \quad k_y = \frac{n}{b} \quad , \tag{7.49}$$

where m and n are two integers. Accordingly, the longitudinal magnetic eld component of the generic transverse electric mode ($\mathrm{TE}_{m,n}$) reads:

$$\mathcal{H}_z = H \ \cos \frac{m\ x}{a} \cos \frac{n\ y}{b} \quad , \tag{7.50}$$

where H is an arbitrary constant. The other eld components, obtained from Eq. (7.50) through Eq. (7.3), are shown in Table 7.1, where, to simplify the notation for the transverse components, which are the most commonly used ones, we set $H = E_o \ ^2/(j\omega \ )$.

The eigenvalue corresponding to Eq. (7.49) is given by:

$$^2_{m,n} = \left(\frac{m}{a}\right)^2 + \left(\frac{n}{b}\right)^2 \quad , \tag{7.51}$$

which lends itself to the simple geometrical interpretation of Figure 7.5. The corresponding *cut-o frequency*, according to Eq. (7.22), is:

$$f_{c_{m,n}} = \frac{1}{2\sqrt{\quad}} \left[\left(\frac{m}{a}\right)^2 + \left(\frac{n}{b}\right)^2 \right]^{1/2} \quad . \tag{7.52}$$

Negative values for m and n should not be taken into account, since a change in sign does not a ect Eq. (7.52), nor Eq. (7.50). In the eld components shown in Table 7.1, changes in sign would not be signi cant, as they can always be considered as part of the arbitrary constant E_o. Zero is a legitimate value

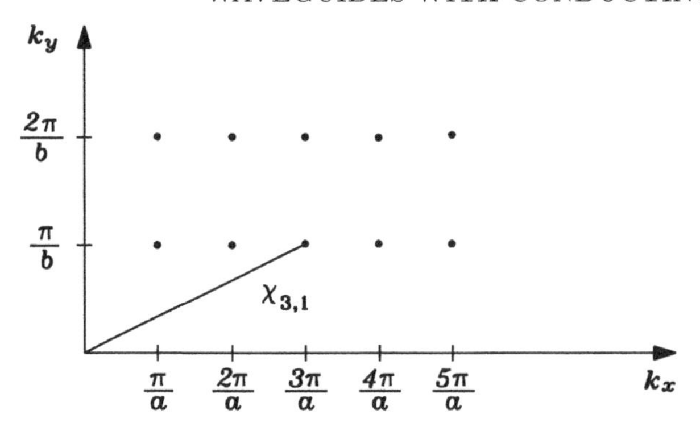

Figure 7.5 *Geometrical construction of the modal eigenvalues in a rectangular waveguide.*

for either m or n, but not simultaneously for both, because in this case in fact the longitudinal component in Eq. (7.50) would become a constant, and consequently all the transverse components of **E** and **H** would vanish. This purely longitudinal magnetic eld, constant with respect to all coordinates, would be the magnetostatic eld of an in nitely long rectangular solenoid, not a traveling guided wave.

As for the TM modes, what we said for the TE modes remains valid until Eq. (7.47). The boundary condition on the wall is now Eq. (7.17), which once the variables are separated reads:

$$X = 0 \qquad \text{for} \qquad x = 0\ ,\ x = a \quad ,$$

$$Y = 0 \qquad \text{for} \qquad y = 0\ ,\ y = b \quad . \tag{7.53}$$

Imposing these conditions on Eqs. (7.47) yields $B = D = 0$, and then Eqs. (7.49) hold again. So, the longitudinal electric eld component of the generic transverse mode $\mathrm{TM}_{m,n}$ is:

$$\mathcal{E}_z = E \sin \frac{m\ x}{a} \sin \frac{n\ y}{b} \quad , \tag{7.54}$$

where E is an arbitrary constant. The corresponding transverse eld components, which can be derived through Eq. (7.3), are also shown in Table 7.1, where, again to simplify the notation for the transverse components, we set $E = \ H_o\ ^2/(j\omega\ )$.

The eigenvalue corresponding to Eq. (7.54) is still given by Eq. (7.51), but now *neither m nor n can take the zero value*, because this would imply $\mathcal{E}_z\ \ 0$ in Eq. (7.54) and thus result in an identically zero eld. Negative values for m and n are also to be left out, for the same reasons as for TE modes.

Eq. (7.51) holding for TM and for TE modes implies that, for $m\ \ n \neq 0$, the $\mathrm{TE}_{m,n}$ and $\mathrm{TM}_{m,n}$ modes are *degenerate*. Any linear combination of two such modes is then a eld which not only satis es the boundary conditions on the waveguide walls, but also depends exponentially on the longitudinal coordinate z and is characterized by the eigenvalue (7.51). Therefore, this eld in turn can be looked at as a mode of the waveguide. In Section 7.9 we will provide an example of a problem — losses in waveguides with non-ideal metallic walls — where special attention has to be paid in the case of degenerate modes and of their linear combinations.

As we said before, for TE modes Eq. (7.51) holds also for $m = 0$ or for $n = 0$. These modes are not degenerate with TM modes. Consequently, *the smallest among all the values that are allowed for* $^{2}_{m,n}$ *corresponds to* $m = 1$ and $n = 0$, i.e., it belongs to the $\mathrm{TE}_{1,0}$ mode. This is, in conclusion, the *fundamental mode* of a rectangular waveguide. Letting $m = 1$ and $n = 0$ in Eq. (7.52), we get its cut-o frequency:

$$f_c = \frac{c}{2a} \quad , \tag{7.55}$$

where $c = (\quad)^{1/2}$ is the speed of light in the dielectric inside the waveguide.

Eq. (7.55) shows that the distance a between the short walls of the waveguide equals a *half wavelength* (measured in the dielectric that lls the guide) at the cut-o frequency of the fundamental mode. This is a hint towards a physical interpretation of the "cut-o " phenomenon in terms of plane waves which impinge on the walls and are re ected on them, giving rise to standing waves in the waveguide. If the distance between two parallel metallic walls is smaller than one half of the plane-wave wavelength, then no standing wave resulting as a superposition of *uniform* plane waves can satisfy the condition of zero tangential eld on both walls, whose distance is shorter than the distance between two nodes of the electric eld, irrespectively of the direction of propagation of the two interfering traveling waves. The conditions on the two walls can be satis ed by the superposition of two *evanescent* waves whose attenuation vectors are parallel to the z axis and whose phase vectors (satisfying $|k| > \omega\sqrt{\quad}$) are orthogonal to the two walls. For these slow waves (see Chapter 4), two consecutive nodes in their interference pattern are less then $/2$ apart from one other. This interpretation recon rms that the eld attenuation in the z direction, which cannot be caused by power dissipation since the structure is lossless by assumption, is indeed an evanescence.

When, on the contrary, a half wavelength is shorter than the distance between the walls, say a, then two nodes of the electric eld at a distance a may indeed result from the superposition of two uniform plane waves, provided their phase vectors have a nonvanishing component in the z direction. The net result is a eld which propagates in the axial direction of the guide. Higher-order modes behave according to the same mechanism, di ering from each other in position and in number of the eld nodes in the interference pattern, which, for $m \neq 0$ and $n \neq 0$, results from re ections on the four walls.

Examples of eld lines, for some of the simplest and yet most important modes of rectangular waveguides, are shown in Figure 7.6.

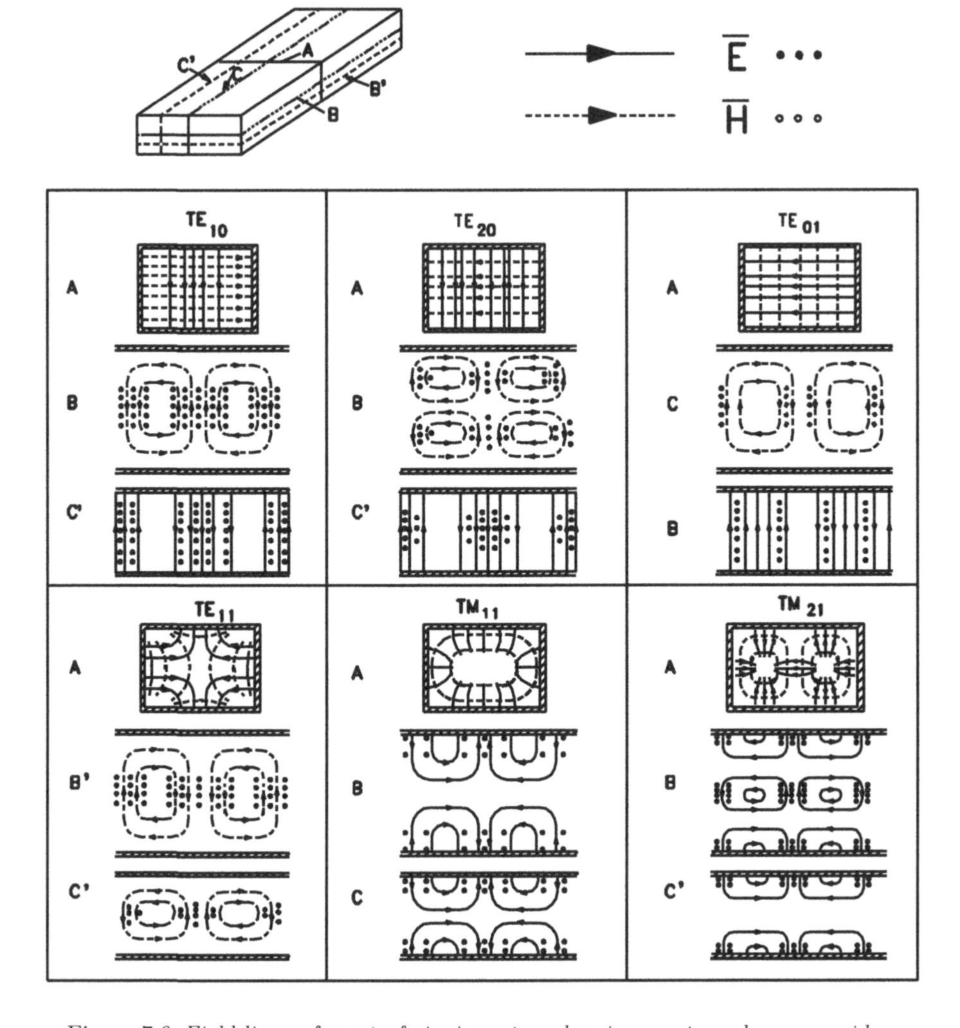

Figure 7.6 *Field lines of a set of signi cant modes, in a rectangular waveguide.*

7.8 Circular waveguides and coaxial cables

The subjects of this section are of technical interest and, at the same time, rather simple as far as calculations are concerned. The mathematical background, although not complicated, is much less dealt with, in standard curricula in electrical engineering and in physics, than that of the previous section. For this reason, Appendix D is devoted to outline it in some detail. Several contents of this section will be useful background points for very important subjects to be dealt with in Chapter 10.

Consider an in nitely long structure with a lossless dielectric either inside a circular-cylinder conductor (cylindrical waveguide, Figure 7.7), or between

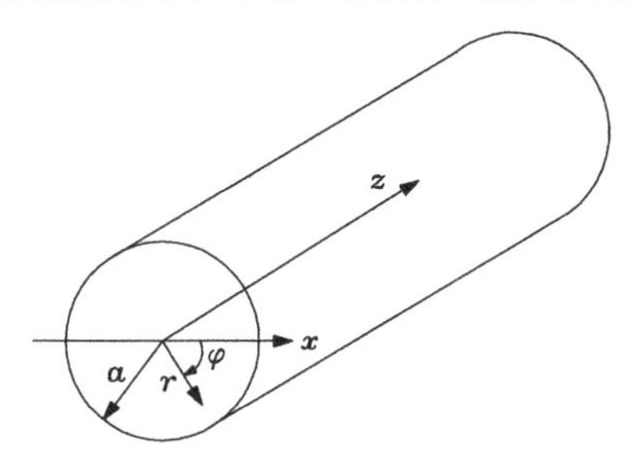

Figure 7.7 *A circular waveguide, and its reference frame.*

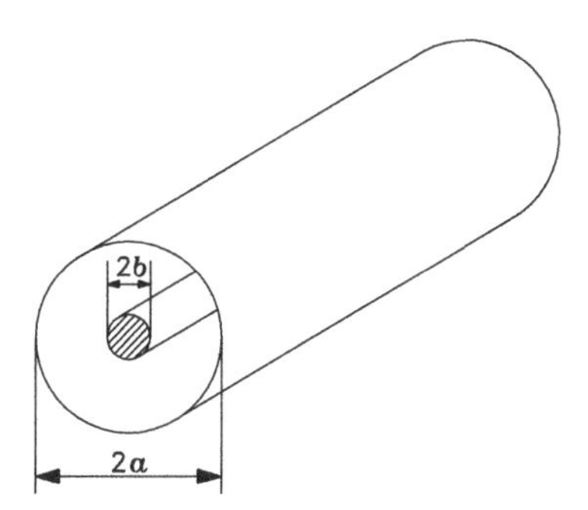

Figure 7.8 *A coaxial cable.*

two circular-cylinder conductors with the same axis (coaxial cable, Figure 7.8). In either case, let us introduce a cylindrical coordinate system (r, φ, z), whose z axis is obviously the axis of the cylinders. The di erential operator of Eq. (7.8) then reads:

$$\nabla_t^2 = \frac{\partial^2}{\partial r^2} + \frac{1}{r}\frac{\partial}{\partial r} + \frac{1}{r^2}\frac{\partial^2}{\partial \varphi^2} = \frac{1}{r}\frac{\partial}{\partial r}\left(r\frac{\partial}{\partial r}\right) + \frac{1}{r^2}\frac{\partial^2}{\partial \varphi^2} \quad . \tag{7.56}$$

To separate the variables, let:

$$\mathcal{F} = R(r) \quad (\varphi) \quad , \tag{7.57}$$

insert Eqs. (7.57) and (7.56) into Eq. (7.8), and then multiply both sides times r^2/R . We obtain the following ODE's, where $\;^2$ is the separation constant:

$$'' = \quad ^2 \quad , \tag{7.58}$$

$$R'' + \frac{1}{r}R' + \left(\,^2_m \quad \frac{^2}{r^2}\right) R = 0 \quad . \tag{7.59}$$

Eq. (7.58) is, once more, the harmonic equation, whose general integral can be written in either of the following forms:

$$(\varphi) \quad = \quad G' \sin \; \varphi + G'' \cos \; \varphi \quad , \tag{7.60}$$

$$(\varphi) \quad = \quad A e^{j \quad \varphi} + B e^{\quad j \quad \varphi} \quad , \tag{7.61}$$

where G', G'' (or A, B) are arbitrary constants. The rst form is more suitable if one wants to stress that the generic solution can be thought of as a superposition of two states of *polarization*, which correspond to $G' = 0$ and to $G'' = 0$, respectively. The second form, on the other hand, represents (φ) as a superposition of two traveling waves which describe a *helical path* around the z axis, one in the clockwise sense ($B = 0$), the other one in the counter-clockwise sense ($A = 0$). This viewpoint becomes particularly attractive in the limit of geometrical optics, i.e., when a ray picture of propagation is associated with these two waves, as we shall see in Chapter 10. The two polarizations in Eq. (7.60) can then be looked at as standing waves which result from interference of two helical waves (Eq. (7.61)) of equal amplitudes.

To be physically meaningful, Eq. (7.57) must be a *single-valued function.* Therefore, it must be:

$$(\varphi + 2l \quad) = \quad (\varphi) \quad , \tag{7.62}$$

where l can be any integer. Consequently, must be an *integer.* To stress this point, we will change the symbol, in agreement with standard conventions in mathematical handbooks and textbooks, and replace with n $(= 0, 1, \ldots)$. Negative values for n are not required, for the same reasons as in Section 7.7.

Under the assumption $\ ^2_m \neq 0$, i.e., if we leave aside the TEM mode, Eq. (7.59) is *the Bessel equation of integer order* $\ = n$. Its general integral can be expressed in several forms (see Appendix D). For the problems discussed in this section, the most convenient one is:

$$R(r) = C\, J_n(\ _m r) + D\, Y_n(\ _m r) \quad , \tag{7.63}$$

where $J_n(x)$ and $Y_n(x)$ are, respectively, the Bessel function of the rst kind, of order n, and the Bessel function of the second kind (also called the Neumann function), of order n.

Some examples are plotted in Figure D.1 and D.2 (see again Appendix D). Note in particular that, for any order n, Y_n tends to $\ \infty$ for $\ \to 0$. An in nite value of the longitudinal eld component $\mathcal{F}$ is unphysical, so the term $Y_n(\ _m r)$ must be disregarded, i.e., we must set $D = 0$ in Eq. (7.63), whenever the z axis (where $r = 0$) belongs to the region where the e.m. eld is de ned. This occurs in cylindrical waveguides (Figure 7.7), but not in coaxial cables (Figure 7.8).

From now on, it is convenient to deal separately with TE and TM modes. For the rst ones, the longitudinal eld component $\mathcal{F} = \mathcal{H}_z$ must satisfy the boundary condition (7.18), which in the present case, keeping track of Eq. (7.57), reads:

$$\left.\frac{dR}{dr}\right|_{r=a} \qquad _m\, C\, J_n'(\ _m a) = 0 \quad , \tag{7.64}$$

where a is the radius of the cylindrical wall, and $J_n'(x)$ indicates the rst derivative of $J_n(x)$ *with respect to its argument.* Let $p'_{n,m}$ be the m-th zero of $J_n'(x)$. Tables of such zeros can be found in several handbooks. Usually, they

Table 7.2 *Field components of TE and TM modes in a circular waveguide.*

$$\nabla_t = \hat{a}_r \frac{\partial}{\partial r} + \hat{a}_\varphi \frac{1}{r} \frac{\partial}{\partial \varphi}$$

TE MODES	
$\mathcal{E}_r = E_o \frac{n}{r} J_n(\ _m r) \begin{cases} \sin n\varphi \\ \cos n\varphi \end{cases}$	$\mathcal{H}_r = \frac{\mathcal{E}_\varphi}{\ _{\text{TE}}}$
$\mathcal{E}_\varphi = E_o \ _m J'_n(\ _m r) \begin{cases} \cos n\varphi \\ \sin n\varphi \end{cases}$	$\mathcal{H}_\varphi = \frac{\mathcal{E}_r}{\ _{\text{TE}}}$
$\mathcal{E}_z = 0$	$\mathcal{H}_z = \frac{\ _m^2}{j\omega\ } E_o J_n(\ _m r) \begin{cases} \cos n\varphi \\ \sin n\varphi \end{cases}$
TM MODES	
$\mathcal{E}_r = \ _{\text{TM}} \mathcal{H}_\varphi$	$\mathcal{H}_r = H_o \frac{n}{r} J_n(\ _m r) \begin{cases} \sin n\varphi \\ \cos n\varphi \end{cases}$
$\mathcal{E}_\varphi = \ _{\text{TM}} \mathcal{H}_r$	$\mathcal{H}_\varphi = H_o \ _m J'_n(\ _m r) \begin{cases} \cos n\varphi \\ \sin n\varphi \end{cases}$
$\mathcal{E}_z = \frac{\ _m^2}{j\omega\ } H_o J_n(\ _m r) \begin{cases} \cos n\varphi \\ \sin n\varphi \end{cases}$	$\mathcal{H}_z = 0$

are labeled also by a second integer, $m = 1, 2, \ldots$ (see, e.g., Abramowitz and Stegun, Eds., 1965). Then, the eigenvalue for the $\text{TE}_{n,m}$ mode is given by:

$$\ _{n,m} = \frac{p'_{n,m}}{a} \quad , \tag{7.65}$$

and the corresponding critical frequency is $f_{c_{n,m}} = p'_{n,m}/(2\ \ a\sqrt{\ \ })$. The longitudinal eld component can be written in the form:

$$\mathcal{H}_z = H\, J_n\left(p'_{n,m}\frac{r}{a}\right) \begin{cases} \cos n\varphi \\ \sin n\varphi \end{cases} \quad , \tag{7.66}$$

where H is an arbitrary constant. Eq. (7.66) stresses that, for $n \neq 0$, the two terms in Eq. (7.60) correspond, as we said before, to two distinct polarizations, giving two modes which are orthogonal (in the sense of Section 7.5) over the range 0 2 and must then be looked at as a *pair of degenerate modes.*

The other eld components, obtained from Eq. (7.66) through Eq. (7.4), are shown in Table 7.2, which also shows the ∇_t operator expressed in cylindrical coordinates. As in the previous section, to simplify the notation for the transverse components, which are the most commonly used ones, we set $H = E_o\ \ ^2/(j\omega\ )$.

For the TM modes, the boundary condition of Eq. (7.17) reads, in the case at hand:

$$R(a) \quad C\, J_n(\ _m a) = 0 \quad . \tag{7.67}$$

Let $p_{n,m}$ be the m-th zero of $J_n(x)$. The eigenvalue of the $\text{TM}_{n,m}$ mode is then given by:

$$\ _{n,m} = \frac{p_{n,m}}{a} \quad , \tag{7.68}$$

where again a is the wall radius. The corresponding critical frequency is $f_{c_{n,m}} = p_{n,m}/(2 \ a\sqrt{\ \ })$, and the longitudinal eld component is:

$$\mathcal{E}_z = E\, J_n\left(p_{n,m}\,\frac{r}{a}\right)\begin{cases}\cos n\varphi \\ \sin n\varphi\end{cases} \quad , \tag{7.69}$$

where E is an arbitrary constant. This expression, like Eq. (7.66), stresses that there are two orthogonal and degenerate polarizations for $n \neq 0$. The transverse eld components are also shown in Table 7.2, where again, for the usual reasons, we set $E = \ \ H_o\ ^2/(j\omega\ )$.

Among all the zeros of the functions J_n and of their derivatives J'_n — incidentally, all of them are irrational numbers — the smallest one is $p'_{1,1} \simeq 1.84$. Hence, the *fundamental mode of the circular guides is the* $\mathrm{TE}_{1,1}$ mode, whose critical frequency is:

$$f_c = \frac{1.84}{2\ \ a\sqrt{\ \ }} \quad . \tag{7.70}$$

This is of the same order of magnitude as the cut-o frequency of the fundamental mode in a rectangular waveguide whose broad side equals in length the diameter of the circular waveguide, given by Eq. (7.55). This indicates that the physical nature of the cut-o phenomenon is not a ected by the waveguide geometry in a very signi cant way.

As for *coaxial cables*, we said before that in the presence of the inner conductor the $r = 0$ axis is out of the region where the eld is de ned, and this implies that the second term in Eq. (7.63) cannot be disregarded. As a matter of fact, it plays a fundamental role in the higher-order modes, but not in the *fundamental mode*, which, as the structure does not have a simply connected cross-section, is de nitely the TEM mode, according to what we saw in Section 7.4.

The TEM mode originates from the Laplace equation, whose solutions, as we said in Section 7.4, obey a well-known uniqueness theorem. We may exploit this fact, and derive the solution of Eq. (7.15) which corresponds to the TEM mode in a heuristic way, without any risk of loss in generality. Let us suppose $\ \ = 0$ in Eq. (7.58). This in turn gives $\ \ =$ constant, and Eq. (7.59) becomes:

$$R'' + \frac{1}{r}\ R' = 0 \quad , \tag{7.71}$$

whose general integral is:

$$R = C_1\ \ln r + C_2 \quad , \tag{7.72}$$

where C_1, C_2 are arbitrary constants. Let us set $\mathcal{F} = \mathcal{T}$ in Eqs. (7.12). Then, we get the following expressions for the eld of the TEM mode:

$$\mathbf{E} \ = \ j\omega\sqrt{\ \ /\ \ }\ C_1\ \frac{1}{r}\ \hat{a}_r\ e^{\ \ j\omega\sqrt{\ \ }z} \quad , \tag{7.73}$$

$$\mathbf{H} \ = \ j\omega\, C_1\ \frac{1}{r}\ \hat{a}_\varphi\ e^{\ \ j\omega\sqrt{\ \ }z} \quad . \tag{7.74}$$

Apart from the propagation factor $\exp(\ j\omega\sqrt{\ }\ z)$, Eq. (7.73) coincides with the electrostatic eld in a cylindrical capacitor where the per-unit-length charge is $\ j\omega(2\ \ )C_1\sqrt{\ }$ on the outer and inner conductors, respectively. Eq. (7.74) coincides — again apart from the propagation factor — with the stationary magnetic eld in a cylindrical inductor where the longitudinal current intensity is $\ 2\ \ \omega C_1$ in the inner and outer conductors, respectively. The wave impedance in the z direction, for the e.m. eld given by Eqs. (7.73) and (7.74), is $\ =\sqrt{\ /\ }$. This point may generate many comments, most of which will be made in Section 8.11.

For what concerns *higher-order modes* in coaxial cables, the boundary conditions must be satis ed on two cylindrical surfaces, $r = a$ and $r = b$, a and b being the radii of the outer and inner conductors, respectively. Consequently, one of the two arbitrary constants in the general integral in Eq. (7.63) can be expressed in terms of the other before one proceeds to nd the eigenvalues. For TE modes, the condition in Eq. (7.18) becomes $dR/dr = 0$ at $r = a$ and $r = b$. Dividing side by side the two equations one gets in this way, we nd the following equation in the unknown $\ _m$:

$$\frac{J_n'(\ _m a)}{J_n'(\ _m b)} = \frac{Y_n'(\ _m a)}{Y_n'(\ _m b)} \quad , \tag{7.75}$$

which is referred to as the *characteristic equation* of the TE modes. One can prove that it has a countable in nity of discrete solutions.

The corresponding longitudinal eld component can be expressed as:

$$\mathcal{H}_z(r,\varphi) = H\Big[Y_n'(\ _m a)\, J_n(\ _m r)\quad J_n'(\ _m a)\, Y_n(\ _m r)\Big]\begin{Bmatrix}\cos n\varphi \\ \sin n\varphi\end{Bmatrix} \quad , \tag{7.76}$$

where H is an arbitrary constant and, once again, two orthogonal polarizations for $n \neq 0$ are written explicitly.

For TM modes, the boundary condition of Eq. (7.17), namely $R(a) = R(b) = 0$, leads to the following *characteristic equation*:

$$\frac{J_n(\ _m a)}{J_n(\ _m b)} = \frac{Y_n(\ _m a)}{Y_n(\ _m b)} \quad , \tag{7.77}$$

and then to the following longitudinal eld component:

$$\mathcal{E}_z(r,\varphi) = E\Big[Y_n(\ _m a)\, J_n(\ _m r)\quad J_n(\ _m a)\, Y_n(\ _m r)\Big]\begin{Bmatrix}\cos n\varphi \\ \sin n\varphi\end{Bmatrix} \quad , \tag{7.78}$$

which lends itself to the same comments as Eq. (7.76). It can be shown that, amongst all the solutions $\ _m \neq 0$ of Eqs. (7.75) and (7.77), the smallest one is the rst solution of Eq. (7.75) for $n = 0$. The corresponding $\mathrm{TE}_{0,1}$ mode is then the rst higher-order mode of the coaxial cable. Its eigenvalue lends itself to an approximation which leads to a physically enlightening expression for its critical frequency, namely:

$$f_c \simeq c/\ \ (a+b) \quad . \tag{7.79}$$

This corresponds to saying that at cut-o the wavelength is equal to the

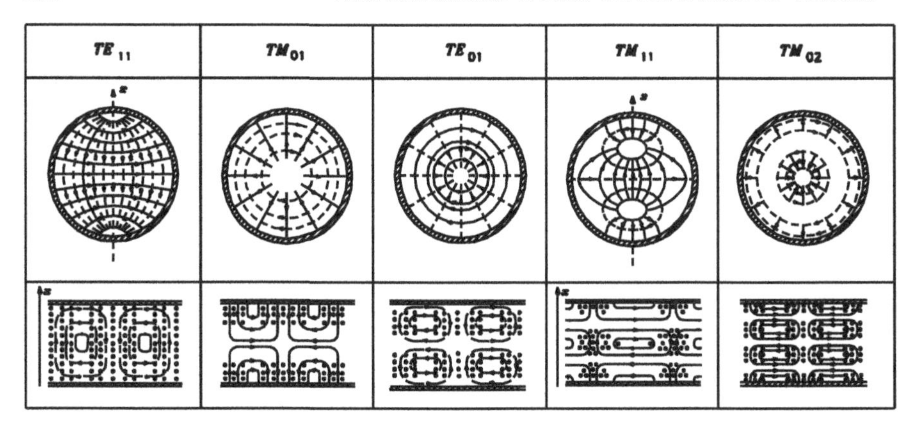

Figure 7.9 *Field lines of a set of signi cant modes, in a circular waveguide.*

length of the mean circle between the conductors. Comments are then similar to those that were given on Eq. (7.55) in Section 7.7.

Examples of eld lines, for some of the simplest and yet most important modes of circular waveguides, are shown in Figure 7.9.

7.9 Waveguides with nonideal walls

An ideal conducting wall is an abstraction, but nevertheless the theory of lossless waveguides is useful in practice. Most of its results apply well to real waveguides whose walls have a large but nite conductivity, ω . The only important quantity of practical interest which is absolutely impossible to evaluate by means of a lossless model is the power loss that a mode propagating in a real waveguide su ers because of dissipation, either in the nonideal conducting walls or in the imperfect dielectric.

This inconvenience can be overcome using a procedure, to be described in this and in the following section, which is part of a broad-scope chapter of Mathematical Physics, known as "perturbation theory" (see, e.g., Morse and Feshbach, 1953). Let us make rst some general comment, which may turn out to be useful also in some of the following chapters.

To nd an e.m. eld in a homogeneous dielectric surrounded by a good conductor in a rigorous way would require the following procedure: rst, to solve Maxwell's equations in the dielectric and in the conductor; then, to apply continuity conditions to the tangential components at the interface between the two media. This is always cumbersome, and often impossible, as we may show with a surprisingly simple example. In a rectangular waveguide, continuity conditions should be imposed on four segments belonging to four di erent coordinate lines. This problem has no analytical solution, as we will see again in the case of dielectric waveguides (Chapter 10).

What we saw in Section 4.12 on plane waves in good conductors provides a good hint towards an approximate solution. Namely, if what we are actually interested in is mainly the eld in the dielectric, then we may characterize the conductor in terms of its *wall impedance*, that is to say in terms of the *boundary condition* it imposes on the eld at its surface. Indeed, one fundamental feature of a wall impedance is that it does not depend on the eld we are looking for in the dielectric.

In a waveguide, it is not always simple to assess which value should be given to the surface impedance of walls which consist of a given material, as the geometrical shape can play an important role also. This statement will be clari ed by some means of examples in the next section. However, once a suitable wall impedance has been identi ed, then to nd the transmission modes of the waveguide means to solve an eigenvalue equation, similar to Eq. (7.8) but with some major changes. In fact, the boundary conditions are no longer Eqs. (7.17) and (7.18). As a consequence, the eigenvalues are not necessarily real and positive.

Still, in most cases this new eigenvalue problem is very hard to solve rigorously. What is much simpler is to use perturbation theory, a procedure which yields a sequence of approximate solutions that converge towards the exact one. To do that, the basic requirement is to know, and to use as the starting point, the exact solution of another eigenvalue problem which is similar enough to the one that one cannot solve exactly. To nd a general criterion to assess what "similar enough" does actually mean, would require a discussion whose size and depth are well beyond the scope of this book. For the problem at hand, we will restrict ourselves to the following statement. In Section 4.12 we saw that the wall impedance of a good conductor tends to zero as $\rightarrow \infty$. Accordingly, the exact solutions of a "similar problem" to start from will be, for us, the modes of the lossless waveguides that we studied in the previous sections.

Furthermore, let us state that, for all practical purposes, attenuation constants of the modes above cut-o are obtained with su cient accuracy as *rst-order corrections* to the eigenvalues of the lossless waveguide modes. This phrase means that calculation of attenuation constants requires only the knowledge of "zero order" elds, i.e., of the modes of the lossless guide. However, let us stress that the procedures to calculate rst-order corrections di er drastically, depending on whether one is dealing with non-degenerate or degenerate modes. There are fundamental reasons behind this di erence, which we will try to clarify by means of an example, later in this section.

For a *non-degenerate mode*, we proceed as follows. We assume that the eld of the (m, n)-th mode of an ideal waveguide ($Z_w = 0$, and lossless dielectric) is present in a waveguide with the same geometry, but with a nite wall impedance, Z_w (later we will discuss also the case of a lossy dielectric). We calculate the per-unit-length active power which, under these conditions, would ow into the walls. By de nition of wall impedance, it equals

$$W_{P(m,n)} = \int_{\ell} \mathrm{Re}[Z_w] \, \frac{|\mathcal{H}_{\tan m,n}|^2}{2} \, \mathrm{d}\ell \quad , \tag{7.80}$$

where ℓ is the contour of the waveguide cross-section, while the su x "tan" denotes, as usual, the tangential component on the wall. The power lost in a piece of waveguide of in nitesimal length dz, $W_{P(m,n)}\, dz$, is proportional to the active power $W_{T(m,n)}$ owing through the waveguide cross-section (the "transmitted" power):

$$W_{T(m,n)} = \int_S Z_{m,n} \, \frac{|\mathcal{H}_{tm,n}|^2}{2} \, dS \quad , \tag{7.81}$$

where $Z_{m,n}$ is the wave impedance in the z direction for the (m,n) mode, and $\mathcal{H}_{tm,n}$ is the *transverse* component of the magnetic eld. As the eld is a solution of homogeneous equations, $\mathcal{H}_t$ is proportional to $\mathcal{H}_{\text{tan}}$.

Proportionality between Eq. (7.80) and Eq. (7.81) entails that, as z increases, the transmitted power decreases exponentially. Hence the e.m. eld also decreases exponentially vs. z. Power being proportional to the squared modulus of the eld, the eld attenuation constant, $_{m,n}$, is one half of the power decay constant, so we nally nd:

$$_{m,n} = \frac{W_{P(m,n)}}{2\, W_{T(m,n)}} \quad . \tag{7.82}$$

This result is accurate enough only if higher-order corrections are much smaller than Eq. (7.82). It can be shown that one *necessary* condition for this to occur is that Eq. (7.82) be *much smaller than unity.* Note that this requirement is violated as the (m,n) mode approaches its cut-o frequency, since then $W_{T(m,n)}$ goes to zero and Eq. (7.82) tends to in nity. We may draw the conclusion that near cut-o it becomes demanding to calculate the attenuation constant accurately enough. Second-order corrections are, in general (see, e.g., Morse and Feshbach, 1953), much more cumbersome to calculate than rst-order ones, as they require summations which extend, at least in principle, over the whole in nity of modes of the lossless waveguide.

The reason why the rst-order procedure, as we described it, can be inadequate for *degenerate modes* is not too di cult to explain at this stage. In fact, in each term of the summations involved in the second-order correction, the denominator contains the di erence between the eigenvalue corresponding to the (m,n)-th mode which we are dealing with, and the eigenvalue of another mode. In the degenerate case, at least one among these di erences vanishes, the corresponding term in the summation becomes in nitely large, and it does not make sense any longer to call it a second-order correction. To avoid this, one has to make sure that the numerator of that term in the summation is also zero. In the case of the problem at hand, it can be shown that the numerator is of the type:

$$W_{P(m,n;q,r)} = \int_\ell Z_w \, \frac{\mathcal{H}_{\tan m,n} \;\; \mathcal{H}_{\tan q,r}}{2} \, d\ell \quad . \tag{7.83}$$

Eq. (7.83) can be looked at as a measure of how much the two modes, (m,n) and (q,r), interact, i.e., loose their orthogonality, in the lossy structure. In precise terms, Eq. (7.83) is the per-unit-length complex power which ows into the imperfect conductor because of the interplay between the two modes, i.e., is the di erence between the total per-unit-length power loss and the sum of those which can be attributed to the two modes individually. If Eq. (7.83) is non-zero, then the two modes are not orthogonal. If, furthermore, they are degenerate, then one of them cannot propagate without being accompanied by the other. It becomes erroneous to calculate its loss as if it were traveling alone.

The correct procedure for two degenerate modes, (m, n) and (q, r), is then the following one. First, one must nd *two linear combinations* of these modes, such that for them the integral in Eq. (7.83) is zero. They (see Section 7.5) are also modes of the ideal lossless guide, but, in contrast to the modes (m, n) and (q, r), they remain orthogonal in the lossy waveguide. One can then apply the procedure in Eq. (7.82) *to each of these new modes.*

An example may help to clarify these points. Take the modes TE_{11} and TM_{11} in a rectangular waveguide, and suppose that Z_w (to be discussed in more detail in the next section) is constant over the waveguide contour and can be taken out of the integral in Eq. (7.83). Let $\mathcal{H}'$ and $\mathcal{H}''$, respectively, be the TE_{11} and the TM_{11} modal elds. From Table 7.1, setting all the arbitrary constants equal to one, as they are irrelevant in these calculations, we have:

$$\mathcal{H}' = \frac{j\ \ _{11}}{a\ \ ^2_{11}}\left\{\frac{1}{b}\sin\frac{\ \ x}{a}\cos\frac{\ \ y}{b}\,\hat{a}_x + \frac{1}{b}\cos\frac{\ \ x}{a}\sin\frac{\ \ y}{b}\,\hat{a}_y\right\} + \cos\frac{\ \ x}{a}\cos\frac{\ \ y}{b}\,\hat{a}_z\ , \tag{7.84}$$

$$\mathcal{H}'' = \frac{j\ k}{\ ^2_{11}}\left\{\frac{1}{b}\sin\frac{\ \ x}{a}\cos\frac{\ \ y}{b}\,\hat{a}_x\quad \frac{1}{a}\cos\frac{\ \ x}{a}\sin\frac{\ \ y}{b}\,\hat{a}_y\right\}\ , \tag{7.85}$$

where:

$$\ ^2_{11} = \left(\frac{\ }{a}\right)^2 + \left(\frac{\ }{b}\right)^2\ , \qquad = \sqrt{\frac{\ }{\ }}\ , \qquad k = \omega\sqrt{\ \ }\ . \tag{7.86}$$

Calculating Eq. (7.83) as:

$$W_P = Z_w \int_0^a \mathcal{H}'_x\,\mathcal{H}''_x\ \ \mathrm{d}x + Z_w \int_0^b \mathcal{H}'_y\,\mathcal{H}''_y\ \ \mathrm{d}y\ , \tag{7.87}$$

we nd:

$$W_P = \frac{Z_w}{2}\ \frac{\ _{11}k\ \ ^2}{\ ^4_{11}ab}\ (a\ \ \ b)\ . \tag{7.88}$$

Therefore, except for a square waveguide ($a = b$), the TE_{11} and TM_{11} modes are not orthogonal, but coupled to each other by the losses in the wall. To nd linear combinations of these modes for which the interaction term in Eq. (7.83) vanishes can be phrased as an eigenvalue problem, as it corresponds to diagonalizing the *loss matrix*:

$$\mathbf{W} = \begin{bmatrix} W' & W_P \\ W_P & W'' \end{bmatrix}\ , \tag{7.89}$$

where W', W'' are Eq. (7.80) calculated for the TE_{11} and the TM_{11} modes, respectively, while W_P is Eq. (7.88). The eigenvalues of the matrix in Eq. (7.89) are, clearly:

$$ _{1,2} = \frac{W' + W''}{2} \quad \frac{1}{2}\sqrt{(W' \quad W'')^2 + W_P^2} \quad . \tag{7.90}$$

These quantities have to be inserted in the numerator of Eq. (7.82) to obtain the correct values for the attenuation constants of the two orthogonalized modes. The magnitude of the error one makes by using Eq. (7.82) without accounting for the coupling between the degenerate modes can be calculated, as an exercise, using for Z_w the value which was found in Section 4.11 for plane waves impinging on a at good conductor. The subject of wall impedance will be dealt with in more detail in the next section.

To conclude this section, let us deal brie y with loss due to an imperfect dielectric inside a waveguide. The concepts remain the same: one assumes that the modal elds of the lossless structure propagate in the lossy one, and evaluates the corresponding per-unit-length power dissipation. Dividing it by the transmitted power, and then by 2 to keep track of the relationship between eld amplitude and power, one nally gets the rate at which the eld decreases exponentially vs. z. Formulas, on the other hand, are much simpler than those of loss due to imperfect conductors. Indeed, the nal result turns out to be the same for TE and for TM modes, and also independent of whether the mode is degenerate or not. It reads:

$$ _d = \frac{\omega \quad ''}{2\sqrt{1 \quad (f_c/f)^2}} \quad , \tag{7.91}$$

where the notation $_c = \ ' \ j \ ''$ has been adopted for the complex permittivity, $ = \sqrt{\ / \ '}$ is the intrinsic impedance of the dielectric in the lossless waveguide, and f_c is the cut-o frequency of the mode we are dealing with. Necessary conditions for the validity of Eq. (7.91) are $'$ $''$, and frequency not too close to the cut-o frequency.

7.10 On wall impedances

The right value for the surface impedance for good conducting walls of a waveguide can always be found using in the right way the same approach as in Section 4.11 for the similar problem of a plane wave impinging on the surface of a half-space good conductor. Namely, we will rst introduce in the Helmholtz equation those approximations which are derived from the de ni- tion of a good conductor (ω). We will see that then, as a consequence, the wave impedance along the normal to the conductor surface becomes in- dependent of the conditions of incidence on the dielectric-conductor interface. Hence, its value may be taken as the wall impedance. We will see that often it is the same as that of Section 4.11; often, but, surprisingly perhaps, not always.

Let us deal rst with a rectangular waveguide (Figure 7.10), and begin from

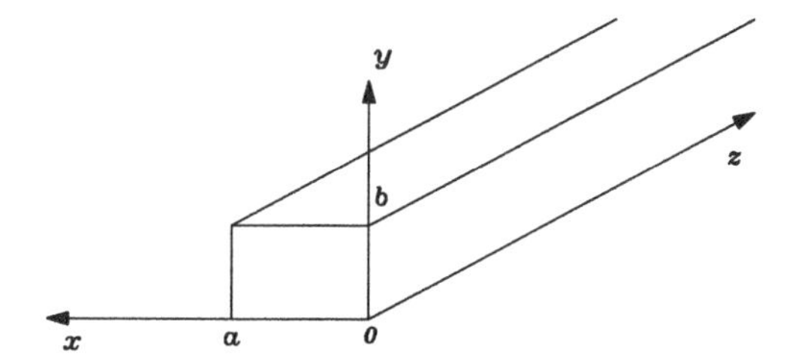

Figure 7.10 *Rectangular waveguide.*

the walls orthogonal to the x axis. Separating variables as in Section 7.7, we get the following equation:

$$\frac{d^2X}{dx^2} = (\omega^2 \quad j\omega \quad + \ ^2 \quad k_y^2)\,X \qquad (k_x^2 \quad j\omega \quad)\,X \quad . \tag{7.92}$$

In a lossless structure, $k_x = m \ /a$ can become large only if ω is large. Hence, good conductor means always $\omega \qquad k_x^2 = |\omega^2 \quad + \ ^2 \quad k_y^2|$. Consequently, the quantity in parenthesis on the right-hand side of Eq. (7.92) can be approximated as $\ j\omega \$. (The same warnings as in Section 4.12 hold, for approximations where real numbers are compared with imaginary ones). From now on, everything ows as in Section 4.12, and obviously the same occurs for the walls orthogonal to the y axis. We conclude that the wall impedance to be used in rectangular waveguide is Eq. (4.107), namely:

$$Z_w = (1+j)\left(\frac{\ f}{\ }\right)^{1/2} \quad . \tag{7.93}$$

For circular waveguides, the appropriate wall impedance is again given by Eq. (7.93). The proof, which requires separation of variables in cylindrical coordinates, is left to the reader as an exercise.

One might then jump to the conclusion that Eq. (7.93) is valid in general, for any shape of the waveguide. This is erroneous. In most cases the error is numerically small, but not insigni cant, as it may a ect, e.g., how the attenuation depends on frequency. As a counter-example, we will deal brie y with waveguides with *elliptical cross-section.* We will show that in such a case the wall impedance Z_w must be taken as a function of position on the waveguide contour, and furthermore, it takes di erent values for TE and TM modes. The rather unexpected conclusion is that the wall impedance is *anisotropic*, although the wall is made of an isotropic conductor.

To prove all these statements, we rst introduce (Figure 7.11) a system of elliptical-cylinder coordinates (, , z), related to cartesian coordinates (x, y, z) as follows:

$$x = q \cosh \quad \cos \quad , \qquad y = q \sinh \quad \sin \quad , \tag{7.94}$$

where $2q$ is the distance between foci (which are the same for all the elliptical coordinate lines $\ = $ constant). Obviously, these foci are also those of the

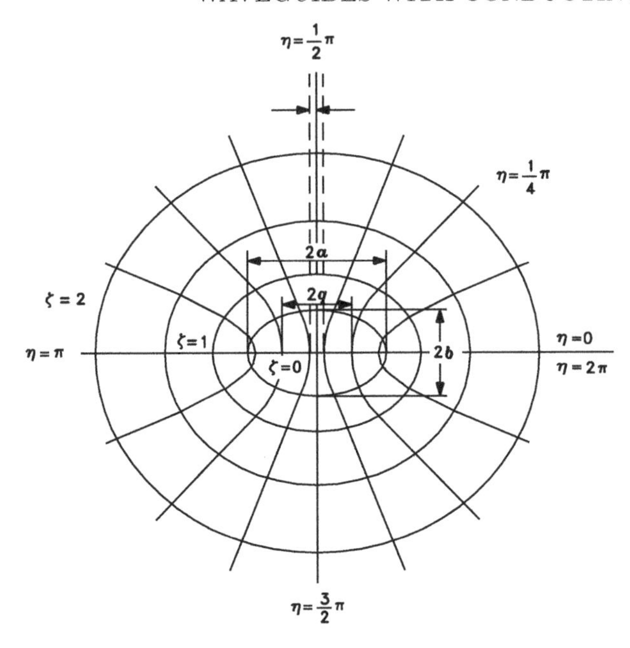

Figure 7.11 *An elliptical coordinate system.*

waveguide contour, which is then a coordinate line, $\;= \;{}_0$. Also, q is related to the ellipticity e, and to the lengths of the major and minor axes of the contour ellipse, $2a$ and $2b$, respectively, as:

$$e = \left[1 \quad \left(\frac{b}{a}\right)^2\right]^{1/2} = \frac{q}{a} \quad . \tag{7.95}$$

To nd the correct wall impedance, one must go from the beginning through the procedure we outlined before. Accordingly, the rst step is to solve Eq. (7.8) under the good-conductor approximation, $\quad \omega$. In elliptical coordinates, Eq. (7.8) reads (see Morse and Feshbach, 1953):

$$\frac{\partial^2 \mathcal{F}}{\partial \;^2} + \frac{\partial^2 \mathcal{F}}{\partial \;^2} + \;{}^2_n\, q^2(\sinh^2 \quad + \sin^2 \quad)\mathcal{F} = 0 \quad . \tag{7.96}$$

We can nd its general integral (see, e.g., Abramowitz and Stegun, Eds., 1965), and then, similarly to what we did in Section 7.8, we can impose on it the restriction that it remains nite for $\;\rightarrow 0$. The net result can be written as:

$$\mathcal{F} = Mc^{(4)}_{(n)}(\;;\;)\left[A\,ce_{(n)}(\;;\;) + B\,se_{(n)}(\;;\;)\right] \quad , \tag{7.97}$$

where the symbols have the following meanings:

A, B = arbitrary constants; n = order of the functions;

$Mc^{(4)}_{(n)}$ = radial Mathieu function of the fourth kind;

$ce_{(n)}$ = even angular Mathieu function of the rst kind;

$se_{(n)}$ = odd angular Mathieu function of the rst kind;

$= \; {}^{2}_{n} q^2/4$.

The assumption of a good conductor implies ${}^{2}_{n} \simeq \; j\omega$ in Eq. (7.96), and also, in most cases, $| \; _n| a \quad 1$. Then, we can make use of an asymptotic expression for $Mc^{(4)}_{(n)}$, which is valid for $| \; _n q \cosh \; | \to \infty$. This is a necessary step in order to obtain wave impedances in the radial direction that are independent of the order n, i.e., independent of the mode under examination. Otherwise, we could not call it a wall impedance.

The expressions one gets in this way are di erent depending on whether we take $\mathbf{E}_{\tan} = E_z \hat{a}_z$ or $\mathbf{E}_{\tan} = E \; \hat{a}$. In detail, it can be shown (Falciasecca *et al.*, 1971; Lewin *et al.*, 1977) that, for $= \; _0$, i.e., on the waveguide wall, it is:

$$Z_{P_{\rm TE}} \quad \frac{E}{H_z} = (1+j)\left(\frac{\;f}{\;}\right)^{1/2} \frac{1}{(1 \quad e^2 \cos^2 \;)^{1/2}} \;, \tag{7.98}$$

$$Z_{P_{\rm TM}} \quad \frac{E_z}{H} = (1+j)\left(\frac{\;f}{\;}\right)^{1/2} (1 \quad e^2 \cos^2 \;)^{1/2} \;. \tag{7.99}$$

It is evident that these expressions are not the same as Eq. (7.93). Changes are due, from a mathematical viewpoint, to the fact that the metric coe - cient of the coordinate changes along the coordinate line ($= \; _0$) where we calculate partial derivatives in order to express the eld components. In a rectangular or circular waveguide, on the contrary, the metric coe cient remains constant along the boundary. From a physical viewpoint, changes with respect to Eq. (7.93) can be explained by looking at the current ow lines inside the conducting wall. We saw in Section 4.12 that, when a uniform plane wave impinges on a conducting half-space, the propagation vector of the transmitted wave is essentially perpendicular to planar interface between the two media. Consequently, the current density $\mathbf{J} = \; \mathbf{E}$ is parallel to the planar interface, at any depth inside the conductor. An equivalent surface current — which is a concept intimately related to that of surface impedance — is easy to de ne, given that all ow lines are parallel. In a rectangular waveguide, if we neglect the corner e ects, the situation remains essentially the same. Current ow lines in the walls are everywhere parallel to the wall surfaces. The same occurs again in a circular waveguide, where the current ow lines have only an axial and an azimuthal component, always parallel to the surface. What we nd in an elliptical waveguide is completely di erent. The current ow lines, at the generic P point, have nonvanishing components along the coordinate elliptical line which passes through P, and consequently (see Figure 7.11) they are not parallel to the waveguide contour. At the same time, the distance between two given coordinate ellipses (which delimit a current ow tube) changes as we move in the azimuthal direction. Also the Poynting vector inside the wall can be shown not to be perpendicular everywhere to the surface of the wall. All these comments justify why Eqs. (7.98) and (7.99)

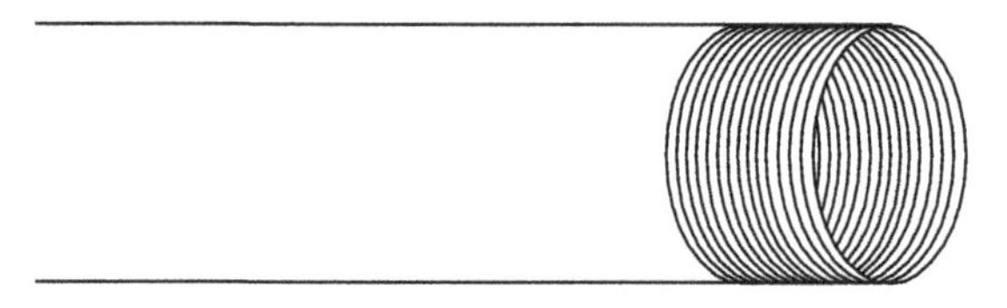

Figure 7.12 *Schematic view of a helical waveguide.*

are surprisingly di erent from the wall impedances we have found so far. In particular, we note that:

none of the wall impedances in an elliptical waveguide is the same as Eq. (7.93);

the wall impedance is anisotropic, since Eqs. (7.98) and (7.99) are not the same;

all the impedances on the elliptical wall change as we move on the waveguide contour.

The di erence between Eqs. (7.98) and (7.99), on one hand, and Eq. (7.93), on the other hand, does not induce any relevant di erence if the purpose is to nd the re ection coe cient of an elliptical object (such as an antenna re ector, for example). On the contrary, with a "small" di erence repeated many times, as in the case of a propagating mode which rebounds again and again on a waveguide wall, then the total di erence on the attenuation constant can become quite signi cant (Lewin *et al.*, 1977). Experimental results on the frequency dependence of loss recon rm that the correct values are those which one obtains with the procedure we outlined in this section.

7.11 Hybrid modes

In the two previous sections, the purpose of the wall impedance was to model more realistically structures which had been analyzed before with an idealized lossless model. The lossless model was a good starting point for the perturbation method. Consequently, we kept on using the terms TE and TM for the modes of waveguides with ohmic losses, although most of them were, strictly speaking, neither TE nor TM.

Wall impedances are used in a slightly di erent sense in modeling structures which do not have an ideal conducting wall, not even as a rst approximation. A typical example in this sense are the *helical waveguides*, structures (see Figure 7.12) whose walls are made of coiled wires or ribbons, the spacing between two coils being usually much smaller than the thickness of the wire or ribbon. These structures strongly discriminate the low attenuation of one family of modes whose currents ow essentially along the wires or ribbons (the $\mathrm{TE}_{0,n}$ family) from the high losses for all other modes. Their behavior can be studied using as a model an anisotropic wall, whose impedance, as a rst approximation, can be taken equal to zero in the azimuthal direction, di erent from zero in the longitudinal direction. Apart from the interest in their practical applications, which is nowadays much less than a few decades ago, what we wish to point out in this section is a concept, namely that, when wall impedances are nite, a complete set of solutions can no longer consist of only TE and TM modes. On the contrary, it is necessary to take into account other elds, which are called *hybrid modes*, are more complicated than TE and TM modes, and will play a dominant role in dielectric waveguides, to be studied in Chapter 10.

To show this, let us start from the general expressions in Eqs. (7.11) and (7.12), and suppose that the anisotropic wall, with a circular cross-section of radius a, sets the following boundary conditions:

$$E_\varphi = 0 \quad , \tag{7.100}$$

$$E_z = ZH_\varphi \quad , \qquad Z \neq 0 \quad . \tag{7.101}$$

First of all, let us verify whether these conditions can both be satis ed by a TM eld, as given by Eq. (7.12). In cylindrical coordinates, after separation of variables, Eq. (7.100) becomes:

$$\left.\frac{\partial \mathcal{T}}{\partial \varphi}\right|_{r=a} \propto R(a) \quad '(\varphi) = 0 \quad , \tag{7.102}$$

where, as usual, the prime indicates the derivative with respect to the argument, while Eq. (7.101) becomes:

$$(\ ^2 \quad \ ^2)\, \mathcal{T}(a) = j\omega \quad Z \left.\frac{\partial \mathcal{T}}{\partial r}\right|_{r=a} \quad . \tag{7.103}$$

From what we saw in Section 7.8 it follows that, for an azimuthal order $n \neq 0$, $\ '(\varphi)$ is not identically zero. Therefore, Eq. (7.102) can be satis ed only for $R(a) = 0$. This condition implies $\mathcal{T}(a) = 0$, and if we insert this into Eq. (7.103) we get $R'(a) = 0$. Nontrivial solutions of the Bessel equation (7.59) do not have any value for their argument such that the function and its rst derivative, both vanish. Thus, Eqs. (7.102) and (7.103) are mutually incompatible, except for $n = 0$ ($\ =$ constant). This means that $\mathrm{TM}_{0,n}$ modes do exist in this type of waveguide, but, on the contrary, there are no TM modes that depend on the azimuthal coordinate.

As for TE modes, we can show in a similar way that they cannot exist, except again for $n = 0$. In fact, assuming $\mathcal{E}_z \quad 0$ implies, via Eq. (7.101), $\mathcal{H}_\varphi = 0$, and because of Eq. (7.11) this entails:

$$\left.\frac{\partial \mathcal{L}}{\partial \varphi}\right|_{r=a} \propto R(a) \quad '(\varphi) = 0 \quad . \tag{7.104}$$

At the same time, the condition Eq. (7.100), again through Eq. (7.11), becomes:

$$\left.\frac{\partial \mathcal{L}}{\partial r}\right|_{r=a} \propto R'(a) \quad (\varphi) = 0 \quad , \tag{7.105}$$

and from now one can proceed as in the previous case.

Evidently, $TE_{0,n}$ and $TM_{0,n}$ modes do not form a complete set, as the waveguide can certainly propagate elds that depend on φ. There must be further solutions of Maxwell's equations. The decomposition theorem of Section 7.2 is quite general, and therefore one should be able to express any solution as a *sum* of a TE eld, of the type in Eq. (7.11), plus a TM eld, of the type in Eq. (7.12). Then, let $\mathcal{L} = A_1 R_1 \;_1$ and $\mathcal{T} = A_2 R_2 \;_2$, with A_1 and A_2 arbitrary constants. The boundary conditions (7.100) and (7.101) become:

$$A_1 \quad R'_1 \;\; _1 + \frac{n}{a} \; A_2 \, R_2 \;\; '_2 = 0 \quad , \tag{7.106}$$

$$^2 \, A_2 \, R_2 \;\; _2 = Z \left\{ j\omega \;\; A_2 \quad R'_2 \;\; '_2 + \frac{j}{\omega \;\; a} \; A_1 \, n \, R_1 \;\; '_1 \right\} , \tag{7.107}$$

where primes indicate, as usual, derivatives with respect to *arguments* (note that the argument of R_i is r, while that of $_i$ is $n\varphi$); furthermore, $^2 =$ ^{2} ^{2}.

From the system of Eqs. (7.106) and (7.107), we can:

eliminate one arbitrary constant A_i, showing that there is only one degree of freedom for amplitude and phase of each mode in this family;

derive an equation in implicit form which links the propagation constant with the angular frequency ω. It is usually referred to as the *characteristic equation* or the *dispersion relationship* of the structure. A simple way to nd it is to set the determinant of the system Eqs. (7.106) and (7.107), in the two unknowns A_i, equal to zero:

$$\frac{j \;\; Z}{\omega \;\; a} \; n \; \frac{R_1 F'_1}{R'_1 F_1} = \frac{^2 R_2 \quad j\omega \quad Z R'_2}{n \;\; R_2} \; a \, \frac{F_2}{F'_2} \quad . \tag{7.108}$$

For this equation in the unknown to admit solutions, it is necessary that the dependence on φ be the same on both sides, i.e.:

$$\frac{F'_1}{F_1} = \frac{F_2}{F'_2} \quad . \tag{7.109}$$

The two choices for the sign in Eq. (7.109) lead to two distinct sets of solutions for Eq. (7.108), resembling what we will see in Chapter 10. There is a need for a new terminology, as we cannot call TE and TM these new modes of helical guides. However, no general agreement on new names has been found throughout the technical literature. In Chapter 10, on the other hand, we will see that there is a general consensus on one nomenclature for the hybrid modes of cylindrical dielectric waveguides.

7.12 Further applications and suggested reading

We said at the beginning of this chapter that one of the reasons why we deal here with waveguides with conducting walls before tackling dielectric waveguides is because they came rst in time. In fact, waveguides surrounded by metal walls were among the key ingredients in the development of radar, which was the rst example of practical use of microwaves, which were also referred to, at that time, as hyperfrequencies. A popular misconception is that those early developments were essentially empirical, based on a cut-and-try strategy. Those who want to see how advanced and sophisticated were, on the contrary, the rst major theoretical models in the area, should take at least a quick tour through the volumes known as the MIT Radiation Laboratory Series, published by McGraw-Hill in the late 1940's. Readers who are familiar with the names of the most prominent theoretical physicists of that time will recognize some of them among the authors of these books. This was because during World War II most physicists in the United States became engaged in applied research for military aims — not only the very famous "Manhattan Project," but also other developments, including of course radar. Accordingly, mathematical tools which were advanced, for that time, and were in use in elds such as quantum mechanics, were promptly and successfully transferred to modeling microwave devices.

Because of this tidy and neat mathematical background, waveguides with ideal conducting walls are an ideal tutorial introduction to guided-wave propagation, much simpler than through any other similar structure. Furthermore, waveguides induce a very bene cial step in the learning process of telecommunication engineers and other students with similar curricula. This step consists of the fact that students who have familiarized themselves with waveguides begin to look at elds, transmission lines, and electrical circuits, in a uni ed way, considering them as three ways to model the same physical reality. This cross-fertilization process can still be appreciated, as a major contribution, in at least one of the books in the previously quoted MIT Series, namely in Montgomery, Dicke and Purcell (1948). Readers who, at this stage, do not yet grasp the meaning and the importance of this comment, should come back to it after studying the next chapter.

For decades, the word *microwaves* has had two meanings. It stood for a frequency range and, at the same time, for a set of technologies, which were then based mainly on waveguides. At present, this is no longer true. At those frequencies which are still called microwaves, up to an upper limit which is not sharply de ned, but is de nitely higher than 10 GHz, and still growing, striplines dominate by far and large over waveguides. As we said when we introduced the TEM mode, striplines, although similar to those structures where the TEM mode propagates, are not easy to model analytically, as they require some approximation. For this reason, they have been deferred to the next chapter. While striplines were displacing waveguides from many applications, most of the electronics gradually became solid-state. Historically, microwave electronics consisted of special tubes which encompassed waveguiding structures, like sequences of cavities, helices, periodic combs, etc. The reader who wants to become familiar with modern microwave technology can be referred to textbooks like Baden Fuller (1990), Gardiol (1984) and Cronin (1995), and to more advanced books like Collin (1966/1991). Those who want to learn about classical microwave tubes, still in use for some applications, may refer to Pierce (1950).

Conducting-wall waveguides have still some practical application, in addition to being pedagogically useful. Compared to striplines, their main advantage is that all the power they convey travels in a well de ned, closed region. This can be crucial in non-telecommunication applications, like industrial heating, domestic cooking, medical diagnosis and therapy, etc., where shielding humans from the eld is essential, to avoid unwanted interaction with biological tissues. Industrial, scienti c and medical (ISM) applications of microwaves are widely described in the literature (e.g., see Roussy and Pearce, 1995, and references therein).

Another case where waveguides are used, because they can convey very high powers with relatively low loss (especially if operated in the TE_{0n} modes of either circular or helical waveguides), is plasma heating. Plasmas, as well known, have been and still are intensively explored, as the way towards generating energy by controlled nuclear fusion, and microwave heating is looked at as one of the possible ways to reach the threshold temperatures that are needed. The eld of plasma diagnostics also relies heavily on microwaves up to about 100 GHz. For these subjects, the reader is referred to Heald and Wharton (1978), Hutchinson (1987), Lochte-Holtgreven (1995) and references therein.

Finally, some kinds of particle accelerators, for high-energy physics experiments, contain parts — typically, pipes where particles gain kinetic energy at the expenses of a radio-frequency or microwave eld — which can be looked at, and designed, as waveguiding structures. An excellent reading for those who are interested in these problems is Panofsky and Phillips (1962).

Problems

7-1 A rectangular waveguide is interrupted on the plane $z = L$ by a at ideal conductor, i.e., is a short-circuited waveguide. There are no sources between $z = 0$ and $z = L$. Assume that the waveguide wall and the dielectric inside it are all lossless. At a frequency such that only the TE_{10} mode is above its cut-o frequency, state whether the real and the imaginary parts of the ux of the Poynting vector through the plane $z = 0$ vanish or not, specifying whether the answers depend on the length L. Repeat the problem assuming that the TE_{10} mode is below its cut-o frequency.

7-2 Derive two relationships which di er from Eqs. (7.38) and (7.39) only because $\mathbf{E}_{tk}$ and $\mathbf{H}_{tk}$ are replaced by $\mathbf{E}_{tk}$ and $\mathbf{H}_{tk}$, respectively. Comment on their physical meanings.

7-3 Estimate the error if Eq. (7.82) is used for a degenerate mode without accounting for its coupling with the other degenerate mode. Use for the wall impedance Z_w the value found in Section 4.12 as the wave impedance for plane waves in a at conductor.

7-4 Return to Problem 1-17, and show that, in general, the eld of any TM mode can be derived from a purely longitudinal magnetic vector potential $\mathbf{A} = A_z \hat{a}_z$.
Derive the constant of proportionality between A_z and E_z. Why is it di erent from $j\omega$?

7-5 Return to Problem 1-18, and show that, in general, the eld of any TE mode can be derived from a purely longitudinal electric vector potential $\mathbf{F} = F_z \hat{a}_z$.
Derive the constant of proportionality between F_z and H_z. Why is it di erent from $j\omega$?

7-6 An important theorem of circuit theory, usually referred to as Foster's theorem, states the following. Let $X(\omega)$ be the input reactance of a perfectly lossless two-terminal lumped-element network. Then, $\partial X/\partial\omega > 0$ for any network, where ω is, as usual, the angular frequency.

a) Verify that the wave impedances of TE and TM modes below cut-o obey this rule, for waveguides of any shape and any size.
b) For any mode of a rectangular waveguide, calculate the magnetic and electric energies stored in a semi-in nite waveguide, $z > 0$, and, using the reactive power term in Poynting's theorem, verify again that these results agree with Foster's theorem.

7-7 Show that in the presence of loss or gain, i.e., when the real parts of the propagation constants of modes above cut-o are nonvanishing, two generic modes are not power-orthogonal, i.e., they do not satisfy Eq. (7.42).

7-8 Design an air- lled rectangular waveguide (i.e., calculate the lengths of its sides, a and b) so that the frequency $f = 3\,\text{GHz}$ is 1.3 times greater than the cut-o frequency of its fundamental mode, and 0.7 times the cut-o frequency of the rst higher-order mode.

7-9 A rectangular waveguide of width $a = 23\,\text{mm}$ and height $b = 11.5\,\text{mm}$ can propagate the TE_{10} mode at a frequency of $6\,\text{GHz}$, only if lled with a suitable dielectric.

a) Find the range of values for the relative permittivity, $_r$, for which the waveguide propagates only the fundamental mode.
b) Study how the phase constant of the fundamental mode, as a function of frequency, varies when the permittivity of the medium varies within this range of values.

7-10 A rectangular waveguide whose dimensions are the same as in the previous problem, lled with air ($_r = 1$), goes from a receiving antenna to a receiver for a total distance $L = 5\,\text{m}$ and conveys a signal whose spectrum is

centered on $f = 1.5\, f_c$, where f_c is the cut-o frequency of the fundamental mode.
Evaluate the di erential phase and group delays (i.e., the di erences in velocities times the distance) over this length, for two frequencies which di er from the center of the spectrum by 5%.

7-11 A standard WR-15 rectangular waveguide (3.76 mm 1.88 mm) ends abruptly in an aperture. Assume that the wave impedance beyond the aperture equals the free-space intrinsic impedance. Find the re ection coe cient for the TE_{10} mode at a frequency of 60 GHz, and the percentage of active power which is re ected back into the waveguide.
(*Comment.* This problem is much simpler to solve after reading the following chapter.)

7-12 Two rectangular waveguides with di erent major sides ($a_1 < a_2$) and equal minor sides ($b_1 = b_2$) are butt-jointed in the $z = 0$ plane. Waveguide 1 propagates the TE_{10} mode only, which travels in the positive z direction and impinges on the plane of discontinuity. Find which modes of waveguide 2 are excited, and which are their amplitudes, if

a) the two waveguides are aligned in a corner ($x = 0$);

b) they are aligned at their centers (the point $x = a_1/2$ of waveguide 1 is the same as the point $x = a_2/2$ of waveguide 2).

(*Hint.* Expand an arc of cosine function into a set of harmonic functions with a di erent period. Make use of orthogonality.)

7-13 Repeat the previous problem for two rectangular waveguides with equal major sides ($a_1 = a_2$) and di erent minor sides ($b_1 < b_2$). Waveguide 1 propagates the TE_{10} mode only.

a) The two waveguides are aligned in a corner ($y = 0$).

b) They are aligned at their centers (the point $y = b_1/2$ of waveguide 1 is the same as the point $y = b_2/2$ of waveguide 2).

7-14 Consider a lossless circular waveguide.

a) Making use of Bessel function identities and integrals given in Appendix D, prove explicitly that its modes are orthogonal.

b) Repeat the same problem for the modes of a coaxial cable.

7-15 Find the modes of a semicircular-waveguide, and those of a waveguide whose cross-section is a quarter of a circle, always assuming that the walls are ideal electric conductors.

7-16 Starting from the wave equation in a good conductor, and using separation of variables in cylindrical coordinates, show that the wall impedance given by Eq. (7.93) is applicable when dealing with circular waveguides and coaxial cables.

7-17 Consider a circular waveguide with a low-loss wall, and use Eq. (7.93) for the wall impedance.

a) Show that the perturbation formula for the attenuation, Eq. (7.80), indicates that the theoretical losses of the $\mathrm{TE}_{0,m}$ modes are monotonically decreasing functions of the frequency, although the wall impedance increases in proportion to the square root of the frequency.

b) Provide a physical explanation for this result.
(*Hint.* Study the current density of these modes in the walls, as a function of frequency, for a constant transmitted power.)

c) Why none of the modes of a rectangular waveguide behaves in the same way?

7-18 A circular waveguide whose modes are all below their cut-o frequencies may be used as an attenuator in a coaxial cable: the internal conductor is removed over a certain length, and short-circuited to the external one at the two extremes of that length.

a) Neglecting backward waves, i.e., assuming a purely exponential z-dependence for the TE_{11} mode below cut-o , write a "calibration chart" of an attenuator whose diameter is 20 mm and whose length is 50 mm, i.e., its attenuation (in decibels) vs. frequency, from dc to the cut-o frequency of the TE_{11} mode.

b) In these calculations, all the structures have been assumed to be lossless. Consequently, attenuation in the device cannot be caused by power dissipation. Where does the lost power go? Is it rigorous to accept, under these circumstances, the previous assumption that backward waves are negligible?

7-19 In coaxial cables, the diameters of the inner and outer conductors which are in current use, as standards, in the design of the so-called Type N connectors, are 0.120 inches and 0.276 inches, respectively. Assuming that the dielectric between them has the same permittivity as air, evaluate (approximately) the uppermost frequency at which only the TEM mode propagates in the cable. Discuss how this cut-o frequency scales vs. relative permittivity of the dielectric.

7-20 A so-called microwave Y-junction is made of three identical rectangular waveguides, as sketched in Figure 3.4. Assume that they are lossless, and that only the TE_{10} mode is above cut-o . When waveguide 1 is connected to a generator, then the two waves at the outputs of waveguides 2 and 3 are equal in amplitude and in phase. By reciprocity, when two waves of equal amplitudes and phases are sent into waveguides 2 and 3, a wave in the TE_{10} mode emerges from waveguide 1.

Using symmetry arguments (see Section 3.9), discuss what happens when:

a) two waves of equal amplitude but phase-shifted by radians are sent into waveguides 2 and 3;

b) one wave is sent into waveguide 2 whereas no wave is sent into waveguide 3.

References

Abramowitz, M. and Stegun, I.A. (Eds.) (1965) *Handbook of Mathematical Functions.* Dover, New York.

Baden Fuller, A.J. (1990) *Microwaves: An Introduction to Microwave Theory and Techniques.* 3rd ed., Pergamon, Oxford, New York.

Collin, R.E. (1966) *Foundations for Microwave Engineering.* McGraw-Hill Physical and Quantum Electronics Series, McGraw-Hill, New York.

Collin, R.E. (1991) *Field Theory of Guided Waves.* IEEE Press, New York.

Cronin, N.J. (1995) *Microwave and Optical Waveguides.* Institute of Physics, Bristol, Philadelphia.

Dicke, R.H. and Wittke, J.P. (1960) *Introduction to Quantum Mechanics.* Addison-Wesley, Reading, MA.

Falciasecca, G., Someda, C.G. and Valdoni, F. (1971) "Wall impedances and application to long-distance waveguides". *Alta Frequenza,* **40**, 426–434.

Gardiol, F.E. (1984) *Introduction to Microwaves.* Artech House, Dedham, MA.

Heald, M.A. and Wharton, C.B. (1978) *Plasma Diagnostics with Microwaves.* R.E. Krieger Pub. Co., Huntington, New York.

Hutchinson, I.H. (1987) *Principles of Plasma Diagnostics.* Cambridge University Press, Cambridge, UK.

Lewin, L., Chang, D.C. and Kuester, E.F. (1977) *Electromagnetic Waves and Curved Structures.* P. Peregrinus, Stevenage, UK.

Lochte-Holtgreven, W. (1995) *Plasma Diagnostics.* American Vacuum Society Classics, AIP Press, New York.

Montgomery, C.G., Dicke, R.H. and Purcell, E.M. (Eds.) (1948) *Principles of Microwave Circuits.* McGraw-Hill, New York. Reprinted by Peter Peregrinus, London, for IEE (1987).

Morse, P.M. and Feshbach, H. (1953) *Methods of Theoretical Physics.* McGraw-Hill, New York.

Panofsky, W.K.H. and Phillips, M. (1962) *Classical Electricity and Magnetism.* 2nd ed., Addison-Wesley, Reading, MA.

Pierce, J.R. (1950) *Traveling-Wave Tubes.* Bell Telephone Laboratories Series, Van Nostrand, New York.

Roussy, M.G. and Pearce, J.A. (1995) *Foundations and Industrial Applications of Microwave and Radio Frequency Fields: Physical and Chemical Processes.* Wiley, New York.

CHAPTER 8

Waves on transmission lines

8.1 Introduction

We will use the phrase "transmission line" for any electrical system which, in one geometrical dimension, i.e., along the axis labeled z, extends signi cantly more than along the others, and more than the wavelength. When we talk of "common" or "ordinary" transmission lines, they consist of two or more conductors, whose cross-sections, orthogonal to the preferential z direction, are separated by dielectrics. Simple examples are parallel wire pairs, and coaxial cables. More sophisticated examples are provided by microwave stripline circuits (see Section 8.11), the main di erence coming from the fact that often their dielectric is not homogeneous.

When the conductors are connected through lumped passive components (resistors, capacitors, inductors, transformers), and fed by lumped voltage or current generators, then usually they can convey an electric current at arbitrarily low frequencies. This is the reason why the classical electrical-engineering approach to transmission lines starts o as an extension of lumped-circuit theory, i.e., begins by postulating Ohm's law and Kirchho 's laws. We will take it for granted that the readers know them from basic physics classes, and follow this standard approach in the rst part of the present chapter. We will shortly be able to show that *waves* may propagate along these lines. They are expressed in terms of voltages and currents, so they must be looked at as being electromagnetic waves.

Our subsequent task will be to reconcile these waves on transmission lines with those e.m. waves that we have studied so far. This will be done also due to the fact that we have used very frequently, for our waves, one of the most typical concepts of circuit theory, that of impedance. Once the link is established between common transmission lines and what we saw in the previous chapters, then the transmission-line *formalism*, developed for actual lines, becomes suitable, and extremely useful, as an *abstract model* for transmission/re ection phenomena with any kind of e.m. wave, no matter whether free or guided. These subjects will be dealt with in the last sections of this chapter.

"Lumped" means that the geometrical size of an element is negligible with respect to wavelength, at all frequencies of interest, so that none of the electrical quantities which can be de ned on that element, like voltage and current, experiences a signi cant phase delay from one point to another within that element.

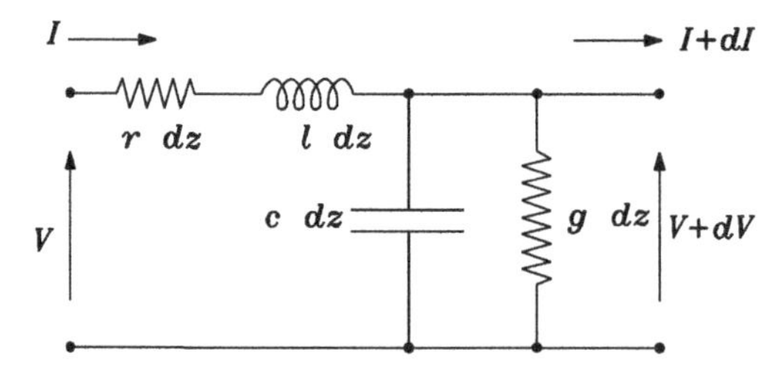

Figure 8.1 *Equivalent circuit for a section of in nitesimal length of a uniform transmission line.*

8.2 Uniform transmission lines

To retain simplicity, we will deal only with transmission lines whose characteristics are constant along the direction of propagation, z. We will refer to them as *uniform* transmission lines.

In any transmission line, energy can be stored along its length under electric and under magnetic form. Some energy can also be dissipated along the line, under the form of ohmic losses in non-ideal conductors and in non-ideal dielectrics. Intuitively, then, it should be possible to model the electromagnetic behavior of any line in terms of *distributed* parameters. Magnetic energy stored in an in nitesimal length dz can be thought of as being stored in an in nitesimal inductance. Ohmic losses in the conductors, over an in nitesimal length, can be attributed to an in nitesimal resistance. The same can be said for the electric energy, stored in an in nitesimal capacitance, and for losses in the dielectric, modeled as an in nitesimal shunt conductance. In a uniform transmission line, these properties must, by de nition, be constant in the z direction. Consequently, we can assume, as a starting point, that the suitable model for an in nitesimal length dz is the "equivalent circuit" shown in Figure 8.1, where the parameters r, l, c, g are independent of z.

We may then de ne a per-unit-length series impedance Z (expressed in ohms per meter), as the series of the per-unit-length inductance l (expressed in henrys per meter) and the per-unit-length resistance r (expressed in ohms per meter), $Z = r + j\omega l$. Similarly, we de ne a per-unit-length shunt admittance $Y = g + j\omega c$ (expressed in inverse ohms per meter) as the parallel of the per-unit-length capacitance c (expressed in farads per meter) and the per-unit-length conductance g (expressed in inverse ohms per meter). Note that this presentation of the "equivalent circuit" is rather phenomenological: at this stage, it should be regarded as a postulate. However, it will be justi ed in more physical terms in Section 8.10. We postpone this justi cation because it is less urgent than the basic results which can be derived from the circuit itself.

Applying Ohm's law and Kirchho 's laws (for voltages around a closed path and for currents in a node, respectively) for time-harmonic voltages and currents at steady state, from the circuit of Figure 8.1 we get the following di erential equations, where the voltage V and the current I are unknown functions of the longitudinal coordinate z:

$$\frac{dV}{dz} = \quad ZI \quad , \qquad \frac{dI}{dz} = \quad YV \quad , \tag{8.1}$$

where:

$$Z = (r + j\omega l) \quad , \qquad Y = (g + j\omega c) \quad . \tag{8.2}$$

Eqs. (8.1) are frequently referred to as the *telegraphist's equations.* Di erentiating them again with respect to z and replacing in one of them the rst derivative by means of the other one, we get:

$$\frac{d^2V}{dz^2} = ZYV \quad , \qquad \frac{d^2I}{dz^2} = ZYI \quad . \tag{8.3}$$

Eqs. (8.3) are sometimes referred to as the *telephonist's equations.* They can also be looked at as one-dimensional forms of the Helmholtz equation. The general integral of either one can be expressed in terms of exponential functions:

$$V(z) = V_+ e^{\quad z} + V \quad e^{\quad z} \quad , \qquad I(z) = I_+ e^{\quad z} + I \quad e^{\quad z} \quad , \tag{8.4}$$

where:

$$= +\sqrt{(r + j\omega l)\,(g + j\omega c)} \quad . \tag{8.5}$$

Since is in general complex, each term in Eqs. (8.4) looks like a wave, traveling either in the forward or in the backward direction along the z axis. Consequently, is referred to as the *propagation constant* of the transmission line. Current symbols are $= \quad + j \quad$, with and real, referred to as the *attenuation constant* and the *phase constant* of the transmission line.

Eqs. (8.4) cannot be looked as the general integral of the system in Eqs. (8.1), as long as the four arbitrary constants are independent. Indeed, the order of the system in Eqs. (8.3) being higher, it has also spurious solutions, i.e., solutions which do not satisfy Eqs. (8.1). To restrict ourselves to the actual solutions, we must introduce Eqs. (8.4) into Eqs. (8.1). So, we nd that the constants must be related among themselves as follows:

$$I_+ = \frac{V_+}{Z_C} \quad , \qquad I \quad = \quad \frac{V}{Z_C} \quad . \tag{8.6}$$

The quantity:

$$Z_C = \sqrt{\frac{(r + j\omega l)}{(g + j\omega c)}} \quad , \tag{8.7}$$

is referred to as the *characteristic impedance* of the transmission line. In general, it is complex, and this indicates that, in a uniform transmission line which is not further speci ed, a voltage and a current waves which travel in the same direction† are not in phase, at the same distance from the origin along the line, z.

† To avoid an unnecessary cumbersome language, on many occasions throughout this chapter we will use the word "direction" with the meaning of "oriented direction," or "sense".

The quantity:

$$P_c(z) = \frac{V(z)\, I\ \ (z)}{2} \quad , \tag{8.8}$$

where the star denotes as usual the complex conjugate, is the *complex power* across the generic cross-section of the line, owing in the *positive* direction of the z axis (i.e., active power is actually owing in the positive z direction if the real part of Eq. (8.8) is positive, in the negative direction if $\mathrm{Re}[P_c] < 0$). Inserting Eq. (8.6) into Eq. (8.8), we see that P_c consists of a term associated with the forward wave, $(V_+ I_+/2)\exp(\ \ 2\ \ z)$, plus a term associated with the backward wave, $(V\ \ I\ \ /2)\exp(2\ \ z)$, plus two "cross" terms, $(V_+ I\ \ /2)\exp(\ \ j2\ \ z)$ and $(V\ \ I_+/2)\exp(j2\ \ z)$.

The remaining two arbitrary constants in the expressions for voltage and current have to be determined by two boundary conditions. Normally, these are imposed by a generator at one end of the line, and a load at the other end. The rest of this chapter will be devoted mainly to how these boundary conditions a ect voltage and current along the entire transmission line.

8.3 Impedance transformation along a transmission line

The ratio between voltage and current, at any value of z along the transmission line, de nes an *input impedance*:

$$Z(z) = \frac{V(z)}{I(z)} = Z_C\, \frac{V_+ e^{\ \ z} + V\ \ e^{\ \ z}}{V_+ e^{\ \ z}\ \ V\ \ e^{\ \ z}} \quad . \tag{8.9}$$

The *input admittance* is, obviously, the inverse of Eq. (8.9).

Suppose that the line is terminated, at $z = 0$, on a linear, passive two-terminal network, whose impedance $Z_L = V(0)/I(0)$ we refer to as the *load impedance*. Incidentally, note that normally we will take the positive z direction from the generator to the load, then, choosing the load terminal section as the origin for the z coordinate implies that the line is in the region $z\ \ 0$. The input impedance at any distance z can be obtained, as a function of z and of the load impedance, imposing $Z(0) = Z_L$ in Eq. (8.9). The result is easily derived as:

$$Z(z) = Z_C\, \frac{Z_L \cosh\ \ z\ \ \ Z_C \sinh\ \ z}{Z_C \cosh\ \ z\ \ \ Z_L \sinh\ \ z} \quad . \tag{8.10}$$

Another useful quantity to de ne at any distance z from the load is the *(voltage) re ection coe cient*, equal to the ratio between the backward and the forward voltage waves at that z. Due, once again, to Eq. (8.9), it is expressed by:

$$\ \ (z) = \frac{V\ \ e^{\ \ z}}{V_+ e^{\ \ \ z}} = \frac{I\ \ e^{\ \ z}}{I_+ e^{\ \ \ z}} = \ \ _L\, e^{2\ \ z} \quad , \tag{8.11}$$

There will be no danger of ambiguity, as the waves considered in this chapter can only travel along the line.

where $\ _L$ is the (voltage) re ection coe cient at $z = 0$, i.e., at the load terminals.

To simplify the expression of $\ (z)$ as a function of the input impedance or admittance, and vice-versa, it is convenient to de ne the so-called *normalized impedance and admittance*, namely:

$$(z) = \frac{Z(z)}{Z_C} \quad , \qquad y(z) = \frac{Y(z)}{Y_C} = \frac{Z_C}{Z(z)} = \frac{1}{\ (z)} \quad . \tag{8.12}$$

The relationships between these adimensional quantities and the re ection coe cient (adimensional also) are easily found by combining the de nitions in Eqs. (8.12) and (8.11), and read:

$$(z) = \frac{1 + \ (z)}{1 \quad (z)} \quad , \qquad y(z) = \frac{1 \quad (z)}{1 + \ (z)} \quad , \tag{8.13}$$

whose inverse is:

$$(z) = \frac{\ (z) \quad 1}{\ (z) + 1} = \frac{Z(z) \quad Z_C}{Z(z) + Z_C} = \frac{1 \quad y(z)}{1 + y(z)} = \frac{Y_C \quad Y(z)}{Y_C + Y(z)} \quad . \tag{8.14}$$

Let us stress how similar these expressions are to Fresnel's re ection formulas, Eqs. (4.92) and (4.93), which we derived for plane wave re ection in Section 4.9. This similarity will induce us to elaborate a broad-scope abstract interpretation for the formalism of transmission lines. This point will be developed in Section 8.10, where, following the same line of thought, we will de ne normalized voltages and currents, quantities which obey Ohm's law with normalized (i.e., adimensional) impedances or admittances and therefore are broadly arbitrary in their physical dimensions.

Coming back to common transmission lines, the previous relationships enable us to express the complex power, Eq. (8.8), in terms of the re ection coe cient, as:

$$P_c(z) = \frac{V_+ I_+}{2} \left[1 \quad (z) \quad (z) + \qquad \right] \quad . \tag{8.15}$$

An exercise which is left to the reader is to separate real and imaginary parts in the last equation, i.e., to learn how to express active and reactive power in terms of the re ection coe cient. This should be done twice, initially assuming that the characteristic impedance is real, and then removing this assumption. The di erence will help the reader to appreciate why the content of the following two sections is very valuable in practice.

8.4 Lossless transmission lines

All the relationships that were de ned or derived in the previous sections become much simpler in the ideal case of a transmission line whose conductors and dielectrics are *lossless*. The reader may easily obtain, setting the per-unit-length series resistance r and shunt conductance g equal to zero, the following results for propagation constant, characteristic impedance, normalized impedance and admittance, re ection coe cient, voltage and current along a lossless transmission line:

$$= j \quad = j\omega\sqrt{lc} \quad , \tag{8.16}$$

$$Z_C = \sqrt{\frac{l}{c}} \quad , \tag{8.17}$$

$$(z) = \frac{{}_L \cos \quad z \quad j \sin \quad z}{\cos \quad z \quad j \,{}_L \sin \quad z} = \frac{{}_L \quad j \tan \quad z}{1 \quad j \,{}_L \tan \quad z} \quad , \tag{8.18}$$

$$y(z) = \frac{1}{(z)} = \frac{y_L \quad j \tan \quad z}{1 \quad j y_L \tan \quad z} \quad , \tag{8.19}$$

$$(z) = {}_L\, e^{j2 \quad z} = \frac{V}{V_+} e^{j2 \quad z} \quad , \tag{8.20}$$

$$V(z) = V_+ e^{\quad j \quad z} + V \quad e^{j \quad z} \quad , \tag{8.21}$$

$$I(z) = \frac{V_+}{Z_C} e^{\quad j \quad z} \quad \frac{V}{Z_C} e^{j \quad z} \quad . \tag{8.22}$$

For a lossless transmission line the characteristic impedance Eq. (8.17) is real, like the intrinsic impedance in a lossless medium (and consequently the wave impedance in any direction in real space, for a uniform plane wave). To insist on this analogy, the quantity $= 2 \; /$ is called the wavelength on the transmission line. Voltage and current of either a forward or a backward traveling wave are in phase, at any z.

Another result which, as the reader may easily check, is peculiar to lossless transmission lines, and holds irrespectively of their loads, is that the re ection coe cient, the normalized input impedance and the normalized input admittance are *periodic* functions of z, with a period equal to $/2$. The modulus of the re ection coe cient is constant along a lossless line. Voltage and current are periodic functions, of period . As for the complex power, Eq. (8.15) shows that along a lossless line its real part (i.e., the active power):

$$P_a = \frac{V_+ I_+}{2} \left[1 \quad (z) \quad (z) \right] \quad , \tag{8.23}$$

is constant vs. z, while the coe cient of its imaginary part (i.e., the reactive power):

$$Q(z) = \quad j \frac{V_+ I_+}{2} [\qquad] \quad , \tag{8.24}$$

varies sinusoidally, with a period equal to $/2$. These properties will be exploited to develop a graphical tool, the Smith chart, to be introduced in Section 8.7.

8.5 Low-loss transmission lines

Strictly lossless transmission lines are not physical entities, in the same sense as the strictly lossless media of the previous chapters. In the real world, losses are inevitable. However, with very rare and special exceptions, like attenuators, transmission lines are of practical interest only if their losses are small enough, at least over a suitable frequency range. Consequently, the lossless model studied in the previous section may be looked at as a "zero-order" approximation for real lines, similarly to what we did for conducting-wall waveguides in Chapter 7. In fact, we will see soon that, for lossy lines, exact solutions are often unnecessarily cumbersome. We will show then how rst-order approximations for real lines can be derived by elaborating on the lossless model. The phrase "small loss," still ambiguous up to this stage, will be used from now on to designate precisely the cases where these rst-order approximations are acceptable.

As we said before, the propagation constant Eq. (8.5) is, in general, a complex quantity, $= \; + j \;$. One may always start from the following expression:

$$^2 = (r + j\omega l)\,(g + j\omega c) = (rg \quad \omega^2 lc) + j\omega(rc + lg) \quad , \tag{8.25}$$

and separate real and imaginary parts, obtaining:

$$\begin{aligned} {}^2 \quad {}^2 &= rg \quad \omega^2\, lc \quad , \\ 2 \quad &= \omega(rc + lg) \quad . \end{aligned} \tag{8.26}$$

Solving this as a system of two equations in two unknowns, we get:

$$\begin{aligned} &= \frac{1}{\sqrt{2}} \sqrt{\sqrt{(r^2 + \omega^2 l^2)\,(g^2 + \omega^2 c^2)} \quad (\omega^2 lc \quad rg)} \quad , \\ &= \frac{1}{\sqrt{2}} \sqrt{\sqrt{(r^2 + \omega^2 l^2)\,(g^2 + \omega^2 c^2)} + (\omega^2 lc \quad rg)} \quad . \end{aligned} \tag{8.27}$$

As anticipated, these expressions are complicated, compared with Eq. (8.16), but become remarkably simpler for small losses, i.e., for $r/\omega l \quad 1$, $g/\omega c \quad 1$. Indeed, truncated binomial expansions for the square roots (neglecting terms of order $rg/\omega^2 lc$ and smaller) give:

$$\begin{aligned} &= \frac{r}{2}\sqrt{\frac{c}{l}} + \frac{g}{2}\sqrt{\frac{l}{c}} = \frac{r}{2}\,Y'_C + \frac{g}{2}\,Z'_C \quad , \\ &= \omega\,\sqrt{lc}\left[1 + \frac{1}{8}\left(\frac{r}{\omega l} \quad \frac{g}{\omega c}\right)^2\right] \quad . \end{aligned} \tag{8.28}$$

From Eq. (8.7) in a similar way we nd:

$$Z_C = Z'_C\left[1 + \frac{r^2}{8\omega^2 l^2} \quad \frac{3g^2}{8\omega^2 c^2} + \frac{rg}{4\omega^2 lc} + j\left(\frac{g}{2\omega c} \quad \frac{r}{2\omega l}\right)\right] \quad , \tag{8.29}$$

where $Z'_C = 1/Y'_C$ is the characteristic impedance of a lossless line having the same per-unit-length inductance and capacitance as the actual line.

Note that, at rst order in $r/\omega l$ and $g/\omega c$, the phase constant remains the same as for an ideal lossless line. The same occurs for the real part of the characteristic impedance. Accordingly, rst-order e ects are only a non-zero attenuation constant, and a reactive part in the characteristic impedance.

As we said in Section 7.9, one of the most important results of perturbation theory in general (see, e.g., Morse and Feshbach, 1953, or Dicke and Wittke, 1960) is that n-th order corrections for an eigenvalue can be calculated by means of the n 1-th order eigenvectors. What this means in our case is that (exactly like in Section 7.9) we can compute the rst-order attenuation constant from the *zero-order* voltages and currents, i.e., from what we found for lossless lines in the previous section. In our case, this statement is easy to prove. Insert the lossless-line voltage and current forward waves (the rst terms in Eqs. (8.21) and (8.22), respectively) into the following formulas, which express the per-unit-length power loss, P_l (in watts per meter) for a forward traveling wave, and the active power P (in watts) carried by that wave through the generic section of the line:

$$P_l = \frac{1}{2}\left(r\,|I_+(z)|^2 + g\,|V_+(z)|^2\right) , \tag{8.30}$$

$$P = \frac{Z_C'\,|I_+(z)|^2}{2} = \frac{Y_C'\,|V_+(z)|^2}{2} . \tag{8.31}$$

On a lossy line, where forward voltage and current decay like $e^{\quad z}$, active power has to vary like $P(z) = P_0\,e^{\ 2\ z}$, and power loss over an in nitesimal length is then $dP = \quad P_l\,dz = \quad 2\ \ P(z)\,dz$. Then, from Eqs. (8.30) and (8.31), the attenuation constant turns out to be:

$$= \frac{rY_C'}{2} + \frac{gZ_C'}{2} , \tag{8.32}$$

which is the same as Eq. (8.28), as we intended to prove.

8.6 Partially standing waves

In this section we will refer to *lossless* lines, unless the opposite is explicitly stated. However, most of the results are not di cult to extend to low-loss lines. The reader is encouraged to do that as an exercise.

The general expression Eq. (8.4) for voltage along a lossless line can be rewritten as follows:

$$V(z) = |V_+|\,e^{j(\ +\quad z)} + |V\ |\,e^{j(\ z+\quad)} , \tag{8.33}$$

where $_+$ and are the phases of the complex numbers V_+ and V , respectively. Adding and subtracting $|V\ |\ e^{\ j\ z}\,e^{j\ +}$, Eq. (8.33) can be rewritten as:

$$V(z) = \Big[|V_+|\quad |V\ |\Big]\,e^{j(\ +\quad z)} + |V\ |\,\Big[e^{j(\ z+\quad)} + e^{\ j(\ z\quad +)}\Big] . \tag{8.34}$$

The rst term on the right-hand side is a forward traveling wave, of amplitude $|V_+| \quad |V \;|$. Comparison with Section 4.5 tells us that the second term is a *pure standing wave*, of maximum amplitude $2\,|V \;|$.

In practice, it occurs often that one can measure either voltage or current (or sometimes a eld related to these) on a transmission line by means of an instrument whose response is *slow* (i.e., it averages over a large number of periods, at the angular frequency ω) and *quadratic* (i.e., the output of the instrument is proportional to the modulus square of either voltage or current). In view of this, let us write explicitly:

$$|V(z)|^2 = |V_+|^2 \left[1 + | \;_L|^2\right] + 2\,|V_+|^2 \;| \;_L| \;\cos(2 \quad z + \quad _L) \quad , \tag{8.35}$$

where $| \;_L|$ and $\;_L$ are modulus and phase, respectively, of the load re ection coe cient, $\;_L = \quad (0) = V \;/V_+$. The voltage modulus reaches its maximum value:

$$|V(z_M)| = |V_+| + |V \;| \quad , \tag{8.36}$$

in those sections where:

$$2 \quad z_M + \quad _L = 2k \qquad (k = 0, 1, 2, \ldots) \quad , \tag{8.37}$$

and its minimum value:

$$|V|_m = |V_+| \quad |V \;| \quad , \tag{8.38}$$

in those sections where:

$$2 \quad z_m + \quad _L = (2k + 1) \qquad . \tag{8.39}$$

It is straightforward to check that the sections where the voltage modulus is maximum, Eq. (8.37), are also those where the current modulus is minimum, and vice-versa for the sections Eq. (8.39), since:

$$|I(z)|^2 = |I_+|^2 \left[1 + | \;_L|^2\right] + 2\,|I_+|^2 \;| \;_L| \;\sin(2 \quad z + \quad _L) \quad . \tag{8.40}$$

Two adjacent minima (or maxima) are separated by a distance:

$$(z_{M+1} \quad z_M) = - = \frac{}{2} \quad , \tag{8.41}$$

whereas the distance between any maximum and either one of the adjacent minima is $\;/4$.

From now on, when using phrases like "voltage maximum" (or minimum) we will refer to the voltage *modulus*, without any danger of ambiguity. The same applies, of course, to the current.

We might have derived the same results also from the following expressions:

$$|V(z)| \quad = \quad |V_+|\;|1 + \quad (z)| \quad , \tag{8.42}$$

$$Z_C|I(z)| \quad = \quad |V_+|\;|1 \quad (z)| \quad , \tag{8.43}$$

which lend themselves to the simple graphical interpretation shown in Figure 8.2, based on the fact that, as Eq. (8.20) shows, when z varies along a lossless line, $\;(z)$ describes a circle.

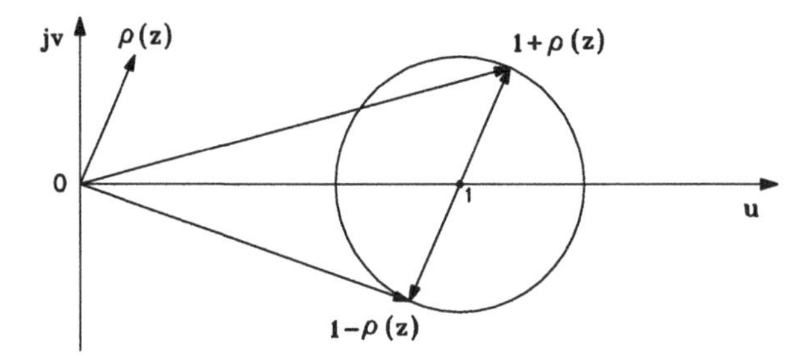

Figure 8.2 *Complex-plane illustration of the behavior of the quantities* 1 *vs. distance, along a lossless transmission line.*

The adimensional quantity:

$$S = \frac{|V|_{\max}}{|V|_{\min}} = \frac{|V_+| + |V\ |}{|V_+| \quad |V\ |} = \frac{1 + |\ |}{1 \quad |\ |} \quad , \tag{8.44}$$

is referred to as the *voltage standing wave ratio*, and frequently designated by its initials, VSWR.

When a transmission line is closed on a *passive* load, i.e., on a two-terminal network such that, at any frequency of interest, active power owing into it is non-negative, then Eq. (8.21) shows that $|\ |$ 1, and consequently S 1.

In the sections where voltage is maximum (and current is minimum), voltage and current are in phase, so the input impedance is real, and indeed it equals:

$$Z(z_{\max}) = \frac{|V(z)|_{\max}}{|I(z)|_{\min}} = \frac{|V_+| + |V\ |}{|V_+| \quad |V\ |} Z_C = S Z_C \quad . \tag{8.45}$$

Similarly, in those sections where voltage is minimum (and current is maximum), voltage and current are again in phase, the input impedance is real, and equals:

$$Z(z_{\min}) = \frac{|V(z)|_{\min}}{|I(z)|_{\max}} = \frac{|V_+| \quad |V\ |}{|V_+| + |V\ |} Z_C = \frac{Z_C}{S} \quad . \tag{8.46}$$

8.7 The Smith chart

In Section 8.3, we derived two formulas, Eq. (8.10) and its reciprocal, which enable us to calculate the input impedance and admittance in any section along a transmission line when the load impedance is speci ed. Those relationships are much more cumbersome than the corresponding transformation rule for the re ection coe cient, Eq. (8.11). Re ection coe cient and input impedance (or admittance) are related to each other by a one-to-one and onto mapping. Hence, it is self-explanatory why it is of interest to exploit this mapping and the simple rules which govern how changes vs. z, to by-pass the cumbersome formulas for $Z(z), Y(z)$.

We derived in Section 8.3 the relationship between normalized input impedance (or admittance) and re ection coe cient:

$$= \frac{\quad 1}{\quad + 1} = \frac{1 \quad y}{1 + y} \quad . \tag{8.47}$$

Relationships of the kind $a \quad + b \quad + c \quad + d = 0$, where a, b, c, d are complex constants, and ,   are two complex variables, are called *bilinear transformations.* It can be shown (Schwerdtfeger, 1961) that any bilinear transformation maps any circle of the   complex plane into a circle on the   complex plane, and vice-versa. Furthermore, any bilinear transformation maps two mutually orthogonal pencils of circles of one of these complex planes into two mutually orthogonal pencils of circles on the other complex plane, and vice-versa. This statement can be extended to the case of two orthogonal sets of parallel straight lines, which can be looked at as degenerate pencils of circles, with base points at in nity. In particular, this can be applied to the straight lines parallel to the real and to the imaginary axis of the   plane (or of the y plane), i.e., to the loci of constant real part of normalized impedance (or admittance), and of constant imaginary part of the normalized impedance (or admittance). In other words, we know *a priori* that these lines will be mapped by Eq. (8.47) into mutually orthogonal pencils of circles on the complex plane of .

To proceed to the calculations, let $\quad = r + jx$ $(y = g + jb)$, and $\quad = u + jv$, where r, x, g, b, u and v are all real.[‡] Starting from:

$$r + jx = \frac{1 + \quad}{1 \quad} = \frac{1 + (u + jv)}{1 \quad (u + jv)} \quad , \tag{8.48}$$

and separating real and imaginary parts, we nd:

$$r = \frac{1 \quad (u^2 + v^2)}{(1 \quad u)^2 + v^2} \quad , \qquad x = \frac{2v}{(1 \quad u)^2 + v^2} \quad . \tag{8.49}$$

These two equations can be rewritten as:

$$\left(u \quad \frac{r}{1 + r}\right)^2 + v^2 = \frac{1}{(1 + r)^2} \quad , \tag{8.50}$$

and:

$$(u \quad 1)^2 + \left(v \quad \frac{1}{x}\right)^2 = \frac{1}{x^2} \quad . \tag{8.51}$$

Eq. (8.50) recon rms that, on the   plane, the loci $r =$ constant are circles. Each of them is centered at a point of coordinates $\left(\frac{r}{1+r}, 0\right)$ and its radius is given by $1/(1 + r)$ (see Figure 8.3). Similarly, Eq. (8.51) recon rms that the

[‡] Several textbooks on transmission lines use the symbols z and x for the normalized impedance and the normalized reactance, respectively. At the same time, they use one of these symbols to denote the longitudinal coordinate along the line. To avoid ambiguity, we call z the longitudinal coordinate,   the normalized impedance. Throughout this section, x means the normalized reactance. Elsewhere, it may indicate a transverse coordinate. The text will always make the symbol unambiguous.

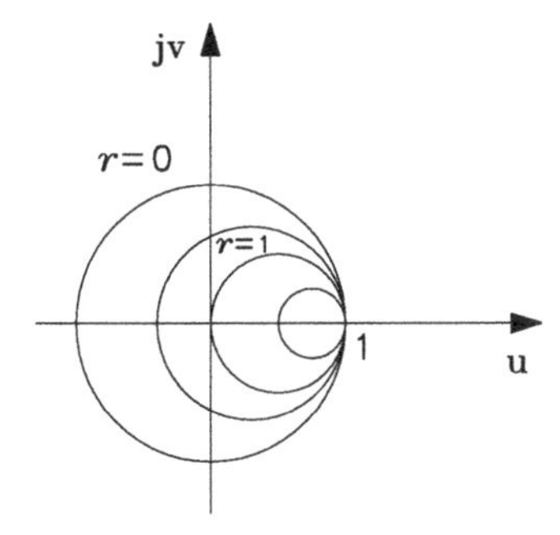

Figure 8.3 *The constant-resistance loci, on the complex plane of the re ection coef- cient.*

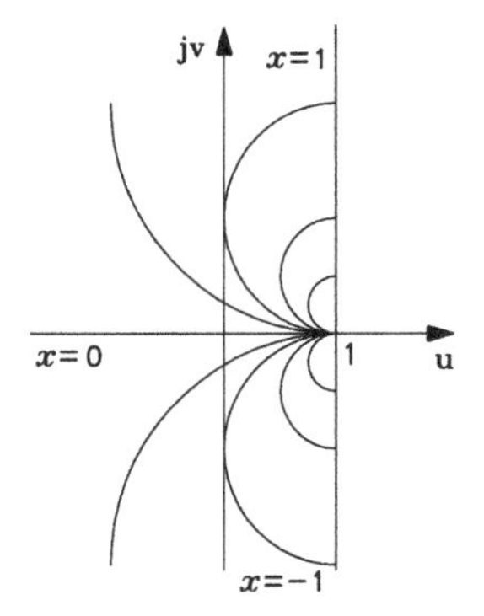

Figure 8.4 *The constant-reactance loci, on the complex plane of the re ection coef- cient.*

loci x = constant on the plane are circles, each of which is centered at a point of coordinates $(1, 1/x)$ and has a radius given by $1/|x|$ (see Figure 8.4). Their sets are indeed two orthogonal pencils. Each pencil is parabolic, i.e., instead of having two distinct base point as the more common elliptic pencils, it has one double base point (the same for both pencils). This base point is located at (1,0). "Double" means that all the circles belonging to the same pencil not only pass through that point, but also have the same tangent in that point. For the r = constant circles, their tangent in common is the line $u = 1$, parallel to the imaginary axis. For the x = constant circles, the tangent in common is the real axis, $v = 0$.

As we saw before, for passive loads we have $| \ |$ 1, and correspondingly, r 0 in Eq. (8.50). Consequently, the region of the plane where passive loads are mapped is the interior of the circle $| \ | = 1$, including the circle itself, which is the locus of the lossless loads. The plot of the r = constant circles and of the portions of the x = constant which belong to this part of the plane is usually referred to as the *Smith impedance chart*, and is shown in Figure 8.5.

An equivalent chart in terms of admittance is easy to build, noting that

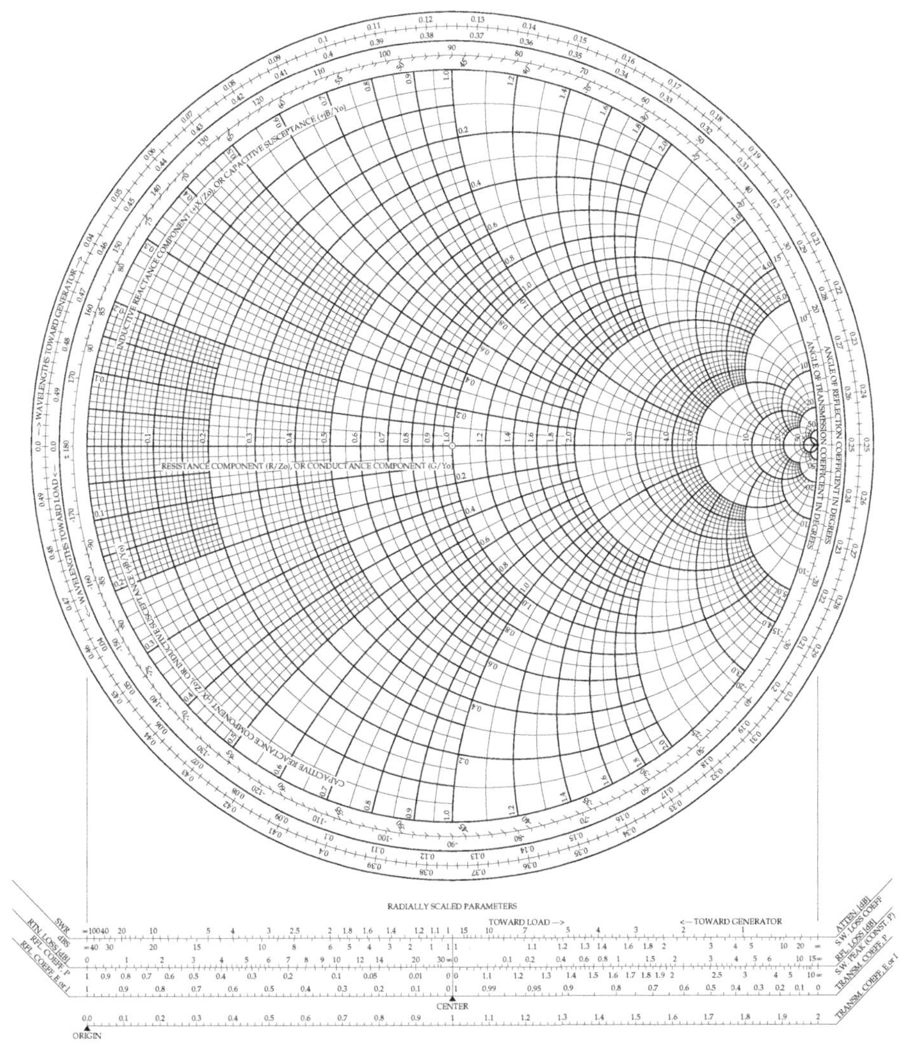

Figure 8.5 *The Smith impedance chart.*

Eq. (8.13) yields:

$$y(\ ) = \frac{1}{(\ )} = (\ )\ . \tag{8.52}$$

Hence, the *Smith admittance chart* can be obtained from the impedance chart merely by changing into , i.e., rotating the plane on which the impedance chart was drawn by radians.

Finally, let us show how the Smith chart can be used when dealing with *active loads*, i.e., for $|\ | > 1$. Let $' = 1/$: in this way, the region outside the circle $|\ | = 1$ is mapped onto the region inside that circle, and vice-versa.

Consequently, plots over Figure 8.5 can be used to represent active loads, provided one reads that plane as that of the inverse of the re ection coe cient. Many commercially available modern instruments for testing active electronic devices, typically microwave transistors, encompass a function which allows passing from the plane to the ^{1} plane on their polar display, by simply pushing a button.

8.8 Remote measurement of the load impedance

An elegant and useful application of the theory that we developed in this chapter comes from a problem that is frequently encountered in actual telecommunication systems. It occurs often that an unknown load impedance, located at the end of a transmission line, is di cult to access. For example, this is typical with antennas installed on the top of buildings, metallic towers, etc., where it is di cult to carry the necessary measuring instrumentation and to operate it in a reliable way. Finding the unknown load impedance from data measured at a distance, along the transmission line, is a remarkable improvement.

We saw in Section 8.3 that if the re ection coe cient is known, in modulus and phase, at a known distance z from the load, then, in principle, it is straightforward to calculate the load impedance or admittance (normalized to the characteristic impedance of the line). We will show here that there is a rather simple experimental recipe to implement this theory. The required instrument is referred to as a voltage standing wave indicator (VSWI), and consists of: a voltage probe that can be moved in a controlled way along the transmission line (usually, it is inserted in a movable mount equipped with a calibrated vernier scale); an RF peak voltmeter (an instrument which measures the modulus of the RF voltage, when connected to the probe); nally, a known load impedance, normally a short circuit.

The rst step is to nd the *modulus* of the re ection coe cient, which we can calculate when we know the VSWR. Moving then the probe along the line, one looks for a maximum, and measures $|V|_{\max}$ by means of the RF peak voltmeter. Next, one looks for an adjacent minimum, and measures $|V|_{\min}$. Assuming, at least during an initial phase, that the line is lossless, we nd:

$$| \; | = | \;_L| = \frac{S \quad 1}{S+1} = \frac{|V|_{\max} \quad |V \;|_{\min}}{|V|_{\max} + |V \;|_{\min}} \quad . \tag{8.53}$$

If we measure the distance between two consecutive minima (or maxima) (equal to /2) by means of the sliding probe, and if the distance between the load and the n-th voltage maximum on the line, z_L, is also known, then in principle it is easy to calculate the *phase* of the load re ection coe cient $\varphi_L = \arg(\;_L)$ as follows:

$$\varphi_L = 2 \; (z_L \quad n \; /2) = 2 \; (2z_L/ \quad n) \quad . \tag{8.54}$$

In practice, however, this procedure is unsuitable, because the result would be a ected by a large error bar. The rst uncertainty is encountered in the

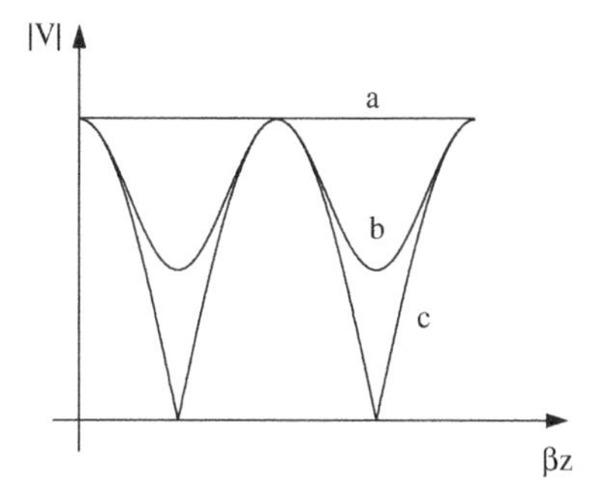

Figure 8.6 *Typical patterns of partially standing waves. a)* VSWR = 1. *b)* VSWR = 2. *c)* VSWR = ∞.

location of a maximum. Indeed, in any partially standing wave, minima (of the modulus, see Figure 8.6) are sharper than maxima. Therefore, locating a minimum is more accurate than locating a maximum. This simply implies that is added to Eq. (8.54). However, another more serious reason why this modi ed procedure is not yet adequate is because, as well known, an unknown quantity measured as the di erence between two quantities of similar magnitude is inevitably a ected by an enlarged relative error bar. In the present case this inconvenience may be extremely signi cant, because quite often it is extremely impractical to measure geometrically the distance z_L. For example, this occurs whenever a line twists or bends, between the load and the measuring equipment.

All these inconveniences are avoided if one measures φ_L by comparison with a reference standing wave, namely the wave which one nds on the same line when the unknown load is replaced by a known impedance. To be a good choice, this known load must yield a large VSWR, with sharply de ned voltage minima. An excellent combination of this requirement and simplicity is a short circuit. Incidentally, note that an open circuit would be equally good in principle, but in practice it is not, because at high frequency a transmission line whose remote terminals are open radiates, similar to an antenna.

If a line is terminated on a short circuit, then the distance between the n-th voltage minimum and the load terminals is $z_{SC} = n\ /2$. If we call z_L the abscissa of the n-th voltage minimum when the line is terminated on the unknown load, then the unknown load phase is simply:

$$\varphi_L = 2\ (z_{SC}\quad z_L)\qquad . \tag{8.55}$$

To perform the tests that were described in this section, there are commercially available VSWI's which encompass a *slotted line*, i.e., a section of either coaxial cable or rectangular waveguide, whose outer wall is slotted in the longitudinal direction, parallel to the current ow in the wall itself. The slot accommodates a movable voltage probe.

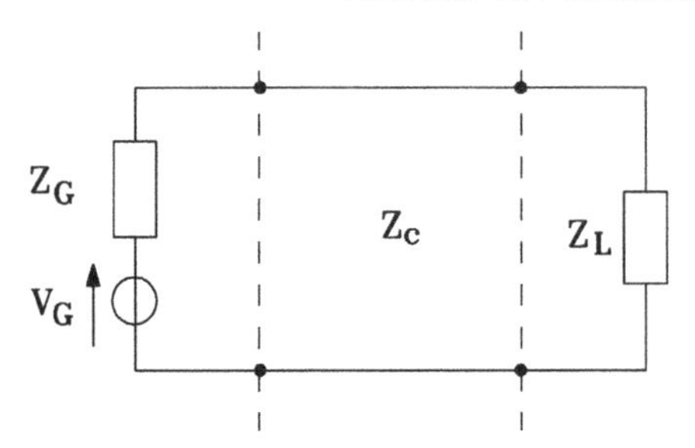

Figure 8.7 *A uniform transmission line, connecting a generator and a load.*

8.9 Impedance matching

Normally, a transmission line of nite length is connected to a generator at one end (the left side in Figure 8.7), and to a load at the other hand. Let Z_G be the internal impedance of the generator (either in series to an ideal voltage generator whose open-circuit voltage is V_G, or in parallel to an ideal current generator whose short-circuit current is $I_G = V_G/Z_G$).

In general, there are re ected waves on the line. As we saw in the previous sections, and also in some of the previous chapters, re ected waves are responsible for unwanted e ects. First, the power delivered to the load becomes smaller than that carried by the forward wave. Second, keeping constant the active power delivered to the load P, the maximum voltage along the line can be expressed as:

$$|V_M| = \sqrt{2Z_C\, PS} \quad , \tag{8.56}$$

where Z_C is, as usual, the characteristic impedance of the line, and S is the VSWR. The larger the re ected waves, the larger S, and, as consequence, $|V_M|$. This must be accounted for, to avoid electrical discharges along the line. Last but not least, in general the re ection is frequency dependent, causing signal distortion in the case of nite spectral width. This intuition is recon rmed by evaluating the voltage transfer function of a lossless transmission line of length l, i.e., the ratio between output and input voltages:

$$H(j\omega) = \frac{V_L}{V_i} = \frac{Z_L}{Z_L \cos \quad l + jZ_C \sin \quad l} \quad . \tag{8.57}$$

It is easy to verify that in general Eq. (8.57) does not satisfy the distortionless transmission conditions ($|H(j\omega)| = 1$, $\angle H(j\omega) \propto \omega$). However, if $Z_L = Z_C$, i.e., when there are no re ections, then $H(j\omega) = e^{\;\; j \;\; l}$ and the signal does not undergo any distortion. An *impedance matching device* is a passive (whenever possible, lossless) two-port network, whose purpose is to eliminate re ections at one of its ports, when the other one is connected to a given load (Figure 8.8). The purpose of this section is to illustrate three of the most commonly used impedance matching networks, which are based on lossless transmission lines.

From the elementary theory of electrical circuits, we know that the active

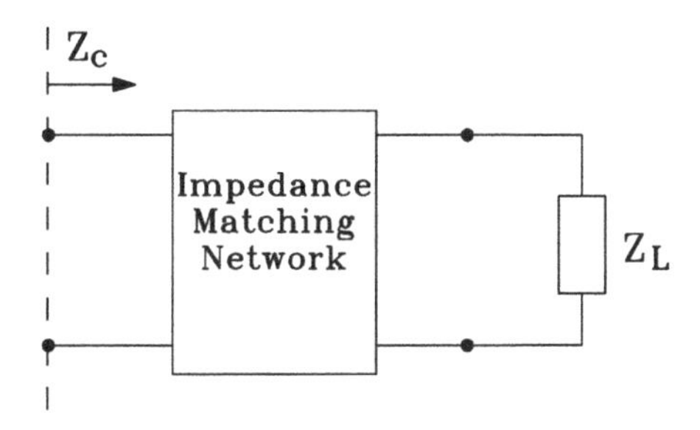

Figure 8.8 *De nition of an impedance-matching network.*

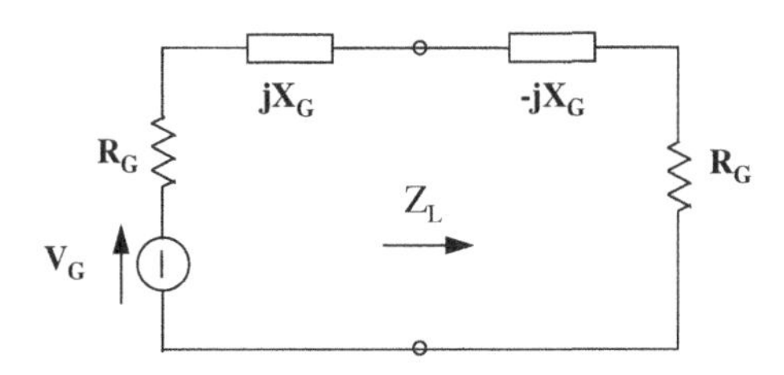

Figure 8.9 *Power matching between a given generator and its load.*

power delivered by a generator of given internal impedance, $Z_G = R_G + jX_G$, reaches its maximum value when the load connected to the generator terminals, Z_i (see Figure 8.9) equals the complex conjugate of Z_G, i.e., for $Z_i = R_G \quad jX_G$. This is referred to as the *power matching* condition. The complex power delivered by the generator under these circumstances is easily calculated from the de nition of complex power (e.g., Eq. (8.8)). For a power-matched load it is $I = V_G/(2R_G)$, so the active power delivered to the load is:

$$P_a = \frac{|V_G|^2}{8R_G} \quad , \tag{8.58}$$

and the reactive power simply equals P_a times the ratio (X_G/R_G). Eq. (8.58) is referred to as the generator's *available power.*

The power matching condition, which is of paramount importance to electrical engineers involved in power systems, in general is not the best answer if we refer to signal transmission problems. In fact, power matching implies that the imaginary parts of the generator and load impedances have opposite signs. Because of the di erent frequency dependence of reactances of opposite signs, their absolute values can be equal just at one frequency, not over a nite bandwidth, except for the case where they are equal to zero. Therefore, in general power matching entails signal distortion. The right recipe in a telecommunication context is not to look for the absolute maximum of the power delivered to the load, but for a *conditional maximum*, namely, under the restriction that it must be frequency-independent (at least over a narrow bandwidth). Again from elementary circuit theory, it can be shown that this corresponds to:

$$Z_i = Z_G \quad , \tag{8.59}$$

which is usually referred to as the *uniformity matching* condition.

Note, however, that in the following we will restrict ourselves to lossless transmission lines, or at most to low-loss lines. It will be taken for granted that their characteristic impedance Z_C is real. As the loads, and often the generators also, will be matched to this real Z_C, the distinction between power matching and uniformity matching disappears.

The purpose of an impedance-matching network is then to provide an input impedance which satis es two independent requirements, if expressed in real numbers: real part equal to Z_C, imaginary part equal to zero. We may then expect that the design of an impedance matching network should involve (at least) two real variables. All the following examples recon rm this expectation. However, we will see that the physical nature of these variables changes completely as we pass from one network to another.

8.9.1 The quarter-wavelength transformer

In Section 4.10 we showed that, for plane waves, a lossless dielectric layer, whose thickness equals a quarter of the wavelength, can in fact cancel the Fresnel re ection at the interface between two semi-in nite lossless media, provided its intrinsic impedance is given a suitable value. The Smith chart of Section 8.7, while on one hand it simpli es the understanding of what we did in Section 4.10, on the other hand it suggests that the same results apply to transmission lines, *provided the load impedance is real.* Indeed, when a lossless line whose length equals an odd integer multiple of /4 is connected to a generic load impedance Z_L, its input impedance equals:

$$Z_i = \frac{Z_C'^2}{Z_L} \quad . \tag{8.60}$$

As the characteristic impedance of the lossless quarter-wavelength line, Z_C', is real, Eq. (8.60) cannot be made equal to a real number (the characteristic impedance of the line to be matched) unless Z_L is itself real. Using a lossy quarter-wavelength line, with a complex Z_C', would mean introducing an unwanted attenuation.

A suitable solution to the problem can be found if we simply accept that the quarter-wavelength line be inserted at a suitable distance d from the load (Figure 8.10), rather than connected directly to its terminals. Indeed, we know from Section 8.3 and the following ones that, as we move along the line, we nd periodically sections where the input impedance is real. They are either maximum-voltage sections, where:

$$Z(z) = SZ_C \quad , \tag{8.61}$$

or minimum-voltage sections, where:

$$Z(z) = Z_C/S \quad , \tag{8.62}$$

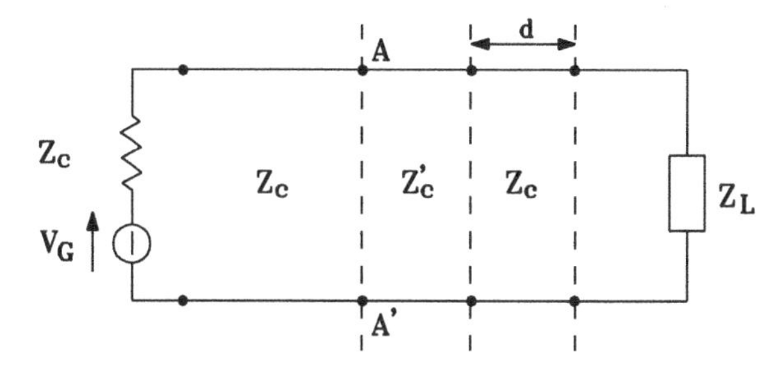

Figure 8.10 *Load matching by means of a quarter-wave line.*

where S is the VSWR. On the Smith chart, these sections correspond to the intersections of the circle $| \ (z)| =$ constant, i.e., the locus of the re ection coe cient along the line, with the real axis of the plane, i.e., the locus of real impedances.

In the rst case, combining Eqs. (8.60) and (8.61), the recipe for the intrinsic impedance of the quarter-wave line is found to be:

$$Z'_C = \sqrt{Z(z)\, Z_C} = Z_C\, \sqrt{S} \quad , \tag{8.63}$$

while in the second case it is found to be:

$$Z'_C = \sqrt{Z(z)\, Z_C} = Z_C / \sqrt{S} \quad . \tag{8.64}$$

These formulas (even better if re-written in terms of normalized impedances, Z'_C/Z_C) are strictly reminiscent of how one calculates the transforming ratio, $n : 1$, for an ideal transformer to convert a resistive impedance R_o into another resistive impedance R_i, $n = \sqrt{R_i/R_o}$. For this reason, in the current terminology of telecommunications, lossless lines whose lengths equal an odd integer multiple of $/4$ are often referred to as *quarter-wavelength transformers*. Note, however, that the analogy disappears when the loads are not purely resistive. Indeed, the sign of the imaginary part of the impedance does not change as one goes through an ideal transformer, while, on the other hand, Eq. (8.60) converts an inductive load into a capacitive input impedance, and vice-versa.

We said in the rst part of this section that we expected impedance-matching network designs to be based on two real parameters. In the case at hand, the two parameters are the characteristic impedance of the quarter-wave section, and its distance from the load, which equals either:

$$d_{(S)} = (\ /2)\,(\varphi_L/2 \ ) + m(\ /2) \quad , \tag{8.65}$$

in the case of Eq. (8.63), and:

$$d_{(1/S)} = d_{(S)} \qquad /4 \quad , \tag{8.66}$$

in the case of Eq. (8.64), where φ_L is the phase angle of the load re ection coe cient, and $m = 0, 1, 2, \ldots$. As for the choice between the solutions in Eqs. (8.63) and (8.64), a suitable criterion is to make the line between the

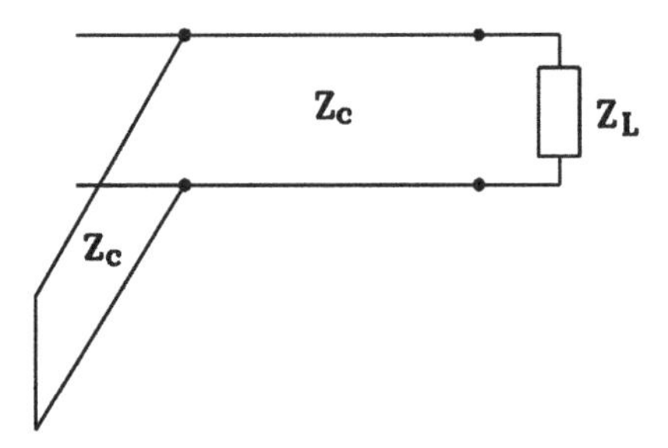

Figure 8.11 *Single-stub impedance-matching network.*

load and the /4-section as short as possible. Indeed, there are re ected waves on this piece of line, and from Eq. (8.64) one may check that the deviation from distortionless transmission grows as the line length grows. Sometimes, however, other more practical criteria become dominant with respect to this. For example, in a coaxial cable to diminish the characteristic impedance is much easier than to increase it, as discussed in one of the problems at the end of this chapter.

8.9.2 Impedance matching with one or two stubs

From the Smith chart, it is easy to check that if a lossless line is either short-circuited, or open-circuited at its end, then its input impedance (and admittance, of course) is purely reactive, irrespectively of its length. Therefore, it is of great interest to study whether any load can be impedance-matched to a lossless transmission line by inserting along the line a pure reactance, either in parallel or in series. In principle, the answer is positive for all these combinations. In practice, however, open-circuited lines are disregarded, because they inevitably radiate an active power from their terminals, thus violating the requirement of being lossless. Insertions in series are also disregarded, normally, being more cumbersome than those in parallel. So, whenever we mention impedance-matching *stubs*, one should refer automatically to short-circuited line sections in parallel to the main line, as shown schematically in Figure 8.11.

From the general expression in Eq. (8.18), with $_L = 0$, we nd immediately that the normalized input impedance of a stub is:

$$_i = 1/y_i = j \tan \ l \quad , \tag{8.67}$$

where l is the stub length. Thus, *any* reactance, from ∞ to $+\infty$, can be realized in this way. Therefore, any normalized admittance lying on the $g = 1$ circle of the plane (i.e., of the Smith admittance chart, see Figure 8.12), can be matched to a lossless transmission line. It is understood that normalization is with respect to the characteristic admittance of this line, $Y_C = 1/Z_C$. So, the only requirement to be satis ed by what is located between the load and

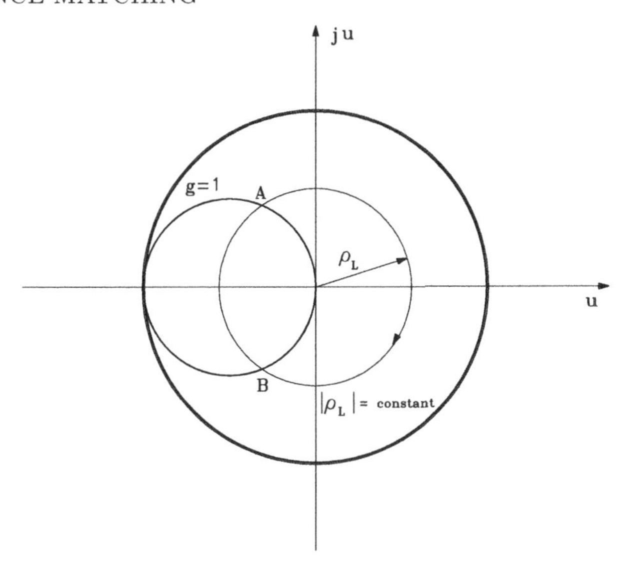

Figure 8.12 *Illustration of the operation principle of a single-stub impedance matching network, on the Smith admittance chart.*

the stub is to transform the point which represents the load into a point on the circle $g = 1$.

Again from Figure 8.12 we see that one simple way to do that, applicable to *any* load, is to leave a suitable length of the main transmission line between the terminals of the load and those of the stub. Indeed, the circle $| \ _L| =$ constant (where $\ _L$ is the re ection coe cient of the load) intersects the circle $g = 1$ at two points (called A and B in Figure 8.12). One of them is in the half-plane where $b > 0$, b being the normalized susceptance; the other one is where $b < 0$. To yield impedance matching, the stub must provide a susceptance which is equal in modulus and opposite in sign to that seen along the main line. Therefore, one solution (point A) entails an inductive stub (with a length $l < \ /4$) and the other one (point B) entails a capacitive stub (with a length $\ /2 > l > \ /4$).

Once again the design of this impedance-matching network relies on two real parameters, namely the length of the piece of line between the load and the stub (to be calculated as $\vartheta/(2 \quad)$, where the angle ϑ is shown in Figure 8.12), and the length of the stub. Note that in the last lines it was taken somewhat for granted that the stub was made with a line with the same characteristic admittance as the main line. This is not necessary at all. When this is not the case, the only change to be made with respect to what we said is the way to calculate the length of the stub, as illustrated in detail in some of the following problems.

Incidentally, the reader may verify as an exercise that designing this type of device (which is currently referred to as a single-stub impedance matching

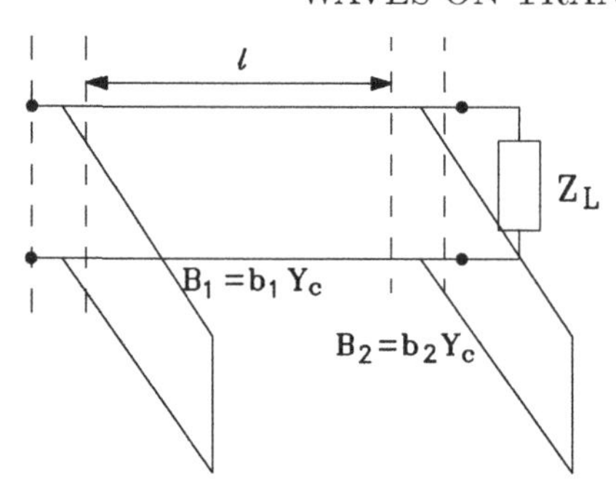

Figure 8.13 *Double-stub impedance-matching network.*

network) using analytical tools, instead of the Smith chart, is possible but much more cumbersome.

Another device which is of common use in practice is the *double-stub* impedance-matching network, schematically illustrated in Figure 8.13. We will call stub 1 the one which is connected to the same terminals as the load, stub 2 the one upstream towards the generator. Contrary to the single stub, distances along the main line are xed. The most commonly used values for the distance between the stubs are either /4, or /8, or 3 /8. The two real variables exploited in the design of this network are the two lengths of the stubs.

To outline how a double-stub network operates, note rst that the stub 2 must play the same role as the stub in the previously discussed single-stub network. Therefore, matching is possible only if the point which represents the input admittance on the main line, as seen from the terminals of the stub 2, lies on the circle $g = 1$ of the admittance Smith chart (see Figure 8.14). Consequently, as nothing can be modi ed on request between the stubs, the point which represents the input admittance immediately upstream of the stub 1 must lay on the curve where the circle $g = 1$ is mapped by the line section between the stubs. Figure 8.14 shows these curves for the three common cases of a distance equal to /4, to /8 and to 3 /8. To avoid confusion, the rest of the procedure is illustrated in detail only for the case /4. The other ones can be completed by the reader as an exercise.

The purpose of the stub 1 is to modify the input admittance, from the given load value to a value lying on the transformed circle, designated as the A-circle. A stub in parallel does not a ect the real part of the admittance. Therefore, we pass from the point y_L to the point on the A-circle along a constant-g line. This intersects the A-circles at two points, indicating that, for a load like the one shown in Figure 8.14, there are two solutions. For each of them, the susceptance of the stub 1 is found as the di erence between that at the point on the A-circle, and that of the load. The susceptance of the stub 2 is then computed in the same way as for a single-stub matching network.

The main advantage of the double stub is to avoid sliding contacts along the

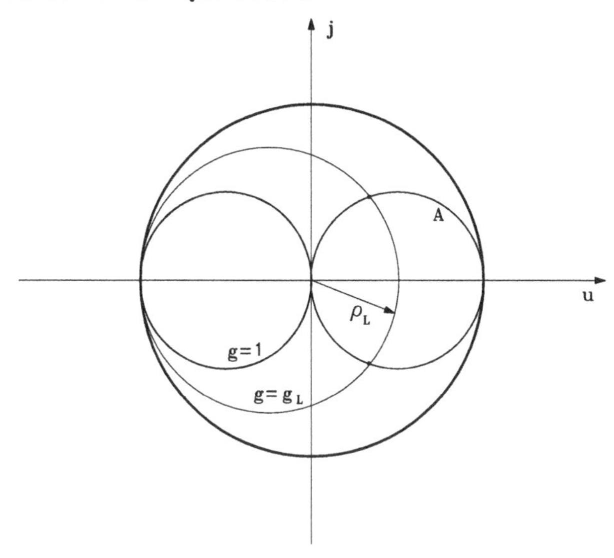

Figure 8.14 *Illustration of the operation principle of a double-stub impedance matching network, on the Smith admittance chart.*

main line. Its main drawback is that it cannot match all possible loads. Indeed, if the normalized load admittance is located in the region whose g = constant lines do not intersect the transformed circle (to be clear, the A-circle in the previous example), then the procedure does not yield any solution. The main reason why double stubs spaced by either /8 or 3 /8 are often preferred to those spaced by /4 is because their "non-match regions" are smaller. The further choice between them is a trade-o : the rst one is shorter, thus giving less distortion, but two closely spaced stubs may tend to interact with each other in a rather unpredictable way.

A powerful solution which makes it possible to match any load is to use a triple stub. In such a case, there are several possible solutions for a generic load to match. The consequent main drawback, usually, is that the device is designed (or made empirically) using a time-consuming cut-and-try procedure.

Double stubs could also be explained and designed in terms of line equations, but the Smith chart turns out to be much simpler, as the reader can verify as an exercise.

8.10 Transmission-line equations: An alternative derivation

In this section we will show how the equations which govern voltage and current waves along transmission lines can be derived from Maxwell's equations, provided some suitable assumptions are satis ed. The rst purpose is to show that what we have seen so far in this chapter, although derived in a rather phenomenological way from the circuit of Figure 8.1, is not in contradiction with the axiomatic approach followed in the rest of the book. Another purpose, may be obscure now, but hopefully not so at the end of this section, is to show that transmission lines provide useful and elegant equivalent models for propagation phenomena in other guiding structures, like the conducting-wall waveguides of the previous chapter, and dielectric waveguides of Chapter 10.

Let us consider a lossless homogeneous cylindrical structure whose cross-section is *not* a simply connected domain, so that it can propagate the TEM mode. Introduce rectangular coordinates where z is the longitudinal axis, and write homogeneous Maxwell's equations by components. *Under the assumptions that the eld is transverse electromagnetic*, i.e.:

$$E_z \quad 0 \quad , \qquad H_z \quad 0 \quad , \tag{8.68}$$

the x and y components read:

$$\frac{\partial E_y}{\partial z} = j\omega \quad H_x \quad , \qquad \frac{\partial E_x}{\partial z} = \quad j\omega \quad H_y \quad , \tag{8.69}$$

$$\frac{\partial H_y}{\partial z} = \quad j\omega \quad E_x \quad , \qquad \frac{\partial H_x}{\partial z} = j\omega \quad E_y \quad . \tag{8.70}$$

The z components are omitted as they are not required in the following steps. The rst of Eqs. (8.69) and the second of Eqs. (8.70) form a system with two unknowns, E_y and H_x. The other two equations form another system (in E_x and H_y), independent of the previous one and formally identical with it, except for a sign. Therefore, it is enough for us to investigate in detail only one of these systems, say the second one. All conclusions could then be repeated for the other one.

Being familiar with separation of variables, we suggest to start o with the following ansatz:

$$\begin{aligned} E_x(x,y,z) &= V(z)\,\mathcal{E}_x(x,y) \quad , \\ H_y(x,y,z) &= I(z)\,\mathcal{H}_y(x,y) \quad . \end{aligned} \tag{8.71}$$

These two equations must be read as if they encompass *two arbitrary multiplicative constants*, A and B, i.e., they can always be replaced by:

$$\begin{aligned} E_x &= V'(z)\,\mathcal{E}'_x(x,y) = \Big[A\,V(z)\Big]\left[\frac{1}{A}\,\mathcal{E}_x(x,y)\right] \quad , \\ H_y &= I'(z)\,\mathcal{H}'_y(x,y) = \Big[B\,I(z)\Big]\left[\frac{1}{B}\,\mathcal{H}_y(x,y)\right] \quad . \end{aligned} \tag{8.72}$$

Note that even the *physical dimensions* of A and B are arbitrary. Therefore, the physical dimensions of the quantities called V, I, $\mathcal{E}_x$ and $\mathcal{H}_y$ in Eqs. (8.71) are not yet de ned at this stage.

Introducing Eqs. (8.71) into (8.69) and (8.70), we get immediately:

$$\frac{dV}{dz}\,\mathcal{E}_x(x,y) = \quad j\omega \quad I\,\mathcal{H}_y(x,y) \quad ,$$

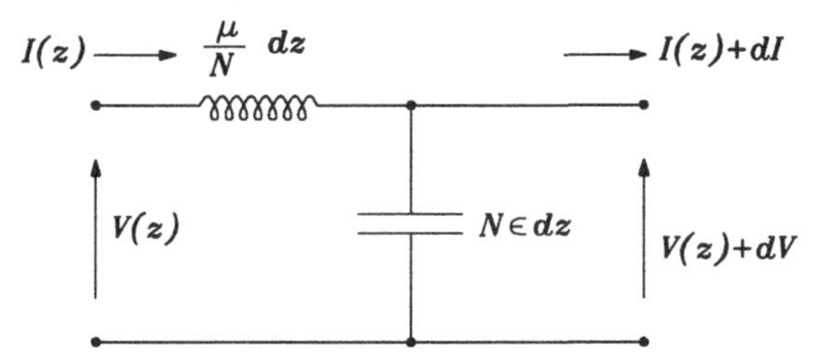

Figure 8.15 *Kirchho 's laws applied to the equivalent circuit are identical to Maxwell's equations for TEM propagation.*

$$\frac{dI}{dz}\,\mathcal{H}_y(x,y) \quad = \quad j\omega \;\; V\,\mathcal{E}_x(x,y) \quad . \tag{8.73}$$

We saw in Section 7.2 that the wave impedance of a TEM eld traveling in the z direction is . The same cannot be said, in general, for a superposition of TEM elds that propagate in both directions along the z axis. Indeed, on several occasions (in Chapter 4 and later) we saw that superimposing two waves is a delicate operation, from the viewpoint of the wave impedance. However, there was one point which was invariably true in all the cases that we examined: namely, the wave impedance did not depend on the transverse coordinates x and y. Consequently, we will assume now that the two functions of the transverse coordinates $\mathcal{E}_x(x,y)$ and $\mathcal{H}_y(x,y)$ in Eqs. (8.71) are *proportional*, i.e.:

$$\mathcal{E}_x(x,y) = N\,\mathcal{H}_y(x,y) \quad . \tag{8.74}$$

Clearly, the proportionality constant N is related to the arbitrary constants A and B of the previous comment. In particular, there is nothing wrong in letting $\mathcal{E}'_x = \mathcal{H}'_y$ in Eq. (8.72), thus obtaining $N = A/B$.

Because of Eq. (8.74), Eqs. (8.73) become:

$$\frac{dV}{dz} \quad = \quad \frac{j\omega}{N}\, I \quad ,$$

$$\frac{dI}{dz} \quad = \quad j\omega \;\; N\,V \quad . \tag{8.75}$$

If we give the arbitrary constants in Eqs. (8.72) physical dimensions such that V and I get, respectively, dimensions of a voltage and of a current, then Eqs. (8.75) become identical with the *telegraphist equations*, Eqs. (8.1). Other choices for A and B do not entail any substantial change. The degrees of freedom in A and B simply mean that the circuit of Figure 8.15, where Eqs. (8.75) are satis ed as Kirchho 's laws, should be looked at as an *equivalent circuit* for any TEM mode propagation, not necessarily as a phenomenological model, as it might have appeared after the initial derivation of the equations from the circuit shown in Figure 8.1.

The circuit shown in Figure 8.15 may look as a simpli ed (or idealized) version of that of Figure 8.1. A natural question to ask then, is whether passing from a lossless medium to a lossy one in Maxwell's equations, Eqs. (8.69) and (8.70), could yield an equivalent circuit like that of Figure 8.1. It is simple to verify that, if we replace by $_c$, only the second of Eqs. (8.75) changes, so that its equivalent circuit encompasses now an in nitesimal *parallel conductance*:

$$g\,dz = \mathrm{Re}\,[j\omega \ _c N]\,dz = N \quad dz \quad . \tag{8.76}$$

No series resistance emerges in this way. The reason is because a series resistance is, strictly speaking, incompatible with the TEM assumption. Indeed, a current owing through conductors with nite conductivity necessarily entails a nonvanishing longitudinal component for the electric eld.

As we saw at the beginning of this chapter, we may elaborate on Eqs. (8.75) so that each unknown, $V(z)$ and $I(z)$, is isolated. The order of the di erential equations increases. Incidentally, note that this procedure is essentially the same as the one used in Chapter 1 to pass from Maxwell's equations to the Helmholtz equation. The equations we get in this way are:

$$\frac{d^2V}{dz^2} = \quad \omega^2 \quad V \quad , \qquad \frac{d^2I}{dz^2} = \quad \omega^2 \quad I \quad . \tag{8.77}$$

Eqs. (8.77) are essentially the same as the *telephonist equations* (8.3). Note that the arbitrary constant N of Eqs. (8.75) disappeared in the last equations. Their general integral is of the type:

$$V = V_+ \, e^{\quad j \quad z} + V \quad e^{+j \quad z} \quad , \tag{8.78}$$

where V_+ and V are arbitrary constants, and:

$$= \omega \sqrt{\quad} \quad . \tag{8.79}$$

Hence, for waves propagating along a TEM transmission line, phase constants are always equal to the intrinsic phase constant of the homogeneous lossless dielectric medium between the (ideal) conductors, independently of the geometrical shape of the structure cross-section.

From Eq. (8.78) and the rst of Eqs. (8.75) we nd:

$$I = \frac{N}{\omega} V_+ \, e^{\quad j \quad z} \quad \frac{N}{\omega} V \quad e^{j \quad z} = I_+ \, e^{\quad j \quad z} + I \quad e^{j \quad z} \quad , \tag{8.80}$$

where now we let:

$$I \quad = \quad V \quad N \sqrt{\frac{\quad}{\quad}} = \frac{V}{(\quad /N)} \quad . \tag{8.81}$$

The physical dimensions of V and I are still unde ned. So are those of the arbitrary constant N, and consequently of $/N$, which appears in Eq. (8.81). Independently of the choice of N, this quantity plays the physical role of an impedance, and is proportional to the intrinsic impedance of the dielectric, $= \sqrt{\ /\ }$. Let us illustrate, by means of some examples, how one may exploit the still arbitrary constants so as to get a set of enlightening results.

1) If we give N the physical dimensions of an impedance, i.e.:

$$[N] = \mathrm{Ohm} \quad , \tag{8.82}$$

then /N is adimensional, while V and I have the same dimensions. *In particular*, V and I can also be taken adimensional. In such a case we have $[\mathcal{E}_x] = \mathrm{V/m}$, $[\mathcal{H}_y] = \mathrm{A/m}$. Everything is ready for applying the so-called scattering-matrix formalism (Corazza, 1974).
An interesting sub-case is to choose $N =$. If we start from it, and extend the formalism of this section to TE and TM modes, as far as possible, then we nd again the results shown in Figure 7.3.

2) If we let N be adimensional, then it is *possible* (although not necessary) to let V have the physical dimensions of voltage, and I those of a current. Then, we have $[\mathcal{E}_x] = [\mathcal{H}_y] = \mathrm{m}^{\ 1}$. If we require that for $\omega \to 0$ our V and I tend to become the same as voltage and current in the time-independent case, then N cannot be chosen arbitrarily, but must be derived from two de nitions of the following types:

$$V = \int_{\ell_1} \mathbf{E} \quad \mathbf{d}\boldsymbol{\ell}_1 \quad , \qquad I = \oint_{\ell_2} \mathbf{H} \quad \mathbf{d}\boldsymbol{\ell}_2 \quad , \tag{8.83}$$

where ℓ_1 is an open path which belongs to the generic $z =$ constant plane, and whose extrema fall on the two conductors of the line, while ℓ_2 is a closed path (also belonging to the generic $z =$ constant plane), concatenated with only one of these conductors. Given that Eqs. (8.71) hold, Eqs. (8.83) lead to:

$$\int_{\ell_1} \mathcal{E}_x\, \hat{a}_x \quad \mathbf{d}\boldsymbol{\ell}_1 = \quad 1 \quad , \qquad \oint_{\ell_2} \mathcal{H}_y\, \hat{a}_y \quad \mathbf{d}\boldsymbol{\ell}_2 = 1 \quad . \tag{8.84}$$

3) If, independently of the physical dimensions one gives to $\mathcal{E}_x$, we choose:

$$N = \int\int_S |\mathcal{E}_x|^2 \, dx \, dy \quad , \tag{8.85}$$

where S is the cross-section of the waveguiding structure, then the complex power which ows through this cross-section is given by:

$$P_c \quad \int\int_S \frac{\mathbf{E} \quad \mathbf{H}}{2} \quad \hat{a_z}\, dS = \frac{VI}{2} \quad . \tag{8.86}$$

This makes the relationship between complex power on one side, and V and I on the other side, *formally* identical with that which holds at the terminals of a one-port electrical network (see Figure 8.16). Note, however, that the dimensions of V and I are still unde ned. To make Eq. (8.86) *really* the same as the usual circuit-theory relationship, it is necessary that, in addition to Eq. (8.85), one of Eqs. (8.84) be valid too.

One can then introduce the scattering-matrix formalism starting from any possible choice of N. To show that, let us rewrite Eqs. (8.75) in the following form:

$$\frac{d}{dz}\begin{pmatrix} V \\ I \end{pmatrix} = \begin{pmatrix} 0 & \dfrac{j\omega}{N} \\ j\omega\ N & 0 \end{pmatrix} \begin{pmatrix} V \\ I \end{pmatrix} \quad . \tag{8.87}$$

To diagonalize the matrix on the right-hand side of Eq. (8.87) (i.e., to nd its

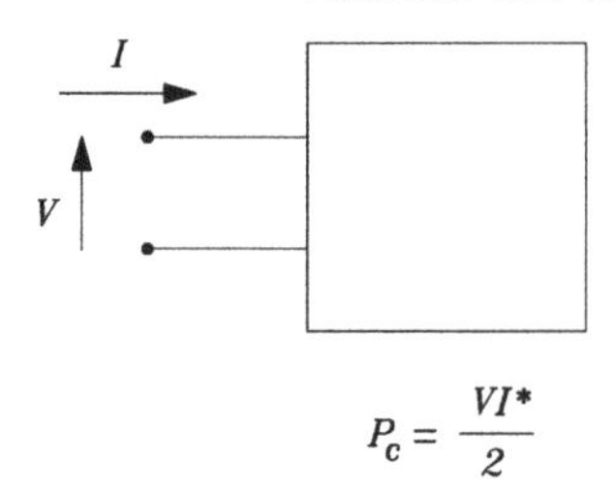

Figure 8.16 *Relationship between voltage, current, and complex power.*

eigenvalues $i = i\omega\sqrt{\ }$ and its eigenvectors) is straightforward algebra, yielding the following equations:

$$\frac{da}{dz} = j\omega\sqrt{\ }\,a \quad ,$$

$$\frac{db}{dz} = j\omega\sqrt{\ }\,b \quad , \qquad (8.88)$$

where:

$$a = \frac{1}{2}\left(\frac{V}{\sqrt{\ /N}} + \sqrt{\ /N}\,I\right) \quad ,$$

$$b = \frac{1}{2}\left(\frac{V}{\sqrt{\ /N}} \quad \sqrt{\ /N}\,I\right) \quad . \qquad (8.89)$$

Their inverse relationships clearly are:

$$V = \sqrt{\frac{\ }{N}}\,(a+b) \quad ,$$

$$I = \sqrt{\frac{N}{\ }}\,(a \quad b) \quad . \qquad (8.90)$$

The new variables Eqs. (8.88) are often referred to as the *normal modes* of the structure. Their practical importance is probably not evident, at this stage, because this whole section su ers from limitations due to some of the simplifying initial assumptions. In fact, after writing Eqs. (8.83), we always mentioned *two* conductors. However, what we proved in Section 7.4 was that the requirement for the existence of the TEM mode were *at least* two conductors, without an upper limit to their number. In a structure with $n > 2$ conductors, it is evident that the TEM mode, in spite of its eigenvalue being unique, has more degrees of freedom than what we assumed when we wrote Eqs. (8.71). Namely, in a *multiwire* transmission line, one can de ne $n \quad 1$ independent voltages (each of them is the voltage between one common reference conductor and the remaining $n \quad 1$ independent conductors) and $n \quad 1$ independent currents (the n-th current in the common reference being the "return" one, equal and

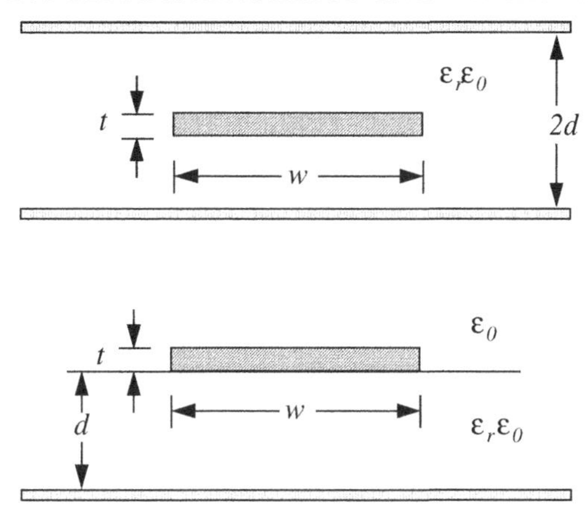

Figure 8.17 *Schematic view of a stripline and a microstrip.*

opposite to the sum of the others). Dealing with such a structure in terms of $2(n - 1)$ independent differential equations would make the formalism unbearably heavy, unless one passes through diagonalizing a matrix like in Eq. (8.87). Further developments on this subject are left to Section 8.12.

8.11 TEM and quasi-TEM propagation in planar lines

For at least three decades, the whole world of electronics evolved in the direction of exploiting planar technologies more and more. Microwaves were not at all an exception to this rule. Waveguides of the types discussed in the previous chapters, coaxial cables, or wire pairs, are not very well matched to this trend. The tendency to build passive transmission lines which consisted of two or more flat conductors, lying on parallel planes and separated by suitable dielectrics (e.g., Itoh, Ed., 1987) was then a natural evolution. Figure 8.17 shows schematically the most commonly used types of planar lines. The conducting planes whose edges are not shown in the drawings are, usually, much wider than all the other transverse geometrical dimensions. In the theoretical analyses they are usually taken as infinitely wide. Unless the opposite is explicitly specified, they are at a common "ground" potential, also when they are more than one. Voltages of interest in the propagation along these structures are those of the other conductors, with respect to ground. The current in the ground electrodes is the opposite of the (algebraic) sum of currents in all the other conductors.

The technological progress, from waveguides to planar lines, is not accompanied by a similar progress in modeling. On the contrary, none of the structures shown in Figure 8.17 is simple to analyze theoretically. Approximations are necessary, almost invariably, often for more than one reason, as we shall explain now.

The theory of guided e.m. waves, as developed in the previous chapter, re-

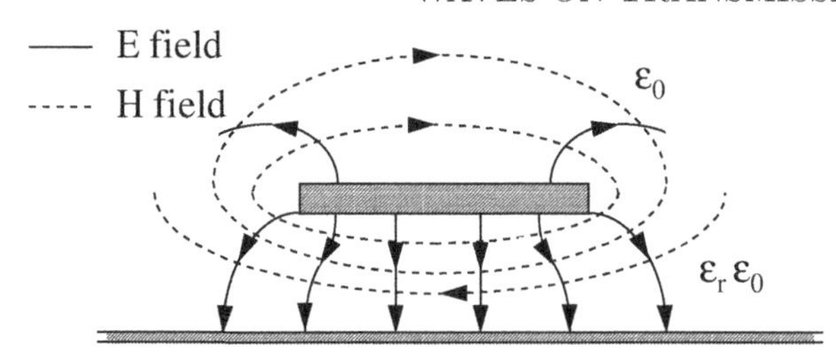

Figure 8.18 *Electric and magnetic eld lines in the cross-section of a typical microstrip.*

quires a homogeneous medium in the region where the guided eld travels. This is the case, for example, for the symmetrical stripline shown in the upper part of Figure 8.17. Then, as the cross-section is not a simply connected domain, the general theory tells us that *the fundamental mode is* TEM. All the contents given so far in this chapter apply. The only di culty that one still has to overcome is calculating the per-unit-length parameters l and c (plus, if required, r and g). Often, though, these entail approximations, because the geometry of the structure does not lend itself to a rigorous analysis. An example will be provided in a short while.

In other cases — typically, the microstrip shown in the lower part of Figure 8.17, which is probably the most practical among all those examples — part of the eld travels in the dielectric between the conductors, and part in the air above them (see Figure 8.18). The theory of the previous chapter no longer applies. Intuition tells us that a structure with more than one conductor can operate at arbitrarily low frequencies, so its fundamental mode must have something in common with the TEM mode. However, it is easy to verify that, if we impose the condition that the longitudinal components of the electric and magnetic elds are zero everywhere, then we do not have enough degrees of freedom left to satisfy all the continuity conditions at the air-dielectric interface, plus the boundary conditions on the conductor surfaces. As a consequence, the fundamental mode is not TEM, but *quasi*-TEM. The whole content of this chapter then applies to it only as an approximation, superimposed upon the di culty of nding the per-unit-length parameters. In the following we will discuss brie y an example, to show what this can mean in practice.

8.11.1 First example: the stripline

For a lossless structure which can propagate the TEM mode, we know from Section 8.4 that its characteristic impedance and its phase constant are given by Eqs. (8.16) and (8.17), respectively, namely $Z_C = \sqrt{l/c}$ and $\ = \omega\sqrt{lc}$. Furthermore, from the previous section we know (see Eq. (8.79)) also that $\ = \omega\sqrt{\ }$. Therefore, when the parameters and of the dielectric medium are known, the phase constant is known, and to determine the characteristic impedance it is enough to determine either c, or l.

As the reader may know from previous classes of static elds, a powerful technique to calculate capacitances, even for apparently complicated geometries, is known as *conformal mapping*. It is explained in several advanced textbooks on static elds (e.g., Plonsey and Collin, 1961), and cannot be dealt with in this book because of space limits. When applied to the geometry of Figure 8.17, part a), under the assumption of negligible thickness t, it yields a per-unit-length capacitance such that the corresponding characteristic impedance is:

$$Z_c \quad \frac{}{4} \frac{K(k)}{K(\sqrt{1 \quad k^2})} \quad , \tag{8.91}$$

where $\quad = (\; / \;)^{1/2}$ is, as usual, the characteristic impedance of the homogeneous dielectric medium, $K(k)$ indicates the so-called complete elliptic integral of the rst kind (Abramowitz and Stegun, Eds., 1965), and:

$$k = \left[\cosh\left(\frac{w}{4d}\right)\right]^{\quad 1} \quad , \tag{8.92}$$

the symbols w and d being explained in Figure 8.17, part a). The practical use of Eq. (8.91) can be simpli ed by using two approximate expressions for the elliptic integral, namely:

$$Z_c \simeq \frac{}{2} \quad \ln\left[2\coth\left(\frac{w}{8d}\right)\right] \quad , \tag{8.93}$$

for $w/d \quad 1.12$, and:

$$Z_c \simeq \frac{}{8[\ln 2 + \quad w/(4d)]} \quad , \tag{8.94}$$

for $w/d > 1.12$.

The assumption of negligible thickness, on the other hand, makes it impossible to calculate the ohmic loss for a nite conductivity and the consequent attenuation constant. Indeed, a conductor of zero cross-section and nite conductivity would entail an in nitely large loss, which is obviously unrealistic. For a nite thickness t, a perturbation approach like that of Section 7.9 yields an attenuation constant:

$$_c = \frac{R_s}{2Z_c d}\left[\frac{w/2d + \ln(8d/ \; t)}{\ln 2 + \quad w/4d}\right] \quad , \tag{8.95}$$

where R_S is, once again, the surface resistance of a conducting half-space, which we found rst in Section 4.12.

8.11.2 *Second example: the microstrip*

It is easy to show that, as we said before, for all the structures shown in Figure 8.17 except for part a), the fundamental mode cannot be rigorously TEM, as they encompass more than one dielectric. Indeed, from the decomposition theorem of Section 3.8 we see that a nontrivial TEM solution (i.e., with nonzero transverse components) requires that the phase constant be $\ = \omega\sqrt{\ _i}$ in each medium ($i = 1, 2$), which is clearly impossible if $\ _1 \neq \ _2$. Approximations are required to begin with. We will outline the way of thinking with reference to the microstrip (Figure 8.17, part b) which is by far and large the most widely used in practice, but the essential features apply also to the other structures.

At relatively low frequency, we may still assume that the longitudinal components are negligible compared to the transverse ones. (The reader may justify this, as an exercise, in quantitative, terms starting from the decomposition theorem). Within this approximation, the only problem we are left with is to calculate a per-unit-length capacitance, to replace Eq. (8.91). However, conformal mapping no longer provides exact results when the dielectric is not homogeneous. Another approximation is needed; the most widely accepted way to proceed is to calculate rst Eq. (8.91) as if the dielectric were air everywhere, and then correct it by means of an "equivalent" relative permittivity, de ned as that of a dielectric medium lling all the half-space above the ground plane, yielding the same per-unit-length capacitance as that of Figure 8.17, part b). Several answers can be found in the literature. The most commonly used one is given in Schneider (1969):

$$\varepsilon_r' = 1 + \frac{(\varepsilon_r \quad 1)}{2}\left[1 + \frac{1}{\sqrt{1 + 10d/w}}\right] \quad . \tag{8.96}$$

At higher frequencies, neglecting the longitudinal components yields unacceptable results, in particular for the phase constant, which is of paramount importance in designing components and circuits. An exact theory, in terms of potentials, can be found in the literature (Collin, 1991), but is too long and demanding to be reported here. A widely used approximation is that of Yamashita *et al.* (1979):

$$\frac{\ }{k_0} = \frac{\sqrt{\varepsilon_r} \quad \sqrt{\varepsilon_r'}}{1 + 4F^{\ 1.5}} + \sqrt{\varepsilon_r'} \quad , \tag{8.97}$$

where $k_0 = 2\ /\ _0$ is the free-space phase constant, and:

$$F = \frac{4d\sqrt{\varepsilon_r \quad 1}}{\ _0}\left\{0.5 + \left[1 + 2\ln\left(1 + \frac{w}{d}\right)\right]^2\right\} \quad . \tag{8.98}$$

One essential feature is that the dispersion in this structure becomes larger when the relative permittivity becomes higher, and when the ratio w/d becomes smaller, i.e., when the structure becomes more compact. Therefore, design of microstrip circuits is inevitably a trade-o between contradictory requirements.

As for attenuation due to nite conductivity, its order of magnitude can be estimated with a rather rough approximation, if a microstrip is considered as a pair of stripes of identical width w. The result of the perturbation approach is then simply:

$$\ _c = R_s/(wZ_c) \quad , \tag{8.99}$$

where R_s is the usual surface resistance of a half-space good conductor, and Z_c is the characteristic impedance of the microstrip. In agreement with intuition, Eq. (8.99) overestimates the attenuation, since, in reality, the width of the conducting backplane is much larger than w. More precise calculations are easily found in modern textbooks on microwaves, and are omitted here because of space limits.

8.12 The coupled-mode equations

Completeness of a set of modes in a *perfect* cylindrical structure was discussed in Section 7.6. The conclusion was that a suitable countable in nity of orthogonal modes could form a basis for expanding any eld in such a structure. One must reconsider completeness, from another point of view, when dealing with an *imperfect* structure, I, which di ers slightly from a perfect one, P, whose modes are known, whilst those of the imperfect structure cannot be found exactly. A typical example in this sense is a waveguide whose metallic walls have a nite conductivity (see Section 7.9).

In such a case, a set of modes $\{\mathbf{E}_i, \mathbf{H}_i\}$ is said to be complete when it satis es the following convergence requirement:

$$\lim_{n\to\infty} \int_{S'} \left[\mathbf{E}_t(z) \quad \sum_{i=1}^{n} V_i(z)\, \mathcal{E}_{ti} \right] \quad \left[\mathbf{H}_t(z) \quad \sum_{i=1}^{n} I_i(z)\, \mathcal{H}_{ti} \right] \quad \hat{a}_z \, dS' = 0 \,, \tag{8.100}$$

where $\{\mathbf{E}_t, \mathbf{H}_t\}$ are the transverse components of any eld which may exist in I; S' is the cross-section of I; $\{\mathcal{E}_{ti}, \mathcal{H}_{ti}\}$ are the (z-independent) transverse components of the i-th mode of the perfect structure P; $\{V_i, I_i \; ; \; i = 1, 2, \ldots\}$ are complex coe cients, which, in general, depend on z. For the time being, we may assume that these coe cients are adimensional.

The condition in Eq. (8.100) means that the two series $(V_i\mathcal{E}_{ti})$ and $(I_i\mathcal{H}_{ti})$ converge towards $\mathbf{E}_t$ and $\mathbf{H}_t$, respectively, at any point on S', except, at most, the points of a zero-measure subset. To state and to prove what are the most general conditions for Eq. (8.100) to be satis ed, is long and complicated, but may be found in the literature (Collin, 1991). We prefer to invest some space and e ort in introducing a formalism which is convenient to use whenever Eq. (8.100) holds. Through it, from Maxwell's equations one obtains a set of ordinary di erential equations (ODE's) for $\{V_i(z), I_i(z)\}$. The derivation is formally reminiscent of that of Section 8.10, and therefore may be phrased in the typical terminology of transmission lines. The only problems we are going to face in this derivation come from the fact that, for the two series in Eq. (8.100), *uniform convergence* is not assured at every point. Thus, we are not guaranteed that they can be di erentiated term by term.

In the structure I, where there are no sources, Maxwell's equations, projected on the z = constant plane, read:

$$\frac{\partial \mathbf{E}_t}{\partial z} = \quad j\omega \quad \hat{a}_z \quad \mathbf{H}_t + \nabla_t E_z \quad ,$$

$$\frac{\partial \mathbf{H}_t}{\partial z} = j\omega \ \hat{a}_z \quad \mathbf{E}_t + \nabla_t H_z \quad . \tag{8.101}$$

When Eq. (8.100) holds, in the previously speci ed sense of this statement, we may set:

$$\mathbf{E}_t = \sum_{i=1}^{\infty} V_i(z)\, \mathcal{E}_{ti} \quad , \qquad \mathbf{H}_t = \sum_{i=1}^{\infty} I_i(z)\, \mathcal{H}_{ti} \quad . \tag{8.102}$$

Let us recall that $\{\mathcal{E}_i, \mathcal{H}_i\}$ is the i-th mode of the structure P. For the sake of generality, we will suppose that it is a hybrid mode (i.e., $\mathcal{E}_z \neq 0$, $\mathcal{H}_z \neq 0$). Then, using Eqs. (7.11) and (7.12), we may write:

$$\mathcal{E}_{ti} = \nabla_t \mathcal{L}_i \quad \hat{a}_z \quad \frac{j \ \ i}{\omega} \nabla_t \mathcal{T}_i \quad ,$$

$$\mathcal{H}_{ti} = \frac{j \ \ i}{\omega} \nabla_t \mathcal{L}_i + \nabla_t \mathcal{T}_i \quad \hat{a}_z \quad . \tag{8.103}$$

The functions $\mathcal{L}_i$ and $\mathcal{T}_i$ satisfy:

$$\nabla_t^2 \mathcal{F}_i = \quad {}_i^2\, \mathcal{F}_i \qquad (\ {}_i^2 = \ {}_i^2 \quad {}^2) \quad . \tag{8.104}$$

To obtain the ODE's in $V_i(z)$ and $I_i(z)$ we are looking for, at rst sight one would think it is enough to express E_z and H_z in terms of V_i and I_i in Eqs. (8.101). Another step, based on mode orthogonality in the perfect structure P, would then allow us to isolate individual coe cients ($i = 1, 2, \ldots$) in Eqs. (8.101). In reality, this procedure would not be correct. As the series do not converge uniformly, $\nabla_t E_z$ and $\nabla_t H_z$ in Eqs. (8.101) cannot be calculated by term-by-term di erentiation of Eqs. (8.102). It is necessary to pass rst from Eqs. (8.101) to equations where E_z and H_z do not go through di erential operators.

To this purpose, let us take the scalar product of the rst of Eqs. (8.101) with $\hat{a}_z \quad \mathcal{H}_{tm}/2$, and of the second one with $\hat{a}_z \quad \mathcal{E}_{tm}/2$. Let us integrate over S', take into account mode orthogonality, and impose on each of them a normalization condition (power through S' equal to 1 W). All these operations result in the following equation:

$$\int_{S'} \frac{\mathcal{E}_{ti} \quad \mathcal{H}_{tm}}{2} \quad \hat{a}_z \, dS' = \quad {}_{i,m} \quad , \tag{8.105}$$

where ${}_{i,m}$ is Kronecker's delta.

Let us write also the following identities, which can be proved by the reader as an exercise using Appendix C and taking into account Eq. (8.104):

$$\nabla_t E_z \quad \nabla_t \mathcal{T}_m = \nabla_t \quad (E_z \nabla_t \mathcal{T}_m) + \ {}_m^2\, E_z \mathcal{T}_m \quad , \tag{8.106}$$

$$\nabla_t E_z \quad \nabla_t \mathcal{L}_m = \nabla_t \quad (E_z \nabla_t \mathcal{L}_m) \quad . \tag{8.107}$$

Let us also consider two identities which di er from the previous ones only because E_z is replaced by H_z, and because $\mathcal{T}_m$ and $\mathcal{L}_m$ have been interchanged. We omit writing them explicitly for the sake of brevity.

In the equations which were derived from Eqs. (8.101), all the terms involved in the latest identities show up inside integrals over the surface S'. Gauss' theorem (A.8) can be applied to the integral containing the rst term on the right-hand side in Eq. (8.106), and to the twin term, containing H_z. Stokes' theorem (A.9) can be applied to the integral containing the right-hand side of Eq. (8.107), and to the twin term containing H_z. The results are the following equations:

$$N_m \frac{dV_m}{dz} = j\omega \quad I_m + \frac{j \quad m}{\omega} \quad {}^2_m \int_{S'} E_z \, \mathcal{T}_m \, dS'$$

$$+ \quad \frac{j \quad m}{\omega} \quad \int_{\ell'} E_z \, \frac{\partial \mathcal{T}_m}{\partial n} \, d\ell' + \int_{\ell'} E_z(\nabla_t \mathcal{L}_m \quad \hat{\ell}) \, d\ell' \, , \qquad (8.108)$$

$$\frac{1}{N_m} \frac{dI_m}{dz} = j\omega \quad V_m \quad \frac{j \quad m}{\omega} \quad {}^2_m \int_{S'} H_z \, \mathcal{L}_m \, dS'$$

$$\frac{j \quad m}{\omega} \quad \int_{\ell'} H_z \, \frac{\partial \mathcal{L}_m}{\partial n} \, d\ell' + \int_{\ell'} H_z(\nabla_t \mathcal{T}_m \quad \hat{\ell}) \, d\ell' \, , \qquad (8.109)$$

where:

$\partial/\partial n$ is the normal derivative;

$\ell' = S' \cap \mathcal{C}$ is the boundary of the section S', oriented clockwise with reference to the unit vector $\hat{a}_z$;

$\hat{\ell}$ is the unit vector tangent to ℓ';

$N_m = |\mathcal{E}_{tm}|/|\mathcal{H}_{tm}|$ is a generalization, for the m-th mode, of the constant N of Eq. (8.74), and is proportional to the wave impedance of the m-th mode.

At this stage, E_z and H_z in Eqs. (8.108) and (8.109) are no longer arguments of di erential operators. Thus, non-uniform convergence of their series expansions does not cause problems any more. These series expansions can be obtained through the longitudinal components of Maxwell's equations, if we let the transverse elds be expressed by Eqs. (8.102), namely:

$$E_z = \frac{1}{j\omega} \nabla \quad \mathbf{H}_t = \sum_{i=1}^{\infty} I_i(z) \, \frac{\nabla_t \quad \mathcal{H}_{ti}}{j\omega}$$

$$= \sum_{i=1}^{\infty} I_i(z) \, \mathcal{E}_{zi} = \quad \frac{1}{j\omega} \sum_{i=1}^{\infty} I_i(z) \quad {}^2_i \, \mathcal{T}_i \quad , \qquad (8.110)$$

$$H_z = \quad \frac{1}{j\omega} \nabla \quad \mathbf{E}_t = \sum_{i=1}^{\infty} V_i(z) \, \frac{\nabla_t \quad \mathcal{E}_{ti}}{j\omega}$$

$$= \sum_{i=1}^{\infty} V_i(z) \, \mathcal{H}_{zi} = \frac{1}{j\omega} \sum_{i=1}^{\infty} V_i(z) \quad {}^2_i \, \mathcal{L}_i \quad . \qquad (8.111)$$

Let us now place these expressions into Eqs. (8.108) and (8.109), and use commutation between the integral over S' and the sum with respect to i. Taking into account further orthogonality relations, of the type in Eq. (7.35), between pre-potentials of the perfect structure P, the only terms among the integrals over S' which survive are those whose arguments contain products of the type $\mathcal{L}_m\mathcal{L}_m$ and $\mathcal{T}_m\mathcal{T}_m$. Because of the normalization in Eq. (8.105), the integrals containing these products are constants:

$$\int_{S'} \mathcal{L}_m \, \mathcal{L}_m \, dS' = A_m \quad , \tag{8.112}$$

$$\int_{S'} \mathcal{T}_m \, \mathcal{T}_m \, dS' = B_m \quad , \tag{8.113}$$

although it is impossible to specify their explicit values as long as the boundary conditions on the walls of P are not speci ed.

In conclusion, Eqs. (8.108) and (8.109) become:

$$\begin{aligned} \frac{dV_m}{dz} &= \frac{j\omega}{N_m} I_m + c_m \, I_m + \sum_{n=1}^{\infty} C_{mn} \, I_n \quad , \\ \frac{dI_m}{dz} &= j\omega \; N_m \, V_m + d_m \, V_m + \sum_{n=1}^{\infty} D_{mn} \, V_n \quad , \end{aligned} \tag{8.114}$$

whose coe cients are shown in Table 8.1.

In the perfect structure P, orthogonality assures that each mode may exist by itself. This implies that in P it must be $C_{mn} = 0$ and $D_{mn} = 0$ for $m \neq n$. Then, Eqs. (8.114) reduce to the standard transmission line equations that we studied for most of this chapter. This makes it self-explanatory why in general Eqs. (8.114), for the imperfect structure I, are referred to as the *coupled-line equations*, or also as the *generalized telegraphist equations*.

Analogy with the previous section entails us to let, for each mode:

$$V_m = \sqrt{\frac{}{N_m}} \, (a_m + b_m) \quad , \qquad I_m = \sqrt{\frac{N_m}{}} \, (a_m \quad b_m) \quad . \tag{8.115}$$

Introducing these new variables into Eqs. (8.104), we get the following equations, usually referred to as the *coupled-mode equations*:

$$\begin{aligned} \frac{da_m}{dz} &= \quad {}_m \, a_m + \sum_{n \neq m} (F_{mn} \, a_n + G_{mn} \, b_n) \quad , \\ \frac{db_m}{dz} &= \quad {}_m \, b_m \quad \sum_{n \neq m} (G_{mn} \, a_n + F_{mn} \, b_n) \quad , \end{aligned} \tag{8.116}$$

where $\;_m$ is the propagation constant of the m-th mode, provided it is alone (i.e., $a_n = b_n = 0$ for $n \neq m$) in the imperfect structure I. In general, it is $\;_m \neq \;_m$ (the propagation constant of the same mode in the perfect structure), as $\;_m$ accounts also for the other terms in Eqs. (8.114) that correspond to $n = m$.

Table 8.1 *Expressions for the coe cients of the coupled-mode equations.*

$C_m = \dfrac{{}^4_m}{\omega^2 \;{}^2}\,\dfrac{m}{N_m}\,B_m$	$d_m = \dfrac{{}^4_m}{\omega^2 \;{}^2}\;{}_m\,N_m\,A_m$
$C_{mn} = \dfrac{m \;{}^2_n}{\omega^2 \;{}^2}\,\dfrac{1}{N_m}\displaystyle\int_{\ell'} \mathcal{T}_n\,\dfrac{\partial \mathcal{T}_m}{\partial n}\,d\ell' + \dfrac{{}^2_n}{j\omega}\,\dfrac{1}{N_m}\displaystyle\int_{\ell'} \mathcal{T}_n(\nabla_t \mathcal{L}_m \;\; \hat{\ell})\,d\ell'$	
$D_{mn} = \dfrac{m \;{}^2_n}{\omega^2 \;{}^2}\,N_m \displaystyle\int_{\ell'} \mathcal{L}_n\,\dfrac{\partial \mathcal{L}_m}{\partial n}\,d\ell' + \dfrac{{}^2_n}{j\omega}\,N_m \displaystyle\int_{\ell'} \mathcal{L}_n(\nabla_t \mathcal{L}_m \;\; \hat{\ell})\,d\ell'$	

The *mode coupling coe cients* are related to the quantities shown in Table 8.1 via:

$$
\begin{aligned}
2F_{mn} &= \sqrt{\frac{N_n}{N_m}}\,C_{mn} + \sqrt{\frac{N_m}{N_n}}\,D_{mn} \quad ,\\
2G_{mn} &= \sqrt{\frac{N_n}{N_m}}\,C_{mn} \quad \sqrt{\frac{N_m}{N_n}}\,D_{mn} \quad . \qquad (8.117)
\end{aligned}
$$

Many examples of solutions of Eqs. (8.116) are available in the literature. Their physical meaning changes from problem to problem. Perhaps the most typical example is that of the directional coupler, which was brie y illustrated in Section 3.9 when we discussed symmetries and their implications. Further suggestions to the reader will be given in the next section.

8.13 Further applications and suggested reading

Solving Maxwell's equations, or using circuit theory, are, as we said on several occasions, two di erent approaches to one physical reality. The di erence is in the axioms. In circuit theory, the axioms are Ohm's law and Kirchho 's laws. In our approach to e.m. waves, the axioms are Maxwell's equations and the charge continuity equation. In this chapter we have shown that, with an additional assumption (namely, that of TEM propagation), the eld approach leads to the same transmission lines equations as the circuit approach. The link between the two approaches is broader, in its scope of validity, than TEM waves, as we outlined brie y in Section 8.10. Hence, one of the most important messages that the reader can extract from this chapter is that, in order to understand more in depth high-frequency e.m. elds, one should have a strong background in circuit theory also. We recommend reading the classical treatise by Newcomb (1966) that we have quoted on several occasions in this book, and at least one other text, more focused on distributed-parameter circuits, such as, for instance, Ghausi and Kelly (1968).

In particular, the Smith chart of Section 8.7 is a crucial intersection between the circuit culture and the e.m. wave culture. On this front, further interest may be stimulated in those readers who like abstraction and mathematical approaches, by saying that bilinear transforms help to solve many other problems related to impedance transformations through various types of linear networks. The short but condensed booklet by Schwerdtfeger (1961), that we quoted when we introduced the Smith chart, is also a good starting point in this direction — see also Libo and Dalman (1985). Another valuable resource is Chapter 7 (by R. Beringer) of Montgomery, Dicke and Purcell (1948). The reader is encouraged to consult it after studying our Chapter 9 on resonant cavities.

A point that might look trivial at rst sight, but in our opinion is not so, is that transmission-line based problems show up in a very large number, and in a very large variety, of occasions in the professional life of anybody in electronic engineering. Solving many problems is a valuable investment for an EE student, in view, for example, of an employment interview. For this reason, the collection of problems at the end of this chapter is wider than the average over the rest of the book. All other textbooks covering this subject are good sources of problems.

The last comment has brought us towards rather practical issues. Continuing in the same direction, let us stress that what has been said in this chapter (and in the whole book) on striplines and microstrips, has to be thought of just as an elementary introduction. All telecommunication engineers should learn much more than that. Many rst-class textbooks cover the area both under the viewpoint of theory and that of applications. Just to quote a few, Collin (1991) is highly recommended in the rst category; Wadell (1991), Gupta *et al.* (1996), and Bhat and Koul (1989) are representative for the second one. Other useful texts include Freeman (1996) and Elliott (1993) (general microwave texts), and Christopoulos (1995), Fache *et al.* (1993), Faria (1993) (texts on mathematical modeling of microwave and strip transmission lines).

On the other hand, the reader must be encouraged not to underestimate the power of the rather abstract approach of Sections 8.10 and 8.12. Without these connections, it would be very hard to understand why the transmission line equations and the coupled-mode equations show up in problems where there are no conductors at all, such as integrated optics, or where it remains hard to visualize two coupled lines, such as microwave tubes based on electron beams, or particle accelerators, or traveling-wave microwave masers. Suggested reading in these areas include Unger (1977) on planar dielectric waveguides, Tamir, Ed., (1979) on integrated optics, Gilmour (1986), Chodorow and Susskind (1964), and Sims and Stephenson (1963) on microwave tubes, Pierce (1950/1954/1974) on traveling wave tubes, electron beams, and waves as well as some excellent popularizations, Louisell (1960) on coupled modes, Roussy and Pearce (1995) on industrial microwave device applications. However, especially in the case of integrated optics, it is preferable for the reader to consult these books only after studying our Chapter 10.

Let us conclude with a comment on the coupled-mode equations (CME's). They are still a popular subject of discussion and investigation, especially in integrated optics. One of the reasons why they are popular is because some important devices (typically, directional couplers and lters) cannot be studied analytically without approximation. However, there is a more subtle reason.

It is because the "perfect" structure P one starts from, in order to study an "imperfect" structure I (see Section 8.12), is *not unique*. For example, to study a directional coupler which encompasses a grating, one can start either from the modes of the two isolated waveguides, or from those (sometimes designated as *supermodes*) of two parallel waveguides. These also are available in exact closed form, for some geometries. Looking back at the general derivation of Section 8.12, the reader, if familiar with perturbation theory, should be able to understand why the accuracy of the perturbation solutions depends on the perfect structure one starts from. For very instructive details on this point, recent research papers by Huang *et al.* (1996) should be consulted, since, to the best of our knowledge, the subject has not yet been published as a book — see Huang (1994) for an overview, and, e.g., Sarangan and Huang (1994), and Huang and Hong (1992) for example applications.

Problems

8-1 Express the active and the reactive power in the generic section of a transmission line in terms of the re ection coe cient, separating real and imaginary parts of Eq. (8.15). Assume rst that the characteristic impedance of the transmission line is real. Then, repeat the exercise assuming a complex characteristic impedance.

8-2 Extend all the results of Section 8.6 to the case of low-loss transmission lines.

8-3 For the lossless transmission lines shown in Figure 8.19, we have $R_G = Z_{c1} = 50$, $Z_{c2} = 100$, $V_G = 10\,\mathrm{V}$. In the line with characteristic impedance Z_{c1}, the VSWR equals 4/3. The voltage modulus goes through a minimum where there is a discontinuity in the characteristic impedance.

a) Evaluate the load impedance Z_L.

b) Calculate the complex power delivered to the load impedance for $X_G = 0$, for $X_G = R_G$, and for $X_G =$ R_G, assuming that the length of the line with characteristic impedance Z_{c1} equals an odd integer multiple of /4.

8-4 In the lossless transmission line shown in Figure 8.20, the phase constant is the same as in free space. The frequency is $f = 100\,\mathrm{MHz}$. We have $R_G = Z_c = 50$, $R_L = 60$, $V_G = 20\,\mathrm{V}$, $C = (1/\ )\,\mathrm{pF}$, $L = 0.1/(4\ )\,\mathrm{mH}$. The line lengths, ℓ_1 and ℓ_2, are both equal to 2 m.

a) Set $R_1 = 40$, and calculate the active power delivered to the load resistance R_L.

b) Find the range of values of R_1 for which the *total* active power, delivered to R_1 and to R_L, is more than one half of the generator's available power.

8-5 For the lossless transmission line shown in Figure 8.21, at the frequency

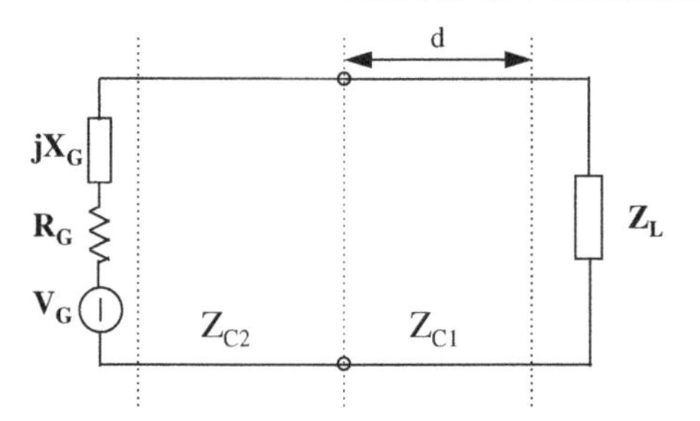

Figure 8.19

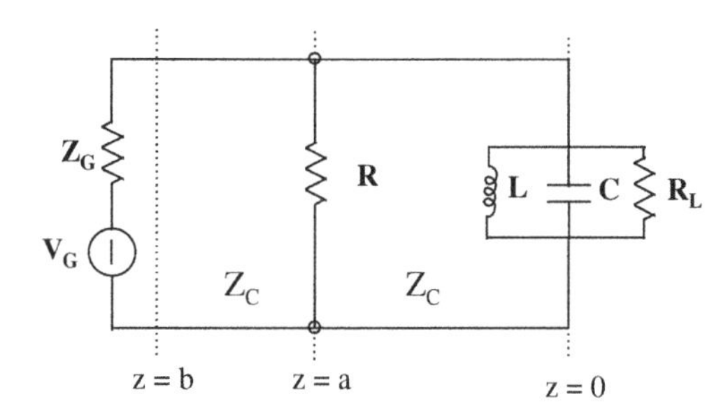

Figure 8.20

$f = 600\,\text{MHz}$, we have $Z_c = 200$, $d_1 = 15\,\text{m}$, $X_L = 100$, $X_C = \quad 400$, $V_G = 10\,\text{V}$. In the Z_{c1} part of the line, the voltage modulus goes through a minimum 0.5 m on the left of the terminals $22'$, while the VSWR is 4. The active power delivered to the load R equals 40 mW.

a) Find all the values of the distance d_2 which are compatible with these data.

b) Find the unknown values of the resistances R and R_G.

c) Find voltage and current at a generic distance from the generator.

8-6 Reconsidering the transmission line shown in the gure of Problem 8-4, and retaining all the other parameter values, nd the internal impedance

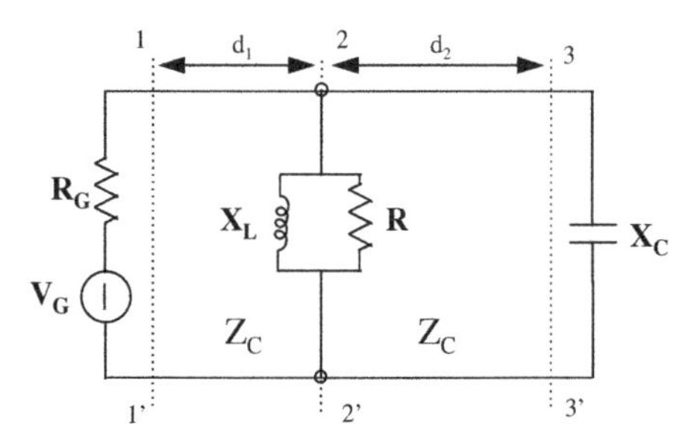

Figure 8.21

of the voltage generator which satis es the power matching condition at the generator terminals.
Repeat the problem modeling the generator as an ideal current generator in parallel with an internal admittance.

8-7 Verify that one can design a single-stub impedance-matching network using an analytical approach, instead of using the Smith chart, but that the procedure becomes quite cumbersome.

8-8 Work out in full detail the procedures to design a double-stub impedance-matching network for the two cases shown in Figure 8.14 but not discussed in the text, i.e., when the distance between the stubs is equal either to /8 or to 3 /8.

8-9 Enumerate relative advantages and disadvantages of single- and double-stub impedance-matching networks. Distinguish between two situations. First, consider a laboratory environment, where the load impedance is changed often, and in general is not known *a priori*. Second, consider a mass production environment, where many identical pieces are fabricated.

8-10 A low-loss transmission line, whose nominal characteristic impedance is 50 , exhibits a loss of 2 dB over a length of 10 m. The loss contributed by the parallel conductance is negligibly small.
Calculate the per-unit-length series resistance r, and the corresponding correction to the phase constant with respect to the lossless case, at 900 MHz, assuming a per-unit-length capacity of 50 pF/m.

8-11 For the network discussed in Problem 8-3, design a single-stub impedance-matching network at the frequency $f = 600$ MHz, to be placed upstream with respect to the discontinuity in the characteristic impedance, but as close as possible to it. Discuss two possibilities, namely a stub with characteristic impedance Z_{c1} and with characteristic impedance Z_{c2}. Calculate the complex power at the load, at the discontinuity, and at the terminals of the stub.

8-12 For the network discussed in Problem 8-4, design a quarter-wavelength impedance-matching network to be placed upstream with respect to the resistor $R = 40$. Find both solutions, one with an increase and the other with a decrease in characteristic impedance. Then, increasing the frequency to 200 MHz, and leaving the whole network unchanged, calculate the VSWR at the generator terminals.

8-13 A lossless transmission line, with phase constant equal to that of free space and characteristic impedance $Z_c = 100$, is closed on a load impedance $Z_L = 100 \quad j50$, and fed by a generator with an open-circuit voltage $V_G = 10$ V and an internal impedance $Z_G = 100 + j50$. The operating

frequency is 300 MHz.
Design a double-stub impedance-matching network connected to the load terminals, and with a spacing $d = 0.25\,\text{m}$ between the stubs. Study how the current and voltage vary with z, from the generator to the load. Calculate the complex power at the terminals of the load and at those of the two stubs.

8-14 For a lossless coaxial cable (whose outer and inner radii are denoted as a and b, respectively), show that if voltage and current are de ned as in the electrostatic case (i.e., according to Eqs. (8.83)), then the characteristic impedance of the cable equals $Z_c = (\ /\ )^{1/2} \ln(a/b)/(2\ )$.

8-15 In an air- lled coaxial cable, the inner conductor is of radius $b = 2.5\,\text{mm}$. Making use of the formula that was proved in the previous problem, adjust the radius of the outer conductor such that $Z_c = 120$.

8-16 The cable of the previous problem is connected to a 50 resistive load impedance, at the frequency of 300 MHz. A metallic sleeve can be used as a /4 impedance-matching device. Find its length, its outer radius, and its distance from the load terminals.
Repeat the same problem for a resistive load impedance of 300 .

8-17 Show, using the decomposition theorem of Section 3.8 or of Section 7.2, that, within each homogeneous dielectric medium present in a microstrip, the longitudinal eld components are negligibly small compared to the transverse ones.

8-18 Using one of the approximations, Eqs. (8.93) or (8.94), design a stripline, in a homogeneous dielectric medium having $_r = 2.2$ and a spacing $d = 10\,\text{mm}$, so that its characteristic impedance is 50 . For copper conductor whose thickness is $t = 0.2\,\text{mm}$, evaluate the attenuation constant vs. frequency between 1 and 2 GHz. (Copper conductivity: 5.8 10^7 ^{1}m ^{1}.)

8-19 A microstrip for which $w/d = 4$ is fabricated with copper on alumina, the dielectric for which $_r = 9.8$. Evaluate the equivalent permittivity, Eq. (8.96), and then the corresponding phase constant, dispersion (i.e., the second derivative of with respect to ω), and attenuation constant, at 3.5 GHz.

8-20 Using Appendix C and taking into account Eq. (8.104), derive Eqs. (8.106) and Eq. (8.107).

References

Abramowitz, M. and Stegun, I.A. (Eds.) (1965) *Handbook of Mathematical Functions.* Dover, New York.

Bhat, B. and Koul, S.K. (1989) *Stripline-Like Transmission Lines for Microwave Integrated Circuits.* Wiley, New York.

Chodorow, M. and Susskind, C. (1964) *Fundamentals of Microwave Electronics.* McGraw-Hill, New York.

Christopoulos, C. (1995) *The Transmission-Line Modeling Method: TLM.* IEEE/OUP Series on Electromagnetic Wave Theory, Institute of Electrical and Electronics Engineers, New York/Oxford University Press, Oxford.

Collin, R.E. (1991) *Field Theory of Guided Waves.* IEEE Press, New York.

Corazza, G.C. (1974) *Fondamenti di Campi Elettromagnetici e Circuiti.* Vol. II, Patron, Bologna.

Dicke, R.H. and Wittke, J.P. (1960) *Introduction to Quantum Mechanics.* Addison-Wesley, Reading, MA.

Elliott, R.S. (1993) *An Introduction to Guided Waves and Microwave Circuits.* Prentice-Hall, Englewood Cli s, NJ.

Fache, N., Olyslager, F. and de Zutter, D. (1993) *Electromagnetic and Circuit Modelling of Multiconductor Transmission Lines.* Oxford Engineering Science Series, Clarendon Press, Oxford/Oxford University Press, New York.

Faria, J.A.B. (1993) *Multiconductor Transmission-Line Structures.* Wiley Series in Microwave and Optical Engineering, Wiley, New York.

Freeman, J.C. (1996) *Fundamentals of Microwave Transmission Lines.* Wiley Series in Microwave and Optical Engineering, Wiley, New York.

Ghausi, M.S. and Kelly, J.J. (1968) *Introduction to Distributed-Parameter Networks: with Application to Integrated Circuits*, Holt, Rinehart and Winston, New York.

Gilmour, A.S., Jr. (1986) *Microwave Tubes.* Artech House Microwave Library, Artech House, Dedham, MA.

Gupta, K.C. *et al.* (1996) *Microstrip Lines and Slotlines.* 2nd ed., Artech House Microwave Library, Artech House, Boston.

Huang, W.-P. and Hong, J. (1992) "A coupled-waveguide grating resonator lter". *IEEE Photonics Technology Letters*, **4** (8), 884–886.

Huang, W.-P. (1994) "Coupled-mode theory for optical waveguides: an overview". *J. Optical Society of America A*, **11** (3), 963–983.

Huang, W.-P., Guo, Q. and Wu, C. (1996) "A polarization-independent distributed Bragg re ector based on phase-shifted grating structures". *J. Lightwave Technology*, **14** (3), 469–473.

Itoh, T. (Ed.) (1987) *Planar Transmission Line Structures.* IEEE Press Selected Reprint Series, IEEE Press, New York.

Libo , R.L. and Dalman, G.C. (1985) *Transmission Lines, Waveguides, and Smith Charts.* Macmillan, New York.

Louisell, W.H. (1960) *Coupled Mode and Parametric Electronics.* Wiley, New York.

Montgomery, C.G., Dicke, R.H. and Purcell, E.M. (Eds.) (1948) *Principles of Microwave Circuits.* McGraw-Hill, New York. Reprinted by Peter Peregrinus, London, for IEE (1987).

Morse, P.M. and Feshbach, H. (1953) *Methods of Theoretical Physics.* McGraw-Hill, New York.

Newcomb, R.W. (1966) *Linear Multiport Synthesis.* McGraw-Hill, New York.

Pierce, J.R. (1950) *Traveling-Wave Tubes.* Bell Telephone Laboratories Series, Van Nostrand, New York.

Pierce, J.R. (1954) *Theory and Design of Electron Beams.* Bell Telephone Laboratories Series, 2nd ed., Van Nostrand, New York.

Pierce, J.R. (1974) *Almost All About Waves.* MIT Press, Cambridge, MA.

Plonsey, R. and Collin, R.E. (1961) *Principles and Applications of Electromagnetic Fields.* McGraw-Hill, New York.

Roussy, M.G. and Pearce, J.A. (1995) *Foundations and Industrial Applications of Microwave and Radio Frequency Fields: Physical and Chemical Processes.* Wiley, New York.

Sarangan, A.M. and Huang, W.-P. (1994) "A coupled mode theory for electron wave directional couplers". *IEEE J. Quantum Electronics*, **30** (12), 2803–2810.

Schneider, M.V. (1969) "Microstrip lines for microwave integrated circuits". *Bell System Technical Journal*, **48** (5), 1421–1444.

Schwerdtfeger, H. (1961) *Introduction to Linear Algebra and the Theory of Matrices.* 2nd ed., Noordho , Groningen.

Sims, G.D. and Stephenson, I.M. (1963) *Microwave Tubes and Semiconductor Devices.* Interscience Publishers, New York.

Tamir, T. (Ed.) (1979) *Integrated Optics.* 2nd corrected and updated ed., Springer-Verlag, Berlin.

Unger, H.-G. (1977) *Planar Optical Waveguides and Fibres.* Clarendon Press, Oxford.

Wadell, B.C. (1991) *Transmission Line Design Handbook.* Artech House Microwave Library, Artech House, Boston.

Yamashita, E., Atsuki, K. and Ueda, T. (1979) "An approximate dispersion formula of microstrip lines for computer-aided design of microwave integrated circuits". *IEEE Trans. Microwave Theory & Techniques*, **MTT-27** (12), 1036–1038.

CHAPTER 9

Resonant cavities

9.1 Introduction

In the previous chapters, we showed what theoretical background is necessary, in order to comply with a self-explanatory technical requirement: how to convey electromagnetic energy along a given path. Similarly, this chapter is devoted to the theoretical basis required to answer another very important technical requirement: how to store electromagnetic energy in a given region in 3-D space.

In elementary circuit theory, as well known, one usually starts o by introducing two ideal components, the capacitor and the inductor. Their purpose is to model, in the most simplistic way, two physical realities, that consist of storing energy, electric and magnetic, respectively, within given regions. The next step is to combine these two elements, and introduce the concept of a resonant circuit. This approach is strongly biased by previous knowledge of electrostatics and magnetostatics. Indeed, the two ideal components are rather easy to de ne as long as electric and magnetic elds are time-invariant. But in a rapidly time-varying regime, electric and magnetic elds are inseparable. Then, as we will see in this chapter, if we start from Maxwell's equations, we may run directly into some objects whose behavior is essentially that of resonant LC circuits; but the main di erence is that it is impossible to split these objects into two separate L and C components.

As we saw in the previous chapter (and also in terms of plane waves, in Chapter 4), the input impedance of a short-circuited single-mode line is purely reactive. It is inductive, if the length ℓ of the line is less than $_g/4$, capacitive if $_g/4 < \ell < _g/2$. Let us consider a transmission line of length $_g/2$, short-circuited at both ends (Figure 9.1). Any plane orthogonal to the z axis (except for the one $z = _g/4$) splits the line in two parts, one of them having an inductive input impedance, the other having a capacitive input impedance. The choice of this plane is, once again, completely arbitrary; this indicates that the resonant behavior is an intrinsic property of the whole system in itself. The de nition of a capacitor and an inductor is an artifact, and was used just as a link with known concepts.

In the example shown in Figure 9.1, a complete round trip, from one of the short-circuit planes to the other and back, along the z axis, has a total length of a wavelength. Consider then a wave that passes through the generic point P with a given phase . When it passes again through P, in the same sense as before, it has the same phase . In the remainder of this chapter we will see

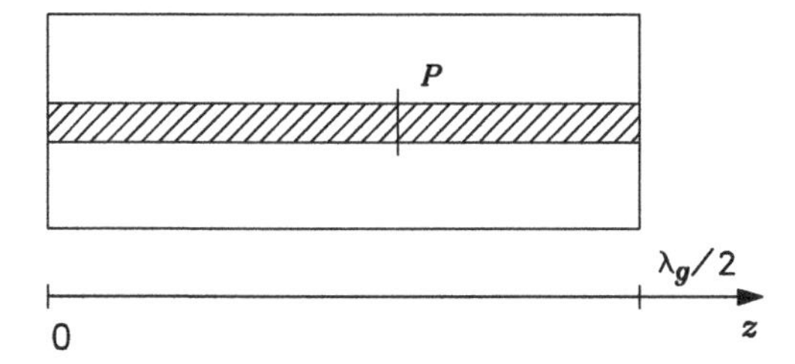

Figure 9.1 *A short-circuited coaxial cable, half wavelength in length, seen as a resonant circuit.*

that this comment has a very broad scope of validity. A resonant behavior always entails that there are closed paths (in general, three-dimensional paths) whose total length is an integer multiple of a wavelength. In Chapter 7, we saw that Maxwell's equations with 2-D boundary conditions admit solutions at any frequency. In this chapter, on the contrary, we will see that 3-D boundary conditions can be satis ed only at *discrete* frequencies; this is a consequence of the phase conservation requirement along closed paths, that we just illustrated. This frequency selection process reinforces the analogy with networks consisting of lumped inductors and capacitors, which also may resonate only at discrete frequencies.

9.2 Separable coordinate systems in three dimensions

In some of the previous chapters, we saw that separation of variables is a powerful method to solve the Helmholtz equation. On the other hand, we saw in Chapter 7 that it su ers from intrinsic limitations, due to the nite number of reference systems in which it can be applied. We will see now brie y how those limitations extend from the 2-D to the 3-D case.

An electromagnetic resonator is, once again, a region in 3-D space, surrounded by a surface S, which can either be completely at a nite distance from an origin (closed resonators), or extend in part to in nity (open resonators). On this surface, we impose *boundary conditions* that the eld must satisfy. When the Helmholtz equation is solved by separation of variables, in a given reference system, boundary conditions are simple to express *only if* the surface S is a *coordinate surface*, or can be subdivided into coordinate surfaces, in that reference system. When this is not the case, then we may still use a complete set of solutions obtained by separation of variables, but the eld must be written as a series expansion. Such a procedure is always a source of nuisance; sometimes, it causes conceptual problems also, as consequences of whether it is legitimate or not, to di erentiate a series term by term.

For these reasons, it is very important to know from the beginning that the 3-D Helmholtz equation is separable only in a *nite number* of coordinate systems. We must omit the proof for the sake of brevity; it is very similar to that of the 2-D case, and can be found in the same references. In any case, we will provide now the list of those coordinate systems (the so-called "separable coordinate systems"), where separation of variables is applicable.

The most general case (in the sense that all the others may be looked at as its degenerate cases) is that of *orthogonal ellipsoidal coordinates* ($_1$, $_2$, $_3$), which are implicitly de ned by the following relations, where (x_1, x_2, x_3) is an orthogonal cartesian system:

$$x_i = \left[\frac{(\ _1^2 \quad a_i^2)(\ _2^2 \quad a_i^2)(\ _3^2 \quad a_i^2)}{a_i^2(a_i^2 \quad a_j^2)} \right]^{1/2} , \qquad \begin{array}{l} i = 1,2 \\ j = 2,1 \quad (\neq i) \end{array}$$

$$x_3 = \frac{\ _1\ _2\ _3}{a_1 a_2} , \qquad _1 > a_1 > \ _2 > a_2 > \ _3 > 0 \quad . \tag{9.1}$$

It can be proved (Morse and Feshbach, 1953) that its coordinate surfaces are *confocal, mutually orthogonal quadric surfaces*, namely:

ellipsoids: $_1$ = constant;

one-fold hyperboloids: $_2$ = constant;

two-fold hyperboloids: $_3$ = constant.

The *degenerate cases*, which complete the set of separable coordinates for the Helmholtz equation, are the following (in brackets, the corresponding coordinate surfaces):

1. *orthogonal cartesian* coordinates (planes);
2. *circular cylindrical* coordinates (circular cylinders; half-planes; planes);
3. *elliptical cylindrical* coordinates (elliptical cylinders; hyperbolic cylinders; planes);
4. *parabolic cylindrical* coordinates (two families of parabolic cylinders; planes);
5. *rotation parabolic* coordinates (two families of rotation paraboloids; planes);
6. *paraboloidal* coordinates (two families of elliptical paraboloids; one family of hyperbolic paraboloids);
7. *spherical* coordinates (spherical surfaces; cones; planes);
8. *prolate spheroidal* coordinates (prolate spheroids; two-fold revolution hyperboloids; planes);
9. *oblate spheroidal* coordinates (oblate spheroids; one-fold revolution hyperboloids; planes);
10. *conical* coordinates (spherical surfaces; one-fold hyperboloids; cones with elliptical cross-section).

9.3 Completeness of resonator modes

Solving the Helmholtz equation by separation of variables is legitimate, similarly to what we saw in Chapters 4 and 7, provided we can prove that the solutions found in this way form a complete set. Completeness of resonator modes can be proved similarly to what we saw in Chapter 4 for plane waves, and in Chapter 7 for waveguide modes.

When we compared Chapter 4 and Chapter 7, we noticed that the dimension of a complete set decreased (from a continuum to a countable in nity), as we passed from the very mild restriction that the eld be square integrable, to boundary conditions at nite distance along two directions. At a given frequency, the component of the propagation vector along a given direction can vary continuously in an unbounded medium, but is discretized in a waveguide. We may wonder now whether in a resonator (for simplicity, a closed one) there is a further reduction in the size of a complete set, as we impose boundary conditions along three directions.

The answer can be given intuitively, going back to the case (see Section 9.1) of a cavity that consists of a nite length of a lossless cylindrical structure, whose transmission modes we assume to know. Two perfect conductors close the structure at its ends. The i-th mode of the cylindrical guide can travel either way; so, we must associate it with two arbitrary complex constants, P_i, Q_i. They represent amplitudes and phases of the two counterpropagating waves, on the plane $z = 0$. The boundary conditions that have been imposed at the ends of the resonator can be expressed as two linear homogeneous equations in P_i, Q_i. This system has non-trivial solutions if and only if its determinant vanishes. From this condition we obtain an equation that admits solutions only for some *discrete* values of the longitudinal propagation constant of the i-th mode $\ _i$, and then, as $\ _i$ is a function of frequency, only for suitable discrete values of ω, $\omega_{i,j} = 2\ f_{i,j}$ $(j = 1, 2, \ldots)$.

In conclusion, a lossless resonator, at a xed frequency, has always a *nite number* of resonant modes. This number is zero, when the frequency we are referring to is not one of those particular $f_{i,j}$'s that were de ned a few lines above, which are obviously called *resonant frequencies*.

The corresponding modes of an ideal resonator form a complete set, in the same sense as those discussed in Chapter 7. If we take a structure whose geometrical shape is identical to that of the ideal resonator, but whose boundary conditions are di erent from those on a perfect electric conductor, then we know that a suitable series of modes of the ideal resonator converges, in the linear mean sense, towards any eld that may exist in the non-ideal structure. This applies not only to resonators with nite conductivity walls, but also to resonators *coupled* to other structures, by means of holes, irises, or similar devices; they are referred to as *loaded* resonators. Actually, a completely closed resonant cavity, without any access, is an easy model to analyze; but it would be useless, if we had not noticed that, as a consequence of mode completeness, the eld in a loaded resonator may be expressed as a series of modes of the isolated resonator.

As we saw in Chapter 7, in a non-ideal waveguide, mode orthogonality collapses; therefore, the eld in general cannot be thought of as being a single mode in a rigorous sense. Modes under their cut-o frequencies are necessary to ful l the boundary conditions, where the real waveguide departs from the ideal one. In the same way, in a loaded resonator, a mode of the corresponding ideal resonator may be present at frequencies which di er from its resonant frequency $f_{n,m}$. Its amplitude coe cient, if all the other parameters are kept constant, decreases as $|f \quad f_{n,m}|$ increases, but in principle it may di er from zero at any frequency. This point will be clari ed, in more quantitative terms, in Section 9.8; but we thought it might be useful to notice now that it is strictly related to the notion of completeness.

9.4 Mode orthogonality in a perfect resonator

Similarly to what was carried out in Section 7.5, we will prove an orthogonality theorem for the modes of a perfectly lossless, completely isolated cavity. The last assumption entails that there are no imposed currents, neither inside the cavity, nor on its walls. The proof holds for two modes *with di erent resonant frequencies*. For two modes with the same resonant frequency (called *degenerate* modes), one can always orthogonalize them, with a procedure based on projections and subtractions (known as *Schmidt's orthogonalization*); this is essentially the same as we saw for transmission modes in a perfect waveguide, and therefore it will not be repeated here.

The proof of the orthogonality theorem is similar to that of Poynting's theorem (Section 3.2); the main di erence comes from the fact that we have to deal with two modes, having resonant frequencies ω_h, ω_k, respectively. Both satisfy homogeneous Maxwell's equations:

$$
\begin{aligned}
&(a) \qquad \nabla \quad \mathbf{E}_h = \quad j\omega_h \quad \mathbf{H}_h \quad , \\
&(b) \qquad \nabla \quad \mathbf{H}_h = j\omega_h \quad \mathbf{E}_h \quad , \\
&(c) \qquad \nabla \quad \mathbf{E}_k = \quad j\omega_k \quad \mathbf{H}_k \quad , \\
&(d) \qquad \nabla \quad \mathbf{H}_k = j\omega_k \quad \mathbf{E}_k \quad .
\end{aligned}
\tag{9.2}
$$

Dot-multiply both sides of equation (a) times $\mathbf{H}_k$, and the complex conjugate of equation (d) by $\mathbf{E}_h$. Then subtract side by side, and use the vector identity (C.4), which we rewrite here for convenience:

$$
\nabla \quad (\mathbf{E}_h \quad \mathbf{H}_k) = \mathbf{H}_k \quad \nabla \quad \mathbf{E}_h \quad \mathbf{E}_h \quad \nabla \quad \mathbf{H}_k \quad . \tag{9.3}
$$

The net result is a formula, similar to one which one nds on the way towards Poynting's theorem, before making a volume integration. We integrate it over the cavity volume V, and then apply Gauss' theorem (A.8) to the integral of the left-hand side of Eq. (9.3) to convert it into a surface integral. It is easy to show that this integral vanishes. In fact, the perfect conductor surrounding the cavity imposes $\mathbf{E}_{\tan} = 0$ at any point. Following again the same path as in Poynting's theorem, we notice that two more terms disappear: the one due to sources (there are none in the cavity, by assumption), and the one due to losses (again zero by assumption). So, we are left with the following relation:

$$\omega_h \int_V \mathbf{H}_h \quad \mathbf{H}_k^* \, dV = \omega_k \int_V \mathbf{E}_h \quad \mathbf{E}_k^* \, dV \quad . \tag{9.4}$$

If we repeat the same steps starting from Eqs. (9.2) (b) and (c), and take the complex conjugate for each term in this new relation, we get:

$$\omega_k \int_V \mathbf{H}_h \quad \mathbf{H}_k^* \, dV = \omega_h \int_V \mathbf{E}_h \quad \mathbf{E}_k^* \, dV \quad . \tag{9.5}$$

Now let us multiply Eq. (9.4) times ω_k, Eq. (9.5) times ω_h, and then subtract; we get the nal result, namely:

$$(\omega_h^2 \quad \omega_k^2) \int_V \mathbf{E}_h \quad \mathbf{E}_k^* \, dV = 0 \quad . \tag{9.6}$$

Multiplying Eq. (9.4) times ω_h and Eq. (9.5) times ω_k, we obtain a similar result for the magnetic elds. Thus if the two modes have di erent resonant frequencies, the following orthogonality relations hold:

$$\int_V \mathbf{E}_h \quad \mathbf{E}_k^* \, dV = 0 \qquad (\omega_h \neq \omega_k) \quad , \tag{9.7}$$

$$\int_V \mathbf{H}_h \quad \mathbf{H}_k^* \, dV = 0 \qquad (\omega_h \neq \omega_k) \quad . \tag{9.8}$$

As we saw in Chapter 7, orthogonality is a strong form of linear independence, with links with the property of completeness. It is also of great help in practical problems, like expanding a generic eld into a series of modes. Just to quote one example of application, it may be useful in calculating the elds excited by a source, either in the cavity, or localized on one of its walls. For more details, the reader should see some of the problems.

9.5 Lossless cylindrical cavities

Let us start, like in Section 7.2, from a cylindrical structure. We call z its axis, and $\hat{a}_z$ its unit vector. Inside the structure there is a homogeneous lossless medium; the lateral surface, $\mathcal{C}$, is a perfect electric conductor. Let us then cut a nite length of this structure, and close it by means of two perfectly conducting planes, orthogonal to the z axis, at $z = 0$ and $z = d$, respectively (Figure 9.2).

We saw in Section 7.2, and in the following ones, that each transmission mode of the in nitely long cylindrical structure can be obtained from a scalar pre-potential (either L or T, depending on whether the mode is TE or TM). It is of the form $\mathcal{F}(u_1, u_2)\ \exp(\quad z)$. But now, we introduced two discontinuities, orthogonal to the z axis. They cause re ections, which couple forward and backward propagating waves. A reasonable ansatz is then to assume that the pre-potentials are of the following type:

$$F = \mathcal{F}(u_1, u_2)\ f(z) \qquad (\mathcal{F} = \mathcal{L}, \mathcal{T}\ ;\ f = \ell, t) \quad . \tag{9.9}$$

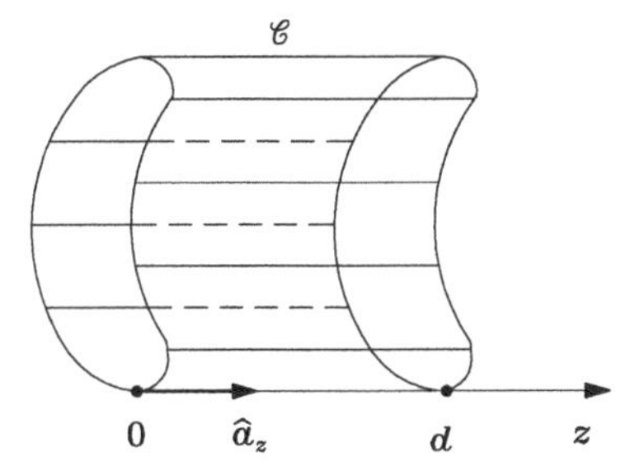

Figure 9.2 *A cylindrical cavity obtained from a waveguide of unspeci ed geometry.*

Here, $\mathcal{L}(u_1, u_2)$ and $\mathcal{T}(u_1, u_2)$, which depend on the transverse coordinates, must satisfy the boundary conditions on $\mathcal{C}$, Eqs. (7.17) and (7.18). Consequently, they remain the same as for TE and TM modes of the in nitely long cylindrical structure. The only quantities which are still unknown are $\ell(z)$ and $t(z)$.

The conducting planes at $z = 0$ and $z = d$ impose the following conditions, to both TE and TM modes:

$$\mathbf{E}_t = 0 \qquad \text{for } z = 0\ ,\ z = d \quad . \tag{9.10}$$

Incidentally, notice that Eq. (9.10) is not the same as $\mathbf{H}_t = 0$: the total eld in the cavity is the sum of forward and backward waves, whose wave impedances have opposite signs. The reader should compare this with what we saw in Chapter 8, and also in Chapter 4 on standing plane waves.

Let us take now Eqs. (3.47) and (3.48), which apply to any TE or TM eld in a homogeneous medium, and introduce Eq. (9.9) into them. We get the following expressions for the eld transverse components:

$$\mathbf{E}_{\mathrm{TE}t} = (\nabla_t \mathcal{L} \quad \hat{a}_z)\, \ell(z) \quad , \qquad \mathbf{E}_{\mathrm{TM}t} = \frac{j}{\omega \ \ c} \nabla_t \mathcal{T} \, \frac{dt(z)}{dz} \quad . \tag{9.11}$$

This shows that an alternative way to express Eq. (9.10) is:

$$\ell(0) = \ell(d) = 0 \qquad \text{for TE modes ;} \tag{9.12}$$

$$\left.\frac{dt}{dz}\right|_{z=0} = \left.\frac{dt}{dz}\right|_{z=d} = 0 \qquad \text{for TM modes .} \tag{9.13}$$

The TE/TM terminology refers, as usual, to the z *axis.* But it is worth noticing that, within a given resonator, modes can be classi ed in di erent ways, depending on which direction the term "transverse" refers to. In particular, this becomes evident in those cavities where the cross-section remains constant along two or three axes.

We know from Section 7.2 that the functions $\mathcal{F}$ satisfy the modi ed Helmholtz equation (7.8), and from Chapter 3 that the functions F satisfy the complete

Helmholtz equation. Consequently, the functions $f(z)$ must satisfy:

$$\frac{d^2 f}{dz^2} = {}^2 f \qquad ({}^2 = {}^2 \quad \omega^2 \quad) \quad , \tag{9.14}$$

where ^{2} is real and positive (see Section 7.4).

Solutions to Eq. (9.14) are compatible with Eqs. (9.12) and (9.13) *only* for ${}^2 < 0$. In this case, letting ${}^2 = \quad k^2$, the solutions we are looking for are:

$$\ell(z) = A \sin kz \quad , \qquad k = p \frac{}{d} \quad (p = 1, 2, \ldots) \quad , \tag{9.15}$$

$$t(z) = B \cos kz \quad , \qquad k = p \frac{}{d} \quad (p = 1, 2, \ldots) \quad , \tag{9.16}$$

where A, B are arbitrary constants.

From the pre-potentials L and T that have been determined in this way, we may now calculate the elds of the cavity *resonance modes*, using the general-purpose formulas in Eqs. (3.47) and (3.48). But the reader can prove, as an exercise, that they can be found in a faster way by changing slightly Eqs. (7.8) and (7.9). These changes consist simply of replacing the exponential functions in Eqs. (7.1) with sine and cosine functions, Eqs. (9.15) and (9.16).

Let us keep the same symbols as in Chapter 7 for the z-independent factors in the waveguide modes; in particular, let $\mathcal{E}_{\mathrm{TE}}, \mathcal{E}_{\mathrm{TM}t}, \mathcal{H}_{\mathrm{TE}t}$ and $\mathcal{H}_{\mathrm{TM}}$ have the same meanings as in Eqs. (7.11) and (7.12). Then, nally we derive the following expressions, where the subscript t stands, as usual, for transverse components with respect to the z axis:

$\mathrm{TE}_{m,n,p}$ mode :

$$E_z \quad 0 \quad , \quad H_z = \mathcal{H}_z \sin p \frac{z}{d} \quad ,$$

$$\mathbf{E}_t = \mathcal{E}_{\mathrm{TE}} \sin p \frac{z}{d} \quad , \quad \mathbf{H}_t = \mathcal{H}_{\mathrm{TE}t} \cos p \frac{z}{d} \quad . \tag{9.17}$$

$\mathrm{TM}_{m,n,p}$ mode :

$$E_z = \mathcal{E}_z \cos p \frac{z}{d} \quad , \quad H_z \quad 0 \quad ,$$

$$\mathbf{E}_t = \mathcal{E}_{\mathrm{TM}t} \sin p \frac{z}{d} \quad , \quad \mathbf{H}_t = \mathcal{H}_{\mathrm{TM}} \cos p \frac{z}{d} \quad . \tag{9.18}$$

Since ^{2} and ${}^2_{m,n}$ are linked by Eq. (9.14), it must be:

$$\omega^2 \quad = {}^2_{m,n} + \left(\frac{p}{d}\right)^2 \quad . \tag{9.19}$$

The results of this section deserve further comments. First of all, one should notice that, to build a resonance mode in a cavity obtained from a waveguide,

The case $p = 0$, in Eq. (9.16), which entails $\mathbf{E}_{\mathrm{TM}t} \quad 0$, is compatible with Eq. (9.13), but it is acceptable *only* for ${}^2 \neq 0$, i.e., if $\mathcal{T}$ is not constant vs. u_1 and/or vs. u_2. If this is not the case, then $p = 0$ must be disregarded, because (like for $p = 0$ in Eq. (9.15)) it would correspond to a eld identical to zero.

a waveguide mode *above its cut-o frequency* is required. Indeed, no non-trivial solution to Eq. (9.14) is compatible with Eqs. (9.12) and (9.13) if we suppose $^2 < 0$ in Eq. (9.14): a resonant mode can never be a superposition of two evanescent waves. As anticipated in Section 9.1, it is a superposition of two *traveling* waves, one forward, the other one backward, and it is straightforward to detect that indeed there are two expressions of the type $\exp(\; jkz)$ $\exp(jkz)$ in Eqs. (9.15) and (9.16).

Resonance occurs when the round-trip length, for the traveling wave that undergoes re ections at both ends of the cavity, is an integer multiple of the wavelength. Notice that, as one can easily infer from Chapter 8, this result depends on the phase shift at the re ections, at both ends, being equal to radians. Should the cavity be terminated by "mirrors" with a di erent phase shift, this should be accounted for. However, if we recall from Chapter 7 that, for the (m, n)-th waveguide mode, $_{m,n} = 2 \; / \; _g$, where $_g$ is the guided wavelength, then the condition $k = p \; /d$ can be rewritten as:

$$2d = p \; _g \quad , \tag{9.20}$$

where p is an integer. This recon rms the previous statement.

Any solution like Eq. (9.17) or (9.18) obeys Eq. (9.19), where $^2_{m,n}$ is determined by the shape and the size of the waveguide. This recon rms that, for a given cavity (i.e., once the transverse eigenvalues are given, and the length d is also given), Eq. (9.9) is satis ed only at *discrete frequencies*, $f_{m,n,p}$. If $f \neq f_{m,n,p}$, the eld is identically zero. This result applies only to a strictly lossless cavity, perfectly isolated from the rest of the universe. What happens in a real cavity has been described in Section 9.3.

9.6 Simple examples

9.6.1 Rectangular resonators

Given a rectangular box, the z axis (with the meaning given to it in the previous sections) can be taken parallel to any edge. Di erent choices for the z axis give rise to di erent nomenclatures for the resonator modes, but such changes do not correspond to any physical di erence.

TE modes. The expressions for $\mathcal{E}_t, \mathcal{H}_t$ and $\mathcal{H}_z$, to be inserted into Eqs. (9.17), come from Eq. (7.50), through Eqs. (7.4) and (7.5). For the $\mathrm{TE}_{m,n,p}$ mode, we get:

$$\begin{aligned}
E_z & \quad 0 \quad , \\
H_z & = H \cos \frac{m \; x}{a} \cos \frac{n \; y}{b} \sin \frac{p \; z}{d} \quad , \\
E_x & = \frac{j\omega}{k_c^2} \; H \frac{n}{b} \cos \frac{m \; x}{a} \sin \frac{n \; y}{b} \sin \frac{p \; z}{d} \quad ,
\end{aligned}$$

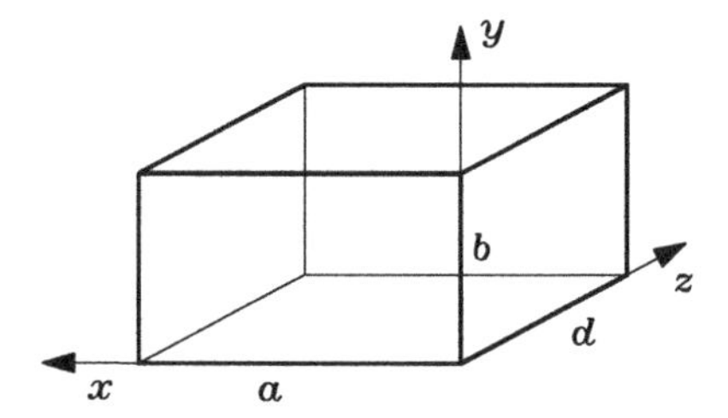

Figure 9.3 *Rectangular resonator.*

$$
\begin{aligned}
E_y &= \frac{j\omega}{k_c^2}\, H\, \frac{m\ }{a} \sin \frac{m\ x}{a} \cos \frac{n\ y}{b} \sin \frac{p\ z}{d} \quad , \\
H_x &= \frac{H}{k_c^2} \frac{p\ }{d} \frac{m\ }{a} \sin \frac{m\ x}{a} \cos \frac{n\ y}{b} \cos \frac{p\ z}{d} \quad , \\
H_y &= \frac{H}{k_c^2} \frac{p\ }{d} \frac{n\ }{b} \cos \frac{m\ x}{a} \sin \frac{n\ y}{b} \cos \frac{p\ z}{d} \quad , \qquad (9.21)
\end{aligned}
$$

where a, b and d are the lengths of the three sides of the cavity (Figure 9.3); H is an arbitrary constant; $k_c^2 = (m\ /a)^2 + (n\ /b)^2$ is a symbol taken from Chapter 7.

TM modes. The expressions for $\mathcal{E}_t, \mathcal{H}_t$ and $\mathcal{E}_z$, to be inserted into Eqs. (9.18), come from Eq. (7.54), through Eqs. (7.4) and (7.5). For the $\mathrm{TM}_{m,n,p}$ mode, we get:

$$
\begin{aligned}
E_z &= E \sin \frac{m\ x}{a} \sin \frac{n\ y}{b} \cos \frac{p\ z}{d} \quad , \\
H_z &\quad\ 0 \quad , \\
E_x &= \frac{E}{k_c^2} \frac{p\ }{d} \frac{m\ }{a} \cos \frac{m\ x}{a} \sin \frac{n\ y}{b} \sin \frac{p\ z}{d} \quad , \\
E_y &= \frac{E}{k_c^2} \frac{p\ }{d} \frac{n\ }{b} \sin \frac{m\ x}{a} \cos \frac{n\ y}{b} \sin \frac{p\ z}{d} \quad , \\
H_x &= \frac{j\omega}{k_c^2}\, E\, \frac{n\ }{b} \sin \frac{m\ x}{a} \cos \frac{n\ y}{b} \cos \frac{p\ z}{d} \quad , \\
H_y &= \frac{j\omega}{k_c^2}\, E\, \frac{m\ }{a} \cos \frac{m\ x}{a} \sin \frac{n\ y}{b} \cos \frac{p\ z}{d} \quad , \qquad (9.22)
\end{aligned}
$$

where the notation is the same as above; E is an arbitrary constant.

The reader can nd, as an exercise, which, among the TE and TM modes, can have one or two integer subscripts equal to zero. No mode can have all three subscripts equal to zero.

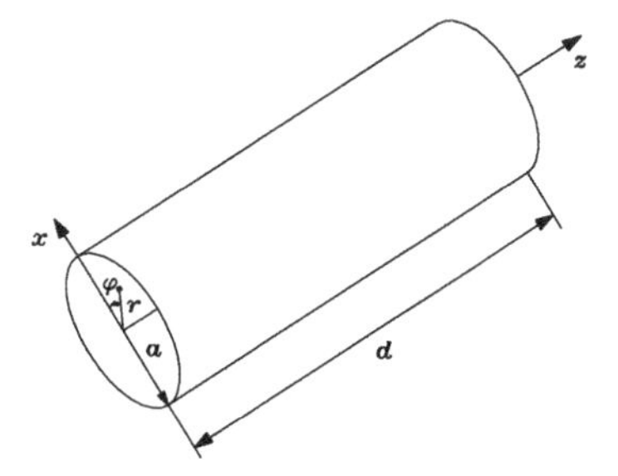

Figure 9.4 *Cylindrical resonator of circular cross-section.*

9.6.2 *Cavities of circular cross-section*

For a cavity shaped like a circular cylinder, of radius a and length d (Figure 9.4), we use of course cylindrical coordinates r, φ, z, the z axis being the cylinder axis. In such a case there is no ambiguity in the mode nomenclature; but this has no physical meaning. Indeed, one might think that the resonator shown in Figure 9.4 is a piece of radial transmission line, short-circuited at $r = a$, or a structure where waves go in loops in the azimuth direction, closing their path at $\varphi = 2$.

TE modes. The expressions for $\mathcal{E}_t, \mathcal{H}_t$ and $\mathcal{H}_z$, to be inserted into Eqs. (9.17), come from Eq. (7.66), through Eqs. (7.4) and (7.5). For the $\mathrm{TE}_{n,m,p}$ mode, we get:

$$\begin{aligned}
E_z & \quad 0 \ , \\
H_z &= H\, J_n\left(p'_{n,m}\frac{r}{a}\right)\begin{Bmatrix}\cos n\varphi \\ \sin n\varphi\end{Bmatrix}\sin\frac{p\ z}{d} \ , \\
E_r &= \frac{j\omega}{p'^2_{n,m}}\, H\,\frac{a^2}{r}\, J_n\left(p'_{n,m}\frac{r}{a}\right) n \begin{Bmatrix}\sin n\varphi \\ \cos n\varphi\end{Bmatrix}\sin\frac{p\ z}{d} \ , \\
E_\varphi & \quad j\omega\ \ H\,\frac{a}{p'_{n,m}}\, J'_n\left(p'_{n,m}\frac{r}{a}\right)\begin{Bmatrix}\cos n\varphi \\ \sin n\varphi\end{Bmatrix}\sin\frac{p\ z}{d} \ , \\
H_r &= \frac{H}{p'^2_{n,m}}\, a\,\frac{p}{d}\, J'_n\left(p'_{n,m}\frac{r}{a}\right)\begin{Bmatrix}\cos n\varphi \\ \sin n\varphi\end{Bmatrix}\cos\frac{p\ z}{d} \ , \\
H_\varphi &= \frac{H}{p'^2_{n,m}}\,\frac{a^2}{r}\,\frac{p}{d}\, J_n\left(p'_{n,m}\frac{r}{a}\right) n \begin{Bmatrix}\sin n\varphi \\ \cos n\varphi\end{Bmatrix}\cos\frac{p\ z}{d} \ ,
\end{aligned} \tag{9.23}$$

where J'_n is the rst derivative of the Bessel function of the rst kind, $J_n(u)$, *with respect to its argument* u (see Appendix D); H is an arbitrary constant; $p'_{n,m}$, the m-th zero of $J'_n(u)$, is linked by Eq. (7.65) to the eigenvalue of the $\mathrm{TE}_{n,m}$ mode in the circular waveguide the cavity was derived from. The functions in brackets stand for the *two possible polarizations* of each mode with $n \neq 0$. Upper or lower rows must be taken together, to get the correct eld components.

TM modes. The expressions for $\mathcal{E}_t, \mathcal{H}_t$ and $\mathcal{E}_z$, to be inserted into Eqs. (9.18), come from Eq. (7.69), through Eqs. (7.4) and (7.5). For the $\mathrm{TE}_{n,m,p}$ mode, we get:

$$
\begin{aligned}
E_z &= E\, J_n\left(p_{n,m}\,\frac{r}{a}\right)\begin{Bmatrix}\cos n\varphi\\ \sin n\varphi\end{Bmatrix}\cos\frac{p\ z}{d} \quad ,\\
H_z &\quad 0 \quad ,\\
E_r &= \frac{E}{p_{n,m}}\, a\,\frac{p}{d}\, J_n'\left(p_{n,m}\,\frac{r}{a}\right)\begin{Bmatrix}\cos n\varphi\\ \sin n\varphi\end{Bmatrix}\sin\frac{p\ z}{d} \quad ,\\
E_\varphi &= \frac{E}{p_{n,m}^2}\,\frac{a^2 p}{rd}\, J_n\left(p_{n,m}\,\frac{r}{a}\right) n\begin{Bmatrix}\sin n\varphi\\ \cos n\varphi\end{Bmatrix}\sin\frac{p\ z}{d} \quad ,\\
H_r &= \frac{j\omega\ E}{p_{n,m}^2}\,\frac{a^2}{r}\, J_n\left(p_{n,m}\,\frac{r}{a}\right) n\begin{Bmatrix}\sin n\varphi\\ \cos n\varphi\end{Bmatrix}\cos\frac{p\ z}{d} \quad ,\\
H_\varphi &= \frac{j\omega\ E}{p_{n,m}}\, a\, J_n'\left(p_{n,m}\,\frac{r}{a}\right)\begin{Bmatrix}\cos n\varphi\\ \sin n\varphi\end{Bmatrix}\cos\frac{p\ z}{d} \quad ,
\end{aligned}
\tag{9.24}
$$

where the symbols are the same as for Eqs. (9.23), except for the arbitrary constant E and for $p_{n,m}$ (m-th zero of $J_n(u)$), linked to the eigenvalue of the $\mathrm{TM}_{n,m}$ mode of the circular waveguide by Eq. (7.68).

Some examples of eld lines, for a few, simple modes of rectangular and circular resonators, are shown in Figure 9.5.

9.6.3 Spherical cavities

In Chapter 11 we will solve the Helmholtz equation using separation of variables in spherical coordinates (r, ϑ, φ). The main reason why this subject is postponed, is because its main applications deal with eld sources (like antennas). After studying Chapter 11, the reader should come back to this section, and impose on each *spherical harmonic* the boundary condition $\mathbf{E}_{\tan} = 0$ on a spherical surface of radius $r = a$. In this way, the modes of a spherical resonator will be found.

Another reasonably simple exercise is to impose the same condition also on coordinate planes $\varphi =$ constant, or on coordinate cones $\vartheta =$ constant. What one obtains in this way are the modes of more complicated structures, which are of some interest in problems related to feeders for aperture antennas (see Chapter 12 and 13), or to tapers between waveguides of di erent cross-section.

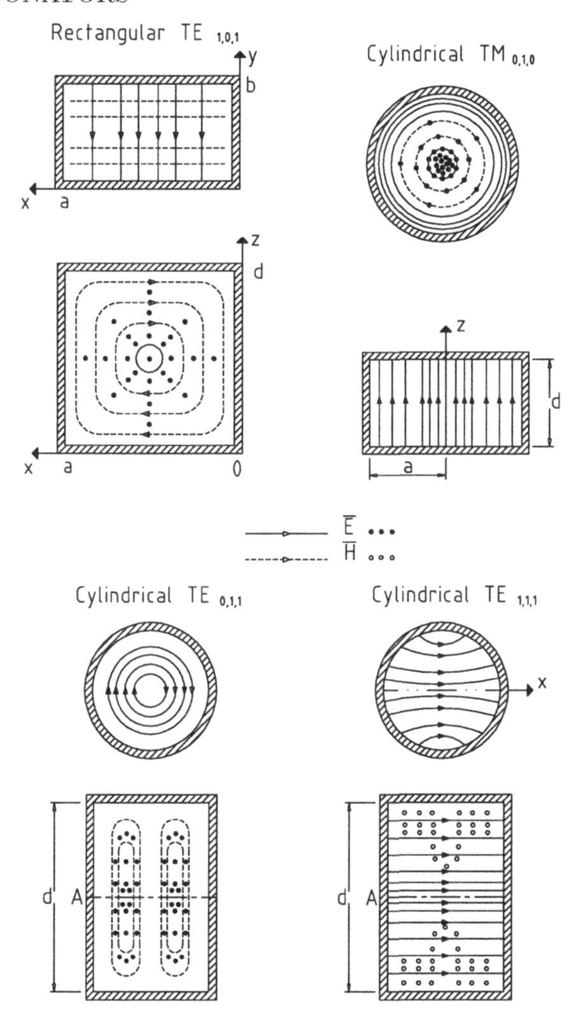

Figure 9.5 *Field lines of a set of signi cant modes, in rectangular and circular resonators.*

9.7 Lossy resonators: Perturbation analysis. Intrinsic Q-factor

From the idealized model of perfectly conducting walls, to a more realistic one where the walls are good conductors (ω , but nite), the step is very similar to that of Chapter 7 about waveguides. On the grounds of what we saw in Section 4.12, the surface which surrounds the resonator can be characterized in terms of its *wall impedance*:

$$Z_0 = (1 + j) \sqrt{\frac{f}{}} \qquad [\ ] \quad , \tag{9.25}$$

which was originally derived for a at *surface*. As we saw in Chapter 7, one can prove that this result is correct also for *cylindrical or spherical walls*. On the other hand, the wall impedance becomes much more complicated (involving metric coe cients of the coordinate system) if the curvature is not constant along the wall itself.

On these premises, the *rst order perturbation theory* is very similar to that of Chapter 7. We start o supposing that a mode of the ideal cavity can exist in the cavity with nite wall impedance, although it cannot be rigorously so. We calculate the ohmic losses under this assumption. As in Chapter 7, special caution is needed in the case of *degenerate modes*: among all the linear combinations of two or more degenerate modes, which are still modes of the ideal cavity, one must identify those that remain orthogonal in the presence of losses.

From lumped-element circuit theory we know that power dissipation in a resonant circuit, W_J, is of little signi cance, unless correlated with the amount of energy which is stored in the circuit, E. The link between these two quantities is, also for a resonant cavity, the adimensional number:

$$Q_0 = \frac{\omega_0 E}{W_J} \quad , \tag{9.26}$$

which is referred to as the *quality factor*, or *factor of merit*, or *gure of merit* (or, for short, the Q-factor) of the resonator mode characterized by the resonance angular frequency ω_0. Let us recall (from Chapter 3) that, for a resonant eld, the time-averaged energies stored in electric and in magnetic form are equal and the total energy, constant in time, is equal to twice those averages.

If one continues along the same examples as in Section 9.6, one sees that, as a general trend, the Q-factor in a given cavity tends to increase with the order of the resonance mode. This can be intuitively justi ed as follows. Consider rst a cavity with lossless walls. In it, the electric eld components of a given mode have nodes, i.e., surfaces on which the tangential components of $\mathbf{E}$ are identically zero. Some of these nodes are located on the resonator walls, but some may be inside. The number of internal nodes grows as the mode order grows. All the nodal surfaces can be metallized with perfect electrical conductors, without a ecting the eld; this operation is equivalent to subdividing the resonator into several cavities, each of which resonates in a lower order mode, compared to that of the original cavity. When we calculate the Q-factor, only the real (external) walls, being non-ideal, contribute to loss. So, the power dissipated in the "large" resonator is less than the sum of the powers that would be dissipated in all the "small" resonators if they would be really separated by the same imperfect conductor the real walls are made of. On the other hand, the total stored energy obviously equals the sum of the stored energies. Consequently, the Q-factor is larger in the large resonator, which is equivalent to saying that (at least within a given "family" of modes) it grows as the order of the mode grows.

From circuit theory we know that in a lumped-element resonator the Q-

factor determines the full width at half maximum of the so-called resonance line, i.e., of the *Lorentzian* function

$$f(\omega) = \frac{1}{1 + jQ\left(\frac{\omega}{\omega_0} \quad \frac{\omega_0}{\omega}\right)} \quad . \tag{9.27}$$

In other words, the Q-factor of a resonant cavity mode is a measure how di cult it is to inject into the cavity a signal whose frequency is di erent from the resonant one.

9.8 Resonators coupled to external loads. Loaded Q-factor

A completely isolated resonant cavity does not bear any practical interest. Any use of a resonator requires coupling it to other structures, that allow to inject energy into it.

Some typical coupling tools at microwave frequencies are:

- connections either with waveguides or with other cavities through holes called *irises* or *diaphragms*;
- connections with transmission lines, or with lumped-element circuits, by means of conductive wires or probes, either straight or loop-shaped;
- connections with other elements (transmission lines or other cavities) by means of electron beams which pass through the resonator.

To evaluate the external coupling of a lossless resonator, suppose that the cavity is at steady state: it is fed by a source which compensates exactly for the power that the cavity loses to the outside world. We de ne as the *external* quality factor of the resonator (or better, of its mode characterized by the resonance frequency ω_0) the adimensional quantity:

$$Q_{\text{ex}} = \frac{\omega_0 E}{W_{\text{ex}}} \quad , \tag{9.28}$$

where E, as in the previous section, is the energy stored in the resonator at steady state, while W_{ex} is the active power leaving the resonator to the outside world.

If we consider a *passive resonator*, as we have always done so far, then the quantity in Eq. (9.28), if inserted into Eq. (9.27), determines, as in the previous section, the spectral width of the Lorentzian function which characterizes the mode under test. It is useful to add that for a resonator *containing sources*, the Q-factor de ned by Eq. (9.28) determines, again through Eq. (9.27), a linewidth which is not that which one measures when the source is running in the large-signal regime. To stress this point, the quantity in Eq. (9.28) is often referred to as the *cold cavity* Q-factor.

In a practical resonator, losses and coupling with the outside world must be accounted for at the same time. In all cases of interest, Q_0 and Q_{ex} are both much greater than one, meaning that the energy stored in the cavity is much larger than the energy lost in the resonator over one period $T = 2 \ /\omega_0$, as well as than the energy delivered to the external load during the same time interval. One consequence is that internal losses and energy given to the external load may be thought of as additive, although in general this is not rigorous. Consequently, it makes sense to define a *loaded quality factor* as:

$$Q_L = \frac{\omega_0 E}{W_J + W_{\text{ex}}} \quad . \tag{9.29}$$

Its link with the other quantities defined in this section and in the previous one is:

$$\frac{1}{Q_L} = \frac{1}{Q_0} + \frac{1}{Q_{\text{ex}}} \quad , \tag{9.30}$$

the loaded Q-factor Eq. (9.39) determines, as usual through Eq. (9.27), the spectral width of the frequency response of a cavity.

9.9 Open resonators

The examples shown in Section 9.6 reconfirm that resonance in a cavity means satisfying a relationship between the geometrical dimensions of the cavity itself and the wavelength at the resonant frequency. When the geometrical shape of the cavity is simple, this relation is simple also. If the shape of the cavity is complicated, this relation becomes complicated but the concept remains the same.

At optical frequencies (of the order of 10^{14}–10^{15} Hz), to make geometrical dimensions comparable with wavelengths becomes a technological challenge. It is necessary to think of resonators which operate in high-order modes, in contrast to what is normally the case for microwave cavities. If we take a cavity of a given shape and given size, and increase the frequency, then the growth in the resonance mode order is accompanied by an increase in the *relative density* of resonances, on the frequency axis (defined simply as the number of modes, n, whose resonance frequencies fall within a unit length on the f axis, $f = 1\,\text{Hz}$, divided by the frequency at which this interval is centered, f_0). Note that the relative density is much more relevant than the absolute one, n, because n/f_0 is the quantity to compare with the cavity Q-factors in order to establish whether in practice it is possible to operate a cavity in a single mode. Single-mode operation requires at the same time a low relative density of modes and a high Q.

In a cavity with metallic walls, of the same type as those we dealt with so far, it is impossible to meet these requirements at optical frequencies. This is because at these frequencies there is no material which satisfies ω . To by-pass the problem, one may proceed as follows:

- replace the metallic walls with *dielectric mirrors*, which are dielectric multilayers (see Section 4.10) designed to yield high reflection coefficient at the wavelength of interest;
- decrease the relative density of modes by eliminating some of the walls, making an *open resonator*, i.e., a 3-D region confined between two highly

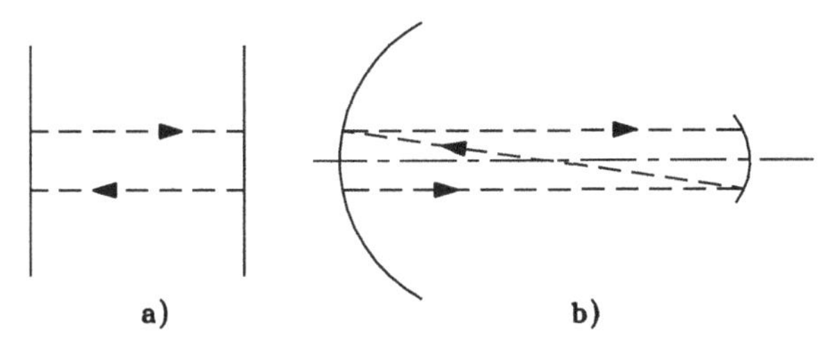

Figure 9.6 *Schematic examples of open resonators.*

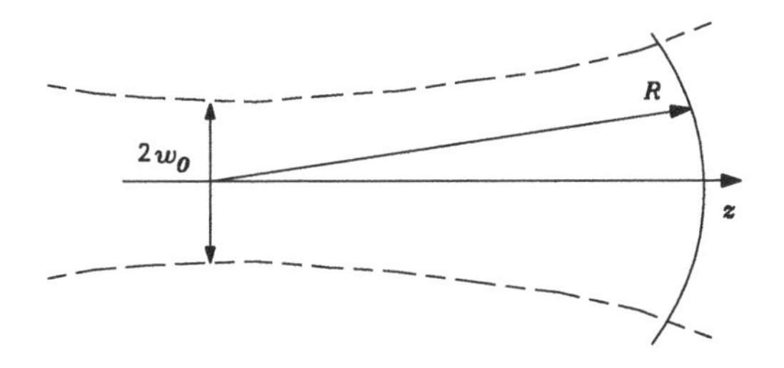

Figure 9.7 *A Gaussian beam in an open resonator.*

re ecting surfaces along one direction, z, but extending *to in nity* in the planes orthogonal to z.

Some examples of open resonators, with plane mirrors (Fabry-Perot cavities, see Section 4.10) and with spherical mirrors, are sketched in Figure 9.6, parts a) and b), respectively. Some rays are also sketched in the same gure, to illustrate how one can justify in intuitive terms high-Q-factor modes in an open resonator. Indeed, for suitable geometries one nds rays that are trapped in the cavity and undergo an in nite number of re ections on the two mirrors in spite of the fact that these have nite transverse dimensions.

As we shall see in the next section, some quantitative information may be grasped from the intuitive ray optics model. However, a much larger amount of information can be extracted from *Gaussian beams.* In fact, we saw in Chapter 5 that, far from the beam waist ($|z| \quad w_0^2/$, see Figure 9.7), the wavefronts of a Gaussian beam are essentially spherical, and their radius of curvature, as given by Eq. (5.96), is:

$$R(z) = z\left[1 + \left(\frac{w_0^2}{z}\right)^2\right] \simeq z \quad . \tag{9.31}$$

If a beam impinges on a spherical mirror whose radius of curvature equals that of the wavefront, re ection gives rise to another Gaussian beam whose parameters are identical to those of the incident beam, and travels in the opposite direction. Intuition indicates that suitable combinations of two mirrors can make the beam go up and down inde nitely in a stable way. Predictions based on this model can be made quantitative, for mirrors having any radius of curvature, by means of the "ABCD law" of Section 5.11. All the main parameters of a mode of an open resonator can be calculated in terms of Gaussian beams: resonance frequency, total stored energy and its spatial distribution, and also, with a few further notions that will be given in Section 9.11, the Q-factor.

In particular, the *resonance frequency* can be calculated ensuring that the total phase shift along a closed path, taking into account phase shifts due to re ections on the two mirrors, has to be an integer multiple of 2 . Making use of the results of Section 5.7, this requirement may be written as:

$$\arg \ _1 + \arg \ _2 + 2kL + 2[\ (z_2) + \ (z_2) \quad (z_1) \quad (z_1)] = 2n \quad , \qquad (9.32)$$

where z_1 and z_2 are the coordinates of the centers of the two mirrors; $L =$ $(z_1 \quad z_2)$ is the cavity length; n is an integer; and have the same meanings as in Sections 5.9 and 5.10; $_1$ and $_2$ are the re ection coe cients of the two mirrors, whose arguments are normally independent of the mode one deals with. The fourth term on the left-hand side of Eq. (9.32) is usually much smaller than the third, so it is neglected in rst approximations; however, it is very important in principle, as it contains the term (z) which, as we saw in Section 5.10, depends on the *transverse order* of the Gaussian beam, labeled in Section 5.10 as (m_1, m_2) for the Hermite-Gauss modes and as (ℓ, p) for the Laguerre-Gauss modes. It is just because of this term that the resonance frequency of the $\mathrm{TEM}_{m_1,m_2,n}$ or $\mathrm{TEM}_{\ell,p,n}$ depends slightly on the transverse order, for a given longitudinal order. This a ects signi cantly the emission spectrum of laser sources, whenever they may oscillate on more than one "transverse mode". Let us stress that such a result would be impossible to obtain using ray optics.

9.10 Stability of open resonators

In the previous section we rst gave the motivations for using open resonators at optical frequencies, and then we saw how they can be studied in terms of Gaussian beams. These beams, as we saw in Chapter 5, have wavefronts that in principle extend to in nity in all directions orthogonal to the z axis. On the contrary, mirrors of open resonators are obviously of nite size. This contrast entails two di erent types of consequences, that will be the subjects of this section and of the next one.

Normally, mirrors are far away from focal planes. In other words, they are located in regions where evanescent waves do not contribute in a signi cant way to the Gaussian beam. Therefore, as a rst approximation we may think in terms of *ray optics*, in order to nd under what conditions a ray remains trapped in the cavity when it undergoes multiple re ections on the two mirrors. Let us consider then what is sketched in Figure 9.8. The cavity consists of two spherical mirrors, with focal lengths $R_1/2$ and $R_2/2$, located at a distance L from each other. Let us restrict ourselves to *meridional rays*, i.e., rays belonging to planes passing through the cavity axis z, like those shown

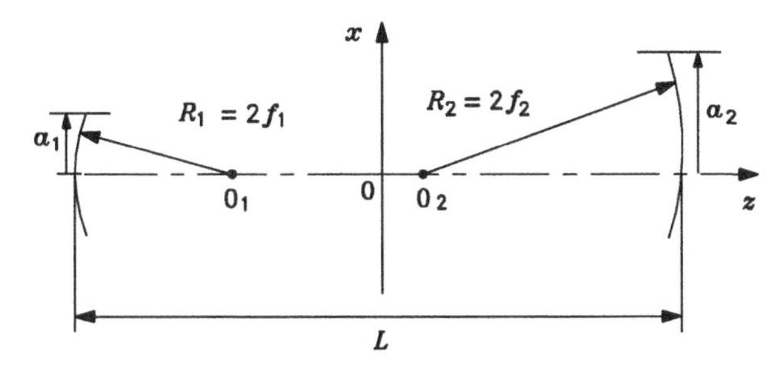

Figure 9.8 *De nitions of the geometrical parameters of an open resonator.*

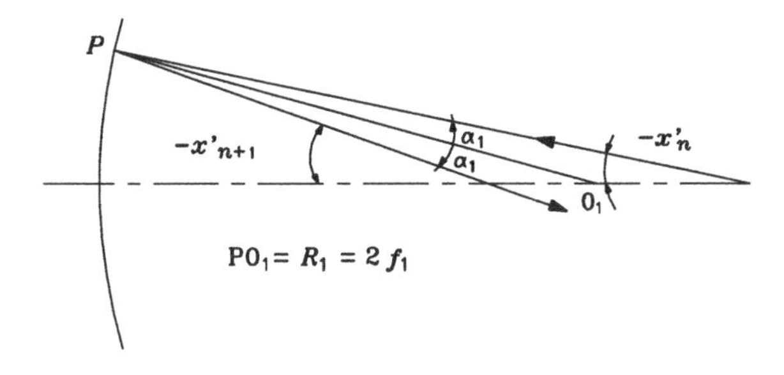

Figure 9.9 *Ray re ection at a mirror at one end of an open resonator.*

in Figure 9.8 which belong to the (x, z) plane. Just to ensure that there is no ambiguity, suppose that the n-th re ection of a given ray takes place on mirror 2, the $(n+1)$-th on mirror 1, and so on. Let x_n be the value of the transverse coordinate x at the site of the n-th re ection, and $x'_i = (dx/dz)_i$ $(i = \ldots, n \quad 1, n, n+1, \ldots)$ be the slope of the ray *after* the i-th re ection, with respect to the positive direction of the z axis.† The law of re ection, if applied on mirror 1 (Figure 9.9), gives the following equation (where the angles are small enough to be approximated by their tangents):

$$x'_{n+1} = \quad x'_n \quad \frac{x_{n+1}}{f_1} \quad . \tag{9.33}$$

Furthermore, as the slope of a ray remains constant between two consecutive re ections, it is straightforward to write that:

$$x_{n+2} \quad x_{n+1} = L\,x'_{n+1} \quad , \qquad x_{n+1} \quad x_n = \quad L\,x'_n \quad . \tag{9.34}$$

Replacing Eq. (9.33) into the rst of Eqs. (9.34) and eliminating x'_n by means of the second of Eqs. (9.34), rearranging the terms, we obtain the

† Note that the signs of x'_i depend on the sense in which the ray propagates: $x'_{n+1} > 0$ means that, traveling from mirror 1 towards mirror 2, x is increasing. On the contrary, $x'_n > 0$ means that, traveling from mirror 2 towards mirror 1, x is decreasing.

following equation, without derivatives:

$$x_{n+2} + x_n = x_{n+1}\left(2 - \frac{L}{f_1}\right) \quad . \tag{9.35}$$

It is self-explanatory why the same equation holds for the previous trips, those in one direction:

$$x_n + x_{n-2} = x_{n-1}\left(2 - \frac{L}{f_1}\right) \quad , \tag{9.36}$$

as well as those in the other direction:

$$x_{n+1} + x_{n-1} = x_n\left(2 - \frac{L}{f_2}\right) \quad . \tag{9.37}$$

Adding Eqs. (9.35) and (9.36) term by term, and eliminating from the right-hand side the term $x_{n+1}+x_{n-1}$ by means of Eq. (9.37), we obtain the following nite di erence equation:

$$x_{n+2} - 2b\,x_n + x_{n-2} = 0 \quad , \tag{9.38}$$

where we set:

$$b = 1 - \frac{L}{f_1} - \frac{L}{f_2} + \frac{L^2}{2f_1 f_2} = 2\left(1 - \frac{L}{2f_1}\right)\left(1 - \frac{L}{2f_2}\right) - 1 \quad . \tag{9.39}$$

The rst thing of interest is to know whether Eq. (9.38) has *stable solutions*, that is whether there are ray trajectories for which the value of the transverse coordinate x does not diverge as the number of re ections increases inde -nitely. For this purpose we let:

$$x_n = x_0\, e^{jn\vartheta} \quad , \tag{9.40}$$

where the unknown ϑ has to be thought of, for the sake of generality, as being complex. Substituting Eq. (9.40) into Eq. (9.38), we nd:

$$e^{j2\vartheta} + e^{-j2\vartheta} = 2b \quad , \tag{9.41}$$

whose solution is:

$$2\vartheta = \arccos b \quad . \tag{9.42}$$

Hence, ϑ is real if $|b| < 1$. In this case, Eq. (9.40) is a periodic solution of Eq. (9.38). As long as a mode of the cavity can be reasonably approximated by plane waves obeying the ray-optics laws, the previous result indicates that these waves remain con ned inside the cavity for $|b| < 1$.

For $|b| > 1$, on the contrary, ϑ becomes imaginary, and so the transverse coordinate Eq. (9.40) diverges exponentially as the number of round trips through the cavity increases. This indicates that, after a nite number of re ections, the generic ray will miss the mirrors and leave the cavity.

The rst case is referred to as a *stable resonator*, the second one as an *unstable resonator*.

It is useful to plot the regions where $|b| > 1$ or $|b| < 1$ on the cartesian

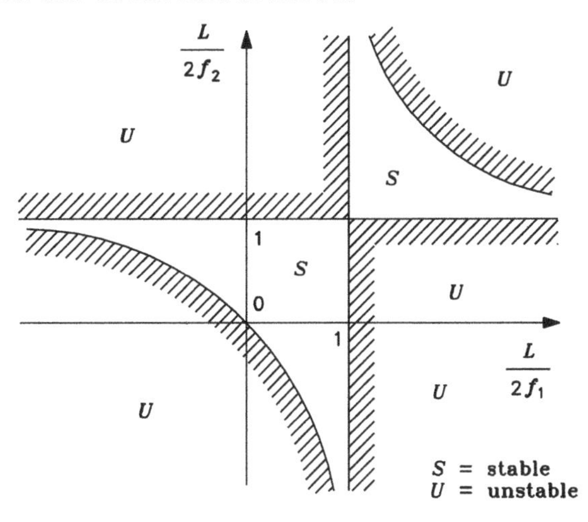

Figure 9.10 *The stability diagram for open resonators.*

plane $\{L/2f_1, L/2f_2\}$. In Figure 9.10, the blank regions correspond to geometries that yield stable resonators, while the dashed regions correspond to unstable resonators. Negative values of one of the coordinates correspond to a convex mirror; so, points whose coordinates are both negative (a pair of convex mirrors) are all in the region of instability, in agreement with intuition. They do not nd any practical application. On the other hand, cavities which are moderately unstable ($|b|$ slightly larger than 1) are used in laser technology, when high-power output beams are required. Indeed, for a given beam power, the energy stored in an unstable cavity is signi cantly smaller than in a stable one, and this can simplify problems related to materials. The price one pays when the stability is lowered (that is, when $|b|$ is increased) is a reduced mode selectivity, that is, a lower quality (both vs. time and vs. spatial coordinates) of the output beam.

9.11 Q-factor of an open resonator

Also for open cavities it is of great interest to de ne an adimensional quantity, called *merit* or *quality factor* or, for short the *Q-factor*. It can still be de ned by multiplying the resonance frequency by the energy stored in the resonator, and dividing it by the time-averaged power loss. There are no di erences with respect to what we saw in other sections from the viewpoint of principles, but now it is more di cult to separate power lost *in the resonator* from that *delivered outside* (see Sections 9.7 and 9.8), a distinction which allowed us to de ne an unloaded Q-factor and a loaded Q-factor for closed resonators.

In an open resonator, both stable and unstable, mirrors have nite diameters, while, as we said before, wavefronts of Gaussian beam extend inde - nitely. Therefore, every re ection is accompanied by loss of that amount of power which corresponds to the part of the beam which travels at distances from the cavity axis larger than the mirror size. This power goes obviously out of the cavity, but must be treated conceptually as *lost* power, because it is not delivered outside in any usable way. Consequently, it is included (together with absorption either within the cavity or on the mirrors) in the *unloaded Q-factor.*

The case of one (or two) partially transparent mirror is completely di erent. If a mirror re ects only a fraction of the incident power, the complementary part can be *delivered* to an external beam. This power passing through the partially transparent mirror has to be accounted for in the de nition of an *external Q-factor.*

When the losses due to nite size of the mirrors and those due to absorption are small with respect to the energy stored in the resonator, the two contributions to the unloaded factor Q_0 can be added, following the rule used in Section 9.8. The low-loss assumption ts always well stable cavities, provided the mirrors are made with low-loss materials. Then, we may write:

$$\frac{1}{Q_0} = \frac{1}{Q_a} + \frac{1}{Q_d} \quad , \tag{9.43}$$

where $Q_a = \omega_0 E / W_a$ is the Q-factor due to absorption (W_a = time-averaged absorbed power), while Q_d is the contribution due to losses at mirrors edges. To nd a way to express these losses, one needs di raction theory, to be dealt with in Chapter 13. Here, we will just quote the formulas we need now, without any proof. All the quantities that we need can be expressed as functions of an adimensional parameter, usually referred to as the *Fresnel number* (Born and Wolf, 1980) which, for two spherical mirrors, can be de ned as:

$$N = \frac{a_1 a_2}{\ L} \quad , \tag{9.44}$$

where is the wavelength, L is the cavity length, a_1 and a_2 are the radii of the cross-sections of the mirrors (see Figure 9.8).

The fraction of the power of a Gaussian beam, , which is lost in one re ection on the mirror, as a function of N, is well approximated by simple expressions for N 1 (di raction losses small compared to the other losses of interest) and for N 1 (that means di raction losses dominant on the others). Respectively, we have:

$$\simeq 16 \ ^2 N e \ ^{4 \ N} \qquad (N \quad 1) \quad , \tag{9.45}$$

and:

$$\simeq 1 \quad ^2 N^2 \qquad (N \quad 1) \quad . \tag{9.46}$$

A simple expression for Q_d as a function of (Ramo *et al.*, 1994) can be easily linked to intuitive properties of Gaussian beams that we saw in Chapter 5. If the radii of curvature of the mirrors are large enough, then the energy distribution in the cavity, on a transverse plane, is almost independent of the longitudinal coordinate, and the ratio between average and maximum values of the energy density depends only on the order of the mode. The

Poynting vector is equal to the product of the energy density times the group velocity, that we can assume equal to the speed of light in the medium which lls the cavity. Therefore, the energy stored in the cavity:

$$E = \quad w_0^2 L U \quad , \tag{9.47}$$

(w_0 = waist width of the Gaussian beam) and the power incident on the mirror:

$$W_i = \quad w_0^2 c U \quad , \tag{9.48}$$

can be easily expressed as functions of the average energy density, U. Taking now into account the fact that power loss at any re ection, W_d, is the product of times Eq. (9.48), we nally get:

$$Q_d = \frac{\omega_0 E}{W_d} = \frac{\omega_0 L}{c} = \frac{2 \quad L}{\quad_0} \quad , \tag{9.49}$$

where $_0 = 2 \quad c/\omega_0$ is the resonance wavelength. So, the quality factor due to di raction losses is, if we keep all other conditions constant, proportional to the number of wavelengths $L/ \quad _0$ contained in the cavity in the axial direction, i.e., to the integer number n of Eq. (9.32) ("longitudinal order" of the resonance mode). This agrees with the statements of Section 9.7, about the Q-factor growing when the order of the mode grows, and justi es why very high values can be obtained without too much di culty for Q-factors at optical frequencies.

To nish this section, it may be useful to recall again that all the de nitions and calculations in the sections on open resonators (Section 9.9 and following) refer to cavities lled with a *passive medium*. In the language of laser technology, these are called *cold cavity* modes. When cavities are lled with an *active medium*, capable of amplifying the resonant wave (the necessary energy comes from other sources), then there are important changes both in the Q-factor, with respect to that de ned in this section, and in the oscillation frequencies, which are "pulled" by the active medium to di erent values with respect to those given by Eq. (9.32). These problems are largely discussed in books on lasers (Svelto, 1989; Yariv, 1976).

9.12 Further applications and suggested reading

We take it for granted that the background in basic physics of all the readers includes resonances in other types of passive systems, typically mechanical ones. The relationship between them and what we have seen in this chapter is not just a vague analogy. The links are very fundamental. Just to quote one of them, a mechanical system can be resonant only if it can store energy under two di erent forms at di erent times in its periodic motion (e.g., kinetic and potential in a pendulum, kinetic and elastic in a spring). Resonance occurs when these two energies balance each other, exactly like electric and magnetic energies in resonant cavities.

It may be of some interest, at least for those readers who have an inclination for the History of Science, to know that the mathematical treatment of resonant cavities, which is rather sophisticated compared to that of simple mechanical systems, has been developed long ago, up to a very advanced level, in the framework of acoustics. The very famous treatise *Theory of Sound* by Lord Rayleigh, now more than one century old, remains a masterpiece of scienti c literature as well as a very important marker in the advancement of science. It is still a highly recommended, very instructive reading.

After these suggestions related to classical physics, we also invite the readers to strengthen their background on resonant systems in Quantum Mechanics, typically, harmonic oscillators (Louisell, 1960). An important analogy in this direction is that between discrete states in a quantized system, and discrete modes in a resonator. The two systems are eigenvalue problems, and it is immediately noted that energy and frequency are proportional, through the Planck constant. Concepts like mode degeneracy and perturbation theory (like the one we used for calculating Q-factors), both in this chapter and in the previous one, may become much more familiar through a cross-correlation with elementary Quantum Mechanics.

Going back to e.m. resonators, it is rather surprising that microwave resonators and optical resonators were studied in completely separate textbooks for decades, in spite of the fact that both are easy to explain, at least qualitatively, in terms of transmission lines (Chapter 8). It is even more surprising since one of the main applications of resonators, ltering, was shared by the two frequency ranges. Probably, this was because most of the experts in microwaves, being electrical engineers, had a strong background in circuit theory, and so they designed their devices using mathematical tools originating from this area, typically those based on topological matrices and on Laplace transforms. On the other side, most of the physicists who worked in optics preferred to use frequency-domain approaches. To see this contrast, one may consult, for example, the sections devoted to ltering properties of Fabry-Perot etalons in Born and Wolf (1980), on one side, and Chapters 4 and 5 of Montgomery, Dicke and Purcell, Eds., (1948) on the other side. Nowadays, thanks to the success of optics in telecommunications, it is much easier to nd the two subjects side by side. However, this comment was not inserted simply as a historical detour. It was intended to remind that, if one has to design resonators for ltering purposes, then it may be advisable, irrespectively of the frequency band, to know at least something in the area of network theory. A sequence of readings in this sense may consist of Desoer and Kuh (1969), Collin (1966) and Howe (1974), to ful l the basic needs and possibly trigger further interest. Expert readers may include dielectric resonators, a subject that was omitted here only because of space limits. One may consult for example Kajfez and Guillon, Eds., (1986). Other readers may wish to be introduced to computer-aided design of ltering devices. At microwave frequencies, one may then consult Gupta *et al.* (1981).

Another practical application of resonant cavities, comparable in importance to ltering, is positive feedback, as required in order to build an oscillator. This extends to microwave and optical frequencies the role played in conventional electronics by lumped-element LC circuits. However, it has been stressed many times in the previous sections that an oscillating device can no longer be modeled as a purely electromagnetic system. Indeed, the eld in it is coupled to something (e.g., an active medium in lasers, or a charged-particle beam in microwave tubes) which provides gain to the eld. One must be warned that this coupling a ects some of the most important characteristics of the eld. Ramo *et al.* (1994) is a recommended source of introductory information on this point, for the microwave case. Textbooks on lasers (e.g., Svelto, 1989) should be consulted for the optical-frequency case.

In this framework, it may be interesting to look again at the relationship between microwave and optical resonators. Section 10.15 of Ramo *et al.* (1994) outlines why it took the work of several outstanding scientists, to grasp the advantages of open resonators, compared to closed ones, when the rst lasers were built. Another important point is that the type and the quality of the resonator a ect both temporal and spatial coherence of the sources. As the subject of coherence will be introduced in Chapter 14, readers are advised to come back to the present chapter after reading that one, and link the two together.

Problems

9-1 Show that for any resonant mode of a closed cavity with perfectly conducting walls, all the eld components can be obtained following the same approach as for the propagation modes in waveguides, provided one replaces the exponential functions in Eqs. (7.1) with either sine or cosine functions, following the indications of Eqs. (9.15) and (9.16).

9-2 Discuss which among the TE and the TM modes of a rectangular resonator can have one or two subscripts equal to zero. Repeat the problem for the modes of a resonator with circular cross-section. Establish a link between these two answers, describing intuitively how the lines of force of the electric and magnetic elds of a given mode in a rectangular resonator deform into those of the corresponding mode in a circular resonator, using a cavity of elliptical cross-section as an intermediate step.

9-3 If one or two ratios between side lengths, in a rectangular resonator, are rational numbers, then we may expect some modes to become degenerate.

a) Find which modes are degenerate when $a = 2b = d/2$.

b) Among these sets of degenerate modes, only some satisfy orthogonality conditions like Eqs. (9.7) and (9.8). Discuss which do and which do not.

9-4 The fundamental mode in a resonator is clearly de ned as the mode with the lowest resonance frequency. In a rectangular cavity, the ordering of the resonant frequencies depends on the inequalities between the side lengths.

a) Keeping $a > b$, discuss which is the fundamental mode as a function of d.

b) Find a design rule that maximizes the spacing between the resonance frequency of the fundamental mode and that of the next higher-order mode.

9-5 In a resonator of circular cross-section, resonant frequencies are ordered in a way which depends on the ratio between the length and the diameter. Repeat, for this case, the two questions of the previous problem.

9-6 In a spherical coordinate frame, impose the boundary condition $\mathbf{E}_{\tan} = 0$ rst on the coordinate planes $\varphi =$ constant, and then on the coordinate cones $\vartheta =$ constant. Find the corresponding solutions of Maxwell's equations. Discuss the physical structures in which these solutions exist as resonant modes or as propagation modes. (*Hint.* Restrict the discussion to TE and TM waves, for which partial derivatives with respect to φ are identically zero).

9-7 Using the contents of Section 7.10, and with the help of a suitable handbook dealing with Mathieu's functions (e.g., Abramowitz and Stegun, 1965), express the longitudinal eld components for the resonant modes of a cylindrical cavity with elliptical cross-section, with an ideal conducting wall.

9-8 Express the intrinsic Q-factors of the generic TE and TM mode of a cubic resonator ($a = b = d$) whose wall impedance is expressed by Eq. (9.25). In particular, discuss how the Q-factor of a given mode depends on the resonant frequency, which in turn depends on the cavity size.

9-9 As sketched in Section 9.1, a piece of coaxial cable whose ends are sort-circuited resonates at those frequencies where its length equals a half wavelength.

a) Express the intrinsic Q-factor due to losses in the good conductor which is used to make the two cylinders and the two at terminal surfaces.

b) What happens to the intrinsic Q-factor if the cavity resonates in a higher-order longitudinal mode?
(*Hint.* Do not forget the loss contributed by the two terminal at walls.)

9-10 Calculate the intrinsic Q-factor of a coaxial cavity, for an outer radius $a = 5\,\mathrm{mm}$, a characteristic impedance of the cable of 120 , a dielectric permittivity inside the cable equal to $_0$, copper walls (conductivity: 5.8 10^7 ^{1}m ^{1}), resonant frequency 600 MHz.

9-11 On some occasions, losses in the dielectric which lls a resonant cavity may contribute signi cantly to the intrinsic Q-factor.

a) Assuming that the conducting walls are ideal, express the intrinsic Q-factor due to a dielectric with a small loss tangent, , for the modes of the cubic resonator of Problem 9-8, and for the TEM modes of the coaxial cavity of Problem 9-9.

b) Which is the expression of the intrinsic Q-factor that accounts for both conductor and dielectric losses, assuming that they are small enough to be additive?

9-12 Tunable resonators are often used to measure frequencies in current microwave practice. They are usually referred to as *wavemeters*. The higher the Q-factor, the sharper the resonance, and so, quite obviously, the more accurate the frequency measurement. This is one of the reasons why cavities of circular cross-section, resonating in a low-loss $\mathrm{TE}_{0,1,p}$ mode (see Section 7.11), are usually preferred to rectangular cavities. To support this statement, calculate the intrinsic Q-factor of the fundamental mode of a cubic cavity (see the previous problem) resonating at 3 GHz, and that of a cylindrical cavity whose radius equals the side length of the previous cavity and resonates in the $\mathrm{TE}_{0,1,1}$ mode at the same frequency. Assume that the wall surfaces are covered with a galvanic silver layer (conductivity) thicker than the skin depth.

9-13 Another reason why cavities of circular cross-section are preferred as wavemeters is because there are no currents owing through the contact between the cylindrical wall and the sliding piston which makes the resonator tunable. To support this statement, nd the current densities in all the walls of a circular-cylinder cavity, in the $\mathrm{TE}_{0,1,1}$ mode. Are there other modes of the same cavity which enjoy the same property of not having currents owing through that contact?

9-14 At microwave frequencies, one method of measuring the permittivity of a dielectric material is to observe the shift in the resonance frequency, ω, when a small sample of the material is placed inside a given cavity. In fact, following a variational approach, Slater proved that $\omega/\omega_0 = U/U_0$, where U is the change in the energy stored in the cavity caused by the presence of the sample, while ω_0 and U_0 are the values which correspond to the cavity without the sample.

a) To enhance the sensitivity of the measurement, is it convenient to place a given dielectric sample near a maximum of the electric eld or near one of the magnetic eld?

b) Assuming that the eld remains unperturbed when the small sample is introduced, evaluate ω for a cubic sample having $_r = 3$, 3 mm in side, placed in the center of a rectangular cavity with $a = 2b = d = 30$ mm, lled with a material having $_r = 1.5$ and resonating in the TE_{101} mode.

9-15 In Chapter 4 we discussed a device, shown in Figure 4.10, which was referred to as the Fabry-Perot interferometer. Actually, this name is usually given to any structure which consists of two at parallel mirrors, separated by a homogeneous medium. It is often used as a frequency (or rather, wavelength) meter at optical frequencies, by making it possible to vary continuously the distance between the two mirrors.

a) To increase the resolution of the frequency measurement, the device is fabricated with dielectrics which exhibit very low loss. What determines then the spectral width of the resonance line, Eq. (9.27) ?

b) Evaluate this external Q-factor of the open resonator.
(*Hint.* Model the device as a piece of lossless transmission line, with a characteristic impedance Z_1, and a length equal to a multiple of /2, sandwiched between two in nitely long transmission lines, with characteristic impedances Z_2.)

c) How does the external Q-factor vary as a function of the di erence between Z_1 and Z_2 ?

9-16 The mirrors which de ne the cavity in a typical semiconductor laser are simply the discontinuities between the high-index semiconductor chip, and air.

a) Assuming a refractive index $n = 3$ (note that in Eq. (9.32) it is $k = nk_0$, where k_0 is de ned in free space), calculate the resonant frequencies and the corresponding free-space wavelengths of the longitudinal modes (i.e., neglect the and terms) for a chip length $L = 300$ m.

b) For the lowest-order answer to the previous question, estimate the frequency separation between two consecutive Hermite-Gauss modes, taking $w_0 =$ 5 m at the beam waist.

9-17 Find one or more physical arguments to justify why (see Figure 9.10):

a) an open resonator with two convex mirrors (f_1 and f_2 both negatives) is always unstable;

b) a convex mirror and a concave one can make a stable resonator, provided their distance L is not too large.

9-18 Consider a Helium-Neon laser, emitting at $= 0.633$ m.

a) Show that, even when the mirrors are so small that di raction losses are extremely high (i.e., for close to unity), the estimated Q-factor, based on Eq. (9.49), is quite high if the cavity length is in the range of 100 mm.

b) Show that mirror radii in the millimeter range are su cient to let Q_d grow by many orders of magnitude.

References

Abramowitz, M. and Stegun, I.A. (Eds.) (1965) *Handbook of Mathematical Functions.* Dover, New York.

Born, M. and Wolf, E. (1980) *Principles of Optics: Electromagnetic Theory of Propagation, Interference and Di raction of Light.* 6th (corrected) ed., reprinted 1986, Pergamon, Oxford.

Collin, R.E. (1966) *Foundations for Microwave Engineering.* McGraw-Hill, New York.

Desoer, C.A. and Kuh, E.S. (1969) *Basic Circuit Theory.* McGraw-Hill, New York.

Gupta, K.C., Garg, R. and Chadha, R. (1981) *Computer-Aided Design of Microwave Circuits.* Artech House, Dedham, MA.

Howe, H. (1974) *Stripline Circuit Design.* Artech House, Dedham, MA.

Kajfez, D. and Guillon, P. (Eds.) (1986) *Dielectric Resonators.* Artech House, Dedham, MA.

Louisell, W.H. (1960) *Coupled Mode and Parametric Electronics.* Wiley, New York.

Montgomery, C.G., Dicke, R.H. and Purcell, E.M. (Eds.) (1948) *Principles of Microwave Circuits.* McGraw-Hill, New York. Reprinted by Peter Peregrinus, London, for IEE (1987).

Morse, P.M. and Feshbach, H. (1953) *Methods of Theoretical Physics.* McGraw-Hill, New York.

Ramo, S., Whinnery, J.R. and Van Duzer, T. (1994) *Fields and Waves in Communication Electronics.* 3rd ed., Wiley, New York.

Svelto, O. (1989) *Principles of Lasers.* 3rd ed., translated from Italian and edited by D.C. Hanna, Plenum, New York.

Yariv, A. (1976) *Introduction to Optical Electronics.* Holt, Rinehart and Winston, New York.

CHAPTER 10

Dielectric waveguides

10.1 Introduction

In all the waveguides dealt with in Chapter 7, the dielectric medium where the guided eld propagates was assumed to be homogeneous. In other structures, which may also be of practical interest, the medium is not homogeneous; for example, this is the case for microwave devices made with ferrite-loaded waveguides, of the same kind as those that were mentioned in Chapter 6. Still, these guides have a boundary de ned by a cylindrical conducting surface. Their analysis is similar, at least with reference to the fundamentals, to what we saw in Chapter 7, although some changes are not trivial. There is still another set of wave guiding structures, whose essential feature is no metallic wall to con ne the eld within a closed region. The dielectric medium does in principle extend to in nity, in the directions orthogonal to the waveguide axis. What drives the eld energy along a given direction is a suitable set of inhomogeneities — either sharp discontinuities, or smooth variations — in the permittivity of the medium. These structures di er radically from those of Chapter 7, and we must approach them in another way.

The theoretical foundations of dielectric waveguides were established in the early decades of the 20th century. However, the space devoted to them in standard Electrical Engineering curricula was extremely marginal until 1970. What made them quite popular, subsequently, was the explosion of ber optics in long-haul telecommunications, and the strictly correlated research e ort in integrated optics. Interest in exploiting optical frequencies for telecommunications had been stimulated, in the early '60's, by the invention of the laser. If one thinks of guiding waves at optical frequencies along a given path, one easily sees that there are at least three very important reasons why the metallic-wall waveguides of Chapter 7 are absolutely unsuitable. Firstly, the size of the guide would be too small, both for manufacturing and handling it. Secondly, conductors would introduce too much loss; actually, the correct physical picture of the interaction between an e.m. eld and charge carriers at the surface of a conductor, at optical frequencies, is not simple, and requires quantum-mechanical methods. Thirdly, for a link length and a bandwidth of technical relevance, dispersion in a metallic waveguide would be prohibitively large.

In this chapter, we will lay down the theoretical foundations of how e.m. waves are guided either by step discontinuities, or by graded inhomogeneities. When this can be done, we will make the analytical approach as similar as possible to that of Chapter 7. In some cases, however, we will be forced to adopt approximate techniques. In these instances, we will follow as closely as possible some ideas that were outlined in Chapter 5, e.g., the WKBJ technique and geometrical optics. However, we will point out that in general all these approximations become accurate only when the transverse dimensions of the waveguides are fairly large compared to the wavelength. Consequently, we will be reasonably satis ed with some analytical solutions only for highly multimode waveguides, not for single-mode or few-mode ones. In these cases, we will face a rather "black or white" situation, where either we are able to nd exact solutions, or we have to resort to numerical methods.

Because of the wide technical interest in optical bers, we will deal explicitly with waveguides where the magnetic permeability is constant throughout the structure, and normally equal to that of free space. However, all the results of this chapter can easily be translated, by duality, into results which apply to structures where the dielectric permittivity is constant and the guiding discontinuities a ect the magnetic permeability .

This chapter is organized as follows. We will commence with an unspeci ed geometry, where the dielectric permittivity exhibits just one step discontinuity. We will show that a structure of this kind supports guided modes (a nite number, at any nite frequency) and unbounded modes (a continuum). With the help of two examples — slab waveguides and cylindrical rods — we will show that only a subset of the guided modes are TE and TM, while the others modes are hybrid. Then, we will outline why the exact analytical modes of step-index circular waveguides are too cumbersome, and how they can be replaced by approximate solutions, whose most interesting feature is that their transverse elds are linearly polarized. Next, we will elaborate further on dispersion in dielectric rods, a very important issue in characterizing optical bers. We will subsequently proceed to graded-index waveguides, show how they can be studied with approximate analytical techniques, and discuss brie y their performance limitations. Finally, we will just touch upon losses in dielectric waveguides.

10.2 Waves guided by a surface of discontinuity. The characteristic equation

The mathematical approach to an inhomogeneous dielectric waveguide varies, depending on whether the structure consists of a nite number of homogeneous media, or of a medium whose permittivity varies continuously as function of transverse coordinates. It is convenient to deal rst with waveguides made of several homogeneous materials, starting from the simplest case, that of two lossless dielectric media, referred to as *core* and *cladding*, separated by a cylindrical surface, C, parallel to the z axis. Suppose that the intersection between C and the plane $z =$ constant is a regular closed curve, ℓ. At each point on ℓ we may de ne in an unambiguous way three mutually orthogonal unit vectors, tangent, normal and binormal to ℓ, denoted by $\hat{t}, \hat{n}, \hat{a}_z$, respectively. The last one is also the unit vector in the direction of the z axis. They are related as

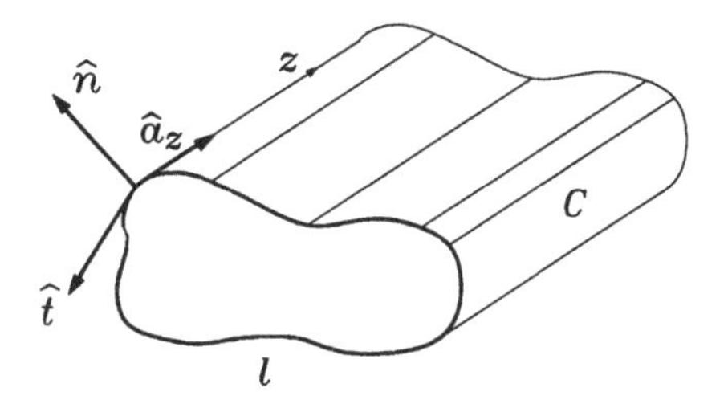

Figure 10.1 *A dielectric waveguide of unspeci ed geometry, consisting of two homogeneous media.*

$\hat{t} \quad \hat{n} = \hat{a}_z$ (see Figure 10.1). Let $_{co}$, $_{cl}$ be the permittivities of the inner and outer medium, with $_{co} > \ _{cl}$, and be the permeability of both media (when not speci ed otherwise, suppose $= \ _0$). We will also use the following symbols, whose meanings are simple extensions of other symbols used in the previous chapters: $^2_i = \omega^2 \quad _i$, $_i = \omega \sqrt{\quad _i}$ $(i = co, cl)$.

Within each of the two regions, since the medium in it is homogeneous and there are no sources, we are entitled to apply the decomposition theorem of a generic eld into a TE and a TM eld, which was proved in Section 3.8 and, in a simpli ed way which is suitable for the case at hand, in Section 7.2. Indeed, we will retain the basic *ansatz* of Chapter 7, that all the unknown functions (pre-potentials and elds) depend on the longitudinal coordinate through an exponential function, in general of complex argument, exp(z). Italic symbols will be used, as in Chapter 7, for the functions of the transverse coordinates where the exponential dependence on z has been factored out. The decomposition theorem also tells us that the longitudinal eld components obey eigenvalue equations like Eq. (7.8), namely:

$$\nabla_t^2 \mathcal{F}_i = \quad ^2_i \mathcal{F}_i \qquad (i = co, cl) \quad , \tag{10.1}$$

where $\mathcal{F}_i$ is either $\mathcal{H}_{iz}$ or $\mathcal{E}_{iz}$, and:

$$^2_i = (\ ^2 + \omega^2 \quad _i) = (\ ^2 \quad \ ^2_i) \quad . \tag{10.2}$$

The logical steps to take from now on are:

to write the total eld, in the core and in the cladding, as the sum of the TE and the TM elds;

to express the continuity conditions of the tangential eld components, Eqs. (1.43) and (1.44), on the core-cladding interface surface, in terms of these elds;

to study whether this system has solutions, and whether they are discrete, or form a continuum.

The rst step is elementary, regardless of the waveguide shape:

$$\mathcal{E}_i = \mathcal{E}_{\mathrm{TE}i} + \mathcal{E}_{\mathrm{TM}i} \quad , \qquad \mathcal{H}_i = \mathcal{E}_{\mathrm{TE}i} + \mathcal{E}_{\mathrm{TM}i} \qquad (i = co, cl) \quad . \tag{10.3}$$

The second step is simple as long as we do it using the general-purpose symbols that have been introduced so far:

$$
\begin{aligned}
\hat{t} \ (\mathcal{E}_{\mathrm{TE}co} + \mathcal{E}_{\mathrm{TM}co}) &= \hat{t} \ (\mathcal{E}_{\mathrm{TE}cl} + \mathcal{E}_{\mathrm{TM}cl}) \quad , \\
\hat{} \ (\mathcal{E}_{\mathrm{TE}co} + \mathcal{E}_{\mathrm{TM}co}) &= \hat{} \ (\mathcal{E}_{\mathrm{TE}cl} + \mathcal{E}_{\mathrm{TM}cl}) \quad , \\
\hat{t} \ (\mathcal{H}_{\mathrm{TE}co} + \mathcal{H}_{\mathrm{TM}co}) &= \hat{t} \ (\mathcal{H}_{\mathrm{TE}cl} + \mathcal{H}_{\mathrm{TM}cl}) \quad , \\
\hat{} \ (\mathcal{H}_{\mathrm{TE}co} + \mathcal{H}_{\mathrm{TM}co}) &= \hat{} \ (\mathcal{H}_{\mathrm{TE}cl} + \mathcal{H}_{\mathrm{TM}cl}) \quad .
\end{aligned}
\tag{10.4}
$$

Comparison between this set of four equations and the boundary conditions in a metallic waveguide, Eqs. (7.17) and (7.18), shows immediately that the mathematical complexity of this problem is much greater than that of Chapter 7. For all the eld components involved in Eqs. (10.4) we might write the general expressions, either Eqs. (7.4) or Eqs. (7.11) and (7.12). This would give us the correct perception that the system in Eqs. (10.4) contains four unknown functions, $\mathcal{H}_{iz}$ and $\mathcal{E}_{iz}$, in addition to the parameter which is still undetermined. However, from a practical viewpoint this is an insigni cant progress, unless the following condition is satis ed: the interface surface between core and cladding, C, is a *coordinate surface in a reference frame where the Helmholtz equation is separable.* To grasp the actual meaning of this point, the reader should return to Section 7.6. It tells us that the geometries which can be studied in closed form are very few: planar slabs, circular-core and elliptical-core waveguides (parabolic waveguides are never found in practice).

Suppose then that we are in one of these fortunate cases, and let (u_1, u_2, z) be the suitable coordinates, $(h_1, h_2, h_3 = 1)$ be their metric coe cients, and $u_1 = a =$ constant be the equation of the coordinate surface C. Then, each of Eqs. (10.1) can be split, as we saw in Chapter 7, into two second-order ordinary di erential equations (ODE's). So, in general every function $\mathcal{F}_i$ is a product of a linear combination of two functions of u_1, multiplied by a linear combination of two functions of u_2, and therefore it contains four arbitrary constants.

At this stage, we must draw a sharp distinction between two possible cases, which will be clari ed by means of examples in the following sections. Let us proceed as follows.

A) In the rst case, purely physical considerations enable us either to break a general integral into two parts (e.g., separating even and odd symmetry solutions), or to eliminate, as unacceptable, some of the functions of u_1 or u_2 which show up in the general integrals (e.g., disregard functions which diverge to in nity within their domain of de nition). On some occasions, as we will see later, a procedure of this kind is capable of cutting the number of arbitrary constants in each function $\mathcal{F}_i$ from four to one, hence the total number of arbitrary constants in the problem from sixteen to four. In those cases, we may write:

$$
\mathcal{H}_{iz}(u_1, u_2) = A_i \, Q_i(u_1) \, M_i(u_2) \qquad (i = co, cl) \quad ,
$$

$$\mathcal{E}_{iz}(u_1, u_2) \quad = \quad B_i\, R_i(u_1)\, N_i(u_2) \qquad (i = co, cl) \quad , \tag{10.5}$$

where A_i, B_i $(i = co, cl)$ are the remaining four arbitrary constants. Using either Eqs. (7.4) or Eqs. (7.11) and (7.12), Eqs. (10.4) can be rewritten in the following form, where it is understood that all the functions of u_1 are evaluated at $u_1 = a$:

$$A_{co}\left[\frac{1}{h_1}\,\frac{dQ_{co}}{du_1}\,M_{co}\right] + B_{co}\left[\frac{}{j\omega \;\; _{co}}\,\frac{1}{h_2}\,R_{co}\,\frac{dN_{co}}{du_2}\right]$$

$$= A_{cl}\left[\frac{1}{h_1}\,\frac{dQ_{cl}}{du_1}\,M_{cl}\right] + B_{cl}\left[\frac{}{j\omega \;\; _{cl}}\,\frac{1}{h_2}\,R_{cl}\,\frac{dN_{cl}}{du_2}\right] ,$$

$$B_{co}\left[\frac{{}^2_{co} \quad {}^2}{j\omega \;\; _{co}}\,R_{co}N_{co}\right] = B_{cl}\left[\frac{{}^2_{co} \quad {}^2}{j\omega \;\; _{cl}}\,R_{cl}N_{cl}\right] ,$$

$$A_{co}\left[\frac{}{j\omega}\,\frac{1}{h_2}\,Q_{co}\,\frac{dM_{co}}{du_2}\right] + B_{co}\left[\frac{1}{h_1}\,\frac{dR_{co}}{du_1}\,N_{co}\right]$$

$$= A_{cl}\left[\frac{}{j\omega}\,\frac{1}{h_2}\,Q_{cl}\,\frac{dM_{cl}}{du_2}\right] + B_{cl}\left[\frac{1}{h_1}\,\frac{dR_{cl}}{du_1}\,N_{cl}\right] ,$$

$$A_{co}\Big[(\;{}^2_{co} \quad {}^2)\,Q_{co}M_{co}\Big] = A_{cl}\Big[(\;{}^2_{cl} \quad {}^2)\,Q_{cl}M_{cl}\Big] . \tag{10.6}$$

For this system of four homogeneous equations in the four variables A_i, B_i to have non-trivial solutions, the determinant of the coe cients (in square brackets) in Eqs. (10.6) must be equal to zero. This yields a *characteristic equation*, where the unknown is the propagation constant . In general, it is a transcendental equation in the complex eld, since in the initial ansatz was supposed to be complex, and other parameters involved in the arguments of the functions M, N, Q, R can be complex too. However, it is *strictly mandatory* to check whether the solutions of the characteristic equation are consistent with the physical arguments that were used in order to arrive at Eqs. (10.6), eliminating some of the mathematically correct solutions of the ODE's. The examples will show that this requirement is satis ed, for any nite value of the angular frequency ω, only by a *nite number* of solutions of the characteristic equation, all of which correspond to *imaginary values* of , and to e.m. elds whose power travels along the z direction. We may express these ndings by saying that the structure has a *nite number of guided modes.* It can also be shown that, quite generally, the values of $= \;\; j$, which correspond to solutions of this kind, fall within the range de ned by the intrinsic phase constants of the cladding, $_{cl}$, and the core, $_{co}$, namely:

$$_{cl} < \quad < \;\; _{co} \quad . \tag{10.7}$$

Clearly, from what we saw in the previous chapters, we may infer that this nite number of solutions cannot be a complete set of modes. To build a complete set, we may account also for solutions of a di erent physical type, to be sketched as the next point.

B) In the second case, on the contrary, one is not able to collect additional physical information, allowing us to write Eqs. (10.5) and so reducing the number of arbitrary constants. To quote just an example, this occurs for a cylindrical rod when one wants to describe elds traveling radially in the cladding. In general, they must be expressed as superpositions of Hankel functions of the rst and the second kind (see Appendix D).

Under these circumstances, Eqs. (10.4) and the separation of variables yield, instead of Eqs. (10.6), a set of equations where the number of unknowns is larger than the number of equations. According to the Rouche-Capelli theorem of algebra, such a set has nontrivial solutions for *any* value of the propagation constant . Therefore, in addition to the nite number of guided modes, any dielectric waveguides has a *continuum* of modes. To grasp their physical nature, one has to check what values of are compatible with Maxwell's equations in the core and in the cladding. It can be shown that, regardless of the waveguide shape, in a lossless structure the net result is as follows:

1. can take imaginary values, provided that $| \ | < \ _{cl} = \omega \sqrt{\ _{cl}}$. This continuum of solutions is referred to as the set of the *radiation modes*. Each of them consists of uniform plane waves, traveling in oblique directions with respect to the waveguide axis;
2. can take real values. This is also a continuum of solutions, referred to as the set of the *evanescent modes*. Each of them consists of evanescent plane waves whose attenuation vector is parallel to the waveguide axis, and consequently whose phase vector is orthogonal to the z axis. Such modes may reach an appreciable amplitude only in the immediate neighborhood of the ends of a waveguide of nite length.

To summarize, at any nite value of the frequency a dielectric waveguide supports:

- a nite number of guided modes;
- a continuum of radiation modes;
- a continuum of evanescent modes.

The number of guided modes depends on the frequency, similarly to the number of propagating modes in a metallic waveguide. However, there is a major di erence between metallic and dielectric waveguides from the viewpoint of cut-o . A mode below cut-o in a metallic waveguide cannot propagate, but does maintain its individual features. When a guided mode goes to its cut-o frequency in a dielectric waveguide, it loses its discrete nature, and merges into the continuum of radiation modes.

Let us point out that in general the characteristic equation of the system (10.6) has other solutions, in addition to the nite number of guided modes, if is permitted to scan the whole complex plane, without restrictions. More precisely, these additional solutions form a countable in nity, and all of them are characterized by complex 's, with nonvanishing real and imaginary parts. At rst sight, each of these solutions might look like a generalized version of a guided mode, the attenuation accounting for some power radiated laterally. For this reason, elds of this kind, satisfying Eq. (10.1) and the following equations, have been called *leaky modes* of the dielectric waveguide. However, it is easy to show that they cannot be true rigorous modes of a lossless structure. Indeed, the characteristic equation can be derived through a crucial step, the selection based on physical requirements among the mathematically correct solutions of Eq. (10.1). Leaky modes violate these physical requirements, and because of this contradiction they cannot be looked at as physically meaningful solutions of the problem. Still, they can be quite useful in practice (Snyder and Love, 1983), to *approximate* the radiation eld (i.e., the total eld minus the guided modes) of a dielectric waveguide as a summation (in general, over a small number of terms) rather than as a line integral in the complex plane. In particular, in a structure a ected by loss, where the cut-o mechanism (i.e., a guided mode becoming radiating) is not simple to describe, leaky modes can be helpful. We will brie y come back to this subject at the end of Section 10.4.

Another important di erence between dielectric and metallic waveguides is that, in general, the set of Eqs. (10.6) does not break into two sets of two equations, one including only $\mathcal{H}_{iz}$ (TE modes), and one including only $\mathcal{E}_{iz}$ (TM modes). Hence, in general dielectric waveguides have *hybrid modes.* TE and TM modes may exist, but are not enough to form a complete set. From this viewpoint, dielectric waveguides may look similar to waveguides surrounded by walls with a non-zero impedance (see Chapter 7). The main di erence is that in a dielectric waveguide the continuity conditions may be expressed as a matching between mode-dependent wave impedances in the direction normal to C, in the core and in the cladding, but not in terms of a eld-independent wall-impedance.

10.3 Guided modes of a slab waveguide

We choose the rst example of application, for the broad-scope approach of the previous section, such that we may use rectangular coordinates, which simplify separation of variables and the following steps. However, we cannot apply the contents of the previous section to a rectangular waveguide, since its contour is not one coordinate surface, as it consists of four segments belonging to di erent coordinates surfaces. In contrast to rectangular metallic waveguides, rectangular dielectric waveguides cannot be solved analytically in exact terms (see Snyder and Love, 1983). Therefore, our rst example will deal with a *slab waveguide*, which extends inde nitely in one direction (say y) orthogonal to that of propagation (say z) (see Figure 10.2). Let $2a$ be the slab width along the x direction.

We will consider only elds which are *independent of the y coordinate.* This assumption will allow us to clarify some basic ideas, but implies a loss in gen-

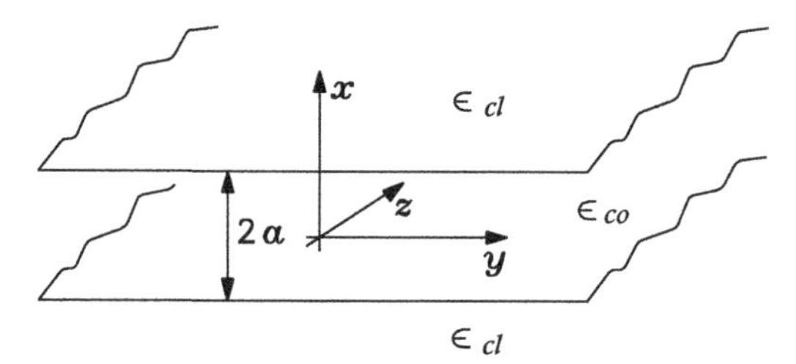

Figure 10.2 *Symmetrical step-index slab waveguide.*

erality, whose consequences might appear to invalidate some of the statements of the previous section (in particular, the point referring to hybrid modes). The overall picture will be fully clari ed only after Section 10.5.

Let us get into the details of the general approach outlined in Section 10.2, writing:

$$\left.\begin{array}{l} Q_i(u_1) \\ R_i(u_1) \end{array}\right\} = X_i(x) \qquad (i = co, cl) \quad , \tag{10.8}$$

and following the same path as in Section 7.7, starting from the Helmholtz equation in the core and in the cladding. One gets that the ODE for X_i reads:

$$X_i'' = \quad k_i^2 \, X_i \quad , \tag{10.9}$$

which is, once again, the harmonic equation, where:

$$k_i^2 = \quad {}_i^2 = \quad {}^2 \quad {}_i^2 = \quad {}^2 + \omega^2 \quad {}_i \qquad (i = co, cl) \quad . \tag{10.10}$$

The general integral of Eq. (10.9) can be written as:

$$X_{co} = C \left\{ \begin{array}{ll} \sin k_{co} x & \text{(even modes)} \\ \cos k_{co} x & \text{(odd modes)} \end{array} \right. \tag{10.11}$$

in the core, and :

$$X_{cl} = D \left\{ \begin{array}{ll} \dfrac{x}{|x|} \, e^{\quad {}_{cl}\,|x|} & \text{(even modes)} \\ e^{\quad {}_{cl}\,|x|} & \text{(odd modes)} \end{array} \right. \tag{10.12}$$

in the cladding, with C and D arbitrary constants. Solutions of the form exp($_{cl}\,|x|$) in the cladding have been disregarded as they diverge when x goes to in nity, therefore are not physically meaningful.

Assuming the eld is independent of the y-coordinate is equivalent to setting

Calling even and odd the modes of a symmetrical slab waveguide is not a uniquely de ned process. In fact, when the longitudinal eld components in Eqs. (10.11) and (10.12) are even functions of x, then the transverse eld components, which descend from the longitudinal ones through a derivative with respect to x, are odd, and vice-versa. This fact is strictly correlated to what we saw about symmetries in Chapter 3. Denominations adopted here are not consistent with those given in Chapter 3, but are the most widely adopted in optical waveguide theory, and re ect the fact that transverse components are more important (and larger) than the longitudinal ones.

M_i' N_i' 0 in Eqs. (10.6). Then, this system breaks down into two sets, of two equations each in two unknowns. One of these sets identi es TE modes, the other one, TM modes. To be more speci c, continuity equations corresponding to *even TE modes* (where the word even is used according to the previous footnote) read:

$$C\,k_{co}\,\cos k_{co}a \quad = \quad D \;\; _{cl}\,e^{\quad _{cl}a} \quad ,$$

$$C\,k_{co}^2\,\sin k_{co}a \quad = \quad D \;\; ^2_{cl}\,e^{\quad _{cl}a} \quad . \qquad (10.13)$$

Setting equal to zero the determinant of the coe cients of Eqs. (10.13) yields the characteristic equation (also called the *dispersion relation*):

$$k_{co}\,\tan k_{co}a = \quad _{cl} \quad , \qquad (10.14)$$

which must be combined with the following equation, which stems from Eq. (10.10):

$$k_{co}^2 + \; ^2_{cl} = \omega^2 \;(\; _{co} \quad _{cl}) \quad . \qquad (10.15)$$

Any solution of Eq. (10.14) corresponds to a nontrivial solution of Eqs. (10.9), which is the longitudinal component, $\mathcal{H}_z$, of an even TE mode. The corresponding transverse eld components can be obtained from it by means of Eqs. (7.4). In detail, if we let $\mathcal{H}_z = X_i$ $(i = co, cl)$, we obtain the following eld in the core:

$$\mathcal{E}_x = 0 \quad , \quad \mathcal{E}_y = \frac{j\omega}{k_{co}}\,C\,\cos k_{co}x \quad , \quad \mathcal{H}_x = \frac{}{j\omega}\,\mathcal{E}_y \quad , \quad \mathcal{H}_y = 0 \quad , \qquad (10.16)$$

and in the cladding:

$$\mathcal{E}_x = 0 \quad , \quad \mathcal{E}_y = \frac{j\omega}{_{cl}}\,D\,e^{\quad _{cl}\,|x|} \quad , \quad \mathcal{H}_x = \frac{}{j\omega}\,\mathcal{E}_y \quad , \quad \mathcal{H}_y = 0 \quad . \qquad (10.17)$$

As we anticipated in the previous footnote, the transverse eld components are even, for the reasons explained in Section 3.9.

The characteristic equation and the eld components of the odd TE modes, and those of the even and odd TM modes, can be obtained in similar ways, and are summarized in Tables 10.1 and 10.2. All the characteristic equations shown in those tables are to be combined with Eq. (10.15).

One of the main points in the theory of the previous section was that at any nite frequency there is a nite number of guided modes. So, we should prove that each of the four characteristic equations has, for a xed nite value of ω, a nite number of solutions. Let us prove it for Eq. (10.14), leaving to the reader as an exercise the very similar proofs for the other three equations.

By means of Eq. (10.15), and imposing that $_{cl}$ is real in order not to lose the physical features of the guided modes, one derives:

$$|k_{co}| \quad \omega\sqrt{\;(\; _{co} \quad _{cl})} = \sqrt{\; ^2_{co} \quad ^2_{cl}} \quad , \qquad (10.18)$$

where the equal sign holds only for $\omega = 0$. Furthermore, by means of Eq. (10.15),

Table 10.1 *TE modes of the symmetrical slab waveguide.*

EVEN TE MODES

Characteristic equation: $k_{co} \tan k_{co} a = \quad _{cl}$

Field components:

Core	Cladding
$\mathcal{H}_z = C \sin k_{co} x$	$\mathcal{H}_z = C \sin k_{co} a \dfrac{x}{\lvert x \rvert} e^{\quad _{cl}(\lvert x \rvert \quad a)}$
$\mathcal{E}_y = \dfrac{j\omega}{k_{co}} C \cos k_{co} x$	$\mathcal{E}_y = \dfrac{j\omega}{\quad _{cl}} C \sin k_{co} a\, e^{\quad _{cl}(\lvert x \rvert \quad a)}$
$\mathcal{H}_x = \quad \dfrac{}{\omega} \mathcal{E}_y$	$\mathcal{H}_x = \quad \dfrac{}{\omega} \mathcal{E}_y$

$$\mathcal{E}_x = \mathcal{H}_y = 0$$

ODD TE MODES

Characteristic equation: $k_{co} \cot k_{co} a = \quad _{cl}$

Field components:

Core	Cladding
$\mathcal{H}_z = C \cos k_{co} x$	$\mathcal{H}_z = C \cos k_{co} a\, e^{\quad _{cl}(\lvert x \rvert \quad a)}$
$\mathcal{E}_y = \quad \dfrac{j\omega}{k_{co}} C \sin k_{co} x$	$\mathcal{E}_y = \quad \dfrac{j\omega}{\quad _{cl}} \dfrac{x}{\lvert x \rvert} C \cos k_{co} a\, e^{\quad _{cl}(\lvert x \rvert \quad a)}$
$\mathcal{H}_x = \quad \dfrac{}{\omega} \mathcal{E}_y$	$\mathcal{H}_x = \quad \dfrac{}{\omega} \mathcal{E}_y$

$$\mathcal{E}_x = \mathcal{H}_y = 0$$

Eq. (10.14) can be rewritten as:

$$\tan k_{co} a = \frac{\sqrt{\quad ^2_{co} \quad ^2_{cl} \quad k^2_{co}}}{k_{co}} \quad . \tag{10.19}$$

The left-hand side of Eq. (10.19) is a periodic function of k_{co}, the right-hand side is a monotonically decreasing function of k_{co} (see Figure 10.3). Within the range de ned by Eq. (10.18), a periodic curve and a monotonical one can intersect each other only at a nite number of points, hence the number of guided modes is nite.

Eq. (10.10) and the inequality in Eq. (10.18) entail that the phase constant of any guided mode, $\quad = \quad j \quad$, satis es, for $\omega \neq 0$, the following constraints:

$$_{cl} < \quad < \quad _{co} \quad , \tag{10.20}$$

which were also anticipated in the general discussion of Section 10.2.

As ω tends to zero, Eq. (10.14) always has one solution, the corresponding phase constant also going to zero, together with k_{co} and $\quad _{cl}$. This result indicates that one guided TE mode may propagate in the slab at an arbitrarily low frequency, which makes a fundamental di erence with respect to metallic waveguides. The same is true for the rst even TM mode as well, but not for any odd mode, neither TE nor TM. Physically, these lowest-order guided

Table 10.2 *TM modes of the symmetrical slab waveguide.*

EVEN TM MODES

Characteristic equation: $\frac{1}{\ _{co}} k_{co} \tan k_{co}a = \frac{1}{\ _{cl}} \ _{cl}$

Field components:

Core	Cladding
$\mathcal{E}_z = C \sin k_{co}x$	$\mathcal{E}_z = C \sin k_{co}a \frac{x}{\lvert x\rvert} e^{\ _{cl}(\lvert x\rvert \ a)}$
$\mathcal{H}_y = \ \frac{j\omega \ _{co}}{k_{co}} C \cos k_{co}x$	$\mathcal{H}_y = \ \frac{j\omega \ _{cl}}{\ _{cl}} C \sin k_{co}a\, e^{\ _{cl}(\lvert x\rvert \ a)}$
$\mathcal{E}_x = \frac{}{\omega \ _{co}} \mathcal{H}_y$	$\mathcal{E}_x = \frac{}{\omega \ _{cl}} \mathcal{H}_y$

$$\mathcal{E}_y = \mathcal{H}_x = 0$$

ODD TM MODES

Characteristic equation: $\frac{1}{\ _{co}} k_{co} \cot k_{co}a = \ \frac{1}{\ _{cl}} \ _{cl}$

Field components:

Core	Cladding
$\mathcal{E}_z = C \cos k_{co}x$	$\mathcal{E}_z = C \cos k_{co}a\, e^{\ _{cl}(\lvert x\rvert \ a)}$
$\mathcal{H}_y = \frac{j\omega \ _{co}}{k_{co}} C \sin k_{co}x$	$\mathcal{H}_y = \frac{j\omega \ _{cl}}{\ _{cl}} \frac{x}{\lvert x\rvert} C \cos k_{co}a\, e^{\ _{cl}(\lvert x\rvert \ a)}$
$\mathcal{E}_x = \frac{}{\omega \ _{co}} \mathcal{H}_y$	$\mathcal{E}_x = \frac{}{\omega \ _{cl}} \mathcal{H}_y$

$$\mathcal{E}_y = \mathcal{H}_x = 0$$

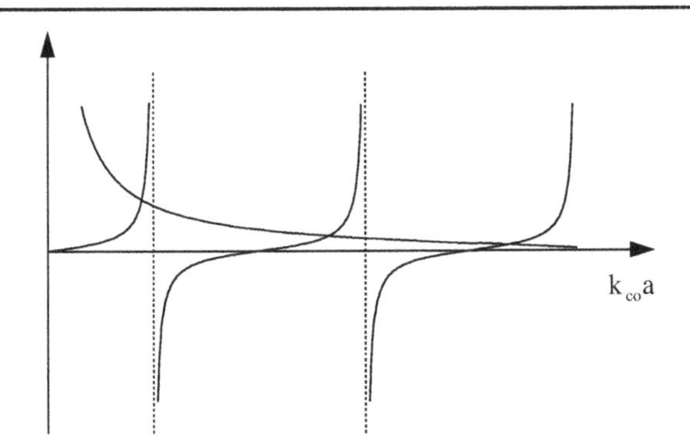

Figure 10.3 *Graphical solution of the characteristic equation of the even TE modes.*

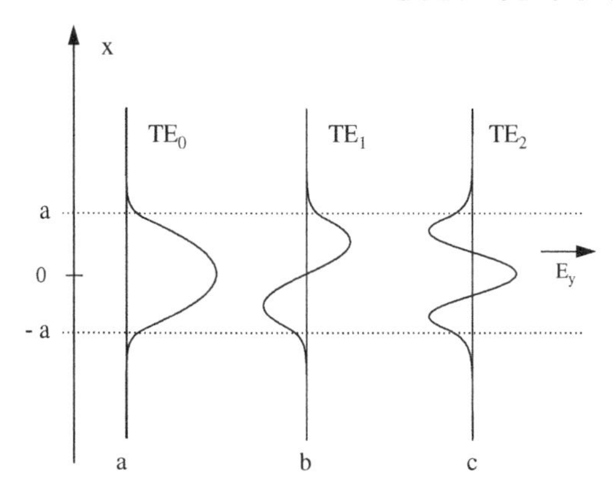

Figure 10.4 *Examples of low-order TE modes in a slab waveguide. a) Lowest-order even mode. b) Lowest-order odd mode. c) First higher-order even mode.*

modes tend to become uniform plane waves in the cladding when ω tends to zero, thus becoming less and less sensitive to the presence of a core. This can never be the case for an odd mode, where the elds in the two half-spaces which form the cladding, have opposite phase. To recon rm that this is the correct physical explanation of the di erence in cut-o behavior between the lowest-order even and odd modes, the reader may prove, as an exercise, that in an asymmetrical slab waveguides where the upper and lower half-spaces have di erent permittivities, all the guided modes have a non-zero cut-o frequency.

Amplitude distributions of the various eld components, and eld lines, for the lowest-order even and odd guided modes of a slab waveguide, are sketched in Figure 10.4.

10.4 Radiation modes of a slab waveguide

For the radiation modes of a slab waveguide, there are no physical grounds to eliminate, in the cladding, any of the oscillatory solutions (or of the two exponential solutions with imaginary argument) of Eq. (10.9). Therefore, we let:

$$X_{cl} = q\,e^{jk_{cl}x} + p\,e^{\ \ jk_{cl}x} \quad , \tag{10.21}$$

where p and q are two arbitrary constants. Eq. (10.15) is replaced by:

$$k_{co}^2 \quad k_{cl}^2 = \omega^2 \ (\ _{co} \quad _{cl}) \quad . \tag{10.22}$$

Continuity of tangential components at $x = \ \ a$ yields, both for TE and for TM modes, *two linear homogeneous equations in three unknowns*, p, q (from Eq. (10.21)) and C (from Eq. (10.11)). The Rouche-Capelli theorem says that such a system has an in nity of nontrivial solutions for any values of the coe cients. Hence, we cannot nd additional conditions on k_{co} and k_{cl}. The *radiation modes are a continuum*, whose transverse eld parameters are related only by Eq. (10.22). They must satisfy only one restriction, coming from Eq. (10.10) as a consequence of $k_{cl}^2 > 0$, which is necessary in order to have, as required, an oscillatory behavior in the cladding. The restriction reads:

$$\quad_{cl} \quad , \qquad (10.23)$$

where is now the phase constant of the generic radiation mode.

We leave it to the reader as an exercise to study the *evanescent modes*, namely the solutions with real , which must satisfy the restriction $k_i \quad {}_i$ ($i = co, cl$), coming again from Eq. (10.10).

Let us strengthen the links between the mathematical description of radiation modes and physical intuition, and try to clarify why Eq. (10.21) contains two terms, one corresponding to a wave traveling away from the slab, the other one to a wave traveling towards the slab. It would be erroneous to believe that radiation modes must be able to describe only elds excited by a source located at $|x| < a$, inferring from this starting point that waves traveling towards the slab are not physically meaningful. There are at least three reasons why this viewpoint would be wrong;

a) To be complete, a set of modes must be able to describe a eld radiated by a source located on one side of the slab, at an arbitrary distance from it. This also requires waves which travel towards the slab;

b) A cladding extending inde nitely in the x direction is a physically acceptable model only if it represents the limit of a cladding with a nite thickness, d, when d goes to in nity. Centripetal waves are needed in order to describe that part of the radiated eld which is re ected back by the external edge of the cladding. This also tells us that it would be erroneous to disregard centripetal waves;

c) The whole centripetal eld, expressed as an integral over the continuum of radiation modes, can carry zero power, without implying zero amplitude for the individual centripetal waves. This is a well known property of the Fourier integral, and the reader is probably familiar with it from time-domain signal analysis.

All these comments recon rm that, as we remarked in Section 10.2, dielectric and metallic waveguides are totally di erent concerning that part of the eld which does not propagate under the form of guided modes. In dielectric guides, each transmission mode loses its individual features as soon as it is cut-o . The "invention" of leaky modes can be thought of as an attempt to eliminate this di erence, using another description for that part of the eld which complements the guided modes.

A non-real solution of the characteristic equation, in the complex eld, could be looked at as an "extension" of a real solution, when a guided mode has gone through a cut-o condition, i.e., when the angular frequency is below a critical value ω_{cn}, similarly to what occurs in Chapter 7. Every complex solution of the characteristic equation would entail Re $\neq 0$, and thus an attenuation which is an increasing function of $\omega_{cn} \quad \omega$. These solutions form

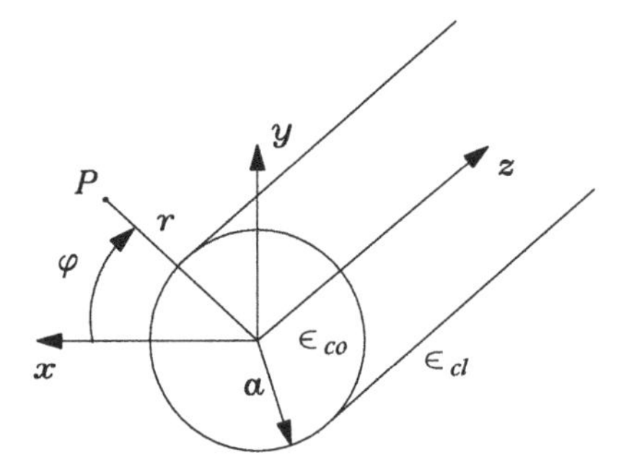

Figure 10.5 *Step-index circular dielectric rod.*

a countable in nity and, as mentioned in Section 10.2, are often referred to as *leaky modes*. However, let us stress once again that the term "modes" might be misleading. As we said in general in Section 10.2, and as we saw in detail in Section 10.3, the characteristic equation was obtained on the grounds of a well de ned physical assumption — a guided mode — and there is no reason why it should hold when the physics changes. To appreciate the usefulness of leaky modes as an approximation, and to appreciate the mathematical subtleties of the procedure, the reader is referred to specialized books (Snyder and Love, 1983).

10.5 The cylindrical rod: Exact modes

This section is devoted to the *guided modes* of a cylindrical dielectric rod, of radius a (Figure 10.5). This example is very important under a practical viewpoint, because of the impact of optical bers in modern telecommunication technology and in other related elds, like sensors. We will show that these guided modes can be identi ed rigorously, but using their exact expressions makes it complicated to express some important quantities, like power density owing along the waveguide. For this reason, we will develop an approximate theory in the following section. Radiation modes of a cylindrical waveguide do not exhibit signi cant di erences with respect to those of a slab waveguide, except for mathematical details (Hankel functions replacing exponentials of imaginary argument) which can be left to the reader as an exercise.

Let us follow the general procedure of Section 10.2, taking into account the contents of Section 7.8 about the Helmholtz equation in cylindrical coordinates. A physical selection criterion is to disregard those solutions of Bessel's equations which diverge, either for $r < a$ or for $r > a$. Accordingly, we can set:

$$\left.\begin{matrix} M_i \\ N_i \end{matrix}\right\} = e^{j\ \varphi} \qquad \begin{matrix} i = co, cl \quad ; \\ \quad = 0, \quad 1, \quad 2, \ldots \quad ; \end{matrix} \tag{10.24}$$

$$\left.\begin{array}{l} Q_{co} \\ R_{co} \end{array}\right\} = J \ (k_{co} r) = J \ \left(\frac{ur}{a}\right) \quad , \tag{10.25}$$

where J is the Bessel function of the rst kind and of integer order , and:

$$\left.\begin{array}{l} Q_{cl} \\ R_{cl} \end{array}\right\} = K \ (\ _{cl} r) = K \ \left(\frac{wr}{a}\right) \quad , \tag{10.26}$$

where K is the modi ed Bessel function of the second kind and of integer order . The following adimensional quantities have been de ned in these equations:

$$u = k_{co} a = a \sqrt{\omega^2 \quad _{co} \quad ^2} = a \sqrt{\omega^2 \quad _{co} + \ ^2} = a \sqrt{\ ^2 \quad ^2_{co}} \quad ,$$

$$w = \ _{cl} a = a \sqrt{\ \omega^2 \quad _{cl} + \ ^2} = a \sqrt{\ (\ ^2 + \omega^2 \quad _{cl})} = a \sqrt{\ ^2_{cl} \quad ^2} \ , \tag{10.27}$$

where $\ = \ j$ is the phase constant of the generic guided mode.

As for the harmonic equation which governs the dependence on the azimuthal coordinate, we chose to express its solutions in terms of exponential functions, Eq. (10.24), rather than in terms of cos(φ) and sin(φ), because the continuity conditions, Eqs. (10.6), involve the functions and also their derivatives, so Eqs. (10.24) end up making the calculations easier.

The metric coe cients are $h_1 = 1$, $h_2 = r$. Following a standard notation, let us use J' and K' for the derivatives of Bessel functions *with respect to their arguments*. Then, Eqs. (10.6) read as follows:

$$\begin{aligned}
A_{co} u \, J' (u) \quad B_{co} \frac{}{j\omega \quad _{co}} J \ (u) \quad &= \quad A_{cl} w \, K' (w) \quad B_{cl} \frac{}{j\omega \quad _{cl}} K \ (w) \quad , \\
B_{co} \frac{u^2 J \ (u)}{j\omega \quad _{co}} \quad &= \quad B_{cl} \frac{w^2 K \ (w)}{j\omega \quad _{cl}} \quad , \\
A_{co} \frac{J \ (u)}{j\omega} + B_{co} u \, J' (u) \quad &= \quad A_{cl} \frac{K \ (w)}{j\omega} + B_{cl} w \, K' (w) \quad , \\
A_{co} \, u^2 \, J \ (u) \quad &= \quad A_{cl} \, w^2 \, K \ (w) \quad .
\end{aligned} \tag{10.28}$$

For $\ = 0$, we can separate Eqs. (10.28) into two sets of two equations, one in the two unknowns (A_{co}, A_{cl}), the other in the two unknowns (B_{co}, B_{cl}). In fact, it is not surprising that these modes, independent of the azimuthal coordinate φ, behave similarly to those independent of the y coordinate in the slab waveguide of Section 10.3. The set which consists of the rst and the fourth of Eqs. (10.28) corresponds to TE modes, and yields the following characteristic equation:

$$\frac{J_0'(u)}{u \, J_0(u)} = \frac{K_0'(w)}{w \, K_0(w)} \qquad (\mathrm{TE}_{0,p}) \quad . \tag{10.29}$$

The set which consists of the second and the third of Eqs. (10.28) corresponds to TM modes, and yields the following characteristic equation:

$$ {}_{co} \frac{J_0'(u)}{u\,J_0(u)} = {}_{cl} \frac{K_0'(w)}{w\,K_0(w)} \qquad (\mathrm{TM}_{0,p}) \quad . \tag{10.30}$$

We see from these characteristic equations that the $\mathrm{TE}_{0,p}$ and the $\mathrm{TM}_{0,p}$ modes are not degenerate (${}_{co} \neq {}_{cl}$), but tend to become degenerate as ${}_{co} \to {}_{cl}$. This comment will be expanded and used in the following section.

Modes whose azimuthal order is not equal to zero are *hybrid modes*, as the set of Eqs. (10.28), for $\neq 0$, does not break down into two separate sets. Their characteristic equation can be written in a compact form if we set:

$$ j'\,(u) = \frac{J'\,(u)}{u\,J\,\,(u)} \quad , \qquad b'\,(w) = \frac{K'\,(w)}{w\,K\,\,(w)} \quad , \tag{10.31}$$

and introduce an adimensional quantity, v, proportional to the angular frequency ω, de ned as follows:

$$ v^2 = u^2 + w^2 = a^2\,\omega^2\,\,({}_{co} \quad {}_{cl}) \quad . \tag{10.32}$$

Dividing the rst and the third of Eqs. (10.28) by $J\,\,K$, eliminating A_{cl} and B_{cl} by means of the second and the forth equation and, nally, setting the determinant of the remaining system of two equations in A_{co}, B_{co} equal to zero, the characteristic equation then reads:

$$ (j'\,+b'\,)\,(\,{}^2_{co} j' + \,{}^2_{cl} b'\,) = \,{}^2\,{}^2 \left(\frac{1}{u^2} + \frac{1}{w^2}\right)^2 \quad , \tag{10.33}$$

where u, w and are related through Eqs. (10.27). Another, perhaps more elegant way to write the characteristic equation is:

$$ (j'\,+b'\,)\,(\,{}_{co} j' + \,{}_{cl} b'\,) = \,{}^2 \left(\frac{1}{u^2} + \frac{1}{w^2}\right) \left(\frac{{}_{co}}{u^2} + \frac{{}_{cl}}{w^2}\right) \quad , \tag{10.34}$$

which must be associated with Eq. (10.32). To reach this result, we divide Eq. (10.33) by v^2, write $1/u^2 + 1/w^2 = v^2/(u^2 w^2)$, replace v^2 with Eq. (10.32), and nally, make use of the following expression, which can be obtained taking the squares of the two Eqs. (10.27), multiplying the rst by ${}_{cl}$ and the second by ${}_{co}$, and summing them:

$$ {}^2\,a^2(\,{}_{co} \quad {}_{cl}) = {}_{co}\,w^2 + \,{}_{cl}\,u^2 \quad . \tag{10.35}$$

Note that Eq. (10.34) breaks into Eqs. (10.29) and (10.30) for $= 0$. For $\neq 0$, Eq. (10.34) is invariant with respect to the change $\to$, since (see Appendix D) $j'\,\, = j'\,\,$, $b'\,\, = b'\,\,$. Therefore, each solution of Eq. (10.34) for $\neq 0$ corresponds to *a pair of degenerate modes*. This agrees with physical intuition: waves propagating helically in an isotropic medium cannot distinguish between traveling clockwise or counterclockwise. This enables us to assume > 0 from now on, to simplify our calculations.

Our next aim is to show that the solutions of the characteristic equation may be separated into two sets, which are physically di erent from each other.

A suitable way to reach this result is simply to rewrite Eq. (10.34) as a second-order algebraic equation in j', and to solve it, while introducing, for future bene ts, two new variables, namely:

$$\bar{\ } = \frac{_{co} + \ _{cl}}{2} \quad , \qquad = \ _{co} \quad _{cl} \quad , \tag{10.36}$$

and using a concise notation, $f^2(u, w)$, for the right-hand side of Eq. (10.34). Make use of the following expressions for derivatives of Bessel functions (see Appendix D):

$$J' \ (u) = \frac{\ J \ (u)}{u} \quad J_{\ +1}(u) \quad , \tag{10.37}$$

when taking the positive square root of the discriminant, and:

$$J' \ (u) = \quad \frac{\ J \ (u)}{u} + J_{\ \ 1}(u) \quad , \tag{10.38}$$

when taking the negative one. The nal results are, respectively:

$$\frac{J_{\ +1}(u)}{J \ (u)} = \frac{\bar{\ }}{_{co}} ub' + \frac{\ }{u} \quad \frac{u}{_{co}} \left[\frac{\ ^2}{4} b'^{\ 2} + \ _{co} f^2(u, w) \right]^{1/2} \tag{10.39}$$

for the plus sign in front of the discriminant, and:

$$\frac{J_{\ \ 1}(u)}{J \ (u)} = \quad \frac{\bar{\ }}{_{co}} ub' + \frac{\ }{u} \quad \frac{u}{_{co}} \left[\frac{\ ^2}{4} b'^{\ 2} + \ _{co} f^2(u, w) \right]^{1/2} \quad , \tag{10.40}$$

for the negative one. Let us emphasize once again that Eqs. (10.39) and (10.40) assume $\ > 0$, and each solution of either one corresponds to a *pair of degenerate modes* having identical behaviors vs. the radial coordinate, given by Eqs. (10.25) and (10.26), while they di er in the sign in front of the azimuthal coordinate, exp($j \ \varphi$). The solutions of Eq. (10.39) are referred to as the *EH* $_{\ ,p}$ *modes*, those of Eq. (10.40) as the *HE* $_{\ ,p}$ *modes*.

Radial and azimuthal components of the e.m. eld may be obtained from the longitudinal components, i.e., from Eqs. (10.24), (10.25) and (10.26), expressing Eqs. (7.4) in cylindrical coordinates. They are summarized in Table 10.3. The reader can verify, as an exercise, that radial components satisfy the following continuity conditions (which, as shown in Chapter 1, are *not* independent of those which hold for tangential components):

$$\begin{aligned} _{co} \mathcal{E}_{rco}(r = a) &= \ _{cl} \mathcal{E}_{rcl}(r = a) \quad , \\ \mathcal{H}_{rco}(r = a) &= \mathcal{H}_{rcl}(r = a) \quad . \end{aligned} \tag{10.41}$$

Table 10.3 shows that the longitudinal components of these exact modal elds are fairly simple but, on the other hand, transverse components are complicated, mainly because Bessel functions of integer order can be written as superpositions of two Bessel functions of di erent orders (to choose between $\ , \ +1$, and $\ \ 1$). Transverse components are fundamental in practice, e.g., to calculate the Poynting vector. Furthermore, it can be shown (see the rest of the chapter) that, in agreement with physical intuition, when the di erence

Table 10.3 *Exact modes of the circular step-index rod.*

	Core (1)	Cladding (1)
$\mathcal{H}_z$	$A J (u) e^{j \varphi}$	$A \frac{J (u)}{K (w)} K (w) e^{j \varphi}$
$\mathcal{E}_z$	$B J (u) e^{j \varphi}$	$B \frac{J (u)}{K (w)} K (w) e^{j \varphi}$
$\mathcal{E}_r$	$\frac{j}{u^2} e^{j \varphi}$	$\frac{j}{w^2} \frac{J (u)}{K (w)} e^{j \varphi}$
	$\left[B \ au J'(u) + A \frac{j\omega \ 0 \ a}{} J (u)\right]$	$\left[B \ aw K'(w) + A \frac{j\omega \ 0 \ a}{} K (w)\right]$
$\mathcal{E}_\varphi$	$\frac{j}{u^2} e^{j \varphi}$	$\frac{j}{w^2} \frac{J (u)}{K (w)} e^{j \varphi}$
	$\left[B \frac{j \ a}{} J (u) \quad A\omega \ _0 aw J'(w)\right]$	$\left[B \frac{j \ a}{} K (w) \quad A\omega \ _0 aw K'(w)\right]$
$\mathcal{H}_r$	$\frac{\omega \ co}{} \mathcal{E}_\varphi$	$\frac{\omega \ cl}{} \mathcal{E}_\varphi$
$\mathcal{H}_\varphi$	$\frac{\omega \ co}{} \mathcal{E}_r$	$\frac{\omega \ cl}{} \mathcal{E}_r$

HE modes: $B = j \sqrt{\frac{0}{co}} A$

EH modes: $B = \quad j \sqrt{\frac{0}{co}} A$

$_{co}$ $_{cl}$ is not too large, the magnitude of the transverse components is much larger than that of longitudinal components. The practical need to simplify transverse eld components is a valid reason why the approximate theory of Section 10.7 is often preferred to the exact theory presented in this section.

10.6 Modal cut-o in the cylindrical rod

A suitable way to clarify the physical di erence between EH and HE modes, is to analyze their *cut-o conditions.* As we found in general in Section 10.2, and saw recon rmed by the example of Section 10.3, a guided mode is cut-o when its eld does not decay, in the radial direction, fast enough to prevent power from owing in the radial direction. This is translated into mathematical terms as $w \to 0$, to be replaced in the previous equations. To proceed further, we make use of the following approximations (Abramowitz and Stegun, Eds., 1965) for small values of w: $b'(w) \simeq \quad /w$ in the case of Eq. (10.39), and $b'(w) \simeq \quad (\ /w) \quad 1/[2(\quad 1)]$ in the case of Eq. (10.40). Keeping also track that Eq. (10.32) implies that at cut-o $(w = 0)$ it is $u = v = v_c$ (critical value of v parameter), after some algebra one nds that the cut-o condition arising

Table 10.4 *Cut-o normalized frequencies for some low-order modes.*

v_c	0	2.405	3.832	5.136	5.520
Modes	HE_{11}	TE_{01}, TM_{01}	EH_{11}, HE_{12}	EH_{21}	TE_{02}, TM_{02}

from Eq. (10.39) reads:

$$J\ (v_c) = 0 \qquad (\text{EH}\ _{,p}\ \text{modes}\ ; \quad = 1, 2, \ldots)\quad , \tag{10.42}$$

while that arising from Eq. (10.40) reads:

$$v_c \frac{J\ (v_c)}{J\ \ _1(v_c)} = \frac{2(\quad 1)}{_{cl}} \qquad (\text{HE}\ _{,p}\ \text{modes}\ ; \quad = 1, 2, \ldots)\quad . \tag{10.43}$$

In particular, the latter equation, for $= 1$ and $v_c \neq 0$, gives $J\ (v_c) = 0$, showing that *at the cut-o* (but *only* at cut-o) $HE_{1,p}$ modes and $EH_{1,p\ \ 1}$ modes are degenerate.

Eqs. (10.42) and (10.43) show that, for any nite value of the normalized frequency v de ned by Eq. (10.32), each family of guided modes is a nite set. This result could also be predicted similarly to what was carried out in Section 10.3 on the characteristic equation of the slab modes, i.e., observing that one side of these equations is monotonic, while the other one is a function which goes, sequentially, through zero and through in nity.

Eq. (10.43) also shows that the *fundamental mode* is the $HE_{1,1}$ mode, which can propagate at an arbitrarily small frequency ($v_c = 0$) like the fundamental mode of a slab. We will see in the following sections that fundamental modes of a slab and of a cylindrical rod are physically very similar, in spite of their labels appearing to be completely di erent from each other. A very practical matter is to establish which are the rst higher-order modes. The question can be answered letting $w = 0$ in Eqs. (10.29) and (10.30), and this yields $J_0(v_c) = 0$. Therefore, the next modes are the $TE_{0,1}$ and the $TM_{0,1}$. They are degenerate only at cut-o . The cut-o value for the normalized frequency is $v_c = 2.405$. Cut-o values of the normalized frequency for other modes are shown in Table 10.4.

For a better understanding of the role played by various parameters in the design of practical single-mode and multimode bers, let us rewrite the normalized frequency, de ned by Eq. (10.32), as follows, in terms of the refractive indices of core and cladding:

$$v = ka\, n_{cl}\, \sqrt{2\ \ } = \frac{2\ \ a}{_0}\, n_{cl}\, \sqrt{2\ \ } \quad , \tag{10.44}$$

where $k = \omega \sqrt{\ _0\ _0} = 2\ \ /\ \ _0$ is the free-space wave number, and:

$$2 \qquad \frac{n_{co}^2 \quad n_{cl}^2}{n_{cl}^2} \tag{10.45}$$

is the relative index di erence. In comparison with Chapter 7, we notice that, while in a conducting-wall waveguide the cut-o frequency of a given mode

Table 10.5 *Typical values of refractive index for glasses in common use.*

Pure silica (SiO_2)	1.4585
GeO_2-doped silica	1.4585 – 1.465
F-doped silica	1.445 – 1.4585
Halide glasses	1.41 – 1.72
Chalcogenide glasses	1.7 – 2.5

depends on the guide size and on the refractive index of the dielectric medium which lls it, in a dielectric waveguide it depends on these parameters plus another one, namely the core-cladding index relative di erence.

For an optical ber to be single-mode, v must be less than 2.405. Wavelengths of highest practical interest (measured in free space) are in the near infrared, more precisely in the range between 0.8 m (the so-called " rst window") and 1.55 m ("third window"; these names are related to historical breakthroughs in the development of optical communications) where silica glass is very transparent (see Section 10.11). The refractive index of pure silica, at those wavelengths, is somewhat less than 1.5 (see Table 10.5). These facts indicate that the product $a\sqrt{2\quad}$ must be smaller than a few units (depending on the exact wavelength) multiplied by $10^{\ 7}$, if a is measured in meters. Actual choices for a and 2 are dictated by other, contradictory requirements (see Snyder and Love, 1983), typically bending loss and coupling loss. However, for all bers which are standard commercial products, a is a few micrometers, and 2 is less than 0.01. These are highly demanding requirements from a technological viewpoint. This point explains why *highly multimode* bers, where a is of the order of tens of micrometers and 2 reaches a few percent so that v 2.405, have attracted extremely serious technical attention for decades, in spite of some major limitation that will be outlined in Section 10.8. On the other hand, few-mode bers, where $v > 2.405$, but not v 2.405, are of interest only for some special applications, like sensors, which are beyond the scope of this book (Dakin and Culshaw, 1988).

Incidentally, let us point out that, as 2 1 in all cases of practical interest, we have $n_{co} + n_{cl} \simeq 2n_{cl}$, so that Eq. (10.45) can be rewritten as

$$\simeq \frac{n_{co} \quad n_{cl}}{n_{cl}} \qquad (\quad 1) \ . \tag{10.46}$$

The practical importance of having 1 will be thoroughly appreciated in the next three sections.

Finally, let us try to explain why the cut-o conditions which were derived from the characteristic equations justify the strange terminology which is in current use for the dielectric waveguide modes. In fact, Eq. (10.42) appears to be the same as the equation which, in a circular waveguide with conducting walls, yields cut-o frequencies for those modes which are most frequently referred to as TM, but are called *E modes* by other authors. As we saw here, modes of a dielectric rod are hybrid; still, as a result of this purely formal similarity, one set was named *EH*, and then, by analogy, the other one was named *HE*.

10.7 Weakly guiding rods: The LP modes

We saw in the previous section that, for most dielectric waveguides used in practice, and in particular for typical optical bres, if we return to the notation of Eq. (10.36), we nd:

$$. \tag{10.47}$$

A dielectric structure where Eq. (10.47) is satis ed is referred to as *weakly guiding*. Introducing this assumption into Eq. (10.34), we nd that *both* characteristic equations, for HE and for EH modes, result in the following approximate form:

$$j' + b' = \left(\frac{1}{u^2} + \frac{1}{w^2}\right) \quad . \tag{10.48}$$

This tells us that HE and EH modes tend to become degenerate as $\rightarrow 0$. Knowing from Section 10.5 that there is a strict two-fold degeneracy related to replacing with , we see that for $\rightarrow 0$ we tend towards a *four-fold degeneracy* of the guided modes, in a dielectric rod.

In mathematical terms, the idea of a "quasi-degeneracy" is not well founded. In any eigenvalue problem, degenerate solutions are critical, in the sense that some di culties show up when two eigenvalues are equal, but disappear as soon as they are not equal, irrespectively of how close they are, provided they are distinct. If we adopt this viewpoint, then two modes which correspond to two di erent eigenvalues can never be linearly combined to yield another mode, even if the two eigenvalues are very close to each other. However, in a weakly guiding dielectric waveguide of circular cross-section, the errors we make if we treat as degenerate EH and HE modes whose eigenvalues are not strictly identical can be shown (Snyder and Love, 1983) to be small, compared to errors that we make neglecting other perturbations, like slight ellipticity, absorption and scattering losses, bends, slight unintentional anisotropy, etc. As a result, the elds that we build up in this section are not exact solutions of Maxwell's equations, and therefore should be called pseudomodes, rather than modes, but they match very well experimental results on actual bers a ected by imperfections, and in this sense can be referred to as modes.

Let us proceed as if the two families of modes were degenerate. It is easy to show that, using the same notation as in the previous section, degeneracy does not occur between the HE $_{,m}$ and the EH $_{,m}$ mode, but *between the HE* $_{,m}$ *and the EH* $_{2,m}$ *mode.* Indeed, using the recurrence formula:

$$\frac{2(\quad 1)\, J \quad _1(u)}{u} = J \ (u) + J \quad _2(u) \quad . \tag{10.49}$$

Eq. (10.43) shows that for $/ \ _{cl} \rightarrow 1$ the cut-o equation for the HE $_{,m}$ mode becomes:

$$J \quad _2(v_c) = 0 \quad , \tag{10.50}$$

which coincides with the cut-o condition for the EH $_{2,m}$ mode, Eq. (10.42), which is insensitive to the approximation $_{cl} \simeq$.

This four-fold degeneracy can be exploited to derive drastically simpler transverse components. The price to pay is to have more complicated longitudinal components, but, as we said before, these are far less used in practical calculations. Furthermore, the reader can prove as an exercise (working, to simplify calculations, on the modes of a slab waveguide or on the TE and TM modes of a cylindrical rod) that the longitudinal components are much smaller than the transverse ones. In a weakly guiding structure, this ratio can be shown to be of order $^{1/2}$, where is still de ned by Eq. (10.45). The procedure which yields the simpli ed transverse elds can be summarized as follows, omitting calculations for the sake of brevity.

To start with, let $\ell = 1, 2, \ldots$. (For $\ell = 0$, there is nothing to add to what was said in Section 10.5). Then:

- take the transverse components of the $\mathrm{HE}_{\ell+1,m}$ mode and those of the $\mathrm{EH}_{\ell \;\, 1,m}$ mode (see Table 10.3);
- use Bessel and modi ed Bessel function recurrence relations to express all these components as linear combinations of Bessel and modi ed Bessel functions of two orders, to be chosen among ℓ, $\ell+1$, $\ell \;\; 1$; for example, let us take those of orders ℓ and $\ell+1$;
- look for a linear combination of the $\mathrm{HE}_{\ell+1,m}$ mode and the $\mathrm{EH}_{\ell \;\, 1,m}$ mode ensuring that it does not contain Bessel functions of one of these remaining orders, say $\ell+1$. Note that this step is not trivial at all, as the $\mathrm{HE}_{\ell+1,m}$ mode and the $\mathrm{EH}_{\ell \;\, 1,m}$ mode have di erent dependences on the azimuthal coordinate, expressed by $\exp[j(\ell \;\; 1)\,\varphi]$, respectively. The reason why one can nd a positive answer to this requirement is precisely the *four-fold mode degeneracy*, i.e., the possibility of combining linearly not only the above mentioned functions of φ, but also $\exp[\;\; j(\ell \;\; 1)\,\varphi]$. The number of independent coe cients one can play with then becomes su cient to eliminate $J_{\ell+1}$ and $K_{\ell+1}$ from all the transverse eld components.

The reader can prove as an exercise that, referring to their dependence on the radial coordinate, the transverse eld components, for the generic mode of this new set, are expressed, respectively, in the core by the upper line and in cladding by the lower line of the following formula:

$$\mathcal{E}_\varphi = E_0'\, F_\ell(\varphi) \begin{cases} J_\ell(ur/a)/J_\ell(u) \\ K_\ell(wr/a)/K_\ell(w) \end{cases} , \qquad \mathcal{H}_r = \;\; \mathcal{E}_\varphi / \;_i \;\; ,$$

$$\mathcal{E}_r = E_0'\, G_\ell(\varphi) \begin{cases} J_\ell(ur/a)/J_\ell(u) \\ K_\ell(wr/a)/K_\ell(w) \end{cases} , \qquad \mathcal{H}_\varphi = \mathcal{E}_r / \;_i \;\; , \quad (10.51)$$

where E_0' is an arbitrary constant, and the pair of functions of the azimuthal coordinate $F_\ell(\varphi), G_\ell(\varphi)$ will be discussed shortly. In Eqs. (10.51) it is $_i =$ $\sqrt{\;\;/\;\;_i} = \;_0/n_i$, n_i being the refractive index of the i-th medium ($i = co, cl$). This result indicates that, when the assumption 1 holds, for any guided

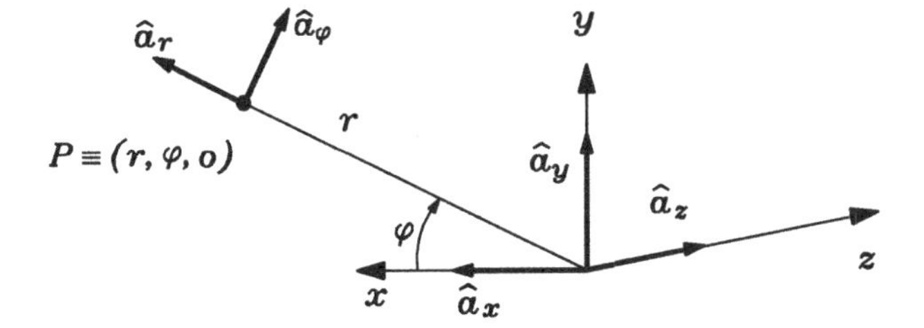

Figure 10.6 *Rectangular and cylindrical reference frames in a circular rod.*

mode expressed by Eqs. (10.51) the *wave impedance* in the direction of propagation is equal to the *intrinsic impedance* of the medium. Consequently, comparing with Chapters 5 and 7, we see that we are dealing with a *quasi-TEM* propagation.

Let us focus now on the azimuthal dependence of the transverse elds. Calculations show that the pair of functions in Eqs. (10.51) can be any among the following four choices (incidentally, note that the four-fold degeneracy has been preserved):

$$\begin{cases} F_\ell = \sin(\ell+1)\,\varphi \\ G_\ell = \quad \cos(\ell+1)\,\varphi \end{cases} ; \qquad \begin{cases} F_\ell = \sin(\ell \quad 1)\,\varphi \\ G_\ell = \cos(\ell \quad 1)\,\varphi \end{cases} ;$$
$$\begin{cases} F_\ell = \cos(\ell+1)\,\varphi \\ G_\ell = \sin(\ell+1)\,\varphi \end{cases} ; \qquad \begin{cases} F_\ell = \cos(\ell \quad 1)\,\varphi \\ G_\ell = \quad \sin(\ell \quad 1)\,\varphi \end{cases} . \tag{10.52}$$

At rst sight, these results may look disappointing, because the φ-dependence is not very simple and is actually not even the same for the $\mathcal{E}_\varphi$ and $\mathcal{E}_r$ components. Fortunately, these problems are very easy to circumvent. Indeed, the eld components in the rectangular coordinate frame shown in Figure 10.6 are related to Eqs. (10.51) as:

$$\mathcal{E}_x = \mathcal{E}_r \cos\varphi \quad \mathcal{E}_\varphi \sin\varphi ,$$
$$\mathcal{E}_y = \mathcal{E}_r \sin\varphi + \mathcal{E}_\varphi \cos\varphi . \tag{10.53}$$

Starting from the four choices expressed by Eqs. (10.52), Eqs. (10.53) yield the following four possibilities:

$$\begin{cases} \mathcal{E}_x \propto \quad \cos\ell\varphi \\ \mathcal{E}_y \propto \sin\ell\varphi \end{cases} ; \qquad \begin{cases} \mathcal{E}_x \propto \cos\ell\varphi \\ \mathcal{E}_y \propto \sin\ell\varphi \end{cases} ; \tag{10.54}$$

$$\begin{cases} \mathcal{E}_x \propto \sin\ell\varphi \\ \mathcal{E}_y \propto \cos\ell\varphi \end{cases} ; \qquad \begin{cases} \mathcal{E}_x \propto \quad \sin\ell\varphi \\ \mathcal{E}_y \propto \cos\ell\varphi \end{cases} . \tag{10.55}$$

Finally, summing and subtracting Eqs. (10.54), and Eqs. (10.55), we get the following four choices:

$$\mathcal{E}_x = 0 , \quad \mathcal{E}_y = E_0\, R(r)\, \sin\ell\varphi ,$$
$$\mathcal{E}_x = 0 , \quad \mathcal{E}_y = E_0\, R(r)\, \cos\ell\varphi , \tag{10.56}$$

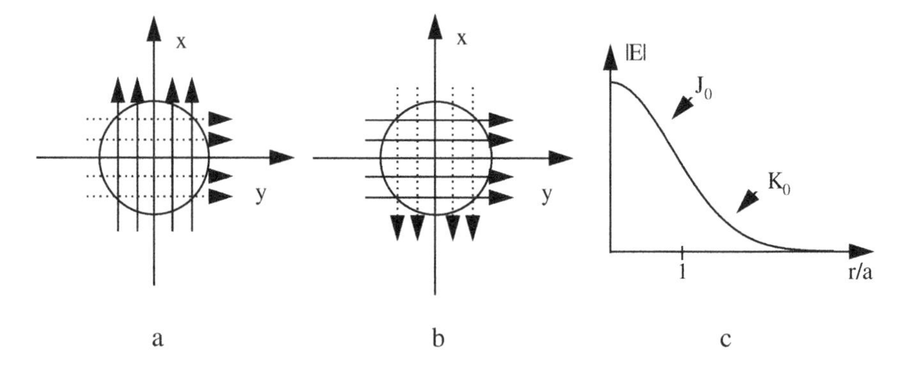

Figure 10.7 *Schematic illustration of the fundamental ($LP_{0,1}$) mode. Part a: x-polarization. Part b: y-polarization (continuous lines = electric eld; dashed lines = magnetic eld). Part c: transverse eld amplitude (arbitrary units) vs. normalized radial coordinate.*

$$\begin{aligned} \mathcal{E}_x &= E_0\ R(r)\ \cos \ell\varphi \quad , \quad \mathcal{E}_y = 0 \quad , \\ \mathcal{E}_x &= E_0\ R(r)\ \sin \ell\varphi \quad , \quad \mathcal{E}_y = 0 \quad , \end{aligned} \tag{10.57}$$

where E_0 is an arbitrary constant and $R(r)$ is the quantity in brackets in Eqs. (10.51).

Each of these electric elds is accompanied by a magnetic eld expressed by:

$$\mathcal{H}_y = \frac{\mathcal{E}_x}{i} \quad , \quad \mathcal{H}_x = \frac{\mathcal{E}_y}{i} \quad . \tag{10.58}$$

Accordingly, we see that the transverse eld of each of these modes consists of one cartesian component for the electric eld, and one cartesian component for the magnetic eld orthogonal to the electric one and with the same radial distribution. This explains why these modes are said to be *linearly polarized.* The corresponding concise notation is $\mathrm{LP}_{\ell,p}$, where $\ell = 0, 1, 2, \ldots$ and $p = 1, 2, \ldots$ Note that:

- the $\mathrm{LP}_{\ell,p}$ mode with $\ell > 1$ is the combination of the $\mathrm{HE}_{\ell+1,p}$ mode and the $\mathrm{EH}_{\ell\ 1,p}$ mode;
- the $\mathrm{LP}_{1,p}$ mode is the combination of the $\mathrm{HE}_{2,p}$ mode, the $\mathrm{TE}_{0,p}$ mode and the $\mathrm{TM}_{0,p}$ mode;
- the $\mathrm{LP}_{0,p}$ mode is the weakly-guiding approximation of the $\mathrm{HE}_{1,p}$ mode;
- any $\mathrm{LP}_{\ell,p}$ mode with $\ell > 0$ exhibits a four-fold degeneracy, while an $\mathrm{LP}_{0,p}$ mode is two-fold degenerate.

Using the new terminology, the *fundamental mode* is the $\mathrm{LP}_{0,1}$ mode. Its eld lines and amplitude distribution are sketched in Figure 10.7. The gure illustrates why the fundamental mode is doubly degenerate: it is a matter of polarization. However, degeneracy is broken when circular symmetry is broken, either because the core cross-section is not circular (*shape birefringence*), or because the medium is anisotropic as a result of internal force distribution

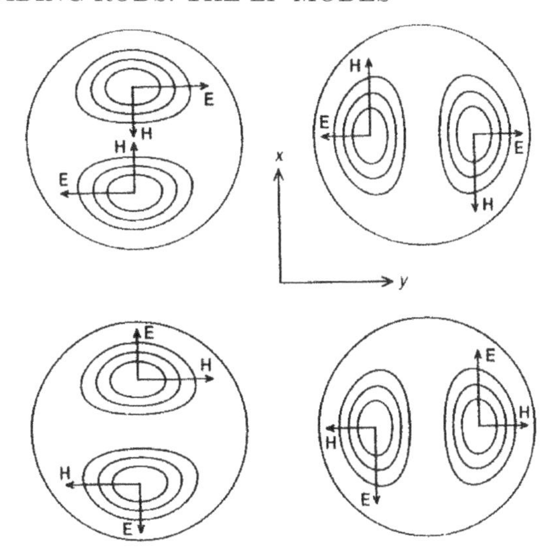

Figure 10.8 *Constant-amplitude contour plots for the rst higher-order ($LP_{1,1}$) mode.*

(*stress birefringence*). Usually, in silica glass, the latter e ect dominates over the rst one. In some cases (high-birefringence bers), stress anisotropy is intentional, and meant to preserve the state of polarization of the guided eld. In typical telecommunication bers, birefringence is unwanted, and varies at random along the ber. It results in distortion of the signal which travels along a single-mode ber, and is usually referred to as *polarization mode dispersion* (Matera and Someda, 1992; Galtarossa and Menyuk, 2005), and probably easier to understand after reading the next section.

The rst higher-order mode, whose normalized cut-o frequency remains $v_c = 2.405$, is now called $\mathrm{LP}_{1,1}$ mode. Its eld distribution is sketched in Figure 10.8, which illustrates the sources of the four-fold degeneracy: one two-fold degeneracy is related to polarization, while the other two-fold degeneracy is related to a rotation of the amplitude distribution around the z axis by an angle $/(2\ell)$.

The eld components of the LP modes are summarized in Table 10.6, which recon rms that, as expected, longitudinal components become complicated while the transverse ones are simpli ed.

The characteristic equation of the LP modes, Eq. (10.48), can be expressed in another way, which will be useful for further developments in the next section. It is also easier to correlate with physical intuition, arising from the elds shown in Table 10.6. Let us take Eq. (10.48) with the plus sign, the expression for j' which results from Eq. (10.37), and the expression for b' which results from:

$$K'\ (w) = \frac{K\ (w)}{w} \quad K_{\ +1}(w) \quad . \qquad (10.59)$$

Table 10.6 *LP modes of the circular step-index rod.*

The expressions shown in this table are related to the second among Eqs. (10.56). The other three cases can be obtained from these with the following replacements:

$$\ell\varphi_1 = \frac{}{2} \quad \ell\varphi \quad ; \quad \varphi_3 = \frac{}{2} \quad \varphi \quad ; \quad \ell\varphi_4 = \ell\varphi \quad (\ell \quad 1)\frac{}{2} \quad .$$

Core

$$\mathcal{E}_y = E_0 \frac{J_\ell(ur/a)}{J_\ell(u)} \cos \ell\varphi$$

$$\mathcal{H}_x = \quad n_{co} \frac{\mathcal{E}_y}{\ \ 0}$$

$$\mathcal{E}_z = \quad j \frac{E_0}{2ka} \frac{u}{n_{co}}$$

$$\left\{ \frac{J_{\ell+1}(ur/a)}{J_\ell(u)} \sin(\ell+1)\,\varphi + \right.$$

$$\left. + \frac{J_{\ell \ \ 1}(ur/a)}{J_\ell(u)} \sin(\ell \quad 1)\,\varphi \right\}$$

$$\mathcal{H}_z = \quad j \frac{E_0}{2ka \ \ _0} u$$

$$\left\{ \frac{J_{\ell+1}(ur/a)}{J_\ell(u)} \cos(\ell+1)\,\varphi \right.$$

$$\left. \frac{J_{\ell \ \ 1}(ur/a)}{J_\ell(u)} \cos(\ell \quad 1)\,\varphi \right\}$$

Cladding

$$\mathcal{E}_y = E_0 \frac{K_\ell(wr/a)}{K_\ell(w)} \cos \ell\varphi$$

$$\mathcal{H}_x = \quad n_{cl} \frac{\mathcal{E}_y}{\ \ 0}$$

$$\mathcal{E}_z = \quad j \frac{E_0}{2ka} \frac{w}{n_{cl}}$$

$$\left\{ \frac{K_{\ell+1}(wr/a)}{K_\ell(w)} \sin(\ell+1)\,\varphi \right.$$

$$\left. \frac{K_{\ell \ \ 1}(wr/a)}{K_\ell(w)} \sin(\ell \quad 1)\,\varphi \right\}$$

$$\mathcal{H}_z = \quad j \frac{E_0}{2ka \ \ _0} w$$

$$\left\{ \frac{K_{\ell+1}(wr/a)}{K_\ell(w)} \cos(\ell+1)\,\varphi + \right.$$

$$\left. + \frac{K_{\ell \ \ 1}(wr/a)}{K_\ell(w)} \cos(\ell \quad 1)\,\varphi \right\}$$

Let us recall that we are now interested in the characteristic equation for the $\mathrm{LP}_{\ell,p}$ mode, corresponding to the $\mathrm{EH}_{\ell \ \ 1,p}$, not to the $\mathrm{EH}_{\ell,p}$ mode; therefore, let $\ = \ell \quad 1$. The nal result is:

$$u \frac{J_{\ell \ \ 1}(u)}{J_\ell(u)} = \quad w \frac{K_{\ell \ \ 1}(w)}{K_\ell(w)} \qquad (\ell = 0, 1, 2, \ldots) \quad , \tag{10.60}$$

to be associated with Eq. (10.32).

Looking carefully at Table 10.6, we see that Eq. (10.60) would be the characteristic equation of the system which we obtain if we stipulate that the *radial components* of the electric and of the magnetic eld are continuous at the core-cladding interface, $r = a$:

$$\begin{aligned} \mathcal{E}_{rco}(r = a) &= \mathcal{E}_{rcl}(r = a) \quad , \\ \mathcal{H}_{rco}(r = a) &= \mathcal{H}_{rcl}(r = a) \quad . \end{aligned} \tag{10.61}$$

The second of Eqs. (10.61) is correct, because there is no discontinuity in magnetic permeability at $r = a$, but the rst one is not rigorous as long as

$_{co} \neq \;_{cl}$. In fact, from Chapter 1 we know that the correct continuity condition at $r = a$ is that the normal components of the electric displacement vector, $\mathbf{D} = \;\mathbf{E}$, are equal in the core and in the cladding. This comment indicates another implicit meaning of the weak-guidance approximation (or, equivalently, LP approximation): the approximation implies that we neglect refraction of the eld lines of the electric vector across the interface between the two media. Consequently, informations concerning *polarization* of the transverse elds, if drawn through LP modes, are not accurate. In reality, the linear polarization ratio of any guided mode is never in nite. In spite of this limitation, LP modes are extremely useful in yielding practical information on optical bers, as we shall see in the following sections.

10.8 Dispersion in dielectric waveguides

In practical applications of dielectric waveguides (among which, optical telecommunication systems are by far and large the most important, nowadays) the main limits to performances come from attenuation and dispersion. Loss in a dielectric waveguide is a rather simple subject to deal with, and will be brie y discussed in Section 10.11. On the contrary, dispersion deserves an accurate discussion, which may sound surprising. In fact, in the previous sections we have established, rstly, that the phase constant of any guided mode, , must be in the range between $_{cl}$ and $_{co}$, the intrinsic phase constants of the two dielectric media; secondly, that the di erence in permittivity between these media is very small in all cases of practical interest. Comparing these results with those concerning guided modes in a conducting-wall waveguide, one might infer that in a dielectric waveguide variations of the phase constant vs. frequency are small and smooth, so that dispersion cannot be large. The reason why this conclusion is unacceptable is easy to explain. The lengths of practical microwave waveguides never exceed a few meters, which means, at most, a few hundreds of wavelengths. Optical bers can be more than hundred kilometers in length, which means more than 10^{10} wavelengths. Accuracy in modeling dispersion must then grow in proportion to these gures.

Dispersion can be dealt with analytically in the weak-guidance approximation, i.e., when:

$$\simeq \frac{n_{co} \quad n_{cl}}{n_{cl}} \quad 1 \quad . \tag{10.62}$$

Given that $_{cl} < \; < \;_{co}$, let us de ne a new quantity whose purpose is to expand this range and so emphasize the variations of vs. frequency. Let:

$$b \quad \frac{\frac{\;^2}{k^2} \quad n_{cl}^2}{n_{co}^2 \quad n_{cl}^2} \simeq \frac{\frac{\;}{k} \quad n_{cl}}{n_{co} \quad n_{cl}} \quad . \tag{10.63}$$

Clearly, b is adimensional and varies between 0 and 1. The last approximate step in the previous equation comes from the fact that $/k + n_{cl} \simeq n_{co} + n_{cl}$. The ratio $/k$ is often referred to as the *e ective index* of the guided mode

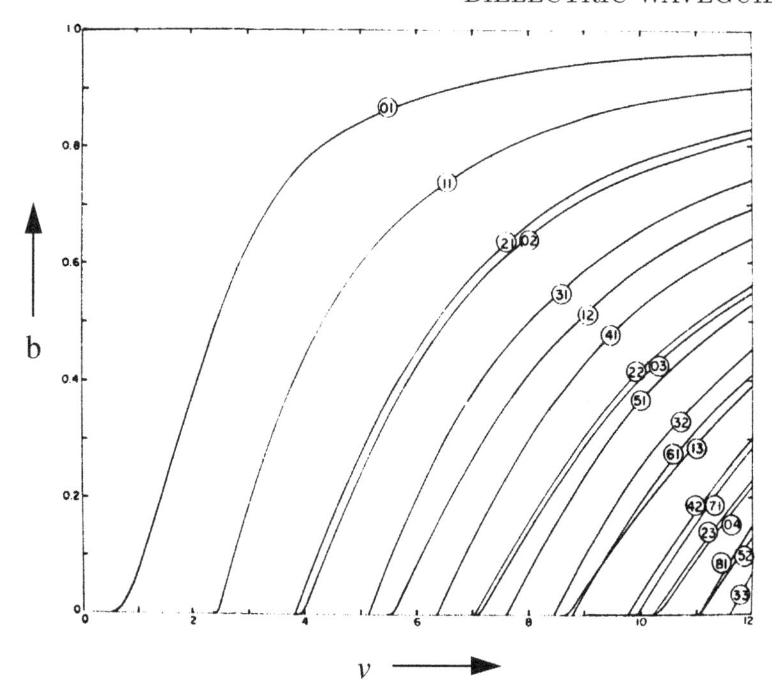

Figure 10.9 *Normalized phase constants vs. normalized frequency. The numbers refer to LP mode su ces.*

characterized by the phase constant . Clearly, its values are con ned in the range between n_{cl} and n_{co}.

Eqs. (10.27) and (10.44) yield another expression for b, namely:

$$b(v) = 1 \quad \frac{u^2}{v^2} \quad , \tag{10.64}$$

where let us stress that u is also a function of v. Plots of Eq. (10.64) for some LP modes are shown in Figure 10.9 (from Gloge, 1971).

Let us calculate, rather than the group velocity v_g of the various LP modes, its inverse, referred to as the (per-unit-length) *group delay,* ${}_g$, following the de nition given in Chapter 5, namely:

$${}_g = \frac{d}{d\omega} = \frac{1}{c_0}\frac{d}{dk} \quad , \tag{10.65}$$

where $c_0 = 1/\sqrt{\ {}_0\ {}_0}$ is the speed of light in free space.

In contrast to the usual situation in conducting-wall waveguides at microwave frequencies, contributions due to *material dispersion*, i.e., to $d(\quad)/d\omega \neq 0$ (see Chapter 5), cannot be neglected in dielectric waveguides at optical frequencies. To account for them without making the notation too cumbersome, let us de ne the following quantities:

$$N_i \quad \frac{d \;\; _i}{dk} = \frac{d(n_i k)}{dk} = n_i + k\,\frac{dn_i}{dk} \qquad (i = co, cl) \quad , \tag{10.66}$$

which are referred to as the *group indices* of the core and cladding materials. Let us de ne also the quantity:

$$\;' \quad \frac{N_{co} \quad N_{cl}}{n_{cl}} = \quad + \frac{k}{n_{cl}}\,\frac{d(n_{co} \quad n_{cl})}{dk} \quad , \tag{10.67}$$

which may di er signi cantly from the relative index di erence, , only in waveguides where the core and cladding materials are completely di erent (e.g., silica-plastic bers). Elaborating on as a function of b, from Eq. (10.63), one nds the following rigorous result:

$$\frac{d}{dk} = \frac{1}{n_{cl}}\left\{n_{cl}N_{cl} + (n_{co}N_{co} \quad n_{cl}N_{cl})\left[b + \frac{1}{2}\,v\,\frac{db}{dv}\right]\right\}(1 + 2 \quad b)^{\;\; 1/2} . \tag{10.68}$$

This expression can be simpli ed, given that 1. Let $(1 + 2 \quad b)^{\;\; 1/2} \simeq 1 \quad b$, then neglect terms of order ^{2}, and use Eq. (10.67). The result is:

$$\frac{d}{dk} \simeq N_{cl} + n_{co} \quad '\,\frac{d(vb)}{dv} \quad \frac{1}{2}\,kn_{co}\,\frac{d}{dk}\,v\,\frac{db}{dv} \quad . \tag{10.69}$$

For a given waveguide, the rst term is independent of the particular mode. Accordingly, in multimode propagation regime, di erences in group delay (referred to as *intermodal dispersion*) are due only to the second and third terms. The last one is essentially negligible for all modes, for which $db/dv \simeq 0$, except, at most, the fundamental mode, but we will return to this later on. The factor $d(vb)/dv$, which governs the second term, is plotted in Figure 10.10 for some low-order LP modes. It may be seen immediately that some of the modes (those of azimuthal order $\ell = 0, 1$) are fast very close to cut-o , when most of their guided power travels in the cladding, where the speed of light is larger than in the core. For the modes of higher azimuthal order, even at cut-o the power fraction in the cladding is too small for the mode to be fast. In any case, as soon as they are su ciently above cut-o all the modes tend to become faster as the normalized frequency increases, for the same reason that the group velocity increases vs. frequency in a conducting-wall waveguide (see Section 7.4).

Figure 10.10 also shows that the di erence between the largest and the smallest normalized group delay depends, in its ne details, on v in a complicated manner, but on the contrary its order of magnitude is unity, and remains essentially constant as soon as v 2.4, as it is in practical multimode bers. Therefore, the intermodal dispersion can be estimated to be:

$$_{\text{int}} = \frac{1}{c_0}\,n_{co} \quad ' \simeq \frac{1}{c_0}(N_{co} \quad N_{cl}) \quad , \tag{10.70}$$

with an approximation $n_{co}/n_{cl} = 1$ in the last step. Essentially, this result tells us that in a multimode step-index ber intermodal dispersion equals the

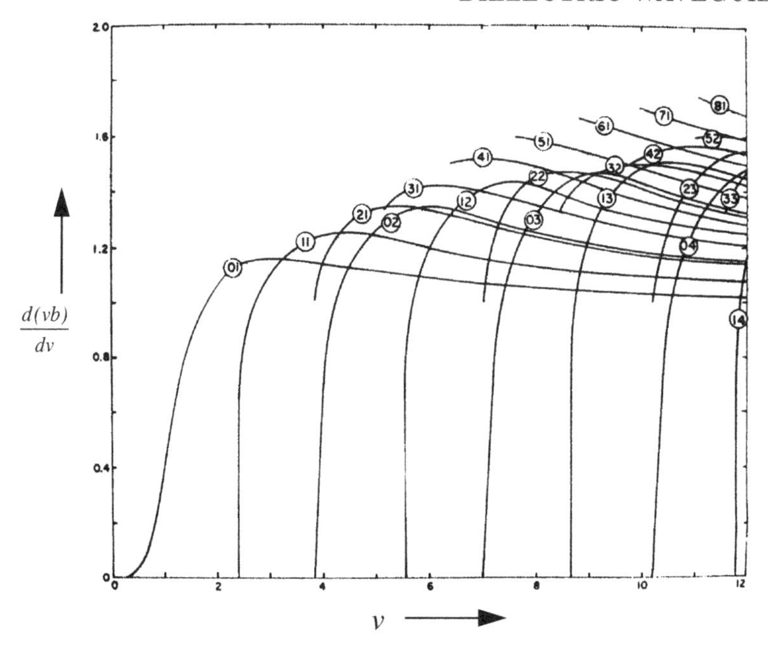

Figure 10.10 *Normalized group delays (waveguide contributions) vs. normalized frequency. The numbers refer to LP mode su ces.*

di erence between the group delays of two uniform plane waves, traveling in the core and in the cladding, respectively.

It is left to the reader as an exercise to prove that an intermodal dispersion of this order of magnitude causes, over a ber length of practical interest (at least 10 km), pulse spreading compatible only with very low bit rates. Actually, experimentally measured intermodal dispersions in step-index bers are somewhat smaller than these predictions, because the modes close to cut-o are ltered out by their losses, signi cantly higher than those of low-order modes. However, they do not allow transmission of attractive bit rates. There are essentially two ways of improving this performance. One is to change the index pro le signi cantly, and will be the subject of the following sections. The other one is to use single-mode bers. Let us now elaborate further on their dispersion properties.

The very small values required for the index di erence (see Section 10.6) implies that core and cladding materials di er at most for a very small amount of dopants. This allows us to let $' \simeq$, i.e., independent of wavelength, thus canceling the third term in Eq. (10.69), and rewrite it (in the presence of just the fundamental mode $LP_{0,1}$) as:

$$\frac{d}{dk} \simeq N_{cl} + (N_{co} \quad N_{cl}) \frac{d(vb)}{dv} \quad . \tag{10.71}$$

In Chapter 5, we saw that a signal is distorted when the group velocity (or

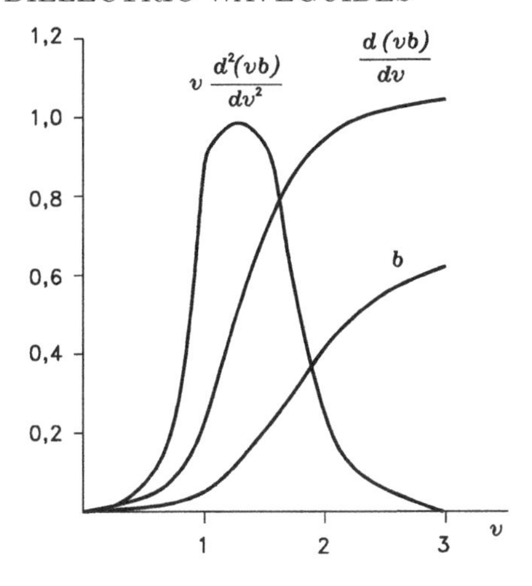

Figure 10.11 *Normalized waveguide dispersion.*

group delay) varies with frequency, within the signal spectral band. In the multimode regime, we overlooked this phenomenon, because it is dominated by the much larger intermodal dispersion. In the single-mode regime, we must account for the source being non-monochromatic, and study how the group delay of the guided eld varies vs. frequency, i.e., for *intramodal dispersion*, sometimes referred to also as *chromatic dispersion*. To calculate the derivative of the group delay with respect to angular frequency:

$$s = \frac{d\ \ _g}{d\omega} = \frac{1}{c_0^2}\,\frac{d^2}{dk^2}\quad, \tag{10.72}$$

starting with Eq. (10.71) we get:

$$\frac{d^2}{dk^2} = \frac{dN_{cl}}{dk} + \frac{d(N_{co}\quad N_{cl})}{dk}\,\frac{d(vb)}{dv} + (N_{co}\quad N_{cl})\,\frac{v}{k}\,\frac{d^2(vb)}{dv^2}\quad. \tag{10.73}$$

Thus, the intramodal dispersion of the $LP_{0,1}$ mode consists of three terms. The rst originates from *material dispersion*. The third term indicates — similar to what we saw in Chapter 7 for conducting-wall waveguides — that a guided wave, being non-TEM, is *inevitably* dispersive, and therefore is referred to as *waveguide dispersion*. The second term is a combination of material and waveguide e ects. Usually, its magnitude is signi cantly less than that of the rst and that of the third term. However, before neglecting it completely, we must investigate whether the rst and third term may cancel each other.

Figure 10.11 shows that in the single-mode regime ($v < 2.405$) the last term in Eq. (10.73) is positive, for any choice of the ber material and geometry. Note that this behavior is opposite to that of a metal-wall waveguide, where the group velocity increases as a function of frequency (anomalous dispersion). This indicates that, in the entire single-mode band, the phenomenon which governs waveguide dispersion is the growth in the power fraction that travels

Table 10.7 *International standards for intramodal dispersion. Source: International Telecommunication Union (formerly CCITT); G.652 and G.653. Unit: ps/(km nm).*

Wavelength range	1285–1330 nm	1270–1430 nm	1525–1575 nm
Fiber optimized at 1300 nm	3.5	6	20
Fiber optimized at 1550 nm	—	—	3.5

in the core, which was outlined previously. On the other hand, the sign of the rst term depends on the speci c waveguide material, and on the wavelength, in a given material. In pure amorphous silica, material dispersion is positive at < 1.27 m, negative at longer wavelengths. Consequently, it is possible to design a weakly guiding glass ber so that its intramodal dispersion Eq. (10.73) goes through zero at one wavelength $_0 > 1.3$ m. The exact zero-dispersion wavelength, $_0$, is determined by the second term in Eq. (10.73), since the sum of the rst and third term is close to zero.

The simplest and most widely used designs give bers whose $_0$ is close to 1.3 m, i.e., in the spectral range which is called, as we said before, "second window". These bers are usually referred to as *standard* single-mode bers. However, the growing interest in exploiting the "third window," where losses pass through their minimum value, and optical ampli ers are commercially available, has encouraged the design of bers whose zero-dispersion wavelength $_0$ is close to 1.5 m. These are referred to as *dispersion shifted* (DS) bers. Typically, their refractive index pro le is more complicated than the single step considered here. The reader who is interested in more details should consult many of the books quoted at the end of this chapter. More complicated bers have been envisaged, where total dispersion is kept below a given upper limit over the whole range from 1.3 m to 1.55 m. They are referred to as *dispersion attened* bers. However, interest in these appears to be low, in recent times.

Units in current technical use, to measure intramodal dispersion of optical bers, are ps/(km nm), i.e., picoseconds (of group delay) divided by kilometers (of ber length) and by nanometers (of source spectral width). Some useful reference data are shown in Table 10.7.

Before we close this section, we would like to remind the reader of the contents of Section 5.2. The quantity expressed by Eq. (10.72) is merely the rst coe cient, in the expansion of the delay as a series of powers of . When the rst coe cient is equal to zero, the signal distortion is governed by the next term, which is proportional to ^{2}. So, at the zero-dispersion wavelength the ber bandwidth is not in nite, but is still orders of magnitude broader than at other wavelengths.

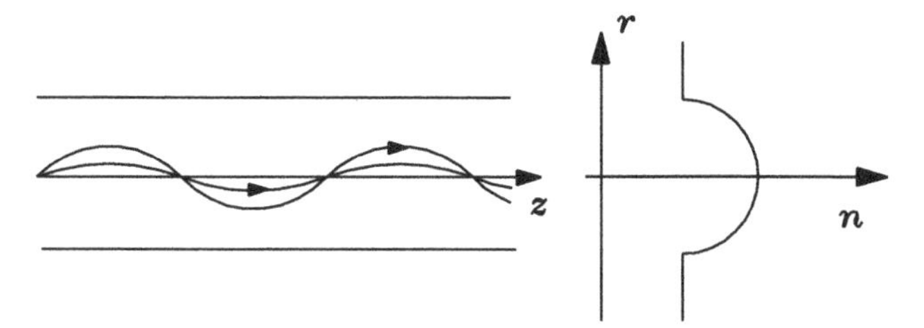

Figure 10.12 *Ray model of propagation in a graded-index waveguide.*

10.9 Graded-index waveguides

As we saw in the previous section, in a multimode step-index ber intermodal dispersion can be too large for long-distance pulse transmission, even at moderate bit rates. A solution, which has been studied in depth theoretically and experimentally, and successfully applied during the early decades of optical communications, is to reduce intermodal dispersion by "equalizing" the speeds of propagation of the various modes. One can do this by acting on the refractive index distribution in the waveguide cross-section, as a function of the radial coordinate. This idea is easy to justify, qualitatively, in terms of ray optics: "times of ight" along guided ray paths of di erent geometrical lengths can be equalized by increasing the speed of propagation (i.e., decreasing the refractive index n) along the longer paths, namely further away from the waveguide axis (see Figure 10.12).

Propagation along graded-index waveguides cannot be studied rigorously, in vector terms. However, most of the results of practical interest can be obtained by means of an analysis based on the WKBJ approximation, which was presented, in general, in Section 5.7. This will be the subject of the next section. Before we proceed in this direction, however, it is worthwhile to sketch brie y, omitting some details, how one can pass from Maxwell's equations to a rather scalar-looking problem, which can then be tackled using the WKBJ method.

Let us see whether Eqs. (7.4), or Eqs. (7.11) and (7.12), which express transverse elds in terms of the longitudinal components $\mathcal{E}_z, \mathcal{H}_z$, are still valid when , and maybe , depend on the transverse coordinates. Using Maxwell's equations, one can show that this is indeed the case. On the contrary, $\mathcal{E}_z$ and $\mathcal{H}_z$ do not satisfy Eq. (7.8), in an inhomogeneous medium. This implies that the theory of guided waves that we developed for conducting-wall waveguides, and then for step-index bers, does not hold for graded index waveguides, because it was based on the pair of scalar functions which were called prepotentials in Section 3.8.

To replace Eq. (7.8) with two new equations, let us go back to Maxwell's equations, calculate the curl of both sides, replace $\nabla \quad \nabla$ with $\nabla^2 + \nabla\,(\nabla\ )$, and account for Eqs. (1.35) and (1.38), which, for $\neq$ constant, $\neq$ constant, yield $\nabla\ \mathbf{H} = (1/\ )\nabla\ \mathbf{H}$ and $\nabla\ \mathbf{E} = (1/\ )\nabla\ \mathbf{E}$, respectively. The nal results are the following equations:

$$[\nabla_t^2 \mathcal{H}_z + {}^2 \mathcal{H}_z] \hat{} + \left[\frac{\nabla_t}{}^{2} \frac{\nabla_t}{}\right] \left[\nabla_t \mathcal{H}_z \quad \hat{}\right] = \frac{}{j\omega} \frac{\nabla_t}{}^{2} \nabla_t \mathcal{E}_z \,, \tag{10.74}$$

$$[\nabla_t^2 \mathcal{E}_z + {}^2 \mathcal{E}_z] \hat{} + \left[\frac{\nabla_t}{}^{2} \frac{\nabla_t}{}\right] \left[\nabla_t \mathcal{E}_z \quad \hat{}\right] = \frac{}{j\omega} \frac{\nabla_t^2}{}^{2} \nabla_t \mathcal{H}_z \,, \tag{10.75}$$

where it is important not to forget that $^2 = {}^2 + \omega^2$ is now a function of the radial coordinate, r. Given our interest in dielectric waveguides, similar to those of Section 10.8, let $= {}_0 =$ constant, $= (r) = {}_0 n^2(r)$, where:

$$n^2(r) = \begin{cases} n_0^2[1 \quad 2 \quad f(r)] & , \quad r < a \quad , \\ n_0^2[1 \quad 2 \quad] \quad n_{cl}^2 & , \quad r \quad a \quad , \end{cases} \tag{10.76}$$

with $\quad 1$. This implies $\nabla \quad = 0$, $\nabla \quad = \quad 2n_0^2 \, {}_0 \quad f'(r) \, \hat{a}_r$, $\nabla \; ^2 = \omega^2 \quad \nabla \;$, where $f' = df/dr$.

We re-de ne the symbols u, v and b so that they keep, as much as possible, the same meanings as in the previous sections, namely:

$$v^2 = k^2 a^2 (n_0^2 \quad n_{cl}^2) = \left(\frac{2 \; a}{0}\right)^2 n_0^2 \, 2 \qquad ,$$

$$u^2 = a^2 (k^2 n_0^2 \quad {}^2) \quad , \qquad b = 1 \quad \frac{u^2}{v^2} \quad ,$$

$$\frac{1}{} = \sqrt{\frac{(r=0)}{0}} = n_0 \sqrt{\frac{0}{0}} \quad . \tag{10.77}$$

Separating variables in cylindrical coordinates (r, φ, z), i.e., setting:

$$\mathcal{E}_z = e(r) \, e^{j \; \varphi} \quad , \qquad \mathcal{H}_z = h(r) \, e^{j \; \varphi} \qquad (\; = 0, \quad 1, \quad 2, \ldots) \quad , \tag{10.78}$$

Eq. (10.74) yields:

$$\frac{d^2 h}{dr^2} + \left(\frac{1}{r} + \frac{f'}{1 \quad b \quad f}\right) \frac{dh}{dr} + \left[\frac{v^2}{a^2} (1 \quad b \quad f) \quad \frac{}{r}^2\right]$$

$$= \quad j \frac{}{r} f' \frac{[1 \quad 2 \quad (1 \quad b)]^{1/2}}{1 \quad b \quad f} \frac{e}{} \quad . \tag{10.79}$$

Similarly, from Eq. (10.75) one obtains an equation which di ers from Eq. (10.79) mainly because the unknown quantities $\{e, h\}$ are interchanged, and is replaced by 1/ . Further di erences, in some of the coe cients, can easily be shown to be negligibly small for $\quad 1$.

For $f' = 0$, i.e., for a step-index waveguide, Eq. (10.79) reduces to the Bessel equation. However, $f' \neq 0$ entails nontrivial complication: both unknowns show up in both equations, Eq. (10.79) and its counterpart following from

Eq. (10.75).[†] To proceed further, we write the system consisting of Eq. (10.79) and its counterpart in the form:

$$Mh' = e' \quad , \qquad Me' = h' \quad , \tag{10.80}$$

where M is the di erential operator on the left-hand side of Eq. (10.79), and the unknowns have been suitably normalized to eliminate an unnecessary coe cient on the right-hand side, and then renamed e', h'. To decouple the equations without increasing their order, take two linear combinations of Eqs. (10.80), with coe cients (1,1) and (1, 1), respectively:

$$\begin{aligned} M(e' + h') &= (e' + h') \quad , \\ M(e' \quad h') &= \quad (e' \quad h') \quad . \end{aligned} \tag{10.81}$$

One conclusion that we may reach now is that, whenever the weakly guiding approximation 1 holds, then, for any index pro le, guided modes exhibit a *four-fold degeneracy*, becoming two-fold for $= 0$. In fact, M being of second order, each part of Eqs. (10.81) has two linearly independent solutions. Therefore, in general, e' and h' are linear combinations of four functions of r, say G_i, $i = 1, \ldots, 4$. However, two of these functions (say G_3 and G_4) diverge, at least at one point in the range 0 r a, similar to Bessel functions of the second kind which diverge in the origin. They must be disregarded for physical reasons. We are left with linear combinations of G_1, G_2, and with the fact that the ansatz expressed by Eqs. (10.78) is invariant for $\rightarrow$, whence the conclusion.

Eqs. (10.81) can be solved rigorously only for a small class of index pro les, some of which are not physically meaningful. For example, the reader may prove as an exercise that for a *parabolic pro le* extending all the way to in nity, i.e., for $f(r) \propto r^2$, $a = \infty$, one nds an equation which is satis ed by the *Laguerre-Gauss modes* which we saw in Chapter 5. However, this example is clearly unacceptable from a strictly physical point of view, as it requires < 0 for su ciently large values of r. Still, it may be looked at as an acceptable approximation, for guided modes well above cut-o , in a ber with parabolic pro le in the core, provided the core radius a is very large compared to the wavelength.

In general, to solve Eqs. (10.81) one has to make approximations, and then it is self-explanatory why there is an interest in seeking rst for equations whose solutions yield directly the *transverse eld components.* As we saw before, these are much larger and much more important in practice than the longitudinal ones, and consequently they lend themselves to clarify better the physical meanings of the approximations. Rather surprisingly, this step becomes simpler if one expands the transverse elds into *circularly polarized components*, using as unit vectors $(\hat{a}_r \quad j\,\hat{a}_\varphi)/\sqrt{2}$, although they change from point to point. In fact, it has been shown (Kurtz and Streifer, 1969; Adams,

[†] There is an exception to this rule, for $= 0$. In this case we nd TE and TM modes, as in Section 10.7.

1981) that the transverse elds can be suitably written as:

$$\boldsymbol{\mathcal{E}}_t^{(i)} = \frac{e^{j\ \varphi}}{2k\ \ n_0^{3/2}}\,[\hat{a}_r \quad j\,\hat{a}_\varphi]\ {}_i \quad , \tag{10.82}$$

$$\boldsymbol{\mathcal{H}}_t^{(i)} = n_0\,\frac{1}{\ _0}\ \hat{}\ \ \boldsymbol{\mathcal{E}}_t^{(i)} \quad , \tag{10.83}$$

where the signs + or are linked to the values $i = 1$, $i = 2$, respectively, and the scalar functions ${}_i$ are related to the G_i's, solutions of Eqs. (10.81), as:

$$_i = \frac{G_i' \quad \frac{\ }{r} G_i}{1 \quad b \quad f} \qquad (i = 1, 2) \quad , \tag{10.84}$$

where the apex stands for derivative with respect to r. Inverting the last formula, we obtain:

$$G_i = \frac{a^2}{v^2}\left[\ {}_i' \quad \frac{1}{r}\ {}_i\right] \qquad (i = 1, 2) \quad , \tag{10.85}$$

and introducing it into Eqs. (10.81), we nally get the di erential equation for ${}_i$, namely:

$$_i'' + \frac{1}{r}\ {}_i' + \left[\frac{v^2}{a^2}(1 \quad b \quad f) \quad \frac{(\quad 1)^2}{r^2}\right]\ {}_i = 0 \qquad (i = 1, 2) \quad . \tag{10.86}$$

In summary:

from Maxwell's equations, in a weakly guiding graded-index rod one extracts a system of two second-order ODE's, each in one scalar unknown function of r;

scalarization is possible because Eq. (10.83) shows that for 1 the eld is quasi-TEM. Locally, it can be looked at as a uniform plane wave;

as in step-index waveguides, degeneracy occurs between modes whose longitudinal components depend di erently on the azimuthal coordinate. Indeed, Eq. (10.86) for $i = 1$ and azimuthal order $= \ell \quad 1$ is the same as that for $i = 2$ and azimuthal order $= \ell + 1$.

The last comment is a straightforward introduction to *linearly polarized modes*, similarly to Section 10.5. Indeed, the two linear combinations $E_{t,\ell\ \ 1}^{(1)} \quad E_{t,\ell+1}^{(2)}$, and those which di er from these because ℓ is replaced by ℓ, give the following modes (for all of which $\boldsymbol{\mathcal{H}}_t$ and $\boldsymbol{\mathcal{E}}_t$ are related by Eq. (10.83)):

$$\boldsymbol{\mathcal{E}}_{t,\ \ \ell\ \ 1}^{(1)} \quad \boldsymbol{\mathcal{E}}_{t,\ \ \ell+1}^{(2)} = e^{\ j\ell\varphi}\,\hat{a}_x \quad , \tag{10.87}$$

$$\boldsymbol{\mathcal{E}}_{t,\ \ \ell\ \ 1}^{(1)} + \boldsymbol{\mathcal{E}}_{t,\ \ \ell+1}^{(2)} = e^{\ j\ell\varphi}\,\hat{a}_y \quad , \tag{10.88}$$

where:

$$= \frac{1}{2j}\ {}_{1,\ell\ \ 1} = \frac{1}{2j}\ {}_{2,\ell+1} \quad . \tag{10.89}$$

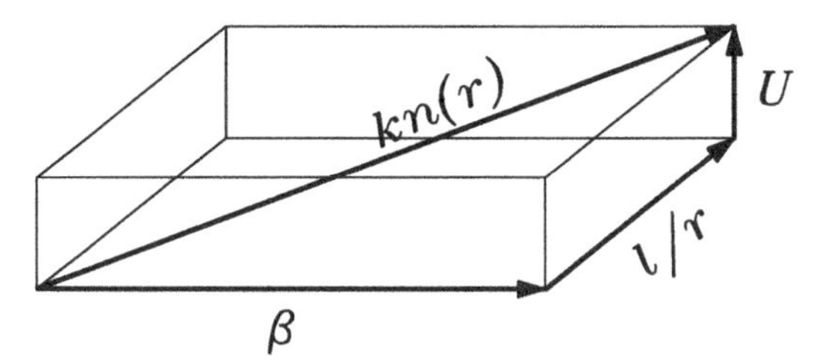

Figure 10.13 *Decomposition of the phase vector into its longitudinal, azimuthal and radial components.*

The nal consequence is that the two Eqs. (10.86) become one equation, namely:

$$\frac{d^2}{dr^2} + \frac{1}{r}\frac{d}{dr} + \left[\frac{v^2}{a^2}(1 \quad b \quad f) \quad \frac{\ell^2}{r^2}\right] \quad = 0 \quad . \tag{10.90}$$

The key feature for deriving Eq. (10.90) has been the assumption of weak guidance, i.e., 1. Hence, we are facing an equation which applies equally well to single-mode, few-mode, or highly multimode bers, but paths for solving it will diverge, from now on. For single-mode and few-mode bers, no broad-scope analytical procedure has been developed, so one must resort to numerical methods. On the contrary, in highly multimode bers Eq. (10.90) can be solved analytically using the WKBJ method outlined in Section 5.7, provided the index pro le function $f(r)$ is slowly varying. The next section will deal with a broad and signi cant class of pro les for which the WKBJ approach may yield elegant closed-form expressions. However, we must rst clarify the meaning of the phrase "slowly-varying pro le". This will automatically explain why the same approach is not suitable for single-mode or few-mode bers.

We know from the previous step-index examples that, in any weakly guiding ber, a guided mode consists of quasi-TEM waves, which travel in a direction almost orthogonal to the radial one, i.e., to the direction along which the index varies. To stress this point, let us go back to Eqs. (10.77) and rewrite the last term in Eq. (10.90) in the following way:

$$\frac{v^2}{a^2}(1 \quad b \quad f) \quad \frac{\ell^2}{r^2} = k^2\, n^2(r) \quad {}^2 \quad \frac{\ell^2}{r^2} \quad U^2(r) \quad . \tag{10.91}$$

This lends itself to the following geometrical interpretation, illustrated in Figure 10.13: the local phase vector (whose magnitude is $k\, n(r)$) can be decomposed into its components in a cylindrical reference frame, i.e., along (r, φ, z). The longitudinal component equals the mode propagation constant, . The azimuthal component is ℓ/r. The radial component, $U(r)$, equals the square root of the left-hand side of Eq. (10.91), and therefore is unknown as long as is unknown.

It was shown in Section 5.7 that the WKBJ method is applicable only if a *local* wavelength, measured in the direction along which the refractive index varies — in our case, the *radial* direction — and therefore expressed by $_r = 2 \;\; /U(r)$, is small compared with the length scale over which f varies, i.e., with f/f'. Consequently, the WKBJ method fails for *any* mode, strictly speaking, near the mode caustics, i.e., to the radii at which $U(r) = 0$. The reader may show, as an exercise, that the loci where $U(r) = 0$ correspond to "turning points" of the rays, in case a medium characterized by an index pro le $f(r)$ is studied in terms of geometrical optics, following the suggestion given by Figure 10.13.

As illustrated by the example given in the next section, errors made by violating this restriction are often negligible for a very large fraction of the guided modes in a highly multimode ber. However, the condition $_r \quad f/f'$ clearly indicates that, for a given pro le function and at a given frequency (i.e., a given k^2), the WKBJ method becomes more accurate as the radial order of the mode increases, i.e., as $U(r)$ grows. This is the reason why it is not suitable for single-mode or few-mode bers.

10.10 The alpha pro les: An important class of multimode graded-index bers

As we remarked in the previous section, there is a class of index pro les for which the WKBJ method yields closed-form results. They are referred to as the r pro les, or sometimes as the Gloge-Marcatili pro les (1973). Their popularity is based on the fact that they form a set which is wide enough to provide an in-depth understanding of propagation in a graded-index medium, as well as to design rules for highly multimode bers. The main reason why such bers are far less used in practice than originally expected is because their performances depend extremely critically on the index pro le. This is predicted, quite accurately, by the theory that we will present in this section. However, this reduced importance of the subject is a good reason for making its presentation somewhat brief, omitting many intermediate steps in the calculations, although this approach is not fully consistent with the rest of this chapter.

In Eq. (10.76), let:

$$f(r) = \left(\frac{r}{a}\right) \quad . \tag{10.92}$$

For $0 < \quad < \infty$, this represents a broad class of index pro les. All of them decrease monotonically from the center towards the periphery (see Figure 10.14). The limiting case $\quad = \infty$ is the step-index pro le which we studied before.

For $\quad 1$, keeping only the linear term in the binomial expansion of the square root of Eq. (10.76), we may write:

$$n(r) = \begin{cases} n_0 \left[1 \quad \left(\frac{r}{a}\right) \; \right] & , \quad r < a \quad , \\ n_0 \left[1 \quad \right] = n_{cl} & , \quad r \quad a \quad . \end{cases} \tag{10.93}$$

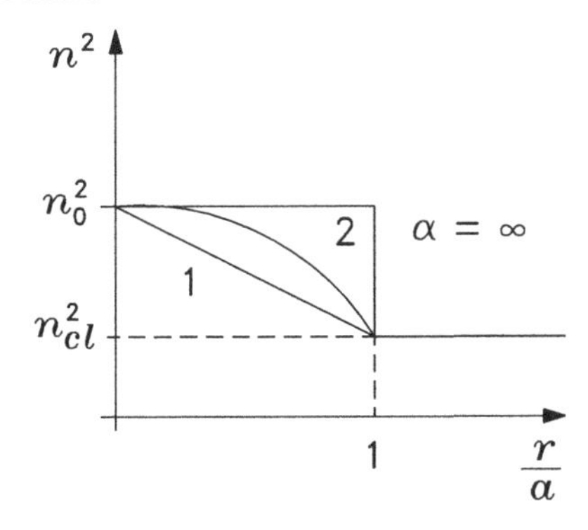

Figure 10.14 *Schematic illustration of the alpha pro les.*

We look for guided modes, so let us recall the basic physical features of their eld dependence on the radial coordinate:

(r) must decrease monotonically for $r > a$, and tend to zero for $r \to \infty$;

(r) must be an oscillating function at least in part of the range $0 \quad r \quad a$.

These features imply that in Eqs. (10.90) and (10.91):

for $r > a$, the radial propagation constant $U(r)$ must be imaginary, i.e., $U^2(r) < 0$;

$U(r)$ must be real, i.e., $U^2(r) > 0$, at least in part of the range $0 \quad r \quad a$.

Consequently, with the help of Figure 10.15 we see that:

$$k\, n_0 > \quad > k\, n_{cl} \quad , \tag{10.94}$$

for any guided mode, as in step-index waveguides. Indeed, for $\quad < kn_{cl}$ the two curves shown in Figure 10.15 have a third intersection, say at $r_3 > a$. Beyond this third caustic (r) oscillates vs. r, and conveys an active power in the radial direction. This behavior is reminiscent of the *leaky modes* mentioned in Section 10.2, and of the electromagneting tunneling discussed in Chapter 4. For these reasons, elds of this kind are called "tunneling leaky modes".

We know that guided modes must form a discrete set. Indeed, solutions of Eq. (10.90) can be discretized on the grounds of the following physical requirement (see again Figure 10.15). Going from the $r = r_1$ caustic to the $r = r_2$ caustic, a wave satisfying Eq. (10.90) must accumulate a total phase shift equal to an integer multiple of 2 , including localized phase shifts at the re ections on the caustics. Using the WKBJ results found in Section 5.7, and the phase corrections at the both caustics, found in Section 5.8,‡ we write this

‡ Modes of azimuthal order $\ell = 0$ do not have an internal caustic. Therefore, in their case the correction of /2 radians should be applied only once. However, we will see very soon that this correction is negligible in most cases.

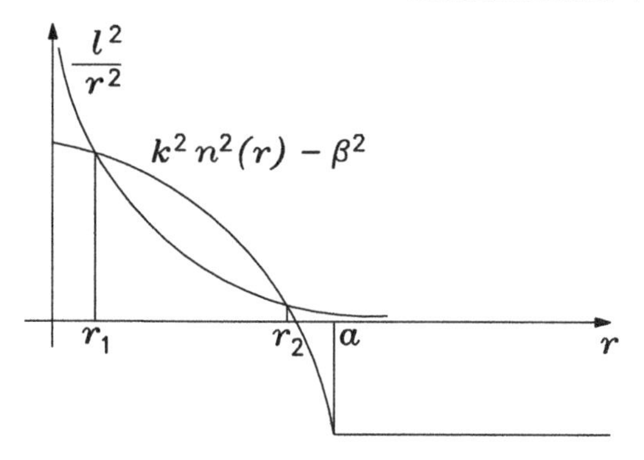

Figure 10.15 *The regions where the modal eld has an oscillatory dependence on the radial coordinate, and those where the eld decays monotonically, in an alpha-pro le waveguide.*

condition as:

$$2\int_{r_1}^{r_2} U(r)\,dr \quad 2\int_{r_1}^{r_2}\left[k^2 n^2(r) \quad {}^2 \quad \frac{\ell^2}{r^2}\right]^{1/2} dr = \quad (2 \quad) \qquad (\quad = 1, 2, \ldots)\,. \tag{10.95}$$

In highly multimode bers (v 1), the correction in Eq. (10.95) is negligible for most of the modes, so we let:

$$\int_{r_1}^{r_2} U(r)\,dr \simeq \quad . \tag{10.96}$$

For v 1, both and ℓ can reach very large values. Hence, although in principle guided modes are discrete, they can be approximated as a continuum, and Eq. (10.96) tells us that their density along the U axis equals $4U/$, the factor 4 accounting for the usual four-fold degeneracy of each solution of the equations we wrote so far. The total number of guided modes, M, is found by integrating their density over their domain of de nition. Interchanging the order of integration, we get:

$$\begin{aligned} M &= \frac{4}{\ } \int_0^{\ell_{\mathrm{MAX}}} d\ell \int_{r_1}^{r_2}\left[k^2 n^2(r) \quad {}^2 \quad \frac{\ell^2}{r^2}\right]^{1/2} dr \\ &= \frac{4}{\ } \int_0^a dr \int_0^{kr[n^2(r) \quad n_{cl}^2]^{1/2}} U(r)\,d\ell \\ &= \int_0^a k^2\left[n^2(r) \quad n_{cl}^2\right] r\,dr \quad . \end{aligned} \tag{10.97}$$

For the pro les de ned by Eq. (10.92), this involves:

$$M = \left(\frac{\ }{\ + 2}\right)\frac{v^2}{2} \quad , \tag{10.98}$$

where v is given by the rst of Eqs. (10.77). Thus, incidentally, guided modes in a parabolic-index ber (= 2) are one half of those in a step-index ber (= ∞), for equal normalized frequency v. This result may a ect coupling loss between a given light source and a ber, as discussed in one of the problems at the end of this chapter.

As we said in Section 10.8, interest in graded-index bers was stimulated mainly by the hope of reducing the intolerable spread among the group delays of guided modes in a step index ber. To calculate the group delay, Eq. (10.65), for any guided mode, we proceed as follows. Di erentiate Eq. (10.95) with respect to k, and account for two facts: the mode indices   and ℓ do not depend on k, while $d\ \ /dk$ is independent of r and can be extracted from the integral. The net result, as the reader can prove as an exercise, is:

$$ _g = \frac{1}{c}\,\frac{\displaystyle\int_{r_1}^{r_2} k\,\frac{n(r)\,N(r)}{U(r)}\,dr}{\displaystyle\int_{r_1}^{r_2} \frac{\ }{U(r)}\,dr} \quad , \tag{10.99}$$

where $U(r)$ is given by Eq. (10.91), and N is the *group index*, which is still de ned like Eq. (10.66), but is now a function of the radial coordinate, i.e.:

$$N(r) = n(r) + k\,\frac{d\,n(r)}{dr} \quad , \qquad N_0 = N(0) \quad . \tag{10.100}$$

For the r pro les de ned by Eq. (10.92), let us make a simple, yet very important assumption. Suppose that, when the frequency changes, the pro le height may vary, but its shape does not. This entails that n_0 and   must be considered k-dependent, but the exponent   is k-independent. After some rather heavy algebra, which we omit for the sake of brevity (see Adams, 1981), one nds:

$$ _g = \frac{N_0}{c}\left[1 \quad 2 \quad \left(\frac{u}{v}\right)^2 \left(\frac{\quad 2}{\ +2}\right)\left(1+\frac{n_0 k}{2N_0}\,\frac{d}{dk}\right)\right]\left[1 \quad 2 \quad \left(\frac{u}{v}\right)^2\right]^{\ 1/2} , \tag{10.101}$$

where u is de ned by Eq. (10.77). Thus, in graded-index bers, as in step-index bers (see Eq. (10.68)), the group delay includes mode-independent terms, which depend on the material, and terms which depend on the mode through u. The mode-dependent terms are sensitive to the index pro le. One of them — the second term in the rst square bracket — depends on the material also through another parameter, namely:

$$y = 2\,\frac{n_0 k}{N_0}\,\frac{d}{dk} \quad , \tag{10.102}$$

which is referred to as the *pro le dispersion parameter*. We will see shortly that y plays a crucial role in optimizing a graded-index pro le.

Using Eq. (10.102), and expanding $[1 \quad 2 \quad (u/v)^2]^{\ 1/2}$ as a power series of $(u/v)^2$, a quantity much smaller than unity, Eq. (10.101) can be rewritten as:

$$
\begin{aligned}
{}_g &= \frac{N_0}{c}\left[1 + \left(\frac{2 \quad y}{+2}\right)\left(\frac{u}{v}\right)^2\right. \\
&+ \frac{^2}{2}\left(\frac{3 \quad 2 \quad 2y}{+2}\right)\left(\frac{u}{v}\right)^4 \\
&+ \left.\frac{^3}{2}\left(\frac{5 \quad 2 \quad 3y}{+2}\right)\left(\frac{u}{v}\right)^6\right] + T(\ ^4) \quad , \qquad (10.103)
\end{aligned}
$$

where $T(\ ^4)$ stands for all the terms containing fourth or higher powers of . All these terms are negligible in any practical case.

Our main task is now to see how the pro le parameter a ects intermodal dispersion. Note that in Eq. (10.103), in addition to where it appears explicitly, also plays a hidden role, because the dispersion relation of any guided mode, $= \ (\omega)$, depends on the index pro le, and consequently the term u/v in Eq. (10.103) is sensitive to . Let $m(\ )$ be the number of guided modes whose phase constants satisfy $\ > \ $. Then, steps that we are forced to omit for brevity (see Adams, 1981) show:

$$
\left(\frac{u}{v}\right)^2 = \left(\frac{m(\ )}{M}\right)^{/(\ +2)} \quad , \qquad (10.104)
$$

where M is the total number of guided modes, given by Eq. (10.98). This tells us that the density of guided modes is not uniform over their domain of de nition. This is a very important point, because it can be shown (Adams, 1981) that the limiting factor of the information carrying capacity of a multimode ber, as transmission channel for a binary digital intensity-modulation, direct-detection system, is the *root-mean-square spread of the group delays*, de ned as:

$$
= L\left\{\langle\ ^2_{gm}(\ _0)\rangle \quad \langle\ _{gm}(\ _0)\rangle^2\right\}^{1/2} \quad , \qquad (10.105)
$$

where L is the ber length, $_{gm}(\ _0)$ is the per-unit-length group delay of the m-th guided mode at the central wavelength of the source spectrum, $_0$, and $\langle\ \rangle$ mean ensemble average over the set of guided modes, which is sensitive, of course, to the mode distribution over their domain of de nition.

Placing Eq. (10.103) into Eq. (10.105), after some algebra one nds:

$$
\begin{aligned}
&= L\,\frac{N_0}{c}\,\frac{}{2}\,\frac{}{+1}\left(\frac{+2}{3 \quad +2}\right)^{1/2} \\
&\left[C_1^2 + \frac{4C_1C_2(\ +1)}{(2\ +1)} + \frac{4C_2^2\ ^2(2\ +2)^2}{(5\ +2)(3\ +2)}\right]^{1/2} \quad , (10.106)
\end{aligned}
$$

where:

$$
C_1 = \frac{2 \quad y}{+2} \quad , \qquad C_2 = \frac{3 \quad 2 \quad 2y}{2(\ +2)} \quad . \qquad (10.107)
$$

The reader may check, as an exercise, that Eq. (10.106) has its minimum, as a function of , for:

$$= {}_{\rm opt} = 2 + y \qquad \frac{(4+y)\,(3+y)}{5+2y} \quad . \tag{10.108}$$

A detailed analysis of Eq. (10.106) shows that this minimum is extremely deep and sharp. In other words, the "optimum pro le" for a multimode graded-index ber is extremely critical. A 1% error in the exponent , which may correspond to an error of $10\ ^{3}$ in the pro le function $f(r)$, can be enough to reduce the information carrying capacity by more than one order of magnitude. Furthermore, after what we said in Section 10.8, the reader should not be surprised if we say that the dispersion pro le parameter y, Eq. (10.102), is usually a function of wavelength. Therefore, a pro le which has been optimized for a given wavelength is far from being optimum at a di erent wavelength. These are, basically, the reasons why interest in multimode graded-index bers decreased rapidly, as we said before, in the early 1980's, leaving more and more room to single-mode bers.

10.11 Attenuation in optical bers

Dielectric waveguides are fundamentally di erent from conducting-wall waveguides, as we have said previously, since they are open structures. This involves a subtle philological di erence, when one speaks of signal attenuation. In a metallic waveguide, the terms "loss" and "attenuation" describe the same phenomenon: a signal traveling along it, and not undergoing any re ection, can become weaker only if a fraction of its power is dissipated, either in an imperfect dielectric or in an imperfect conducting wall. In a dielectric waveguide, on the other hand, guided modes can lose part of their power because this is transferred, by some kind of waveguide imperfection which breaks mode orthogonality, into *radiation modes* of the waveguide. This conversion can be a lossless process in a strictly thermodynamic sense, as it can preserve the total power of the e.m. eld. However, as radiation modes in most cases do not reach the eld detector at the end of the waveguide, this process results in a signal attenuation. With this in mind, terminology in current use for optical waveguides distinguishes between *absorption losses* (i.e., conversion of e.m. energy into other forms of energy, typically heat) and *radiation losses.*

In dielectric waveguides of practical interest, such as optical bers and planar guides in integrated optics,[§] absorption losses are so small, that a rst-order perturbation theory, similar to those we saw in Chapters 7 (metallic waveguides) and 9 (resonators) is certainly adequate. Let the imperfect dielectric media be characterized by their complex permittivity, $_c = {}' \quad j\,''$. Then, it is left to the reader as an exercise to prove that rst-order perturbation theory yields the following expression for the attenuation constant of the

[§] The only case which may, occasionally, o er an exception to this rule, is that of waveguide lasers, when their active medium is not pumped, and therefore is heavily absorbing.

m-th guided mode:

$$_m = \frac{1}{2} \frac{\displaystyle\int_S \omega \;'' \frac{\boldsymbol{\mathcal{E}}_m \;\; \boldsymbol{\mathcal{E}}_m}{2} \, dS}{\displaystyle\int_S \frac{\boldsymbol{\mathcal{E}}_{tm} \;\; \boldsymbol{\mathcal{H}}_{tm}}{2} \; \hat{} \; dS} \,, \tag{10.109}$$

where S is the waveguide cross-section, $\boldsymbol{\mathcal{E}}_m$ is the electric eld of the m-th mode, $\boldsymbol{\mathcal{E}}_{tm}$ is its transverse component. So, whereas the integral in the denominator is the Poynting vector ux, the integral in the numerator, strictly speaking, is not proportional to the Poynting vector, as it also contains a term related to the longitudinal eld component. However, we saw on several occasions in this chapter that longitudinal components are of order $^{1/2}$, so their square is of order and therefore may be neglected. The integrals in the numerator and in the denominator become proportional, so that, if $''$ is constant throughout the cross-section S (core and cladding), we have:

$$_m \simeq \frac{1}{2} \omega \;'' \sqrt{\frac{}{_{co}}} \,, \tag{10.110}$$

which is independent of the mode under consideration. On the contrary, if $''$ in the core di ers from that in the cladding, $_m$ is mode-dependent, and mode ltering may occur. Actually, in the early multimode step-index bers absorption in the cladding was larger than in the core. This implied that the modes barely above cut-o experienced signi cantly higher loss, compared to the others. The net result was that, as we said brie y in Section 10.8, experimental values of intermodal dispersion were smaller than theoretical predictions, because the slowest modes did not reach the detector with a signi cant power level. In a single-mode ber, on the other hand, where a signi cant fraction of the guided power travels in the cladding, this result implies that the quality of the cladding material must be as good as that of the core.

Absorption losses can be subdivided into two sets: those which characterize the waveguide material (usually referred to as *intrinsic* absorption losses), and those which are induced by unwanted *impurities* in the host material. The latter ones are, clearly, technology dependent. The dramatic reduction in the amounts of impurities contained in silica glass (now as low as a few parts per billion in weight) was probably the main cause of the impressive progress in ber optics in the 1970's and '80's. In modern bers their contribution to total attenuation is negligible, except for a peak, centered at a wavelength of about 1.4 m, caused by hydroxyl ions (see Figure 10.16). Intrinsic absorption in silica glass is very small throughout the visible range and the very near infrared, but begins to grow rapidly vs. beyond 1.6 m (see again Figure 10.16). Combining this fact with what we are going to see soon about radiation loss, one understands why the wavelengths of highest interest for optical telecommunications are, as we said in previous sections, between 1.3 and 1.5 m.

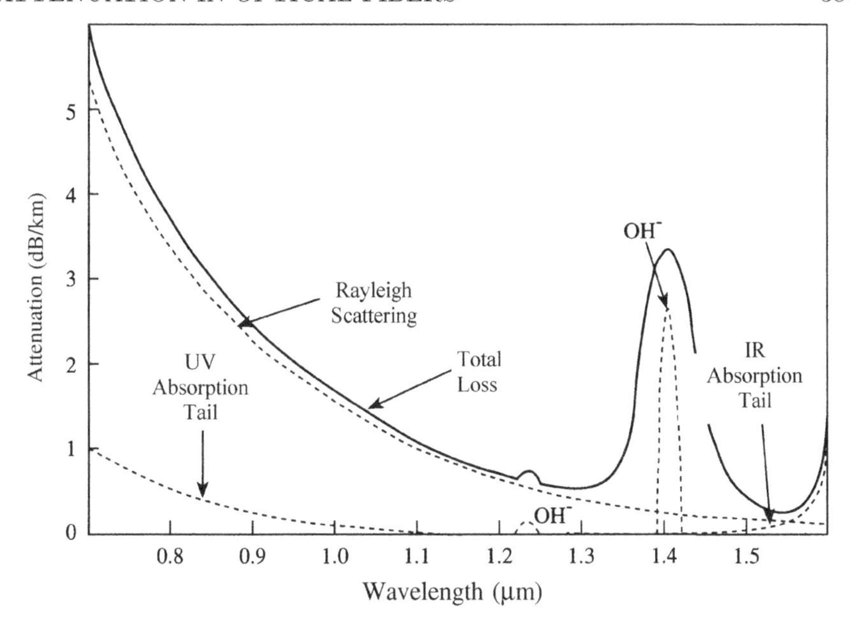

Figure 10.16 *Attenuation vs. wavelength in a typical optical ber for telecommunication or sensor application.*

Referring now to radiation losses, we can also draw a distinction between an intrinsic part, related to basic and unavoidable phenomena in the host material, and extrinsic contributions. The intrinsic part of radiation loss is due to *scattering phenomena* which originate from the fact that, on a microscopic scale, in an amorphous material such as glass there are small density uctuations, which in turn translate into small uctuations of the local refractive index. This phenomenon, well modeled since the late 19th century, is referred to as *Rayleigh scattering* (Born and Wolf, 1980). It occurs uniformly in the entire volume of the dielectric waveguide, in proportion to the local eld amplitude, and therefore the attenuation induced by it is essentially the same for all guided modes. It sets the theoretical lower limit to the attenuation of a dielectric waveguide. Since it decreases strongly as the wavelength increases (in proportion to ^{4}), it induces us to use, in optical communications, the longest wavelengths before running into a region of high intrinsic absorption, as illustrated, once again, by Figure 10.16. Incidentally, it may be of interest to the reader to know that Lord Rayleigh developed his theory as an explanation of the blue color of the sky, following an intuition by Leonardo da Vinci, who was, apparently, the rst who correlated this color with light scattering.

Extrinsic radiation losses may consist of several contributions. One comes from scattering caused by macroscopic inhomogeneities. In early multimode bers, scattering centers were mainly air bubbles at the core-cladding interface, which introduced a selective modal attenuation, very strong for modes

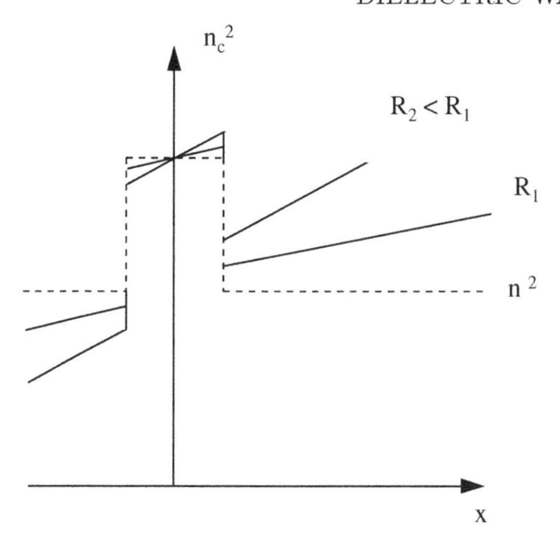

Figure 10.17 *Equivalent index pro le for a bent ber.*

near cut-o . Presently, this phenomenon has been essentially eliminated by technological progress in bers, but scattering centers due to technological processes (e.g., dopant di usion or ion exchange) can still be a serious drawback in some types of planar waveguides for integrated optics.

Another very important type of extrinsic radiation losses are *bending losses.* In an open waveguide, the eld of any guided mode extends all the way to in nity in the transverse direction. Therefore, if the waveguide axis is bent, the tangential phase velocity of the wave front should grow inde nitely, as we go away from the core. The net result is that the waveguide radiates laterally. More precisely, it can be shown (by means of a classical mathematical procedure called *conformal mapping*, used frequently in static eld problems) that a bent waveguide is equivalent to a straight waveguide with an e ective pro le of the kind shown in Figure 10.17. This pro le cannot propagate a strictly guided mode, because, according to what we saw in Section 10.9, the eld beyond a certain distance from the axis is again an oscillating function of the transverse coordinate. Incidentally, note that this entails that the modes of a curved ber form a continuum.

Demanding calculations (see Tosco, Ed., 1990), based on expanding the eld as a continuous superposition of Hankel functions (see Appendix D), yield an analytical expression for the per-unit-length eld attenuation, $_b$, in a single-mode ber bent with a constant radius of curvature, R. The steps are omitted here rst because of space limits, and also because their evaluation requires subjects to be discussed in Chapters 12 and 13. The nal result is:

$$_b = F R^{\ \ 1/2} \exp(\ A R) \quad , \tag{10.111}$$

where:

$$F = \frac{1}{2} \left(\frac{}{aw^3}\right)^{1/2} \left[\frac{u}{v\,K_1(w)}\right]^2 ,$$

$$A = \frac{4}{3} \quad \frac{w^3}{av^2} , \tag{10.112}$$

with the usual meanings of the symbols u, w (normalized transverse propagation constants in core and cladding, respectively), v (normalized frequency), (relative index di erence), K_1 (modi ed Bessel function). The very strong (actually, exponential) dependence of the bending loss on some of these parameters has a paramount impact on optical cable design.

Actually, Eq. (10.112) gives the per-unit-length loss when the eld propagating around the bend has reached a steady-state distribution, as a function of the transverse coordinates. However, more often than not, there is an additional *transition loss*, localized at the beginning and at the end of the bend, and caused by mismatch between modal eld distributions in the straight and bent waveguides. Many more details on loss calculation in bent bers and in bent planar structures can be found in the literature (e.g., Unger, 1977).

What we said so far about bending loss refers to a single curve, with a constant radius of curvature. There is a completely di erent situation which is also encountered often in practice, typically when an optical ber experiences a lateral force pressing it against a surface which is not perfectly at (e.g., inside a cable, or on a drum). The surface roughness will induce a random sequence of small changes in the direction of the waveguide axis. This phenomenon is usually referred to as *microbending*, and clearly, is another source of extrinsic radiation loss, which can play a role of paramount importance in the design of optical cables (e.g., in choosing the ber coating materials). Theoretical expressions for this type of loss, which match very well experimental data, are available in the literature. However, their derivation requires us to treat the departure of ber axis from a straight line as a random phenomenon, whose statistical model is beyond the scope of this book. In the next section, the interested reader will nd indications on where this subject is covered.

10.12 Further applications and suggested reading

As we said at the beginning of this chapter, and again on other occasions, engineering interest in dielectric waveguides has increased enormously over slightly more than two decades. Technical and economical advantages of optical bers, with respect to all the previously exploited transmission media based on conductors (wire pairs, coaxial cables, metallic-wall waveguides) have been, by far and large, the main driving force in this process. This entails two consequences, which might look rather obvious to the reader who has followed us throughout the chapter. One is that an overwhelming fraction of the practical applications of dielectric waveguides is strongly associated to *optical frequencies*; more precisely, a signi cant part of the work is in the visible range, a minute fraction is in the ultraviolet, and the large majority is in the near infrared. The second consequence is that the largest market for dielectric waveguides, as well as the main driving force for ongoing research in the eld, is in telecommunications. Many books which focus on dielectric waveguides at optical frequencies for telecommunications have already been quoted in previous sections.

An inexperienced reader might draw, from this rather narrow-looking range of applications, the erroneous conclusion that there is not much to learn, at least in terms of fundamentals, beyond the contents of this chapter. Let us quote at least one reason why such a statement is untenable. We focused almost exclusively on *ideal* structures: perfect geometrical shape, in nite length without longitudinal discontinuities, perfectly isotropic materials. Yet, performances of actual waveguides are limited, on many occasions, by their imperfections. In order to have at least a general idea of how many problems related to imperfections can be addressed, the reader should rst look again at those parts of Chapter 7 and Chapter 8 which deal with imperfections (collapse of mode orthogonality, coupled-mode equations, etc.), and then expand on this subject, reading, e.g., the *ad-hoc* chapters of Snyder and Love (1983), and those of Unger (1977). For a more in-depth approach to some problems related to one type of "imperfection," anisotropy, one may consult the rst chapters of Someda and Stegeman, Eds., (1992).

One of the reasons — certainly not a minimal one — why imperfect optical waveguides are an important subject, is because a fast developing technique exploits their imperfections in order to monitor external physical variables which a ect deviations of dielectric waveguides — typically, optical bers — from the ideal behavior. So, we have *ber sensors* of temperature, strain, vibration frequencies of mechanical structures, etc., in the research stage as well as already on the market. The reader should at least examine a book covering the fundamentals of this subject, e.g., Culshaw (1984/1996), Dakin and Culshaw (1988). Those who wish to learn more about testing those properties of bers which may become limiting factors for telecommunication applications, but can be exploited for sensor applications, may consult Cancellieri, Ed., (1993).

A few decades after the rst proposal to exploit bers for telecommunications (Kao and Hockham, 1966), and after a history of success beyond any expectation (see Neumann, 1988, for an extensive bibliography), optical waveguides are still a fascinating eld for advanced research. One of the main reasons why this occurs, is because in dielectric waveguides it is rather easy to have electromagnetic waves reaching the *nonlinear propagation regime.* Let us emphasize that dealing with *waveguides* is not a marginal detail, in this respect, but, on the contrary, plays a fundamental role. In fact in the 1960's, when the invention of the laser stimulated the rst interest in nonlinear optics, silica glass was never considered as a suitable material for performing nonlinear optics experiments. Indeed, nonlinear coe cients of glass are very small, compared with other solid, liquid and gaseous media. Yet, in an optical ber two phenomena concur to yield a gigantic enhancement of nonlinear phenomena: con nement of radiation within a very small area, and extremely long interaction lengths, for waves propagating in the same direction, given the very low losses they experience in bers. Note that even in telecommunications, where the main concern is never a high power level, nonlinearities set limits to system performances, or, on the contrary, can be exploited to yield extremely good results. The reader is strongly encouraged to read many papers and books in this area, which consists, in turn, of many di erent phenomena. The last chapters in Someda and Stegeman, Eds., (1992) are a good starting point, and contain large lists of references.

An up-to-date blend of ber imperfections and nonlinearities can be found in Galtarossa and Menyuk, Eds. (2005), under the common heading of polarization mode dispersion.

For those readers who wish, on the other hand, to familiarize themselves with a larger number of examples of application of the theory outlined in this chapter, and to nd a larger list of problems, a suitable reading is van Etten and van der Plaats (1991).

Finally, let us remark that even an in-depth knowledge of optical waveguides is of limited use and bene t, if it is not complemented by an adequate background in Quantum Electronics, the discipline which studies interaction between radiation and matter and is the basis to designing lasers. The reader should consult at least a few among the many excellent textbooks on lasers. Let us quote, for example, Yariv (1989), Svelto (1989), and Siegman (1986).

This nal statement becomes more and more appropriate, nowadays, as ber-based optical ampli ers and all- ber lasers gain industrial momentum. Understanding them requires, as we just said, a well balanced, broad-scope view of Quantum Electronics and dielectric waveguides.

Problems

10-1 Prove that the characteristic equations of the odd TE modes, the even TM modes, and the odd TM modes of a slab waveguide, are analogous to Eq. (10.14), in the sense that they have a nite number of solutions at any nite frequency. Which ones, among these families of modes, include a solution which propagates as a guided mode at an arbitrarily small frequency ? Discuss the physical reasons why some do, and the others do not.

10-2 Show that Maxwell's equations in a slab waveguide are veri ed by a continuous in nity of modes which satisfy the restriction $k_i \quad {}_{co} \quad {}_{cl}$. They are the *evanescent modes* of the slab waveguide.

10-3 Derive the characteristic equation of the even TE modes of an asymmetrical slab waveguide, a planar structure where the core (refractive index n_{co}) is sandwiched between two half-spaces whose refractive indices n_1, n_2 are di erent, still lower than n_{co}.
Show that the fundamental mode of this structure has a nite cut-o fre-

quency. Provide a physical explanation for the di erent behaviors of this mode, in a symmetrical slab and in an asymmetrical one.

10-4 The most schematic model of an integrated-optics *directional coupler* is a pair of parallel dielectric slab waveguides, spaced by a gap of width $2g$ in the x-direction (the reference frame is parallel to that of Figure 10.2, but the origin is mid way between the slabs). The refractive index in the region between the cores is equal to that in the outer half-spaces. Assume that the coupler is *symmetrical*: the two cores are identical in width, $2a$, and in index, n_{co}. Applying continuity of tangential components at the various planar interfaces, and assuming either harmonic (in the cores) or exponential (elsewhere) eld dependence on x, derive the characteristic equation of the even and odd modes of the complete structure (sometimes referred to as "supermodes" to distinguish them from the modes of the individual guides). Use the terms "even" and "odd" in the same senses as they were used when discussing spatial symmetries (Section 3.9).

10-5 Return to the previous problem and suppose now that the two slabs di er either in width ($a_1 \neq a_2$), or in index ($n_1 \neq n_2$), or in both.

a) Write the characteristic equation of the TE modes whose elds in the two cores are in phase, and that of the TE modes whose elds in the two cores are out of phase by radians.
b) Discuss why none of these modes can be even or odd, with respect to any plane orthogonal to the x axis.

10-6 Write explicitly the eld components of the radiation modes of a dielectric rod of circular cross-section, with core radius a and refractive indices n_{co}, n_{cl}.

10-7 Verify that the continuity conditions for the radial components, Eqs. (10.41), are satis ed by the elds of the exact guided modes of a circular dielectric rod.

10-8 Consider the guided modes of a weakly guiding slab waveguide.

a) Show that their longitudinal eld components are of order $^{1/2}$ with respect to the transverse ones, where is de ned by Eq. (10.45).
b) Repeat the previous calculation for the TE and TM modes of a circular cylindrical rod.

10-9 Prove that Eqs. (10.51) hold in the weakly-guiding approximation.

10-10 A *ring-shaped* dielectric waveguide is a structure whose guiding core (refractive index n_{co}) is a cylindrical tube, of inner and outer radii a and b, respectively. The refractive index equals n_{cl} (smaller than n_{co}) for $r < a$ and for $r > b$. Applying the continuity conditions at $r = a$ and at $r = b$, derive the characteristic equations of TE, TM and hybrid modes of this structure. (*Hint.* In the region $r < a$, the suitable solutions of Bessel's equation to represent guided modes are the modi ed Bessel functions of the rst kind).

10-11 A *multiple-core* optical ber is a structure where the refractive index equals n_{co} in a central region of radius a_1 *and* in a cylindrical tube of inner and outer radii a_2 and a_3, respectively; it equals $n_{cl} < n_{co}$ for $a_1 < r < a_2$ and for $r > a_3$.

a) Consider the range of single-mode propagation. Provide a qualitative, physical argument to explain why in this structure the peak of the normal waveguide dispersion (i.e., the peak of $vd^2(vb)/dv^2$ in Figure 10.11) is shifted towards longer wavelengths, compared with a simple step-index ber whose core radius is nearly equal to a_3.

b) Applying the continuity conditions at all the cylindrical interfaces, derive the characteristic equations of TE, TM and hybrid modes of this structure.

10-12 Evaluate the maximum number of pulses per unit time which can be transmitted over a 10 km long, highly multimode ber, where the relative di erence between the core and cladding group indices is 0.01. Assume, as an order of magnitude, that $N_{co} = 1.5$.

10-13 In a so-called "standard" single-mode ber, chromatic dispersion goes through zero at a wavelength close to 1.3 m and its typical value at $=$ 1.55 m is 17 ps/(km nm). Evaluate the maximum number of pulses per unit time which can be transmitted over 100 km, when the spectral linewidth of the optical source (centered at 1.55 m) equals 1 GHz, 10 GHz, or 100 GHz, respectively.

10-14 At the wavelength of zero chromatic dispersion, the distortion of a pulse traveling along a single-mode ber is governed by the next higher-order term in the Taylor series expansion of the phase constant vs. frequency (see Eq. (5.5). Find an expression for $d^3 \ /dk^3$, starting from Eq. (10.73) and neglecting the last term in this expression.

10-15 A widely used numerical approach to graded-index bers, which respects the vector nature of the modal elds, is based on the so-called "staircase approximation": the continuous index pro le in the ber core is replaced by a discrete multiple step one, where the index is constant within each concentric layer. Continuity conditions of tangential eld components are applied at the various interfaces, to obtain the characteristic equation of guided modes, which is solved numerically. Prepare a ow-chart for a computer program based on this method. Follow two approaches: (a) the radial widths of all layers are equal, (b) for a monotonically decreasing pro le the index steps at all interfaces are equal.

Note. Matrix inversion requires a processing time which increases approximately as the third power of the matrix order. Therefore, avoid large matrices. Eliminate some unknowns at each interface, before proceeding to the next one.

10-16 Prove that for a parabolic pro le which extends all the way to in nity, i.e., for $f(r) \propto r^2$, $a = \infty$, Eqs. (10.81) are satis ed by the *Laguerre-Gauss modes*, which were addressed in Chapter 5.

10-17 Show that the loci of $U(r) = 0$ ($U(r)$ is de ned by Eq. (10.91)) become "turning points" of the rays, if a medium characterized by an index pro le $f(r)$ is studied in terms of geometrical optics.

10-18 Derive Eq. (10.99).

10-19 Prove that the minimum of Eq. (10.106), as a function of , is where Eq. (10.108) is satis ed.

10-20 Using Eqs. (10.111) and (10.112), evaluate the per unit length bending loss for a radius of curvature $R = 0.1\,\mathrm{m}$, for a single-mode ber with a core radius $a = 5.5$ m, at a normalized frequency $v = 2.2$, at a wavelength of 1.33 m, with pure silica cladding ($n_{cl} = 1.4585$).
(*Hint.* Use the asymptotic form $K_1(w) \simeq \sqrt{\ /(2w)}\ \exp(\ \ w)$.)

References

Abramowitz, M. and Stegun, I.A. (Eds.) (1965) *Handbook of Mathematical Functions.* Dover, New York.

Adams, M.J. (1981) *An Introduction to Optical Waveguides.* Wiley, New York.

Born, M. and Wolf, E. (1980) *Principles of Optics: Electromagnetic Theory of Propagation, Interference and Di raction of Light.* 6th (corrected) ed., reprinted 1986, Pergamon, Oxford.

Cancellieri, G. (Ed.) (1993) *Single-Mode Optical Fiber Measurement: Characterization and Sensing.* Artech House, Boston-London.

Culshaw, B. (1984) *Optical Fibre Sensing and Signal Processing.* Peter Peregrinus, London.

Culshaw, B. (1996) *Smart Structures and Materials.* Artech House, Norwood, MA.

Dakin, J.K. and Culshaw, B. (1988) *Optical Fiber Sensors: Principles and Devices.* Vols. 1-2, Artech House, Norwood, MA.

Etten, W., van der and Plaats, J., van der (1991) *Fundamentals of Optical Fibers Communications.* Prentice-Hall, New York.

Galtarossa, A. and Menyuk, C.R., Eds. (2005) *Polarization Mode Dispersion.* Springer New York Inc. (Optical and Fiber Communication Reports, Volume 1), New York.

Gloge, D. (1971) "Weakly guiding bers". *Applied Optics,* **10**, 2252–2258.

Gloge, D. and Marcatili, E.A.J. (1973) "Multimode Theory of Graded-Core Fibers". *Bell System Technical Journal,* **52**, 1563–1578.

Kao, K.C. and Hockham, G.A. (1966) "Dielectric- ber surface waveguides for optical frequencies". *Proc. IEE,* **113**, 1151–1158.

Kurtz, C.N. and Streifer, W. (1969) "Guided Waves in Inhomogeneous Focussing Media. Part I: Formulation, Solution for Quadratic Inhomogeneity". *IEEE Trans. on Microwave Theory and Techniques,* **17**, 11–15.

Neumann, E.-G. (1988) *Single-Mode Fibers: Fundamentals.* Springer Series in Optical Sciences, Tamir Ed., Springer-Verlag, Berlin.

Petersen, J.K. (2003) *Fiber Optics Illustrated Dictionary.* CRC Press, Boca Raton.

Saleh, B.E.A. and Teich, M.C. (1991) *Fundamentals of Photonics.* Wiley, New York.

Siegman, A.E. (1986) *Lasers.* University Science Books, Mill Valley, California.

Snyder, A.W. and Love, J.D. (1983) *Optical Waveguide Theory.* Chapman & Hall, London.

Matera, F. and Someda, C.G. (1992) "Random Birefringence and Polarization Dispersion in Long Single-Mode Optical Fibers". In C.G. Someda and G. Stegeman, Eds., *Anisotropic and Nonlinear Optical Waveguides,* Elsevier, Amsterdam.

Svelto, O. (1989) *Principles of Lasers.* 3rd ed., translated from Italian and edited by D.C. Hanna, Plenum, New York.

Tosco, F. (Ed.) (1990) *Fiber Optic Communications Handbook.* TAB Professional and Reference Books, Blue Ridge Summit, PA.

Unger, H.-G. (1977) *Planar Optical Waveguides and Fibres.* Clarendon Press, Oxford.

Yariv, A. (1989) *Quantum Electronics.* 3rd ed., Wiley, New York.

CHAPTER 11

Retarded potentials

11.1 Introduction

As we said at the beginning of Chapter 4, the most common ways of solving partial di erential equations (PDE's) are separation of variables, and the so-called Green's functions method. In many of the previous chapters we used separation of variables to solve either the Helmholtz equation or other PDE's derived from it. This chapter will be an introduction to the Green's function technique, and illustrate some basic results which stem from solving Maxwell's equations in this new way. Chapters 12 and 13 will be devoted to more speci c applications of these broad-scope results.

It was brie y mentioned in Chapter 4 that separation of variables is very convenient when the region of interest does not contain eld sources. Green's functions are more convenient in the presence of sources. However, let us stress from the beginning that there is no sharp boundary between the scopes of application of the two techniques. In fact, a sharp contrast would not even match the physical world. When we deal, in a given region of the 3-D space, with a eld without sources, it must be understood that this eld has been generated by sources which are located elsewhere. There is complete agreement between this obvious physical fact and the mathematical approaches, in the sense that there are links between Green's functions and what we saw in the previous chapters. Sections 11.7, 11.8 and 11.9 will clarify this statement by means of some examples.

We will assume here that the reader knows already at least the fundamentals of distribution theory. More precisely, the so-called Dirac function will be widely used throughout the rest of this book. In contrast to what we do for other items of the mathematical background, which are sketched or summarized in the appendices, we do not deal at all with distributions, whose fundamentals are very di cult to compress into an appendix. Readers who feel that they should refresh their memories on this subject should consult textbooks (e.g., Zemanian, 1965). Before we proceed, let us just warn the reader that, in the following, the function will be involved in formulas (e.g., Green's theorem) that have been proven for *continuous and di erentiable* functions. There are two ways to justify such steps:

proofs given for regular functions are replaced by new *rigorous* proofs, in the framework of distribution theory. This approach is quite demanding. To start with, it requires new de nitions, e.g., the *convolution product* among distributions (Donoghue, 1969). Although more rigorous than the second one, this approach will not be followed here, for the sake of simplicity;

we make use of a so-called *regular model* for the function. This procedure, as we brie y mentioned in Chapter 1, begins with a regular function, (P), that depends on a parameter, d, so that $(P) \to (P)$ for $d \to 0$. For example, this occurs for a Gaussian function, $G(x)$, normalized so that its integral remains equal to unity, when its width at $G(0)/e$ amplitude tends to zero. Following this approach, all the formulas containing a (P)-function must be interpreted as if they had been written rst with a regular (P), and *later* sent to the limit for $d \to 0$. These intermediate steps will not be written explicitly, to save space and to avoid very cumbersome presentations.

11.2 Green's functions for the scalar Helmholtz equation

The Green's functions method can be used, in principle, to solve any *linear* PDE with constant coe cients. It can be looked at as a practical way to implement the superposition principle. Those readers who are interested in a more general overview of this subject should consult a treatise (e.g., Morse and Feshbach, 1953; Barton, 1989; Greenberg, 1971); still, it may be useful to outline here the basic features of the method. The vector nature of e.m. elds entails some complications which might make this outline less clear, so let us deal rst with the *scalar Helmholtz equation*, which governs propagation of other types of waves (acoustic, elastic, etc.) (DeSanto, 1992). It may also be of interest in some problems in electromagnetism; indeed, electric and magnetic scalar potentials in a homogeneous medium, if we follow the Lorentz gauge (Chapter 1), obey this equation. The same is true for the pre-potential functions L and T of Chapter 3.

The *non-homogeneous* scalar Helmholtz equation reads:

$$\nabla^2 \quad {}^2 \quad = S(\mathbf{r}) \quad . \tag{11.1}$$

The function $S(\mathbf{r})$ — in general, complex — is the source of the scalar eld $(\mathbf{r})$, whose physical dimensions are irrelevant to what we want to discuss here. Let us use another symbol, G, for *any* solution (i.e., irrespectively of the boundary conditions) of the same equation when the right-hand side is a 3-D Dirac function, $(\mathbf{r} \quad \mathbf{r}') = (x \quad x') \ (y \quad y') \ (z \quad z')$, centered at the point $P = O + \mathbf{r}' = O + x'\hat{a}_x + y'\hat{a}_y + z'\hat{a}_z$:

$$\nabla^2 G \quad {}^2 G = \ (\mathbf{r} \quad \mathbf{r}') \quad . \tag{11.2}$$

This $G(\mathbf{r}, \mathbf{r}')$ is referred to as a *Green's function* of the problem described by Eq. (11.1). Note that the operator ∇^2 in Eq. (11.2) operates on the coordinates of $Q = O + \mathbf{r}$. Multiply now Eq. (11.1) by G, and Eq. (11.2) by , and subtract side by side. Integrate the result over a region V which includes the point $P = O + \mathbf{r}'$ where $(\mathbf{r} \quad \mathbf{r}')$ is centered, and is bounded by a closed surface C.

By de nition of $(\mathbf{r} \quad \mathbf{r}')$, we obtain:

$$(\mathbf{r}') = \int_V G(\mathbf{r}, \mathbf{r}')\, S(\mathbf{r})\, \mathrm{d}V + \int_V (\quad \nabla^2 G \quad G \nabla^2 \quad) \, \mathrm{d}V \quad . \tag{11.3}$$

Apply Green's theorem (A.11) to the second term on the right-hand side of Eq. (11.3), yielding:

$$(\mathbf{r}') = \int_V G(\mathbf{r}, \mathbf{r}')\, S(\mathbf{r})\, \mathrm{d}V + \int_C \left(\quad \frac{\partial G}{\partial n} \quad G \frac{\partial}{\partial n} \right) \mathrm{d}C \quad . \tag{11.4}$$

Note that $P = O + \mathbf{r}'$ can be *any point* inside the region V. Therefore, Eq. (11.4) represents a formal solution to the problem in Eq. (11.1). It indicates that, in order to compute $(\mathbf{r}')$ in the whole region V, all we need to know, in addition to the source $S(\mathbf{r})$, is:

and its normal derivative everywhere on the closed surface C that surrounds V;

one solution $G(\mathbf{r})$ of Eq. (11.2). Saying *one* we stress that it is not obliged to satisfy given boundary conditions on the surface C. Its domain of de nition can be wider than V, it can even be the whole 3-D space.

For $S(\mathbf{r}) = \quad (\mathbf{r} \quad \mathbf{r}')$, and assuming that and G satisfy the same boundary conditions on C, Eq. (11.4) gives $(\mathbf{r}) = G(\mathbf{r}, \mathbf{r}')$. Thus, Eq. (11.4) is, as we anticipated, just the superposition principle: the function $(\mathbf{r}')$ is the overlap of the in nitesimal contributions due to an in nite number of point sources, $d \quad = S(\mathbf{r}) \quad (\mathbf{r}' \quad \mathbf{r})\, dV$, provided that all these contributions obey identical boundary conditions. If, on the contrary, they obey di erent boundary conditions, the last term in Eq. (11.4) accounts for these di erences.

The result deserves further comments. Solutions of the scalar Helmholtz equation obey a *uniqueness theorem*, very similar to the classical theorem of electrostatics (i.e., the theorem holding for solutions of the Laplace equation $\nabla^2 \quad = S(\mathbf{r})$). In particular, one can show that the values of and those of its normal derivative $\partial \quad / \partial n$ *cannot be pre-assigned independently of each other* at all points of a closed contour C. Consequently, Eq. (11.4) is just a *formal* solution for Eq. (11.1). In more rigorous terminology, it is what mathematicians call an *integral representation* of the unknown . An expression like Eq. (11.4) becomes actually a solution only in well-speci ed cases, that is, only when and $\partial \quad / \partial n$ have carefully chosen, mutually compatible values on C.

With this in mind, we may proceed now to determine the Green's function. We are, by assumption, in a homogeneous medium, and have complete freedom to choose the conditions G must satisfy on C. This allows us to suppose that G has *spherical symmetry* around the point P where the -function source is located. Using then a spherical coordinate frame $(\ , \vartheta, \varphi)$ whose origin is in P (Figure 11.1), partial derivatives of G with respect to ϑ and to φ are identically zero by symmetry.

The Laplacian in spherical coordinates, Eq. (B.14), reduces itself to the only

On this point, see again the last part of Section 11.1.

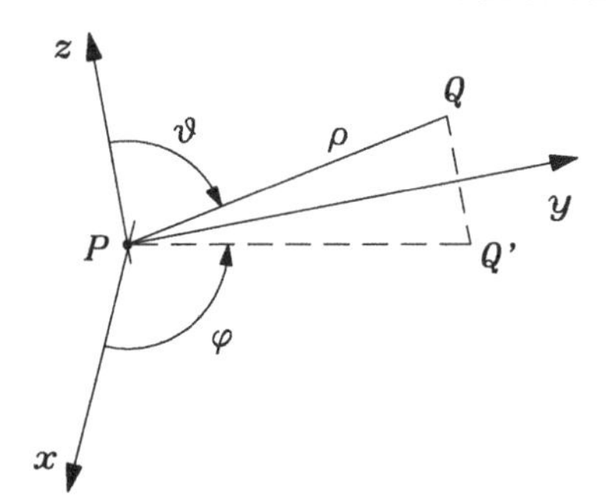

Figure 11.1 *Reference frame used in the derivation of the spherically symmetrical Green's function.*

term $\ ^{1}(\partial^2/\partial\ ^2)\ $. Eq. (11.2) becomes the following ordinary di erential equation:

$$\frac{1}{\ }\frac{d^2}{d\ ^2}\left[\ G(\ )\right]\quad ^2 G(\ ) = \ (\ )\ \ . \tag{11.5}$$

A solution of Eq. (11.5) can be found heuristically in the following way. Leave aside, for the time being, the point P ($= 0$) where there is a singularity. Anywhere else, Eq. (11.5) is a homogeneous equation, whose general integral can be found immediately:

$$G(\ ) = a\,\frac{e}{\quad} + b\,\frac{e}{\quad}\ , \tag{11.6}$$

where a and b are two arbitrary constants. If we pass to the time domain, it is easy to see that the rst term on the right-hand side of Eq. (11.6) is an *expanding* spherical wave, whose wavefront radius grows linearly with time. The second term is a *contracting* spherical wave, whose wavefront radius decreases linearly with time. So far, we have assumed that the source of the wave G is located at $= 0$; therefore, causality suggests that the rst term is physically suitable, and, exploiting our freedom to choose *one* solution of Eq. (11.2), we set $b = 0$ in Eq. (11.6). However, we will show in Section 11.6 that also the second term in Eq. (11.6) can lend itself to a physically sensible interpretation, at least in the lossless case.

The nal detail we are left with is to give the constant a a value such that the rst term in Eq. (11.6) satis es Eq. (11.5) even for $\to 0$. Let us integrate both sides of Eq. (11.2) over the volume of a sphere, , centered at P. The integral of the right-hand side equals unity, by de nition of the Dirac delta-function. On the left-hand side, the integrand is identically zero as soon as $\neq 0$, and therefore:

$$\int (\nabla^2 G \quad ^2 G)\,\mathrm{d}\ = \lim_{\to 0}\int \nabla^2 G\,\mathrm{d}\quad \lim_{\to 0}\int\ ^2 G\,\mathrm{d}\quad . \tag{11.7}$$

As d is proportional to $^2 d$, the last term in Eq. (11.7) vanishes. Apply then Gauss' theorem (A.8) to the rst term on the right-hand side (with

$\hat{n}$ = unit vector normal to the surface, u, which surrounds ; $du = \ {}^2 d\ $), and make use, for G, of Eq. (11.6) with $b = 0$. We obtain:

$$\lim_{\to 0} \int \nabla^2 G \, d \ = \lim_{\to 0} \int_u \nabla G \ \hat{n} \, du = \lim_{\to 0} \left\{ \ ^2 \int_4 \frac{dG}{d} d \ \right\} = \ 4 \ a \, . \quad (11.8)$$

The integral of the right-hand side of Eq. (11.2) is equal to unity, and must be equal to Eq. (11.8) as well. This yields $a = \ 1/(4\ )$, so nally we get:

$$G = \ \frac{1}{4} \frac{e}{} = \ \frac{1}{4} \frac{e^{\ |\mathbf{r} \ \mathbf{r}'|}}{|\mathbf{r} \ \mathbf{r}'|} \ . \quad (11.9)$$

The next sections will show that the function in Eq. (11.9) is extremely relevant and useful also to solve the vector Helmholtz equation.

When using the spherical-symmetry Green's function, Eq. (11.9), it may be convenient to write a di erent expression for the surface integral which includes $(\partial G/\partial n)$ in Eq. (11.4). From Eq. (11.9) it is straightforward to get:

$$\frac{\partial G}{\partial n} = (\nabla G) \ \hat{n} = \frac{dG}{d} \ \hat{} \ \ \hat{n} = \left(\ \ \ \frac{1}{\ } \right) G \ \hat{} \ \ \hat{n} \ , \quad (11.10)$$

where:

$$\hat{} = \frac{\mathbf{r} \ \ \mathbf{r}'}{|\mathbf{r} \ \ \mathbf{r}'|} \quad (11.11)$$

is the unit vector in the direction from P toward Q. Thus, Eq. (11.4) can be rewritten as:

$$\begin{aligned} (\mathbf{r}') \ &= \ \int_V G(|\mathbf{r} \ \ \mathbf{r}'|) \, S(\mathbf{r}) \, \mathrm{d}V \\ &\ \int_C \left[\ \left(\ + \frac{1}{\ } \right) \hat{} \ \ \hat{n} + \frac{\partial}{\partial n} \right] G(|\mathbf{r} \ \ \mathbf{r}'|) \, \mathrm{d}C \ . \end{aligned} \quad (11.12)$$

We will make use of this on a few occasions in the next section, and again in Chapter 13.

11.3 Lorentz-gauge vector potentials in a homogeneous medium

Having solved the scalar Helmholtz equation, the Green's function method will enable us to deal now with vector problems. Indeed, it was shown in Section 1.8 that, if we are in a homogeneous medium and adopt the *Lorentz gauge*, the magnetic vector potential, $\mathbf{A}$, and the electric vector potential, $\mathbf{F}$, satisfy a vector Helmholtz equation, which is not homogeneous when there are sources $\mathbf{J}_0, \mathbf{M}_0$:

$$\nabla^2 \mathbf{A} \ \ \ {}^2 \mathbf{A} \ = \ \ \mathbf{J}_0 \ , \quad (11.13)$$

$$\nabla^2 \mathbf{F} \ \ \ {}^2 \mathbf{F} \ = \ \ {}_c \mathbf{M}_0 \ . \quad (11.14)$$

In the following, we will concentrate on Eq. (11.13), unless the opposite is stated. Eq. (11.14) becomes useful in case the sources are magnetic currents, and can be approached by applying the duality theorem of Section 3.7 to the results that we will nd about Eq. (11.13).

As well known, in rectangular coordinates the components of the Laplacian of a vector, $\mathbf{A}$, are the (scalar) Laplacians of the vector components, A_k $(k = 1, 2, 3)$. The vector equation (11.13) reduces to three scalar Helmholtz equations, independent of each other, and these can be solved using the Green's functions approach of Section 11.2. Usually, however, what is not easy is to write, on the closed surface which surrounds the region where Eq. (11.13) is to be solved, boundary conditions that apply separately to the three cartesian components of the vector $\mathbf{A}$. The way to circumvent this problem is the *equivalence theorem* of Section 3.5, whose content is worth recalling here.

Let S be a regular closed surface, and $\hat{n}$ be the unit vector orthogonal to S towards outside S (note that this makes a di erence with respect to Eqs. (3.24) and (3.25), where $\hat{n}$ was inwards). Let the following surface current densities:

$$\mathbf{J}_S = \mathbf{H} \quad \hat{n} \ , \qquad \mathbf{M}_S = \hat{n} \quad \mathbf{E} \ , \tag{11.15}$$

circulate on S. Do not modify the sources that are located in the region, V, which is inside S. On the contrary, let all sources outside S become equal to zero. Then:

at any point in V, the eld is identical to that generated by *all* the *actual* sources, located inside and outside S;

at any point *outside* S, *the eld is zero*; as such, it can be derived from potentials that are identically zero outside S.

As the sources can be grouped into three sets, the superposition principle entails that the total eld, at any point Q in V, can be calculated as the sum of three e.m. elds. One is due to the vector potential $\mathbf{A}'$ which satis es Eq. (11.13). The other two contributions are generated by the *equivalent sources*, Eq. (11.15). Also these elds can be found via two vector potentials, $\mathbf{A}''$ and $\mathbf{F}''$, which satisfy the following equations:†

$$\nabla^2 \mathbf{A}'' \quad {}^2\mathbf{A}'' = \begin{cases} \quad \mathbf{H} \quad \hat{n} & \text{on } S \\ 0 & \text{in } Q \notin S \end{cases} , \tag{11.16}$$

$$\nabla^2 \mathbf{F}'' \quad {}^2\mathbf{F}'' = \begin{cases} \quad c\,\hat{n} \quad \mathbf{E} & \text{on } S \\ 0 & \text{in } Q \notin S \end{cases} . \tag{11.17}$$

We are now ready to address the question of the boundary conditions, i.e., to nd what replaces the surface integral of the scalar formula (11.4). When we invoke the equivalence theorem, we have the freedom to choose a surface S which is entirely *within* the region where Eq. (11.13) (or Eq. (11.14), in the dual case) holds. This, in turn, allows us to choose the boundary-condition closed surface C (i.e., the surface which made it possible to pass from Eq. (11.3) to

† For physical dimensions to be consistent, on the right-hand side of Eqs. (11.16) and (11.17) the functions centered on S must be thought of as having dimensions of an inverse length.

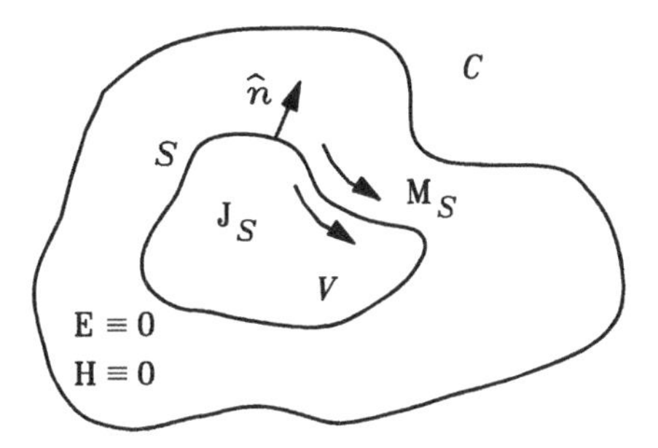

Figure 11.2 *Application of the equivalence theorem.*

Eq. (11.4)) completely located *outside* S (see Figure 11.2). Then C lies where the potentials and their derivatives are *identically* zero. With this choice of C, *no contribution to the eld comes from integrals over C itself*, as all terms in the integral over C are zero.

To summarize, we wrote four vector Helmholtz equations. Each of them has a solution which consists of only one integral: an integral over the volume V for the vectors involved in Eqs. (11.13) and (11.14), an integral over the surface S for the vectors that are generated by the equivalent sources in Eq. (11.15):

$$\mathbf{A}' = \frac{}{4} \int_V \mathbf{J}_0 \frac{e}{} \, \mathrm{d}V \quad , \quad \mathbf{F}' = \frac{c}{4} \int_V \mathbf{M}_0 \frac{e}{} \, \mathrm{d}V \, , \quad (11.18)$$

$$\mathbf{A}'' = \frac{}{4} \int_S \mathbf{H} \quad \hat{n} \frac{e}{} \, \mathrm{d}S \quad , \quad \mathbf{F}'' = \frac{c}{4} \int_S \hat{n} \quad \mathbf{E} \frac{e}{} \, \mathrm{d}S \, , \quad (11.19)$$

where $= |P \quad Q|$.
Let us stress that:

to apply this procedure, the medium cannot be strictly lossless, as this was a requirement of the equivalence theorem. However, Eqs. (11.18) and (11.19) behave continuously as $\mathrm{Re}(\quad) \to 0$ and therefore they can be used also in a lossless medium, provided that "lossless" means *the limiting case as losses tend to zero*;

the medium must be homogeneous in a region which *contains entirely* the surface S over which the integrals (11.19) are calculated. However, S can be arbitrarily near to the boundaries of other media. Furthermore, note that only the tangential components of $\mathbf{E}$ and of $\mathbf{H}$ are involved in Eq. (11.19), and, as known from Chapter 1, these components are continuous even at sharp discontinuities between media. In practice, S may coincide with the surface which surrounds a homogeneous medium, with the only constraint that there cannot be imposed surface currents circulating on this surface.

11.4 Field vectors in terms of dyadic Green's functions

From the previous section one can infer, without any conceptual di culty, how to calculate an e.m. eld in a homogeneous medium, given its sources and the boundary conditions. Indeed, the eld $\{\mathbf{E}, \mathbf{H}\}$ can be found from the potentials through the relationships that are derived in Chapter 1. If the potentials found in the previous section are inserted in the general formulas, we get:

$$
\begin{aligned}
\mathbf{E}(P) \;=\;& \frac{1}{j\omega 4\quad c} \nabla_P \quad \nabla_P \quad \left(\int_V \mathbf{J}_0 \,\frac{e}{\quad}\, \mathrm{d}V_Q + \int_S \mathbf{H} \quad \hat{n} \,\frac{e}{\quad}\, \mathrm{d}S_Q \right) \\
& \frac{1}{4} \nabla_P \quad \left(\int_V \mathbf{M}_0 \,\frac{e}{\quad}\, \mathrm{d}V_Q + \int_S \hat{n} \quad \mathbf{E} \,\frac{e}{\quad}\, \mathrm{d}S_Q \right) \\
& \frac{\mathbf{J}_0}{j\omega \quad c} \quad , \qquad (11.20)
\end{aligned}
$$

$$
\begin{aligned}
\mathbf{H}(P) \;=\;& \frac{1}{4} \nabla_P \quad \left(\int_V \mathbf{J}_0 \,\frac{e}{\quad}\, \mathrm{d}V_Q + \int_S \mathbf{H} \quad \hat{n} \,\frac{e}{\quad}\, \mathrm{d}S_Q \right) \\
+\;& \frac{1}{j\omega \quad 4} \nabla_P \quad \nabla_P \quad \left(\int_V \mathbf{M}_0 \,\frac{e}{\quad}\, \mathrm{d}V_Q + \int_S \hat{n} \quad \mathbf{E} \,\frac{e}{\quad}\, \mathrm{d}S_Q \right) \\
+\;& \frac{\mathbf{M}_0}{j\omega} \quad , \qquad (11.21)
\end{aligned}
$$

where the symbol ∇_P is meant to stress that the ∇ operator must operate *on the coordinates of the "potentiated" point P.*

In the volume and surface integrals of Eqs. (11.20) and (11.21) the integration variables are the coordinates of the "potentiating" point Q. Therefore, all integrals commute with the operator ∇_P. Moreover, neither the sources nor the elds on the surface S depend on P, so ∇_P operates only on the function exp( )/ . Then, taking into account the vector identities of Appendix C, Eqs. (11.20) and (11.21) can be replaced by the following expressions:

$$
\begin{aligned}
\mathbf{E}(P) \;=\;& \frac{1}{j\omega \quad c 4} \int_V \left[\mathbf{J}_0 \,\nabla_P^2 \left(\frac{e}{\quad} \right) \quad (\mathbf{J}_0 \quad \nabla_P)\, \nabla_P \left(\frac{e}{\quad} \right) \right] \mathrm{d}V_Q \\
& \frac{1}{j\omega \quad c 4} \int_S \left[(\mathbf{H} \quad \hat{n}) \nabla_P^2 \left(\frac{e}{\quad} \right) \quad (\mathbf{H} \quad \hat{n} \quad \nabla_P) \nabla_P \left(\frac{e}{\quad} \right) \right] \mathrm{d}S_Q \\
& \frac{1}{4} \int_V \nabla_P \left(\frac{e}{\quad} \right) \quad \mathbf{M}_0 \mathrm{d}V_Q \quad \frac{1}{4} \int_S \nabla_P \left(\frac{e}{\quad} \right) \quad \hat{n} \quad \mathbf{E}\, \mathrm{d}S_Q \\
& \frac{\mathbf{J}_0}{j\omega \quad c} \quad , \qquad (11.22)
\end{aligned}
$$

$$\begin{aligned}
\mathbf{H}(P) &= \frac{1}{4}\int_V \nabla_P\left(\frac{e}{\quad}\right)\ \ \mathbf{J}_0 \mathrm{d}V_Q + \frac{1}{4}\int_S \nabla_P\left(\frac{e}{\quad}\right)\ \ \mathbf{H}\ \ \hat{n}\,\mathrm{d}S_Q \\
&+ \frac{1}{j\omega\ \ 4}\int_V \left[\mathbf{M}_0 \nabla_P^2\left(\frac{e}{\quad}\right)\ \ (\mathbf{M}_0\ \ \nabla_P)\nabla_P\left(\frac{e}{\quad}\right)\right]\mathrm{d}V_Q \\
&+ \frac{1}{j\omega\ \ 4}\int_S \left[(\hat{n}\ \ \mathbf{E})\nabla_P^2\left(\frac{e}{\quad}\right)\ \ (\hat{n}\ \ \mathbf{E}\ \ \nabla_P)\nabla_P\left(\frac{e}{\quad}\right)\right]\mathrm{d}S_Q \\
&+ \frac{\mathbf{M}_0}{j\omega}\ \ .
\end{aligned} \tag{11.23}$$

Eqs. (11.22) and (11.23) express, once again, the superposition principle, giving the elds as sums of an in nite number of contributions due to in nitesimal point sources and to the boundary conditions. They play essentially the same logical role as Eqs. (11.4), (11.18), (11.19), and di er from these only because the linear relations between the vectors on the left-hand side, and the source and boundary terms (which are vectorial too) on the right-hand side, are expressed by *dyadic operators.*

Reordering the terms and introducing new symbols, whose meanings will be explained shortly, Eqs. (11.22) and (11.23) can be rewritten as:

$$\begin{aligned}
\mathbf{E}(P) &= \frac{1}{4}\int_S \mathbf{G}_{11}\ \ \mathbf{E}\,\mathrm{d}S_Q\ \ \frac{1}{j\omega\ {}_c 4}\int_S \mathbf{H}\ \ \mathbf{G}_{12}\,\mathrm{d}S_Q \\
&\quad \frac{1}{j\omega\ {}_c 4}\int_V \mathbf{J}_0\ \ \mathcal{G}_{11}\,\mathrm{d}V_Q\ \ \frac{1}{4}\int_V \mathcal{G}_{12}\ \ \mathbf{M}_0\,\mathrm{d}V_Q\ \ \frac{\mathbf{J}_0}{j\omega\ \ {}_c}\ \ ,
\end{aligned} \tag{11.24}$$

$$\begin{aligned}
\mathbf{H}(P) &= \frac{1}{4}\int_S \mathbf{G}_{22}\ \ \mathbf{H}\,\mathrm{d}S_Q + \frac{1}{j\omega\ \ 4}\int_S \mathbf{E}\ \ \mathbf{G}_{21}\,\mathrm{d}S_Q \\
&+ \frac{1}{j\omega\ \ 4}\int_V \mathbf{M}_0\ \ \mathcal{G}_{22}\,\mathrm{d}V_Q + \frac{1}{4}\int_V \mathcal{G}_{21}\ \ \mathbf{J}_0\,\mathrm{d}V_Q + \frac{\mathbf{M}_0}{j\omega}\ \ .
\end{aligned} \tag{11.25}$$

The *dyadic Green's functions* $\mathbf{G}_{lj}$ and $\mathcal{G}_{lj}$ are de ned by the following relations, that the reader can elaborate further, as an exercise (remembering in particular how the cross product of three vectors is expressed in terms of their inner products):

$$\begin{aligned}
\mathbf{G}_{11}\ \ \mathbf{E} &= \nabla_P\left(\frac{e}{\quad}\right)\ \ \hat{n}\ \ \mathbf{E}\ \ , \\
\mathbf{H}\ \ \mathbf{G}_{12} &= (\mathbf{H}\ \ \hat{n})\ \left\{\nabla_P^2\left(\frac{e}{\quad}\right)\mathbf{i}\ \ \nabla_P\left[\nabla_P\left(\frac{e}{\quad}\right)\right]\right\}\ \ , \\
\mathbf{J}_0\ \ \mathcal{G}_{11} &= \mathbf{J}_0\ \left\{\mathbf{i}\nabla_P^2\left(\frac{e}{\quad}\right)\ \ \nabla_P\left[\nabla_P\left(\frac{e}{\quad}\right)\right]\right\}\ \ ,
\end{aligned}$$

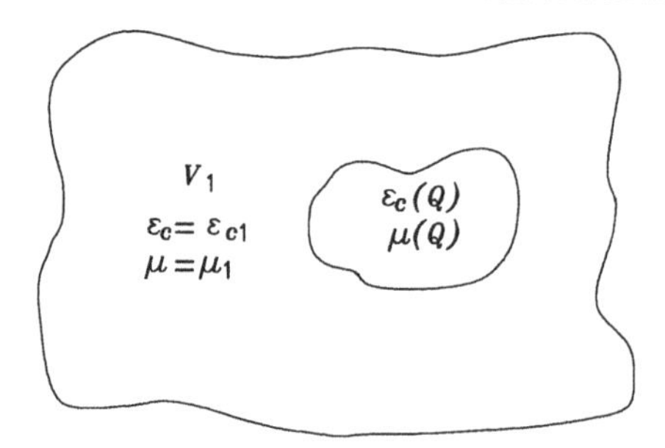

Figure 11.3 *De nition of the polarization currents.*

$$\mathcal{G}_{12} \ \mathbf{M}_0 \quad = \quad \nabla_P \left(\frac{e}{} \right) \quad \mathbf{M}_0 \quad . \tag{11.26}$$

The symbol $\mathbf{i}$ indicates the identity dyadic. The remaining operators $\mathbf{G}_{lj}$ and $\mathcal{G}_{lj}$ can be calculated from Eq. (11.26) by duality.

Eqs. (11.22) and (11.23) express the eld $\{\mathbf{E}, \mathbf{H}\}$ *directly*, without passing through potentials, as a function of the sources and of the values the elds themselves take on the contour. However, this is only an apparent simpli ca- tion, as Eq. (11.24) entails very complicated calculations. For this reason, in the following we will always make use of the vector potentials, in the form we found for them in the previous section. For more details on dyadic Green's functions, the interested reader may consult Tai (1971).

11.5 Inhomogeneous media: Polarization currents

We stressed several times that the results of this chapter are valid only if the region where the eld is de ned contains a *homogeneous medium* (i.e., if , , are constant throughout this region). There is no general and rigorous procedure to extend these results to inhomogeneous media. Approximations are unavoidable. The purpose of this section is to illustrate the one which is used most frequently to solve practical problems, such as, for example, a dielectric lens or dome in front of an aperture antenna.

Let V be the 3-D region where the eld is de ned. Suppose that $_c = {}_{c1}$ and $= {}_1$ are constant in a subset, (V_1), of V, but do not keep the same values in the complementary subset, (V_2), where, in general, $_c(Q)$ and (Q) may change from point to point (see Figure 11.3). Let us de ne the following vectors, which di er from zero only at points $Q \in V_2$:

$$\begin{aligned} \mathbf{J}_P(Q) &= j\omega\,[\ _c(Q) \quad _{c1}]\,\mathbf{E}(Q) \quad , \\ \mathbf{M}_P(Q) &= j\omega\,[\ (Q) \quad _1]\,\mathbf{H}(Q) \quad . \end{aligned} \tag{11.27}$$

Their physical dimensions are those of an electric current density and of a magnetic one, respectively. They are referred to as *polarization current densi- ties.* The name stems from the formal analogy between Eqs. (11.27) and the quantities which are in common use in microscopic-scale models of electric and magnetic properties of matter (Jackson, 1975; Panofsky and Phillips, 1962).

Through Eqs. (11.27), Maxwell's equations, accounting for "real" sources $(\mathbf{J}_0, \mathbf{M}_0)$ as well, can be written anywhere in V in the following way:

$$\begin{aligned} \nabla \quad \mathbf{E} &= \quad j\omega \; {}_1 \mathbf{H} \quad \mathbf{M}_p \quad \mathbf{M}_0 \quad , \\ \nabla \quad \mathbf{H} &= \quad j\omega \; {}_{c1} \mathbf{E} + \mathbf{J}_p + \mathbf{J}_0 \quad . \end{aligned} \tag{11.28}$$

Formally, Eqs. (11.28) are identical to Maxwell's equations in a homogeneous medium ($_{c1}$, $_1$ constants), with the addition of two extra terms, $\mathbf{J}_p$ and $\mathbf{M}_p$, which look like sources at rst sight. In this way, the new problem has been transformed into the one we solved in the previous sections, paying only the price of introducing two additional sources. In reality, however, the step we took is only a formality. Quantities like Eqs. (11.27) are not sources, in fact, as they depend on *unknown* quantities, $\mathbf{E}(Q), \mathbf{H}(Q)$. In mathematical terms, we may say that the Green's functions method applied to Eq. (11.28) does not solve Maxwell's equations in inhomogeneous media, but just transforms them into integral equations.

Anyway, this procedure entails remarkable bene ts in all those cases where a zero-order approximation is easily guessed, for $\{\mathbf{E}, \mathbf{H}\}$, in the whole region V. From this, approximate values can be calculated for the polarization currents $\mathbf{J}_p$ and $\mathbf{M}_p$, and then the e.m. eld can evaluated more accurately, dealing with these approximate $\mathbf{J}_p$ and $\mathbf{M}_p$ as if they were sources, and using Green's functions. The procedure can be iterated, until the di erence between two successive approximations becomes smaller than a pre-selected value. A trade-o between accuracy of the results and computational load is always necessary, but progress in computers and in numerical methods have made these approaches more and more attractive. These statements will be corroborated by other results to be outlined in Chapter 12.

11.6 Time-domain interpretation of Green's functions

We said several times in the previous sections that the Green's function method is an operational implementation of the superposition principle. We will outline this again in this section from a di erent standpoint, after passing from the complex-vector space for time-harmonic elds to the time domain. The time-domain version of the scalar Green's function in Eq. (11.9) is:

$$\mathrm{Re}\left\{G\, e^{j\omega t}\right\} = \mathrm{Re}\left\{ \frac{e}{4} \; e^{j(\omega t \quad)} \right\} = \frac{e}{4} \; \cos(\omega t \qquad) \, , \tag{11.29}$$

where, as usual, $+ j \; = \; = (\; \omega^2 \; _c)^{1/2}$, with and real and non-negative, while (see Figure 11.1) $= |P \quad Q|$, where P is the point where the source is located. The wave radiated by the point source, Eq. (11.29), is clearly a spherical wave, centered at P, as the constant-phase surfaces are the spherical surfaces $=$ constant. These spheres are, at the same time, constant-amplitude surfaces. The 1/ factor, which a ects only the amplitude, accounts for energy conservation: as the wavefront surface increases as 4 ^{2}, the power density must decrease. The phase velocity of the wave in Eq. (11.29) is $v_f = \omega /$, equal to the speed of light, if the medium is lossless. To avoid inconsistencies, some cautions are necessary while analyzing Eq. (11.29). First, by de nition of spherical coordinates, it is 0; consequently, a wavefront (i.e., a constant value of the argument of the cosine function) cannot be tracked back at times t before the instant, $t_0 = /\omega$, when $= 0$. This disagrees with the basic assumption of the time-harmonic regime, that all quantities are sinusoidal functions of time over the whole range $\infty < t < +\infty$. A possible way out of this inconsistency is not to make a gross use of Steinmetz's method, but rather to look at the Green's function Eq.(11.9) as a spectrum in the frequency domain, and to reconstruct the time-domain behavior through an inverse Fourier transform. For the sake of simplicity, let us deal with the lossless case, i.e., with $G = (\ 1/4 \ )(e^{\ j} \ / \ )$ where $= \omega/c$. We have:

$$\mathcal{G} \ (t) = \mathcal{F}^{\ 1}\left(\frac{e^{\ j\omega \ /c}}{4} \right) = \frac{1}{4} \int_{\ \infty}^{+\infty} e^{\ j\omega \ /c} e^{j\omega t} \, \mathrm{d}\omega$$

$$= \frac{1}{4} \left(t \quad \frac{}{c} \right) , \tag{11.30}$$

where is Dirac's function, as usual. The inverse transform of a product is the convolution of the inverse transforms of the two factors; hence, to get the equivalent of Eqs. (11.18) and (11.19) in the time domain, Eq. (11.30) must be convolved with the time-domain source terms. An integration with respect to time has to be added to those over the spatial coordinates, namely:

$$\int_{\ \infty}^{t} \left(t' \quad \frac{}{c} \right) S \ (t \quad t', \mathbf{r}) \, \mathrm{d}t' \quad . \tag{11.31}$$

Eq. (11.31) points out that the source value at the time t is sensed, at a distance , at a later time $t' = t + \ /c$, that is, the *delay* equals the distance divided by the speed of light. This explains why the scalar and vector potentials calculated through Green's functions are commonly referred to as *retarded potentials.*

The interpretation of Green's function as a spherical wave radiated from a point source is also applicable to Eq. (11.30) and to what follows from it. Hence, the linear operator:

$$\int_{V_Q} \int_{\ \infty}^{t} \mathcal{G} \ (t', \ ) [\quad] \, \mathrm{d}t' \, \mathrm{d}V_Q \quad , \tag{11.32}$$

where the operand (in square brackets) is the whole set of the eld sources, can be looked at as a set of spherical expanding waves, each of them leaving its source at the right instant so that it arrives at P at the time t.

What we said so far is also an appropriate approach to the second term

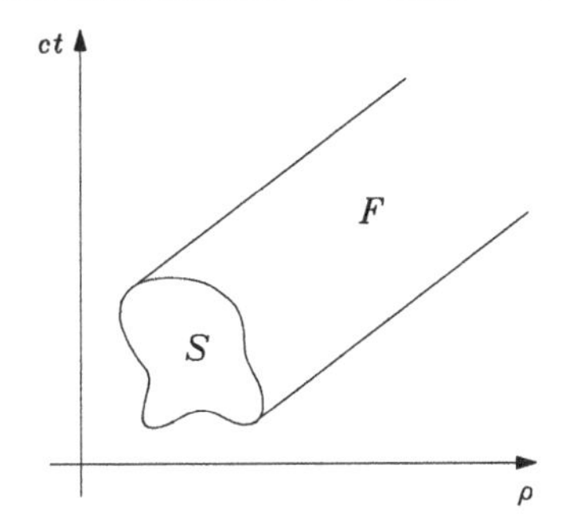

Figure 11.4 *A Minkowsky diagram.*

in Eq. (11.6), which, as we anticipated in Section 11.2, has an acceptable physical meaning in a lossless medium. Choosing $a = 0$, instead of $b = 0$, in Eq. (11.6), means to replace $t' \quad \frac{}{c}$ with $t' + \frac{}{c}$. The linear operator which takes the place of Eq. (11.32) represents a contracting spherical wave which, as time grows, shrinks, with phase velocity c, to the potentiated point P, and collects all the contributions that the sources, at distances from P, supply at the instant $t' = t \quad (\,/c)$, t being the instant when the wave reaches the point P. However, this picture collapses as soon as the medium is lossy, because then the argument of exponential function of has a positive real part which entails — in contrast to basic physics — the farther the source, the stronger its e ect.

The graphical representations of the content of this section are the so-called *Minkowsky diagrams*, whose meaning can be explained as follows. On the cartesian plane ($, ct$), where is the distance in space from an arbitrarily chosen origin, let the set S be the locus of those sources which are located at distance , and active at the instant t (Figure 11.4). From what we said so far, it follows that the potentials given by this set of sources, and consequently the elds, can di er from zero only inside the locus, on the ($, ct$) plane, (F), which consists of the unit-slope half-lines originating in S. In other words, given a point $A \quad (\,, ct)$ on this space-time cartesian plane, the only "e ects" which can be sensed at A are those due to "causes" that are located along the backward going half-line with slope +1. For example, the S_1 source of Figure 11.5 can generate a nonvanishing eld in A, but this is not possible for the source S_2 of the same gure.

If we introduce rectangular coordinates (x_1, x_2, x_3) in 3-D space so that $= (x_1^2 + x_2^2 + x_3^2)^{1/2}$, then the half-lines with slope 1 of Figures 11.4 and 11.5 satisfy:

$$x_1^2 + x_2^2 + x_3^2 = c^2\, t^2 + \text{constant} \quad , \tag{11.33}$$

which, in an orthogonal 4-D space (x_1, x_2, x_3, t), is the equation of a conical surface, whose generatrices are tilted by c with respect to the t axes. This is the reason why the name *future-light cone* is given to any unit-slope half-line of Figure 11.4 originating at a source point. The vertex of the cone is located on

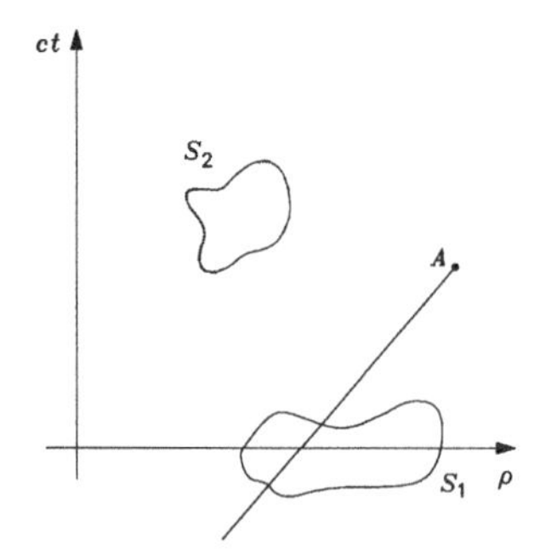

Figure 11.5 *An example of application of the Minkowsky diagrams.*

the source itself. Similarly, the name *past-light cone* is given to any unit-slope half-line like the one shown in Figure 11.5.

11.7 Green's function expansion into orthogonal eigenfunctions

In Chapter 12, and in a large fraction of Chapter 13, the broad-scope results that were derived in the previous sections will be applied to sources of practical relevance. On those occasions, most of our attention will be focused on properties of the sources themselves. This will induce us to look for boundary conditions whose in uence on the eld generated by a given source is as small as possible. This attitude might induce the reader to think, erroneously, that the Green's function method is not applicable to problems where boundary conditions become of primary importance, such as, for example, a eld source placed inside a waveguide. To avoid this misunderstanding, we will devote this section to establish links between the Green's function method and the approaches explained in the previous chapters, whose key features are complete sets of eigenfunctions of a linear di erential operator. First, we will tackle the subject in abstract terms; we will come to actual problems in the following sections and in the problems.

For the sake of simplicity and without loosing generality, let us start o with the scalar case, since many problems in electromagnetism can be reduced to scalar PDE's, making use of the decomposition theorem of Section 3.8. Assume that we know a complete set of (orthogonal and normalized) solutions $_i$ of the 2-D or 3-D *homogeneous* scalar Helmholtz equation, satisfying the boundary conditions imposed by a given structure. For simplicity, suppose that this set, J, is discrete; we will come back to this point later on. Because of this assumption, the following relations hold:

$$\nabla^2 \quad _i(\mathbf{r}) + \left[f(\mathbf{r}) + \quad ^2_i \right] \quad _i(\mathbf{r}) = 0 \quad , \tag{11.34}$$

$$\int_D \quad _i(\mathbf{r}) \quad _l(\mathbf{r}) \, d\mathbf{r} = \quad _{i,l} \quad , \tag{11.35}$$

where D is a suitable integration domain, $\ _i^2$ (eigenvalue) is a constant — in general a complex one — and $\ _{i,l}$ is Kronecker's symbol (0 for $i \neq l$, 1 for $i = l$). The symbol $f(\mathbf{r})$ is meant to underline that the formalism is applicable to piecewise homogeneous media, like, for example, a step-index dielectric waveguide.

A Green's function $G(\mathbf{r}, \mathbf{r}')$ (not necessarily the spherical-symmetry solution used in the previous sections) is, by de nition, a solution of the following equation:

$$\nabla_Q^2 G(\mathbf{r}, \mathbf{r}') + \left[f(\mathbf{r}) + \ ^2 \right] G(\mathbf{r}, \mathbf{r}') = \ (\mathbf{r} \quad \mathbf{r}') \quad , \tag{11.36}$$

where the subscript Q recalls, as usual, that the partial derivatives of G are with respect of the coordinates of $Q = O + \mathbf{r}$. Suppose that the quantity $\ ^2$ in Eq. (11.36) *does not equal any of the eigenvalues* $\ _i^2$ of Eq. (11.34). The more di cult (and far less general) case where $\ ^2$ equals one of the eigenvalues will be discussed in Section 11.9. Thanks to completeness of the set J, the function G can be expressed as a linear combination of the functions $\ _l$:

$$G(\mathbf{r}, \mathbf{r}') = \ _l\, a_l(\mathbf{r}') \ _l(\mathbf{r}) \quad . \tag{11.37}$$

The symbols stress that, in general, the coe cients $a_l(\mathbf{r}')$ do depend on the point $P = O + \mathbf{r}'$ where the elementary source $\ (\mathbf{r} \quad \mathbf{r}')$ is centered. Moreover, note that Eq. (11.37) can be written without even thinking of boundary conditions on the functions $\ _l(\mathbf{r})$, as long as one is not required to impose boundary conditions on $G(\mathbf{r}, \mathbf{r}')$. This means that each function $\ _l$, while it must satisfy Eq. (11.34), *is not obliged to be a mode* of the structure we are dealing with.

At this stage, the unknowns in the problem are the coe cients $a_l(\mathbf{r}')$. If we are able to determine them, then the scalar eld due to any source $S(\mathbf{r})$ can be calculated through Eq. (11.4), using Eq. (11.37) as Green's function. Substituting Eq. (11.37) into Eq. (11.36), and taking Eq. (11.34) into account, we get:

$$\ _l\, a_l(\mathbf{r}') \ _l(\mathbf{r}) \left(\ ^2 \quad \ _l^2 \right) = \ (\mathbf{r} \quad \mathbf{r}') \quad . \tag{11.38}$$

Multiplying both sides of Eq. (11.38) by $\ _i(\mathbf{r})$ (for $i = 1, 2, \ldots$), integrating over the eld domain D, and applying orthogonality, i.e., Eq. (11.35), we nd for each unknown coe cient an expression in terms of an eigenfunction $\ _i$, namely:

$$a_i(\mathbf{r}') = \frac{\ _i(\mathbf{r}')}{\ ^2 \quad \ _i^2} \quad . \tag{11.39}$$

So, nally, the Green's function decomposition in Eq. (11.37) as an eigenfunction series reads:

$$G(\mathbf{r}, \mathbf{r}') = \ _l \frac{\ _l(\mathbf{r}') \ _l(\mathbf{r})}{\ ^2 \quad \ _l^2} \quad . \tag{11.40}$$

In the case of a *continuous set* of eigenfunctions, the summation in Eq. (11.40) is replaced by an integral. In this case one must properly account for the *density* of the eigenfunctions, that is not necessarily uniform throughout the do-

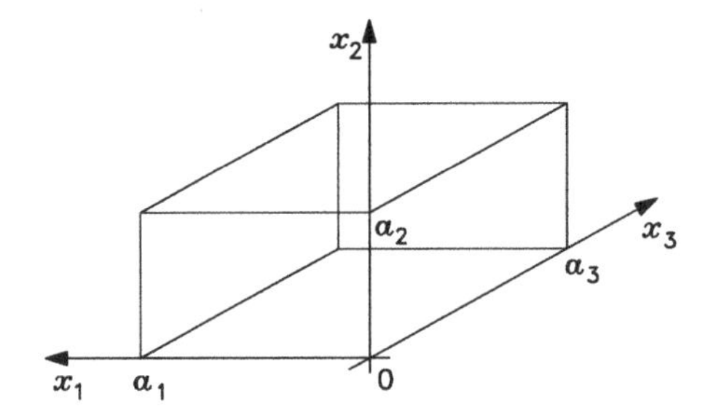

Figure 11.6 *A rectangular box and its reference frame.*

main of integration. The procedure remains fairly simple as long as the denominator ^{2} ^{2}( ) does not vanish. When this happens, caution is necessary, and the interested reader can nd the answer to this question in advanced textbooks (e.g., Morse and Feshbach, 1953).

11.8 An example: Field in a rectangular box

Let us test the formalism that was developed in the previous section with an example where simple geometry can make calculations easy. Let us take a rectangular box, completely lled by a homogeneous medium, and do not specify, as long as we are not obliged to, the boundary condition on its surface. The reference frame is shown in Figure 11.6, and the side lengths are a_1, a_2, a_3, respectively. The eld is not de ned outside the box and this allows us to extend it periodically in three dimensions. Consequently, thanks to well-known properties of Fourier series, we may assess that the following functions form a complete set, to be used as a basis in Eq. (11.40):

$$ {}_{mnp}(x_1, x_2, x_3) = C_{mnp}\, F_1\left(\frac{m}{a_1} x_1\right) F_2\left(\frac{n}{a_2} x_2\right) F_3\left(\frac{p}{a_3} x_3\right) \quad , \tag{11.41} $$

where every function F_i can be written in either one of the following forms:

$$ F_i\left(\frac{l}{a_i} x_i\right) = \begin{cases} \cos \dfrac{l}{a_i} x_i \quad , & (i = 1, 2, 3) \quad , \\ \sin \dfrac{l}{a_i} x_i \quad , & (l = 0, 1, 2, \ldots) \quad . \end{cases} \tag{11.42} $$

The constants C_{mnp} can be evaluated imposing the normalization condition, Eq. (11.35), where the domain of integration is the rectangular volume. It can be shown as an exercise that:

$$
\begin{aligned}
C_{mnp} &= \sqrt{\frac{8}{a_1 a_2 a_3}} && \text{for } m \quad n \quad p \neq 0 \ , \\
C_{mnp} &= \frac{2}{\sqrt{a_1 a_2 a_3}} && \text{when a subscript is null} \ , \\
C_{mnp} &= \sqrt{\frac{2}{a_1 a_2 a_3}} && \text{when 2 subscripts are null} \ , \\
C_{mnp} &= \frac{1}{\sqrt{a_1 a_2 a_3}} && \text{for } m = n = p = 0 \ .
\end{aligned}
\tag{11.43}
$$

Since $\ ^2 = \omega^2 \quad$, then:

$$
{}^2_{mnp} = \ {}^2 \left[\left(\frac{m}{a_1} \right)^2 + \left(\frac{n}{a_2} \right)^2 + \left(\frac{p}{a_3} \right)^2 \right] \quad . \tag{11.44}
$$

To write explicitly Eq. (11.40), the reader should rst multiply Eq. (11.41) by its copy where the three coordinates are replaced by x_1', x_2', x_3', respectively. Then, divide by $\omega^2 \qquad {}^2_{mnp}$. Finally, make the triple sum with respect to m, n and p, from 0 to ∞.

Comparing Eq. (11.41) to what we found in Chapter 9 about a rectangular resonator with conducting walls, we see that the terms of the basis in Eq. (11.41) are much more numerous than the resonator modes. This stems from the fact that here the *boundary conditions* on the walls of the box have not yet been speci ed. If they are made to coincide with those of Chapter 9, then, since the resonator mode set is, as well known, a complete one, the basis in Eq. (11.41) is redundant. To decide which terms survive and which do not when those boundary conditions are enforced, one may proceed in the following way.

In Chapter 7, two problems dealt with the proof that a TM [a TE] mode, in a rectangular waveguide, can be deduced from a magnetic [an electric] vector potential which is purely longitudinal (i.e., parallel to the waveguide axis) and proportional to the longitudinal component of the electric [the magnetic] eld. Along the same lines, as an exercise, the reader can prove that, in a metallic-wall resonator, a mode which is TM with respect to the x_3 axis can be deduced from a magnetic vector potential parallel to the x_3 axes, given by:

$$
\mathbf{A} = A_3 \, \hat{a}_3 \, \sin\left(\frac{m}{a_1} x_1 \right) \sin\left(\frac{n}{a_2} x_2 \right) \cos\left(\frac{p}{a_3} x_3 \right) = A \, \hat{a}_3 \quad . \tag{11.45}
$$

Suppose now that the eld sources are electric currents owing in the x_3 direction, with a density $\mathbf{J} = J \, \hat{a}_3$, located inside the resonator. Then, the vector integral:

$$
\mathbf{A} = \qquad \int_V \mathbf{J}_0 \, G \, \mathrm{d}V \tag{11.46}
$$

can be reduced to the following scalar relationship:

$$A = \int_V J\, G\, \mathrm{d}V \quad . \tag{11.47}$$

Therefore, the Green's function to be used in this case (which, as we remarked before, is TM with respect to the x_3 axes) is:

$$G_{\mathrm{TM}_3} = \sum_{m,n,p} C_{mnp} \sin\left(\frac{m}{a_1} x_1'\right) \sin\left(\frac{n}{a_2} x_2'\right) \cos\left(\frac{p}{a_3} x_3'\right)$$

$$\sin\left(\frac{m}{a_1} x_1\right) \sin\left(\frac{n}{a_2} x_2\right) \cos\left(\frac{p}{a_3} x_3\right) \quad . \tag{11.48}$$

Dually, in the case of magnetic currents owing in the x_3 direction we have

$$G_{\mathrm{TE}_3} = \sum_{m,n,p} C_{mnp} \cos\left(\frac{m}{a_1} x_1'\right) \cos\left(\frac{n}{a_2} x_2'\right) \sin\left(\frac{p}{a_3} x_3'\right)$$

$$\cos\left(\frac{m}{a_1} x_1\right) \cos\left(\frac{n}{a_2} x_2\right) \sin\left(\frac{p}{a_3} x_3\right) \quad . \tag{11.49}$$

If the sources are electric [magnetic] currents owing in a generic direction, they can be written by components. In the volume shown in Figure 11.6, there is no fundamental di erence between the x_3 axis and the other ones. Therefore, the x_1 and x_2 components can be dealt with in the same way as the previous one. The whole eld can be calculated by superimposing the three contributions. Note that, because of the completeness, a mode which is either TE or TM with respect to another axis can also be expressed as a superposition of TE and TM modes with respect to the x_3 axes.

Sources located on the walls (which are meant to model the coupling of the resonator with a waveguide or with another resonator through one or more holes, called irises in microwave terminology) are dealt with in depth in other books (Collin, 1992).

11.9 Spherical harmonics

It may occur that the condition $^2 \neq {}_i^2$, which enabled us to write the Green's function in the form in Eq. (11.40), is not satis ed by a complete orthonormal set of eigenfunctions. A series expansion of Green's function is still possible, but more delicate. It is not simple to give an abstract and general formulation to this problem, like in Section 11.7. It is more advisable to concentrate on two examples of high practical relevance, namely *spherical and cylindrical harmonics*. We will deal in full detail with the rst example (MacRobert, 1967). In this section we will study the orthonormal basis functions, and in the next one we will expand the Green's function in this basis. For cylindrical harmonics, the reader should be able to develop the subject following the hints that will be given in Section 11.11.

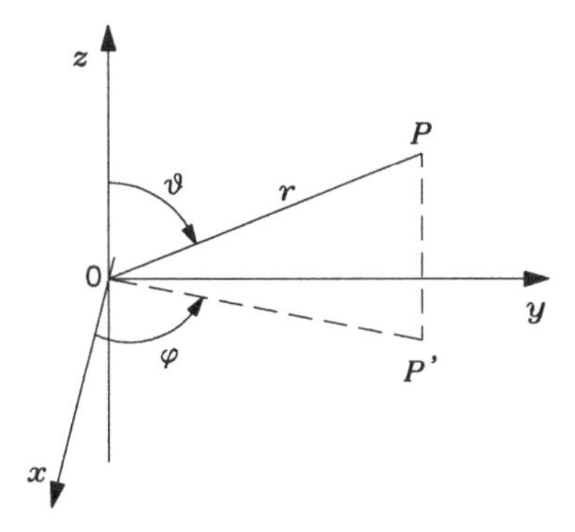

Figure 11.7 *Reference frame for spherical harmonics.*

Suppose that the region, lled by a homogeneous medium, where we need to solve the Helmholtz equation, is the *whole 3-D space*. Then, as we will see now, to nd a complete set of orthonormal eigenfunctions requires only to impose *Sommerfeld's radiation conditions* (see Section 3.3) as boundary conditions at in nity, i.e.:

$$\lim_{r\to\infty}\Big[r\,|\mathbf{E}|\Big]=0\quad,\qquad\lim_{r\to\infty}\Big[r\,|\mathbf{H}|\Big]=0\quad,\tag{11.50}$$

r being the distance from an arbitrarily chosen origin (not at in nity). The orthonormal base will consist of a *countable in nity* of functions, which are referred to as *spherical harmonics*. For the sake of brevity, we will only sketch the proof of completeness.

Let us start solving the scalar Helmholtz equation by separation of variables in a *spherical coordinate frame* (r,ϑ,φ) (Figure 11.7), similarly to what we did in previous chapters in cartesian and in cylindrical coordinates. The Laplacian (see Appendix B) is expressed in spherical coordinates as:

$$\begin{aligned}\nabla^2 &= \frac{1}{r^2}\frac{\partial}{\partial r}\left(r^2\frac{\partial}{\partial r}\right)+\frac{1}{r^2\sin\vartheta}\left[\frac{\partial}{\partial\vartheta}\left(\sin\vartheta\,\frac{\partial}{\partial\vartheta}\right)+\frac{1}{\sin\vartheta}\frac{\partial^2}{\partial\varphi^2}\right]\\ &= \frac{1}{r^2}\frac{\partial}{\partial r}\left(r^2\frac{\partial}{\partial r}\right)+\nabla_\perp^2\quad,\end{aligned}\tag{11.51}$$

where the symbol $\nabla_\perp^2$ summarizes all the terms which include a derivative with respect to the angular coordinates ϑ,φ.

Let $\;\; = R(r)\,Y(\vartheta,\varphi)$, where R is a function only of the radial coordinate and Y is a function of the angular coordinates. Divide all terms in the Helmholtz equation by RY/r^2 and write them in the following order:

$$r^2\,\frac{R''}{R}+2r\,\frac{R'}{R}\qquad{}^2\,r^2=r^2\,\frac{\nabla_\perp^2\,Y}{Y}\quad.\tag{11.52}$$

The two sides of Eq. (11.52) depend on di erent independent variables, therefore, to be equal they must be constant.‡ For later convenience, let us call $l(l+1)$ the separation constant, and use a subscript to underline that the unknown function R depends on the parameter l. We get:

$$\frac{d^2R_l}{dr^2}+\frac{2}{r}\frac{dR_l}{dr}\quad\left[\ {}^2\quad\frac{l(l+1)}{r^2}\right]R_l=0\quad . \tag{11.53}$$

Incidentally, when $\ {}^2=0$, i.e., when we deal with the Laplace equation instead of the Helmholtz one, the general integral of Eq. (11.53) is:

$$R_l\big|_{\ =0}=C_1\,r^l+C_2\,r^{\ (l+1)}\quad , \tag{11.54}$$

where C_1, C_2 are arbitrary constants. The function $Y(\vartheta,\varphi)$, that we will nd soon, is insensitive to any change in $\ {}^2$. These comments may be useful if one wants to decompose G as a series of solutions of the Laplace equation (taking Section 11.7 as a hint).

Coming back to the general case $\ {}^2\neq 0$, it may be useful (restricting our scope to lossless media) to set $\ {}^2=\ \ k^2$, and to make another change of variable, namely:

$$R_l(r)=r^{\ 1/2}\,u_l(r)\quad . \tag{11.55}$$

Eq. (11.53) then becomes:

$$\left[\frac{d^2}{dr^2}+\frac{1}{r}\frac{d}{dr}\right]u_l+\left[k^2\quad\frac{\left(l+\frac{1}{2}\right)^2}{r^2}\right]u_l=0\quad , \tag{11.56}$$

which is (see Appendix D) the *Bessel equation of order* $l+1/2$. We will show very soon that l is an integer; the order of Eq. (11.56) then is clearly *half-integer*. The general integral of Eq. (11.53) can be written as:

$$R_l(r)=C_1\,r^{\ 1/2}\,J_{l+(1/2)}(kr)+C_2\,r^{\ 1/2}\,N_{l+(1/2)}(kr)\quad , \tag{11.57}$$

where J and N are Bessel functions of the rst and second kind, respectively, of half-integer order, and C_1 and C_2 are arbitrary constants.

To simplify the notation, it is convenient to use the so-called *spherical Bessel and Hankel functions*, de ned as follows:

$$j_l(x)=\left(\frac{}{2x}\right)^{1/2}J_{l+(1/2)}(x)\quad ,\qquad n_l(x)=\left(\frac{}{2x}\right)^{1/2}N_{l+(1/2)}(x)\quad ,$$

$$\left.\begin{matrix}h_l^{(1)}(x)\\ h_l^{(2)}(x)\end{matrix}\right\}=\left(\frac{}{2x}\right)^{1/2}\left[J_{l+(1/2)}(x)\quad j\,N_{l\ \ (1/2)}(x)\right]\quad . \tag{11.58}$$

They satisfy interesting relations, namely:

$$j_l(x)\ =\ (\ \ x)^l\left(\frac{1}{x}\frac{d}{dx}\right)^l\left(\frac{\sin x}{x}\right)\quad ,$$

‡ Note that the coordinate r on the right-hand side of Eq. (11.52) is an artifact: indeed, all the terms of $\nabla_\perp^2$ are proportional to $r^{\ 2}$.

$$n_l(x) \quad = \quad (\ \ x)^l \left(\frac{1}{x}\frac{d}{dx}\right)^l \left(\frac{\cos x}{x}\right) \quad ,$$

$$\left.\begin{array}{l} h_l^{(1)}(x) \\ h_l^{(2)}(x) \end{array}\right\} \quad = \quad (\ \ x)^l \left(\frac{1}{x}\frac{d}{dx}\right)^l \left(\frac{e^{\ \ jx}}{x}\right) \quad . \tag{11.59}$$

The last relationship, in detail, shows that the l-th expanding or contracting scalar spherical wave can be derived by applying l times the operator $(1/x)\, d/dx$ to the spherical-symmetry Green's function $e^{\ \ x}/(4\ \ x)$.

Let us now nd $Y(\vartheta, \varphi)$, i.e., the solution of:

$$\frac{1}{\sin\vartheta}\frac{\partial}{\partial\vartheta}\left(\sin\vartheta\,\frac{\partial Y}{\partial\vartheta}\right) + \frac{1}{\sin^2\vartheta}\frac{\partial^2 Y}{\partial\varphi^2} + l(l+1)\,Y = 0 \quad . \tag{11.60}$$

Apply, once again, separation of variables, setting $Y(\vartheta,\varphi) = Z(\vartheta)\,T(\varphi)$ and isolating on one side the term involving the derivative with respect to φ. Arising from the periodicity $T(\varphi + 2\ \ ) = T(\varphi)$ which comes from being single-valued at any point in real space, we get:

$$\frac{d^2T}{d\varphi^2} = \quad m^2 T \qquad (m \text{ integer}) \quad . \tag{11.61}$$

The general integral of Eq. (11.61) is easily found to be:

$$T = T_1\, e^{jm\varphi} + T_2\, e^{\ \ jm\varphi} \quad , \tag{11.62}$$

where T_1, T_2 are arbitrary constants. So, we are left with one ODE to solve, namely:

$$\frac{d}{dx}\left[(1 \quad x^2)\,\frac{dZ}{dx}\right] + \left[l(l+1) \quad \frac{m^2}{1 \quad x^2}\right] Z = 0 \quad , \tag{11.63}$$

where the change of variable $x = \cos\vartheta$ has been used. This equation is known as the *generalized Legendre equation.* Its general integral (Abramowitz and Stegun, Eds., 1964) is usually written as follows:

$$Z(x) = D_1\, P_l^{(m)}(x) + D_2\, Q_l^{(m)}(x) \quad , \tag{11.64}$$

where $P_l^{(m)}$ and $Q_l^{(m)}$ are known as the *associated Legendre functions*, of the rst and second kind, respectively, of degree l and of order m. To discuss brie y their main properties, we can start from the case $m = 0$ (i.e., azimuthal symmetry, $T(\varphi)$ = constant). In this case, Eq. (11.63) simply becomes:

$$\frac{d}{dx}\left[(1 \quad x^2)\,\frac{dZ}{dx}\right] + l(l+1)\,Z = 0 \quad . \tag{11.65}$$

This is known as the *ordinary Legendre equation.* We must look only for solutions which are bounded over the whole range $|x| \quad 1$: this is a necessary condition for the solutions to be physically meaningful, as $x = \cos\vartheta$. It can be shown that this requirement is satis ed *only if l is an integer.* After a suitable normalization procedure, these bounded solutions can be shown to be the so-called *Legendre polynomials.* For a few examples, see Table 11.1. They can be

Table 11.1 *Example of Legendre polynomials.*

$P_0(x) = 1$	$P_3(x) = (1/2)\,(5x^3 \quad 3x)$
$P_1(x) = x$	$P_4(x) = (1/8)\,(35x^4 \quad 30x^2 + 3)$
$P_2(x) = (1/2)\,(3x^2 \quad 1)$	

summarized with the following expression, referred to as the *Rodriguez rule*:

$$P_l(x) = \frac{1}{(2)^l\, l!}\,\frac{d^l}{dx^l}\,(x^2 \quad 1)^l \quad . \tag{11.66}$$

When m is a positive integer, then it can be shown that the associated Legendre function of the rst kind can be written as:

$$P_l^{(m)}(x) = (\quad 1)^m\,(1 \quad x^2)^{m/2}\,\frac{d^m}{dx^m}\,P_l(x) \quad , \tag{11.67}$$

where $P_l(x)$ is the l-th degree Legendre polynomial, solution of Eq. (11.65). The last expression is bounded over the range $|x| \quad 1$ when $m \quad l$. On the contrary, the associated functions of the second kind are not bounded over the whole range, so they must be rejected in our framework because they do not obey an elementary physical restriction.

Inserting Eq. (11.66) into Eq. (11.67), we obtain the following expression for the associated Legendre functions of the rst kind:

$$P_l^{(m)}(x) = \frac{(\quad 1)^m}{2^l\, l!}\,(1 \quad x^2)^{m/2}\,\frac{d^{l+m}}{dx^{l+m}}\,(x^2 \quad 1)^l \quad , \tag{11.68}$$

and we notice that also for $m = \quad l, \quad l+1, \ldots, \quad 1$ this expression is well de ned, satis es Eq. (11.65), and is bounded for $|x| \quad 1$. Therefore, for any given l, Eq. (11.63) has $2l+1$ distinct and physically meaningful solutions, of the type $e^{jm\varphi}\,P_l^{(m)}(x)$, where $P_l^{(m)}(x)$ is given by Eq. (11.68) with $|m| \quad l$.

In order to normalize these solutions, i.e., to have:

$$\int_0^{2} d\varphi \int_0 Y_{l,m}(\vartheta,\varphi)\,Y_{l,m}(\vartheta,\varphi)\,\sin\vartheta\,d\vartheta = 1 \quad , \tag{11.69}$$

the reader can prove, as an exercise, that it must be:

$$Y_{l,m}(\vartheta,\varphi) = \left[\frac{2l+1}{4}\,\frac{(l \quad m)!}{(l+m)!}\right]^{1/2} P_l^m(\cos\vartheta)\,e^{jm\varphi} \quad . \tag{11.70}$$

Table 11.2 shows a few examples of *normalized spherical harmonics*, Eq. (11.70), with the explicit use of $\cos\vartheta = x$ and $\sin\vartheta = (1 \quad x^2)^{1/2}$ in the associated Legendre functions.

It can also be shown that solutions of the type in Eq. (11.70) obey the following orthogonality relation between:

$$\int_0^{2} d\varphi \int_0 Y_{l,m}\,Y_{l',m'}\,\sin\vartheta\,d\vartheta = \;_{ll'}\;\;_{mm'} \quad , \tag{11.71}$$

Table 11.2 *Examples of spherical harmonics.*

l	
$l = 0$	$Y_{00} = \frac{1}{\sqrt{4}}$
$l = 1$	$Y_{11} = \quad \sqrt{\frac{3}{8}} \sin \quad e^{j\varphi}$
	$Y_{10} = \sqrt{\frac{3}{4}} \cos$
$l = 2$	$Y_{22} = \frac{1}{4}\sqrt{\frac{15}{2}} \sin^2 \quad e^{2j\varphi}$
	$Y_{21} = \quad \sqrt{\frac{15}{8}} \sin \quad \cos \quad e^{j\varphi}$
	$Y_{20} = \sqrt{\frac{5}{4}} \left(\frac{3}{2} \cos^2 \quad \frac{1}{2}\right)$
$l = 3$	$Y_{33} = \quad \frac{1}{4}\sqrt{\frac{35}{4}} \sin^3 \quad e^{3j\varphi}$
	$Y_{32} = \frac{1}{4}\sqrt{\frac{105}{4}} \sin^2 \quad \cos \quad e^{2j\varphi}$
	$Y_{31} = \quad \frac{1}{4}\sqrt{\frac{21}{4}} \sin \quad (5\cos^2 \quad 1)\, e^{j\varphi}$
	$Y_{30} = \sqrt{\frac{7}{4}} \left(\frac{5}{2} \cos^3 \quad \frac{3}{2} \cos \quad \right)$

where $_{hk}$ is Kronecker's symbol. Finally, the following relation can be proved as well:

$$\sum_{l=0}^{\infty} \sum_{m= \quad l}^{l} Y_{l,m}(\vartheta', \varphi')\, Y_{l,m}(\vartheta, \varphi) = \quad (\cos\vartheta \quad \cos\vartheta') \quad (\varphi \quad \varphi') \quad , \qquad (11.72)$$

where is the Dirac delta-function. Eq. (11.72) says that the orthonormal set of functions de ned by Eq. (11.70) is *complete*. Indeed, multiplying both sides of Eq. (11.72) by any function $f(\vartheta, \varphi)$, and integrating over the whole 3-D space, we nally get:

$$F(\vartheta, \varphi) = \sum_{l=0}^{m} \sum_{m= \quad l}^{l} Y_{l,m}(\vartheta, \varphi) \int_0^{2} d\varphi' \int_0 F(\vartheta', \varphi')\, Y_{l,m}(\vartheta', \varphi') \sin\vartheta'\, d\vartheta' . \qquad (11.73)$$

It is worth mentioning that this set of functions is suitable for solving not only the problem described at the beginning of this section (eld in the whole 3-D space), but also for problems where boundary conditions are imposed on coordinate surfaces of the spherical reference frame. Among these, we may mention:

spherical resonators, that were quoted in Chapter 9;

conical waveguides, like tapers connecting circular waveguides of di erent diameters;

conical horn antennas.

The reader who is interested in some of these subjects can nd in-depth discussions in other textbooks (e.g., Schelkuno , 1948; Ramo *et al.*, 1994).

11.10 Multipole expansion

Our next task is to use spherical harmonics, i.e., the results of the previous section, to make a series expansion of the type in Eq. (11.40) for the Green's function. After this, we will be able to make a few broad-scope remarks on the elds radiated by any nite-size sources, in a homogeneous medium extending to in nity. The procedure begins very similarly to that of Section 11.8. The general results of Section 11.7 tell us that a series expansion of the type:

$$G(\mathbf{r},\mathbf{r}') = \quad {}_{l,m}\, g_l(r,r')\, Y_{l,m}(\vartheta',\varphi')\, Y_{l,m}(\vartheta,\varphi) \quad , \tag{11.74}$$

is certainly possible, the $Y_{l,m}$'s being given by Eq. (11.70). All we know about the functions $g_l(r,r')$'s, to be determined here, is that they must be related to the expressions in Eq. (11.57). Inserting then Eq. (11.74) into Eq. (11.2) (and taking it for granted that all the regularity requirements, to di erentiate the series term by term, are met), keeping track of Eqs. (11.71) and (11.72) we get:

$$\left[\frac{d^2}{dr^2} + \frac{2}{r}\frac{d}{dr} + k^2 \quad \frac{l(l+1)}{r^2}\right] g_l(r,r') = \frac{1}{r^2} \quad (\mathbf{r} \quad \mathbf{r}') \quad . \tag{11.75}$$

This di ers from Eq. (11.53) only at $r = r'$, where its right-hand side is di erent from zero. So, similarly to what we did in Section 11.2, we can solve Eq. (11.75) matching two solutions of Eq. (11.53) in such a way that at $r = r'$ there is a discontinuity in the rst derivative, and hence a peak in the second derivative, equal to the peak on the right-hand side. One among the solutions of Eq. (11.75) satis es these requirements and also Sommerfeld's condition ($rg_l \to 0$ for $r \to \infty$); furthermore, it is bounded at any nite r, including $r = 0$. It is:

$$g_l(r,r') = \begin{cases} jk\, j_l(kr)\, h_l^{(2)}(kr') & \text{for } r < r' \quad , \\ jk\, j_l(kr')\, h_l^{(2)}(kr) & \text{for } r > r' \quad . \end{cases} \tag{11.76}$$

The arbitrary constant which appears in the general integral of the associated homogeneous equation has been given a value, in Eq. (11.76), such that the left-hand side of Eq. (11.75), multiplied by r^2, and integrated over an interval that encompasses $r = r'$, equals one, as required by equality with the right-hand side. Eq. (11.76) can be rewritten in a more compact way as:

$$g_l(r, r') = \quad jk\, j_l(kr_m)\, h_l^{(2)}(kr_M) \quad , \tag{11.77}$$

where $r_m = \min(r, r')$, and $r_M = \mathrm{MAX}(r, r')$.

The boundary conditions on spherical harmonics at in nity are the same as those satis ed by the Green's function in Eq. (11.9). Therefore, the following relation holds:

$$\frac{1}{4} \frac{e^{\quad jk|\mathbf{r} \quad \mathbf{r}'|}}{|\mathbf{r} \quad \mathbf{r}'|} = \quad jk \sum_{l=0}^{\infty} j_l(kr_m)\, h_l^{(2)}(kr_M) \sum_{n= \quad l}^{+l} Y_{ln}(\vartheta', \varphi')\, Y_{ln}(\vartheta, \varphi) \,. \tag{11.78}$$

The series expansion in Eq. (11.78) is suitable for calculating elds only when the sources satisfy all the regularity requirements which must be met in order to:

commute integration over the volume where the sources are located, and summation of the various terms in Eq. (11.78);

calculate term by term all the partial derivatives of the series which are needed in order to derive the elds $\{\mathbf{E}, \mathbf{H}\}$ from the potentials $\{\mathbf{A}, \mathbf{F}\}$.

When this is indeed the case, then each term in the series in Eq. (11.78), passing through Eqs. (11.20) and (11.21) (where the surface integrals equal zero), generates an e.m. eld which is referred to as the *multipole eld of order* (l, n). The physical meaning of the two integer indices will be illustrated soon.

Some general information about multipole elds can make an appropriate introduction to Chapter 12. For this purpose, it may be convenient to express the symbolic vector ∇ in spherical coordinates, as follows:

$$\nabla = \hat{r}\, \frac{\partial}{\partial r} + \frac{1}{r^2}\, \mathbf{r} \quad \nabla \quad \mathbf{r} \quad \hat{r}\, \frac{\partial}{\partial r} + \frac{j}{r^2}\, (\mathbf{L} \quad \mathbf{r}) \quad , \tag{11.79}$$

where a new symbolic vector, called *rotator*, is de ned as:

$$\mathbf{L} = \quad j\, \mathbf{r} \quad \nabla \quad , \tag{11.80}$$

and can be written by components as follows:

$$\mathbf{L} \qquad (L_x; L_y; L_z)$$
$$\left(j\, \sin\varphi\, \frac{\partial}{\partial\vartheta} + j\, \frac{\cos\varphi}{\sin\vartheta}\, \frac{\partial}{\partial\varphi}\, ; \quad j\, \cos\varphi\, \frac{\partial}{\partial\vartheta} + j\, \frac{\sin\varphi}{\sin\vartheta}\, \frac{\partial}{\partial\varphi}\, ; \quad j\, \frac{\partial}{\partial\varphi} \right) . \tag{11.81}$$

An outer or an inner product of $\mathbf{L}$ by a vector which is separable in spherical coordinates is an operation on the angular coordinates only, and leaves the dependence on the radial coordinate r unchanged.

Let us invoke Eq. (C.6), which, when the vector $\mathbf{a}$ does not depend on the coordinates on which ∇ operates, reads simply:

$$\nabla \quad (\mathbf{a}\, f(P)) = \nabla f(P) \quad \mathbf{a} \quad . \tag{11.82}$$

We are now able to comment on multipoles of any order l, i.e., on the l-th term of Eq. (11.78) and on what can be derived from it. For the sake of simplicity, we will con ne ourselves to the lossless case, i.e., $k = 2\ \ /\ \ $, a real quantity. The following comments are independent of whether the sources are electric or magnetic currents, and restricted only by the assumption that the sources are localized (i.e., there are no sources at in nity).

Near the the origin of the coordinate system (i.e., for r), but outside of the source region (i.e., for $r_M = r$), we may use the following approximation, which holds for x l (Abramowitz and Stegun, Eds., 1964):

$$h_l^{(2)}(x) \simeq j\, \frac{1\ \ 3\ \ 5\ \ \ldots\ \ (2l\ \ \ 1)}{x^{l+1}} \quad . \tag{11.83}$$

If we calculate the electric [the magnetic] eld of the magnetic [the electric] multipole of order l with this approximation, then all its components, radial and transverse, are proportional to $r^{\ \ (l+2)}$. The electric [the magnetic] eld, to be found in turn as the curl of the magnetic [the electric] one, has all its components proportional to $r^{\ \ (l+3)}$. Eq. (11.83) also shows that the phase of the vector potential (and consequently that of the eld) does not depend on the distance from the origin. Hence, in the region r where the approximation in Eq. (11.83) holds, the nite speed of propagation of the eld is not appreciable. The reader can check, as an exercise, that the elds calculated according to this approximation are formally identical to those of the *static multipoles*, i.e., the solutions of the Laplace equation (Jackson, 1975). This is the reason why the region r is referred to as the *induction zone*, either electric or magnetic depending on which eld is the dominant one.

Far from the origin of the reference frame, i.e., for r , we may use an approximation, which holds for x l (Abramowitz and Stegun, Eds., 1964):

$$h_l^{(2)}(x) = (\ \ j)^{l+1}\, \frac{\exp(\ \ jx)}{x} \quad . \tag{11.84}$$

Therefore, in this region (which is referred to as the *wave zone*) the vector potential and the eld of *any* multipole behave like a *spherical wave*, of the kind $e^{\ \ jkr}/r$. The operator $\partial/\partial r$, applied to Eq. (11.84), gives rise to two terms, namely:

$$\frac{\partial}{\partial r}\left(\frac{e^{\ \ jkr}}{r}\right) = \left(\ \ jk\ \ \ \frac{1}{r}\right)\left(\frac{e^{\ \ jkr}}{r}\right) \quad . \tag{11.85}$$

The rst one is of order $r^{\ \ 1}$, the second one is of order $r^{\ \ 2}$. The operator $(j/r^2)\,\mathbf{L}\ \ \ \mathbf{r}$, applied to Eq. (11.84), gives rise to several terms, all of which are of order $r^{\ \ 2}$. For r , the term of order $r^{\ \ 1}$ dominates over all the other ones. Therefore, at any point P the magnetic [the electric] eld of the electric [the magnetic] multipole of order l is essentially transverse to the line which joins P with the origin of the reference frame. However, let us point out that this is a comparison between *amplitudes*, neglecting completely the phase information. To see unexpected consequences of this fact, the reader should wait until Section 12.4. Furthermore, the comparisons we have made among

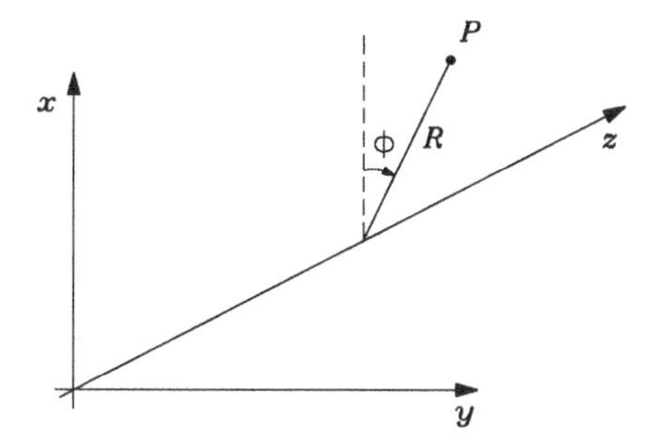

Figure 11.8 *Reference frame for cylindrical harmonics.*

the various terms may fail in the neighborhood of those directions along which $Y_{l,n}(\vartheta, \varphi) = 0$ while $\partial Y_{l,n}/\partial \varphi \neq 0$ or $\partial Y_{l,n}/\partial \vartheta \neq 0$. Indeed, in those directions one term generated by the rotator, although small, dominates over the term of order $r^{\ 1}$, because this goes to zero where $Y_{l,n}(\vartheta, \varphi)$ vanishes.

To calculate the electric [the magnetic] eld, as the curl of the magnetic [the electric] eld found in this way, let us apply once again the identity in Eq. (11.82), making the approximation that the vector **a** be normal to **r**, and expressing ∇ as in Eq. (11.79). A term proportional to $e^{\ jkr}/r$ shows up in **E** [in **H**] as well. With the same caution as in the previous paragraph, we may say that **E**(P) [**H**(P)] is transverse to the line which joins P with the origin of the reference frame. The proof of the following statements is left to the reader as an exercise:

E and **H**, if linearly polarized, are orthogonal to each other, in the region r ;

$|\mathbf{E}|/|\mathbf{H}| =$, the intrinsic impedance of the medium;

in the neighborhood of any point $P = O + \mathbf{r}$ at r , the eld can be approximated with a uniform plane wave, which travels in the sense of increasing **r**.

All these remarks indicate that, in the wave region (r), multipole elds are not very sensitive to the order of the multipole itself. This is in contrast to what occurs for r , where the order of the multipole plays a role in the eld formulas. We will show in Section 12.2 that, at a distance which is also *large with respect to the source size*, the eld can be derived, within a very reasonable approximation, passing through just one vector function of ϑ and φ, which we will call *equivalent moment* of the source. This is true irrespectively of the source details, and is a consequence of what we have now pointed out on multipole expansions, namely that the far elds are essentially the same for all multipole orders.

11.11 An introduction to cylindrical harmonics

The procedure outlined in Section 11.9 can be completely translated from spherical to *cylindrical coordinates*, which we denote here as $R,\ \ ,z$ to avoid any misunderstandings with previous symbols (Figure 11.8). The results obtained in this way are suitable for problems with an *axial rotational symmetry*, such as, for example, index pro le perturbations in circular dielectric waveguides, elds radiated or guided by straight in nitely long conductors, etc. The radiation condition at in nity must be modi ed, with respect to Eq. (11.50), to account for the fact that the area of a cylindrical surface increases in proportion to R. Accordingly, Sommerfeld's conditions in a cylindrical frame read:

$$\lim_{R\to\infty}\left(|\mathbf{E}|\,\sqrt{R}\right)=0 \quad , \qquad \lim_{R\to\infty}\left(|\mathbf{H}|\,\sqrt{R}\right)=0 \quad . \tag{11.86}$$

For reasons which stem from Chapters 7 and 10, let us impose that all eld components are proportional to $\exp(\ \ j\ \ z)$. Furthermore, let:

$$w^2=(\ \ \omega^2\ \ \ +\ \ ^2)\,a^2 \quad , \tag{11.87}$$

where a is a normalizing value for the radial coordinate R; for the time being, it can be chosen arbitrarily.

Let us proceed now, like in Section 11.9, to select those among the Bessel functions which satisfy Eq. (11.86), and do not diverge over the domain where they must be used. The net result of such a selection is that the cylindrical harmonic expansion, playing the same role as Eq. (11.78) did in spherical coordinates, is the following one:

$$G(\mathbf{r},\mathbf{r}')=\frac{1}{2\ \ }\sum_{l=\ \ \infty}^{+\infty} I_l\left(w\,\frac{R_m}{a}\right)K_l\left(w\,\frac{R_M}{a}\right)e^{jl(\ \ \ \ ')} \quad , \tag{11.88}$$

where $\mathbf{r},\mathbf{r}'$ are position vectors, as usual; I and K are the modi ed Bessel functions of the rst and second kind, respectively (Appendix D). Furthermore, we set:

$$R_m=\min(R,R') \quad , \qquad R_M=\mathrm{MAX}(R,R') \quad . \tag{11.89}$$

Note that, to get the correct result, one must remember that the two-dimensional Dirac function, in polar coordinates $(R,\ \ )$, reads:

$$(\mathbf{r}\ \ \ \mathbf{r}')=\frac{1}{R}\ \ (R\ \ \ R')\ \ (\ \ \ \ \ ') \quad . \tag{11.90}$$

By similarity with Sections 11.9 and 11.10 it is easily understood that Eq. (11.88) holds only when the whole 3-D space is lled by a homogeneous medium.

If, on the other hand, we want to apply it to problems of great practical interest where has a *step* discontinuity in the radial direction (see Figure 11.9), then we are obliged to use the polarization currents de ned in Section 11.5. This may entail a trade-o between accuracy and computational load. To by-pass this obstacle, we can make use of the *step pro le* Green's function, instead of the *free-space* Green's function Eq. (11.9). It is an expansion where

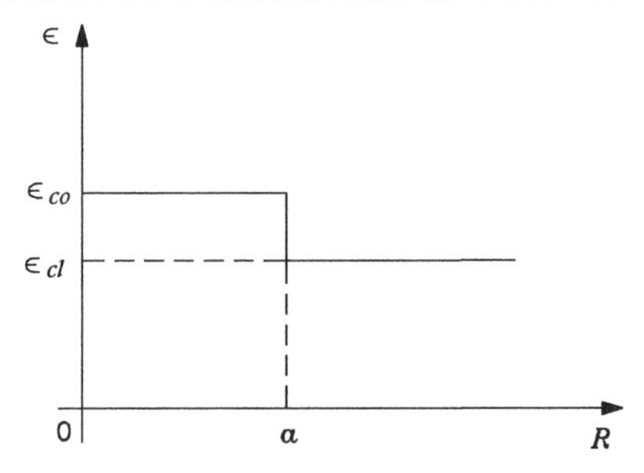

Figure 11.9 *A dielectric permittivity pro le exhibiting a step discontinuity.*

each series term satis es the continuity conditions on the cylindrical surface, $R = a$, where has its step discontinuity. It reads:

$$G(\mathbf{r}, \mathbf{r}') = \frac{1}{2} \sum_{l= \ \infty}^{+\infty} g_l(R, R')\, e^{jl(\quad ')} \quad , \tag{11.91}$$

with:

$$g_l(R, R') = A\, y_1(R_m)\, y_2(R_M) \quad ,$$

$$y_1(R) = \begin{cases} \dfrac{J_l(uR/a)}{J_l(u)} & R \quad a \quad , \\ \dfrac{K_l(wR/a) + a_l\, I_l(wR/a)}{K_l(w) + a_l\, I_l(w)} & R > a \quad , \end{cases}$$

$$y_2(R) = \begin{cases} \dfrac{J_l(uR/a) + b_l\, Y_l(uR/a)}{J_l(u) + b_l\, Y_l(u)} & R \quad a \quad , \\ \dfrac{K_l(wR/a)}{K_l(w)} & R > a \quad , \end{cases}$$

where J_l, Y_l, K_l, I_l indicate Bessel functions, w is de ned by Eq. (11.87) setting $\ = \ _m$, whilst u^2 can also be obtained through Eq. (11.87) if we let $\ = \ _n$ and $u^2 = \ w^2$. Furthermore:

$$A = \ K_l(w)\, I_l(w) \left[1 + (J \quad I)/(K \quad J)\right] \quad ,$$

$$a_l = \frac{K \quad J}{J \quad I} \frac{K_l(w)}{I_l(w)} \quad , \qquad b_l = \frac{J \quad K}{K \quad Y} \frac{J_l(u)}{Y_l(w)} \quad ,$$

$$J = u\, J_l'(u)/J_l(u) \quad , \qquad K = w\, K_l'(u)/K_l(u) \quad ,$$

$$Y = u\, Y_l'(u)/Y_l(u) \quad , \qquad I = w\, I_l'(u)/I_l(u) \quad ,$$

where the primes denote derivatives with respect to the arguments.

Clearly, the price we paid to avoid using polarization currents was a high level of analytical complexity, to start with. In real cases, which way to proceed depends often on the tastes and the experience of the person who is tackling the problem.

11.12 Further applications and suggested reading

Retarded potentials have very broad scopes of application. In this sense, the present chapter is similar to the rst chapters of this book. It is not surprising, therefore, that many points that we dealt with in Sections 1.11 and 3.10 are relevant, again, in this new context. Consequently, the same holds for the reference books which were quoted in those chapters. Another consequence that should not surprise the reader is that, to a large extent, the contents of this chapter can be completely by-passed in a strongly telecommunication-oriented curriculum, where on the contrary many other chapters of this book are instrumental.

We do recommend to the reader to consult Jackson (1975) — not only for what refers to retarded potentials — because that book adopts the so-called *CGS Gaussian system of units*, where dielectric permittivity and magnetic permeability of vacuum are both chosen equal to unity. It seems that most of the students in electronic engineering, and often those in physics also, have lost any familiarity with this system of units. Undoubtedly, it will take them a substantial e ort to start reading Jackson, or any other book written in CGS units. However, we believe that their e ort will be rewarded by the pleasure of appreciating the formal elegance of the treatise. It will also entail some useful consequences, for example a better understanding of the duality theorem (which looks awkward in MKS units) and of the so-called Babinet's principle, to be covered in Section 13.3.

As illustrated in Jackson (1975), retarded potentials nd applications, among many other cases, in magnetohydrodynamics and in calculating the elds which are radiated by moving charged particles. For the latter problem, the reader should give at least a glimpse to the so-called *Lienard-Wiechert potentials*, which are treated by Jackson in its Section 14.1.

In this chapter, we ran into another very important item of mathematical physics, namely, the Green's functions. We already stressed, on several occasions, that (i) this is an extremely powerful and broad-scope method; (ii) there are no reasons why one should use it *only* to solve non-homogeneous partial di erential equations, i.e., problems with sources; (iii) Eq. (11.9) is just an *example* of Green's function, though a particularly simple and useful one. Experience of instructors indicates that many students have a tendency to think of Eq. (11.9) as *the* Green's function, and of what we saw in Section 11.3 as *the* application of the Green's function method. To convince them that this is not true, it should be enough to have them consult Felsen and Marcuvitz (1973), a powerful, very rigorous treatise where Green's functions are used in relationship with all the subjects covered in it, underlying very well relationships between similar scalar and vector problems, and providing evidence of how powerful Green's functions are in tackling problems like those which we saw in the frameworks of plane waves and of guided waves.

At this stage, some readers might be concerned that, little by little, Green's functions have brought us very far from the telecommunication-oriented applications of e.m. waves. The contrary is true: consult, for example, Chapter 34 of Snyder and Love (1983), which illustrates how in optical waveguides Green's functions can be used to solve very practical problems, such as estimating launching e ciency from a given source into a given ber, or studying non-ideal index pro les. Specialists may proceed further and link these subjects with those of Chapter 6, studying the Green's function approach to anisotropic dielectric waveguides.

Problems

11-1 Show that, in cylindrical coordinates r, φ, z, a function of the type exp(r) r , with real , cannot be a Green's function. Comment on this result in view of the non-periodicity of Bessel function oscillations, pointed out in Appendix D.

11-2 Elaborate on Eqs. (11.26), writing explicit expressions for the gradient and the Laplacian of exp( )/ . Write the two remaining dyadic Green's functions by duality.

11-3 A small cubic sample of a dielectric material whose relative permittivity is 1.2 is placed in an air- lled rectangular cavity resonating in the $\mathrm{TE}_{1,0,1}$ mode. The lengths a, b and d of the resonator sides are much longer than that of the cubic sample side, s. Assuming that at rst order the electric eld in the sample remains the same as in the empty resonator, express the polarization currents de ned by Eqs. (11.27), as a function of the sample position.
Repeat the same exercise for a small sample of magnetic material of relative permeability of 1.1.

11-4 A "microlens" is a half sphere of radius a, large compared to the wavelength, of relative permittivity $_r = 1.5$, surrounded by air. Suppose that when a plane wave impinges on it, polarization currents at rst order can be evaluated from an unperturbed plane-wave eld.

a) Calculate the polarization current density when the plane wave impinges on the at side of the half sphere.
b) Calculate the polarization current density when the plane wave impinges at a generic angle on the semi-spherical surface.

11-5 Write explicitly all the analytical steps which are required to derive Eqs. (11.43) and Eq. (11.44).

11-6 Show that, in a metallic-wall rectangular resonator, a mode which is TM with respect to the x_3 axis can be derived from a magnetic vector potential which is parallel to the x_3 axis and given by Eq. (11.45).

11-7 Using Green's function expansions like Eqs. (11.48) and (11.49), together with the results of Problem 11-3, nd the magnetic or electric vector potentials of the generic mode of the rectangular resonator, generated by the polarization currents which describe the small dielectric or magnetic samples.

11-8 Prove that Eq. (11.70) must hold, in order to satisfy the normalization condition expressed by Eq. (11.69). Also show that Eq. (11.71) holds, and provide comments on its physical meaning.

11-9 Show that the normalized spherical harmonics given by Eq. (11.70) satisfy the following "sum rule":

$$P_\ell(\cos \ ) = \frac{4}{2\ell+1} \sum_{m=\ l}^{l} Y_{l,m}(\vartheta', \varphi')\, Y_{l,m}(\vartheta, \varphi) \quad ,$$

where $\cos \ = \cos\vartheta \, \cos\vartheta' + \sin\vartheta \, \sin\vartheta' \, \cos(\varphi \quad \varphi')$.

11-10 Write Maxwell's equations in spherical coordinates in a region without sources. Show that if there is rotational symmetry around the polar axis (i.e., for $\partial/\partial\varphi = 0$), their solutions split into TE and TM types, i.e., into elds with $E_r = 0$ and elds with $H_r = 0$.
Which eld components are nonvanishing for each of these classes of modes? Correlate the answer to this question with features of TE and TM elds in waveguides.

11-11 From the previous problem, derive a second-order di erential equation in spherical coordinates where the only unknown is rE_φ, for a TE mode, and one where the only unknown is rH_φ, for a TM mode. Compare it with Eq. (11.51), for $\partial/\partial\varphi = 0$: why are they di erent?
(*Hint.* Is it correct to say that in spherical coordinates the components of the Laplacian of a vector are the Laplacians of its components?)

11-12 Elaborating on the TM elds of Problem 11-10, show that a set of resonant frequencies of the φ-independent modes of a spherical resonator surrounded by an ideal conducting wall of radius a (imposing the boundary condition $E_\vartheta = 0$ at $r = a$) are solutions of the characteristic equation $\tan ka = ka/[1 \quad (ka)^2]$.

11-13 Which are the modes of a semi-spherical resonator, i.e., the modes of a spherical resonator compatible with an ideally conducting sept in the coordinate plane $\vartheta = \ /2$? And which are the modes of a quarter-of-a-sphere resonator, with conducting walls in the coordinate planes $\vartheta = \ /2$ and $\varphi = 0$?

11-14 A wave , satisfying the scalar Helmholtz equation propagates in a conical horn which imposes the boundary condition $= 0$ on the coordinate surface $\vartheta =$ /8, going outwards from a radial coordinate a to a radial coordinate b. Choose among the spherical harmonics those which satisfy these requirements.

11-15 After reading Chapter 12, show that the rigorous eld radiated by a short electric current element is a rst-order TM spherical harmonic. Which source radiates its dual, namely the rst-order TE spherical harmonic?

11-16 Show that the elds calculated according to the approximation in Eq. (11.83), i.e., in the region r , are formally identical to those of static multipoles, i.e., solutions of the Laplace equation (which can be found, e.g., in Jackson, 1975).

11-17 Prove that, in the region r , the elds discussed in Section 11.10 satisfy the following statements:

a) **E** and **H**, if linearly polarized, are orthogonal to each other;

b) $|\mathbf{E}|/|\mathbf{H}| =$, the intrinsic impedance of the medium;

c) locally, any of these elds can be approximated by a uniform plane wave which travels towards increasing values of **r**.

11-18 Outline explicitly all the analytical steps which are required to derive Eq. (11.88) and Eqs. (11.91).

11-19 A surface current, at frequency $f = 9.55\,\mathrm{GHz}$, with a peak value density of $1\,\mathrm{mA/mm}$, ows in the azimuthal direction on the surface of a cylinder, 10 mm in diameter. Using Eqs. (11.91), and accounting properly for the phase shift of the current density as a function of φ, derive the magnetic vector potential in the region $r > a$, assuming that permittivity and permeability are those of free space.

11-20 Find the e.m. eld generated in the region $r > a$ by the current distribution of the previous problem.

References

Abramowitz, M. and Stegun, I.A., (Eds.) (1964) *Handbook of Mathematical Functions.* Dover, New York.

Barton, G. (1989) *Elements of Green's Functions and Propagation: Potentials, Diffusion, and Waves.* Clarendon Press, New York/Oxford.

Collin, R.E. (1992) *Foundations for Microwave Engineering.* 2nd ed., McGraw-Hill, New York.

DeSanto, J.A. (1992) *Scalar Wave Theory: Green's Functions and Applications.* Springer, Berlin/New York.

Donoghue, W.F. (1969) *Distributions and Fourier Transforms.* Academic Press, New York.

Felsen, L.B. and Marcuvitz, N. (1973) *Radiation and Scattering of Waves.* Prentice-Hall, Englewood Cli s, NJ.

Greenberg, M.D. (1971) *Application of Green's Functions in Science and Engineering.* Prentice-Hall, Englewood Cli s, NJ.

Jackson, J.D. (1975) *Classical Electrodynamics.* 2nd ed., Wiley, New York.

MacRobert, T.M. (1967) *Spherical Harmonics: An Elementary Treatise on Harmonic Functions, with Applications.* Pergamon, Oxford.

Morse, P.M. and Feshbach, H. (1953) *Methods of Theoretical Physics.* McGraw-Hill, New York.

Panofsky, W.K.H. and Phillips, M. (1962) *Classical Electricity and Magnetism.* 2nd ed., Addison-Wesley, Reading, MA.

Ramo, S., Whinnery, J.R. and van Duzer, T. (1994) *Fields and Waves in Communication Electronics.* 3rd ed., Wiley, New York.

Schelkuno , S.A. (1948) *Electromagnetic Waves.* Van Nostrand, New York.

Snyder, A.W. and Love, J.D. (1983) *Optical Waveguide Theory.* Chapman & Hall, London.

Tai, C. (1971) *Dyadic Green's Functions in Electromagnetic Theory.* Intext Educational Publishers, Scranton.

Zemanian, A.H. (1965) *Distribution Theory and Transform Analysis.* McGraw-Hill, New York.

CHAPTER 12

Fundamentals of antenna theory

12.1 Introduction

A frequent situation, in telecommunication engineering, is that an e.m. field is either determined experimentally, or given as a design specification, while its sources $\mathbf{J}_0$ and $\mathbf{M}_0$, on the other hand, are not known. How to determine these sources is a trivial problem if the field $\{\mathbf{E}, \mathbf{H}\}$ is known everywhere in a region, V, which contains the sources. In that case, $\mathbf{J}_0$ and $\mathbf{M}_0$ can be computed using Maxwell's equations. On the contrary, if one knows the field $\{\mathbf{E}, \mathbf{H}\}$ only in a part of the region V where the field is defined, and wishes to extract from this partial knowledge some information on sources located *elsewhere*, the problem is not trivial at all. For example, to design a transmitting antenna, the typical specification one starts from is the field intensity at the receiving antennas. Starting from these data one must design a distribution of imposed currents which sustains the specified field at a distance. In order to be able to proceed in this way, we must first go through the opposite path. At least, this will give us the ability to distinguish physically acceptable specifications from those which can never be satisfied. This is in fact the purpose of this chapter.

In general, a field in a given region V depends not only on its sources, but also on *the boundary conditions* on the surface which encloses V. The relationship between field and sources becomes one-to-one only if suitable boundary conditions are specified, so that the uniqueness theorem of Section 3.3 applies. The relationship between field and sources becomes straightforward when the *surface integrals* (11.19) *vanish.* This is compatible with the uniqueness theorem, if the region V is the whole 3-D space, filled by a homogeneous medium, and if the field satisfies, at infinity, Sommerfeld's radiation conditions in Eq. (3.17). Throughout this chapter, we will always give it for granted that these conditions hold, except where the contrary is explicitly stated.

These assumptions entail implications which may affect engineering practice quite profoundly. Indeed, the field generated by an actual source will become more and more similar to the theoretical expressions derived in this chapter as the medium inhomogeneities are removed further and further away from the source and from the points where the field is tested. In any case, however, it is required that the field "reflected" by inhomogeneities be weak. Therefore, as a general rule, variations in parameters of the medium can be tolerated only if small and smooth.

For the same reasons, antennas are normally tested in the so-called *anechoic chambers*, i.e., in rooms where reflections from the walls, including the floor

Figure 12.1 *Anechoic Chamber (12 14 16 meters) containing the Spherical Near Field Antenna Test Facility at the Technical University of Denmark. The picture illustrates a micro satellite model being made ready for antenna tests. The front of the picture shows the L-band measurement probe.*

and the ceiling, are made as small as possible by means of suitable absorbing elements. An example is shown in Figure 12.1. Again for similar reasons, in real radio links the e ects of an inhomogeneous medium are not always negligible. In some cases they can be accounted for using methods which, in general, are further elaborations on the contents of this chapter. Some of them will be brie y described in the following chapter. The interested readers will be given further references, for in-depth approaches to these problems, in the nal sections of this chapter and of the next one.

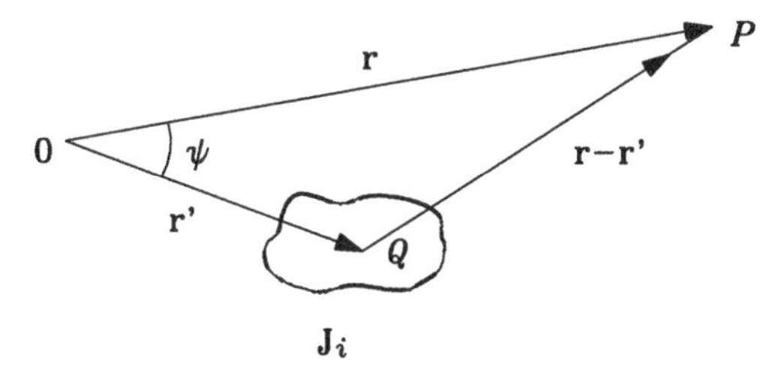

Figure 12.2 *The geometry of the equivalent-moment problem.*

12.2 Equivalent dipole moment of an extended source

Suppose that, in an unbounded homogeneous medium, all the eld sources are located at nite distances from an arbitrarily chosen origin. Let us deal explicitly with the case where these sources are electric current densities, $\mathbf{J}_0$. Imposed magnetic currents can be treated by duality.

Under these assumptions, the eld $\{\mathbf{E}, \mathbf{H}\}$ can be computed through the magnetic vector potential, Eq. (11.18):

$$\mathbf{A}(\mathbf{r}) = \frac{}{4} \int_V \mathbf{J}_0(\mathbf{r}') \, \frac{e^{\quad |\mathbf{r} \quad \mathbf{r}'|}}{|\mathbf{r} \quad \mathbf{r}'|} \, d\mathbf{r}' \quad , \tag{12.1}$$

where, as always, $\ ^2 = \ \omega^2 \ \ _c$, while the meanings of the position vectors $\mathbf{r}$ and $\mathbf{r}'$ are illustrated in Figure 12.2.

As we saw in Section 11.10, the Green's function in Eq. (12.1) can be expanded as a series of spherical harmonics. For r , using the asymptotic form for spherical Hankel functions of the second kind, Eq. (11.84), we can write:

$$\frac{1}{4} \frac{e^{\quad |\mathbf{r} \quad \mathbf{r}'|}}{|\mathbf{r} \quad \mathbf{r}'|} \simeq \frac{e^{\quad r}}{r} \left\{ \sum_{\ell=0}^{\infty} j_\ell(\ j \quad r) (\ j)^{\ell+1} \sum_{m= \ \ell}^{\ell} Y_{\ell m}(\vartheta', \varphi') \, Y_{\ell m}(\vartheta, \varphi) \right\} . \tag{12.2}$$

The term $e^{\quad r}/r$ is a common factor in the summation, and can be taken out of the integral in Eq. (12.1). This step underlines that far from the origin the eld is like one *spherical wave*, for any source distribution, provided sources do not extend to in nity.

Let us look now for an approximation (to hold for $r \quad r'$) to that part of Eq. (12.2) which depends on the integration point $Q = O + \mathbf{r}'$, namely:

$$\left(\frac{e^{\quad |\mathbf{r} \quad \mathbf{r}'|}}{|\mathbf{r} \quad \mathbf{r}'|} \right) \Big/ \left(\frac{e^{\quad r}}{r} \right) = \frac{r}{|\mathbf{r} \quad \mathbf{r}'|} \, e^{\quad (|\mathbf{r} \quad \mathbf{r}'| \quad r)} \quad . \tag{12.3}$$

The aim is to simplify the calculation of the vector potential, Eq. (12.1), far from the sources. Then, apply Carnot's theorem (see again Figure 12.2) and write the binomial expansion of the square-root term. We get:

$$|\mathbf{r} \quad \mathbf{r}'| = (r^2 + r'^2 \quad 2rr' \cos \quad)^{1/2} = r \left[1 \quad \frac{r'}{r} \cos \quad + T \left(\frac{r'}{r} \right)^2 + \ldots \right] , \tag{12.4}$$

where $T(r'/r)^2$ denotes a quantity proportional to $(r'/r)^2$, through a coe - cient the reader may evaluate as an exercise. For $r \quad r'$, we are entitled to write simply $r/|\mathbf{r} \quad \mathbf{r}'| \simeq 1$, but a more accurate procedure is necessary for the *di erence* between r and $|\mathbf{r} \quad \mathbf{r}'|$, rst because it is not correct to neglect a quantity with respect to itself, second because in Eq. (12.3) this di erence gets into the argument of a rapidly varying function, an exponential.

If we truncate the expansion of Eq. (12.4) at the term of order r'/r, i.e., if we let:

$$|\mathbf{r} \quad \mathbf{r}'| \simeq r \quad r' \cos \quad = r \quad \mathbf{r}' \quad \hat{r} \quad , \tag{12.5}$$

the corresponding error in Eq. (12.3) can be estimated by rewriting Eq. (12.5) as:

$$\exp \left[\quad r T \left(\frac{r'}{r} \right)^2 + \ldots \right] = 1 \quad r T \left(\frac{r'}{r} \right)^2 + \ldots \simeq 1 \quad . \tag{12.6}$$

To give a simple physical interpretation for the approximation, let us restrict ourselves to lossless media ($\quad = j2 \quad / \quad$). Then, we see that Eq. (12.6) is acceptable whenever:

$$\frac{2 \quad r'^2}{\quad r} \quad 1 \quad . \tag{12.7}$$

So, although Eq. (12.5) was obtained using a purely geometrical argument (i.e., comparing r and r'), still its scope of validity is restricted by a condition which entails also the *wavelength*. In practice, this comment pertains only to sources whose size is large compared to . For them, the condition in Eq. (12.7) is satis ed farther away from the sources than the condition $r \quad r'$.

When all these restrictions are satis ed, Eq. (12.1) can be replaced by the following approximation:

$$\mathbf{A}(\mathbf{r}) = \frac{\quad}{4} \frac{e^{\quad r}}{r} \int_V \mathbf{J}_0(\mathbf{r}') \, e^{\quad \mathbf{r}' \quad \hat{\mathbf{r}}} \, d\mathbf{r}' \quad , \tag{12.8}$$

which contains the quantity:

$$\boldsymbol{\mathcal{M}}(\vartheta, \varphi) \quad \int_V \mathbf{J}_0(\mathbf{r}') \, e^{\quad \mathbf{r}' \quad \hat{\mathbf{r}}} \, d\mathbf{r}' \quad , \tag{12.9}$$

referred to as the source *equivalent dipole moment* (or, for short, equivalent moment). To avoid misunderstandings, let us point out that Eq. (12.9) does not coincide with the coe cient of the electric-dipole term in the multipole expansion of Section 11.10. $\boldsymbol{\mathcal{M}}$ depends only on the *angular coordinates* of

In other textbooks (e.g., Ramo *et al.*, 1994), Eq. (12.9) is called the *radiation vector* of the source.

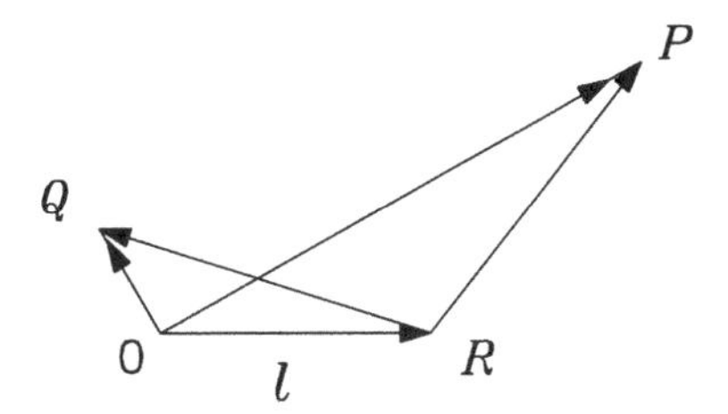

Figure 12.3 *Translation of the reference origin.*

the "potentiated point" $P = O + \mathbf{r}$. In other words, it depends on the direction of the vector $\mathbf{r}$, but is *independent of the distance* $r = |P \quad O|$. Let us stress that the direction of $\mathbf{r}$ is that of the straight line from the origin of the reference frame, O, to the point P. Consequently, the equivalent moment of a given source *depends on the mutual position between the origin and the source.* Very important practical consequences of this fact will be illustrated in Section 12.8. For further clari cation of this point, let $\boldsymbol{\mathcal{M}}_O(P)$ and $\boldsymbol{\mathcal{M}}_R(P)$ be the equivalent moments of a given source, at the same potentiated point P but referred to two distinct origins, O and R, respectively. Let:

$$\boldsymbol{\ell} = R \quad O \quad , \tag{12.10}$$

be the origin translation vector (Figure 12.3). Assume that the inequality in Eq. (12.7) is satis ed in P with reference to *both* origins. Then, the reader may prove that, if:

$$|\boldsymbol{\ell}| \quad |P \quad O| \quad , \qquad |\boldsymbol{\ell}| \quad |P \quad Q| \quad , \tag{12.11}$$

then, within the same order of accuracy as Eq. (12.9), it is:

$$\boldsymbol{\mathcal{M}}_O(P) \simeq e^{\quad \boldsymbol{\ell}\,\hat{r}}\, \boldsymbol{\mathcal{M}}_R(P) \quad . \tag{12.12}$$

Eqs. (12.9) and (12.12) allow us to comment on the previous approximation from a physical point of view. The assumption $r/|\mathbf{r} \quad \mathbf{r}'| \simeq 1$, which eliminates the factor $1/r$ from Eq. (12.9), means that, far from the sources, small di erences in the distance covered by individual in nitesimal spherical waves, radiated by di erent points within the source, cause negligible di erences in *amplitudes.* On the contrary, the exponential function in Eq. (12.9) — with an imaginary argument in the lossless case — indicates that *phases* of the contributions, coming from di erent points of the source, have crucial in uence on the far eld, so that in this respect di erent path lengths have to be evaluated more carefully.

12.3 Far- eld approximations

Let us show that in the region where Eq. (12.8) applies for the magnetic vector potential (keeping the assumption of a homogeneous medium in the whole 3-D space), simple approximate expressions for the e.m. eld can be obtained from Eq. (12.8) itself.

Place Eq. (12.8) into the general expression Eq. (1.60) for the magnetic eld. The symbol ∇_P for the curl stresses, like in the previous chapter, that it operates only on the coordinates of the potentiated point $P = (r, \vartheta, \varphi)$. We get:

$$\mathbf{H} = \frac{1}{\ } \nabla_P \quad \mathbf{A} = \frac{1}{4\ } \nabla_P \left(\frac{e^{\quad r}}{r} \right) \quad \boldsymbol{\mathcal{M}} + \frac{1}{4\ } \frac{e^{\quad r}}{r} (\nabla_P \quad \boldsymbol{\mathcal{M}}) . \quad (12.13)$$

The rst term is simply

$$\nabla_P \left(\frac{e^{\quad r}}{r} \right) = \left(\qquad \frac{1}{r} \right) \hat{r} \frac{e^{\quad r}}{r} \quad . \quad (12.14)$$

In the second term, ∇_P can be brought inside the integral in Eq. (12.9). Then, using the Eq. (B.11) for the gradient in spherical coordinates, we get:

$$\nabla_P \quad \left[\mathbf{J}_0(\mathbf{r}') \, e^{\quad r' \cos} \right] = \quad \mathbf{J}_0(\mathbf{r}') \quad \nabla_P(e^{\quad r' \cos} \quad)$$

$$= \qquad r'(e^{\quad r' \cos} \quad) \, \mathbf{J}_0(\mathbf{r}') \quad \left[\frac{1}{r} \frac{\partial(\cos \quad)}{\partial \vartheta} \hat{\vartheta} + \frac{1}{r \sin \vartheta} \frac{\partial(\cos \quad)}{\partial \varphi} \hat{\varphi} \right] . \quad (12.15)$$

All the terms in Eq. (12.13) contain a factor $e^{\quad r}/r^2$, except for one contribution, in the rst term, which is proportional to $e^{\quad r}/r$. Let us refer, as usual, to the lossless case, where $\quad = j2 \ / \ $. Then, for $r \qquad$, all the other terms are negligible in comparison with that which is proportional to $r^{\ 1}$. (Incidentally, note that we are comparing real quantities with imaginary ones. Consequently, our approximations are correct for the moduli, but may yield unacceptable consequences on phases. Another point to be heeded is the di erent dependence of the various terms on the *angular coordinates*. We will go in depth through all these points in Section 12.4). Then, we have:

$$\mathbf{H} \simeq \quad - \hat{r} \quad \mathbf{A} = \frac{\ }{4\ } \hat{r} \quad \boldsymbol{\mathcal{M}} \frac{e^{\quad r}}{r} \quad . \quad (12.16)$$

In the lossless case ($\quad = j\omega \sqrt{\ }$), Eq. (12.16) can be written also as follows:

$$\mathbf{H} \simeq \quad j\omega \frac{1}{\ } \hat{r} \quad \mathbf{A} \quad , \quad (12.17)$$

where $\quad = \sqrt{\ / \ }$ is the *intrinsic impedance of the medium*, de ned in Chapter 4.

We can nd the electric eld from Eq. (12.16) using the second Maxwell's equation and the vector identity (C.7):

$$\mathbf{E} = \frac{\nabla_P \quad \mathbf{H}}{j\omega \ _c} \simeq \frac{\ }{j\omega \ _c} \left(\hat{r} \nabla_P \quad \mathbf{A} \quad \frac{\partial \mathbf{A}}{\partial r} \right) \quad . \quad (12.18)$$

For $r \qquad$, but with the warnings given before, higher-order terms in $1/r$

can be neglected with respect to low-order ones, yielding:

$$\nabla_P \ \mathbf{A} \simeq \frac{1}{r^2}\frac{\partial}{\partial r}\left(r^2 A_r\right) \simeq \qquad \mathbf{A} \ \hat{r} \quad ,$$

$$\frac{\partial \mathbf{A}}{\partial r} \simeq \qquad \mathbf{A} \quad . \tag{12.19}$$

We get in the end:

$$\mathbf{E} = \ j\omega\Big[\mathbf{A} \quad (\mathbf{A} \ \hat{r})\,\hat{r}\Big] = \ j\omega\,\hat{r} \quad \mathbf{A} \quad \hat{r} = \ \frac{j\omega}{4}\,\hat{r} \quad \boldsymbol{\mathcal{M}} \quad \hat{r}\,\frac{e^{\quad r}}{r} \, . \tag{12.20}$$

It can be proved as an exercise that the last expression holds even in a lossy medium.

Eqs. (12.20) and (12.16) show that, at a distance which is large compared to both and the source size, the eld of *any* source not at in nity, in a homogeneous unbounded medium, is an out owing spherical wave, $(e^{\quad r}/r)$. In general it is not a *uniform* wave, i.e., $\boldsymbol{\mathcal{M}}(\vartheta,\varphi) \neq$ constant. *Locally*, i.e., in the neighborhood of a xed point, on a length scale comparable with the wavelength, this eld is very well approximated by a uniform plane wave. Its constant-phase planes are orthogonal to the unit-vector $\hat{r}$, therefore they are tangent to the wavefronts of the spherical wave. This agrees also with the fact that, for a linearly polarized eld, the three vectors $\mathbf{E}, \mathbf{H}, \hat{r}$ are mutually orthogonal. It matches also with the fact that the wave impedance in the radial direction is equal to the intrinsic impedance of the medium, .

The expressions found in this section show that the far eld is easy to compute once the source equivalent moment, Eq. (12.9), is known. For this reason, to design an antenna essentially means to nd a distribution of imposed current densities, $\mathbf{J}_0(Q)$, such that the corresponding equivalent moment $\boldsymbol{\mathcal{M}}$ has a pre-speci ed pattern as a function of the two angular coordinates ϑ and φ.

12.4 First example: Short electric-current element

In the general expression of the equivalent moment, Eq. (12.9), the integral over the source volume may be read as a superposition of the e ects due to an in nite number of elementary sources $\mathbf{J}_0\,dV$. This pushes us to study rst a *point source*, i.e., a source for which we can let $\mathbf{r}' = 0$ in Eq. (12.9). An elementary source can be envisaged in three, in two, or in one dimension, by means of the following relationships, respectively:

$$\boldsymbol{\mathcal{M}} = \left\{ \begin{array}{c} \mathbf{J}_0\,dV \\ \mathbf{J}_{0s}\,dS \\ I\,\mathbf{d}\boldsymbol{\ell} \end{array} \right\} = \text{constant} \quad . \tag{12.21}$$

A small magnetic current element can be de ned and treated by duality.

For simplicity, without losing in generality, we may assume that $\boldsymbol{\mathcal{M}}$ is real. Then, let us introduce two reference systems, a rectangular one and a spherical one, whose z axis is parallel to $\boldsymbol{\mathcal{M}}$ and whose origin is where the elementary

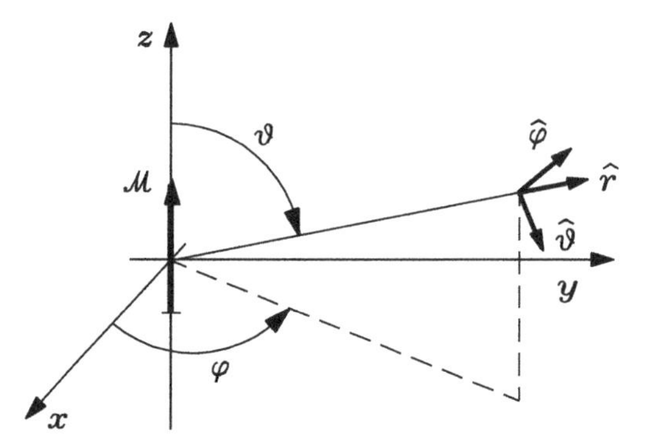

Figure 12.4 *The reference system for the elementary source.*

source is located (see Figure 12.4). Then, the magnetic vector potential in Eq. (12.8) can be written as:

$$\mathbf{A} = A_z\,\hat{a}_z = \frac{}{4}\,\mathcal{M}\,\frac{e^{\quad r}}{r}\,\hat{a}_z \quad . \tag{12.22}$$

In general, this expression holds, as we said in Section 12.2, for $r \quad r'$ and $r \quad r'^2$. Now, however, $r' = 0$ implies that Eq. (12.22) is valid in the whole 3-D space. From the potential, Eq. (12.22), we can compute the eld $\{\mathbf{E}, \mathbf{H}\}$ everywhere, using exact relationships derived in Chapter 1. This will be done later on; rst, let us proceed to calculate the far eld ($r \quad$) using the approximations in Eqs. (12.20) and (12.16).

In the spherical coordinate system that we introduced above (see again Figure 12.4), the following relationships hold:

$$\hat{r} \quad \hat{a}_z \quad \hat{r} = \quad \sin\vartheta\,\hat{\vartheta} \quad , \qquad \hat{\vartheta} \quad \hat{r} = \quad \hat{\varphi} \quad . \tag{12.23}$$

Then, from Eq. (12.22), in a lossless medium we get immediately:

$$\mathbf{E} \simeq j\,\frac{\mathcal{M}}{2}\,\sin\vartheta\,\frac{e^{\quad r}}{r}\,\hat{\vartheta} \quad ,$$

$$\mathbf{H} \simeq j\,\frac{\mathcal{M}}{2}\,\sin\vartheta\,\frac{e^{\quad r}}{r}\,\hat{\varphi} \quad . \tag{12.24}$$

At any point where these expressions are applicable $\mathbf{E}$ and $\mathbf{H}$ are linearly polarized. This is a consequence of our assumption that the vector $\boldsymbol{\mathcal{M}}$ is real. The ow lines of $\mathbf{E}$ (of $\mathbf{H}$) are the meridians (the parallels) of the spherical coordinate system. The Poynting vector is real, radial, and oriented towards growing r's. Its ux through the spherical surface of radius r centered at the origin (the surface element, as well known, is $dS = r^2 \sin\vartheta\, d\vartheta\, d\varphi$) equals:

$$W_a = \int_0^{2} d\varphi \int_0 \frac{|\mathcal{M}|^2}{8\ ^2} \sin^2\vartheta\, e^{\ 2\,\mathrm{Re}(\ \ )\,r} \sin\vartheta\, d\vartheta$$

$$= \quad \frac{|\mathcal{M}|^2}{3 \qquad 2} \quad e^{\;2\,\mathrm{Re}(\;\;)\,r} \quad . \tag{12.25}$$

As this quantity is real, the far eld carries only an *active power* in the radial direction. The reader can prove as an exercise (using Section 12.3) that essentially the same result applies to *any source* which does not extend to in nity. This real power is *positive*, meaning that it is radiated by the current element.

In a lossless medium ($= j$), Eq. (12.25) is independent of the sphere radius r. So, total power is conserved while it ows away from the source. This agrees, obviously, with Poynting's theorem. However, this theorem implies that in a lossless medium power through a spherical surface centered in the source must be independent of the radius even when the restriction r , which we used to derive Eqs. (12.24), is violated. This implies that the two vector functions Eq. (12.24) yield *everywhere* the correct active power through a sphere centered at the origin, although they do not satisfy rigorously Maxwell's equations. To express this, they are referred to as the *radiation eld* (or the radiative eld) of the electric current element.

It is enlightening to nd now the rigorous eld expressions which descend from the vector potential, Eq. (12.22), and compare it with the approximate eld, Eq. (12.24). The vector identity (C.6) entails $\nabla \quad (A_z \hat{a}_z) = \nabla A_z \quad \hat{a}_z$. Taking then into account that A_z depends only on the r coordinate, and that $\hat{r} \quad \hat{a}_z = \quad \sin\vartheta\,\hat{\varphi}$, we nd:

$$\mathbf{H} = \frac{1}{\;}\nabla \quad \mathbf{A} = \frac{\mathcal{M}}{4}\left(\quad + \frac{1}{r}\right)\frac{e^{\quad r}}{r}\sin\vartheta\,\hat{\varphi} \quad . \tag{12.26}$$

Through the second Maxwell's equation and using the curl in spherical coordinates (Eq. (B.13)), from Eq. (12.26) it follows:

$$\mathbf{E} = \frac{1}{j\omega \quad c}\frac{\mathcal{M}}{2}\left(\quad + \frac{1}{r}\right)\frac{e^{\quad r}}{r^2}\cos\vartheta\,\hat{r}$$
$$+ \frac{1}{j\omega \quad c}\frac{\mathcal{M}}{4}\left(\;^2 + \frac{\;}{r} + \frac{1}{r^2}\right)\frac{e^{\quad r}}{r}\sin\vartheta\,\hat{\vartheta}\,. \tag{12.27}$$

To go back from Eqs. (12.26) and (12.27) to Eqs. (12.24) we must neglect:

the term $1/r$ with respect to in the magnetic eld;

the component $E_r \propto 1/r^2$ with respect to the component $E_\vartheta \propto \quad /r$ in the electric eld;

the terms $/r$ and $1/r^2$ with respect to ^{2} in the E_ϑ component.

These approximations are certainly not acceptable if the condition $r \quad 1/$ ($\simeq \quad /2$ in low-loss media) is not satis ed. Still, they must be handled with care even when this condition is satis ed. Indeed:

$1/r$ is real but is complex (purely imaginary in the lossless case). If we neglect the rst term with respect to the second one, the modulus of the complex number $+1/r$ is well approximated, but its phase is inaccurate. The same comment applies to the comparison between ^{2}, $/r$, $1/r^2$;

the E_r and E_ϑ components depend on ϑ in di erent ways so that it is not correct to consider $|E_r| \quad |E_\vartheta|$ near $\vartheta = 0, \quad$, where $\sin\vartheta \simeq 0$. Therefore, the approximation leading to Eqs. (12.24) is not justi ed inside a thin double cone, whose axis is in the direction of the source moment $\boldsymbol{\mathcal{M}}$.

One of the most important results for practical purposes, the ux of the Poynting vector through a sphere centered at the origin, is not a ected at all by the second objection. In fact, the radial component of the electric eld does not contribute to this ux. On the other hand, a rigorous evaluation of the radial component of the Poynting vector entails enlightening remarks on what the rst objection really means. Using Eqs. (12.26) and (12.27), in the lossless case we obtain:

$$W_c = \int_S \frac{\mathbf{E} \quad \mathbf{H}}{2} \quad \hat{r}\, dS = \frac{}{3} \quad \frac{|\boldsymbol{\mathcal{M}}|^2}{2} \quad \left(1 \quad j \frac{3}{8 \quad ^3 r^3}\right) \quad . \qquad (12.28)$$

Comparing Eq. (12.28) and Eq. (12.25) we see that, when we use the approximations in Eqs. (12.24), we obtain everywhere the correct active power, but we miss completely the *reactive power*, which is given by:

$$W_q = \quad \frac{3}{8 \quad ^3 r^3} W_a = \quad \frac{}{3} \quad \frac{|\boldsymbol{\mathcal{M}}|^2}{8 \quad ^3 r^3} \quad . \qquad (12.29)$$

Eq. (12.29) gives a negative (i.e., capacitive) reactive power, which for r is certainly very small compared to the active power, W_a, but never rigorously equal to zero. On the other hand, the reactive power dominates over W_a for $r \quad$, i.e., in the *near eld.*

Looking in more detail at the region $r \quad$, we nd that the near eld matches perfectly with the theory of quasi-static elds (e.g., Plonsey and Collin, 1961; Kupfmuller, 1973). Indeed, with the cautions which, as we said before, are necessary when comparing real and imaginary quantities, for r and assuming $e^{\ j \ \ r} \simeq 1$, from Eq. (12.27) we get:

$$\mathbf{E} = \frac{\mathcal{M}}{j\omega}\left[\frac{1}{2 \quad r^3} \quad \cos\vartheta\, \hat{r} + \frac{1}{4 \quad r^3} \quad \sin\vartheta\, \hat{\vartheta}\right] = \quad \nabla\left[\frac{q\, d\ell}{4 \quad r^2} \quad \cos\vartheta\right] \quad , \qquad (12.30)$$

where we set:

$$\frac{\mathcal{M}}{j\omega} = \frac{I\, d\ell}{j\omega} = q\, d\ell \quad , \qquad (12.31)$$

while $q = I/(j\omega)$ is the complex number representing the time-harmonic electric charge equal to the time integral of the current which ows through the dipole length $d\ell$. In Eq. (12.30) one recognizes immediately the static eld of an *electric dipole* whose moment equals $qd\ell$, i.e. (see Figure 12.5), two point charges q, located at the extremes of a segment $d\ell$. This is the reason why the elementary source de ned by Eq. (12.22) is called, sometimes, an *elementary electric dipole.*

In contrast to an electrostatic dipole, the source characterized by the magnetic vector potential given by Eq. (12.22) generates, in the near eld, also a *magnetic* eld. Indeed, with the approximations which hold for $r \quad$,

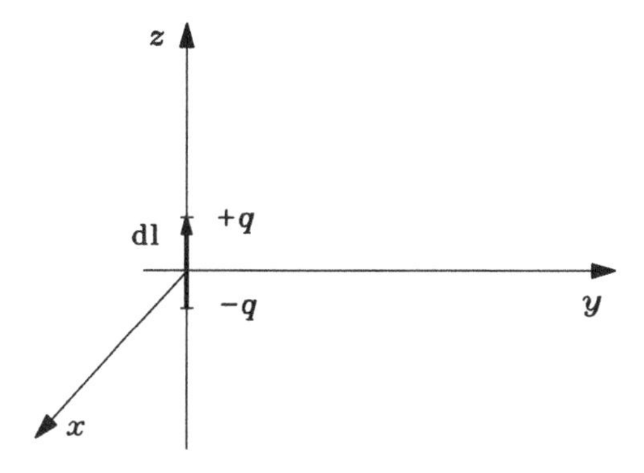

Figure 12.5 *Schematic representation of an electrostatic dipole.*

Eq. (12.26) becomes:

$$\mathbf{H} = \frac{I\,d\ell}{4\ \,r^2}\,\sin\vartheta\,\hat{\varphi} = \frac{I\,\mathbf{d}\boldsymbol{\ell}\quad \mathbf{r}}{4\ \,r^3}\quad . \tag{12.32}$$

It is easy to recognize this as the so-called *Ampere's law of elementary action* (Ramo *et al.*, 1994; Plonsey and Collin, 1961) which expresses the magnetic eld generated by a steady-state electric current I owing through a line element $\mathbf{d}\boldsymbol{\ell}$.

The elds given by Eqs. (12.30) and (12.32) *do not satisfy Maxwell's equations.* This should not come as a surprise: it is well known that static or slowly-varying electric elds appear to be independent of magnetic ones, and vice-versa. Our approach shows that this apparent independence is, in reality, the result of an approximation which is acceptable only if one can neglect delays due to the nite velocity of propagation of e.m. waves.

Starting from Eqs. (12.30) and (12.32), let us compute the ux of Poynting vector through a spherical surface of radius r, centered at the origin. We nd only a *reactive power*, equal to Eq. (12.29). No active power is an error, whose source is the rough approximation for the phase of the complex factors in $\mathbf{E}$ and in $\mathbf{H}$. The correct result would be $W_a \quad W_q$ but $W_a \neq 0$, in the near eld.

Often, when is very large with respect to the size of the eld sources, elds have a detectable level only at distances r . Under these conditions, active power is negligibly small *everywhere*. The practical meaning of this result is that at low frequencies (typically, at industrial frequencies) it is practically impossible for an e.m. eld *in a homogeneous medium* to convey an active power. To carry power, it is strictly necessary to exploit suitable inhomogeneities in the medium, and these are called transmission lines.

12.5 Characterization of antennas

Those terms in Maxwell's equations which we call "field sources" are mathematical models of practical objects, which radiate e.m. fields according to some specifications and are usually called "antennas". Theory and practice are consistent when a mathematical "source" describes accurately an actual antenna. In this context, let us stress again the following assumptions that must be satisfied if one wants to exploit the results of previous sections:

- the medium must be homogeneous *in the whole 3-D space*. Usually, in practice, an antenna is built with a conducting material surrounded by a dielectric, but the mathematical source is an imposed current density *in a homogeneous dielectric*, equal to the medium which surrounds the actual antenna;
- to apply the far-field approximations, the distance from the source must be much larger than the antenna size *and* the field wavelength.

In this section, unless the opposite is explicitly stated, the medium will be assumed to be *lossless*.

From Section 12.3, and from some of the problems listed at the end of this chapter, we know that under these assumptions the Poynting vector $\mathbf{P}$, at any point $R = (r, \vartheta, \varphi)$, is real, is oriented as $\mathbf{r} = R - O$, and is proportional to $1/r^2$. These features are independent of the source of the field, so they do not yield any information that may characterize an individual antenna. On the contrary, peculiarities of each antenna are expressed by how the Poynting vector $\mathbf{P}$ depends on the *angular coordinates* ϑ, φ. In other words, directional properties characterize an antenna. In this framework, it turns out to be useful to define the following quantity, which is referred to as the *radiation intensity* in the direction (ϑ, φ):

$$I_r(\vartheta, \varphi) = |\mathbf{P}(R)| \; |R - O|^2 = \frac{\mathbf{E}(R) \; \mathbf{E} \; (R)}{2} \; r^2 \quad . \tag{12.33}$$

Eq. (12.33) is independent of the distance between the point of observation R and the origin O. However, useful information on the antenna directional properties is still mixed with the trivial dependence of I_r on the source equivalent moment, through its modulus squared. In turn, this quantity is proportional to the square of the modulus of the current which feeds the antenna. In other words, to specify unambiguously Eq. (12.33) one must measure the current (or the power) which enters the antenna. To eliminate this often unwanted dependence, let us define the *normalized radiation intensity*:

$$i = \frac{I_r(R)}{I_r(M)} = \frac{\mathbf{E}(R) \; \mathbf{E} \; (R)}{\mathbf{E}(M) \; \mathbf{E} \; (M)} \quad , \tag{12.34}$$

where M is a fixed reference point, and R is a generic point on the spherical surface of radius $r = |M - O|$. If nothing else is stated, it must be taken for granted that M is located along the direction (or one of the directions) along which the radiation intensity in Eq. (12.33) reaches its *absolute maximum*

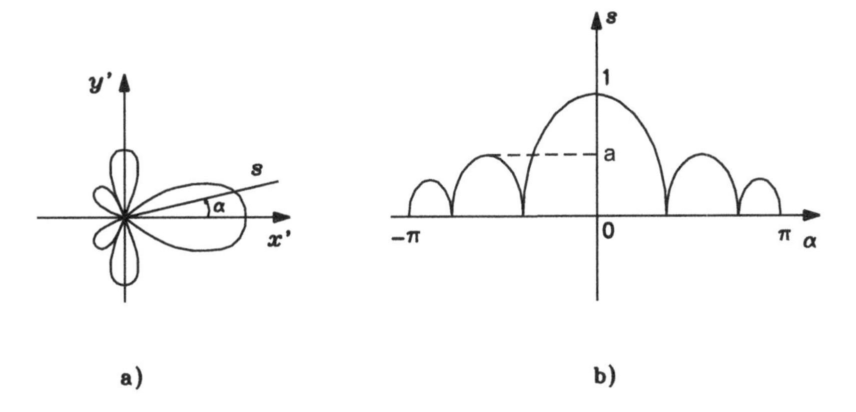

Figure 12.6 *An example of radiation pattern.*

value. Then, we have i 1 everywhere. This i also o ers the advantage, with respect to Eq. (12.33), to make it needless to measure the distance r.

The quantity:

$$f(\vartheta, \varphi) = \sqrt{i(\vartheta, \varphi)} \quad , \tag{12.35}$$

proportional to $|\mathbf{E}|$ and to $|\mathbf{H}|$ (while the radiation intensity is proportional to their squares, and so to the power density), is referred to as the *radiation function* of the antenna.

Let us introduce another spherical coordinate system (s, ϑ, φ), whose origin and whose polar axis coincide with those of the reference system (r, ϑ, φ) which we use to express the eld and the other quantities de ned so far. Then, the equation:

$$s = f(\vartheta, \varphi) \tag{12.36}$$

de nes a surface which is referred to as the *radiation surface* of the antenna. This geometrical representation of the function (12.35) is a very helpful hint towards an intuitive understanding of the properties of an antenna.

The curve where the radiation surface intersects a plane, p, passing through the origin, is called the *radiation diagram* of the antenna in the p plane. An example is shown in Figure 12.6, whose part a) shows a polar diagram, as we have just de ned it, while part b) is just a cartesian plot of the same diagram.

An antenna is called *omnidirectional* in a plane $\mathbf{p}$ if its radiation diagram in $\mathbf{p}$ is a circle. Omnidirectional antennas are interesting for broadcasting and similar applications, where transmitted signals must reach an extended region. For point-to-point links, interest is focused on highly directional antennas. However, let us stress that even in principle (i.e., leaving aside technological di culties) any physically realizable source has a radiation function $f(\vartheta, \varphi)$ which vanishes only at *discrete values* of ϑ and φ. In other words, any physically realizable antenna has *null directions*, but there are no sources for which $f(\vartheta, \varphi) = 0$ over nite intervals of either ϑ or φ. This result can be proved as a remote consequence of the theory of analytic functions of complex variables, which entails that a function which vanishes over a nite interval is zero everywhere. Another way to explain it is to start from the equivalence theorem of Chapter 3. This implies that, if **E** and **H** are equal to zero over a nite interval of either ϑ or φ (i.e., over an angle in the 3-D space), then they are *identically zero* in the whole 3-D space.

The part of a radiation diagram which lies between two consecutive null directions is called a *radiation lobe.* Lobes whose maximum equals unity are called *main lobes,*† while all the other lobes are *secondary* ones.

The *angular width* of a radiation lobe may be de ned in two ways. One is simply the angle between the null directions which encompass the lobe. The other one is the so-called *half-power* (or 3 decibel) width, i.e., the angle between the directions along which the radiation intensity equals its maximum value divided by two. Both de nitions are popular and in current use. The rst one is useful for calculations that one makes while designing an antenna, but its experimental measurements can be a ected by a large error bar. The converse is true for the second de nition.

The *lobe ratio* of an antenna is de ned as the ratio, $1/a$, between the absolute maximum of $f(\vartheta, \varphi)$ (which is unity) and the largest, a, among the secondary lobe amplitudes (see Figure 12.6).

A problem which has been dealt with in depth is whether it is possible to make a *super-gain antenna,* de ned as a source whose lobe ratio is larger than an arbitrarily large pre-selected value, and whose main lobe angular width is smaller than an arbitrarily small pre-selected value. On paper, the answer to this question is yes, but it yields recipes which cannot be implemented in practice because they would be too complicated to build. We will come back to this subject, shortly, in Section 12.8.

The de nitions which we gave so far about the directional properties of an antenna were all local, i.e., had to do with quantities along one or more discrete directions. It turns to be useful to de ne adimensional numbers which summarizes the directional properties of an antenna, averaged over all directions. The most commonly used quantity is the inverse of the average of the normalized radiation intensity, namely:

$$d = \frac{1}{\frac{1}{4} \int_0^{4} i \, d} \quad , \tag{12.37}$$

where $d \;\; = \sin\vartheta \, d\vartheta \, d\varphi$ is the elemental solid angle. Eq. (12.37) is referred to as the *directivity gain* (or simply the *directivity*) of the antenna. Given two antennas whose directivities are d_1 and d_2, respectively, the relationship:

$$d_{12} = \frac{d_1}{d_2} \quad , \tag{12.38}$$

† Clearly, in this de nition we are assuming that the point M of Eq. (12.34) has been chosen so that $f(\vartheta, \varphi) \quad 1$ everywhere.

de nes the relative gain (in directivity) of antenna 1 with respect to antenna 2. This enables us to note that Eq. (12.37) can be interpreted as the relative gain of the antenna we are dealing with, say A, with respect to an antenna having $i = 1$ in all directions, which is called *isotropic* antenna. We may indeed recon rm this interpretation if we insert Eq. (12.34) into Eq. (12.37) and let $dS = r^2\, d$ (element of the spherical surface of radius r centered at the origin). So, we get:

$$d = \frac{4 \quad \frac{\mathbf{E}(M) \quad \mathbf{E}\ (M)}{2}\, r^2}{\int_S \frac{\mathbf{E}(R) \quad \mathbf{E}\ (R)}{2}\, dS} = \frac{P_0}{P_i} \quad . \tag{12.39}$$

In Eq. (12.39), d plays the role of ratio between two powers. The numerator, P_0, is the power that an isotropic antenna (placed at the same point, 0, where the antenna A is located) should radiate in order to generate, at the reference point M, the same power density as A. The denominator, P_i, is the power which is actually radiated by A. This viewpoint suggests that it might be more appropriate to call the directivity a "saving," rather than a gain, since an antenna is a passive component. However, the term gain has been in use for decades and can be accepted without any danger of misunderstandings.

For several reasons — rst of all, because it is di cult to measure the radiated power P_i — another adimensional quantity, called the antenna *power gain*, is used in practice at least as much as d. It is de ned as:

$$g = \frac{P_0}{P_f} \quad , \tag{12.40}$$

where P_0 is the same as in Eq. (12.39), while P_f is the active power measured at the antenna input terminals, and referred to as the *antenna feeding power*. Because of energy conservation, P_f equals the sum of radiated power, P_i, and power loss in the antenna, P_d, due to nite conductivity of the materials it is made of. Therefore, Eqs. (12.39) and (12.40) are related as:

$$g = \frac{P_0}{P_i + P_d} = d\,\frac{P_i}{P_i + P_d} = d \quad , \tag{12.41}$$

where the ratio:

$$= \frac{P_i}{P_i + P_d} \quad , \tag{12.42}$$

is called the antenna *e ciency*.

Another quantity of great technical interest is the so-called *radiation resistance* of the antenna, R_i. It is de ned *assuming* that the relationship between the radiated power, P_i, and the current at the antenna input terminals, I, is formally identical to Joule's law which yields the amount of power dissipated in a resistor, i.e.:

$$P_i = R_i\,\frac{|I|^2}{2} \quad . \tag{12.43}$$

The resistance de ned in this way coincides with the real part of the antenna input impedance if the power losses in the antenna are negligible, com-

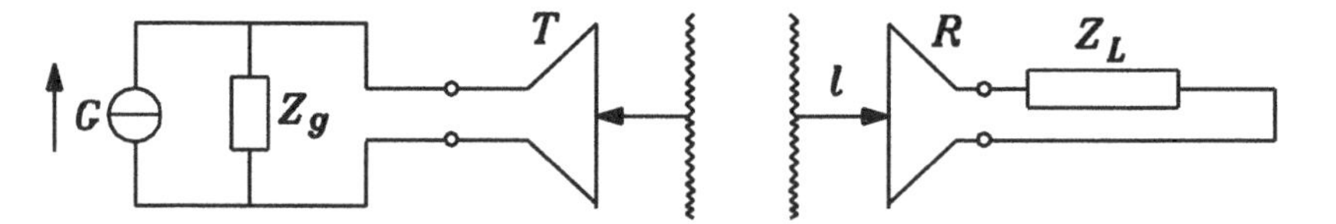

Figure 12.7 *A system consisting of two antennas, a transmitter and a receiver.*

pared with radiated power ($\simeq 1$). The imaginary part of the antenna input impedance is also important in practice, for load matching (see Chapter 8) at the antenna terminals. It will be one of the subjects of the next section.

12.6 Behavior of receiving antennas. Reciprocity

We saw on many occasions that, when we study the transmitting function, we model an antenna as an imposed current density. This enables us to avoid complications we would encounter if we modeled an antenna as an electrical two-port network, connected to a generator on one side, and delivering a fraction of the generator power to the other port. If we adopted this approach, the imposed quantity should be the generator short-circuit current, not the current owing in the antenna.

We pass now to the antenna *receiving* function. One might think that it could also be modeled in terms of an imposed current, which absorbs power from a eld impinging on it. This would not be the right approach. Indeed, the current owing through a receiving antenna is neither known *a priori*, nor independent of the impinging eld. A receiving antenna *must* therefore be modeled as a two-port network, which delivers to a load (the receiver) some power, depending on the impinging eld and also on the load impedance.

Let us consider a system consisting of two antennas. One of them is connected to a generator, the other one to a load, as shown schematically in Figure 12.7. The symbol G stands for the load-independent current of the ideal generator which is part of the transmitter model. Suppose that the medium surrounding the antennas is linear, isotropic, lossless, and homogeneous throughout the whole 3-D space. The only inhomogeneities are the antennas themselves, and their distance ℓ is much larger than their size and than the wavelength . Under all these assumptions, it is reasonable to state that the eld in the neighborhood of the transmitting antenna T is not affected by the presence of the receiving antenna R. So, it remains the same as the eld we calculated under the assumptions of Section 12.5.

On the other side, for a *given* receiving antenna R, the active power delivered by R to the load Z_L, which we refer to as the *received power*, depends on:

the load impedance, Z_L;

the direction towards which R is aiming;

the polarization of the eld which impinges on R.

Let P_R be the maximum value that the received power can reach, when we act on all the three variables we just mentioned, i.e., when:

the load impedance is matched to the antenna impedance;‡

R aims at the direction which maximizes the power received from a given transmitter;

polarization of the received signal is optimal, in a sense to be clari ed soon.

Under these assumptions, we de ne the *e ective area* of the antenna R as:

$$A_e = \frac{P_R}{\frac{\mathbf{E} \quad \mathbf{E}}{2}} , \tag{12.44}$$

where $\mathbf{E}$ is the electric eld radiated by the transmitting antenna, measured *in the absence* of the receiving antenna R in the same place where R is to be located. The last remark is intended to stress that the presence of R modi es the eld, locally, in a rather unpredictable way, so that any eld measurement in the presence of R is unreliable. On the other hand, as ℓ , the far eld $\mathbf{E}$ of the transmitting antenna looks like a uniform plane wave, and so $\mathbf{E} \quad \mathbf{E}$, in the absence of R, is essentially constant over the whole region to be occupied by R. Under the same assumptions, it is sometimes convenient to use another de nition, namely, the *e ective height* of the antenna R. It is de ned as:

$$h_e = \frac{V}{|\mathbf{E}|} , \tag{12.45}$$

where V is the modulus of the open-circuit voltage at the antenna terminals, while $\mathbf{E}$ has the same meaning as in the previous de nition. It is left as an exercise to the reader to show that:

$$A_e = h_e^2 \frac{\quad}{4R} , \tag{12.46}$$

where R is the real part of the impedance at the antenna terminals. As we said before, in this section we deal only with *isotropic media.* Therefore, the behavior of a receiving antenna is not independent of the behavior of the same antenna when used as a transmitting one, as a consequence of the reciprocity theorem of Section 3.4. To nd the link between the two functions, we must deal with what remains between the terminals of the two antennas, and model it as a two-port network. It is linear and passive, because of the assumptions we made about the antennas and about the surrounding medium. Suppose that both ports are connected to their impedance-matched loads. As we said

‡ We will see soon that calculations in this section are based on assuming *power matching*, i.e., a load impedance equal to the complex conjugate of the equivalent generator internal impedance. It is well known (see Chapter 8) that (except for a pure resistance) this condition can be satis ed only at a single frequency. So, calculations in this chapter might appear not to be consistent with antennas transmitting or receiving signals with a nite bandwidth. To avoid inconsistencies, all the quantities and the relationships introduced in this section must refer to the *carrier frequency*, and the spectra of modulated signals must be narrow enough, so that the impedances involved in our calculations remain essentially constant over the whole bandwidth.

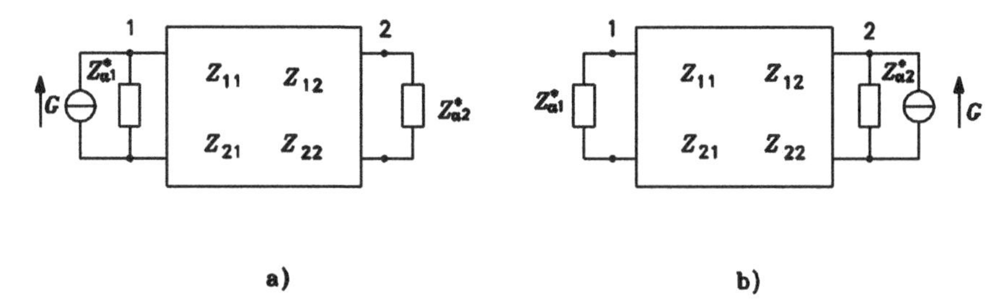

Figure 12.8 *A circuit model for the system shown in Figure 12.7.*

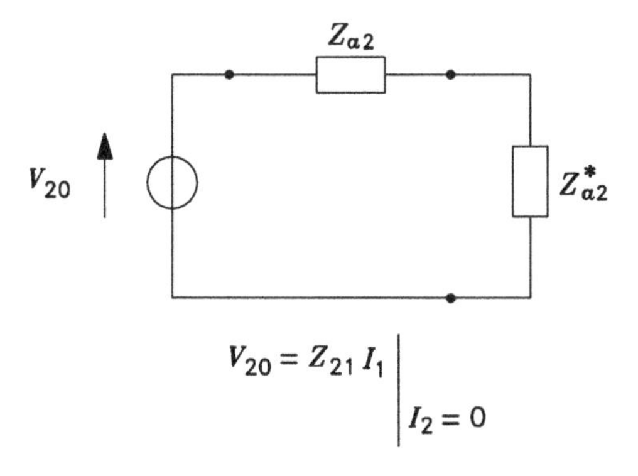

Figure 12.9 *The equivalent-generator circuit.*

previously, the eld near the transmitter is not a ected by the presence of the receiver; hence, the two matched loads are equal to Z_{a1} and Z_{a2}, respectively, Z_{a1} and Z_{a2} being the input impedances measured at the terminals of each antenna in the *absence* of the other one.

Consider now the two schemes shown in Figure 12.8, where, like in Figure 12.7, transmitters are modeled as current generators. Powers through ports 1 and 2 can be evaluated using Thevenin's equivalent-generator theorem (Figure 12.9), and taking into account that, again because of the very large distance between R and T, the input impedance at port 1 does not change when the load at port 2 is an open circuit, instead of Z_{a2}.

In the case a) of Figure 12.8, if we call $P_R^{(a)}$ (received power) and $P_T^{(a)}$ (transmitted power) the active powers owing through ports 2 and 1, respectively, we nd:

$$P_R^{(a)} = \frac{1}{8} \frac{|Z_{21}|^2 \, |I_1|^2}{\mathrm{Re}(Z_{a2})} \quad , \qquad P_T^{(a)} = \frac{1}{2} \, \mathrm{Re}(Z_{a1}) \, |I_1|^2 \quad ,$$

from which we get:

$$\frac{P_R^{(a)}}{P_T^{(a)}} = \frac{1}{4} \, |Z_{21}|^2 \, \frac{1}{\mathrm{Re}(Z_{a1}) \, \mathrm{Re}(Z_{a2})} \quad . \tag{12.47}$$

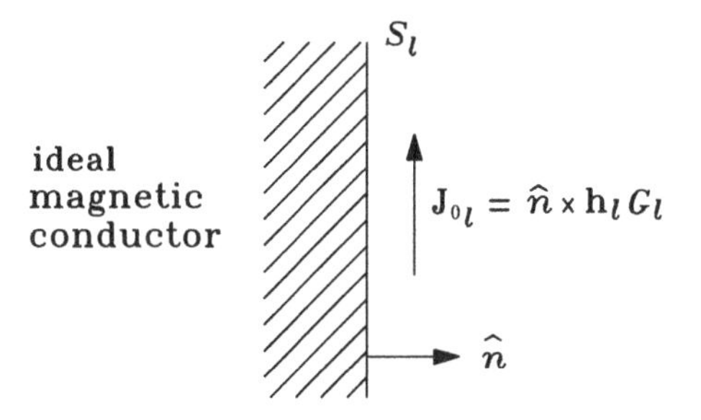

Figure 12.10 *Representation of an ideal current generator in terms of elds.*

In the case b) of the same gure, similarly we get:

$$\frac{P_R^{(b)}}{P_T^{(b)}} = \frac{1}{4} |Z_{12}|^2 \frac{1}{\mathrm{Re}(Z_{a1}) \, \mathrm{Re}(Z_{a2})} \quad . \tag{12.48}$$

To reach the conclusion that the two antennas are interchangeable, what we must still prove is that $Z_{12} = Z_{21}$. This equality could be borrowed from network theory, where it is usually taken as the de nition reciprocity. However, it is interesting to show that we are able to derive it from properties of the elds, and that is what we will do next.

Suppose that the antenna 1 is connected to the port 1 through a *single-mode* waveguide, or transmission line. This assumption is instrumental to de ne an impedance in an unambiguous way. In the transverse eld $\{\mathbf{E}_t, \mathbf{H}_t\}$, let us write two factors (see Chapter 8) which are mode functions, $\{\mathbf{e}, \mathbf{h}\}$, normalized over the cross-section S_1 of the line itself, i.e.:

$$\mathbf{E}_{t1} = V_1 \, \mathbf{e}_1 \quad ,$$

$$\mathbf{H}_{t1} = I_1 \, \mathbf{h}_1 \quad ,$$

$$\int_{S_1} \mathbf{e}_1 \quad \mathbf{h}_1 \;\; \hat{n} \, dS_1 = 1 \quad . \tag{12.49}$$

Note that $\mathbf{h}_1$, not $\mathbf{h}_1$, appears in the normalization condition. This will be useful in the following steps; in any case, it is not a restriction, given that $\mathbf{e}_1$ and $\mathbf{h}_1$ can be supposed to be real in Eqs. (12.49). The unit vector $\hat{n}$, normal to S, is oriented towards the antenna. Similar assumptions can be made, of course, on the line connected to the antenna 2.

Let us model the two ideal current generators, connected to these lines, in terms of two imposed current densities:

$$\mathbf{J}_{01} = \hat{n} \quad \mathbf{h}_1 \, G_1 \quad , \qquad \mathbf{J}_{02} = \hat{n} \quad \mathbf{h}_2 \, G_2 \quad , \tag{12.50}$$

on two cross-sections (S_1, S_2) which are immediately downstream with respect to two open circuits, i.e., two plane ideal magnetic conductors (see Figure 12.10). Now let us proceed and calculate Z_{12} and Z_{21}.

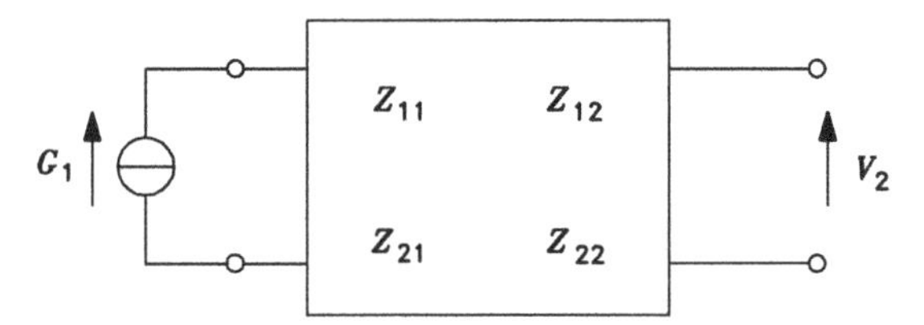

Figure 12.11 *Scheme of the electrical network used to calculate the antenna mutual impedances.*

From the scheme shown in Figure 12.11 we get:

$$Z_{21} = \frac{V_2}{G_1} = \frac{V_2 G_2}{G_1 G_2} \int_{S_2} \mathbf{e}_2 \quad \mathbf{h}_2 \ \hat{n}\, dS_2 \quad , \tag{12.51}$$

where G_2 is, clearly, an arbitrary quantity, and therefore can be chosen equal to the quantity indicated with the same symbol in Eq. (12.50). Hence, we have:

$$Z_{21} = \frac{1}{G_1 G_2} \int_{S_2} (V_2\, \mathbf{e}_2) \quad (\hat{n} \quad G_2\, \mathbf{h}_2)\, dS_2 = \frac{1}{G_1 G_2} \int_{S_2} \mathbf{E}_{t2} \quad \mathbf{J}_{02}\, dS_2 \,, \tag{12.52}$$

where $\mathbf{E}_{t2}$ is the electric eld on section 2 due to the antenna 1, when only the antenna 1 is fed by its generator. Repeating the calculation for Z_{12}, and applying Lorentz's reciprocity theorem (see Section 3.4) to the integral over S_1, we get:

$$Z_{12} = \frac{1}{G_1 G_2} \int_{S_1} \mathbf{E}_{t1} \quad \mathbf{J}_{01}\, dS_1 = \frac{1}{G_1 G_2} \int_{S_2} \mathbf{E}_{t2} \quad \mathbf{J}_{02}\, dS_2 = Z_{21} \,, \tag{12.53}$$

which is exactly what we intended to prove.

An important corollary of antenna reciprocity is that, at any given frequency, *power gain and e ective area are proportional.* Indeed, let us express the power received by antenna 2 using the de nition of e ective area, Eq. (12.44), and then multiply and divide by 4 ℓ^2, where ℓ is the distance between the two antennas. We get:

$$P_R = A_{\mathrm{e}} \ _2 \left\{ \frac{\mathbf{E} \quad \mathbf{E}}{2} \ 4 \ \ell^2 \right\} \frac{1}{4 \ \ell^2} \quad . \tag{12.54}$$

The quantity in parenthesis in Eq. (12.54) equals the power which should be radiated by an isotropic antenna, in place of antenna 1, to get the same radiation intensity as that given by antenna 1 where antenna 2 is located. Thanks to Eq. (12.40) it can be expressed as:

$$P_0 = g_1 \ P_{f1} = g_1 \ P_T \quad , \tag{12.55}$$

where g_1 is the power gain of the antenna 1, and $P_{f1} = P_T$ is the power feeding the antenna 1. Combining Eqs. (12.54) and (12.55), we get:

$$P_R = \frac{1}{4 \quad \ell^2} \; A_{\mathrm{e}} \;\; {}_2 \, g_1 \, P_T \quad . \qquad (12.56)$$

Interchanging now the roles of the two antennas, and remembering that Eq. (12.53) assures that the ratio between received and transmitted powers remain the same, one gets $A_{\mathrm{e}} \;\; {}_2 \, g_1 \;\; = \;\; A_{\mathrm{e}} \;\; {}_1 \, g_2$. As these two antennas are completely arbitrary, it is therefore:

$$A_{\mathrm{e}} \quad = g \qquad \text{constant} \quad , \qquad (12.57)$$

as long as *the frequency remains constant.* The ratio Eq. (12.47) can be evaluated for any speci c example. As shown in the following section, one nds:

$$A_{\mathrm{e}} \quad = g \, \frac{\;\;{}^2}{4} \quad , \qquad (12.58)$$

where is the wavelength in free space.

This result allows us to express Eq. (12.56) in the following form, which is often referred to as the *Friis formula*, and is the starting point in the design of radio links:

$$P_R = g_2 (g_1 \, P_T) \left(\frac{}{4 \quad \ell} \right)^2 \quad . \qquad (12.59)$$

Let us emphasize, once again, that this result was derived under the following assumptions: each antenna is in the *far eld* of the other (ℓ); the medium is *homogeneous* in the whole 3-D space (there are no re ections from other objects, neither guided waves, nor "multiple paths"), and is also *lossless.*

The factor ($/4 \quad \ell)^2$ is called *free-path attenuation*, to underline these assumptions. Being proportional to $1/\ell^2$, it reminds us that, for any source, waves in the far eld are spherical and therefore their wavefront surfaces increase in proportion to ℓ^2.

The factor $(g_1 \, P_T)$, whose physical dimensions are those of power, has been written in parenthesis to underline that it summarizes all the roles played by the transmitting side, for what refers to the received power. In other words, a receiver cannot discriminate between a case where the transmitted power is large, but the transmitting antenna gain is small, and a case where the transmitted power is smaller, but the transmitting antenna gain is larger so that the products of these factors remain the same. For this reason, the factor $g_1 \, P_T$ is referred to as the *e ective radiated power* (ERP) of the transmitting plant. It is in current use in international standards and speci cations relating to services based on radio waves. For example, receivers on board of vehicles like ships or aircraft are designed starting from a given ERP for the transmitters which are located either on the ground or on satellites.

A realistic model of a radio receiver must account for the fact that the receiving antenna captures noise, in addition to the wanted signal. Indeed, the main speci cation for a radio communication system is always to insure a signal-to-noise ratio above a given minimum level at the antenna terminals. *Independently of the physical source of the noise* which is actually captured

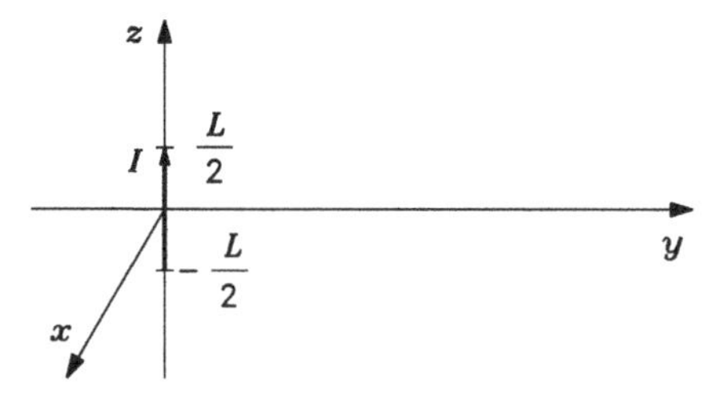

Figure 12.12 *A short electric dipole.*

by the antenna, it is in current use to characterize this noise by means of the following formula, which is, as well known, the standard formula for thermal (Johnson-Nyquist) noise generated by a matched resistor:

$$N = k\,T_a\,B \quad , \tag{12.60}$$

where N is the time-averaged noise power measured at the receiver input terminals, $k = 1.38 \times 10^{-23}$ J/°K is Boltzmann's constant, and B is the receiver bandwidth. Since N is determined experimentally, Eq. (12.60) can be taken as the *definition* of the absolute temperature T_a, which is referred to as the *noise temperature* of the antenna. Typical values of this noise temperature in terrestrial radio links are about 300 K, i.e., are comparable to ambient temperatures. This happens because, in those links, the main contribution to the received noise is black-body radiation from the earth, as the antenna main lobe is tangent to the ground. A black body is indeed quite similar to a matched resistor in its physical essence, whence the result. On the other hand, antenna noise temperatures can decrease substantially when antennas are oriented towards the sky, in directions where there are no radiation sources near the earth. This is the typical situation in satellite communication systems. However, Nobel Laureates Penzias and Wilson discovered, in the early '60's, that the noise temperature experiences a lower limit, even when an antenna is oriented towards the deepest space. This lower limit, due to background black-body radiation traveling throughout the universe, equals 4 K.

12.7 Examples

Let us compute now the quantities defined in the previous sections for some antennas of simple shape and of practical interest, in a lossless medium. Several calculations are left to the reader as exercises. On the contrary, the first example (short electric dipole) goes into the details of how one evaluates the effective area, in order to find, on this occasion, the proportionality constant in Eq. (12.58).

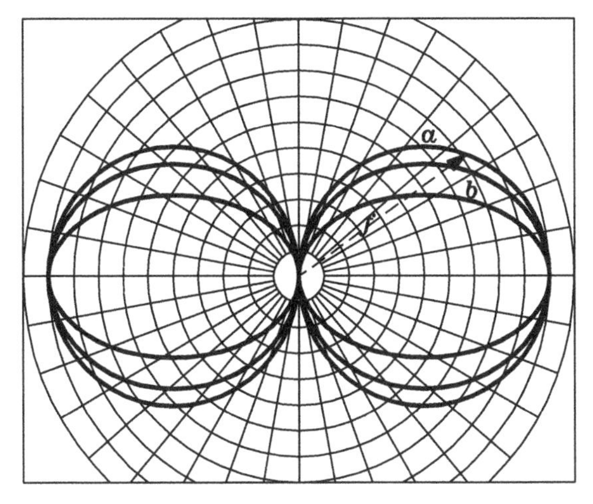

Figure 12.13 *Radiation diagrams for three thin-wire linear antennas.*

12.7.1 Short electric dipole

In a straight conductor of length L , the current I can be assumed to be constant for $|z|$ $L/2$ (see Figure 12.12). The equivalent moment and the far eld are well modeled by what we saw in Section 12.4, setting $\mathcal{M} = IL$. The following results are easy to derive.

The *radiation intensity* and its *normalized* version are, respectively:

$$I_r = \frac{\ }{8} \left(\frac{L}{\ } \right)^2 |I|^2 \sin^2 \vartheta \quad , i = \sin^2 \vartheta \quad , \tag{12.61}$$

and so the *radiation function* is:

$$f = |\sin \vartheta| \quad . \tag{12.62}$$

Hence, a short dipole is omnidirectional in the (x, y) plane. In any plane passing through the z axis the radiation diagram is that shown in Figure 12.13 (curve a), with null directions at $\vartheta = 0$ and $\vartheta =$.

The *directivity* is:

$$d = \frac{4}{\int_0 d\vartheta \int_0^2 d\varphi \sin^2 \vartheta \sin \vartheta} = \frac{3}{2} \quad . \tag{12.63}$$

As for the *radiation resistance*, the radiated power is given by Eq. (12.25) (with Re( ) = 0 as the medium is lossless), so we get:

$$R_i = \frac{2}{3} \left(\frac{L}{\ } \right)^2 \quad . \tag{12.64}$$

To nd the e *ective area*, let us compute the power delivered to a matched load. The open-circuit voltage between the two ends of a segment of length L

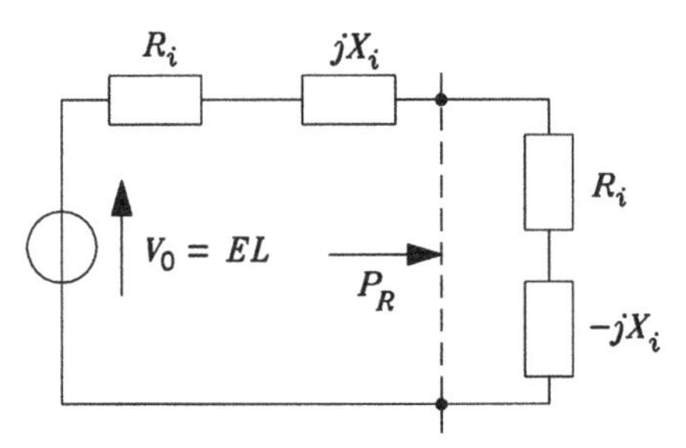

Figure 12.14 *Equivalent circuit used to evaluate the received power.*

( ) is obviously maximum, for a given eld strength, when the electric eld is linearly polarized and parallel to the segment itself. In this case we have $V_0 = EL$. This is then the open-circuit voltage of the generator equivalent to the receiving antenna (Figure 12.14). Its internal impedance has a real part R_i given by Eq. (12.64). Its imaginary part is not relevant to the present problem, since the load must be power-matched and then the reactances of the generator and of the load cancel each other.

Thus, we get:

$$P_R = \frac{|V_0|^2}{8R_i} = \frac{3}{16} |E|^2 \frac{\ ^2}{\ } \quad . \tag{12.65}$$

Then, from the de nition in Eq. (12.44) of e ective area, it is:

$$A_{\rm e} \quad = \frac{3}{8} \ ^2 \quad . \tag{12.66}$$

Comparing it with Eq. (12.63) we get:

$$A_{\rm e} \quad = d \, \frac{\ ^2}{4} \quad , \tag{12.67}$$

a result which was anticipated by Eq. (12.58). Note that the conclusion relies on the fact that we used a lossless model for the short dipole, so its directivity gain equals its power gain.

12.7.2 Thin linear antennas

When the imposed current density $\mathbf{J}_0$ has the same direction over the whole region where it is de ned, then Eq. (12.9) shows that the equivalent moment $\boldsymbol{\mathcal{M}}$ and the magnetic vector potential $\mathbf{A}$ are also parallel to this direction. If we take it as the z axis of a spherical reference frame, then Eqs. (12.17) and (12.20) imply that in the far eld the vector $\mathbf{H}$ is parallel to φ and the vector $\mathbf{E}$ is parallel to ϑ, independently of the antenna details.

For an antenna which consists of a single thin wire, of length $2L$ (Figure 12.15), fed by an ideal current generator at its center, then, on the ground

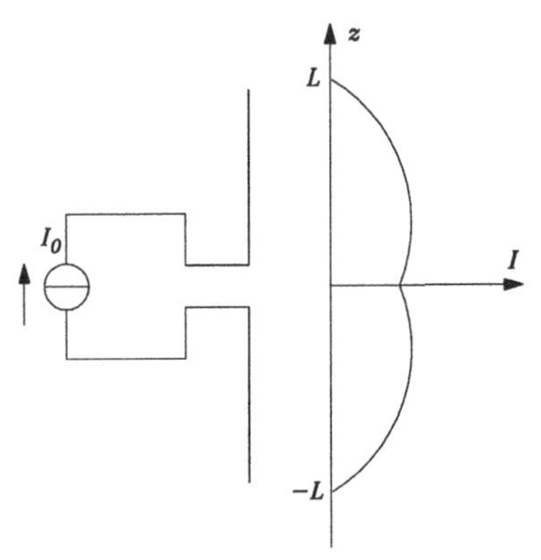

Figure 12.15 *Thin-wire linear antenna, of a generic length.*

of transmission line theory (Chapter 8), we assume that the current varies as:

$$I = I_M \sin\Big[\ (L \quad |z|)\Big] \quad , \tag{12.68}$$

where $= \omega \sqrt{\ _0 \ _0}$, and I_M is the largest value of the current in the antenna, related to the generator current as $I_M = I_0/\sin \ L$. One of the most interesting cases is when $I_M = I_0$, i.e., $L = \ /4$ (the so-called *half-wavelength* resonant dipole). In the following we will list the results for an arbitrary length L as well as those for $L = \ /4$.

The *radiation intensity* is:[§]

$$I_r = \frac{|I_M|^2}{8 \ ^2} \left[\frac{\cos(\ L \cos\vartheta) \quad \cos \ L}{\sin\vartheta} \right]^2 \qquad (L \text{ generic}), \tag{12.69}$$

$$I_{\ /2} = \frac{|I_0|^2}{8 \ ^2} \left[\frac{\cos\left(\frac{\ \cos\vartheta}{2} \right)}{\sin\vartheta} \right]^2 \qquad (L = \ /4). \tag{12.70}$$

The radiation intensity is simple to normalize if we take as the reference its value for $\vartheta = \ /2$, in spite of this not being its maximum value for any length L (it is indeed the absolute maximum for $L = \ /4$, but not for a generic L). We get:

$$i = \left[\frac{\cos(\ L \cos\vartheta) \quad \cos \ L}{(1 \quad \cos \ L) \sin\vartheta} \right]^2 \qquad (L \text{ generic}), \tag{12.71}$$

§ When proving, as an exercise, this result, the reader may nd it useful to exploit the following identity:

$$\int e^{ax} \sin(bx + c)\, dx = \frac{e^{ax}}{a^2 + b^2} \Big[a \sin(bx + c) \quad b \cos(bx + c) \Big] + D \quad ,$$

where D is an arbitrary constant.

$$i_{\ /2} \quad = \quad \frac{\cos^2\left(\dfrac{\ \cos\vartheta}{2}\right)}{\sin^2\vartheta} \qquad (L = \ /4)\,. \tag{12.72}$$

Hence, *any* straight wire fed by a steady-state time-harmonic current is omnidirectional in the plane orthogonal to the wire. Radiation diagrams ($f = \sqrt{i}$) in a plane passing through the z axis are shown in Figure 12.13, where curve a), as we said before, refers to a short dipole, and curve b) is for $L = \ /4$. For $L < \ /4$ the diagram falls between a) and b). Hence, under the viewpoint of directivity the length of a thin wire does not play a major role. Radiation resistance, on the contrary, will be shown to depend very strongly on the wire length.

To compute the *directivity* and the *radiation resistance* of a thin-wire linear antenna, one must evaluate:

$$\int_0 \frac{[\cos(\ L\cos\vartheta) \quad \cos\ L]^2}{\sin\vartheta}\,d\vartheta \quad . \tag{12.73}$$

With a change of variable $u = \cos\vartheta$, after some algebra this integral can be reduced to a combination of $\log(2\ \ L)$ and the following functions, whose tables and plots can be found in the literature (Kraus, 1950; Ramo *et al.*, 1994):

$$Si(x) \quad = \quad \int_0^x \frac{\sin x}{x}\,dx \quad ,$$

$$Ci(x) \quad = \quad \int_x^\infty \frac{\cos x}{x}\,dx \quad . \tag{12.74}$$

In particular, for $2L = \ /2$ (, as usual, is the intrinsic impedance of the medium which surrounds the antenna) one gets:

$$R_{i(\ /2)} = 2.44\,\frac{}{4} \quad , \quad d_{\ /2} = 1.64 \quad . \tag{12.75}$$

What is in current use is to refer the radiation resistance of a linear antenna, R_i, to the *maximum current*, I_M, not to the generator current, I_0. Clearly, this distinction is meaningless for a half-wave dipole, but in the generic case the di erence can be quite substantial, since the two currents are related as:

$$I_0 = I_M \cos\ \ x \quad , \tag{12.76}$$

x being the distance between the antenna input terminals and the nearest current maximum. What must be independent of how one de nes the radiation resistance is, obviously, the radiated power; therefore, the radiation resistance referred to the input current is:

$$R_0 = \frac{R_i}{\cos^2\ \ x} \quad . \tag{12.77}$$

So, an antenna whose length equals the wavelength might look attractive at rst sight, but in reality it is a poor choice, because its input impedance is very large.

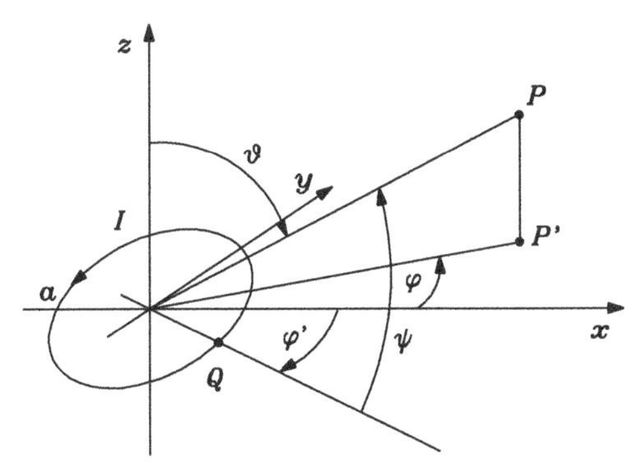

Figure 12.16 *Circular-loop thin-wire antenna.*

At those frequencies at which the ground (or the sea water) can be looked at as an almost ideal conductor, the results which we found now for an antenna of length $2L$, fed at its center and placed in an homogeneous medium, apply, thanks to symmetries discussed in Section 3.9, to a thin vertical wire of length L, fed at its lower end and close to the ground (or water), which provides its mirror image. This antenna is sometimes referred to as the *Marconi aerial.*

12.7.3 Circular loop

When the imposed current density $\mathbf{J}_0$ exhibits rotational symmetry around an axis, it is convenient to use a spherical coordinate system whose polar axis, z, is the symmetry axis. Then, Eq. (12.9) shows that the equivalent moment $\boldsymbol{\mathcal{M}}$ (and so $\mathbf{A}$ too) has to be parallel to the $\hat{\varphi}$ unit-vector. It follows, through Eqs. (12.17) and (12.20), that in the far eld $\mathbf{H}$ is parallel to $\hat{\vartheta}$ and $\mathbf{E}$ is parallel to $\hat{\varphi}$. So, the directions of $\mathbf{E}$ and $\mathbf{H}$ are interchanged with respect to the previous example. This did not occur by accident. Indeed, it can be shown that the eld generated by a wire loop and the eld generated by a short electric dipole satisfy the *duality theorem* of Section 4.7. Hence, similarly to what happens in magnetostatics (i.e., analogous to the so-called *Ampere magnetic sheets*), even in electrodynamics an electric-current loop may be considered as equivalent to a short *magnetic dipole.*

For a circular loop of radius a, assuming that the current I is constant around the loop, in a reference frame as in Figure 12.16 we have:

$$\cos \quad = \sin\vartheta \, \cos(\varphi \quad \varphi') \quad . \tag{12.78}$$

If we insert this into the integral in Eq. (12.9), then, due to spatial symmetry (see Figure 12.16 again), what actually contributes to the equivalent moment in P is not $I(Q)$ but $I(Q)\cos(\varphi \quad \varphi')$. Therefore, it is:

$$\boldsymbol{\mathcal{M}} = \hat{\varphi}\, I a \int_0^{2} e^{jka \sin\vartheta \cos(\varphi \quad \varphi')} \cos(\varphi \quad \varphi')\, d\varphi' \quad . \tag{12.79}$$

To evaluate this integral, set $\cos(\varphi \quad \varphi') = [e^{j(\varphi \quad \varphi')} + e^{\quad j(\varphi \quad \varphi')}]/2$, and make use of the following identity, which stems from the so-called Bessel-Fourier decomposition (Abramowitz and Stegun, Eds., 1965):

$$\int_0^{2} e^{j(x \sin\varphi \quad m\varphi)}\, d\varphi = 2 \quad J_m(x) \qquad (m \text{ integer}) \quad ,$$

where J_m is Bessel's function of the rst kind of order m. Recalling the identity (see Appendix D) $J_{\ 1}(x) = \quad J_1(x)$, one nally gets:

$$\boldsymbol{\mathcal{M}} = j\,\hat{\varphi} I a\, 2 \quad J_1(ka \sin\vartheta) \quad . \tag{12.80}$$

From Eq. (12.80) one may proceed without any further approximation. The reader may treat this as an exercise. The following results, on the other hand, refer to the case (a very common one, in the practical applications of loop antennas) $ka \quad 1$, where the approximation:

$$J_1(ka \sin\vartheta) \simeq (ka/2) \sin\vartheta \quad , \tag{12.81}$$

applies. Within the limits of this approximation, the radiation intensity is:

$$I_r = \frac{}{32}\,(ka)^4\,|I|^2 \sin^2\vartheta = \frac{|I|^2}{2}\left(\frac{a}{}\right)^4 \sin^2\vartheta \quad . \tag{12.82}$$

The corresponding normalized radiation intensity is

$$i = \sin^2\vartheta \quad , \tag{12.83}$$

which coincides with that of the short electric dipole (the rst example in this section) and yields, consequently, the same directivity, $d = 3/2$. This stems, of course, from duality. The loop is equivalent, in the far eld, to a *short magnetic dipole* whose moment equals:

$$\boldsymbol{\mathcal{N}} = j\omega \quad I \quad a^2 \hat{a}_z \qquad (\text{V} \quad \text{m}) \quad . \tag{12.84}$$

Incidentally, it is worth writing in detail the far eld of an imposed magnetic current which results in an equivalent moment:

$$\boldsymbol{\mathcal{N}} = \int_V \mathbf{M}_0(r')\, e^{\quad \hat{\mathbf{r}}\ \mathbf{r}'}\, dV_{\mathbf{r}'} \quad , \tag{12.85}$$

and thus in an electric vector potential $\mathbf{F} = (\ _c/4\ )\boldsymbol{\mathcal{N}}(e^{\quad r}/r)$. By duality, from Eqs. (12.17) and (12.20) we get:

$$\mathbf{E} = \frac{}{4}\,\hat{r} \quad \boldsymbol{\mathcal{N}}\,\frac{e^{\quad r}}{r} \quad , \tag{12.86}$$

$$\mathbf{H} = \frac{j\omega}{4}\,\hat{r} \quad \boldsymbol{\mathcal{N}} \quad \hat{r}\,\frac{e^{\quad r}}{r} \quad . \tag{12.87}$$

Coming back to the loop of Figure 12.16, its e ective area is equal to Eq. (12.66), so it has no relation whatsoever with the geometrical area of the loop itself. Its radiation resistance is:

$$R_i = \frac{}{6}\left(\frac{2 \quad a}{}\right)^4 \quad , \tag{12.88}$$

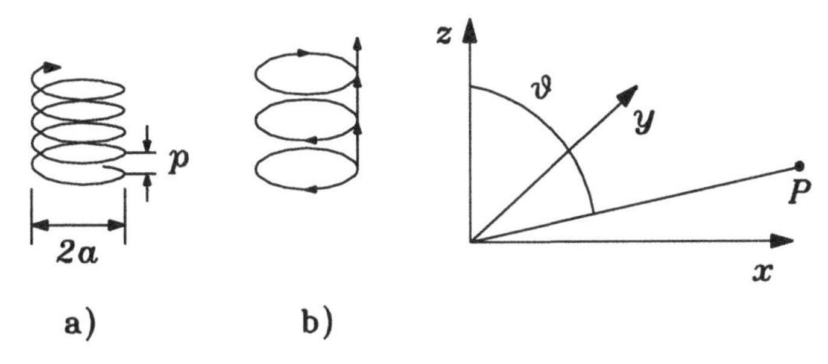

Figure 12.17 *A helical antenna and its simpli ed models.*

which, for the same conductor lengths ($2\ \ a = L$) and at the same wavelength , may di er considerably from that of the electric dipole, Eq. (12.64).

12.7.4 Helical antennas

A thin wire wrapped as a cylindrical helix, of radius a and pitch p, (Figure 12.17a), is an interesting antenna, whose properties depend strongly on the pitch angle, $\ = \arctan(p/2\ \ a)$, and on the ratio between its geometrical parameters and the wavelength . For p and a , the phase of the current is essentially constant along the whole helix. In this case, the previous examples which dealt with the short electric dipole and with the loop provide results which can be taken, within a reasonable approximation, as starting points, if we decompose the helix as sketched in Figure 12.17b. By superposition, we see that the radiation intensity is maximum in the plane orthogonal to the z axis ($\vartheta = \ /2$). The radiation function does not di er much from Eq. (12.62). The helix is then said to operate in the *transverse mode.*

On the contrary, for a, and if the pitch angle is not too large, the helix becomes similar to a circular radiating aperture, to be dealt with in Chapter 13. Its main radiation lobe is then along the z axis, and the helix is said to operate in the *longitudinal mode.* In between, as we may see from the de nition in Eq. (12.9), the features of the radiation function depend strongly on the phase delays experienced by the current as it ows along the conductor. Therefore, if all the other parameters are kept constant, they are very sensitive to the value of the pitch angle (Kraus, 1950).

12.7.5 Huygens source

Dealing with the previous example, we brie y touched upon the superposition of short electric and magnetic dipoles, which were parallel to each other. The reader may write, as an exercise, the eld due to two moments, say $\boldsymbol{\mathcal{M}}$ and $\boldsymbol{\mathcal{N}}$, whose directions are completely arbitrary. This can be done writing $\boldsymbol{\mathcal{M}}$ and $\boldsymbol{\mathcal{N}}$ by components, and recalling the relationships between unit vectors of an orthogonal cartesian reference frame $(\hat{a}_1, \hat{a}_2, \hat{a}_3)$, and those of a spherical

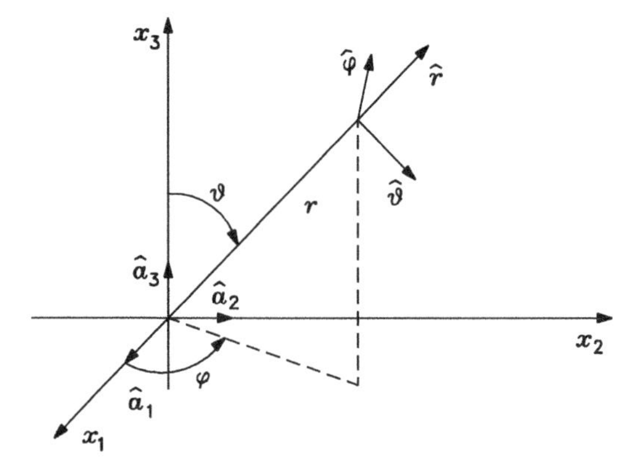

Figure 12.18 *The reference frame for the Huygens source.*

frame (Figure 12.18):

$$\begin{aligned}\hat{a}_1 &= \sin\vartheta\,\cos\varphi\,\hat{r} + \cos\vartheta\,\cos\varphi\,\hat{\vartheta} \quad \sin\varphi\,\hat{\varphi} \quad , \\ \hat{a}_2 &= \sin\vartheta\,\sin\varphi\,\hat{r} + \cos\vartheta\,\sin\varphi\,\hat{\vartheta} + \cos\varphi\,\hat{\varphi} \quad , \\ \hat{a}_3 &= \cos\vartheta\,\hat{r} \quad \sin\vartheta\,\hat{\vartheta} \quad . \end{aligned} \tag{12.89}$$

The most interesting case is probably that where the electric and magnetic dipoles are *orthogonal* in space. The equivalence theorem of Section 3.5 suggests (and this point will be dealt with in more detail in the next chapter) that, when the two vectors $\boldsymbol{\mathcal{M}}$ and $\boldsymbol{\mathcal{N}}$ are orthogonal, and the ratio between their moduli takes a suitable value, their combination may represent very well a surface element illuminated by a known e.m. wave. In particular, if $|\boldsymbol{\mathcal{N}}| = \quad |\boldsymbol{\mathcal{M}}|$, where is the intrinsic impedance of the medium, then the illuminating eld $\{\mathbf{E}, \mathbf{H}\}$ is an element of a uniform plane wave. As we saw in Section 12.3, this is a good local approximation for the far eld of *any* primary source. To express this fact, the combination:

$$\boldsymbol{\mathcal{N}} = \mathcal{N}\,\hat{\imath} \quad , \qquad \boldsymbol{\mathcal{M}} = \frac{\mathcal{N}}{\quad}\,\hat{\jmath} \quad , \tag{12.90}$$

(where $\hat{\imath}, \hat{\jmath}$ are a generic pair of real orthogonal unit vectors) is called the *Huygens source.*

Let us use now, in Figure 12.18, a reference frame such that $\hat{\imath} = \hat{a}_1$, $\hat{\jmath} = \hat{a}_2$ (see Figure 12.19). From the previous results, we nd that the far eld is:

$$\mathbf{E} = j\,\frac{\mathcal{N}}{2}\,\frac{e^{\ j\ r}}{r}\,(1 + \cos\vartheta)\,(\sin\varphi\,\hat{\vartheta} \quad \cos\varphi\,\hat{\varphi}) \quad . \tag{12.91}$$

The corresponding radiation intensity is:

$$I_r = \frac{|\mathcal{N}|^2}{4\ ^2}\,(1 + \cos\vartheta)^2 = \frac{|\mathcal{N}|^2}{^2}\,\cos^4\frac{\vartheta}{2} \quad . \tag{12.92}$$

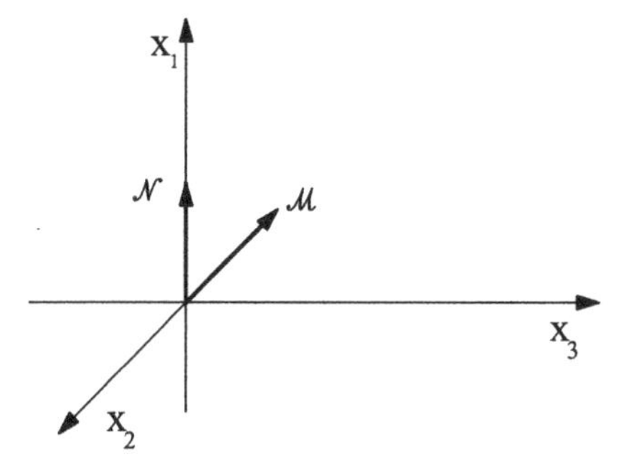

Figure 12.19 *The Huygens source.*

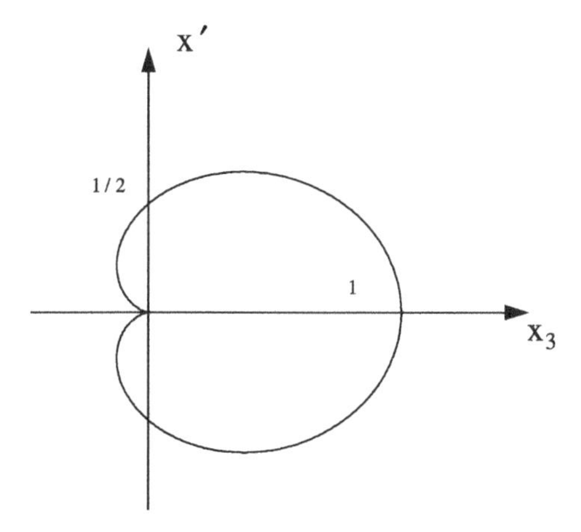

Figure 12.20 *Radiation diagram of a Huygens source.*

Once again, I_r is independent of the φ coordinate, i.e., this source is omnidirectional in the (x_1, x_2) plane, which is now the plane of the two dipoles.

The radiation function

$$f = \frac{1}{2}\,|1 + \cos\vartheta| = \cos^2\frac{\vartheta}{2}\quad, \tag{12.93}$$

is described, in the generic plane passing through the x_3 axes, by the radiation diagram shown in Figure 12.20, referred to in geometry as a *cardioid.*

The directivity, to be computed by the reader as an exercise, is $d = 3$. It should not cause any surprise if combining two elements, each of which, separately, has a directivity $d = 3/2$, yields a directivity larger than those of the individual elements. On the other hand, what is not guaranteed in general is that the directivities of the elements add up, as they do in this example. Directivity is derived from the eld through a nonlinear process, since the radiation intensity is proportional to the square of the electric eld modulus; so, directivities add up only in very special cases.

12.8 Antenna arrays

It was pointed out in Section 12.2 that the equivalent moment of a given source depends on the position of the source with respect to the origin of the reference frame. Then, let us take n identical sources, close to each other but mutually independent. Each of them can be described, if we aim at its far field, in terms of its equivalent moment (although, to be rigorous, this quantity has been defined when just one source is present, in a homogeneous medium which extends to infinity). As the equivalent moments of these n sources are not equal to each other, the total far field will not be, trivially, n times the field of a single source. The practical fallout of this statement comes from the fact that combinations of mass-produced modular elements are, in general, far less expensive than a single piece, made for a specific purpose. Furthermore, it will be shown in this section that even the design of a composite source which consists of identical and known elements can be simpler than that of a single source of an unknown kind. We will use a notation which refers to 3-D current distributions, but the content of this section applies equally well to surface currents or to linear ones.

Consider a set of n equal and equally oriented sources, i.e., n current distributions which differ from each other only by a complex constant factor, a_h ($h = 0, 1, \ldots, n-1$), which accounts for differences in amplitude ($|a_h|$) and in phase ($\angle a_h$) among the currents feeding the n antennas:

$$\mathbf{J}_0(Q_h) = \mathbf{J}_0(Q_0)\, a_h \quad , \qquad a_0 = 1 \ , \ h = 0, 1, \ldots, n-1 \quad , \tag{12.94}$$

where Q_h is the point which scans the region where the h-th current density flows. Their numbers go from 0 to $n-1$ for future simplicity. Suppose that each of these n regions can be superimposed onto the $h = 0$ region by translation along a given direction, which we call the *direction of alignment.* In such a case, the whole source is called a *linear array*. In the sketch of Figure 12.21, O is the origin of the reference frame and $\mathbf{r}'_h = Q_h - O$, $\boldsymbol{\ell}_h = Q_h - Q_0$ (therefore, $\boldsymbol{\ell}_0 = 0$).

In the far field (i.e., at distances much larger than *the size of the array*) field computation may start, as we saw in Section 12.3, from the equivalent moment of the whole array:

$$\begin{aligned}\boldsymbol{\mathcal{M}} &= \sum_{h=0}^{n-1} \int_{V_h} \mathbf{J}_0(Q_h)\, e^{\ \mathbf{r}'_h\ \hat{\mathbf{r}}}\, dV_h \\ &= \left(\sum_{h=0}^{n-1} a_h\, e^{\ \boldsymbol{\ell}_h\ \hat{r}} \right) \int_{V_0} \mathbf{J}_0(Q_0)\, e^{\ \mathbf{r}'_0\ \hat{\mathbf{r}}}\, dV_0 = M\, \boldsymbol{\mathcal{M}}_0 \quad ,\end{aligned} \tag{12.95}$$

where $\boldsymbol{\mathcal{M}}_0$ is the equivalent moment of the $h = 0$ antenna. The scalar function:

$$M = \sum_{h=0}^{n-1} a_h\, e^{\ \boldsymbol{\ell}_h\ \hat{r}} \quad , \tag{12.96}$$

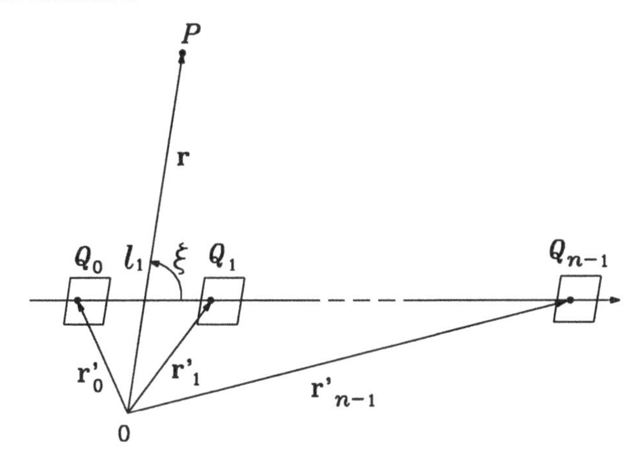

Figure 12.21 *The basic scheme of an antenna array.*

is called the *complex composition factor*; the quantity $|M|$, much more useful than M, is called simply the *composition factor*. We see that M depends on how the n individual sources are fed and on their mutual positions, but does not depend on the *type of antennas* that are used to form the array.

Eq. (12.96) means that an array can be analyzed or designed without using integrals, only with algebra. This point becomes even more evident as soon as we assume that the elements of an array are *equally spaced*, i.e.:

$$\boldsymbol{\ell}_h = h\,\boldsymbol{\ell} \qquad (h = 0, 1, \ldots, n \quad 1) \quad . \tag{12.97}$$

The vector $\boldsymbol{\ell}$ is called the *pitch* of the array. If we set:

$$e^{\quad \boldsymbol{\ell}\,\hat{r}} = w \quad , \tag{12.98}$$

the complex composition factor (12.96) can be rewritten as:

$$M = \sum_{h=0}^{n \quad 1} a_h\, w^h \quad , \tag{12.99}$$

which is a *polynomial* of degree $n \quad 1$ in the complex variable de ned by Eq. (12.98). If the medium is lossless, then ($\quad = j \quad$), and Eq. (12.98) implies $|w| = 1$. So, the complex composition factor is a polynomial whose independent variable is bound on the unit-radius circle of the complex plane. Incidentally, this entails us to borrow several results from neighboring elds, like control systems and lter theory, to design an antenna array.

We will deal now in more detail with two special problems. The rst one is mainly of theoretical importance, the second one has relevant practical fallouts.

12.8.1 Super-gain antennas

The "super-gain antenna" problem can be phrased as follows. We ask ourselves whether it is possible to design a source satisfying the following speci cations:

its lobe ratio is larger than an arbitrarily large pre-selected value, ;

the width of its main lobe is smaller than an arbitrarily small pre-selected value, ϑ.

The theory of polynomials tells us that, as the independent variable w spans the unit-radius circle on the complex plane, the modulus of a polynomial in w can indeed exhibit a major peak whose width is as small as one wishes, and secondary maxima which stay lower than a speci ed fraction of the absolute maximum. However, to reach such performances one must be free to make the degree of the polynomial as large as one wishes. Moreover, one can show that the tolerances on the coe cients of the polynomial (i.e., the margins within which the source performances remain good) become extremely narrow when becomes very large and ϑ becomes very small. In summary, to build a super-gain antenna array one needs an arbitrarily large number of elements, and must be able to comply with arbitrarily tight speci cations on modulus and phase of the current which feeds each element. This entails that super-gain antennas exist on paper but are not a technical reality.¶

12.8.2 Uniform arrays

An array is *uniform* if its n equally spaced elements are fed by currents whose n amplitudes are all equal, and whose phases shift from one element to the next by an amount which is independent of the particular pair of elements, i.e., if:

$$a_h = e^{\ jh} \qquad (h = 0, 1, \ldots, n \quad 1) \quad , \tag{12.100}$$

being the phase shift between the current in the h-th antenna and that in the $(h \quad 1)$-th one. Taking into account also Eq. (12.97) (equally spaced elements), and considering, in order to simplify calculations, a lossless medium, we get:

$$M = \sum_{h=0}^{n \quad 1} \left[e^{j(\ \boldsymbol{\ell} \ \hat{r} \quad)} \right]^h \quad . \tag{12.101}$$

This quantity is easy to evaluate as the sum of the rst n terms in a geometrical series, of ratio $q = \exp[j(\ \boldsymbol{\ell} \ \hat{r} \quad)] = \exp(j2v)$, so:

$$M = \frac{1 \quad q^n}{1 \quad q} = e^{j(n \quad 1)\,v} \, \frac{\sin nv}{\sin v} \quad , \tag{12.102}$$

where we set

$$v = \frac{1}{2} (\ \boldsymbol{\ell} \ \hat{r} \quad) = \frac{\ell}{\ } \cos \quad \frac{\ }{2} \quad , \tag{12.103}$$

¶ To avoid possible misunderstandings, let us mention that some authors (e.g., Ramo *et al.*, 1994) use the term "super-gain array" to designate simply an array whose gain is larger than that of a uniform array, the subject of the next subsection.

and is the angle between the vectors P O (P is the generic observation point) and $\boldsymbol{\ell}$ (which is in the alignment direction, see Figure 12.21). Eqs. (12.102) and (12.103) show that, for a uniform array, the e ect of composing several elements depends only on one angular coordinate of the point P. Hence, we have a *rotational symmetry* around the alignment axis. However, Eq. (12.95) shows that this symmetry breaks down in the radiation surface of a speci c array, because of the factor $\boldsymbol{\mathcal{M}}_0$, the equivalent moment of the single array element, which in general is not omnidirectional in the plane normal to $\boldsymbol{\ell}$.

Let us leave aside, for the time being, the physical meaning of the variable v in Eq. (12.103), and study the function:

$$|M| = \left| \frac{\sin nv}{\sin v} \right| \quad . \tag{12.104}$$

For any integer n, Eq. (12.104) is periodic and its period equals $v =$. It is also an even function, so its knowledge over the range 0 v /2 is su cient to reconstruct it everywhere. Figure 12.22 shows some diagrams, for values of n which are of practical interest. Let us stress once more that in practice these diagrams must be handled with care, as the actual range spanned, in reality, by the variable in Eq. (12.103) is not always simple to estimate at a glance. In fact, v depends on the angular coordinate and also on the ratio between the array pitch and the wavelength ($\ell/$), and nally on the phase shift ( ). Some of the problems at the end of this chapter will outline in more detail this point and its implications.

Going back to Eq. (12.104), let us make the following remarks. Its absolute maximum (at $v = 0$, $,\ldots$) equals n. The meaning is that, in the directions of the maxima, elds generated by the n elements of the array add up in phase. Consequently, the power density radiated in those directions (proportional to $|E^2|$) grows *as the square of the number of antennas.*

As long as the radiation resistance of each array element can be assumed to remain the same as that of an isolated element, acting as the only eld source in an unbounded medium, then, clearly, the total power fed to the n elements equals n times the power fed to the individual antenna. The conclusion is that the array gain equals n times the gain of the individual element. The assumption of constant radiation resistance applies well to aperture antennas, but not always to thin wires. In these cases, the gain remains proportional to n, but through other factors smaller than unity, which depend on the array geometry. Enlightening examples can be found, for instance, in Collin (1985).

The null directions which delimit the main lobe correspond to the following values of v:

$$v = \frac{\quad}{n} \quad . \tag{12.105}$$

So, the angular width of the main lobe tends to zero as the number of elements increases. To nd the directions of the secondary lobes, and then their peak values, suppose that n is large enough to neglect the derivative of the denominator in Eq. (12.104) with respect to that of the numerator, so that

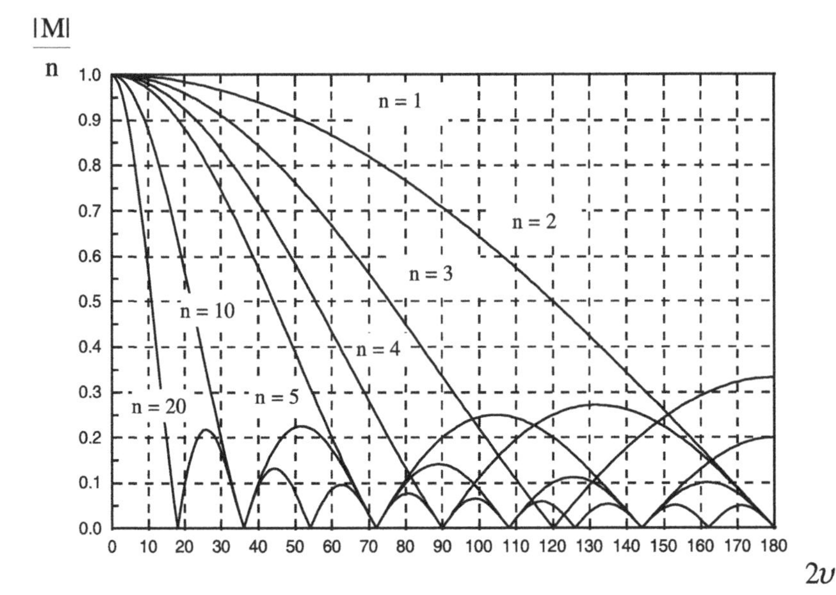

Figure 12.22 *Normalized chart of the composition factor, for uniform arrays.*

we may write:

$$\frac{d\,|M|}{dv} = 0 \qquad \text{for} \qquad nv \simeq \ \frac{\ }{2}\,(2p+1) \quad , \qquad p = 1, 2, \ldots \qquad (12.106)$$

The solution which corresponds to $p = 0$ in Eq. (12.106) is a spurious one, introduced erroneously by the previous approximation. This is easy to check, noting that values $v = \quad (\ /2n)$ are smaller (in modulus) than Eq. (12.105), so they correspond to directions which are still within the main lobe. Therefore, the largest secondary lobes correspond to $v = \quad (3\ /2n)$. For a su ciently large n, their amplitude can be approximated as follows:

$$b = \left| M\left(v = \ \frac{3\ }{2n}\right)\right| = \frac{1}{\sin\dfrac{3\ }{2n}} = \frac{2n}{3\ } \quad . \qquad (12.107)$$

Thus the ratio between lobes, as n increases, increases too, but tends to a nite upper bound, given by:

$$\frac{n}{b} \simeq \frac{3\ }{2} \quad . \qquad (12.108)$$

This is the reason why a uniform array cannot become a super-gain antenna, in spite of the fact that the width of its main lobe has no lower bound, as we saw earlier.

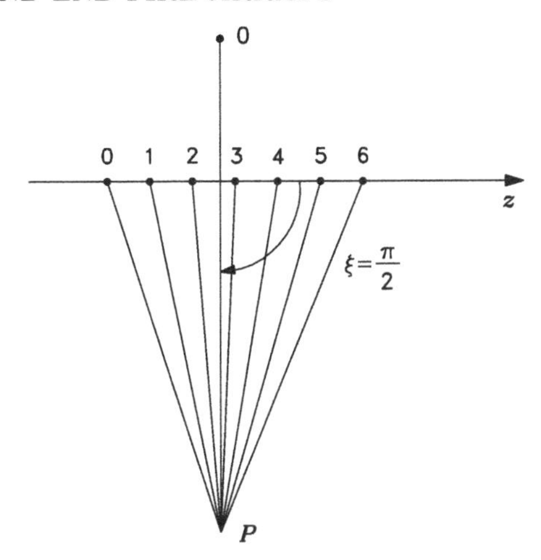

Figure 12.23 *The basic scheme of a broad-side array.*

12.9 Broad-side and end- re arrays

Quite frequently, a speci cation for a uniform array is that its main lobe be orthogonal to the direction of alignment. This is referred to as a *broad-side* array. Another frequent speci cation is that the main lobe be along the direction of alignment. This is referred to as an *end- re* array.

a) *Broad-side arrays.* Since the composition factor exhibits, as we said before, a rotational symmetry around the alignment axis z, the absolute maximum of $|M|$ is in the whole plane orthogonal to z. In general, the single elements of a broad-side array are not omnidirectional in this plane, so the rotational symmetry is not a general feature of the array far eld.

As we saw in the previous section, $|M|$ reaches its maximum value for $v = 0$. Therefore, the requirement that the main lobe be oriented along the direction $= /2$ entails the following recipe for Eq. (12.103) (see also Figure 12.23):

$$v\left(= \frac{}{2}\right) = 0 \quad . \qquad (12.109)$$

Eq. (12.109) is satis ed if

$$= 0 \quad , \qquad (12.110)$$

i.e., if the all the array elements are fed *in phase*. Recalling that $|M|_{\mathrm{MAX}} = n$, where n is the number of elements in the array, the result is simple to explain in physical terms. Consider a point P lying in the far eld and located on a straight line orthogonal to the alignment, like in Figure 12.23. Spherical waves radiated in phase by the n antennas add up in phase in P because di erences in the distance they traveled are negligible. This makes the total eld n times larger than that radiated by a single source.

The null directions which delimit the main lobe can be found letting $= 0$

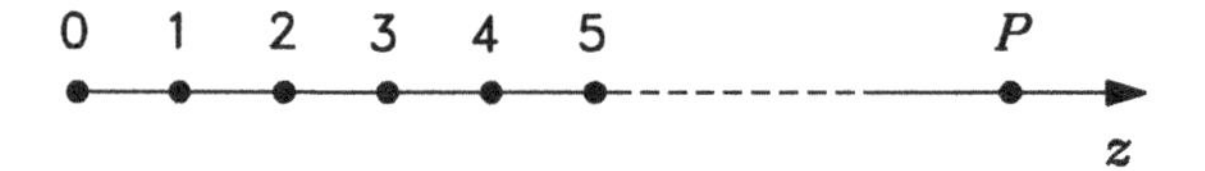

Figure 12.24 *The basic scheme of an end- re array.*

in Eq. (12.105), and are:

$$_1 = \arccos \left(\frac{}{n\ell} \right) \quad . \tag{12.111}$$

If n is large enough, it is simple to check (letting cos $=$ sin(/2), and approximating the sine function with its argument) that $_1$ $_1$ is essentially equal to 2 $/n\ell$.

Further details on these arrays can be found solving some of the problems at the end of this chapter.

b) *End- re arrays.* The speci cation says that the absolute maximum of $|M|$ is in the alignment direction, i.e., either at $= 0$ or at $=$. As the antenna numbers, from 0 to n 1, can go either way, it is obvious how to reduce one of these cases to the other one. Let us then consider only the case:

$$v(= 0) = 0 \quad . \tag{12.112}$$

Eq. (12.103) shows that this requirement is satis ed if:

$$= \frac{2 \ \ell}{} = \ \ell \quad . \tag{12.113}$$

So, the absolute maximum of $|M|$ is along the alignment direction if each feeding current is phase shifted, with respect to the next array element, by an amount equal to the phase shift experienced by a uniform plane wave which travels the distance between these two elements.

This result lends itself to a simple physical explanation, similar to the previous one on broad-side arrays. To make the spherical waves radiated by n antennas arrive in phase at a point P, placed in the far eld along the alignment direction (see Figure 12.24), it is necessary to compensate for the path differences, and this can be done by shifting the phases of the feeding currents, according to the recipe expressed by Eq. (12.113).

This comment is consistent with what we carried out so far only if we refer to the far eld. Actually, each array element is in the near eld of the other elements. In some end- re arrays — among them, the very popular *Yagi antenna*, which consists of 4 half-wave dipoles — only one element is connected to a generator. The currents in the other elements are induced by the near eld of the "active" element. In such a case, a precise calculation of the geometrical distances between elements must account for the fact that, as soon as ℓ is less than (i.e., in the near eld), eld phase delays can no longer be expressed as simply as $e^{\ j \ \ell}$. For more details, see, e.g., Jordan, (1986).

To calculate the width of the main lobe of an end- re array, we set $=$ ℓ

in Eq. (12.105). We get:

$$1 \quad \cos \ _1 \quad 2\left(\sin \frac{\ _1}{2}\right)^2 = \frac{}{n\ell} \quad . \tag{12.114}$$

If n is large enough, then the full angular width of the main lobe can be written as:

$$2 \ _1 = 4\sqrt{\frac{}{2n\ell}} \quad . \tag{12.115}$$

Comparing with the previous results, we see that the main lobe width of an end- re array decreases, as n increases, more slowly than that of a broad-side array.

Also for end- re arrays more details can be learned solving some of the problems at the end of this chapter.

12.10 Further applications and suggested reading

Radio systems have played a fundamental role in the whole history of telecommunications, and antennas have always been instrumental ingredients in such systems. Among several competitors who, at the end of the 19th century, shared the purpose to exploit Hertzian waves as signal carriers, Marconi succeeded rst, and that may be ascribed, to a large extent, to his physical intuition which suggested to him how to use "aerials". Readers should not be surprised, henceforth, if many excellent treatises on antennas can be found in all good engineering libraries. It was certainly not a coincidence, when the title we chose for this chapter was *Fundamentals* of antenna theory.

Anyway, it might be unfair to let the readers believe that the recent spectacular success of some radio-based systems was mostly due to dramatic progress in antennas. Satellite communications, beyond any doubt, make use of some the best antennas which exist today, both on board and on earth. However, the key to their technical and commercial success has not been antennas, but electronics, above all low-noise front-ends for the receivers — rst, parametric ampli ers and microwave masers, later, GaAs microwave integrated circuits.

As a second example, let us take mobile cellular telephony systems. Their big leap in one decade — from a few, small, privately-operated networks, serving small numbers of customers, to a public service which covers continents and reaches millions of subscribers — was made possible by gigantic progress, in performance and in cost, of computers and networks. Consistently with these remarks, we suggest to the readers, especially to those whose main interests are in telecommunications, to broaden their knowledge of all ingredients of modern radio systems, and make it well balanced. An excellent, up-to-date coverage in this sense is o ered, whit reference to mobile systems, by the book edited by Steele (1992) and the handbook edited by Gibson (1996).

The invitation to broaden the knowledge of the scenario in which radio systems operate applies not only to the two examples which we mentioned explicitly, but to others as well, for instance radar systems (see, e.g., Kingsley and Ouegan, 1992). Also in this eld, advances during the last decades were driven mainly by progress in electronics, in signal processing, and in real-time computing. However, it was remarkable how all these ingredients mixed with electromagnetic theory. A very signi cant example in this sense is the so-called *synthetic-aperture radar*: one antenna, on board an aircraft or spacecraft moving at constant speed, followed by suitable signal processing circuits, "simulates" an antenna array. However, some of the basic features of how such a system operates can be understood only after reading the next chapter.

The next suggestion we give to our readers is to delve more deeply into the contents of this chapter, which, on purpose, was focused on basic theory. As for more practical information, let us say that for an antenna specialist, experience is at least as important as the theoretical background. It seems to be more so in this eld than in other branches of applied electromagnetism, because with antennas many situations which are unavoidable in practice are very di cult to model in mathematical terms. Just to quote an example, mutual in uence between antennas which are located in the near eld of each other is largely governed by obstacles in their neighborhood, like buildings or metallic towers on which they are installed. It is very hard to learn these subjects from textbooks, and to a substantial extent, practical experience is proprietary know-how of companies.

Yet, some readings may help to bridge the gap between basic theory and advanced professional experience. One very famous book, which has been a daily work tool for generations of professional engineers, has been entitled, throughout many editions, *Reference Data for Radio Engineers*, and was written by the Technical Sta of IT&T. A revised — and signi cantly broadened in scope — evolution, now on the market, is quoted in the reference list under the name of its Editor in Chief, E.C. Jordan (1986). It is, in fact, a handbook, but at the same time it encompasses fully adequate theoretical background items. Chapter 32 is entirely devoted to antennas, and at least two more chapters deserve to be quoted here. Chapter 34 is entitled "Radio Noise and Interference," and is subdivided into sections dealing with natural noise, man-made radio noise, precipitation static, thermal noise calculations, and noise measurements. Chapter 35 is entitled "Broadcasting, Cable Television, and Recording System Standards". In addition to very valuable practical information, this reading conveys a pedagogical message. It reminds future radio engineers that they are going to "share the ether" with many other users, therefore they *must* comply with standards and regulations.

For those who wish to proceed with the theory of antennas, among many excellent textbooks we recommend the classic treatise by Kraus (1950), and that by Fradin (1961). Further suggestions, more speci cally oriented towards aperture antennas, will be provided at the end of the next chapter. Finally, a subject whose importance is rapidly growing, and might become fundamental in the near future, is microstrip antennas, or, more generally, printed antennas. These objects are becoming popular, not so much because of their performances, which are not particularly outstanding, but because of their low cost. A good introductory reading is James and Hall (1989).

Problems

12-1 Derive the coe cient of proportionality between $T(r'/r)^2$ and $(r'/r)^2$ in Eq. (12.4).

12-2 Prove that Eq. (12.20) still holds in a lossy dielectric medium.

12-3 Using the results of Section 12.3, show that for any source which does not extend to in nity, the Poynting vector derived from the far eld is always real and in the outward radial direction.
Show that in a lossless medium its ux through a spherical surface centered in the origin of the reference frame is independent of its radius.

12-4 Do the expressions of the far eld, as found in Section 12.3, satisfy Maxwell's equations *exactly*? If not, justify the missing terms as consequences of the approximations that were made.

12-5 Compare the restrictions $r \quad r'$, $r \quad r'^2/ \ $, $r \quad$, nding the most restrictive one in each of the following cases:

a) the source is a thin wire, 1 m in length, at the frequency $f = 500\,\mathrm{kHz}$ (medium-wave radio frequency);
b) the source is a circular aperture, 1 m in diameter, at $f = 1\,\mathrm{GHz}$ (microwave frequency);
c) the source is a round hole in a cardboard screen, at $\ = 0.5 \ \mathrm{m}$ (visible light).

12-6 Show that the eld of a short dipole, of length L, in a lossless medium, can be derived from a Hertz vector potential (see Chapter 1) given by:

$$= \frac{IL}{j\omega} \ \frac{e^{\ j \ r} 4 \ r}{\hat{a}_z} \ .$$

Discuss the physical meaning of the relationship between this result and the approach to the short dipole, Section 12.4.

12-7 Find the radius of the spherical surface, centered in the origin, such that, for a short dipole placed at the origin, the real and the imaginary parts of the Poynting vector ux through it are equal (in modulus). Find the exact values, in modulus and phase, of each eld component (i.e., E_r, E_ϑ and H_φ) on this surface.

12-8 A mobile telephone is equipped with an antenna which can be modeled as a short dipole, 50 mm in length. It radiates a power of 0.3 W at a frequency

of 980 MHz. Find the current which circulates in the antenna.
What would be the maximum current, if the same power was radiated by a half-wave dipole?
Assuming that the e ciency of the short dipole is 0.3, how much power is dissipated in the antenna?

12-9 Derive the expression of the radiation intensity for a thin wire antenna, Eq. (12.69).

12-10 Analyze in detail the normalized radiation intensity of a thin wire antenna, Eq. (12.71), for the case where its total length $2L$ equals $3\ /2$.

a) Show that this antenna has two main lobes, symmetrical with respect to the $z = 0$ plane, at an angle ϑ with respect to the wire axis satisfying the equation $\tan y = y/(A^2 \quad y^2)$, where $y = A \cos\vartheta$, $A = 3\ /2$.
b) Show that it has a secondary lobe in the $z = 0$ plane.
c) Find its null directions.

12-11 Calculate the directivity of the Huygens source, discussed in Section 12.7.

12-12 Derive the eld which follows from Eq. (12.80) without any further approximation, and compare it with the expressions found in the text.

12-13 A non-ideal (i.e., lossy) parabolic-re ector antenna, with an aperture diameter of 1 m, has a power gain of 43 dB at $f = 15$ GHz. Find its e ective area and its e ciency.

12-14 Comparing a terrestrial radio-relay link with a geostationary satellite system, we notice that the link lengths di er by about 3 orders of magnitude. Noting the di erence in antenna noise temperatures, estimate the order of magnitude of the antenna gain improvement in decibels which is necessary in order to preserve comparable signal-to-noise ratios at the receiver input.

12-15 An antenna for satellite communication is pointed in a direction such that its noise temperature is $T_a = 40$ K. Asume $B = 10$ MHz, $g_R = 1o^6$. What is the minimum received power in order to guarantee a signal-to-noise ratio $S/N = 63$ dB?
What is the corresponding ERP if the distance between transmitter and receiver is 30,000 km and the frequency is 6 GHz?

12-16 Show that the e ective area of an array of n identical antennas can never exceed n times the e ective area of the individual antenna, and reaches that upper limit only in the idealized case where the radiation resistance of each antenna is insensitive to the presence of the other $n \quad 1$ elements of the array.

12-17 The three-half-wavelength antenna of the Problem 12-10 can be looked at as an array of three half-wavelength antennas, the array pitch being also $/2$, and the phase shift between two consecutive sources being . Show that the product of the composition factor (squared) times the radiation intensity of the half-wave antenna yields the same radiation intensity as found in the previous problem.

12-18 A transmitting array consists of 6 thin-wire elements, parallel to the z axis, aligned along the y axis, and fed in phase. Its total power gain at $f = 600\,\text{MHz}$ is $10\,\text{dB}$, the transmitted power is $100\,\text{W}$. The receiving antenna is also a thin wire, and has a gain $g_r = 1.5$. Assuming an inde nitely extended homogeneous medium, calculate the received power under the following circumstances:

a) the receiving antenna is at P $(0, 10^4\,\text{m}, 0)$, and oriented parallel to the z axis;

b) the receiving antenna is at P $(0, 10^4\,\text{m}, 0)$, and oriented parallel to the y axis;

c) the receiving antenna is at P $(0, 0, 10^4\,\text{m})$, and its orientation is arbitrary;

d) the receiving antenna is at P $(10^4\,\text{m}, 0, 0)$, oriented parallel to the z axis.

Assume that the pitch of the transmitting array equals rst $/4$, then $/2$, and nally .

12-19 Find the restrictions on $\ell/$ and which a uniform array must satisfy in order to:

a) exhibit at least one absolute maximum of the composition factor (i.e., with $v = 0$ corresponding to a value of between 0 and);

b) exhibit only one absolute maximum of the composition factor (i.e., are out of the interval spanned by v when varies from 0 to);

c) have secondary lobes which decrease monotonically as the angle with the direction of absolute maximum increases.

12-20 An "array of arrays" consists of 12 elements, aligned on 3 "rows" parallel to the x axis and 4 "columns" parallel to the y axis. The pitch equals $/3$ in both directions. The four elements in each row are fed in phase. The phase shift between consecutive rows equals $2\ /3$ radians. Take as composition factor the product of those along the two axes.

a) Show that the composition factor has only one absolute maximum. Find its direction.

b) In which coordinate plane, (x, y), (y, z) or (x, z), do we nd the minimum angular width of the main lobe?

c) Find the null directions which delimit the main lobe in that plane.

12-21 Show that the eld of a short electric dipole is compatible with an "electric wall" (i.e., an ideal electric conductor) in the plane $z = 0$. Also show that the same is true for any thin-wire antenna whose current distribution is even with respect to the point $z = 0$. Infer a comment on vertical antennas over a perfectly conducting ground.

12-22 Consider a loop antenna, as discussed in the text.

a) Show that its eld is compatible with a "magnetic wall" (i.e., an ideal magnetic conductor) in the plane $z = 0$.

b) Show that the total eld of two identical and parallel loop antennas, fed with a phase shift of radians, is compatible with an "electric wall" (i.e., an ideal electric conductor) in the plane mid-way between them.

References

Abramowitz, M. and Stegun, I.A. (Eds.) (1965) *Handbook of Mathematical Functions.* Dover, New York.

Collin, R.E. (1985) *Antennas and Radiowave Propagation.* McGraw-Hill, New York.

Fradin, A.Z. (1961) *Microwave Antennas.* Pergamon, Oxford.

Gibson, J.D. (Ed.) (1996) *The Mobile Communications Handbook.* CRC/IEEE Press, New York.

James, J.R. and Hall, P.S. (1989) *Handbook of Microstrip Antennas.* P. Peregrinus Ltd., London.

Jordan, E.C. (Editor in Chief) (1986) *Reference Data for Engineers: Radio, Electronics, Computer, and Communications.* H.W. Sams & Co., Indianapolis.

Kingsley, S. and Ouegan, S. (1992) *Understanding Radar Systems.* McGraw-Hill, London.

Kraus, J.D. (1950) *Antennas.* McGraw-Hill, New York.

Kraus, J.D. and Marhefka, R.J. (2003) *Antennas for All Applications,* 3rd ed.. McGraw-Hill, New York.

Kupfmuller, K. (1973) *Fondamenti di Elettrotecnica.* Utet, Torino.

Plonsey, R. and Collin, R.E. (1961) *Principles and Applications of Electromagnetic Fields.* McGraw-Hill, New York.

Ramo, S., Whinnery, J.R. and van Duzer, T. (1994) *Fields and Waves in Communication Electronics.* 3rd ed., Wiley, New York.

Steele, R. (Ed.) (1992) *Mobile Radio Communications.* Pentech Press, London.

Stutzman, W.L. and Thiele, G.A. (1997) *Antenna Theory and Design.* 2nd ed., Wiley, New York.

CHAPTER 13

Di raction

13.1 Introduction

The behavior of an electromagnetic wave impinging on a sharp discontinuity was studied in Chapters 4 and 10, under the assumption that the surface of discontinuity extends to in nity, at least in one direction. In the part of Chapter 9 where we dealt with open resonators, we pointed out that the mirrors which delimit them do not satisfy an assumption of this kind. However, until now we did not consider the consequences induced by surfaces of discontinuity limited by rims.

In the presence of obstacles, i.e., of abrupt variations in the parameters of the medium where they propagate, e.m. waves (and in general, 2-D or 3-D waves of any kind) are very sensitive to the ratio between their wavelength, , and the size of the obstacles. When the geometrical dimensions involved in the phenomenon are not large with respect to , or in general at a distance from an obstacle rim comparable to , the experimentally observed eld cannot be expressed in terms of a small number of plane waves, along the guidelines of the re ection-refraction formalism. It can neither be modeled in terms of guided waves, nor by means of the approaches that we introduced in Chapter 5 for slowly-varying media, like ray optics and the WKBJ method. This tells us that we may still adopt Sommerfeld's de nition for di raction, as the set of those wave phenomena which cannot be explained in terms of re ected and refracted rays. Originally, the de nition referred to optics, but, as we said, the concept applies to other waves (in acoustics, in the theory of elasticity in solids, in hydrodynamics, in the ondulatorial approach to Quantum Mechanics, etc.). In electromagnetism, di raction is not a peculiarity of optical frequencies, although that was the spectral range where it was rst observed and modeled. Typical wavelength-to-size ratios in modern aperture antennas are such that analysis and design of these objects are deeply based on the theory of di raction, i.e., on analytical approaches that were rst developed to study optical phenomena.

13.2 The di raction integral: The vector formulation

The equivalence theorem of Section 3.5, and the induction theorem of Section 3.6, are the most suitable background items for the theory of di raction. Suppose that all the eld sources, as well as all the inhomogeneities in the medium, belong to a 3-D region surrounded by a closed regular surface S

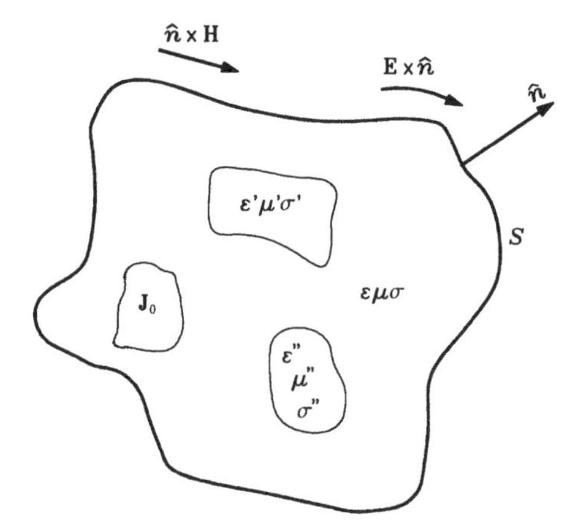

Figure 13.1 *The starting point for the theory of di raction: the equivalence theorem.*

(Figure 13.1). To evaluate the eld outside S, the equivalence theorem tells us that it is enough to know the tangential components of **E** and **H** at every point on S: two surface integrals over S provide the solution to the problem. So, we can state that knowing the eld in the vicinity of the obstacles enables us to determine the eld (the di racted eld, in fact) in all the remainder of the 3-D space, provided the latter is lled by a homogeneous medium.

If taken literally, this statement does not yield any practical result. Indeed, the eld in the neighborhood of the di racting objects is also unknown. Consequently, exact solutions are not known in general. However, the same statement is a powerful hint towards an approximation, which we shall justify quantitatively later on. Suppose one is able to identify a closed surface S, such that, for given sources inside S, the values of the elds on S are not very sensitive to the details of the medium outside S. Also suppose that, introducing simplifying assumptions on S itself — for example, perfect re ection on an ideal conductor — one is able to nd a solution for Maxwell's equations in the region inside S. Then, clearly, the next step is to take this eld as an approximate solution for the case where the simplifying assumption is removed, introduce it into the surface integrals of the equivalence theorem, and nd in this way an approximate solution (quite often, an acceptable approximation) for the eld outside S.

The classical theory of di raction is just a translation of these qualitative ideas into quantitative relationships. Let (see Figure 13.2) all the sources be inside a closed surface S which is the union of two open surfaces. Suppose that the rst one, A', is absolutely not transparent to electromagnetic energy, i.e., that the normal component of the Poynting vector is identically zero on A'. Assume that the medium is homogeneous in the neighborhood of the remaining open surface, A, which we call the *aperture*. Incidentally, note that on A' there is necessarily an inhomogeneity in the medium. Also, suppose

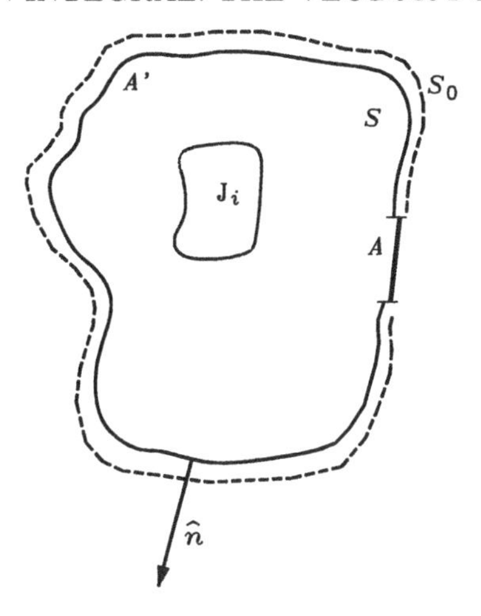

Figure 13.2 *Introduction to the concept of di raction through an aperture.*

that the medium in the region outside S is homogeneous. Then, applying the equivalence theorem we evaluate its surface integrals over a closed surface, S_0, which consists of the aperture A plus another surface, arbitrarily close to the external face of A' (see Figure 13.2). Eqs. (11.18) and (11.19) hold at any point P outside S_0. Let us re-write them in the following way, which is commonly used in di raction theory, with the unit-vector $\hat{n}$ towards the region where we compute the potentials:

$$\mathbf{A} = \frac{}{4} \int_{S_0} \hat{n} \quad \mathbf{H} \frac{e}{} \, \mathrm{d}S \quad , \tag{13.1}$$

$$\mathbf{F} = \frac{c}{4} \int_{S_0} \mathbf{E} \quad \hat{n} \frac{e}{} \, \mathrm{d}S \quad . \tag{13.2}$$

Here, is the distance between P and the point Q that scans the surface S_0. Incidentally, in view of the next sections let us point out that in the previous equations the factor $(1/4\ )(e \quad / \ )$ could be replaced by any other Green's function, G.

Eqs. (13.1) and (13.2) require that one knows **E** and **H** at all points of S_0, but when Q does not belong to A it is outside S, in the region where the eld is unknown and we are aiming at nding it. This, as we said before, forces us to make approximations.

The so-called "classical" theory of di raction begins with an approximation which, at this stage, can be justi ed only qualitatively. Quantitative checks of its validity rely mainly on comparisons with experimental results. Let us introduce it in the following way.

We assumed the normal component of the Poynting vector equal to zero at any point on A'. If A' were a closed surface (i.e., if the aperture A disappeared), the uniqueness theorem of Section 3.3 would tell us that the eld outside S

would be identically zero. Hence, we say that, if the aperture is small enough compared to the rest of S, we can approximate Eqs. (13.1) and (13.2) *assuming the tangential components of both* **E** *and* **H** *equal to zero on the external face of* A'. Eqs. (13.1) and (13.2) thus become two integrals over the aperture A, namely:

$$\mathbf{A} \;=\; \frac{}{4} \int_A \hat{n} \quad \mathbf{H} \frac{e}{\qquad} \, \mathrm{d}S \quad , \tag{13.3}$$

$$\mathbf{F} \;=\; \frac{c}{4} \int_A \mathbf{E} \quad \hat{n} \frac{e}{\qquad} \, \mathrm{d}S \quad , \tag{13.4}$$

which are usually referred to as the *di raction integrals.*

In rigorous terms, this statement is not acceptable. To prove it wrong, let us take a closed surface S_1 which coincides with S_0 on the aperture A, but is completed by an arbitrary surface S' on the outer side of S_0. The equivalence theorem says that the eld outside S_1 can be evaluated either by means of Eqs. (13.3) and (13.4), or by means of similar integrals over S_1. Consequently, it must be:

$$\int_{S'} \hat{n} \quad \mathbf{H} \frac{e}{\qquad} \, \mathrm{d}S = 0 \quad , \tag{13.5}$$

$$\int_{S'} \mathbf{E} \quad \hat{n} \frac{e}{\qquad} \, \mathrm{d}S = 0 \quad , \tag{13.6}$$

and hence, as S' is arbitrary, we may infer:

$$\mathbf{E} \quad \mathbf{H} \quad 0 \quad , \tag{13.7}$$

everywhere in the region outside S. This is a very important point as a matter of principle. However, we shall see soon that, nonetheless, this "classical" approach to di raction yields results which are quite satisfactory for most practical purposes.

The e.m. eld di racted by the aperture A can then be computed at any point in the region outside S from the potentials, Eqs. (13.3) and (13.4), by means of Eqs. (11.20) and (11.21), where we let $\mathbf{M}_i = \mathbf{J}_i = 0$. One gets:

$$\mathbf{E}(P) \;=\; \frac{1}{j\omega \;\; c} \frac{1}{4} \nabla_P \quad \nabla_P \quad \int_A \hat{n} \quad \mathbf{H} \frac{e}{\qquad} \, \mathrm{d}S$$

$$\frac{1}{4} \nabla_P \quad \int_A \mathbf{E} \quad \hat{n} \frac{e}{\qquad} \, \mathrm{d}S \quad , \tag{13.8}$$

So far, we supposed that S_0 could be arbitrarily close to S, but could not coincide with it. In fact, the "screening" surface is surely an inhomogeneity in the medium: if it were in the region where we apply the equivalence theorem, we should replace it with polarization currents. These, in turn, because of the assumptions that were made before, should be located inside the region surrounded by S_0. Now, however, after justifying the steps from Eqs. (13.1) and (13.2) to Eqs. (13.3) and (13.4), we can consider $\mathbf{E}_{\tan} = \mathbf{H}_{\tan} = 0$ as the boundary conditions on S. One of the consequences would be that Eq. (13.7) holds in the entire region outside S.

$$\begin{aligned}\mathbf{H}(P) &= \frac{1}{4}\nabla_P \quad \int_A \hat{n} \quad \mathbf{H}\,\frac{e}{\quad}\,\mathrm{d}S \\ &+ \frac{1}{j\omega}\,\frac{1}{4}\,\nabla_P \quad \nabla_P \quad \int_A \mathbf{E} \quad \hat{n}\,\frac{e}{\quad}\,\mathrm{d}S \quad . \end{aligned} \tag{13.9}$$

The electric and magnetic elds at P are determined by the tangential components of the elds on the aperture (to be considered, in the framework of this approximation, known quantities obtained solving Maxwell's equations in the region within S), and by partial derivatives of $e \quad / \quad$ with respect to the coordinates of P. As P does not belong to the aperture A, derivatives and integration in Eqs. (13.8) and (13.9) commute. After changing their order, one can also account for the fact that the coordinates of P are present only in the Green's function, not in $\hat{n} \quad \mathbf{H}$ and $\mathbf{E} \quad \hat{n}$ which are de ned on A. So, using the vector identities (C.6) and (C.9) in the integrals, and accounting for the fact that $e \quad / \quad$ satis es the homogeneous Helmholtz equation for $\quad \neq 0$, one nally gets:

$$\nabla_P \quad \left[(\mathbf{E} \quad \hat{n})\,\frac{e}{\quad}\right] = (\mathbf{E} \quad \hat{n}) \quad \left[\nabla_P\left(\frac{e}{\quad}\right)\right] \quad , \tag{13.10}$$

$$\begin{aligned}\nabla_P \quad \nabla_P \quad \left[(\hat{n} \quad \mathbf{H})\,\frac{e}{\quad}\right] &= \\ &= (\hat{n} \quad \mathbf{H})\,\nabla_P^2\left(\frac{e}{\quad}\right) + \left[(\hat{n} \quad \mathbf{H}) \quad \nabla_P\right]\nabla_P\,\frac{e}{\quad} \\ &= (\hat{n} \quad \mathbf{H}) \quad {}^2\,\frac{e}{\quad} + \left[(\hat{n} \quad \mathbf{H}) \quad \nabla_P\right]\nabla_P\,\frac{e}{\quad} .\end{aligned} \tag{13.11}$$

The terms which appear in Eq. (13.9) can be handled in a very similar way.

At this stage, it may be useful to exploit symmetry in the function $e \quad / \quad$, with $\quad = |\mathbf{r} \quad \mathbf{r}'|$. In fact, the coordinates of P and of Q (which scans the aperture A) play mutually symmetrical roles. Hence, now (but not before) the operator ∇_P can be replaced by ∇_Q. As the net result, instead of Eqs. (13.8) and (13.9) we may write:

$$\begin{aligned}\mathbf{E}(P) &= \frac{1}{j\omega \quad c}\,\frac{1}{4}\int_A \left[(\hat{n} \quad \mathbf{H}) \quad \nabla_Q\right]\nabla_Q\left(\frac{e}{\quad}\right)\mathrm{d}S_Q \\ &\quad j\omega \quad \frac{1}{4}\int_A \hat{n} \quad \mathbf{H}\,\frac{e}{\quad}\,\mathrm{d}S_Q \\ &+ \frac{1}{4}\int_A \mathbf{E} \quad \hat{n} \quad \nabla_Q\left(\frac{e}{\quad}\right)\mathrm{d}S_Q \quad ,\end{aligned} \tag{13.12}$$

$$\mathbf{H}(P) = \quad \frac{1}{4}\int_A \hat{n} \quad \mathbf{H} \quad \nabla_Q\left(\frac{e}{\quad}\right)\mathrm{d}S_Q$$

$$+ \quad \frac{1}{j\omega} \frac{1}{4} \int_A \Big[(\mathbf{E} \quad \hat{n}) \quad \nabla_Q \Big] \nabla_Q \left(\frac{e}{\quad} \right) \mathrm{d}S_Q$$

$$j\omega \;_c \frac{1}{4} \int_A (\mathbf{E} \quad \hat{n}) \frac{e}{\quad} \mathrm{d}S_Q \quad . \tag{13.13}$$

Let us point out once again that in Eqs. (13.12) and (13.13) the quantity $(1/4\ )(e \quad / \ )$ could be replaced by *any* Green's function G, as de ned in Section 11.2, under the only restriction that the coordinates of the source point and those of the potentiated point play symmetrical roles in G. The last part of this statement should be memorized in view of the second part of Section 13.4.

The main advantages o ered by this new formulation of the di raction integrals are:

all calculations are made on the aperture A;

consequently, some approximations, to be seen in Sections 13.5 and 13.6, are easier to justify;

comparison between the vector theory, that we begun to outline in this section, and the scalar theory, to be outlined in Section 13.4, is easier.

In view of this last comparison (Section 13.4) it may be useful to include here an alternative to Eq. (13.12), whose derivation requires, after Eq. (13.12), Maxwell's equations and vector identities, and is left to the reader as an exercise (see Rush and Potter, 1970). For the sake of brevity, set $\quad = e \quad / \ .$ Then, one obtains:

$$\mathbf{E}(P) \quad = \quad \frac{1}{4} \int_A \left(\mathbf{E} \frac{\partial}{\partial n} \quad \frac{\partial \mathbf{E}}{\partial n} \right) \mathrm{d}S$$

$$\frac{1}{4} \int_A \Big\{ (\mathbf{E} \quad \nabla) \hat{n} + \mathbf{E} \quad \nabla \quad \hat{n} \Big\} \quad \mathrm{d}S$$

$$+ \quad \frac{1}{4} \oint \ (\mathbf{E} \quad \mathrm{d}\boldsymbol{\ell}) \quad \frac{1}{4} \frac{1}{j\omega \;_c} \oint \nabla \ (\mathbf{H} \quad \mathrm{d}\boldsymbol{\ell}) \quad , \tag{13.14}$$

where the closed curve is the edge of the aperture A. A similar expression can be obtained for $\mathbf{H}(P)$, replacing Eq. (13.13). These expressions are more complicated to compute than the previous ones, so they are not much in use. However, as we said before, they will be useful in the vector/scalar comparison of Section 13.4.

13.3 Illumination conditions. Babinet's principle

In order to compute the di raction integrals, Eqs. (13.12) and (13.13), one needs to know $\mathbf{E} \quad \hat{n}$ and $\hat{n} \quad \mathbf{H}$ on the aperture. As mentioned before, these quantities would be known exactly only if the whole eld itself was known exactly, and in practice it is necessary to proceed by approximations. The simplest ones assume that one can break down the problem into two stages.

The rst one is to evaluate the eld that would exist if there was no aperture, i.e., the so-called "illuminating" eld. *As a second step*, one evaluates the di racted eld, replacing the generic $\hat{n} \quad \mathbf{H}$ and $\mathbf{E} \quad \hat{n}$ in the di raction integrals of the previous section with the illuminating elds. If this does not yield a result with su cient accuracy, then one may use iterative procedures, for which the reader is invited to consult specialized treatises.

Coming now to the illuminating eld, let us discuss in detail two cases which di er remarkably from each other. Both of them match well with suitable experimental conditions.

A) Suppose that the aperture A is an opening in a *perfectly absorbing* screen, i.e., in a surface whose wave impedance equals the intrinsic impedance of the surrounding medium, $\quad = \sqrt{\quad / \quad}$. In this case, an obvious choice is to take as illuminating eld $\{\mathbf{E}, \mathbf{H}\}$ that eld which would be generated by the same sources if they were located in an unbounded homogeneous medium. We will denote this eld as $\{\mathbf{E}_i, \mathbf{H}_i\}$. As the intrinsic impedance of the medium equals the wave impedance in the far eld (see Chapter 12), this model only is acceptable provided that the distance between the screen and the primary sources is much larger than *and* than the size of the sources. In that case, the following relationship holds:

$$\mathbf{E}_i = \hat{n} \quad \mathbf{E}_i \quad \hat{n} = \quad \mathbf{H}_i \quad \hat{n} \quad . \tag{13.15}$$

Note that whereas the illuminating eld is a solution of Maxwell's equations (a plane wave), in general we cannot be sure that the same is true for the di racted eld. We will come back to this point in Sections 13.4 and 13.6.

For several reasons — rst of all, because a dark opaque surface is easily made at optical frequencies — this illuminating eld is usually referred to as the *geometrical-optics approximation.*

B) Let us suppose now that the aperture is an opening in a screen made with an *ideal electric conductor.* Taking again for granted that the aperture is so small that it does not perturb the impinging eld, then on the entire screen, including the aperture, we may set:

$$\mathbf{E} \quad \hat{n} = 0 \quad . \tag{13.16}$$

Consequently, three of the six terms in the di raction integrals in Eqs. (13.12) and (13.13) vanish. As for the tangential magnetic eld, we set it equal to *twice* the incident eld $\mathbf{H}_i$, since Eq. (13.16) implies that the re ection coe cient on the surface equals 1 (see Chapters 4 and 8):

$$\hat{n} \quad \mathbf{H} = 2\hat{n} \quad \mathbf{H}_i \quad . \tag{13.17}$$

The di raction integrals can be simpli ed as:

$$\begin{aligned} \mathbf{E}(P) &= \frac{1}{j\omega \quad _c} \frac{1}{2} \int_A \left[(\hat{n} \quad \mathbf{H}_i) \quad \nabla_Q\right] \nabla_Q \left(\frac{e}{\quad}\right) \mathrm{d}S_Q \\ & \quad j\omega \quad \frac{1}{2} \int_A \hat{n} \quad \mathbf{H}_i \frac{e}{\quad} \mathrm{d}S_Q \quad , \end{aligned} \tag{13.18}$$

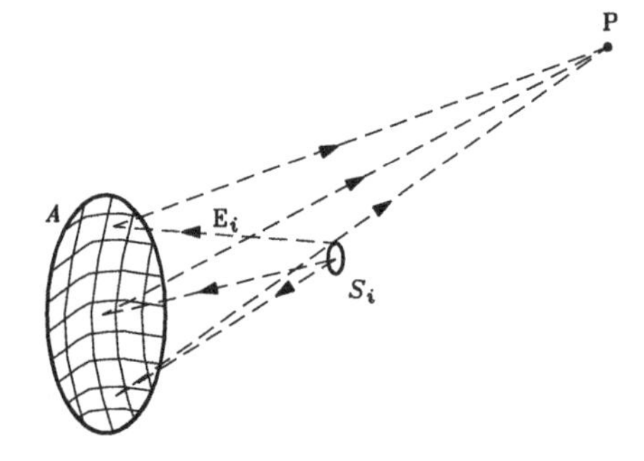

Figure 13.3 *Introduction to the concept of di raction by a re ecting object.*

$$\mathbf{H}(P) \quad = \quad \frac{1}{2} \int_A \hat{n} \quad \mathbf{H}_i \quad \nabla_Q \left(\frac{e}{\quad} \right) \mathrm{d}S_Q \quad . \tag{13.19}$$

These expressions t well also for the case (a quite frequent one, in microwave antennas) where the di racting aperture is a metallic re ector (see Figure 13.3) rather than a hole in a screen. At rst sight, there is a signi cant di erence between the problems: the source of the illuminating eld, S_i, is now on the same side as the di racted eld, with respect to the aperture. We will show in the following that actually, thanks to symmetries in the eld (see Section 3.9), this di erence is irrelevant.

Approaching the problem through Eqs. (13.16) and (13.17) is equivalent to considering di raction as a "local" problem, in the sense that:

we do not account for possible multiple re ections between the primary source and the screen;

we assume implicitly that those parts of the aperture (or of the re ector) which are not illuminated directly by the primary source do not contribute to the di racted eld.

These conditions are acceptable only if the size of the aperture (or re ector), as well as its radius of curvature, are both large compared with the wavelength. To bear this in mind, and also to distinguish the present situation from the previous one (Eqs. (13.15)), Eqs. (13.16) and (13.17) are referred to as the *physical-optics approximation.* An important example of application will be provided in Section 13.7.

There may be cases where di racting objects have shapes and/or sizes such that the previous assumptions are violated. For example, this occurs for re ectors which encompass sharp corners or edges. The choice of the illuminating eld becomes more critical. A short comment on this point will be given in Section 13.9, and the reader should refer to specialized treatises for more details.

C) Let us go back to the previous comment on the side where the primary source is located with respect to the screen. It is quite simple to write relationships between elds on the two sides of the aperture, in the case of a

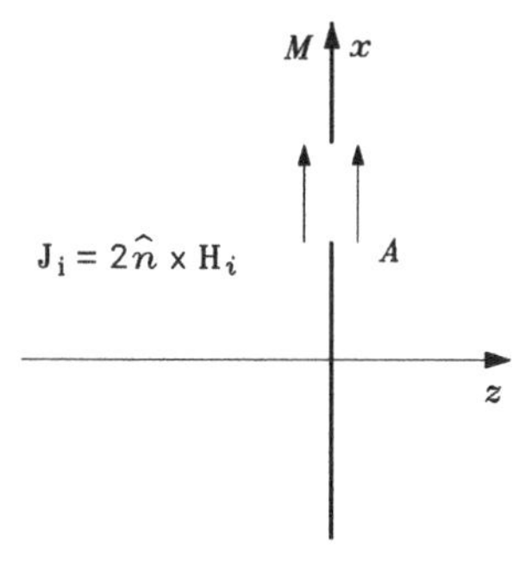

Figure 13.4 *An aperture in an in nitely thin conducting screen.*

plane in nitely thin conducting screen. Indeed, the problem can be modeled as shown in Figure 13.4, in terms of an imposed surface current density on the aperture A, given by Eq. (13.17). Thanks to the geometrical symmetry with respect to the $z = 0$ plane, solutions of Maxwell's equations (see Section 3.9) must be of *odd symmetry*, i.e.:

$$\begin{aligned} \mathbf{E}_t^{(d)}(x,y,z) &= \quad \mathbf{E}_t^{(d)}(x,y,\quad z) \quad , \\ E_z^{(d)}(x,y,z) &= E_z^{(d)}(x,y,\quad z) \quad , \\ \mathbf{H}_t^{(d)}(x,y,z) &= \mathbf{H}_t^{(d)}(x,y,\quad z) \quad , \\ H_z^{(d)}(x,y,z) &= \quad H_z^{(d)}(x,y,\quad z) \quad , \end{aligned} \tag{13.20}$$

where the subscript t denotes the transverse components with respect to the z axis, which is orthogonal to the screen, and the superscript (d) remains for "di racted". In order to nd the total eld, one has to sum the di racted eld and the illuminating eld $\{\mathbf{E}^{(i)}, \mathbf{H}^{(i)}\}$. Note, however, that the latter is not symmetrical with respect to the screen surface, because it was assumed to exist only in the left half-space.†

Another point which deserves a short discussion is the relationship between elds di racted by two *complementary screens*, M_1 and M_2. We mean that the conducting parts on the plane of M_1 coincide with apertures in the plane of M_2, and vice-versa (see Figure 13.5).

From above, we know that in either case $\{\mathbf{E}^{(d)}, \mathbf{H}^{(d)}\}$ in $z \quad 0$ must satisfy the following boundary conditions on the $z = 0$ plane:

$$0 = \mathbf{E}_{\tan}^{(d)}(x,y,0+) = \mathbf{E}_{\tan}^{(i)}(x,y,0\quad) + \mathbf{E}_{\tan}^{(d)}(x,y,0\quad) \quad , \tag{13.21}$$

† For non-planar screens or apertures, the problem becomes more complicated. Nonetheless, interesting results can be found using the equivalence theorem. For example, one can replace a parabolic re ector with equivalent surface currents on its planar aperture. One can then decompose any eld into its even and odd parts, as discussed in Section 3.9.

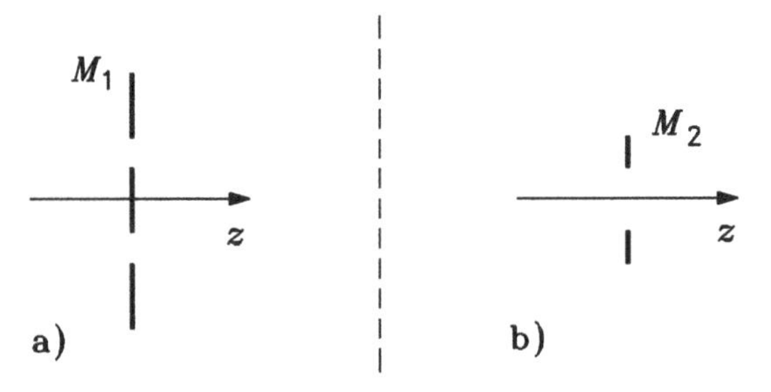

Figure 13.5 *An example of complementary screens.*

on the metallic parts of the plane, and:

$$\mathbf{H}_{\tan}^{(d)}(x,y,0+) = \mathbf{H}_{\tan}^{(i)}(x,y,0\ \ ) + \mathbf{H}_{\tan}^{(d)}(x,y,0\ \ )\quad , \tag{13.22}$$

on the apertures. Passing from M_1 to the complementary screen M_2, the regions where Eqs. (13.21) hold become regions where Eqs. (13.22) hold, and vice-versa. Applying the duality theorem and the uniqueness theorem of Chapter 3, we reach the following conclusions, which are currently referred to as *Babinet's principle*: two complementary screens made of *ideal conductors* and illuminated by dual elds, give rise to di racted elds which are the dual of each other. Note that when the illuminating eld is a linearly polarized plane wave (a frequent case in practice), two "dual" elds simply have orthogonal polarizations.

In the case of two *perfectly absorbing* complementary screens, the two di racted elds result from integrals, Eqs. (13.3) and (13.4), over two complementary regions of the $z = 0$ plane. Then, let us recall that, *when there is no screen*, using the equivalence theorem, the eld in the half-plane $z > 0$ can be obtained from the same integrals *over the whole plane* $z = 0$. So, we may conclude that the eld di racted by a screen equals the *di erence* between the eld in the absence of the screen, and the eld di racted by the complementary screen. This statement is also often quoted as Babinet's principle. Note its close connections with the concept of absorbing currents, which were de ned in Section 3.5.

13.4 The scalar theory of di raction

Scalar functions which, in a homogeneous medium, satisfy the scalar Helmholtz equation (like acoustic waves, or wave functions in quantum mechanics) are also subject to di raction. These phenomena can be dealt with analytically, starting from Green's theorem in its scalar version (see Section 11.2) and proceeding along the guidelines that we followed in Sections 13.2 and 13.3 for vector waves. We saw in previous chapters that for e.m. waves it is convenient to pass from a vector formalism to a scalar one, rigorously only in a few cases (e.g., for guided waves), as an approximation in many other instances.

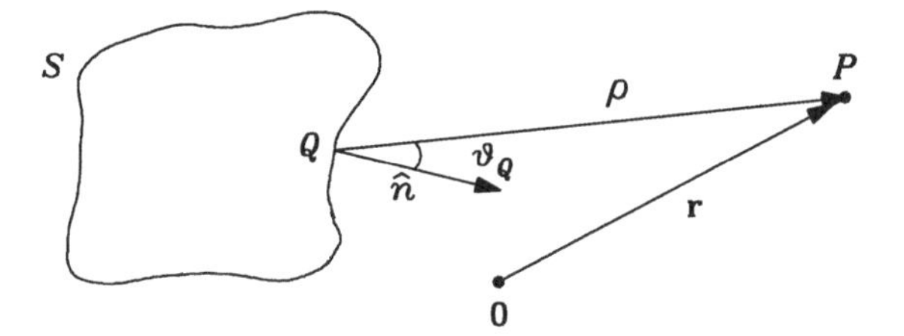

Figure 13.6 *Illustration of the symbols used in Eqs. (13.23) and (13.25).*

In this section, we will see that di raction belongs to the second set: strictly speaking, the scalar formulation will turn out to be just an approximation. On the other hand, its mathematics is much simpler than that of the rigorous vector approach. Another point in favor of the scalar theory is that it helps to link some simplifying approximations, which are very important in practical applications and will be discussed in the following sections, to physical intuition.

The starting point for the scalar theory is Eq. (11.4), namely:

$$(P) = \int_S \left[\frac{\partial \ (Q)}{\partial n} G \qquad (Q) \frac{\partial G}{\partial n} \right] \mathrm{d}S_Q \quad , \tag{13.23}$$

where G can be *any* Green's function, S is a closed surface, and P is a point belonging to the region towards which we orient the normal direction along which the derivatives are taken. If P is on the outer side of S (see Figure 13.6), we can imagine that the region P belongs to is surrounded by another surface at in nity, whose contribution to Eq. (13.23) is zero because of boundary conditions on it, similar to Sommerfeld's radiation conditions of Section 3.3, namely:

$$\lim_{r \to \infty} r \left| \frac{\partial}{\partial r} \right| = \lim_{r \to \infty} r \left| \ \right| = 0 \quad , \tag{13.24}$$

where r is the distance between P and an origin, which can be chosen arbitrarily as long as it is not at in nity.

13.4.1 The Helmholtz-Kirchho formulation

The conventional way to formulate the scalar theory, and at the same time that which is most similar to the vector theory, begins with the choice of the elementary spherical wave, Eq. (11.9), as the Green's function:

$$G = \quad \frac{1}{4} \frac{e}{\quad} \quad , \tag{13.25}$$

where $\ = P \quad Q = \mathbf{r} \quad \mathbf{r}'$. If we let (see Figure 13.6):

$$\cos \ _Q = \frac{\hat{n}}{| \ |} \quad , \tag{13.26}$$

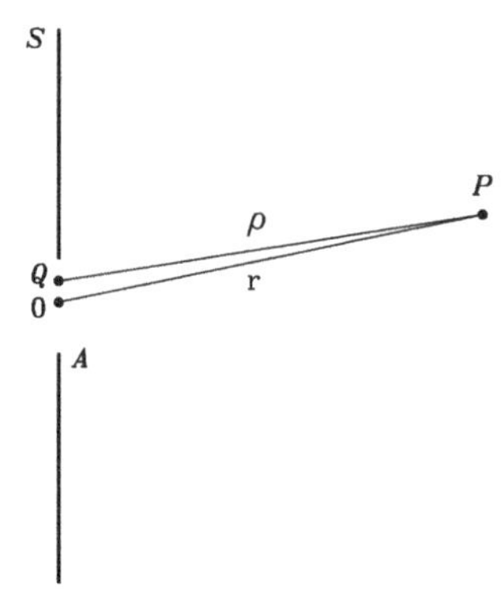

Figure 13.7 *An aperture A on a screen S.*

and note that ∇ forms an angle ϑ_Q with $\hat{n}$, we get:

$$\frac{\partial G}{\partial n} = \frac{\partial G}{\partial} \cos \ _Q = \frac{1}{4} \left(+ \frac{1}{} \right) \frac{e}{} \cos \ _Q \quad . \tag{13.27}$$

Consequently, Eq. (13.23) can be rewritten as:

$$(P) = \frac{1}{4} \int_S \frac{e}{} \left[(Q) \left(+ \frac{1}{} \right) \cos \ _Q \quad \frac{\partial}{\partial n} \right] \mathrm{d}S_Q \quad . \tag{13.28}$$

So far, this approach has been rigorous, as S is a closed surface. Passing now to an aperture A on a screen S (Figure 13.7), it is in current use to introduce the so-called *Kirchho 's boundary conditions*, which are essentially the same as those which lead us to Eqs. (13.3) and (13.4), namely:

$$(Q) \quad 0 \quad , \quad \frac{\partial \ (Q)}{\partial n} \quad 0 \qquad \text{for} \qquad Q \notin A \quad . \tag{13.29}$$

The scalar di raction integral reads then:

$$(P) = \frac{1}{4} \int_A \frac{e}{} \left[(Q) \left(+ \frac{1}{} \right) \cos \ _Q \quad \frac{\partial \ (Q)}{\partial n} \right] \mathrm{d}S_Q \quad . \tag{13.30}$$

Rigorously, the two conditions in Eq. (13.29) cannot hold together. Indeed, following exactly the same guidelines as in Section 13.2, one can prove that they would imply 0 everywhere. Nevertheless, Eq. (13.30) is very widely used in practice, because it yields results which agree well with experiments.

For further comparison with the vector theory, let us go back to Eq. (13.14), recall Eq. (13.27), and compare the result with Eq. (13.30). Then, we see that Eq. (13.30) may be looked at as one cartesian component of Eq. (13.14), *provided* that the second, third and fourth integrals in Eq. (13.14) vanish, so that we can rewrite it as:

$$\mathbf{E}(P) = \frac{1}{4} \int_A \frac{e}{} \left[\mathbf{E} \left(+ \frac{1}{} \right) \cos \ _Q \quad \frac{\partial \mathbf{E}}{\partial n} \right] \mathrm{d}S \quad . \tag{13.31}$$

It can be proved (see Fradin, 1961) that those three integrals vanish only if A

is a *closed surface*, which obviously has no contour . Henceforth, we see that using the scalar theory to evaluate a di racted e.m. eld entails an additional approximation with respect to the vector approach. This statement can be recon rmed by the following observation. If we use Eqs. (13.3) and (13.4) to calculate the far eld, then, as $\to \infty$ this eld tends to become TEM with respect to its direction of propagation. This result is correct, according to the general results of the previous chapter on the far eld. On the contrary, if we take Eq. (13.30) as a cartesian component of a vector eld, the implications are that — at least in some cases, for example when $\partial \ /\partial n = 0$ — the eld preserves the polarization it has on the aperture everywhere. This is not correct.

We will recall the last comment in Section 13.7, when we discuss under what conditions we can accept the scalar approximation in the so-called Fraunhofer region.

13.4.2 The Rayleigh-Sommerfeld formulation

One reason why in the previous approach we had to impose two conditions, Eqs. (13.29), was because we retained the Green's function we had found in Section 11.2, when we were dealing with a point source in free space. Two boundary conditions are more than the degrees of freedom in the partial differential equation we want to solve. As a consequence, we were forced to make a rather rough approximation. These di culties can be by-passed using a Green's function that satis es some boundary conditions which account for *spatial symmetries* in the di racting object.

This procedure can be applied in several cases. The simplest one is an aperture A in a planar screen (Figure 13.8). In this situation it is easy to de ne an odd Green's function and an even one, which are (with the upper and the lower sign, respectively):

$$G^{(i)}(Q) = \quad \frac{1}{4} \left(\frac{e}{\quad} \quad \frac{e \quad '}{\quad '} \right) \qquad (i = o, e) \quad , \qquad (13.32)$$

where the su ces o and e stand for odd and even, respectively. One can easily verify that the two functions in Eq. (13.32) satisfy the Helmholtz equation with *two point sources*, located at two points which are symmetrical with respect to the screen, P and P', and are in counterphase for $i = o$, in phase for $i = e$.

Reconsidering the procedure of Section 11.2, we see that P' does not belong to the domain of integration where we apply Green's theorem. Therefore, the source at P' does not yield any additional contribution, so the two functions in Eq. (13.32) are perfectly suitable for insertion into Eq. (13.23).

The reader can prove as an exercise that, since $| \ | = | \ '|$ and $'_Q = 2 \qquad _Q$

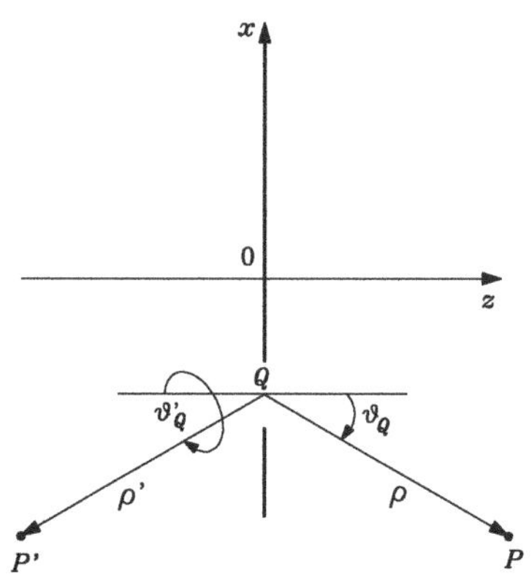

Figure 13.8 *Schematic illustration of the geometry involved in the de nition of even and odd Green's functions.*

at any point Q belonging to the $z = 0$ plane, then on that plane it is:

$$G^{(o)} = 0 \quad , \quad \frac{\partial G^{(o)}}{\partial n} = \frac{\partial G^{(o)}}{\partial z} = \frac{1}{2} \cos {}_Q \left(\; + \frac{1}{\;} \right) \frac{e}{\quad} ,$$

$$G^{(e)} = \frac{1}{2} \frac{e}{\quad} \quad , \quad \frac{\partial G^{(e)}}{\partial n} = 0 \quad . \tag{13.33}$$

Therefore, if we insert either one of Eqs. (13.32) into Eq. (13.23), one of the two terms in the integral disappears. Consequently, we may proceed further imposing, at any point on the screen except for the aperture, *either* $\;= 0$, *or* $\partial \;/\partial n = 0$. In conclusion, this procedure is not a ected by the inconsistencies that we outlined for the previous one.‡

There are no logical reasons why one should prefer either $G^{(o)}$, or $G^{(e)}$. However, it is more customary to use $G^{(o)}$, probably because intuition suggests to think of the screen as an ideal electric conductor, and of the point sources at P and P' as two electric currents owing on $z =$ constant planes. This is a purely conventional matter, especially in the framework of the scalar theory where the physical dimensions of the wave function are irrelevant.

With this choice for the Green's function, the di raction integral gets the following shape, known as the *Rayleigh-Sommerfeld* formulation:

$$(P) = \frac{1}{2} \int_A (Q) \left(\; + \frac{1}{\;} \right) \frac{e}{\quad} \cos {}_Q \, \mathrm{d}S_Q \quad . \tag{13.34}$$

Thinking of as one cartesian component of a vector eld, Eq. (13.34) recon- rms our previous comments on the polarization of the incident eld, preserved by the di racted eld.

‡ One should not confuse self-consistency of an approximate solution with its *precision*. The latter depends on the number of terms which are kept when the expansion as a power series in is truncated.

13.5 Di raction formulas and Rayleigh-Sommerfeld

Henceforth, for the sake of simplicity we will refer only to the scalar theory, unless the opposite is explicitly stated. In most of their applications, the expressions obtained in the previous sections are used under the assumption that the di racted eld propagates in a *lossless medium*, letting:

$$= jk = j\omega\sqrt{\ } = j\,\frac{2}{\ } \quad . \tag{13.35}$$

In most instances, one is interested in the di racted eld only *far from the aperture*, on the legth scale of the wavelength. Consequently, in Eqs. (13.27) and (13.33), or more in general in all relationships involving normal derivatives of Green's functions, the following approximation can be used, with the usual warning about its validity for the moduli but not for the phases of the complex numbers:

$$k = \frac{2}{\ } \quad \frac{1}{\ } \quad . \tag{13.36}$$

This simpli es Eqs. (13.28) and (13.34) in a rather obvious way.

Another condition which is frequent in practice, especially at optical frequencies, is that the aperture A is located *in the far eld of the primary source.* Then, arising from what we said in Section 13.3, the illuminating eld on the aperture is completely characterized by the *equivalent moment* of its source. On the other hand, we must remember from Chapter 11 that the equivalent moment was de ned for a source in an unbounded homogeneous medium. However, for (see Figure 13.9):

$$|\mathbf{d}| \qquad , \qquad |\mathbf{d}| \quad |\mathbf{d}'|_{\mathrm{MAX}} \quad , \tag{13.37}$$

the illuminating eld on A is of the type:

$$(Q) = F(\ ',\ ')\,\frac{e^{\ jkd}}{d} \quad , \tag{13.38}$$

where $'$ and $'$ are the angular coordinates of the point Q in a spherical coordinate system which we use as reference for the source of the illuminating eld, S_i. As for the function $F(\ ',\ ')$, it is a volume integral over S_i and is proportional to the equivalent moment. The proportionality constant depends on the assumptions one makes about the screen S, in the same sense as we said in Section 13.3.

Eqs. (13.37) allow us to make also the following approximation:

$$\left.\frac{\partial}{\partial n}\right|_Q = \frac{\partial}{\partial d}\cos\ _0 = F(\ ',\ ')\,(\ jk)\cos\ _0\,\frac{e^{\ jkd}}{d} \quad . \tag{13.39}$$

Therefore, when all the assumptions that we introduced so far in this section hold, the scalar di raction integral (13.30) can be approximated as:

$$(P) = \frac{jk}{2}\int_A F(\ ',\ ')\,\frac{e^{\ jk(\ +d)}}{d}\left[\frac{\cos\ _0 + \cos\ _Q}{2}\right] dA \quad . \tag{13.40}$$

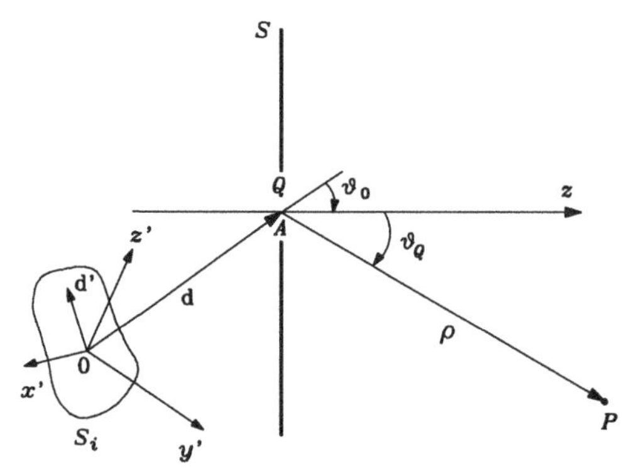

Figure 13.9 *Illustration of the symbols used in the equations of this section.*

Often, from the origin O to which we refer the primary source the aperture A is seen under an angle small enough to allow us to let:

$$F(\ ',\ ') = F = \text{constant} \quad , \tag{13.41}$$

in Eq. (13.40), i.e., to take *locally* a uniform spherical wave, centered at O, as the illuminating eld. Now the constant F can be taken out of the integral, and so Eq. (13.40) becomes what is known in the literature as the *Fresnel-Kirchho formula.*

If the same approximations are introduced in Eq. (13.34), then it is easy to prove that one obtains:

$$(P) = \frac{jk}{2} \int_A F(\ ',\ ') \frac{e^{\ jk(\ +d)}}{d} \cos\ _Q \, \mathrm{d}A \quad . \tag{13.42}$$

When Eq. (13.41) holds, Eq. (13.42) becomes what is known as the *Rayleigh-Sommerfeld formula.* It di ers from Eq. (13.40) only in the term which accounts for the angles ϑ_0 and ϑ_Q, usually referred to as the *obliquity factor.*

Eqs. (13.40) and (13.42) can be interpreted as two of the numerous quantitative formulations of what goes under the generic name of "Huygens' principle". Indeed, let us consider:

$$F(\ ',\ ') \cos\ _0 \, e^{\ jkd} \tag{13.43}$$

(or the same quantity without the factor $\cos\ _0$, depending on the formulation) as a *secondary source*, whose intensity is proportional to that of the primary source S_i. The di racted eld can be considered as being radiated by these secondary sources.

When $F(\ ',\ ') = \text{constant}$ can be taken out of the integral, either Eq. (13.40) or Eq. (13.42), then the di racted eld depends mainly on the *shape* of the di racting aperture. The exponential functions in these integrals suggest that this shape dependence may be linked to Fourier transforms. We will return to this point in Section 13.7, after introducing further approximations.

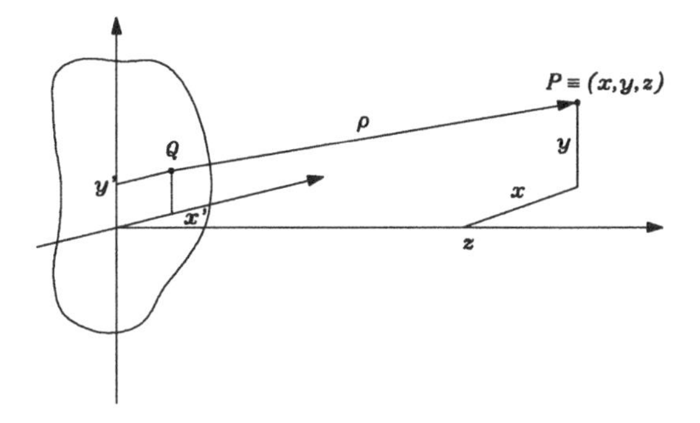

Figure 13.10 *The reference frame used in the de nition of the Fresnel region.*

13.6 The Fresnel di raction region

In the previous section, we saw how the di raction integrals can be remarkably simpli ed when the distance between the primary source and the di racting object is large. Now, we will assume that the distance between the di racting object and the *observation point* is large, with respect not only to the wavelength but also to *all the other geometrical dimensions* involved in the problem.

Let us introduce a rectangular coordinate frame whose origin is on the aperture A, and whose (x, y) plane is tangential to A in the origin (Figure 13.10). As usual, $P \quad (x, y, z)$ is a generic observation point, while $Q \quad (x', y', z' = 0)$ is any point on the aperture A.

In many instances, the far eld is of practical interest only within a *paraxial region*; therefore, we are allowed to suppose:

$$z \quad |x| \quad , \quad |x'| \quad , \quad |y| \quad , \quad |y'| \quad . \tag{13.44}$$

Consequently, in Eqs. (13.30) and (13.34), and henceforth in Eqs. (13.40) and (13.42), we can set:

$$\cos \quad_Q = 1 \quad . \tag{13.45}$$

The exact expression for the distance between P and Q, for these formulas, would be:

$$= \left[z^2 + (x \quad x')^2 + (y \quad y')^2\right]^{1/2} \quad . \tag{13.46}$$

However, since z is independent of the point Q on A, functions of z only can be brought out of the integral (13.42). This entails that we are looking for an approximation for Eq. (13.46), and we will use the binomial expansion of the square root. The number of terms to save in this expansion determines the accuracy of the approximation, therefore we must discuss it following the same lines as in Section 12.2 where we dealt with the equivalent moment of an antenna. Let us deal, for the sake of simplicity, with Eq. (13.40) or Eq. (13.42). In their denominator, we are certainly allowed to let:

$$= z \quad . \tag{13.47}$$

For the factor exp(jk) we must make comparisons not only among the quantities which appear in Eq. (13.44), but also with the wavelength. The reader can prove as an exercise that, if at any point $Q \quad (x', y', 0)$ located on the aperture the following condition is satis ed:

$$z^3 \quad \left[(x \quad x')^2 + (y \quad y')^2\right]^2 \quad , \tag{13.48}$$

then it is legitimate to limit the expansion to the rst term, i.e., to set:

$$= z \left[1 + \frac{1}{2}\left(\frac{x \quad x'}{z}\right)^2 + \frac{1}{2}\left(\frac{y \quad y'}{z}\right)^2\right] \quad . \tag{13.49}$$

We will come back shortly to Eq. (13.49), and show that its range of validity is, in fact, much broader than Eq. (13.40).

The region where Eq. (13.49) holds is referred to as the *Fresnel di raction region* (or Fresnel zone). In it, Eq. (13.40) can be simpli ed even further. If the constant factor $e^{\quad jkd}$ is incorporated into F, that equation becomes:

$$\begin{aligned} (P) \quad &= \quad \frac{j \; e^{\quad jkz}}{z} \int\!\!\int_A \frac{F(1 + \cos \quad _0)}{2d} \; e^{\quad j(k/2z)\,[(x \quad x')^2 + (y \quad y')^2]}\, \mathrm{d}x' \, \mathrm{d}y' \\ &= \quad \frac{j \, e^{\quad jkz}}{z} \; e^{\quad j(k/2z)\,(x^2+y^2)} \int\!\!\int_A \frac{F(1 + \cos \quad _0)}{2d} \\ & \qquad e^{\quad j(k/2z)\,(x'^2+y'^2)} \quad e^{j(k/z)\,(xx'+yy')}\, \mathrm{d}x' \, \mathrm{d}y' \quad . \end{aligned} \tag{13.50}$$

When we use Eq. (13.50) to compute the di racted eld, we speak of *Fresnel di raction.*

Eq. (13.50) lends itself to a comment, in connection with other comments made in Chapter 4 about completeness of plane waves. Ignoring, for the time being, the factor in front of the integral, which represents only a phase delay and a bending of the wave fronts, we see that the eld in the Fresnel region appears as a two-dimensional *Fourier transform* of the function:

$$(Q) = \frac{F(1 + \cos \quad _0)}{2d} \; e^{\quad j(k/2z)\,(x'^2+y'^2)} \quad , \tag{13.51}$$

if the quantities:

$$f_x = \frac{kx}{2 \quad z} = \frac{x}{\quad z} \quad , \qquad f_y = \frac{ky}{2 \quad z} = \frac{y}{\quad z} \quad , \tag{13.52}$$

are looked at as *spatial frequencies.* Hence, the plane-wave spectrum of the di racted eld is linked to the illuminating eld on the aperture through a proportionality factor $F(1 + \cos \quad _0)/2d$. Further implications of this comment will be outlined in the next section, and then again in Section 13.12.

The spatial frequencies in Eq. (13.52) can be interpreted as *angular apertures* normalized with respect to the wavelength. Thanks to the assumptions in Eq. (13.44), all angles in the problem at hand can be considered small, so we may write:

$$\vartheta_x = \frac{x}{z} = f_x \quad , \qquad \vartheta_y = \frac{y}{z} = f_y \quad . \tag{13.53}$$

A well-known theorem about Fourier transforms tells us that a change in scale of the independent variables in Eq. (13.51), expressed by a factor D ($x'' = Dx'$, $y'' = Dy'$), causes a change D^{-1} in the corresponding scale of the spatial frequencies (13.52). This factor D^{-1} shows up in Eq. (13.53) too, if the scale is changed. This proves the following fundamental property of diffraction phenomena: the *angular aperture* of the lobes diffracted by an aperture is *inversely proportional to the linear size* of the aperture itself, measured using the wavelength as a unit. This property is correlated to what we said in Chapter 4 about the fact that sizes and angles are conjugate variables. Also, it is correlated to what we said in Chapter 5 on Gaussian beams, and in Chapter 12 on the effective area of an antenna. Additional correlated subjects will be pointed out in Chapter 14.

Let us go back now to the question that we left aside for some time, about the approximations which define the Fresnel region. The condition (13.48) is certainly a sufficient one for Eq. (13.49) to hold, but is not necessary for Eq. (13.50) to hold. To prove this, suppose, for simplicity, that the factor $F(1 + \cos\theta_0)$ is constant. Let the next higher-order term in the expansion of the distance be accounted for, and call it t for brevity. Suppose that t is small enough so that:

$$e^{-jkt} = 1 - jkt \quad . \tag{13.54}$$

Then, the consequent correction on Eq. (13.50) is given by:

$$-jk \int\!\!\int_A t\, e^{-j(k/2z)\,[(x-x')^2+(y-y')^2]}\, \mathrm{d}x'\, \mathrm{d}y' \quad . \tag{13.55}$$

Thanks to the averaging effect of the exponential functions with an imaginary argument, which phase-mismatch contributions coming from different points on A, the quantity (13.55) can be neglected globally even when t is not negligible locally.

The same line of thinking can help us to answer the following question. What is the *minimum size* of an aperture, in order that the diffracted field at a given point of observation be essentially the same as if there were no screen? This problem will be discussed in Section 13.10.

13.7 The Fraunhofer diffraction region

Assume now that another condition, involving the wavelength and more restrictive than Eq. (13.44), is also satisfied, in addition to Eq. (13.44) itself. Namely, suppose that:

$$z \gg \frac{k}{2}\,(x'^2 + y'^2) \quad , \tag{13.56}$$

for any point $Q \quad (x', y', 0)$ belonging to the aperture A.[§]

The region where Eq. (13.56) is satis ed is referred to as the *Fraunhofer di raction region* (or zone). Strictly speaking, the Fraunhofer region is therefore a subset of the Fresnel region; however, normally the phrase "Fresnel region" is used to indicate the region where Eq. (13.48) is satis ed but Eq. (13.56) is not.

Eq. (13.56) entails that the factor $\exp[\ \ jk(x'^2 + y'^2)/2z]$ in Eq. (13.50) can be replaced with unity. So, we get:

$$(P) = \frac{j\, e^{\ \ jkz}}{z}\, e^{\ \ j(k/2z)\,(x^2+y^2)} \int\!\!\int_A F\ \frac{1+\cos\ \ _0}{2d}\ e^{j(k/z)\,(xx'+yy')}\, \mathrm{d}x'\, \mathrm{d}y' \,. \tag{13.57}$$

Comparing with Eq. (13.50), we see that the di racted eld in the Fraunhofer region is the *Fourier transform of the illuminating eld on the aperture*, i.e.:

$$(Q) = F\ \frac{1+\cos\ \ _0}{2d} \quad . \tag{13.58}$$

One fallout of this result, of paramount practical importance, is *holography*. This subject cannot be included in this book because of space limitation, but the reader should consult the broad, high-level literature which is available on it, e.g., Goodman (1968).

In the Fraunhofer region, the scalar theory and the vector theory agree well, without any need to postulate awkward conditions like those that were mentioned in Section 13.4. To prove this statement, take Eqs. (13.12) and (13.13), assume that the medium is lossless ($\ \ = jk$), use the approximation $\ \ = z$ in the denominator, and let:

$$= (\mathbf{r} \quad \mathbf{r}') \quad \hat{r} = z + \frac{x^2+y^2}{2z} \quad \left(x'\,\frac{x}{z} + y'\,\frac{y}{z} \right) \quad , \tag{13.59}$$

in the exponential factors. To simplify the calculation of the gradient of $e^{\ \ jk}\ /\ $, use spherical coordinates centered at O (see Figure 13.11).

Using those arguments which allowed us, in Section 12.3, to neglect all terms in comparison with the term jk which comes out from the radial derivative, we get:

$$\nabla\ \frac{e^{\ \ jk}}{} = \frac{jk}{z}\ e^{\ \ jkz}\ e^{\ \ j(k/2z)\,(x^2+y^2)}\ e^{j(k/z)\,(xx'+yy')}\ \hat{r} \quad , \tag{13.60}$$

and then, with a few more steps:

$$(\mathbf{a}\ \nabla)\,\nabla\ \frac{e^{\ \ jk}}{} = \quad k^2(\mathbf{a}\ \hat{r})\ \frac{e^{\ \ jkz}\ e^{\ \ j(k/2z)\,(x^2+y^2)}\ e^{j(k/z)\,(xx'+yy')}}{z}\ \hat{r}\,, \tag{13.61}$$

where either $\mathbf{a} = \hat{n} \quad \mathbf{H}$, or $\mathbf{a} = \mathbf{E} \quad \hat{n}$. From this, it follows that Eqs. (13.12) and

[§] The reader should compare this condition with what we said in Section 12.2 when we de ned the equivalent moment of a source. Note, in particular, the exponential factor in the integral in Eq. (13.57), and compare it with Eq. (12.9).

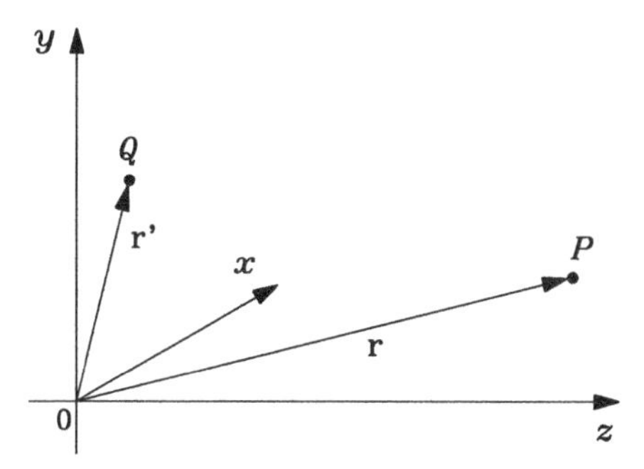

Figure 13.11 *Geometrical illustration of the symbols used in the approximation (13.59).*

(13.13) can be written as follows (we let, as usual, $k^2 = \omega^2 \quad , \quad = (\ / \ )^{1/2}$):

$$\mathbf{E}(P) = \frac{jk}{4 \ z} e^{\ jkz} e^{\ j(k/z)\,(x^2+y^2)}$$

$$\left[\int_A \hat{r} \quad (\mathbf{E} \quad \hat{n}) \, e^{j(k/z)\,(xx'+yy')} \, \mathrm{d}A \right.$$

$$\left. + \int_A \hat{r} \quad [\hat{n} \quad \mathbf{H}] \quad \hat{r} \, e^{j(k/z)\,(xx'+yy')} \, \mathrm{d}A \right] , \quad (13.62)$$

$$\mathbf{H}(P) = \frac{jk}{4 \ z} e^{\ jkz} e^{\ j(k/z)\,(x^2+y^2)}$$

$$\left[\frac{1}{\ } \int_A \hat{r} \quad [\mathbf{E} \quad \hat{n}] \quad \hat{r} \, e^{j(k/z)\,(xx'+yy')} \, \mathrm{d}A \right.$$

$$\left. + \int_A (\hat{n} \quad \mathbf{H}) \quad \hat{r} \, e^{j(k/z)\,(xx'+yy')} \, \mathrm{d}A \right] . \quad (13.63)$$

Clearly, these expressions look very similar to Eq. (13.57). The similarity can be enhanced even further, by decomposing the total eld $\{\mathbf{E}, \mathbf{H}\}$ in Eqs. (13.62) and (13.63) as the sum of two elds: the rst one, $\{\mathbf{E}_m, \mathbf{H}_m\}$, is generated by magnetic currents on A, while the second one, $\{\mathbf{E}_e, \mathbf{H}_e\}$, is generated by electric currents on A. Since the unit-vector $\hat{r}$ does not depend on the integration point on A, if we set, for brevity, $\ = k[z + (x^2/z) + (y^2/z)]$, we nally get:

$$\mathbf{E}_m = \frac{j \, e^{\ j}}{2 \ z} \hat{r} \quad \int \int_A (\mathbf{E} \quad \hat{n}) \, e^{j(k/z)\,(xx'+yy')} \, \mathrm{d}A \quad ,$$

$$\mathbf{H}_m = \frac{1}{\ } \mathbf{E}_m \quad \hat{r} \quad , \quad (13.64)$$

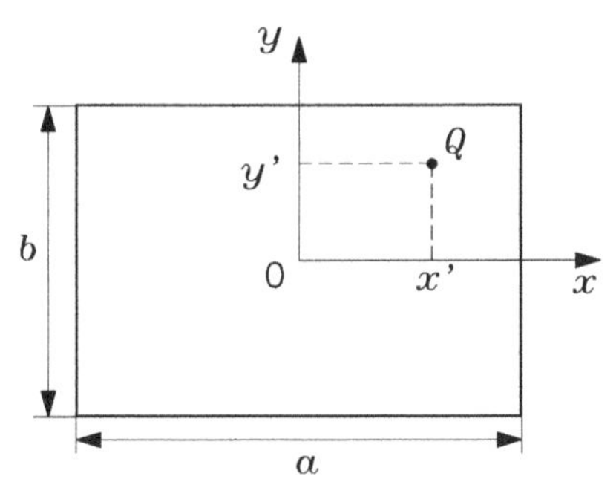

Figure 13.12 *A rectangular aperture and its reference frame.*

$$\mathbf{H}_e = j\,\frac{e^{\ \ j}}{2\ \ z}\,\hat{r}\ \int\!\!\int_A (\hat{n}\quad \mathbf{H})\, e^{j(k/z)\,(xx'+yy')}\,\mathrm{d}A \quad ,$$

$$\mathbf{E}_e = \quad \mathbf{H}_e \quad \hat{r} \quad . \tag{13.65}$$

So, in the Fraunhofer region the vector theory gives an electric and a magnetic eld which are both *transverse* to the direction of propagation (de ned by the unit-vector $\hat{r}$). Furthermore, the wave impedance in the direction of $\hat{r}$ equals the intrinsic impedance of the medium, as it has to be for a TEM eld. These results match with those of Chapter 12 for the far eld of any source, as well as with those of Chapter 5 concerning the consequences of the scalar approximation. The agreement can be extended even further, assuming that the aperture is planar. In this case, from Eqs. (13.44), which are still valid, it follows that:

$$\hat{r} = \hat{n} \quad . \tag{13.66}$$

Comparing Eq. (13.64) and the rst of Eqs. (13.65), we see that in the Fraunhofer region the eld *polarization* is the same as that of the *transverse components of the illuminating eld on the aperture.* We conclude that in these conditions the scalar theory is satisfactory. Comparison between two examples, 13.8.2 and 13.8.3 in the next section, will make these concepts easier to understand.

13.8 Examples

The contents of this section will refer, unless the opposite is stated, to the Fraunhofer region. We will use mainly the scalar theory, and invoke the vector theory in just a few cases where comparison between theories may be enlightening.

13.8.1 Rectangular aperture

De ne the aperture (Figure 13.12) as the set of those points $Q \quad (x', y', 0)$ which satisfy:

$$|x'| \quad a/2 \quad , \qquad |y'| \quad b/2 \quad . \tag{13.67}$$

Assume that the illuminating eld is a plane wave, traveling along the z axis. Henceforth, irrespectively of whether the screen is perfectly absorbing or perfectly conducting, the eld on the aperture is *constant.* Therefore we may replace Eq. (13.58) with:

$$\begin{cases} \ (Q) = F \dfrac{1+\cos \ _0}{2d} = U = \text{constant} & \text{for } |x'| \quad a/2\ ,\ |y'| \quad b/2 \quad , \\ \ (Q) = 0 & \text{otherwise} \quad . \end{cases} \tag{13.68}$$

Then, Eq. (13.57) gives the following di racted eld:

$$\begin{aligned} \ (P) &= Uj \frac{e^{\ jkz}}{z} e^{\ j(k/2z)(x^2+y^2)} \int_{\ (a/2)}^{a/2} \mathrm{d}x' \int_{\ (b/2)}^{b/2} \mathrm{d}y' \, e^{j(k/z)(xx'+yy')} \\ &= Uj \frac{e^{\ jkz}}{z} e^{\ j(k/2z)(x^2+y^2)} \, ab \, \frac{\sin x}{x} \frac{\sin y}{y} \quad , \end{aligned} \tag{13.69}$$

where, for our own convenience, we introduced the *normalized* coordinates:

$$x = \frac{ax}{z} \quad , \qquad y = \frac{by}{z} \quad . \tag{13.70}$$

For the reasons that were outlined in Chapter 12 when dealing with general properties of antennas, also here it is of interest to compute the *normalized radiation intensity.* In the scalar case, it is de ned, by a straightforward analogy with Eq. (12.34), as:

$$i = \frac{\ (P) \quad (P)}{\ (M) \quad (M)} \quad , \tag{13.71}$$

where M is a xed point, chosen in the direction along which $| \ |$ reaches its maximum value; the point P, on the other hand, spans the entire spherical surface of radius $r = |M \quad O|$. Inserting Eq. (13.69) into Eq. (13.71), we get:

$$i = \frac{\sin^2 x}{x^2} \frac{\sin^2 y}{y^2} \quad . \tag{13.72}$$

Figure 13.13 shows a plot of this quantity, as a function of one of the normalized coordinates, calculated where the other normalized coordinate equals zero. The angular width of the main lobe — between zero- eld directions — is then given by:

$$\frac{2 \ ax}{z} = 2 \quad , \tag{13.73}$$

and, as the approximations in Eq. (13.44) allow us to replace the angle $\ _x$ with its tangent (x/z), we may also write:

$$\ _x = \frac{x}{z} = \frac{\ }{a} = \frac{2}{ka} \quad . \tag{13.74}$$

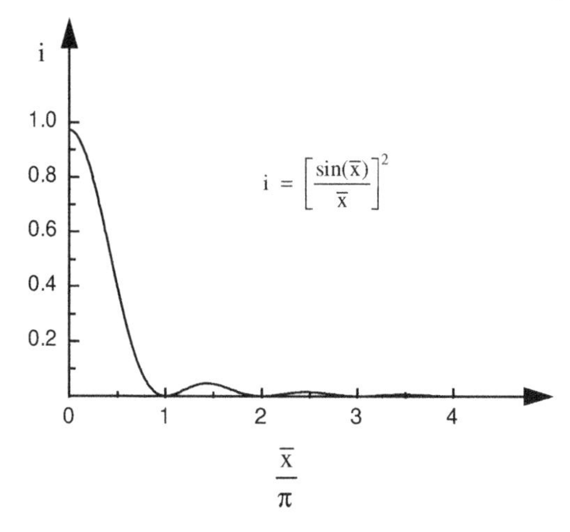

Figure 13.13 *Normalized radiation intensity in the Fraunhofer region of a rectangular aperture.*

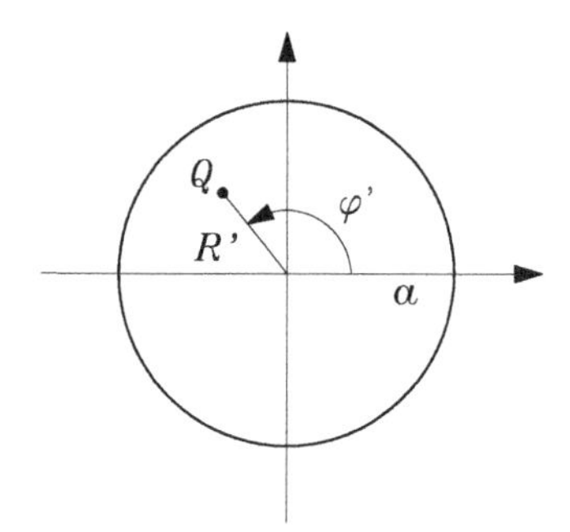

Figure 13.14 *A circular aperture and its reference frame.*

Let us go back to the comments on completeness of plane waves (see Chapter 4), and to the interpretation of the eld in the Fraunhofer region as the spatial Fourier transform of the eld on the aperture. Then, the reader can prove as an exercise that Eq. (13.74) is the *minimum* angular width for a plane-wave beam which passes through an aperture of width a. For this reason, a set of plane waves satisfying Eq. (13.74) is referred to as a *di raction-limited* beam.

13.8.2 Circular aperture: the scalar approach

De ne the aperture as the set of those points Q which satisfy (see Figure 13.14):

$$R' \quad a \; . \tag{13.75}$$

If the illuminating eld is a plane wave, then Eq. (13.68) holds again. To simplify the computation of Eq. (13.57), it is useful to adopt cylindrical coordinates,¶ setting:

$$x = R\cos\ \ , \quad y = R\sin\ \ , \quad x' = R'\cos\ ' \ , \quad y' = R'\sin\ ' \ . \tag{13.76}$$

Taking out of the integral whatever does not depend on Q, we are left with the following expression to compute:

$$\int_0^a \mathrm{d}R' R' \int_0^{2} \mathrm{d}\ ' \, e^{j(k/z)\, RR' \cos(\ \ \ ')} \ . \tag{13.77}$$

Let us recall a Bessel function identity, which was already used in Subsection 12.7.3, namely:

$$J_0(u) = \frac{1}{2} \int_0^{2} e^{ju\cos\ '} \mathrm{d}\ ' \ . \tag{13.78}$$

We obtain then from Eq. (13.77) the following expression for the di racted eld, known as the Bessel-Fourier transform of the eld on the aperture:

$$(P) = jU \frac{e^{\ jkz}}{z} e^{\ j(kR^2/2z)} \ 2 \int_0^a R' J_0\left(\frac{k}{z} RR'\right) dR' \ . \tag{13.79}$$

If we exploit also the following identity (see Appendix D):

$$\int_0^x J_0(\ )\, \mathrm{d}\ = x J_1(x) \ , \tag{13.80}$$

the nal result is:

$$\begin{aligned} (P) &= jU \frac{e^{\ jkz}}{z} e^{\ j(kR^2/2z)} \left(\frac{z}{kR}\right)^2 \frac{k}{z} Ra\, J_1\left(\frac{k}{z} Ra\right) 2 \\ &= jU\, e^{\ jkz} e^{\ j(kR^2/2z)} \frac{a}{R} J_1\left(\frac{k}{z} Ra\right) \ . \end{aligned} \tag{13.81}$$

The corresponding normalized radiation intensity can be obtained inserting Eq. (13.81) into Eq. (13.71). It is then advisable to use the following normalized radial coordinate:

$$R = \frac{2a\, R}{z} \ , \tag{13.82}$$

where, to emphasize similarity with Eqs. (13.70), we used the diameter of the aperture as the normalization reference. Recalling that the function $J_1(u)/u$ reaches its maximum, $(1/2)$, for $u = 0$, we may write:

$$i = 4 \left| \frac{J_1(R)}{R} \right|^2 \ . \tag{13.83}$$

¶ Do not confuse the cylindrical coordinate R, that we are going to introduce now, with the distance r between the observation point P and the origin O, which is a spherical coordinate. In some of the following examples, we will be forced to use both cylindrical and spherical coordinates, superimposed on each other.

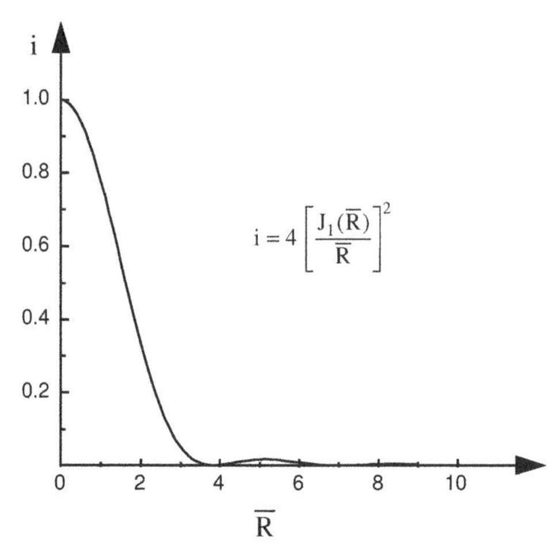

Figure 13.15 *Normalized radiation intensity in the Fraunhofer region of a circular aperture.*

The corresponding plot, shown in Figure 13.15, is often referred to as the *Airy diagram.*

13.8.3 Circular aperture: the vector approach

The eld di racted by a circular aperture is of major technical interest since the "mouth" of the widely used *parabolic antenna* is in fact a circular aperture. Comparison between this subsection and the previous one will outline that the theoretical discussion of Section 13.7 has practical implications. The next subsection, in turn, will underline the important role of the assumptions about the eld on the aperture. For the time being, let us take an illuminating eld as simple as possible: a plane wave traveling in the z direction, orthogonal to a perfectly absorbing screen. Polarization is assumed to be linear.

All this means that in Eqs. (13.64) and (13.65), with a suitably chosen reference frame, we may let:

$$\begin{aligned} \mathbf{E} &= \mathbf{E}_0 = \text{cost} = E_0\,\hat{a}_x \quad , \\ \mathbf{H} &= \frac{E_0}{\ _0}\,\hat{a}_y \quad . \end{aligned} \tag{13.84}$$

The approximation in Eq. (13.66), if accepted over the whole region of interest, entails $\hat{r} = \hat{a}_z$. Then, it is easy to test that nothing fundamental changes, in the vector approach, compared with the previous example 13.8.2.

If, on the contrary, we account for the true direction of the unit vector $\hat{r}$, speci ed by the observation point P, then, using the spherical reference frame shown in Figure 13.16, we get:

$$\hat{r} \quad \mathbf{E} \quad \hat{n} = E_0(\cos\varphi\ \hat{}\quad \cos\quad \sin\varphi\hat{\varphi}) \quad ,$$

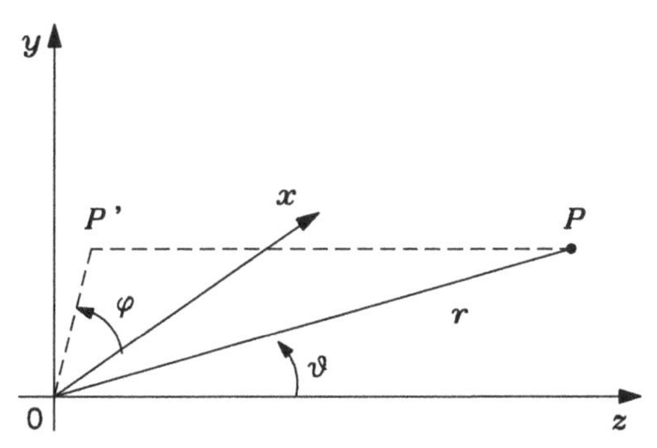

Figure 13.16 *The reference frame used in the calculation of the vector approach to a circular aperture.*

$$\hat{r} \quad (\hat{n} \quad \mathbf{H}) \quad \hat{r} \;=\; E_0(\cos \quad \cos\varphi\,\hat{} \quad \sin\varphi\hat{\varphi}) \quad . \tag{13.85}$$

Inserting these expressions into Eqs. (13.64) and (13.65), we get:

$$\mathbf{E} \;=\; \mathbf{E}_m + \mathbf{E}_e = j\,\frac{e \quad j}{2 \quad z}\;E_0(1+\cos \quad)\;(\cos\varphi\,\hat{} \quad \sin\varphi\hat{\varphi})$$

$$\int_0^a \mathrm{d}R'R' \int_0^{2} \mathrm{d} \quad ' \; e^{j(k/z)\,RR'\,\cos(\quad ')} \quad ,$$

$$\mathbf{H} \;=\; \frac{1}{-}\,\hat{r} \quad \mathbf{E} \quad . \tag{13.86}$$

The integral in Eqs. (13.86) is exactly the same as in Eq. (13.77), that we computed a short time ago for the scalar case. So, changes with respect to the scalar theory arise only because Eqs. (13.86) contain the factors $(1+\cos \quad)$ and $(\cos\varphi\,\hat{} \quad \sin\varphi\hat{\varphi})$. The latter a ects the eld *polarization*, and shows that the di racted eld does not remain parallel to the incident eld, for $\hat{r} \neq \hat{a}_z$. This agrees with the fact that the far eld *must be TEM* with respect to its direction of propagation. On the other hand, this factor does not a ect the spatial distribution of the normalized radiation intensity. Indeed, a straightforward calculation yields:

$$(\cos\varphi\,\hat{} \quad \sin\varphi\hat{\varphi}) \;\; (\cos\varphi\,\hat{} \quad \sin\varphi\hat{\varphi}) = \cos^2\varphi + \sin^2\varphi = 1 \quad . \tag{13.87}$$

The factor $(1+\cos \quad)$ does a ect the behavior of the normalized radiation intensity i, since:

$$|1+\cos \quad |^2 = 4\cos^2 \frac{}{2} \quad . \tag{13.88}$$

Hence, for the vector case, the Airy diagram of Figure 13.15 has to be redrawn, as Eq. (13.83) must be multiplied by Eq. (13.88). This correcting factor is close to unity as long as $\quad (\simeq R/z)$ is much smaller than $\quad$. So, the main lobe of Figure 13.15 is almost una ected by vector correction, provided that the aperture diameter $2a$ is not too small compared to the wavelength. We may conclude that under these circumstances it makes sense to compute the directivity of a parabolic antenna from Eqs. (13.83). The result is:

$$d = \left(\frac{2\ \ a}{\quad}\right)^2 = \frac{4}{\ ^2}\,(\ \ a^2) \quad . \tag{13.89}$$

Eqs. (13.89) and (12.56) show that under these conditions the *e ective area* of a lossless parabolic antenna coincides exactly with the *geometrical area* of its mouth.

Let us stress that, on the other hand, the correcting factor in Eq.(13.88) a ects in a signi cant way the radiation pattern (even within the main lobe) if the condition a is not satis ed.

13.8.4 Parabolic antenna: a comment on the illuminating eld

We noted several times that the main reason why we deal often with circular apertures is because parabolic re ectors are popular antennas. In the two previous examples, the illuminating eld was assumed to be a uniform plane wave. A point source located at the focus F of a parabolic re ector generates a plane wave on the aperture only within the *ray optics approximation.* This entails that this approach yields approximate results, inevitably, as it neglects:

- the nite size of the actual primary source, called *feeder*;
- the nite size of the parabolic re ector surface;
- the wavelength, which in most cases is not negligibly small compared to the geometrical dimensions involved in the problem, because there are obvious practical limits to the antenna size and weight.

An additional drawback of ray optics is that it does not provide any hint on how to design the feeder. Modeling the illuminating eld with improved accuracy is a problem which has been tackled in many ways. Among them, those which are easier to link with the contents of this book are:

- spherical-harmonics expansion (Chapter 11);
- Gaussian-beam expansion (Chapter 5).

To learn more about both methods, as well as about other approaches, the reader should consult specialized texts.

13.8.5 Sinusoidal gratings

Until now, we assumed always that the eld was zero at all points of the screen which do not belong to the aperture, and proportional to the incident eld on the aperture. Let us deal now with a completely di erent example. Suppose that the features of the screen vary continuously and periodically as a function of one coordinate: the screen induces a spatial modulation on the eld transmitted through it. We will study in detail the case of an *amplitude modulation*, taking for simplicity a purely *sinusoidal* modulation. The reader can nd in the literature many other examples, e.g., phase modulation (Goodman, 1968). Many technological processes can be used to fabricate objects whose characteristics match more or less these assumptions: holography, selective depositions followed by di usion, etching, surface acoustic waves, etc. The object itself is referred to as a *di raction grating.*

Let us deal rst, for simplicity, with a grating which extends to in nity, although the approximations in Eq. (13.44) break down in this situation. We will bypass this obstacle a little later. Let $Q \quad (x', y', 0)$ be any point on the grating, and suppose that the grating is illuminated by a plane wave, traveling normal to the grating plane. We can write:

$$(Q) = U\left[\frac{1}{2} + \frac{1}{2} m \cos(2 \quad f_0 x')\right] \quad , \tag{13.90}$$

where the modulation index m falls between 0 and 1.

As we said in Section 13.7, the eld in the Fraunhofer region is proportional to the Fourier transform of the illuminating eld. Hence, it is:

$$\mathcal{F}\left[\frac{1}{2} + \frac{1}{2} m \cos(2 \quad f_0 x')\right] = \frac{1}{2} \quad (f_x) + \frac{1}{4} m \quad (f_x + f_0) + \frac{1}{4} m \quad (f_x \quad f_0) \, , \tag{13.91}$$

where $f_x = (x/ \quad z)$ and stands for the Dirac delta function. This shows that the eld transmitted through the grating is the sum of a plane wave of amplitude $U/2$, traveling in the z direction, plus two plane waves (each of amplitude $mU/4$), traveling along the directions de ned by the following equations:

$$f_x \quad f_0 = 0 \quad , \tag{13.92}$$

corresponding to the angles:

$$_B = \quad \arctan \frac{x}{z} = \quad \frac{x}{z} = \quad \frac{2 \quad f_0}{k} = \quad f_0 \quad . \tag{13.93}$$

The angle $_B$ de ned by Eq. (13.93) is called the *Bragg angle*, and the phenomenon we are analyzing here is referred to as *Bragg's di raction.* The central plane wave, traveling along the z direction, is said to be of *order zero*, the remaining waves are said to be of *order plus and minus one.*

Using the Bessel-Fourier series expansion (see Appendix D), which is in common use to describe phase-modulated signals and was already mentioned in Section 12.7, the reader can prove as an exercise that, in the case of a *phase grating*, one also gets a countable in nity of discrete *higher-order* di racted waves. This subject should be cross-correlated by the reader with what we saw in Chapter 12 when dealing with antenna arrays.

We may now take care of the inconsistency of the grating extending from ∞ to $+\infty$, which forces us to violate one of the far- eld assumptions. Assume that Eq. (13.90) applies only over a *rectangular aperture*, $|x'| \quad a$, $|y'| \quad b$ (Figure 13.12). The eld on the aperture is then equal to the product of Eq. (13.90) and Eq. (13.68). The *convolution theorem* for Fourier transforms (Papoulis, 1968) tells us that the eld in the Fraunhofer region equals the convolution of Eqs. (13.69) and (13.91), i.e.:

$$(P) \;=\; Uj\,\frac{e^{\;jkz}}{z}\,e^{\;j(k/2z)\,(x^2+y^2)}\;2ab\,\frac{\sin y}{y}$$

$$\left[\frac{\sin x}{x}+\frac{m}{2}\,\frac{\sin(x+x_B)}{x+x_B}+\frac{m}{2}\,\frac{\sin(x\quad x_B)}{x\quad x_B}\right] \quad , \tag{13.94}$$

where x, y are given by Eqs. (13.70), whilst the quantity x_B, referred to as the *Bragg shift*, given by:

$$x_B = \quad f_0\, a \quad , \tag{13.95}$$

is proportional to the number of grating modulation periods which fall within the aperture.

If we compute the square of Eq. (13.94), we see immediately that all the cross products are negligible everywhere *provided* that $x_B \quad 1$. In that case, we can speak of "orders" of di raction, similarly to Eq. (13.91).

13.9 The eld near a focus: First example of Fresnel di raction

Examples of Fraunhofer di raction were given before those of Fresnel di raction because these entail more demanding mathematics. These complications will restrict us to the scalar theory.

We know from Chapter 5 that the equations of geometrical optics break down near a focus. The transport equation would yield an in nite amplitude for the eld where a ux tube collapses to an in nitesimal cross-section. To model in a reasonably correct way how the eld behaves near a focus, we must use Fresnel di raction theory.

Let us start (Figure 13.17) from an incoming spherical wave, centered at the focal point, where we place the origin of the reference frame. For simplicity the wave is assumed to be uniform, i.e., its amplitude depends only on the distance from the focus. The wave is transmitted through a circular aperture of radius a. This is a suitable model for a converging lens or for a receiving parabolic antenna, assuming that a plane wave impinges on either one.

If the aperture radius a is small compared with the focal length ℓ, then the obliqueness factor in Eq. (13.50), $(1 + cos\ \ _0)/2$, is very close to unity. Thanks to the assumption of a uniform spherical wave, we may write the amplitude factor F/d as $\ /\ell$, where stands, consequently, for the amplitude of the spherical wave one meter before the focus.

Let us de ne the following normalized coordinates of the observation point P, which is close to the focus 0:

$$u = \frac{2}{\ }\left(\frac{a}{\ell}\right)^2 z \quad , \qquad v = \frac{2}{\ }\left(\frac{a}{\ell}\right) R = \frac{2}{\ }\left(\frac{a}{\ell}\right)\sqrt{x^2+y^2} \quad . \tag{13.96}$$

Also let $\ ,\ '$ be, respectively, the azimuthal coordinates of $P \quad (x, y, z)$ and of $Q \quad (x', y', z')$ in the cylindrical reference frame $(R, \ , z)$ superimposed on the cartesian frame shown in Figure 13.17. Inserting these new variables into

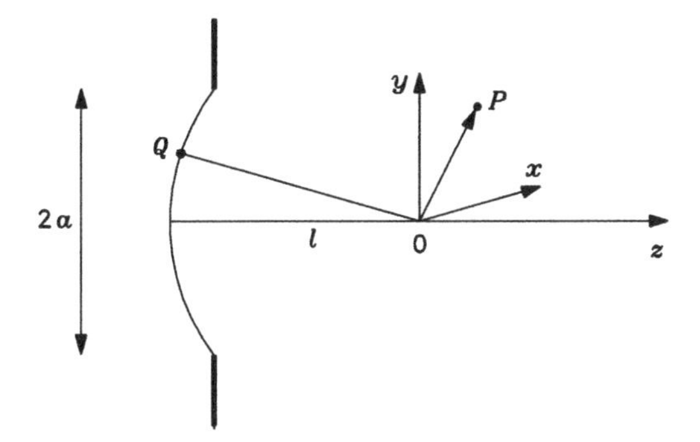

Figure 13.17 *Fresnel di raction of a circular aperture illuminated by a spherical wave: the geometrical layout.*

Eq. (13.50), after some algebra we arrive at:

$$(P) = \frac{j}{\ell^2}\, e^{\; j[\ell/a]^2 u} \int_0^a \int_0^2 e^{\; j[v(R'/a)\cos(\quad ')\quad (1/2)\, u(R'/a)^2]}\, R'\,\mathrm{d}R'\,\mathrm{d}\quad ' . \tag{13.97}$$

Comparing this expression with Eq. (13.77) — which holds only in the Fraunhofer region — we nd the same azimuthal dependence; accordingly, integration over $'$ can again make use of the identity in Eq. (13.78). If we de ne a normalized radial coordinate $q = R'/a$ for the point that scans the wave front, we get:

$$(P) = j\,\frac{2}{\quad}\,\frac{a^2}{\ell^2}\, e^{\; j(\ell/a)^2 u} \int_0^1 J_0(vq)\, e^{j(uq^2)/2}\, q\, dq \quad . \tag{13.98}$$

The integral in Eq. (13.98) is di cult to compute, as it requires integration by parts, plus some Bessel function recurrence formulas, which can be found in an *ad hoc* appendix of Born and Wolf (1980). Let us go directly to the nal result, which can be written as:

$$(P) = j\,\frac{2}{\quad}\,\frac{a^2}{\ell^2}\, e^{j(\ell/a)^2 u}\,\frac{1}{2}\left\{C(u,v) + j\,S(u,v)\right\} \quad , \tag{13.99}$$

where:

$$\begin{aligned} C(u,v) &= \frac{\cos\frac{u}{2}}{\frac{u}{2}}\,U_1(u,v) + \frac{\sin\frac{u}{2}}{\frac{u}{2}}\,U_2(u,v) \\ &= \frac{2}{u}\sin\frac{v^2}{2u} + \frac{\sin\frac{u}{2}}{\frac{u}{2}}\,V_0(u,v) \quad \frac{\cos\frac{u}{2}}{\frac{u}{2}}\,V_1(u,v) \quad , \end{aligned} \tag{13.100}$$

$$
\begin{aligned}
S(u,v) &= \frac{\sin\frac{u}{2}}{\frac{u}{2}}\,U_1(u,v) \quad \frac{\cos\frac{u}{2}}{\frac{u}{2}}\,U_2(u,v) \\
&= \frac{2}{u}\cos\frac{v^2}{2u} \quad \frac{\cos\frac{u}{2}}{\frac{u}{2}}\,V_0(u,v) \quad \frac{\sin\frac{u}{2}}{\frac{u}{2}}\,V_1(u,v) \quad , \qquad (13.101)
\end{aligned}
$$

while U_n, V_n stand for the following expressions, currently referred to as the *Lommel functions* of order n:

$$
\begin{aligned}
U_n(u,v) &= \sum_{i=0}^{\infty} (\ \ 1)^i \left(\frac{u}{v}\right)^{n+2i} J_{n+2i}(v) \quad , \\
V_n(u,v) &= \sum_{i=0}^{\infty} (\ \ 1)^i \left(\frac{v}{u}\right)^{n+2i} J_{n+2i}(v) \quad . \qquad (13.102)
\end{aligned}
$$

The reason why we gave two di erent expressions for C and S in Eqs. (13.100) and (13.101) is because the series in Eq. (13.102) do not converge everywhere at the same rate. The functions U converge faster when:

$$
\left|\frac{u}{v}\right| < 1 \quad , \qquad (13.103)
$$

while the functions V converge faster when $|u/v| > 1$. If we go back to the de - nition of the normalized coordinates, we see that the condition in Eq. (13.103) is the same as:

$$
\frac{R}{z} > \frac{a}{\ell} \quad , \qquad (13.104)
$$

so it means that the observation point P is in the *geometrical shadow*, namely that region where ray optics would predict zero eld. Obviously, then, $|u/v| >$ 1 states that P is in the "illuminated" region.

The radiation intensity $|\ \ |^2$, normalized with respect to the value it reaches in the geometrical focus, equals:

$$
I_0 = |\ \ (0)|^2 = \left(\frac{a^2}{\ell^2}\right)^2 \quad , \qquad (13.105)
$$

and is illustrated, as a contour plot in the (u,v) plane, in Figure 13.18. The reader can compare these results with what we saw in Chapter 5 about Gaussian beams. Discrepancies are large enough to indicate that tightly focused Gaussian beams are not the exact solutions of Maxwell's equations in free space (see Marcatili and Someda, 1987).

13.10 Di raction from a straight edge: Second example of Fresnel di raction

Another theoretical problem with important practical fallouts is di raction caused by the edge of either an absorbing or a re ecting half-plane. For exam-

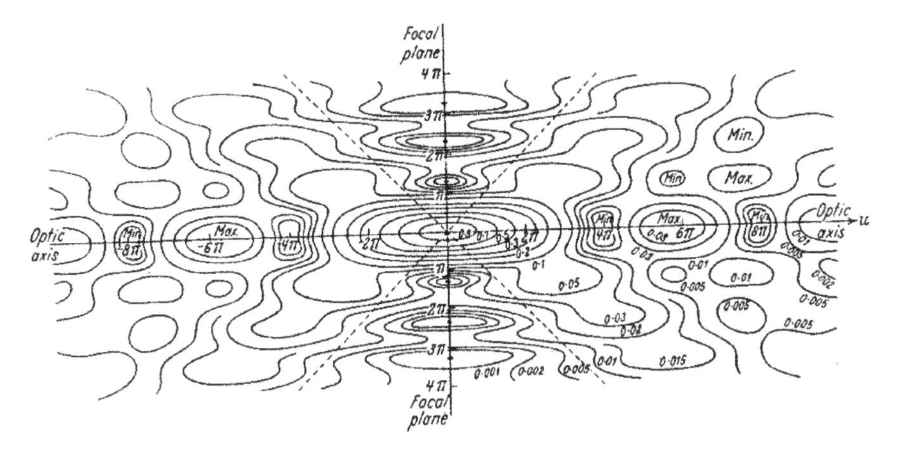

Figure 13.18 *Contour plot of the eld intensity near the geometrical focus, according to Fresnel di raction theory.*

ple this can be an adequate model for radio waves propagating just above a mountain crest. The aperture is the complementary half-plane, which extends to in nity so that strictly speaking it does not match our initial assumptions in Eq. (13.44). Nevertheless, theoretical results are in good agreement with experimental data if one starts from a rectangular aperture, studies its Fresnel di raction, and then takes a limit for the size of the rectangle going to in nity. The relationship between this problem and the one which we discussed in the rst example of Section 13.8 is the same as that between the previous section and the second example in Section 13.8.

Let us start again from Eq. (13.50), and assume that the complex amplitude of the incident wave is constant over the aperture. Then, the term $F(1+\cos\ _0)/2d = A$ can be factored out of the di raction integral. The rst formulation of Eq. (13.50) o ers the advantage that, within the exponential function in the integral, there is a perfect square. Let us then introduce two new normalized variables u, v, de ned as:

$$\begin{aligned} u^2 &= \frac{k}{z}\,(x \quad x')^2 \quad , \\ v^2 &= \frac{k}{z}\,(y \quad y')^2 \quad . \end{aligned} \tag{13.106}$$

Using these, we get:

$$\mathrm{d}x'\,\mathrm{d}y' = \frac{z}{2}\,\mathrm{d}u\,\mathrm{d}v \quad . \tag{13.107}$$

Separating real and imaginary parts in the argument of the exponential, we nd:

$$(P) = j\,A\,e^{\ jkz}\,(C + j\,S) \quad , \tag{13.108}$$

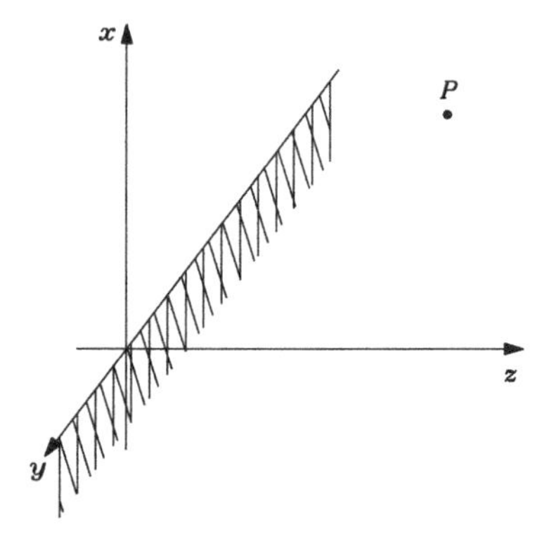

Figure 13.19 *A di racting straight edge: the geometry of the problem.*

where the symbols C, S stand for:

$$
\begin{aligned}
C &= \int\!\!\int_{A'} \cos\left\{\frac{}{2}\,(u^2+v^2)\right\}\, du\, dv \quad , \\
S &= \int\!\!\int_{A'} \sin\left\{\frac{}{2}\,(u^2+v^2)\right\}\, du\, dv \quad . \qquad (13.109)
\end{aligned}
$$

In Eqs. (13.109), A' is the map of A on the (u, v) plane using the transformation (13.106). By means of well-known trigonometric identities, each of Eqs. (13.109) breaks down into four integrals of the following types:

$$
\begin{aligned}
\mathcal{C}(a) &= \int_0^a \cos\left(\frac{}{2}\, W^2\right) dW \quad , \\
\mathcal{S}(a) &= \int_0^a \sin\left(\frac{}{2}\, W^2\right) dW \quad , \qquad (13.110)
\end{aligned}
$$

which are referred to as the *Fresnel integrals*, and are discussed thoroughly in many texts and handbooks (e.g., Abramowitz and Stegun, Eds., 1965). In particular, in the following, for $a \to \infty$ we will make use of the fact that $\mathcal{C}(\infty) = \mathcal{S}(\infty) = 1/2$. Furthermore, from symmetry arguments, we have:

$$
\begin{aligned}
\int_{\infty}^{0} \cos\left(\frac{}{2}\, W^2\right) dW &= \mathcal{C}(\infty) = \frac{1}{2} \quad , \\
\int_{\infty}^{0} \sin\left(\frac{}{2}\, W^2\right) dW &= \mathcal{S}(\infty) = \frac{1}{2} \quad . \qquad (13.111)
\end{aligned}
$$

Coming back to our original problem, introduce a reference frame whose y axis is along the edge (Figure 13.19), and whose x axis is oriented so that x is positive at a point P in the illuminated region, negative in the geometrical shadow. The aperture A is then de ned as $0 < x' < +\infty$, $\infty < y' < +\infty$. With a suitable choice for the sign of the square root in Eq. (13.106), the

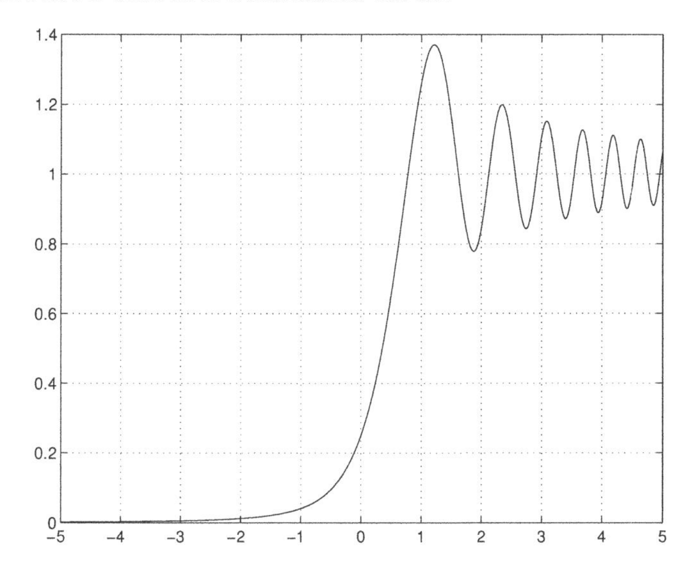

Figure 13.20 *Normalized radiation intensity in the Fresnel region, vs. normalized transverse coordinate of the observation point.*

domain A' turns out to be de ned as:

$$\infty < u < a = x \sqrt{\frac{1}{\ } \frac{k}{z}} \quad . \tag{13.112}$$

As we have noted before, the assumptions (13.44) are not satis ed, strictly speaking. Nonetheless, using Eqs. (13.111) the di raction integrals give the following results:

$$\int_{\infty}^{a} \cos\left(\frac{\ }{2} W^2\right) \mathrm{d}W = \frac{1}{2} + \mathcal{C}(a) \quad , \tag{13.113}$$

$$\int_{\infty}^{a} \sin\left(\frac{\ }{2} W^2\right) \mathrm{d}W = \frac{1}{2} + \mathcal{S}(a) \quad , \tag{13.114}$$

$$\int_{\infty}^{+\infty} \cos\left(\frac{\ }{2} W^2\right) \mathrm{d}W = \int_{\infty}^{+\infty} \sin\left(\frac{\ }{2} W^2\right) \mathrm{d}W = 1 \quad . \tag{13.115}$$

The corresponding *radiation intensity* is:

$$I = \frac{|\ |^2}{2} = \frac{|A|^2}{2} \left[\left(\frac{1}{2} + \mathcal{C}(a)\right)^2 + \left(\frac{1}{2} + \mathcal{S}(a)\right)^2 \right] \quad . \tag{13.116}$$

This function is plotted in Figure 13.20. In the geometrical shadow ($a < 0$), the intensity grows monotonically as the observation point gets closer to the illuminated region. In the illuminated region ($a > 0$), the intensity exhibits dumped oscillations, and, for $a \to \infty$, tends to become the same as the intensity of a uniform plane wave when there is no edge. At the border line

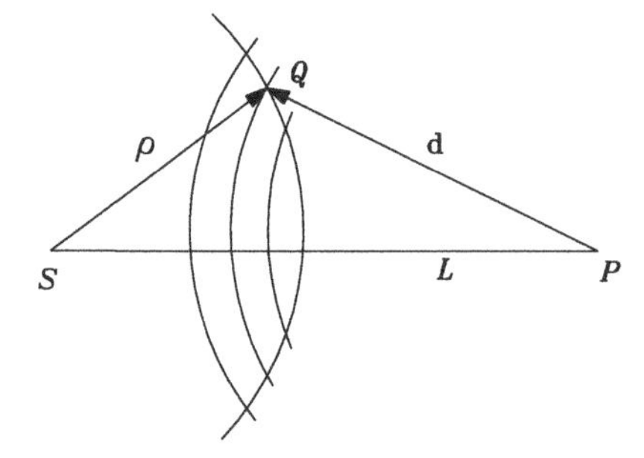

Figure 13.21 *Schematic illustration of the Fresnel ellipsoides.*

of the geometrical shadow, the intensity equals one fourth of its asymptotic value. Its absolute maximum is located slightly inside the geometrical shadow.

From these results we may infer a rule of thumb, for a rough estimate of the received signal intensity in a point-to-point radio link. Obstacles do not signi cantly a ect the received signal when their edges are at least half a wavelength apart from the "straight ray," i.e., from the trajectory of the ray passing through transmitter and receiver, calculated by means of geometrical optics.

Most likely, the reader has been exposed to similar (but not identical) results in classes on general physics, and has heard about a process known as *Fresnel zones*, in use to model wave propagation from a point source S to an observation point P *in the absence of any obstacle.* The process starts with the de nition of the so-called *Fresnel ellipsoids* (for the symbols, see Figure 13.21), which are the loci of those points Q that satisfy the following relationships:

$$+ d = L + m \frac{\ }{2} \qquad (m = 1, 2, 3, \ldots \; ; \; L = |S \quad P|) \quad . \tag{13.117}$$

Consider now a spherical wave, radiated from the source. Its intersections with the Fresnel ellipsoids divide the wave front into the so-called *Fresnel zones.* Each zone contributes to the total eld at P by an amount that can be computed as a di raction integral. Their sum must coincide with the spherical-wave eld in P. This allows us to eliminate one proportionality constant, and therefore to complete the evaluation of the individual contributions of the zones to the total eld. One may conclude that neighboring zones yield contributions of opposite signs, their moduli decreasing monotonically as we move outwards. Their sum exhibits dumped oscillations as we add more terms, and so it is qualitatively similar to Figure 13.20. Still, convergence of this process towards the asymptotic value is signi cantly slower than what we might infer from Figure 13.20. It can be shown that this slow convergence comes from the fact that the analysis based on Fresnel zones is less rigorous than that based on edge di raction. Yet, its nal results are in good agreement with experiments. To summarize, the rather empirical rule "leave the rst Fresnel ellipsoid free from obstacles" is well justi ed for obstacles with straight edges.

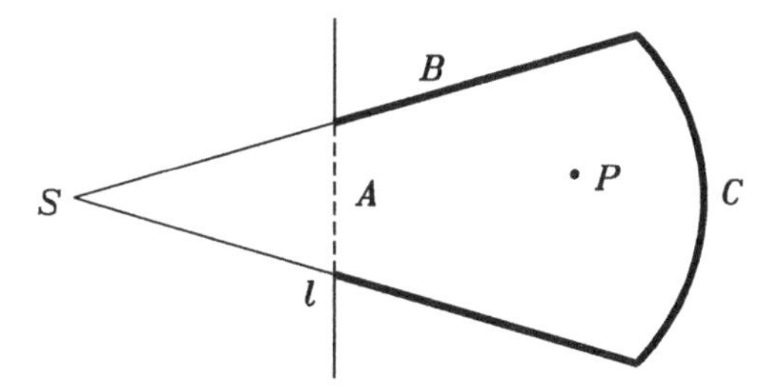

Figure 13.22 *Schematic layout used to introduce the geometrical theory of di raction.*

It becomes much less accurate for circular holes, due to the slow convergence that we pointed out previously.

13.11 A short note on the geometrical theory of di raction

In the rst sections of this chapter we remarked that, to be rigorous, di raction integrals should be calculated over a closed surface. In practice, all the classical approaches are approximate, as the integral is computed over an open surface. However, even within these approximations di raction still appears like a "global" phenomenon, not like a "local" one: the eld at any observation point depends upon the *whole* di racting object. Just to quote an impressive example, this is instrumental for the far eld in the Fraunhofer region to be the Fourier transform of the eld on the aperture: as well known, the Fourier transform of any function senses the whole function over its entire domain of de nition.

This viewpoint is di cult to reconcile with another intuitive idea. When the wavelength becomes small compared to the geometrical size of all the objects invested by a given wave, there must be a way to link di raction with ray optics, which is typically a "local" phenomenon. Hence, di raction must tend to become a local e ect when the wavelength becomes very small. Indeed, it is possible to change this intuition into a rigorous idea.

To justify it qualitatively, consider an aperture A in a planar screen, and for simplicity suppose that it is illuminated by a monochromatic point-source S (Figure 13.22). At any point P on the other side of the screen, the eld can be expressed *rigorously* in terms of integrals over the closed surface which is the union of A, B and C in Figure 13.22. Over A, the integral contains the eld due to the source. C can go to in nity, thus yielding an arbitrarily small contribution. As for B, it can be proved (Born and Wolf, 1980; Rush and Potter, 1970) that this surface integral can be transformed into a *line integral* over ℓ, the contour of the aperture A, i.e., the intersection between the surfaces A and B. Consequently, the eld at P can be computed as the sum of a eld transmitted through the aperture (which, if the aperture is large compared to , can be computed using ray optics), plus a eld originating from a line integral, referred to as the *edge-di racted eld*. This is the rst step towards considering di raction as a local phenomenon.

The next step is to nd under what conditions one is allowed to split the line integral into contributions due to localized point sources, each of which emits rays. These conditions are discussed in depth in the specialized literature (Kouyoumjian, 1975). To give an ultra-simpli ed idea, we may say that, when they hold, integrals of complex functions can be replaced either by sums or by integrals of real functions, neglecting completely the phase. Such a procedure is intimately related to the following argument, which is known in optics as the "principle of stationary phase". Let us consider a di raction integral; if the *derivatives of the phase* of the integrand, with respect to the coordinates, are zero at a certain point on the aperture A, then the region around that point plays a major role in that integral, because elementary contributions from various points in that region add in phase. On the contrary, regions where the phase of the integrand varies quickly contribute negligibly to the integral, as the elementary contributions average out.

When the conditions we are referring to are satis ed (and this can happen only if the aperture is broad enough compared to), then it can be shown that the di racted eld can be computed in terms of rays also in the geometrical shadow. This is what is currently meant when one speaks of geometrical theory of di raction.

13.12 Further applications and suggested reading

Many important applications of di raction theory have already been quoted, throughout this chapter. Let us summarize them, skipping mathematical details, and without repeating quotations which can be found in previous sections.

Holography and *Fourier optics* (see Goodman, 1968) are very important consequences of Fraunhofer di raction, and are potentially spectacular too. For a better comprehension of their details — for example, to understand why it is possible to make white-light holograms — the readers are advised not to attempt the problem as long as they are not familiar with coherence, which is the subject of the next chapter. For a broad and balanced overview of their applications, one should not restrict oneself to the theory of electromagnetic phenomena, but also read about other disciplines like signal processing, image processing, pattern recognition, and so on. Concepts borrowed from system theory and communication theory are also extremely important background items, for those who want to enhance their knowledge in this eld.

Theory and modern design of *aperture antennas* is another subject that has been mentioned several times throughout this chapter, as a still lively product of di raction theory. In fact, even some signi cant theoretical advances beyond the inaccuracies of the classical Helmholtz-Kirchho and Rayleigh-Sommerfeld formulations were triggered by the technical interest in designing very advanced, complicated high-performing spaceborne antennas.

Another advice, similar to what was mentioned at the end of other previous chapters, is to remind the reader that the basics of di raction theory remain the same in all those areas of physics which pivot around the wave equations. Di raction phenomena are, therefore, very important in quantum mechanics. In this eld, as well as in electromagnetism, they share many features with *scattering* theory, where a known object (or a given potential distribution) is illuminated by a primary beam, and the question is to nd how the impinging waves or particles are de ected in space. In the so-called *Born approximation*, i.e., if multiple scattering can be neglected, conceptual similarity with di raction phenomena is quite evident because scattering also reduces to a surface integral.

The next subject we come across, if we follow this line of thinking, is the so-called theory of *inverse scattering*, a phrase which originally meant that one wants to reconstruct the shape of an unknown scattering center from the scattered eld. The reason why we stress that this was the *original* meaning, is because this question turned out to trigger an extremely powerful technique for solving wide classes of nonlinear partial di erential equations, which apply to many di erent physical contexts. The reason why the same name, inverse scattering, is kept in use, is because an essential feature shared by all these techniques is a change of variables, which may di er from one problem to another but always evokes reconstructing an object from a scattered eld. A recommended starting point for newcomers into this eld is, for example, Ablowitz and Segur (1981).

Finally, let us point out that some very popular numerical approaches to wave propagation problems are indebted to di raction theory, which provided part of the background. This is the case, for example, of the so-called Beam Propagation Method (BPM), in its original implementation, where di raction theory stays behind the use of the Fast-Fourier transform, at any step, to pass from the eld distribution in the 3-D real space to plane-wave spectra, and vice-versa. The BPM is now one of the most popular numerical methods in integrated optics. There are profound links between di raction and propagation in dielectric waveguides. The interested reader may understand them thoroughly by reading, e.g., the relevant chapters in Snyder and Love (1983). However, the BPM was originally developed to study underwater propagation of acoustic waves: another indication that di raction theory is ubiquitous, throughout wave propagation phenomena. For further examples of such ubiquity, see, e.g., Berry (1966), and Erko *et al.* (1996).

Problems

13-1 Show that Eq. (13.14) is an alternative to Eq. (13.12). Follow the hints given in the text, and make use of the following identity (which should also be proved as part of the exercise):

$$\int_S \left[(\hat{n} \quad \mathbf{H}) \quad \nabla \right] \nabla \quad dS = \int_S j\omega \, (\hat{n} \quad \mathbf{E}) \nabla \quad dS \quad \oint_\ell \nabla \quad (\mathbf{H} \quad d\boldsymbol{\ell}) \quad ,$$

where S is an open surface with two faces and ℓ is its contour.

13-2 Derive an expression for the magnetic eld similar to Eq. (13.14).

13-3 Prove Eqs. (13.33).

13-4 Show that whenever Eq. (13.48) is satis ed, Eq. (13.49) is an acceptable approximation.

13-5 Prove that Eq. (13.74) is the *minimum* angular width for a beam of plane waves which passes through an aperture of width a.

13-6 Show that the function $F(\vartheta', \ ')$ of Eq. (13.38) is the equivalent moment of the source S_i, if the screen is perfectly absorbing.

13-7 In the Fraunhofer region, test whether the vector di racted elds, Eqs. (13.62) and (13.63), are a true e.m. eld, i.e., an exact solution of Maxwell's equations. If not, justify the approximations involved.

13-8 A circular aperture with a 2 mm diameter is illuminated by a Helium-Neon laser beam, whose wavelength is 633 nm. How far from the aperture does the Fraunhofer region begin?
Repeat the same calculation for the same aperture and a CO_2 laser beam (= 10 m).

13-9 The so-called active region of a semiconductor laser, i.e., that part where the laser radiation is generated, ends at the two terminal facets as a rectangular strip. Suppose that its width and height are $w = 30$ m (in the x, z plane) and $h = 5$ m (in the y, z plane) and that the wavelength equals 1.5 m. Calculate the 3 dB angular widths, in the x, z and y, z planes, of the radiated beam, in the Fraunhofer region, under the following assumptions:

a) a uniformly illuminated rectangular aperture;

b) Gaussian distributions (see Chapter 5) of $1/e$ widths w and h, in the two planes.

Compare the two results and draw some comments.

13-10 Reconsider, in the scalar approach, the Fraunhofer di raction of a rectangular aperture illuminated by a plane wave whose phase vector, lying in the (x, z) plane, forms an angle ϑ_0 with respect to the z axis (i.e., introducing a phase factor exp($jk \sin\vartheta_0 x'$) in the function F).

13-11 Study in detail the case of a *phase modulated grating*, and compare the results with those of Problem 13-6.
(*Hint.* Make use of the Bessel-Fourier series expansion, see Section 12.7.3.)

13-12 In the Fraunhofer region, write the details of the vector approach to the eld di racted by a uniformly illuminated rectangular aperture, equal to that of the rst example of Section 13.8. For the illuminating eld, adopt the same assumptions as for the circular aperture discussed in the same section, except for the following detail: the incident electric eld is linearly polarized along the direction which forms an angle with the x axis.
Discuss the polarization properties of the di racted eld as a function of the angle and compare them with the scalar approach.

13-13 Discuss how the vector correction to the Airy diagram, Eq. (13.88), a ects the angular width of the main lobe of a circular aperture, as a function of its normalized radial size $a/$. Consider both de nitions of the angular width, i.e., the rst direction of null as well as the 3 dB one.

13-14 A radiation source consists of two rectangular apertures, identical to that discussed in Section 13.8, parallel to each other, and with their centers at a distance $2d > 2a$, along the x direction. Each of them is illuminated by a uniform plane wave traveling in the z direction. In the scalar approach, study how the total eld in the Fraunhofer region varies as a function of:

a) the phase shift between the illuminating plane waves of equal amplitude;

b) the distance $2d$, for equal phases and amplitudes;

c) the amplitude ratio, for equal phases and xed distance.

Compare the answers to the rst two questions with what was seen on antenna arrays in Chapter 12.

13-15 Referring to the previous problem, it is noted that the di racted eld in the Fraunhofer region vanishes along the z axis if the two apertures are illuminated with equal amplitudes and a radian phase shift.
Could this result be predicted using symmetry properties of the Fourier transform?

13-16 Consider a rectangular aperture, illuminated by the TE_{10} mode of a rectangular waveguide. Using the same coordinate system and the same symbols as in Figure ??, the illuminating eld is $F = E_0 \cos(x'/a)$.

a) In the scalar approach, evaluate the di racted eld in the Fraunhofer region.

b) Show that one of the reasons why waveguides are often terminated with ared ends is to reduce the angular width of the main radiation lobe.

13-17 Show that in the ray-optics focal plane ($u = 0$) the radiation intensity of the circular aperture in the Fresnel region, discussed in Section 13.9, equals:

$$I_r(0, v) = I_0 \left[2J_1(v)/v^2\right]^2 \quad .$$

Provide an intuitive physical justi cation of why this distribution is identical

to that which is found in the Fraunhofer region.
(*Hint.* Pay attention to the shape of the phase fronts.)

13-18 Show that along the symmetry axis ($v = 0$) the radiation intensity of the circular aperture in the Fresnel region, discussed in Section 13.9, is expressed by:

$$I_r(u, 0) = I_0 \left[\frac{\sin(u/4)}{u/4} \right]^2 \quad .$$

Find the positions of its nulls and of its relative maxima.

13-19 A coaxial cable, where only the TEM propagates (outer and inner diameters $2a$ and $2b$, respectively), is cut perpendicularly to the z axis. Assume that the mismatch between cable and free space does not generate any higher-order mode, so that the electric and magnetic elds are proportional to $1/r$. Find the di racted eld and the normalized radiation intensity in the Fraunhofer region.
(*Hint.* Apply the identity (13.78) and then, since kRR'/z 1 in the whole region of interest, use a small-argument approximation for J_0.)

References

Ablowitz, M.J. and Segur, H. (1981) *Solitons and the Inverse Scattering Transform.* Siam, Philadelphia.

Abramowitz, M. and Stegun, I.A. (Eds.) (1965) *Handbook of Mathematical Functions.* Dover, New York.

Berry, M.V. (1966) *The Di raction of Light by Ultrasound.* Academic, London.

Born, M. and Wolf, E. (1980) *Principles of Optics.* 6th (corrected) ed., reprinted 1986, Pergamon, Oxford.

Erko, A.I., Aristov, V.V. and Vidal, B. (1996) *Di raction X-Ray Optics.* Translated by S. Chomet, Institute of Physics Pub., Philadelphia, PA.

Fradin, A.Z. (1961) *Microwave Antennas.* Pergamon, Oxford.

Goodman, J.W. (1968) *Introduction to Fourier Optics.* McGraw-Hill, New York.

Kouyoumjian, R.G. (1975) "The Geometrical Theory of Di raction and Its Applications". In R. Mittra, Ed., *Numerical and Asymptotic Techniques in Electromagnetics*, Springer-Verlag, Berlin.

Marcatili, E.A.J. and Someda, C.G. (1987) "Gaussian beam are fundamentally different from free-space modes". *IEEE J. of Quantum Electronics*, **23** (2), 164–167.

Papoulis, A. (1968) *Systems and Transforms with Applications in Optics.* McGraw-Hill, New York.

Rush, W.U.T. and Potter, P.D. (1970) *Analysis of Re ector Antennas.* Academic, New York/London.

Snyder, A.W. and Love, J.D. (1983) *Optical Waveguide Theory.* Chapman & Hall, London.

Stutzman, W.L. and Thiele, G.A. (1997) *Antenna Theory and Design.* 2nd ed., Wiley, New York.

CHAPTER 14

An introduction to radiation statistics and to the theory of coherence

14.1 Background and purpose of the chapter

In the previous chapters, we have been dealing mostly with monochromatic elds. Whenever it was necessary to account for nite spectral width (for instance, to introduce concepts like group velocity and dispersion, in Chapter 5), we considered only functions whose behavior was fully deterministic, both in time and in space. On the other hand, when we mentioned brie y in Chapter 12 that receiving antennas inevitably pick up noise, we went all the way to the other extreme, assuming implicitly that we were then dealing with completely random variables.

In reality, physical e.m. elds can never be perfectly deterministic, and it happens very seldom that they vary completely at random. The purpose of this chapter is to introduce some concepts which may help describe any situation in between. One reason why this chapter is short, in comparison with the extent of the whole book, is because, when dealing with these problems, the readers can exploit the most classical notions on random process analysis, which they are supposed to learn elsewhere. Consequently, here we may restrict our analysis to a small number of problems which are strictly relevant to electromagnetic waves. Relevance comes mainly from the fact that the best way to focus on these problems is to describe experiments which exploit optical instruments.

14.2 The analytical signal

Throughout this book we used the well-known Steinmetz representation for deterministic time-harmonic variables. We can generalize the complex representations and cover a broader scope. This section is devoted to de nitions, introduction of suitable symbols, and other background items.

Let us start from a *real* time-dependent vector (for instance, an electric eld $\mathbf{E}(t)$), at a given point in 3-D space. Suppose that, in a rectangular coordinate frame, all components of $\mathbf{E}(t)$ can be Fourier-transformed. Then, the vector $\mathbf{E}(t)$ can be written as a 3-D complex Fourier integral, namely:

$$\mathbf{E}(t) = \frac{1}{4} \int_{\infty}^{+\infty} \mathbf{E}(\omega)\, e^{j\omega t}\, d\omega \quad , \tag{14.1}$$

where ω is, obviously, the angular frequency.

Using Euler's formula $e^{j\omega t} = \cos\omega t + j \sin\omega t$, and simple considerations on parity of these functions, it is easy to show that, for $\mathbf{E}\ (t)$ to be real, the complex vector $\mathbf{E}(\omega)$ must satisfy the constraint:

$$\mathbf{E}(\ \ \omega) = \mathbf{E}\ (\omega) \quad , \tag{14.2}$$

i.e., its modulus must be an even function of ω, and its phase must be an odd function of ω.

As a result of Eq. (14.2), Eq. (14.1) can be rewritten in the following form, where "c.c." stands for the complex conjugate of the rst term within the same square bracket:

$$\begin{aligned} \mathbf{E}\ (t) &= \frac{1}{2}\left[\frac{1}{2}\ \int_0^{+\infty} \mathbf{E}(\omega)\ e^{j\omega t}\ d\omega + \text{c.c.}\right] \\ &= \text{Re}\left[\frac{1}{2}\ \int_0^{+\infty} \mathbf{E}(\omega)\ e^{j\omega t}\ d\omega\right] \quad \text{Re}\left[\ \mathbf{E}^{(a)}\ (t)\right] \quad . \end{aligned} \tag{14.3}$$

The complex function of time:

$$\mathbf{E}^{(a)}\ (t) = \frac{1}{2}\ \int_0^{+\infty} \mathbf{E}(\omega)\ e^{j\omega t}\ d\omega \quad , \tag{14.4}$$

is referred to as the *analytical signal* associated with the real signal $\mathbf{E}\ (t)$. It can be shown (Papoulis, 1968) that the imaginary part, $\mathbf{E}^{(i)}$, of the analytical signal and its real part are related to each other by a *Hilbert transformation*, i.e., satisfy:

$$\text{Im}\left[\ \mathbf{E}^{(a)}\ (t)\right] \quad \mathbf{E}^{(i)} = \frac{1}{\ }\ \int_{\ \infty}^{+\infty} \frac{\mathbf{E}\ (t')}{t\ \ t'}\ dt' \quad , \tag{14.5}$$

whose inverse is also a Hilbert transformation, namely:

$$\mathbf{E}\ (t) = \frac{1}{\ }\ \int_{\ \infty}^{+\infty} \frac{\mathbf{E}^{(i)}\ (t')}{t\ \ t'}\ dt' \quad . \tag{14.6}$$

Incidentally, Eqs. (14.4), (14.5) and (14.6) imply that the three instantaneous vectors $\mathbf{E}$, $\mathbf{E}^{(a)}$, $\mathbf{E}^{(i)}$ are *parallel.*

By means of the analytical signal we can give simple de nitions for two quantities, which we will refer to as the *instantaneous amplitude* and the *instantaneous phase* of the real eld $\mathbf{E}\ (t)$. Indeed, without any restriction, we

In most textbooks on the Fourier integral and its applications, the relationship between a time-domain function and its Fourier transform di ers from Eq. (14.1), because $\exp(j\omega t)$ is replaced by $\exp(\ \ j\omega t)$. We prefer Eq. (14.1) because it is consistent with our previous chapters, where, for time-harmonic elds, we set $\mathbf{E} = \text{Re}(\mathbf{E}\, e^{j\omega t}) = (\mathbf{E}\, e^{j\omega t} + \mathbf{E}\ \ e^{\ \ j\omega t})/2$. Also in order to be consistent with our previous chapters the Fourier integral was de ned here with an unusual factor $1/(4\ \ )$. Indeed, by doing so we get $E = \cos\omega_0 t$ when the frequency-domain spectrum consists of two in nitely narrow lines, i.e., for $E(\omega) = \ \ (f\ \ f_0) + \ \ (f + f_0) = (1/2\ \ )[\ \ (\omega\ \ \ \omega_0) + \ \ (\omega + \omega_0)]$.

may write:

$$\mathbf{E}\,(t) = \mathbf{A}\,(t)\,\cos\left(\omega_0 t + \varphi(t)\right) \quad , \tag{14.7}$$

where the constant angular frequency ω_0 is, for the time being, completely arbitrary, while $\mathbf{A}\,(t)$ is a time-dependent real vector. We have then:

$$\mathbf{E}^{(a)}\,(t) = \mathbf{A}\,(t)\,e^{j(\omega_0 t + \varphi(t))} = \mathbf{C}(t)\,e^{j\omega_0 t} \quad , \tag{14.8}$$

and the quantity $\mathbf{C}(t) = \mathbf{A}\,(t)\,e^{j\varphi(t)}$ is referred to as the *complex amplitude* of the analytical signal. The following relationships are rather obvious consequences of these de nitions:

$$|\mathbf{C}(t)| \;=\; |\,\mathbf{A}\,(t)| = \left[\,\mathbf{E}^{(a)} \quad \mathbf{E}^{(a)}\,\right]^{1/2} = \left|\mathbf{E}^{(a)}\right| \quad , \tag{14.9}$$

$$\varphi(t) \;=\; \arctan\left(\frac{\mathbf{E}^{(i)}}{\mathbf{E}}\right) \quad \omega_0(t) \quad . \tag{14.10}$$

These statements and comments are always true in principle, but become useful in practice only when the instantaneous eld $\mathbf{E}\,(t)$ is a "band-pass" one, i.e., when its spectrum is narrow enough. In that case, it is advisable to choose as ω_0 in Eq. (14.7) the central frequency in the eld spectrum, and suppose that amplitude and phase, $\mathbf{A}\,(t)$ and $\varphi(t)$, are *slowly varying*:

$$\left|\frac{d\mathbf{A}}{dt}\right| \quad \omega_0\,|\,\mathbf{A}\,| \quad , \qquad \left|\frac{d\varphi}{dt}\right| \quad \omega_0 \quad . \tag{14.11}$$

In that case, $\mathbf{E}\,(t)$ is a "quasi time-harmonic" vector, whose envelope has an instantaneous amplitude given by Eq. (14.9), and a phase given by Eq. (14.10).

In the following, we will always assume that the conditions in Eq. (14.11) are satis ed. Experimentally, this corresponds to passing the eld through a su ciently narrow band-pass lter, centered at ω_0.

An extension of this formalism, that we are going to exploit in the rest of this chapter, is the superposition of two or more analytical signals, namely:

$$\begin{aligned} \mathbf{E}\,(t) \;&=\; {}_m \operatorname{Re}\left[\,\mathbf{E}_m^{(a)}(t)\right] = \operatorname{Re}\left[\;{}_m\;\mathbf{E}_m^{(a)}(t)\right] \\ &=\; \operatorname{Re}\left[\;{}_m\;\mathbf{C}_m(t)\,e^{j\omega_0 t}\right] = \operatorname{Re}\left[\;{}_m\;\mathbf{A}_m(t)\,e^{j(\omega_0 t + \varphi_m(t))}\right] \quad . \end{aligned} \tag{14.12}$$

Such a model may describe many situations of practical interest, ranging from simultaneous presence of several sources, to multipath or multimode propagation of elds generated by a single source, in free space as well as in waveguides. Details will be clari ed in the following sections.

The band-pass assumption and its formalism re ect the fact that the key experiments to be discussed in this chapter were made initially at optical frequencies, using detectors all of which (from photodiodes to video cameras, from photographic plates to the human eye) have the following features:

a square-law response, i.e., an output signal (either a current or a voltage) proportional to the square of the electric eld, $\mathbf{E}\ (t)$, impinging on the detector;

a long response time $_R$, compared with the inverse of any frequency in the spectrum of $\mathbf{E}\ (t)$. Detectors can then be looked at as low-pass lters, which average over time slots of length $_R$ and so eliminate all terms, in the square of the incident eld, whose angular frequencies are comparable with ω_0, or with its harmonics.

At radio frequencies, and at microwave frequencies, it is possible (but not necessary) to reproduce the same conditions experimentally. Accordingly, henceforth we will always assume that the band-pass assumptions are satis ed, regardless of whether we deal with radio signals or with optical ones.

14.3 Complex degree of coherence

In this section, as well as in the following ones, unless the opposite is explicitly stated we will deal only with random processes which are *ergodic and stationary*. That is to say, we will consider elds whose time average (as seen by a slow detector) remains constant in time, and is identical to the statistical *ensemble average.*

Let us take a detector which satis es the assumptions of Section 14.2, and expose it to the sum of *two elds* whose spectra are centered at the same ω_0, and whose spectral widths are narrow compared with ω_0. The two elds may be, at least on some occasions, two replicas, generated by the same source either at di erent times or in di erent directions in space. To avoid needless complication, suppose that in any case the two elds have *identical polarization.*

The detector is sensitive to the total *intensity*, i.e.:

$$I = \overline{\mathbf{E}\ (t)\ \ \mathbf{E}\ (t)} = \langle \mathbf{E}\ (t)\ \ \mathbf{E}\ (t) \rangle \quad , \qquad (14.13)$$

where the overline denotes a time average over the detector response time, and the symbol $\langle\ \ \rangle$ stands for an ensemble average. Let:

$$\mathbf{E}\ (t) = \mathbf{E}_1(t) + \mathbf{E}_2(t) = \mathrm{Re}\left[\ \mathbf{A}_1 e^{j(\omega_0 t + \varphi_1(t))} + \mathbf{A}_2 e^{j(\omega_0 t + \varphi_2(t))}\right] . \qquad (14.14)$$

We easily obtain:

$$I = \langle \mathbf{E}_1\ \ \mathbf{E}_1 \rangle + \langle \mathbf{E}_2\ \ \mathbf{E}_2 \rangle + 2\,\langle \mathbf{E}_1\ \ \mathbf{E}_2 \rangle = I_1 + I_2 + 2\,\langle \mathbf{E}_1\ \ \mathbf{E}_2 \rangle \quad . \qquad (14.15)$$

The terms at $2\omega_0$ which appear in the product $\mathbf{E}\ (t)\ \ \mathbf{E}\ (t)$ are cancelled by the slow detector. So, in general the total intensity is the sum of three terms. Two of them are the intensities that each individual eld would generate if it were present separately from the other. The magnitude of the third term depends on the amount of correlation between the two elds, and is referred to as the *mutual coherence term* between $\mathbf{E}_1$ and $\mathbf{E}_2$.

Using the analytical signal, and recalling that, when only one of the two elds impinges on the detector, Eq. (14.13) gives $|\ \mathbf{A}_m\ | = |\mathbf{C}_m| = \sqrt{2I_m}$ $(m = 1, 2)$, we may rewrite Eq. (14.15) as:

$$I = I_1 + I_2 + 2\sqrt{I_1 I_2}\ \mathrm{Re}(\ _{12}) = \frac{|\ \mathbf{A}_1|^2}{2} + \frac{|\ \mathbf{A}_2|^2}{2} + |\ \mathbf{A}_1||\ \mathbf{A}_2|\ \mathrm{Re}(\ _{12})\,. \tag{14.16}$$

In this way we de ne a new quantity, namely:

$$_{12} = \frac{\langle \mathbf{C}_1\ \ \mathbf{C}_2\rangle}{\sqrt{I_1 I_2}} = \frac{\langle \mathbf{C}_1\ \ \mathbf{C}_2\rangle}{\langle|\mathbf{C}_1|\rangle\ \langle|\mathbf{C}_2|\rangle}\quad , \tag{14.17}$$

which is referred to as the *complex degree of mutual coherence* between the elds. The corresponding quantity Re($_{12}$) is called their *degree of mutual coherence.*

We can look at Eq. (14.15) as the result of an *interference* experiment: at a given point in space, where the detector is placed, and at a given instant in time, the total intensity is the sum of the intensities of the individual elds, plus a third term, which depends on the phase shift between the interfering elds. This term is responsible, as well known from basic physics (see, e.g., Born and Wolf, 1980), for the so called *interference fringes.* What Eq. (14.16) tells us is that the *visibility* of these fringes, de ned as:

$$v = \frac{I_{\mathrm{MAX}}\quad I_{\mathrm{min}}}{I_{\mathrm{MAX}} + I_{\mathrm{min}}}\quad , \tag{14.18}$$

depends on the degree of mutual coherence between the two interfering elds. In the next sections we will focus on those cases, very important in practice, where the two elds are, as we have said previously, replicas generated by the same source.

14.4 Temporal coherence of a source

We will apply now the general de nitions given in the previous section to the interference between elds which are generated by the same source at di erent times, in the same direction in space. This subject is referred to as *temporal coherence.* The case of elds generated at the same time in di erent directions, referred to as *spatial coherence*, will be dealt with in the next section.

To investigate temporal coherence we need an instrument which is capable of splitting a eld in two, and introducing a variable delay in one part of it. The scheme shown in Figure 14.1 actually does this. It is known in classical optics as a *Michelson interferometer.* S stands for a 3 dB beam splitter (i.e., a semitransparent mirror), M_1 and M_2 are two at mirrors, ℓ_1 and ℓ_2 are the lengths of the so-called arms of the interferometer. The symbol of a diode stands for a square-law slow photodetector.

Suppose that, on the length scale which is set by the size of this instrument, the incident eld $\mathbf{E}_0$ does not depend on the transverse coordinates, so any transverse displacement of the detector does not a ect the output signal.

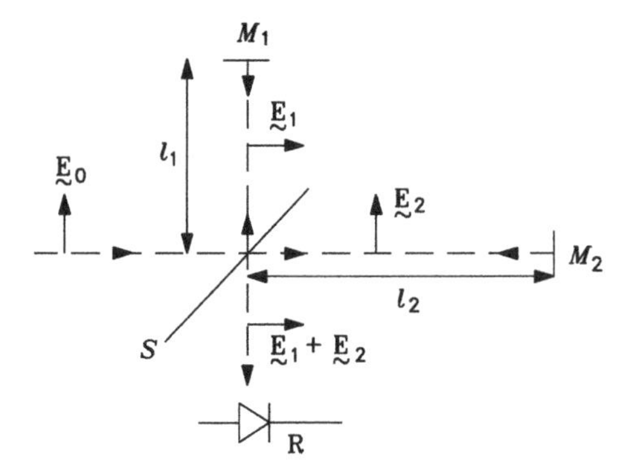

Figure 14.1 *Schematic diagram of a Michelson interferometer.*

Let:

$$= 2(\ell_2 \quad \ell_1)/c \quad , \tag{14.19}$$

be the di erential delay which corresponds to the di erence in length between the arms (c = speed of light in free space). Then, Eq. (14.15) can be rewritten as:

$$I = I_1 + I_2 + 2\overline{\mathbf{E}_1(t) \;\; \mathbf{E}_1(t \qquad)} = 2\left[I_1 + \overline{\mathbf{E}_1(t) \;\; \mathbf{E}_1(t \qquad)}\right] \quad , \tag{14.20}$$

where, as usual, the overline means time average over the detector response time. For $= 0$, the incident eld interferes with itself: intuition and experience say that, under such conditions, the degree of mutual coherence must be very high. This can be con rmed rigorously using the analytical signal in the form (14.12), and assuming that is smaller than the inverse of the spectral width of the two elds, so that we may take $\mathbf{C}_1(t \qquad) = \mathbf{C}_1(t)$. Hence, Eq. (14.20) can be rewritten as:

$$\begin{aligned} I &= 2I_1 + 2\left[\frac{\mathbf{C}_1(t) \;\; \mathbf{C}_1(t \qquad)}{4} e^{j\omega_0} + \text{c.c.}\right] \\ &= 2I_1 + 2I_1 \;\cos\omega_0 \quad = 2I_1 \left\{1 + \cos\left[\frac{2}{\;\;_0}\, 2(\ell_2 \quad \ell_1)\right]\right\} \end{aligned} \quad , \tag{14.21}$$

with $_0 = 2 \;\; c/\omega_0$.

Eq. (14.21) yields $I_{\text{min}}[= I(\omega_0 \quad = \quad)] = 0$. So, under these circumstances, the fringe visibility, Eq. (14.18), is:

$$v = \frac{I_{\text{MAX}}}{I_{\text{MAX}}} = 1 \quad , \tag{14.22}$$

reaching its highest possible value. This result is independent of the source, except for the assumption that Eqs. (14.11) are satis ed.

What depends on the statistical properties — order or disorder — of an individual source, is how long it maintains memory of its past, or, in other words, how the fringe visibility behaves vs. the di erential delay in the interferome-

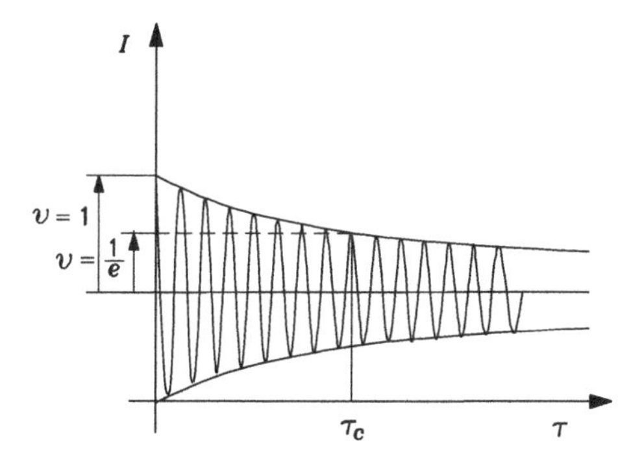

Figure 14.2 *Interference fringes vs. di erential delay in the interferometer.*

ter, $v(\ )$. We will restrict ourselves to *free-running* sources (i.e., sources which are not externally modulated).† For free-running sources, v is a *monotonically decreasing* function of . Its typical behavior is sketched in Figure 14.2. The value of for which $v = 1/e$ is referred to as the source *coherence time*, i.e.:

$$v(\ _c) = 1/e \quad . \tag{14.23}$$

For any function of time which has all the necessary requirements to be Fourier-transformable, there is a lower bound to its time-bandwidth product, the so-called uncertainty relationship. With our de nition of the Fourier transform, the time duration of a signal, , and the bandwidth of its source, B, it reads:

$$B \quad 1 \quad . \tag{14.24}$$

We may think of the coherence time as a time slot after which the source has lost memory of what it was at the beginning, so that nothing changes if it is turned o , and then on again. Comparing this viewpoint with Eq. (14.24), we conclude that it must be:

$$_c \quad _{\min} = \frac{1}{B} \quad . \tag{14.25}$$

Since the bandwidth B cannot be in nite, there are no sources which are *completely* incoherent in time, i.e., able to loose instantaneously their memory. *As a convention*, we will refer to a source which obeys Eq. (14.25) as an equality as *temporally incoherent*. On the contrary, we will call temporally coherent a source for which $_c \quad _{\min}$.

14.5 Spatial coherence of a source

We can pass from temporal coherence to spatial coherence without too much trouble, if we recall from Chapters 4 and 13 that there is a Fourier-transform

† Modulation of any kind introduces further correlations, over longer time scales. The interested reader should consult more advanced books.

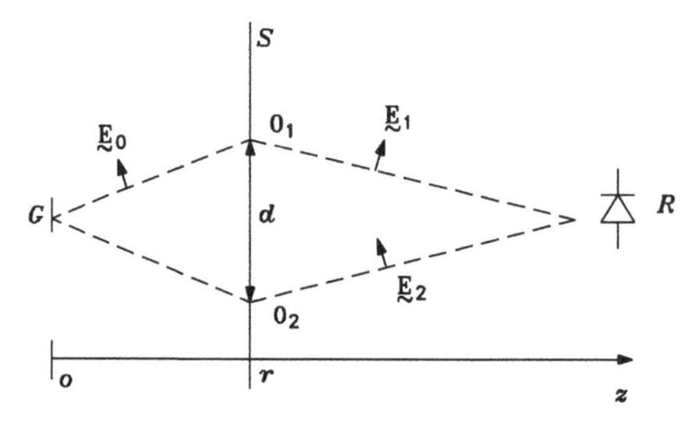

Figure 14.3 *Schematic diagram of Young's interferometer.*

relationship between any square-integrable eld in 3-D space, and its plane-wave spectrum. The subsequent hint is to investigate spatial coherence following the same path as in the previous section.

To investigate spatial coherence we need an instrument which is capable of generating two replicas of a eld generated by the same source at the same time, but in two di erent directions in space. The scheme shown in Figure 14.3 does it. This device is known in classical optics as *Young's interferometer*. G stands for the primary eld source, S is an opaque screen with two holes, O_1 and O_2, which for simplicity are dealt with as point sources, R stands for a square-law photodetector. For the sake of clarity, in Section 14.4 we ignored spatial coherence, assuming that the eld was independent of the transverse coordinates on the length scale given by the size of the instrument. Similarly, we will suppose now that the eld $\mathbf{E}_0$ which illuminates the interferometer has a coherence time much longer than all the delays due to di erent path lengths, and concentrate on spatial coherence.

The detector R is large in size, compared to the wavelength. Accordingly, the detection process averages in space and cancels those terms which depend on spatial coordinates with periods of order /2, similarly to the averaging of temporal oscillations at angular frequency $2\omega_0$ in the previous section.

Interference fringes are observed when the detector R is moved in a plane parallel to the screen S. They result from the phase shift between the two contributions $\mathbf{E}_1$ and $\mathbf{E}_2$, which depends on the position of R. With suitable geometries, their period can be made much larger than ${}_0$; its exact value depends on the angle between the incident beams (see Chapter 4). Their *visibility* reaches its maximum (ideally, $v = 1$), when the two holes, O_1 and O_2, are at an in nitesimal distance from each other. It would remain $v = 1$ as the distance between the holes, d, increases, if at any point on the screen S the eld $\mathbf{E}_0$ had perfect memory of what its complex amplitude is at any other point of S. In reality (leaving aside "spatially modulated" sources, such as antenna arrays or di raction gratings, which make spatial correlation more complicated), the fringe visibility decreases monotonically as d increases. The value of d at which $v = 1/e$, d_c, de nes what is usually referred to as the *coherence area*:

$$A_c = \quad d_c^2 \quad . \tag{14.26}$$

The reader may reconsider now, or more rewardingly after reading the rest of this section, the relationship between antenna gain and e ective area, which we derived in Chapter 12.

To nd what the smallest possible coherence area can be, in analogy with the shortest coherence time, Eq. (14.25), one may reconsider either Chapter 13, or Section 5.7 on Gaussian beams. If we call $\quad x$ the source size along one direction, x, then the smallest angular aperture of the emitted beam in the (x, z) plane (where z is the direction normal to the source surface) is given by:

$$\vartheta_x = \frac{\quad}{\quad x} \quad . \tag{14.27}$$

Let r be the distance between the source and the plane over which we measure its coherence (in our example, the plane of the screen S), and suppose, for simplicity, that the problem is symmetrical around the z axis. Then, the minimum coherence area is:

$$A_{\min} = \quad (r \quad \vartheta_x)^2 = \quad \left(\frac{r}{\quad x}\right)^2 \quad . \tag{14.28}$$

This shows that, in free propagation, the smallest coherence area grows in proportion to the square of the distance from the source, irrespectively of the physical nature of the source itself. One of the consequences is a phenomenon that is very well known in astronomy: from the earth, stars look as spatially coherence sources. Readers who are familiar with Shannon's sampling theorem in signal theory should not be surprised by this result: two samples taken on the earth cannot be mutually independent, because they are so close to each other, in comparison with the distance from the source.

Eq. (14.28) enables us to another comment: *spatial ltering*, i.e., introduction of an iris which makes $\quad x$ smaller, enlarges the minimum coherence area, downstream after the iris.

We may conclude this section with a comment very similar to what we said at the end of Section 14.4 when we discussed temporal coherence. There are no sources which are completely spatially incoherent. *Conventionally*, we call spatially incoherent those sources whose coherence area coincides with Eq. (14.28), spatially coherent those for which, on the contrary, $A_c \quad A_{\min}$.

14.6 Higher-order coherence: An introduction

In the previous two sections we gave "classical" de nitions of coherence, as they can be found in all the textbooks on optics written before the '60's. De ned in this way, coherence is a way to measure what is called, in the framework of statistics and probability, the *autocorrelation function* of a random process. For example, Eq. (14.20) can be written in the form:

$$I = 2\left[I_1 + G^{(1)}(\ )\right] \quad , \tag{14.29}$$

where $G^{(1)}(\ )$ is precisely the autocorrelation function of the eld $\mathbf{E}\ (t)$.

Those readers who are familiar with the fundamentals of random processes know that an autocorrelation function does not identify completely, in an unambiguous way, the corresponding probability distribution. The autocorrelation function identi es only the second-order moment of the distribution, or in other words its mean-square deviation. Consequently, in general an interferometry experiment is not enough to characterize completely the probability distribution of an e.m. eld. It is su cient whenever one knows *a priori* that the eld in question obeys Gaussian statistics, since a Gaussian probability distribution is fully characterized by its mean value and its variance. Gaussian statistics were in fact the only case of practical interest for classical optical sources, so this explains why the classical theory of coherence is built around what we have seen so far (Born and Wolf, 1980). With the advent of the laser, and with the extension of the theory of coherence to radio frequencies, it became clear that further experimental observations were needed, to measure higher-order moments of the probability distribution or, as an equivalent alternative, higher-order correlation functions. The moment approach will be discussed in Section 14.7. Here we will describe the rst experiment which allowed measurements of higher-order correlation. It was conceived by two radioastronomers, Hanbury-Brown and Twiss, who aimed at increasing the angular resolution of their radiotelescope to study twin stars.

Suppose that, in addition to the term $\langle \mathbf{E}_1\ \ \mathbf{E}_2\rangle$ which appears in Eq. (14.15), we want to known the quantity:

$$\langle (\mathbf{E}_1\ \ \mathbf{E}_2)^2\rangle = \overline{(\mathbf{E}_1\ \ \mathbf{E}_2)^2} \quad . \tag{14.30}$$

It cannot be measured using two detectors of the same kind as those used in classical interferometry, whose response time is long not only compared with $T = 2\ \ /\omega_0$, but also on the time scale which characterizes the random uctuations in the complex amplitudes $\mathbf{C}_i$. These detectors would average out these uctuations, whereas they are exactly what we want to see in order to measure Eq. (14.30). If one multiplies the outputs of two *slow* detectors (Figure 14.4a), the output signal would be proportional to:

$$\overline{(\mathbf{E}_1\ \ \mathbf{E}_2)}\ \ \overline{(\mathbf{E}_1\ \ \mathbf{E}_2)} = (\langle \mathbf{E}_1\ \ \mathbf{E}_2\rangle)^2 \quad\ \ \langle (\mathbf{E}_1\ \ \mathbf{E}_2)^2\rangle \quad , \tag{14.31}$$

where the equal sign between the second and the third expressions holds *only* if the two random processes $\mathbf{E}_1$ and $\mathbf{E}_2$ are *completely correlated.*

To measure Eq. (14.30), we may use the scheme shown in Figure 14.4b, where the symbols R'_A, R'_B stand for two *fast* photodetectors, ltering out the terms at angular frequency $2\omega_0$ but not averaging over the characteristic times of uctuations in the incident eld. Average on this time scale is performed downstream, by a low-pass lter at the output of the multiplier. The *variable*

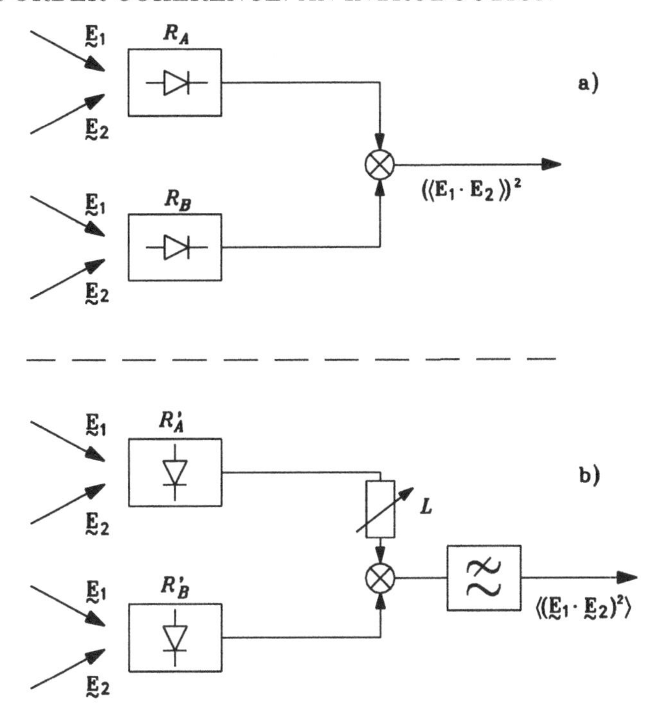

Figure 14.4 *Schematic diagram of a set-up to measure second-order correlation functions.*

delay line L, inserted between one of the fast detectors and the multiplier, plays a role which is similar to the variable arm length in the Michelson interferometer, as it allows us to measure:

$$I^{(2)}(\ ) = \overline{[\mathbf{E}\ (t)]^2\ [\mathbf{E}\ (t \qquad)]^2} \quad . \tag{14.32}$$

If we let $\mathbf{E} = \mathbf{E}_1 + \mathbf{E}_2$ then we see that Eq. (14.32) contains, like Eq. (14.29), terms which do not depend on and other terms which depend on . The last ones can be looked at as a *second-order time-domain correlation function.*

Switching to *spatial correlation*, suppose now that the two detectors are located at two points de ned by the position vectors $\mathbf{r}_A$ and $\mathbf{r}_B$, respectively. Then, Eq. (14.32) is replaced by:

$$I^{(2)}(\mathbf{r}_A \quad \mathbf{r}_B) = \overline{[\mathbf{E}\ (\mathbf{r}_A)]^2\ [\mathbf{E}\ (\mathbf{r}_B)]^2} \quad . \tag{14.33}$$

This equation lends itself to comments which are very similar to those we made after Eq. (14.32).

In contrast to what might appear at rst sight, the results of the Hanbury-Brown and Twiss experiment cannot be guessed from those of standard interferometry. To outline this point, let us deal with three examples.

A) Let the two interfering elds be:

$$\mathbf{E}_\ell(\mathbf{r},t) = 2\,\mathrm{Re}\left[C_\ell\;\hat{v}\;e^{j(\omega_0 t\quad \mathbf{k}_\ell\;\mathbf{r})}\right] \qquad (\ell = 1,2) \quad , \tag{14.34}$$

where $\hat{v}$ is a constant unit-vector. We may suppose that the propagation vector $\mathbf{k}_\ell$ is constant over the whole spectrum of the eld. C_1 and C_2 (complex scalar quantities) are two *statistically independent*, stationary and ergodic random variables, whose real and imaginary parts have Gaussian probability distributions, with average values equal to zero. The factor 2 was introduced merely to simplify the following calculations.

Remembering that the detectors lter out the terms at angular frequency $2\omega_0$, we see that Eq. (14.33) gives:

$$\begin{aligned} I^{(2)}(\mathbf{r}_A\quad \mathbf{r}_B) \;=\; & \langle(|C_1|^2 + |C_2|^2 + C_1 C_2\,e^{j(\mathbf{k}_2\quad \mathbf{k}_1)\,\mathbf{r}_A} + C_1 C_2\,e^{j(\mathbf{k}_2\quad \mathbf{k}_1)\,\mathbf{r}_A}) \\ & (|C_1|^2 + |C_2|^2 + C_1 C_2\,e^{j(\mathbf{k}_2\quad \mathbf{k}_1)\,\mathbf{r}_B} + C_1 C_2\,e^{\quad j(\mathbf{k}_2\quad \mathbf{k}_1)\,\mathbf{r}_B})\rangle\,. \end{aligned} \tag{14.35}$$

If each of the factors in brackets were averaged separately from the other, then, given the previous assumptions on the statistics of C_1 and C_2, the result would consist only of the term $\langle|C_1|^2 + |C_2|^2\rangle$, and so the experiment would be trivial. As averaging occurs after their multiplication, we get:

$$\begin{aligned} & I^{(2)}(\mathbf{r}_A\quad \mathbf{r}_B) = \\ & = \langle(|C_1|^2 + |C_2|^2)^2\rangle + \langle|C_1|^2\;|C_2|^2\rangle\left[e^{j(\mathbf{k}_2\quad \mathbf{k}_1)\,(\mathbf{r}_B\quad \mathbf{r}_A)} + e^{j(\mathbf{k}_2\quad \mathbf{k}_1)\,(\mathbf{r}_A\quad \mathbf{r}_B)}\right] \\ & = \langle(|C_1|^2 + |C_2|^2)^2\rangle + 2\,\langle|C_1|^2\;|C_2|^2\rangle\,\cos\left[(\mathbf{k}_2\quad \mathbf{k}_1)\;(\mathbf{r}_A\quad \mathbf{r}_B)\right]. \end{aligned} \tag{14.36}$$

The second term in the last expression is the second-order correlation function we are looking for. Its ratio to the rst term can be taken as the de nition of "visibility of power density fringes," distinct from the "amplitude fringes" that one sees in ordinary interferometry.‡ Note that the result turned out to be Eq. (14.36) because we supposed that the functions C_1 and C_2 describe the behavior of the eld *on both detectors*, R'_A and R'_B. This assumption implies that the two detectors must be located *within the same coherence area* of the incident eld. A very similar comment can be made for the corresponding experiment in the time domain.

B) Suppose now that the elds on detector R'_A are still described by Eq. (14.34), but the elds on R'_B are:

$$\mathbf{E}_\ell(\mathbf{r}_B,t) = 2\,\mathrm{Re}\left[D_\ell\;\hat{v}\;e^{j(\omega_0 t\quad \mathbf{k}_\ell\;\mathbf{r}_B)}\right] \qquad (\ell = 1,2) \quad . \tag{14.37}$$

Suppose that the Gaussian variables D_1, D_2 are *completely uncorrelated* with

‡ Eq. (14.36) contains a term which is sensitive to the di erence between the propagation vectors of the two incident waves, $\mathbf{k}_1$ and $\mathbf{k}_2$. Measuring this di erence was, in fact, the original purpose of the Hanbury-Brown and Twiss experiment.

C_1, C_2 of Eq. (14.34). This happens if each detector is located *outside the coherence area* centered on the other one. With these new assumptions, Eq. (14.33) yields:

$$I^{(2)}(\mathbf{r}_A, \mathbf{r}_B) = \langle (|C_1|^2 + |C_2|^2)\,(|D_1|^2 + |D_2|^2) \rangle \quad . \tag{14.38}$$

All the terms which depend on the mutual position of the detectors disappeared. In other words, we do not see "power density fringes" any more.

C) It would be erroneous to draw from the previous examples the conclusion that, in the Hanbury-Brown and Twiss experiment, we see fringes if there is coherence (in the classical sense of the word) and we do not see them if there is no coherence. If it were so, the experiment would be a rather trivial duplicate of standard interferometry. To show that this conclusion is wrong, let us deal with two detectors illuminated by a *monochromatic uniform plane wave*, i.e., by an ideally coherent field:

$$\mathbf{E} = \mathbf{E}_0 \, e^{j(\omega_0 t - \mathbf{k}\cdot\mathbf{r})} \quad , \tag{14.39}$$

where $\mathbf{E}_0$ is a constant vector. Eq. (14.33) gives then:

$$I^{(2)} = \langle |\mathbf{E}_0|^2 \, |\mathbf{E}_0|^2 \rangle = |\mathbf{E}_0|^4 \quad . \tag{14.40}$$

In this case, again, there are absolutely no terms depending on spatial coordinates. The result is the same as in case B), in spite of the statistical properties of the fields being completely different in the two examples.

To summarize, we have examined here:
A) a random field, staying inside its coherence area;
B) a random field, going out of its coherence area;
C) an ideally coherent field.

While in classical interferometry A) and C) give the same results, and B) is different, in the Hanbury-Brown and Twiss experiment B) and C) give identical results, and differ from those of A).

14.7 An introduction to photocount distributions

The second-order correlation function that was defined in the previous section can be generalized even further. In fact, one may define quantities which contain products of more than two fields. One can also imagine measuring them using square-law detectors, but such a procedure is awkward for more than two detectors, so it can be at most of theoretical interest. Alternatively, one may start from another classical description of any probability distribution, namely, that in terms of its *moments*:

$$\mu_n = \int_{-\infty}^{+\infty} (x - m)^n \, p(x)\, d(x) \qquad (n = 1, 2, \ldots) \quad , \tag{14.41}$$

where m is the statistical average and $p(x)$ is the probability density function of the random variable x. This approach is simple to convert into an experimental technique at optical frequencies. This is the reason why it is often referred to as *photocount distribution*, or the photon statistics method (Arecchi, 1969). Here, being space-limited, we can only provide a short introduction to this subject.

What we would like to measure is the probability density, $P(C)$, of a random complex variable C de ned by Eq. (14.8).§ However, we are bound to use square-law detectors, so the phase of C cannot be measured directly. It could be measured letting C beat with another signal ("heterodyne" detection process), but often this reference signal is a ected by strong phase noise; this was an almost unavoidable problem with lasers, until recently. In this case, the result would be a convolution of two statistics, and might be di cult to interpret.

In the photocount method, this problem is circumvented by measuring the probability distribution of the *power density*, which is proportional to $|C|^2$. In quantum-optics terms, we may call it the probability distribution of the number of per-unit-time photons which impinge on the photodetector, $p(n)$. From it one can infer $P(C)$ by means of suitable theoretical models.

In order to extract $p(n)$ from the numbers of photons detected in various time slots of equal length, T_R, the following conditions must be satis ed, to be consistent with what we said in the previous sections:

T_R 2 $/\omega_0$, where ω_0 is the central angular frequency in the spectrum of the detected eld;

$T_R <$ $_c$ ($_c$ = coherence time of the detected eld), to avoid a premature time averaging, which would make the probability distribution narrower and "squeeze" it around the average photon number;

the time gap between two consecutive measurement slots, T, must be larger than $_{\min} = 1/B$, to ensure that the samples are statistically independent of each other.

Consequently, one needs a fast and sensitive photodetector (typically, a high-gain, low-noise photomultiplier), followed by a fast sampling circuit, which must operate over time intervals of length T_R, separated by time gaps of length T. The sampled signal (see Figure 14.5) is then converted to a digital format and sent to a multichannel amplitude analyzer, i.e., to an electronic "abacus" which counts the number of times, during the whole data acquisition time, the signal corresponds to $1, 2, \ldots, n$ photons in each time slot.

Figure 14.6 shows some plots, corresponding to some of the typical probability distributions which are encountered in photon counting. Curve L is typical for a single-mode laser eld, and corresponds to a *Poisson distribution*, namely:

$$p(n) = e^{\quad m} \frac{m^n}{n!} \quad , \qquad (14.42)$$

where m is the average per-unit-time number of detected photons.

Curve G is an example of what is measured for a standard thermal light

§ According to our de nition, C is a complex scalar quantity, as, to avoid needless complication, we still suppose that the eld has a constant polarization.

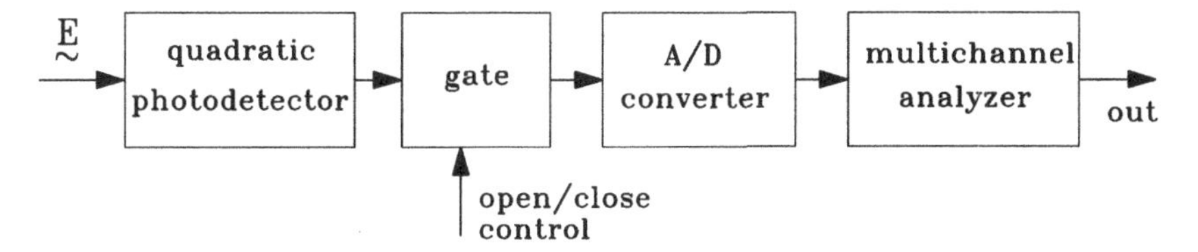

Figure 14.5 *Block diagram of a set-up to measure photocount distributions.*

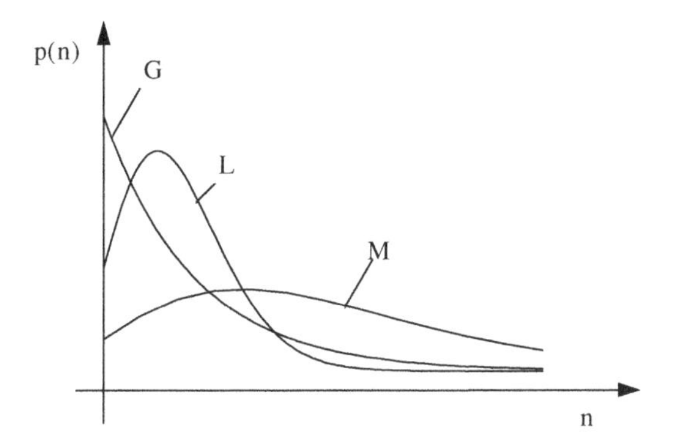

Figure 14.6 *Three examples of probability densities for photocount distributions.*

source, and corresponds to the so-called *geometrical* or *Bose-Einstein distribution*, very popular in atomic physics, namely:

$$p(n) = \frac{1}{1+m} \left(\frac{m}{1+m} \right)^n \quad , \tag{14.43}$$

where m is again the average number of detected photons per unit time.

Finally, curve M refers to a superposition of elds of the two previous types, and corresponds to the following probability distribution:

$$p(n) = \frac{N^n}{(1+N)^{n+1}} L_n \left(\frac{S}{(1+N)\,N} \right) e^{\ S/(1+N)} \quad , \tag{14.44}$$

where S and N are the average numbers of per-unit-time detected photons belonging to the laser eld and to the random one, respectively; L_n is the Laguerre polynomial of order n (see Chapter 5).

It can be shown that a photocount distribution (14.43) corresponds to a so-called *normal* or *Gaussian* probability distribution, for the complex amplitude C, on the complex plane, i.e.:

$$P_g(C) = \frac{1}{m} e^{\ (|C|^2)/m} \quad . \tag{14.45}$$

The corresponding probability density for the variable $|C|$, on the positive real

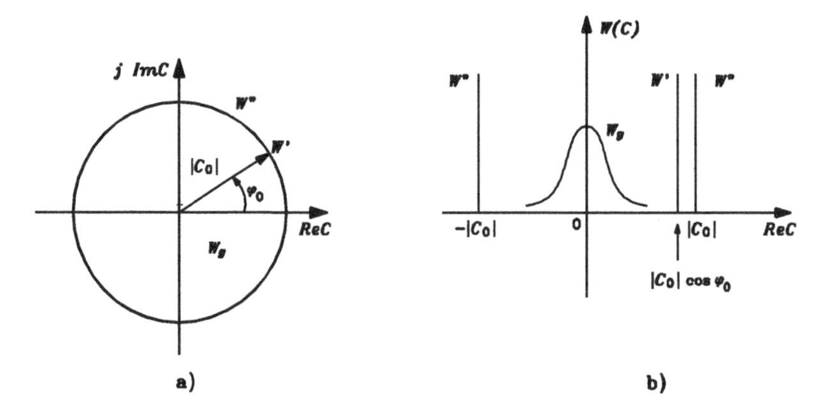

Figure 14.7 *Complex-plane representation of the eld statistics which correspond to the three probability densities shown in the previous gure.*

right-half axis, is a so-called *Rayleigh distribution*, namely:

$$P_R(|C|) = \frac{|C|}{\sqrt{m}}\, e^{\ (|C|^2)/2m} \quad . \tag{14.46}$$

The Poisson distribution (14.42) may correspond to two di erent probability densities $P(C)$. One of them is given by:

$$P'(C) = \ (C \quad C_0) \quad , \tag{14.47}$$

where $\ (z)$ is a Dirac delta function of the complex variable z, and C_0 is an arbitrary complex constant. The other one is given by:

$$P''(C) = \frac{\ (|C| \quad |C_0|)}{2\ |C_0|} \quad . \tag{14.48}$$

Eq. (14.47) represents a perfectly deterministic, ideally coherent time-harmonic eld with constant complex amplitude C_0, while Eq. (14.48) represents an amplitude-stabilized eld whose phase varies with time in a completely arbitrary way. Physical considerations beyond the scope of this book show that in the large-signal regime a free-running laser, oscillating in a single mode without external feedback loops, emits a eld which corresponds to Eq. (14.48), i.e., is a ected by a large amount of phase noise (see also the previous comment on heterodyne detection).

Plots of the three distributions, Eqs. (14.45), (14.47) and (14.48), on the complex plane are shown in Figure 14.7. Let us recall that C was de ned as the complex amplitude of a eld whose spectrum is centered on the angular frequency ω_0. Accordingly, Figure 14.7a can be thought of as the picture of a generalized Steinmetz procedure.

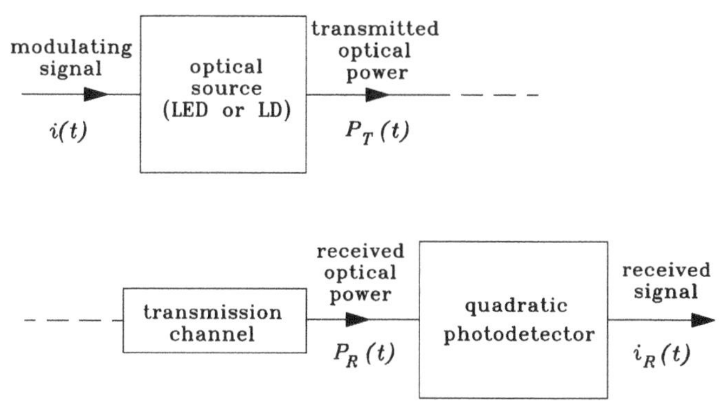

Figure 14.8 *Block diagram of a direct-detection optical- ber communication system.*

14.8 Modal noise in optical- ber transmission systems: A short outline

Performance of optical communication systems, especially those based on multimode bers, can be impaired, under some circumstances, by phenomena which can be explained using concepts introduced in this chapter and in Chapter 10, on dielectric waveguides.

The block diagram shown in Figure 14.8 describes part of an optical- ber direct-detection intensity-modulated transmission system, from the electrical-to-optical converter to the converse operation. The block labelled "transmission channel" is an optical ber, which propagates N transmission modes.¶ Within the framework of our model, the channel is *linear with respect to the e.m. eld.*

The optical-source instantaneous *power* is proportional to the driving current. The square-law photodetector gives an output current proportional to the optical *power* incident on it. Therefore, to have an output signal $i_R(t)$ proportional to the modulating signal $i(t)$, the transmission channel should be *linear with respect to the power* which travels through it. For this to be true, it is necessary that the total power P_{tot} through any cross-section of the channel be equal the sum of the powers carried by the individual modes, P_ℓ ($\ell = 1, 2, \ldots, N$), namely:

$$P_{\text{tot}} = \sum_{\ell=1}^{N} P_\ell \quad . \tag{14.49}$$

This requirement must hold for quantities which carry information, i.e., for powers which are time-averaged over a period of the optical carrier frequency,

¶ In a single-mode ber, it is known from Chapter 10 that two orthogonal states of polarization may propagate, and that they are degenerate only in an ideal structure. Furthermore, radiation modes can play an important role also, over short distances (for example, the length of a ber-based passive device, such as an attenuator or a directional coupler). This comment means that the rest of this section applies also to single-mode systems.

$T_0 = 2\ \ /\omega_0$, but not averaged over times comparable with the inverse of the modulating signal frequencies.

Let us discuss under which assumptions Eq. (14.49) can be considered as being veri ed in an actual link. From Chapter 7, we know that, in general, in a waveguide complex powers carried by guided modes can be added, yielding the total power, as a result of to their *orthogonality* relationships, of the type:

$$\int_S \mathcal{E}_{ti} \ \ \mathcal{E}_{tj}\, dS = 0 \qquad (i \neq j) \quad . \tag{14.50}$$

These relationships hold when S is the *ideal* cross-section of the waveguide. Orthogonality breaks down, in general, when a cross-section, say S', di ers from S. This can actually happen in real bers, for instance in a misaligned connector or in a photodetector which is not properly coupled to the ber output section.‖ Terms like Eq. (14.50), if they do not vanish, cause signal *distortion*, because two interfering modes (say i and j) in general have *di erent group delays* ($_i \neq \ _j$), so that the two factors in Eq. (14.50) correspond to two values that the modulating signal has taken at two di erent times. The combination of all these distorting terms (which, in the multimode case, are many and hard to correlate to one another) is called *modal noise.*

To complete this introductory analysis, let us compare the di erential group delays with the times that we de ned in Section 14.4 to characterize the source. In fact, terms like Eq. (14.50) may disappear not only because of mode orthogonality, but also for other reasons. Let $\ _c$ be the coherence time of the optical source. Then, if:

$$|\ _i \quad \ _j| > \ _c \quad , \tag{14.51}$$

products like Eq. (14.50) involve two elds, of the i-th and the j-th mode, respectively, which are mutually uncorrelated and therefore have *zero time-average value.* Hence, if a relationship like Eq. (14.51) is satis ed by all pairs of waveguide modes, and if the largest modulation frequency f_{MAX} satis es:

$$f_{\mathrm{MAX}} \qquad \frac{1}{\ _c} \quad , \tag{14.52}$$

then Eq. (14.49) can be considered satis ed even over non-ideal cross-sections, and the transmission channel is linear with respect to power, as required. Note, in particular, that Eq. (14.51) must have been already satis ed at the *rst* imperfect joint. Thus, connectors close to the optical source are critical points referring to modal noise.

14.9 Further applications and suggested reading

For a long time, whose beginning is not easy to identify, but whose length is measured in decades, and whose end is rather sharply peaked around 1960,

‖ An iris whose cross-section is smaller than S can be looked at as a spatial lter. Therefore (see Section 14.5), it increases the coherence area, and destroys the peculiar form of partial spatial incoherence which is expressed by Eq. (14.50).

optics has been looked at as a chapter of physics which had reached its complete maturity, and was then enjoying something comparable to an honorable retirement. The invention of the laser made a most dramatic change, not only because a broad range of new technical applications of light became possible and convenient, but also because some fundamentals of physics had to be re-examined in depth. Coherence is certainly one of the most impressive examples of this development. As we said in Section 14.7, the readers can easily verify this statement, consulting rst a classical treatise on optics (e.g., Born and Wolf, 1980), and then moving on to modern books, beginning with those which were written in the late 60's, just a few years after the invention of the laser.

The classical theory of coherence occupies the entire Chapter X (about 65 pages) in Born and Wolf (1980). Consulting the tables of contents, the reader may grasp why such a theory is intimately related not only to interference, as outlined by our approach in this chapter, but also to other subjects of classical optics, like di raction and polarization (especially if one uses the Stokes parameters which we de ned in Chapter 2).

Even from the earliest presentations of the re-examined theory (e.g., Arecchi, 1969) one may easily grasp that in the modern framework, coherence becomes intimately related to quantum theory. Those who wish to proceed further should, rst of all, familiarize themselves with the fundamentals of quantum statistics, typically with the density-matrix formalism (Louisell, 1960). We believe this to be extremely useful to those readers who are interested in optical communications and in photonics. They will nd close — possibly, unexpected — relationships between these subjects and the main source of concern in telecommunications and in electronics: noise. They will face — also possibly for the rst time — a physical world where the dominant enemy is not thermal noise, but quantum noise. Statistics become completely di erent. The challenge of detecting signals "beyond the quantum limit," which deserves to be monitored very carefully by modern engineers, requires coherence as a background item. The same is true in order to understand the so-called "laser without inversion".

Quantum electronics is a discipline dealing with interaction of e.m. waves and matter. In its framework, coherence properties are of paramount importance to model correctly not only elds, but also systems which consist of several particles. A highly recommended reading, to this purpose, is Meystre and Sargent (1991).

Modeling interactions between radiation and matter is not too dissimilar, in many respects, from modeling interactions between di erent waves in a nonlinear medium. So, we may say that knowledge about coherence is not complete unless the reader has at least a rudimentary background in nonlinear optics. It is interesting to note how even the earliest books on nonlinear optics (e.g., Bloembergen, 1965) touch often upon coherence. Many phenomena, discovered and exploited more recently in guided-wave nonlinear optics (see Agrawal, 1995), are impossible to explain without coherence. The idea of "deterministic chaos" — a new frontier in basic physics, after the deterministic mechanics of the 18th century and the intimately statistical thermodynamics of the 19th century — originates from many background items; coherence and nonlinear optics are certainly among the most prominent ones.

To conclude, those readers who enjoy puzzles should consult the literature (e.g., Mandel and Wolf, 1995) about mysteries of single-photon interferometry, a branch of optics which is still challenging our very fundamental ideas on the world in which we live.

Problems

14-1 Consider the Gaussian pulse expressed by Eq. (5.23), dealt with in Section 5.2, where dispersion was de ned and discussed. Find the corresponding analytical signal at the receiver, without and with dispersion. Show that its real and imaginary parts satisfy the Hilbert-transform relationships.

14-2 Refer again to the Gaussian pulses of Section 5.2, and assume no dispersion ($k'' = 0$). Evaluate the complex degree of mutual coherence, Eq. (14.17), between the pulse at a xed distance z_0 and the pulse at a generic distance z, as a function of z.
Repeat the problem in the presence of dispersion.

14-3 The spectrum of a white-light source extends from 400 to 700 nm.

a) Evaluate its minimum coherence time.
b) Suppose next that radiation emitted by this source is passed through an optical passband lter whose transfer function has a Gaussian shape, is centered at = 500 nm, and has a $1/e$ spectral width of 10 nm.
   Evaluate the coherence time after the lter.

14-4 The radiation emitted by the source of the previous problem is injected into an interference microscope: a system where the incoming light beam is split into two parts of equal amplitudes, one travels through a sample to be tested, the other goes through a reference arm; nally, the two channels are recombined in front of an eyepiece, where they interfere with each other. Suppose that the geometrical length of the two arms is the same, but that there is air in the reference arm (unit refractive index), while the sample under test has a refractive index of about 1.5.

a) What is the maximum thickness of the sample which enables a fringe visibility higher than $1/e$, using the white-light source?
b) Which value does this maximum thickness reach if one inserts the lter of the previous problem?
c) Describe, qualitatively, what happens if the reference arm is also equipped with a layer of a material whose refractive index is close to that of the test sample.

14-5 The light emitted by a continuous-wave (cw) laser source, whose emission line is peaked at $_0 = 633$ nm, and whose coherence time is of several seconds, is modulated by an intensity modulator, driven by a square-wave signal whose frequency is 10 GHz. The output of the modulator is coupled to a Michelson interferometer.
Draw qualitative diagrams of the fringe visibility as a function of the unbalance (i.e., the relative delay) between the interferometer arms, under the following assumptions:

a) the modulator is an ideal device, i.e., in the "o " state its output signal equals zero;

b) the extinction ratio of the modulator is only 3 dB, i.e., the output power in the "o " state is one half of that in the "on" state.

14-6 The radius of the Sun is of the order of 700,000 km, and its mean distance from the Earth is about 150 million kilometers.
Estimate the coherence area of the solar radiation when it reaches the Earth atmosphere, in the visible range (let $= 500$ nm), and then for a radioastronomy experiment at 100 MHz.

14-7 The eld emitted from any real source, observed at a given distance d from the source, has a nite coherence time as well as a nite coherence area. Exploiting the de nitions of coherence time and coherence area, de ne a *coherence volume*, as the volume of that region in the real 3-D space within which the eld in question has a signi cant degree of mutual coherence.

14-8 Study the dependence of the Poisson probability distribution, Eq. (14.42), on the average number of photons per unit time, m. In particular, discuss how the abscissa of the peak depends on m, and how a point in the "tail" (e.g., the point where the probability equals $10^{\ 3}$) depends on m.

14-9 A two-mode optical ber is designed in such a way that, at $= 1.3$ m, the per unit length group delays of the two modes di er by 1 ps/km.
Estimate, as an order of magnitude, the ber length after which the two modes, as a result of their mutual incoherence, do not yield any cross term in the output power when their spatial orthogonality is broken by a misaligned connector. Consider the following radiation sources feeding the ber input:

a) a light-emitting diode, with a spectral width of 7.5 nm;

b) a semiconductor laser, with a linewidth of 100 MHz;

c) a ber laser, based on an erbium-doped ber, with a linewidth of 5 MHz.

14-10 In a single-mode ber a ected by a gentle bend, a part of the power of the guided mode is radiated into the cladding, but then, after the bend, is guided along the ber by the index di erence between the nite cladding and the surrounding medium. Suppose that the per unit length group delay of the guided mode is proportional to the core group index (de ned in Chapter 10), while that of the "cladding mode" is proportional to the cladding group index.

a) Let us measure the ber output power placing a photodetector of nite size in front of the ber end. Why does the detected signal exhibit an oscillatory behavior as a function of wavelength?

b) Will the amplitude of these oscillations be sensitive to loss in the medium surrounding the cladding? In which sense?

c) How far from the bend shall the detector be for these oscillations to become negligible, in the lossless case, if the two group indices di er by 0.5%, and the emission line of the source used in the experiment, centered at 1.5 m, has a spectral width of 1 nm?

References

Agrawal, G.P. (1995) *Nonlinear Fiber Optics.* 2nd ed., Academic, San Diego.

Arecchi, F.T. (1969) "Photocount distribution and eld statistics". In R.J. Glauber, Ed., *Proc. Int. School of Physics "E. Fermi," XLII Course*, Academic, New York/London.

Bloembergen, N. (1965) *Nonlinear Optics.* Benjamin, New York/Amsterdam.

Born, M. and Wolf, E. (1980) *Principles of Optics.* 6th ed., reprinted 1986, Pergamon, Oxford.

Louisell, W.H. (1960) *Coupled Mode and Parametric Electronics.* Wiley, New York.

Mandel, L. and Wolf, E. (1995) *Optical Coherence and Quantum Optics.* Cambridge University Press, Cambridge, UK.

Meystre, P. and Sargent, M. (1991) *Elements of Quantum Optics.* 2nd ed., Springer-Verlag.

Papoulis, A. (1968) *Systems and Transforms with Applications in Optics.* McGraw-Hill, New York.

Appendices

APPENDIX A

Vector calculus: De nitions and fundamental theorems

Unless otherwise indicated, the meanings of the symbols are:

, = two scalar functions of space coordinates, de ned and regular in a three-dimensional domain D;

$\mathbf{A}$ = a vector function of space coordinates, de ned and regular in D;

S = a closed surface, contained in D and regular;

V = a three-dimensional region, surrounded by the surface S (sometimes, the volume of that region);

$\hat{n}$ = the unit-vector orthogonal to S, oriented towards the external region (see Figure A.1).

The *gradient of a scalar eld* is the following vector:

$$\nabla \quad \text{grad} \quad = \lim_{V \to 0} \frac{1}{V} \int_S \hat{n}\, \mathrm{d}S \quad . \tag{A.1}$$

The *divergence of a vector eld* is the following scalar:

$$\nabla \quad \mathbf{A} \quad \text{div}\, \mathbf{A} = \lim_{V \to 0} \frac{1}{V} \int_S \mathbf{A} \quad \hat{n}\, \mathrm{d}S \quad . \tag{A.2}$$

The *curl of a vector eld* is the following vector:

$$\nabla \quad \mathbf{A} \quad \text{curl}\, \mathbf{A} = \lim_{V \to 0} \frac{1}{V} \int_S \hat{n} \quad \mathbf{A}\, \mathrm{d}S \quad . \tag{A.3}$$

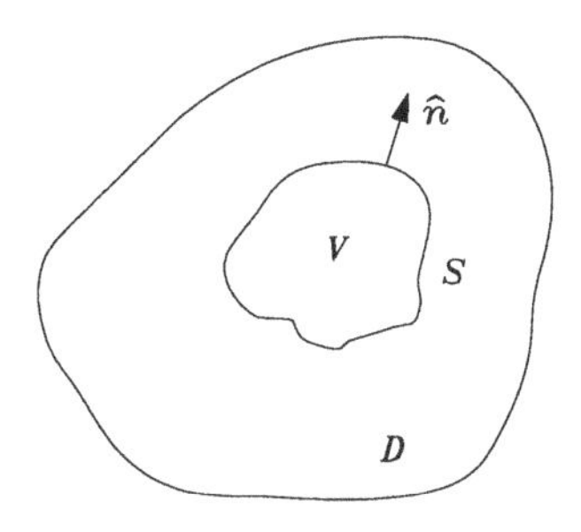

Figure A.1 *Illustration of the symbols used in the de nitions of the vector calculus operators.*

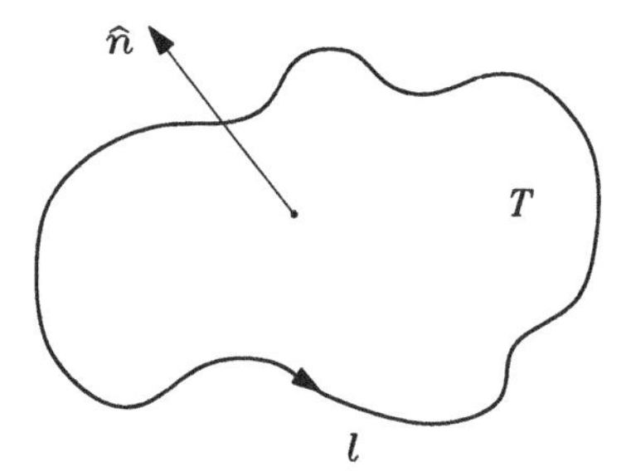

Figure A.2 *Illustration of the symbols used in Stokes' curl theorem.*

The *gradient of a vector eld* is the following dyadic:

$$\nabla \mathbf{A} \quad \text{grad}\,\mathbf{A} = \lim_{V \to 0} \frac{1}{V} \int_S \hat{n}\,\mathbf{A}\,\mathrm{d}S \quad . \tag{A.4}$$

The *symbolic vector* ∇ can be de ned as follows:

$$\nabla(\;) = \lim_{V \to 0} \frac{1}{V} \int_S \hat{n}(\;)\,\mathrm{d}S \quad , \tag{A.5}$$

where the pair of parentheses stands for any of the quantities on which ∇ operates in the previous four de nitions.

The *Laplacian of a scalar eld* is the following scalar:

$$\nabla^2 \quad = \nabla \quad \nabla \qquad \text{div grad} \quad . \tag{A.6}$$

The *Laplacian of a vector eld* is the following vector:

$$\nabla^2\mathbf{A} = \nabla(\nabla \quad \mathbf{A}) \quad \nabla \quad \nabla \quad \mathbf{A} \quad \text{grad}(\text{div}\,\mathbf{A}) \quad \text{curl}(\text{curl}\,\mathbf{A}) \quad . \tag{A.7}$$

Gauss' divergence theorem: with the previously de ned symbols, it can be proved that:

$$\int_V \nabla \quad \mathbf{A}\,\mathrm{d}V = \oint_S \mathbf{A} \quad \hat{n}\,\mathrm{d}S \quad . \tag{A.8}$$

Stokes' curl theorem: let T be a two-faced regular surface, ℓ be its contour (a connected and oriented line), $\hat{n}$ be the unit-vector orthogonal to T whose orientation is related to that of ℓ by the so-called right-hand rule (see Figure A.2). Then, it can be shown that:

$$\int_T \nabla \quad \mathbf{A} \quad \hat{n}\,\mathrm{d}T = \int_\ell \mathbf{A} \quad d\boldsymbol{\ell} \quad . \tag{A.9}$$

First Green's identity: with the symbols whose meanings were de ned at the beginning of this appendix, it can be shown that:

$$\int_V (\quad \nabla^2 \quad + \nabla \quad \nabla \quad)\,\mathrm{d}V = \int_S \quad \nabla \quad \hat{n}\,\mathrm{d}S \quad \int_S \quad \frac{\partial}{\partial n}\,\mathrm{d}S \quad , \tag{A.10}$$

where $\partial/\partial n$ stands for the direction derivative along the normal to S.

Second Green's identity: it can be shown that the following relationship follows from Eq. (A.10):

$$\int_V (\ \nabla^2 \qquad \nabla^2\ )\,\mathrm{d}V = \int_S (\ \nabla \qquad \nabla\ )\ \hat{n}\,\mathrm{d}S$$

$$\int_S \left(\ \frac{\partial}{\partial n} \qquad \frac{\partial}{\partial n}\right)\mathrm{d}S\ . \tag{A.11}$$

APPENDIX B

Vector di erential operators in frequently used reference systems

Rectangular cartesian coordinates (x_1, x_2, x_3):

$$\nabla f = \sum_{i=1}^{3} \hat{a}_i \frac{\partial f}{\partial x_i} \quad , \tag{B.1}$$

$$\nabla \ \mathbf{A} = \sum_{i=1}^{3} \frac{\partial A_i}{\partial x_i} \quad , \tag{B.2}$$

$$\begin{aligned} \nabla \ \ \mathbf{A} &= \hat{a}_1 \left(\frac{\partial A_3}{\partial x_2} \quad \frac{\partial A_2}{\partial x_3} \right) + \hat{a}_2 \left(\frac{\partial A_1}{\partial x_3} \quad \frac{\partial A_3}{\partial x_1} \right) + \hat{a}_3 \left(\frac{\partial A_2}{\partial x_1} \quad \frac{\partial A_1}{\partial x_2} \right) \\ &= \begin{vmatrix} \hat{a}_1 & \hat{a}_2 & \hat{a}_3 \\ \dfrac{\partial}{\partial x_1} & \dfrac{\partial}{\partial x_2} & \dfrac{\partial}{\partial x_3} \\ A_1 & A_2 & A_3 \end{vmatrix} \quad , \end{aligned} \tag{B.3}$$

$$\nabla^2 f = \sum_{i=1}^{3} \frac{\partial^2 f}{\partial x_i^2} \quad , \tag{B.4}$$

$$\nabla^2 \mathbf{A} = \sum_{i=1}^{3} \hat{a}_i \, \nabla^2 A_i \quad . \tag{B.5}$$

Note that sometimes, in the text, the base unit vectors are denoted as $\hat{x}_1, \hat{x}_2, \hat{x}_3$.

Cylindrical coordinates (r, φ, z):

$$\nabla f = \hat{a}_r \frac{\partial f}{\partial r} + \hat{a}_\varphi \frac{1}{r} \frac{\partial f}{\partial \varphi} + \hat{a}_z \frac{\partial f}{\partial z} \quad , \tag{B.6}$$

$$\nabla \ \mathbf{A} = \frac{1}{r} \frac{\partial}{\partial r} (r A_r) + \frac{1}{r} \frac{\partial A_\varphi}{\partial \varphi} + \frac{\partial A_z}{\partial z} \quad , \tag{B.7}$$

$$\nabla \ \ \mathbf{A} = \hat{a}_r \left(\frac{1}{r} \frac{\partial A_z}{\partial \varphi} \quad \frac{\partial A_\varphi}{\partial z} \right) + \hat{a}_\varphi \left(\frac{\partial A_r}{\partial z} \quad \frac{\partial A_z}{\partial r} \right) +$$

$$+ \hat{a}_z \left(\frac{1}{r} \frac{\partial (r A_\varphi)}{\partial r} \quad \frac{1}{r} \frac{\partial A_r}{\partial \varphi} \right) \quad , \tag{B.8}$$

$$\nabla^2 f = \frac{1}{r} \frac{\partial}{\partial r} \left(r \frac{\partial f}{\partial r} \right) + \frac{1}{r^2} \frac{\partial^2 f}{\partial \varphi^2} + \frac{\partial^2 f}{\partial z^2} \quad , \tag{B.9}$$

$$\begin{aligned} \nabla^2 \mathbf{A} &= \hat{a}_r \left(\nabla^2 A_r \quad \frac{2}{r} \frac{\partial A_\varphi}{\partial \varphi} \quad \frac{A_r}{r^2} \right) + \\ &+ \hat{a}_\varphi \left(\nabla^2 A_\varphi + \frac{2}{r^2} \frac{\partial A_r}{\partial \varphi} \quad \frac{A_\varphi}{r^2} \right) + \hat{a}_z (\nabla^2 A_z) \quad . \end{aligned} \tag{B.10}$$

Note that sometimes, in the text, the base unit vectors are denoted as $\hat{r}, \hat{\varphi}, \hat{z}$.

Spherical coordinates (r, ϑ, φ):

$$\nabla f = \hat{a}_r \frac{\partial f}{\partial r} + \hat{a}_\vartheta \frac{1}{r} \frac{\partial f}{\partial \vartheta} + \hat{a}_\varphi \frac{1}{r \sin \vartheta} \frac{\partial f}{\partial \varphi} \quad , \tag{B.11}$$

$$\nabla \quad \mathbf{A} = \frac{1}{r^2} \frac{\partial}{\partial r} (r^2 A_r) + \frac{1}{r \sin \vartheta} \frac{\partial}{\partial \vartheta} (\sin \vartheta \, A_\vartheta) + \frac{1}{r \sin \vartheta} \frac{\partial A_\varphi}{\partial \varphi} \, , \tag{B.12}$$

$$\begin{aligned} \nabla \quad \mathbf{A} &= \frac{\hat{a}_r}{r \sin \vartheta} \left[\frac{\partial}{\partial \vartheta} (A_\varphi \sin \vartheta) \quad \frac{\partial A_\vartheta}{\partial \varphi} \right] + \\ &+ \frac{\hat{a}_\vartheta}{r} \left[\frac{1}{\sin \vartheta} \frac{\partial A_r}{\partial \varphi} \quad \frac{\partial}{\partial r} (r A_\varphi) \right] + \frac{\hat{a}_\varphi}{r} \left[\frac{\partial}{\partial r} (r A_\vartheta) \quad \frac{\partial A_r}{\partial \vartheta} \right] , \end{aligned} \tag{B.13}$$

$$\nabla^2 f = \frac{1}{r^2} \frac{\partial}{\partial r} \left(r^2 \frac{\partial f}{\partial r} \right) + \frac{1}{r^2 \sin \vartheta} \frac{\partial}{\partial \vartheta} \left(\sin \vartheta \frac{\partial f}{\partial \vartheta} \right) + \frac{1}{r^2 \sin^2 \vartheta} \frac{\partial^2 f}{\partial \varphi^2} \, , \tag{B.14}$$

$$\begin{aligned} \nabla^2 \mathbf{A} &= \hat{a}_r \left[\nabla^2 A_r \quad \frac{2}{r^2} \left(A_r + \frac{1}{\mathrm{tg}\, \vartheta} A_\vartheta + \frac{1}{\sin \vartheta} \frac{\partial A_\varphi}{\partial \varphi} + \frac{\partial A_\vartheta}{\partial \vartheta} \right) \right] + \\ &+ \hat{a}_\vartheta \left[\nabla^2 A_\vartheta \quad \frac{1}{r^2} \left(\frac{1}{\sin^2 \vartheta} A_\vartheta \quad 2 \frac{\partial A_r}{\partial \vartheta} + \frac{2}{\mathrm{tg}\, \vartheta \sin \vartheta} \frac{\partial A_\varphi}{\partial \varphi} \right) \right] + \\ &+ \hat{a}_\varphi \left[\nabla^2 A_\varphi \quad \frac{1}{r^2} \left(\frac{1}{\sin^2 \vartheta} A_\varphi \quad \frac{2}{\sin \vartheta} \frac{\partial A_r}{\partial \varphi} \quad \frac{2}{\mathrm{tg}\, \vartheta \sin \vartheta} \frac{\partial A_\vartheta}{\partial \varphi} \right) \right] . \end{aligned} \tag{B.15}$$

Note that sometimes, in the text, the base unit vectors are denoted as $\hat{r}, \hat{\vartheta}, \hat{\varphi}$.

APPENDIX C

Vector identities

List of symbols:

f, g = scalar functions of space coordinates;

$\mathbf{A}, \mathbf{B}, \mathbf{C}$ = vector functions of space coordinates;

= scalar (or internal) product;

= vector (or external) product;

$(\mathbf{A}\ \nabla)$ = direction derivative, along the direction of the vector $\mathbf{A}$, multiplied by $|\mathbf{A}|$;

$\hat{v}$ = real unit-vector.

$$\nabla(fg) = f\,\nabla g + g\,\nabla f \quad , \tag{C.1}$$

$$\nabla(\mathbf{A}\ \ \mathbf{B}) = (\mathbf{A}\ \ \nabla)\,\mathbf{B} + (\mathbf{B}\ \ \nabla)\,\mathbf{A} + \mathbf{A}\quad (\nabla\quad \mathbf{B}) + \mathbf{B}\quad (\nabla\quad \mathbf{A})\,, \tag{C.2}$$

$$\nabla\quad (f\mathbf{A}) = f\,\nabla\quad \mathbf{A} + \nabla f\quad \mathbf{A} \quad , \tag{C.3}$$

$$\nabla\quad (\mathbf{A}\quad \mathbf{B}) = \mathbf{B}\ \ \nabla\quad \mathbf{A}\quad \mathbf{A}\ \ \nabla\quad \mathbf{B} \quad , \tag{C.4}$$

$$\nabla\quad (\nabla\quad \mathbf{A}) = 0 \quad , \tag{C.5}$$

$$\nabla\quad (f\mathbf{A}) = f\,\nabla\quad \mathbf{A} + \nabla f\quad \mathbf{A} \quad , \tag{C.6}$$

$$\nabla\quad (\mathbf{A}\quad \mathbf{B}) = \mathbf{A}(\nabla\ \ \mathbf{B})\quad \mathbf{B}(\nabla\ \ \mathbf{A}) + (\mathbf{B}\ \ \nabla)\,\mathbf{A}\quad (\mathbf{A}\ \ \nabla)\,\mathbf{B}\,, \tag{C.7}$$

$$\nabla\quad (\nabla f) = 0 \quad , \tag{C.8}$$

$$\nabla^2\mathbf{A} = \nabla(\nabla\ \ \mathbf{A})\quad \nabla\quad \nabla\quad \mathbf{A} \quad , \tag{C.9}$$

$$\mathbf{A} = (\mathbf{A}\ \ \hat{v})\,\hat{v} + \hat{v}\quad \mathbf{A}\quad \hat{v} \quad , \tag{C.10}$$

$$\mathbf{A}\quad (\mathbf{B}\quad \mathbf{C}) = (\mathbf{A}\ \ \mathbf{C})\,\mathbf{B}\quad (\mathbf{A}\ \ \mathbf{B})\,\mathbf{C} \quad . \tag{C.11}$$

APPENDIX D

Fundamentals on Bessel functions

D.1 Bessel, Neumann and Hankel functions

In many di erent chapters of Mathematical Physics, one nds di erential equations which can be brought to the following form:

$$\frac{d^2y}{dz^2} + \frac{1}{z}\frac{dy}{dz} + \left(1 \quad \frac{{}^2}{z^2}\right) y = 0 \quad , \tag{D.1}$$

where the parameter , in general, can be real or complex. Eq. (D.1) is called *Bessel's equation of order* and its solutions are called *Bessel functions* or *cylindrical functions.*

It can be shown (Gatteschi, 1973; Watson, 1958) that all the properties of Eq. (D.1) could be derived from those of the so-called con uent equations and, similarly, the properties of its solutions could be derived from those of the hypergeometric con uent functions. But, given its popularity, Eq. (D.1) has been studied directly, and the properties of its solutions can be found in detail in the mathematical literature (Abramowitz and Stegun, Eds., 1965; Watson, 1958; Erdelyi *et al.*, 1953).

Like for any homogeneous ordinary di erential equation of order 2, the general integral of Eq. (D.1) can be expressed as a linear combination of two linearly independent particular integrals. Assuming just the existence of a solution that admits a power series expansion, then one particular integral is found to be expressed by:

$$y_1 = J \ (z) = \left(\frac{z}{2}\right) \sum_{k=0}^{\infty} \frac{(\ 1)^k}{k! \ (\ + k + 1)} \left(\frac{z}{2}\right)^{2k} \quad , \tag{D.2}$$

where denotes Euler's gamma function:

$$(z) = \int_0^{\infty} e^{\ t} t^{z \ 1} \, dt \quad , \qquad \mathrm{Re}\, z > 0 \quad , \tag{D.3}$$

which, for integer values of its argument, becomes a factorial:

$$(n + 1) = n! \quad . \tag{D.4}$$

Eq. (D.2) is called *Bessel function of the rst kind* of order .

If is not an integer, then a second linearly independent integral of Eq. (D.1) is simply found by changing into in Eq. (D.2):

$$y_2 = J \quad (z) = \left(\frac{z}{2}\right) \quad \sum_{k=0}^{\infty} \frac{(\ 1)^k}{k! \ (\quad + k + 1)} \left(\frac{z}{2}\right)^{2k} \quad . \tag{D.5}$$

When becomes an integer, n, then Eqs. (D.2) and (D.5) are no longer linearly independent, as they satisfy:

$$J_{\ n}(z) = (\ \ 1)^n J_n(z) \quad . \tag{D.6}$$

The case of integer order is of particular importance in physical applications. Therefore, one has to nd for the second integral an expression which remains linearly independent of the rst one for $= n$. This is the case for:

$$Y\ (z) = \frac{J\ (z) \cos \quad J\ \ (z)}{\sin} \quad , \tag{D.7}$$

which, for $\to n$, becomes undetermined, but can be calculated via de l'Hospital rule, and found to be:

$$Y_n(z) = \lim_{\to n} Y\ (z) = \frac{1}{\ } \left[\frac{\partial J_n(z)}{\partial n} \quad (\ \ 1)^n \frac{\partial J_{\ n}(z)}{\partial n} \right] \quad , \tag{D.8}$$

from which one nally gets:

$$\begin{aligned} Y_n(z) \quad &= \quad \frac{2}{\ } \left(C + \log \frac{z}{2} \right) J_n(z) \quad \frac{1}{\ } \sum_{k=0}^{n\ \ 1} \frac{(n \quad k \quad 1)!}{k!} \left(\frac{z}{2} \right)^{\ \ n+2k} \\ &\quad \frac{1}{\ } \sum_{k=0}^{\infty} \frac{(\ \ 1)^k}{k!\,(n+k)!} (H_k + H_{n+k}) \left(\frac{z}{2} \right)^{n+2k} \quad , \end{aligned} \tag{D.9}$$

where $C = 0.5772156649\ldots$ is the so-called *Euler-Mascheroni's constant* and $H_m = 1 + \frac{1}{2} \ldots + \frac{1}{m}$ for $m = 1, 2, \ldots$; $H_0 = 0$.

Eq. (D.9) is called *Bessel function of the second kind* or *Neumann function* of order n. It is easily veri ed that it satis es the following relationship, with an obvious correspondence to Eq. (D.6):

$$Y_{\ n}(z) = (\ \ 1)^n Y_n(z) \quad . \tag{D.10}$$

Let us warn the reader that in the literature one can nd several de nitions for Bessel functions of the second kind, which di er from each other just by multiplicative constants. It is legitimate to use any of them, but it is important to be self-consistent.

Some examples of the behavior of J and of Y, for integer orders and real argument, are shown in Figure D.1.

Bessel's equation (D.1) is reminiscent of the harmonic equation. It actually tends to become a harmonic equation for $z \to \infty$. Similarly, its solutions J and Y are seen in Figure D.1 to be reminiscent of the oscillatory-type solutions of the harmonic equation, sine and cosine. Hence, one wonders what information one may extract from linear combinations of Eqs. (D.2) and (D.7) with simple complex coe cients, in analogy with Euler's formulas. In fact, the two functions:

$$H^{(1)}(z) = J\ (z) + jY\ (z) \quad , \qquad H^{(2)}(z) = J\ (z) \quad jY\ (z) \tag{D.11}$$

obviously satisfy Bessel's equation (D.1) for any value of , and can easily be

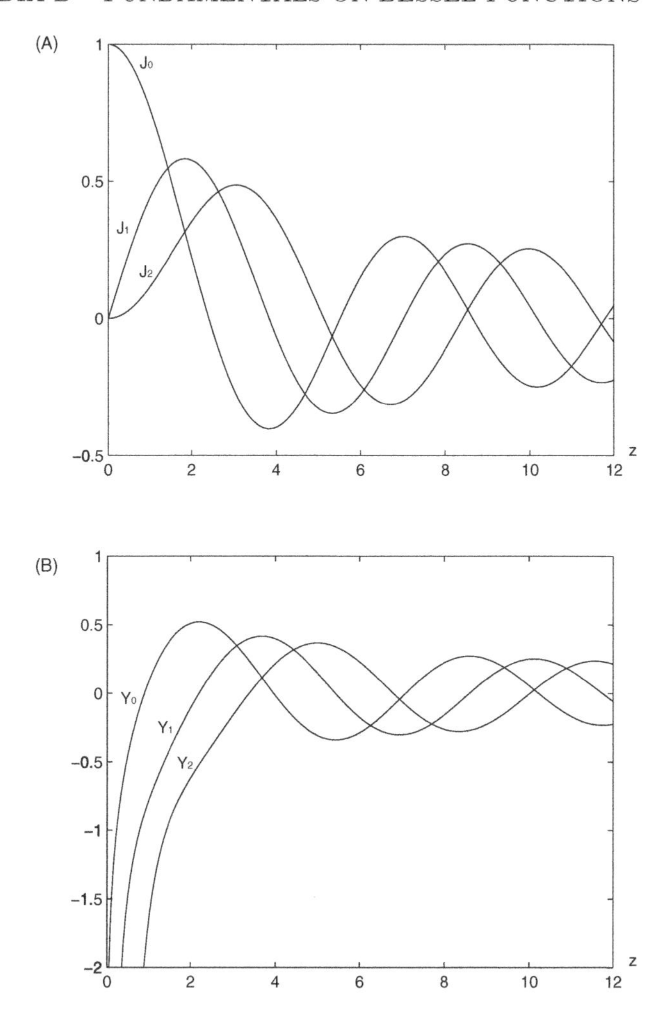

Figure D.1 *Examples of Bessel functions of integer orders (A) of the rst kind and (B) of the second kind*

shown to be linearly independent. Thus, they can be used as a base, alternative to Eqs. (D.2) and (D.7), to express the general integral of Eq. (D.1). They are called *Hankel functions of the rst and second kind*, respectively (or sometimes *Bessel functions of the third kind*).

The inverse of Eqs. (D.11) are:

$$J\ (z) = \frac{H^{(1)}(z) + H^{(2)}(z)}{2} \quad , \qquad Y\ (z) = \frac{H^{(1)}(z) \quad H^{(2)}(z)}{2j} \quad . \tag{D.12}$$

The analogy between Hankel functions and exponential functions of imaginary argument helps in understanding the behavior of waves having cylindrical wavefronts. This approach yields for Bessel functions of the rst and second kind a role of standing waves.

From a purely mathematical point of view, Hankel functions are considered to be very useful because their asymptotic forms (see Section D.3) are easier to derive than those of Eqs. (D.2) and (D.7).

D.2 Modi ed Bessel functions

The analogy between Bessel's equation (D.1) and the harmonic equation has been a useful hint, in the previous section, in view of de ning Hankel functions. It may also remind us of the major interest that one has, in physical applications of the harmonic equation, in studying what happens when the sign in front of the second-derivative term changes. As well known, this change leads to solutions that are easily expressed in terms of hyperbolic functions. Let us study now what occurs for a similar change in Bessel's equation.

Let the complex independent variable z in Eq. (D.1) be replaced by jz, yielding the following equation:

$$\frac{d^2y}{dz^2} + \frac{1}{z}\frac{dy}{dz} \quad \left(1 + \frac{\ ^2}{z^2}\right) y = 0 \quad . \tag{D.13}$$

The general integral of Eq. (D.13) can be written as a linear combination of any two, linearly independent, among the solutions of Eq. (D.1) which we found in the previous section, with z replaced by jz. But, instead of that, it is preferable to de ne two new functions, which are called *modi ed Bessel functions*, respectively *of the rst kind*:

$$\begin{aligned} I\ (z) &= \sum_{k=0}^{\infty} \frac{(z/2)^{\ +2k}}{k!\ \ (\ +k+1)} \quad , \\ I\ (z) &= e^{\ \ j\ /2}\, J\ \left(z\, e^{\ j\ /2}\right) \quad , \end{aligned} \tag{D.14}$$

and *of the second kind*:

$$K\ (z) = \frac{j\ }{2}\, e^{\ j\ /2}\, H^{(1)}\ \left(z\, e^{\ j\ /2}\right) \quad ,$$

$$\begin{cases} K\ (z) = \frac{\ }{2}\,\frac{I\ \ (z)\quad I\ (z)}{\sin\ \ } \quad , & \ \neq 0, \quad 1, \quad 2, \ldots \\ K_n(z) = \lim_{\ \to n} K\ (z) \quad , & n = 0, \quad 1, \quad 2, \ldots \end{cases} \tag{D.15}$$

The last one is also known as *Macdonald function.*

Some examples of the behavior of I and of K, for integer orders and real argument, are shown in Figure D.2.

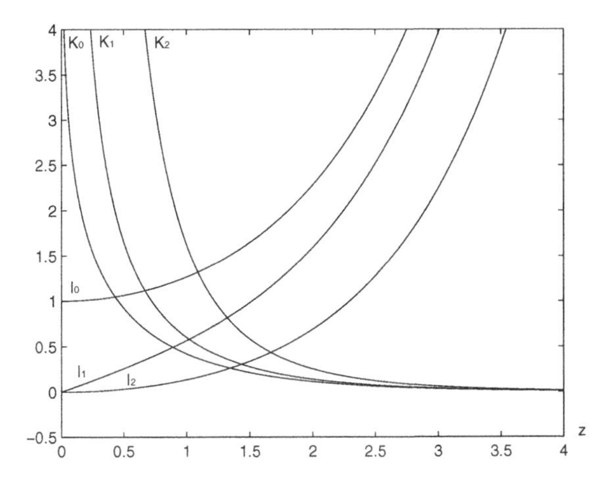

Figure D.2 *Examples of modi ed Bessel functions of integer orders of the rst and second kind*

D.3 Bessel function formulas

Asymptotic forms

Starting from their power series de nitions, it can be shown (Gatteschi, 1973) that for large *real* values of their arguments, the functions that were de ned in the two previous sections tend (for $v \to \infty$) to the following forms:

$$H^{(1)}(v) \to \sqrt{\frac{2}{v}}\, e^{j[v\ \ (\ /4)\ \ (\ \ /2)]} \quad , \tag{D.16}$$

$$H^{(2)}(v) \to \sqrt{\frac{2}{v}}\, e^{\ \ j[v\ \ (\ /4)\ \ (\ \ /2)]} \quad , \tag{D.17}$$

$$J\ (v) \to \sqrt{\frac{2}{v}} \cos\left(v \quad \frac{}{4} \quad \frac{}{2}\right) \quad , \tag{D.18}$$

$$Y\ (v) \to \sqrt{\frac{2}{v}} \sin\left(v \quad \frac{}{4} \quad \frac{}{2}\right) \quad , \tag{D.19}$$

$$j \quad J\ (jv) = I\ (v) \to \sqrt{\frac{1}{2\ \ v}}\, e^{v} \quad , \tag{D.20}$$

$$j^{\ \ +1} H^{(1)}(jv) = \frac{2}{\ }\, K\ (v) \to \sqrt{\frac{2}{v}}\, e^{\ \ v} \quad . \tag{D.21}$$

From these expressions, and remembering the analogies with harmonic equations that were pointed out in Section (D1), the reader can easily draw simple comments on the physical meaning of propagation of waves described by Bessel functions of the various kinds.

Derivatives

Derivatives with respect to the argument can be found by direct di erentiation of the power series. The following relationships apply to any of the four functions J, Y, $H^{(1)}$ and $H^{(2)}$ that were de ned in Section D.1:

$$R_0'(z) = R_1(z) \quad , \tag{D.22}$$

$$R_1'(z) = R_0(z) \quad \frac{1}{z} R_1(z) \quad , \tag{D.23}$$

$$z\, R' \,(z) = R \,(z) \quad z\, R_{\ +1}(z) \quad , \tag{D.24}$$

$$z\, R' \,(z) = R \,(z) + z\, R_{\ \ 1}(z) \quad , \tag{D.25}$$

$$\frac{d}{dz}\left[z \quad R \,(z)\right] = z \quad R_{\ +1}(z) \quad , \tag{D.26}$$

$$\frac{d}{dz}\left[z \ R \,(z)\right] = z \ R_{\ \ 1}(z) \quad . \tag{D.27}$$

For the modi ed Bessel functions, the following rules, which can be derived via their relationships with the previous functions, apply:

$$z\, I' \,(z) = I \,(z) + z\, I_{\ +1}(z) \quad , \tag{D.28}$$

$$z\, I' \,(z) = I \,(z) + z\, I_{\ \ 1}(z) \quad , \tag{D.29}$$

$$z\, K' \,(z) = K \,(z) \quad z\, K_{\ +1}(z) \quad , \tag{D.30}$$

$$z\, K' \,(z) = K \,(z) \quad z\, K_{\ \ 1}(z) \quad . \tag{D.31}$$

Derivatives with respect to *order* may be of interest, on some occasions. They can easily be found in the mathematical literature (e.g., Abramowitz and Stegun, Eds., 1965).

Recurrence formulas

For any given value of the independent variable z, Bessel functions whose orders di er by integers are related among themselves. The basic recurrence formula for $J, Y, H^{(1)}$ and $H^{(2)}$ can be obtained simply by subtracting Eq. (D.25) from Eq. (D.24):

$$\frac{2}{z} R \,(z) = R_{\ +1}(z) + R_{\ \ 1}(z) \quad . \tag{D.32}$$

Similarly, for the modi ed Bessel functions one obtains:

$$\frac{2}{z} I \,(z) = I_{\ \ 1}(z) \quad I_{\ +1}(z) \quad , \tag{D.33}$$

$$\frac{2}{z} K\ (z) \quad = \quad K_{\ +1}(z) \quad K_{\ 1}(z) \quad . \qquad \text{(D.34)}$$

Integrals

Many integrals involving Bessel functions can be calculated analytically, and are easily found in the mathematical literature (e.g., Abramowitz and Stegun, Eds., 1965). We quote here just two integrals, which are of particular relevance to waveguide problems dealt with in this book. Once again, R can be J, Y, $H^{(1)}$ or $H^{(2)}$:

$$\int z\, R\ (\ z)\, R\ (\ z)\, dz =$$

$$= \frac{z}{\ ^2 \quad ^2} \Big[\ R\ (\ z) R_{\ 1}(\ z) \quad R_{\ 1}(\ z) R\ (\ z)\Big], \quad \neq \quad . \qquad \text{(D.35)}$$

This expression enables us to prove orthogonality of modes having the same azimuthal order but di erent radial orders, e.g., in metallic-wall waveguides. The second expression is:

$$\int z\, R^2(\ z)\, dz \quad = \quad \frac{z^2}{2}\Big[R^2(\ z) \quad R_{\ 1}(\ z)\, R_{\ +1}(\ z)\Big]$$

$$= \quad \frac{z^2}{2}\left[R'^{\,2}(\ z) + \left(1 \quad \frac{^2}{^2 z^2}\right) R^2(\ z)\right] \quad . \qquad \text{(D.36)}$$

These expressions are useful in order to evaluate the Poynting vector ux through the cross-section of a circular waveguide. For modi ed Bessel functions, the second expression is no longer correct, because of changes in signs in recurrence formulas and in derivatives, which we pointed out before. The rst expression still holds. However, given that its main use, in the context of this book, is to evaluate the Poynting vector ux through the cladding of a circular dielectric waveguide, it may be convenient to rewrite it explicitly as:

$$\int_z^{\infty} z\, K^2(\ z)\, dz = \frac{z^2}{2}\Big[K_{\ 1}(\ z)\, K_{\ +1}(\ z) \quad K^2(\ z)\Big] \quad . \qquad \text{(D.37)}$$

References

Abramowitz, M. and Stegun, I.A. (Eds.) (1965) *Handbook of Mathematical Functions.* Dover, New York.

Erdelyi, A., Magnus, W., Oberhettinger, F. and Tricomi, F.G. (1953) *Higher Transcendental Functions*, Vol. II, McGraw-Hill, New York.

Gatteschi, L. (1973) *Funzioni speciali.* Utet, Torino.

Watson, G.N. (1958) *A Treatise on the Theory of Bessel Functions.* 2nd ed., Cambridge University Press, Cambridge, UK.

Further Suggested Reading

Adams, M.J. and Henning, I.D. (1990) *Optical Fibres and Sources for Communications.* Plenum, New York.

Balanis, C.A. (1997) *Antenna Theory: Analysis and Design.* 2nd ed., Wiley, New York.

Beran, M.J. and Parrent, G.B. (1964) *Theory of Partial Coherence.* Prentice-Hall, Englewood Cli s, NJ (reprinted 1974 by SPIE, Bellingham, WA).

Desurvire, E. (1994) *Erbium-Doped Fiber Ampli ers: Principles and Applications.* Wiley, New York.

Dyott, R.B. (1995) *Elliptical Fiber Waveguides.* Artech House, Boston, 1995.

Elliott, R.S. (1981) *Antenna Theory and Design.* Prentice-Hall, Englewood Cli s, NJ.

Iga, K. (1994) *Fundamentals of Laser Optics.* Technical editor: R.B. Miles, Plenum, New York.

Kapany, N.S. and Burke, J.J. (1972) *Optical Waveguides.* Academic, New York.

Lo, Y.T. and Lee, S.W. (Eds.) (1988) *Antenna Handbook. Theory, Applications and Design.* Van Nostrand Reinhold, New York.

Louisell, W.H. (1973) *Quantum Statistical Properties of Radiation.* Wiley, New York.

Marcuse, D. (1982) *Light Transmission Optics.* 2nd ed., Van Nostrand Reinhold, New York.

Marcuse, D. (1991) *Theory of Dielectric Optical Waveguides.* 2nd ed., Academic, New York.

Roach, G.F. (1982) *Green's Functions: Introductory Theory with Applications.* 2nd ed., Cambridge University Press, Cambridge, UK.

Saleh, B.E.A. and Teich, M.C. (1991) *Fundamentals of Photonics.* Wiley, New York.

Schelkuno , S.A. (1952) *Advanced Antenna Theory.* Wiley, New York.

Silver, S.S. (Ed.) (1984) *Microwave Antenna Theory and Design.* P. Peregrinus, London.

Stamnes, J.J. (1986) *Waves in Focal Regions: Propagation, Di raction and Focusing of Light, Sound and Water Waves.* Adam Hilger, Boston.

Suematsu, Y. and Iga, K. (1982) *Introduction to Optical Fiber Communications.* Translated by H. Matsumura, edited and revised by W.A. Gambling, Wiley, New York.

Toraldo di Francia, G. (1956) *Electromagnetic Waves.* translated from Italian (Onde Elettromagnetiche), Interscience, New York.

Vendelin, G.D., Pavio, A.M. and Rohde, U.L. (1990) *Microwave Circuit Design using Linear and Nonlinear Techniques.* Wiley, New York.

Index

Page numbers appearing in *italic* refer to tables.
Page numbers appearing in **bold** refer to gures.

B

C

F

N

O

P

For Product Safety Concerns and Information please contact our EU representative GPSR@taylorandfrancis.com Taylor & Francis Verlag GmbH, Kaufingerstraße 24, 80331 München, Germany

Printed and bound by CPI Group (UK) Ltd, Croydon, CR0 4YY
01/05/2025
01858546-0006